The Search for Mathematical Roots, 1870–1940

LOGICS, SET THEORIES AND
THE FOUNDATIONS OF MATHEMATICS
FROM CANTOR THROUGH
RUSSELL TO GÖDEL

I. GRATTAN-GUINNESS

PRINCETON UNIVERSITY PRESS

PRINCETON AND OXFORD

Copyright © 2000 by Princeton University Press
Published by Princeton University Press, 41 William Street,
Princeton, New Jersey 08540
In the United Kingdom: Princeton University Press, 3 Market Place,
Woodstock, Oxfordshire, OX20 1SY

Library of Congress Cataloging-in-Publication Data

Grattan-Guinness, I.
The search for mathematical roots, 1870–1940 : logics, set theories and the foundations
of mathematics from Cantor through Russell to Gödel / I. Grattan-Guinness.
p. cm.
Includes bibliographical references and index.
ISBN 0-691-05857-1 (alk. paper) — ISBN 0-691-05858-X (pbk. : alk. paper)
1. Arithmetic—Foundations—History—19th century. 2.
Arithmetic—Foundations—History—20th century. 3. Set theory—History—19th century. 4.
Set theory—History—20th century. 5. Logic, Symbolic and mathematical—History—19th
century. 6. Logic, Symbolic and mathematical—History—20th century. I. Title.

QA248 .G684 2000
510--dc21   00-036694

This book has been composed in Times Roman

The paper used in this publication meets the minimum requirements of
ANSI/NISO Z39.48-1992 (R1997) (*Permanence of Paper*)

www.pup.princeton.edu

Printed in the United States of America

10 9 8 7 6 5 4 3 2 1

CONTENTS

CHAPTER 4
Parallel Processes in Set Theory, Logics and Axiomatics, 1870s–1900s

CHAPTER 6
Russell's Way In: From Certainty to Paradoxes, 1895–1903

CHAPTER 7
Russell and Whitehead Seek the *Principia Mathematica*, 1903–1913

CHAPTER 11
Transcription of Manuscripts

The Search for Mathematical Roots,
1870–1940

CHAPTER 1

Explanations

1.1 SALLIES

Language is an instrument of Logic, but not an indispensable instrument.
                                                          Boole *1847a*, 118

We know that mathematicians care no more for logic than logicians for mathematics. The two eyes of exact science are mathematics and logic; the mathematical sect puts out the logical eye, the logical sect puts out the mathematical eye; each believing that it sees better with one eye than with two.
                                                          De Morgan *1868a*, 71

That which is provable, ought not to be believed in science without proof.
                                                          Dedekind *1888a*, preface

If I compare arithmetic with a tree that unfolds upwards in a multitude of techniques and theorems whilst the root drives into the depths [...]
                                                          Frege *1893a*, xiii

Arithmetic must be discovered in just the same sense in which Columbus discovered the West Indies, and we no more create numbers than he created the Indians.
                                                          Russell *1903a*, 451

1.2 SCOPE AND LIMITS OF THE BOOK

1.2.1 *An outline history.* The story told here from §3 onwards is re-garded as well known. It begins with the emergence of set theory in the 1870s under the inspiration of Georg Cantor, and the contemporary development of mathematical logic by Gottlob Frege and (especially) Giuseppe Peano. A cumulation of these and some related movements was achieved in the 1900s with the philosophy of mathematics proposed by Alfred North Whitehead and Bertrand Russell. They claimed that "all" mathematics could be founded on a mathematical logic comprising the propositional and predicate calculi (including a logic of relations), with set theory providing many techniques and various other devices to hand, especially to solve the paradoxes of set theory and logic which Russell discovered or collected. Their position was given a definitive presentation in the three volumes of *Principia mathematica* (1910–1913). The name 'logicism' has become attached to this position; it is due (in this sense of the word) to Abraham Fraenkel (§8.7.6) and especially Rudolf Carnap (§8.9.3) only in the late 1920s, but I shall use it throughout.

Various consequences followed, especially revised conceptions of logic and/or logicism from Russell's followers Ludwig Wittgenstein and Frank Ramsey, and from his own revisions of the mid 1920s. Then many techniques and aims were adopted by the Vienna Circle of philosophers, affirmatively with Carnap but negatively from Kurt Gödel in that his incompletability theorem of 1931 showed that the assumptions of consistency and completeness intuitively made by Russell (and by most mathematicians and logicians of that time) could not be sustained in the form intended. No authoritative position, either within or outside logicism, emerged: after 1931 many of the main questions had to be re-framed, and another epoch began.

The tale is fairly familiar, but mostly for its philosophical content; here the main emphasis is laid on the logical and mathematical sides. The story will now be reviewed in more detail from these points of view.

1.2.2 *Mathematical aspects.*   First of all, the most pertinent parts of the prehistory are related in §2. The bulk of the chapter is given over to developments of new algebras in France in the early 19th century and their partial adoption in England; and then follow the contributions of George Boole and Augustus De Morgan (§2.4–5), who each adapted one of these algebras to produce a mathematicised logic. The algebras were not the same, so neither were the resulting logics; together they largely founded the tradition of algebraic logic, with some adoption by others (§2.6). By contrast, the prehistory of mathematical logic lies squarely in mathematical analysis, and its origins in Augustin-Louis Cauchy and extension led by Karl Weierstrass are recalled in §2.7, the concluding section of this chapter, to lead in to the main story which then follows. A common feature of both traditions is that their practitioners handled collections in the traditional way of part-whole theory, where, say, the sub-collection of Englishmen is part of the collection of men, and membership to it is not distinguished from inclusion within it.

The set theory introduced in §3 is the '*Mengenlehre*' of Georg Cantor, both the point set topology and transfinite arithmetic and the general theory of sets. In an important contrast with part-whole theory, an object *was* distinguished from its unit set, and belonged to a set S whereas sub-sets were included in S: for example, object a belongs to the set $\{a, b, c\}$ of objects while sets $\{a\}$ and $\{a, b\}$ are subsets of it. The appearance of both approaches to collections explains the phrase 'set theor*ies*' in the sub-title of this book.

Next, §4 treats a sextet of related areas contemporary with the main themes outlined above, largely over the period 1870–1900. Firstly, §4.2 records the splitting in the late 1890s of Cantor's *Mengenlehre* into its general and its topological branches, and briefly describes measure theory and functional analysis. Next, §4.3–4 outlines the extension of algebraic logic by Ernst Schröder and Charles Sanders Peirce, where in particular the contributions of Boole and De Morgan were fused in a Boolean logic of relations; Peirce also introduced quantification theory, which Schröder developed. All this work continued within part-whole theory. §4.5 outlines the creation of a version of mathematical logic by Frege, highly regarded today but (as will be explained) modestly noted in his own time; it included elements of set theory. Then follows §4.6 on the first stages in the development of phenomenological logic by Edmund Husserl. Finally, §4.7 notes the early stages of David Hilbert's proof theory (not yet his meta-mathematics), and of American work in model theory influenced by E. H. Moore.

Then §5 describes the work of Peano and his followers (who were affectionately known as the 'Peanists'), which gained the greatest attention of mathematicians. Inspired by Weierstrass's analysis and *Mengenlehre*, this 'mathematical logic' (Peano's name) was used to express quite a wide range of mathematical theories in terms of proportional and predicate calculi with quantification (but the latter now construed in terms of members of sets rather than part-whole theory). The period covered runs from 1888 to 1900, when Russell and Whitehead became acquainted with the work of the Peanists and were inspired by it to conceive of logicism.

Russell's career in logic is largely contained within the next two chapters. First, §6 begins with his début in both logic and philosophy in the mid 1890s, and records his progress through a philosophical conversion inspired by G. E. Moore, and the *entrée* of Whitehead into foundational studies in 1898. Next comes Russell's discovery of Peano's work in 1900 and his paradox soon afterwards, followed by the publication in 1903 of *The principles of mathematics*, where his first version of logicism was presented. Then §7 records him formally collaborating with Whitehead, gathering further paradoxes, discovering an axiom of choice in set theory, adopting a theory of definite descriptions, and trying various logical systems before settling on the one which they worked out in detail in *Principia mathematica* (hereafter '*PM*'), published in three volumes between 1910 and 1913. Some contemporary reactions by others are recorded, mainly in §7.5.

In §8 is recorded the reception and use of *PM* and of logicism in many hands of various nationalities from the early 1910s to the late 1920s. Russell's own contributions included applications of logical techniques to philosophy from the 1910s, and a new edition of *PM* in the mid 1920s (§8.2–3). His most prominent successors were Wittgenstein and Ramsey, and interest continued in the U.S.A. (§8.3–5). Considerable concern with foundational studies was shown among German-speaking philosophers and mathematicians (§8.7), including the second stage of Hilbert's 'meta-mathematics' and the emergence of the 'intuitionistic' philosophy of mathematics, primarily with the Dutchman L. E. J. Brouwer. Two new groups arose: logicians in Poland, led by Jan Łukasiewicz and Stanisław Leśniewski, and soon joined by the young Alfred Tarski (§8.8); and the group of philosophers which became known as the 'Vienna Circle', of whom Moritz Schlick, Carnap and Gödel are the most significant here (§8.9).

In briefer order than before, §9 completes the story by reviewing the work of the 1930s. Starting with Gödel's incompletability theorem of 1931, other contemporary work is surveyed, especially by members of the Vienna Circle and some associates. The returns of both Whitehead and Russell to logicism are described, and some new applications and countries of interest are noted. Finally, with special attention to Russell, the concluding §10 reviews the myriad relationships between logics, set theories and the foundations of mathematics treated in this book; the concluding §10.3 contains a flow chart of the mathematical developments described in the book and stresses the lack of an outright "winner". Ten manuscripts, mostly letters to or from Russell, are transcribed in §11. Then follow the bibliography and index.

1.2.3 *Historical presentation.* This book is intended for mathematicians, logicians, historians, and perhaps philosophers and historians of science who take seriously the concerns of the other disciplines. No knowledge of the history is assumed in the reader, and numerous references are given to both the original and the historical literature. However, it does not serve as a textbook for the mathematics, logic or philosophy discussed: the reader is assumed to be already familiar with these, approximately at the level of an undergraduate in his final academic year.

From now on I shall refer to the 'traditions' of algebraic and of mathematical logic; the two together constitute 'symbolic logic'. Occasionally mention will be made of other traditions, such as syllogistic logic or Kantian philosophy. By contrast of term, logicism will constitute a 'school', in contention with those of metamathematics, intuitionism and phenomenology.

Inter-disciplinary relationships were an important part of the story itself, for symbolic logic was usually seen by mathematicians as too philosophical and by philosophers as too mathematical. De Morgan's remark quoted in §1.1 is especially brilliant, because not only was he both mathematician and logician but also he had only one eye! Thus the title of this book, 'The search for mathematical roots', is a *double entendre*: whether mathematics (or at least some major parts of it) could be founded in something else, such as the mathematical logic of Whitehead and Russell; or the inverse stance, where mathematics itself could serve as the foundation for something else. A third position asserted that mathematics and logic were overlapping disciplines, with set theories occupying some significant place which itself had to be specified; it was upheld by the Peanists, and gained more support after Gödel, especially with W. V. Quine (§9.4.4).

The final clause of the sub-title of this book would read more accurately, but also a little too clumsily, as 'inspired in different ways by Lagrange and

Cauchy, and pursued especially but not only from Cantor and Peano through Whitehead and Russell to Carnap and Gödel', with some important names still missing. Its story differs much from the one in which Frege dominates, the details of the mathematics are at best sketched, and everything is construed in terms of analytic philosophy. For example, the discussion here of *Principia mathematica* does not stop after the first 200 pages but also takes note of the next 1,600, where the formulae are presented. The quality and merits of Whitehead and Russell's logicism should then become clearer, as well as its well-known (and important) confusions and limitations. Again, most histories of these topics are of the 'great man' variety; but here many other people play more minor but significant roles—either as minor figures in the tale or as major ones in some related developments.

Another novelty is that much new information is provided from about 50 archival sources which have been examined. Russell left an enormous *Nachlass*, known as the 'Russell Archives' and cited in this book as 'RA'; so did some other figures (for example, Hilbert, Peirce and Carnap). For several more, valuable collections are available (Boole, Cantor, Dedekind and Gödel); for some, sadly, almost nothing (Peano and Whitehead). Important information has come from the manuscripts of many other figures (including several named earlier), and from some university and publishers' archives. Normally a collection is cited as, say, 'Cantor Papers', followed by an identifying clause or code of a particular document appropriate for its (dis-)organisation. Its location is indicated at the head of the list of his cited works in the bibliography. The main archive locations are recorded in the front matter there, and are also named in the acknowledgements in §1.4.

1.2.4 *Other logics, mathematics and philosophies.* To temper the ambitions just outlined, some modesty is required.

1) A few concurrent developments outside mathematical logic are described, though not in much detail. The limited coverage of algebraic logic was mentioned in §1.2.2: its own relationships with other algebras are treated lightly. An integrated history of post-syllogistic and algebraic logic from the 1820s to the 1920s is *very* desirable.

Again, in §6.2–3 notice is taken of the influential but very non-mathematical neo-Hegelian tradition in logic only in connection with the young Russell, who started out with it but then rejected it at the end of the century.[1] Similarly, phenomenological logic is noted just to the extent of §4.5 on Husserl and §8.7.2,8 on a few followers; and §8.8 and §9.6.7 contain only some of the work of the Polish community of logicians.

[1] Since those kinds of philosophy have fallen out of favour (apart from centres where Germanic influences remain active), the history has become quite mis-remembered. It is thought, even by some historians, that they died very quickly, especially in Anglo-Saxon countries, after the rise of Russell and his associates in the 1900s; however, a different course will be revealed in §9.5.

2) An important neighbour is metamathematics, which in this period was created and dominated by Hilbert with an important school of followers. The story of his search for mathematical roots from Cantor to Gödel is very important; but it is rather different from this one, more involved with the growth of axiomatisation in mathematics and with metamathematics and granting a greater place to geometry, and less concerned with mathematical analysis and the details of Cantor's *Mengenlehre*. So only some portions of it appear here, mainly in §4.4, §8.8 and §9.6.2. Similarly, no attempt is made here to convey other foundational studies undertaken in mathematics at that time, such as the foundations of geometry and of mechanics, or the development of abstract algebras and of quantum mechanics.

3) Another neighbouring discipline to logic is linguistics, which during our period was concerned not only with grammar and syntax but also with traditional questions such as the origins of language in humans and the classification of languages. One would assume that links to logics, especially mathematicised ones, were strong, in particular through the common link of semiotics, the science of signs, for which common algebra was the supreme case; indeed, we shall note in §2.2.1 that in the 17th century John Locke had used 'semiotics' and 'logic' as synonyms. However, with the exception of Peirce (§4.3.8) the connections were slight—indeed, already so in the 18th century when linguistics was well developed while logic languished. More work is needed on this puzzling situation, which is largely side-stepped here.

4) Almost all of the logics described here were 'finitary'; that is, both formulae and proofs were finitely long. From time to time we shall come across an 'infinitary' logic, usually "horizontal" extensions to infinitely long formulae while in §9.2.5 appears a "vertical" foray to infinitely long proofs; but their main histories lie after our period.

5) A few modern versions of logicism have been proposed in recent years, and also various figures in our story have been invoked in support or criticism of current positions in epistemology and the philosophy of science. I have noted only a few cases in a footnote in §10.2.3, since modernised versions of the older thought are involved. More generally, I have made no attempt to treat the huge literature which comments without originality on the developments described in this book. Logicism has inspired many opinions about logic and the philosophy of mathematics from Russell's time to today, but often offered with little knowledge of the technical details or applications of his logic.

6) The story concentrates upon the research level of work: its (non-)diffusion into education is touched upon only on opportune occasions. The impact upon teaching during the period under consideration seems to have been rather slight, but the matter merits more investigation than it receives here.

1.3 Citations, terminology and notations

1.3.1 *References and the bibliography.* The best source for the original literature is the German reviewing journal *Jahrbuch über die Fortschritte der Mathematik*, where it was categorised in amusingly varied distributions over the years between the sections on 'Philosophie', 'Grundlagen', 'Mengenlehre' and 'Logik'. Among bibliographies, Church *1936a* and *1938a* stand out for logic, and Risse *1979a* and Vega Reñon *1996a* are also useful; for set theory Fraenkel *1953a* is supreme. Toepell *1991a* provides basic data on German mathematicians, including several logicians. My general encyclopaedia *1994a* for the history and philosophy of mathematics has pt. 5 devoted to logics and foundations, and each article has a bibliography of mostly secondary sources; some articles in other parts are also relevant. Among philosophical reference works, note especially Burkhardt and Smith *1991a*.

Most works are cited by dating codes in italics with a letter, such as 'Russell *1906a*'; the full details are given in the bibliography, which also conveys dates of all authors when known. When a manuscript is cited, whether or not it has been published on some later occasion, then the reference is prefaced by '*m*' as in 'Russell *m1906a*', in which case there is no '*1906a*'. Collected or selected editions or translations of works and/or correspondence are cited by words such as '*Works*' or '*Letters*'; if a particular volume is cited in the text, then the volume number is added also in italics, as in 'Husserl *Works 12*'. Different editions of a work are marked by subscript numbers. '*PM*' is cited wherever possible by the asterisk number of the proposition or definition; if page numbers are needed, they are to the second edition. A few works on a figure without named author or editor are cited under his name with a prime attached; for example, 'Couturat *1983a'*' is a volume on his life and work.

This strategy of avoiding page numbers has been followed whenever possible for works which have received multiple publication—original appearance (maybe more than once), re-appearance in an edition of the author's works and/or anthologies, and maybe a translation or two. In such cases, article or even theorem or equation numbers have been used instead. Where a page number is necessary, an accessible and reliable source has been chosen, and its status is indicated in the bibliography entry by the sign '‡'.

Finally, '§' is used to indicate chapters and their sections and sub-sections; no chapter has more than nine sections, and no section has more than nine sub-sections. Equations or expressions are numbered consecutively within a sub-section; for example, (255.3) is the third equation in §2.5.5.

1.3.2 *Translations, quotations and notations.*   All non-English texts have been translated into English; usually the translations are my own. Several of our main authors have been translated into English, but not always with happy results—too free, and often not drawing upon the correct philosophical distinctions in the original language (especially German). Occasionally issues of translation are discussed. Apart from in §11, my own insertions into quotations, of any kind, are enclosed within square brackets.

As far as possible, I have followed the terms and symbols used by the historical figures, and in quotations they are preserved or translated exactly. But several ordinary words, in any language, were used as technical terms (for example, 'concept' and 'number'). Quite often I have used quotation marks or quoted the original word alongside the translation; and I use 'notion' as a neutral all-purpose word to cover concepts and general ideas. In addition, a variety of terms, or changes in terms, has occurred over time, and the most modern version is often *not* adopted here. In particular, I use 'set' when in Cantor's *Mengenlehre* but follow Russell in speaking of 'classes', which was his technical term with 'sets' as informal talk. Some further terms in Russell are explained in §6.1.1.

From 1904 the word 'logistic' was adopted to denote the new mathematical logic (§7.5.1), but it covered both the position of the Peanists and that of Russell. I try to make clear its sense in each context, and use 'Peanism' or 'logicism' where possible.

Related problems arise from our custom of distinguishing a theory, language or logic from its metatheory, metalanguage or metalogic; for it clearly emerged only during the early 1930s (§9.2–3, §9.6.7). Apart from some tantalising partial anticipations in the 1920s, especially in the U.S.A. (§8.5), earlier it was either explicitly avoided (by Russell, for example) or observed only in certain special cases, such as distinguishing a descriptive phrase from its possible referent. In particular, the conditional connective ('if . . . then') between propositions was muddled with implication between their names, and propositions themselves with (well-formed) sentences in languages. I have tried to *follow* these kinds of conflation, in order to reconstruct the muddles of the story; the logic is worse, but the history much better. So I have not distinguished name-forming single quotation marks from quasi-quotes; however, I use double quotation marks as scare-quotes for special uses of terms. Lastly, the reader should bear in mind that often I *mention* an historical figure *using* some quoted term or notation.

In quotations from and explanations of original work, the original symbols are used or at least described. However, for my own text I have had to make choices, since various notations have been entertained in logic and set theories over the decades. Several of them have their origins in Peano or in Whitehead and Russell, and they serve as my basic lexicon here (including some conflations discussed above):

$\sim$ *or* $^{-}$	not	$\vee$ (inclusive) or	$\cdot$	and $\vdash$ assertion
$\supset$	if … then *or* implication	$\equiv$		if and only if *or* equivalence
(x) *or* $_x$	… for all x …	$(\exists x)$		there is an x such that …
$=$	identity *or* equality	$:=$ *or* $=$ Df.^{2}		equality by definition
$\ni$	such that	$\in$ *or* ε		is a member of
$\cup$	union of classes	$\cap$		intersection of classes
$\subseteq$	improper inclusion of classes	$\subset$ *or* $\supset$		proper inclusion of classes
$-$	difference of classes	ι'		unit class of
V	universal class *or* tautology	Λ		empty class *or* contradiction
$\{a, b, \dots\}$	unordered class	$(a, b, \dots)$		ordered class
$\hat{x}\phi(x)$	the x's such that (class abstraction)	$(\iota x)(\phi x)$		the x such that (definite description)

In addition, to reduce the density of brackets I have made some use of Peano's systems of dots: the larger their number at a location, the greater their scope. Dots indicating logical conjunction take the highest priority, and there the scope lies in both directions; then come dots following expressions which use brackets for quantifiers; and finally there are dots around connectives joining propositions.

I use the usual Roman or Greek letters for mathematical and for logical functions, distinguishing the two types by enclosing the argument variable of a mathematical function within brackets (such as '$f(x)$'). Relations are normally represented by upper case Roman letters. Further explanations, such as Russell's enthusiastic use of '!', are made in context.

1.4 PERMISSIONS AND ACKNOWLEDGEMENTS

Over the three decades of preparing this book, I have enjoyed many valued contacts. Among people who have died during that period, I recall especially Jean van Heijenoort, Alonzo Church and Sir Karl Popper. The most constant and continuing obligations lie to Kenneth Blackwell, the founder

---

^{2} ' := ' has become popular in recent years: De Morgan had used it to define 'singular identity' between individual members of classes (*1862a*, 307). ' = Df.' belongs to Russell: according to Chwistek *1992a*, 242, the variant ' $=_{Df}$ ' (not employed here) was introduced by W. Wilcosz; but it was already presented in the form ' $=_{Def}$ ' in Burali-Forti *1894b*, 26 (§5.3.7).

Russell Archivist at McMaster University, Canada; Albert Lewis, long-time member of the Russell Edition project (an appointment which I gladly recall as instigating) and now with the Peirce Project; Joseph Dauben, the best biographer of Georg Cantor; and Volker Peckhaus, the leading student of German foundational studies for our period. In addition, I acknowledge advice of various kinds from Liliana Albertazzi, Gerard Bornet, Umberto Bottazzini, John Corcoran, Tony Crilly, John Crossley, John and Cheryl Dawson, O. I. Franksen, Eugene Gadol, Massimo Galuzzi, Nicholas Griffin, Leon Henkin, Larry Hickman, Claire Hill, Wilfrid Hodges, Nathan Houser, Ken Kennedy, Gregory Landini, Desmond MacHale, Saunders Mac Lane, Corrado Mangione, Elena Anne Marchisotto, Daniel D. Merrill, Gregory Moore, Eduardo Ortiz, Maria Panteki, Roberto Poli, W. V. Quine, Francisco Rodriguez-Consuegra, Adrian Rice, Matthias Schirn, Gert Schubring, Peter Simons, Barry Smith, Gordon Smith, Carl Spadoni, Christian Thiel, Michael Toepell, Alison Walsh, George Weaver, Jan Wolenski, and the publishers' anonymous referees. As publishers' reader, Jennifer Slater carried the spirit of the infinitesimal into textual preparation.

Some writing of this book, and much archival research, were supported by a Fellowship from the Leverhulme Foundation for 18 months between 1995 and 1997. I express deep gratitude for their provision of money and, as an even more precious commodity, time. Further archival research in 1997 was made possible by a Research Grant from the Royal Society of London.

The main archives and their excellent archivists are housed as follows. In Britain, East Sussex Record Office; Cambridge University Library; Churchill College, King's College, and Gonville and Caius College, Cambridge; Victoria University of Manchester; Royal Holloway College and University College, University of London; Reading University; and The Royal Society of London. In Ireland, Cork University. In Germany, Erlangen, Freiburg and Göttingen Universities. In Austria, Vienna University. In the Netherlands, the State Archives of North Holland, Haarlem. In Switzerland, the Technical High School, Zürich; and the University of Lausanne. In Sweden, the Institut Mittag-Leffler, Djursholm. In the U.S.A., Indiana University at Indianapolis and at Bloomington; the University of Chicago; the University of Texas at Austin; Southern Illinois University at Carbondale; Columbia University, New York; Pittsburgh University; Harvard University; Massachusetts Institute of Technology; Smith College; and the Library of Congress, Washington. In Canada, McMaster University (which holds especially the Russell Archives). In Israel, the late Mrs. M. Fraenkel.

For permission to publish manuscripts by Russell I thank the McMaster University Permissions Committee. Similar sentiments are offered to Quine and to Leon Henkin, for their correspondence with Russell published in §10.8–9; and to Cambridge University Press for the diagram used in §9.5.3.

All efforts have been made to locate copyright holders of a few other quoted texts.

Finally, much gratitude is due to my wife Enid for secretarial help, to Humphrey for all his attention during the actual writing, and to his brother Monty for usually realising that one cat in the way at a time was enough.

January 2000

CHAPTER 2

Preludes: Algebraic Logic
and Mathematical Analysis
up to 1870

2.1 PLAN OF THE CHAPTER

The story begins in French mathematics and philosophy in the late 18th century: specifically the semiotic 'logique' of Condillac and Condorcet and the connections with the algebraic theories, especially the calculus, developed by Lagrange (§2.2). Then it moves to England, for both topics: the adoption of Lagrangian mathematics by Babbage and Herschel, and the revival of logic (although not after the French model) in the 1820s (§2.3). Next come the two principal first founders of algebraic logic, De Morgan and Boole (§2.4–5). The main initial reactions to Boole are described in §2.6.

In a change of topic, §2.7 also starts with the French, but charts a rival tradition in the calculus: that of Cauchy, who inaugurated mathematical analysis, based upon the theory of limits and including a radical reformulation of the calculus. Then the refinements brought about from the 1860s by Weierstrass and his followers are noted; the inspiration drawn from a doctoral thesis by Riemann is stressed. Thereby the scene is set for Cantor in §3.

While two important philosophers, Bolzano and Kant, are noted (§2.8.2), the chapter does not attempt to cover the variety of approaches adopted in logic in general during the period under study. For a valuable survey of the teaching of logic internationally, see Blakey *1851a*, chs. 14–22. A pioneering revision of the history of linguistics for this period and later is given in Aarsleff *1982a*.

2.2 'LOGIQUE' AND ALGEBRAS IN FRENCH MATHEMATICS

2.2.1 The 'logique' and clarity of 'idéologie'. Supporters of the doctrine of 'ideology' became engaged in the political life of France in the mid 1790s, including collision with the young General Napoléon Bonaparte; and the word 'ideology' has carried a political connotation ever since. However, when Antoine Destutt de Tracy introduced the word 'idéologie' in 1796, it referred not to a political standpoint but to an epistemological position: namely, to ideas, their reference and the sign used to represent them. It exemplified the strongly semiotic character of much French philosophy of the time, especially following certain traits of the Enlightenment.

This was already marked in the hands of the Abbé Condillac, the father-figure of the Idéologues. His treatise *La logique* was published in 1780, soon after his death in that year.[1] The 'logic' that it espoused was the method of 'analysis' of our ideas as originating in simple sensory experiences, followed by the process of 'synthesis' in which the ideas were reconstructed in such a way that the relations between them were clearly revealed (Rider *1990a*). To us the book reads more like a work in semiotics than logic: both words had been used by Condillac's father-figure, John Locke (1632–1704), in his *Essay concerning human understanding*, and he took them as synonyms because words were the most common kind of sign (Locke *1690a*, book 4, ch. 21: this seems to be the *origin* of the word 'semiotics'). For Condillac the procedure of analysis and synthesis followed nature: 'the origin and generation both of ideas and of the faculties of the soul are explained according to this method' (Condillac *1780a*, title of pt. 1). When the *Ecole Normale* was opened in Paris in 1795 for its short run of four months as a teacher training college,[2] a copy of this book was given to every student.

Condillac did not present logical rules in his doctrine: instead, broadly following views established in Port-Royal logic and Enlightenment philosophy, he laid great emphasis on language. In order that the ideas could indeed be clearly stated and expressed, the language of which the signs were elements had to be well made, so that indeed 'the art of reasoning is reduced' to it (title of pt. 2, ch. 5). He did not discuss syllogistic logic, where the rules were assumed to apply to reasoning independently of the language in which it was expressed. In showing this *degree* of uninterest in tradition, his approach was rather novel. But he gained attention from *savants* in various fields of French science. For example, the chemist Antoine-Laurent Lavoisier was influenced by Condillac to improve the notation of his subject, even to the extent of writing down chemical equations. Similarities between logics and chemistry were to recur at times later (Picardi *1994b*).

2.2.2 *Lagrange's algebraic philosophy.* Obviously mathematics was the apotheosis of a clear science, and within mathematics algebra gained a preferred place. Condillac himself wrote a treatise on algebra entitled 'The

---

[1] On the political significance of Condillac's thought, see Albury *1986a*; his edition/translation of the *Logique* has a very useful introduction. On the general background in Port-Royal logic and Enlightenment philosophy, see Auroux *1973a* and *1982a*.

[2] Bad planning and poor financing caused the early demise of the *Ecole Normale*. The current institution carrying this name was founded in 1810 as the elite establishment of the new *Université Impériale*, which despite its name was basically the school-teaching organisation for the Empire. On the French educational structure of the time, see my *1988a*.

language of calculation' which was published posthumously as his *1798a*, in which the formal rules of ordinary arithmetic and algebra were explained, the legitimacy of the negative numbers *as* numbers was stressed (§2.4.2), and so on.

Some mathematicians of the time were drawn to the doctrine. The most prominent was the Marquis de Condorcet (1743–1794), although his emphasis on the mathematical rather than the linguistic features inevitably made his position less well appreciated. Much of his work in probability and the calculus was heavily algebraic in character (for example, he esteemed closed-form solutions to differential equations over any other kind). But the master of algebras of the time was Joseph Louis Lagrange (1736–1813), who had come to Paris from Berlin in 1787. He popularised his position in teaching both at the *Ecole Normale* and especially at the *Ecole Polytechnique*. This latter was a preparatory engineering school which opened in 1794 (the year of Condorcet's suicide, incidentally); in contrast to the failure of the other school, it ran successfully.

Lagrange had formed his preferences for algebraic mathematical theories in his youth in the late 1750s, quite independently of Condillac or the Idéologues (indeed, rather prior to them). But he found a congenial philosophical climate within which his views could be propounded. He tried (unsuccessfully, but that is another matter) to ground all mechanics in principles such as that of least action, which could be stated entirely in algebraic terms, without resource to either geometrical theories or the intuition of experience: 'One will not find Figures in this work' is a famous quotation from his *Méchanique analitique* (*1788a*, preface).

The algebras involved are not the common ones of Condillac but the differential and integral calculus and the calculus of variations, of which Lagrange had proposed algebraic versions (see Dickstein *1899a* and Fraser *1985a* respectively). As the former calculus is of some importance for our story, a little detail is in order.

According to Lagrange, every mathematical function $f(x + h)$ could be expanded in a power series in the increment variable h on the argument variable x; and the 'derived functions' $f'(x), f''(x), \ldots$ (these were his terms and notations) were *definable* in terms of the coefficients of the appropriate powers of h. These definitions, and the manner of their determination, were held by him to be obtainable by purely algebraic means, without resource to limits or infinitesimals, common procedures of the time but unrigorous in his view. The integral was also defined algebraically, as the inverse of the derived function. The whole approach was extended to cover functions of several independent variables. The only exceptions to be allowed for were 'singular values' of x, where $f(x)$ was undefined or took infinite values; even multi-valued functions were allowed. Other theories, such as the manipulation of functions and of finite and infinite series, were also to be handled only by algebraic means.

Lagrange gave his theory much publicity in connection with the courses which he taught in some of the early years of the *Ecole Polytechnique*, and his textbook *Théorie des fonctions analytiques* (*1797a*) was widely read both in France and abroad. The next section contains a few of the new results to which it led. However, the standpoint lacked a measure of conviction; was it actually possible to define the derived function and the integral in every case, or even to produce the Taylor-series expansion of a function in the first place, or to manipulate series and functions, without admitting the dreaded limits or infinitesimals? These alternative approaches, particularly the latter, continued to maintain a healthy life; and we shall see in §2.7.3 that in the 1820s Cauchy was to give the former its golden age.

2.2.3 *The many senses of 'analysis'.* One further link between 'logique' and mathematics merits attention here: the use in both fields of the word 'analysis'. We saw it in Condillac's philosophy, and it occurred also in the titles of both of Lagrange's books. In both cases the method of reducing a compound to its constituent parts was involved: however, one should not otherwise emphasise the common factor too strongly, for the word was over-used in both disciplines. Among mathematicians the word carried not only this sense but also the 'analytic' type of proof known to the Greeks, where a result was proved by regressing from it until apparently indubitable principles were found; the converse method, of starting from those principles and deriving the result, was 'synthetic'. Neither type of proof is necessarily analytic or synthetic in the senses of decomposition or composition. Further, during the 17th and 18th centuries 'analytic' proofs were associated with algebra while 'synthetic' ones were linked with geometry (Otte and Panza *1997a*). However, developments in both these branches of mathematics made such associations questionable; for example, precisely around 1800 the subject called 'analytic geometry' began to receive textbook treatment.

Thus the uses of these terms were confusing, and some of the more philosophically sensitive mathematicians were aware of it. One of these was Sylvestre-François Lacroix (1765–1843), disciple of Condorcet and the most eminent textbook writer of his day. In an essay *1799a* written in his mid thirties, he tried to clarify the uses to which these two words should be put in mathematics and to warn against the two associations with branches of mathematics. However, his battle was a losing one, as Joseph-Diez Gergonne (1771–1859) pointed out in a most witty article in his journal *Annales de mathématiques pures et appliquées*; for example, 'an author who wants to draw the regards and the attention of the public to his opus, hardly neglects to write at its head: "Analytical treatise" ' (*1817a*, 369)! His joke was to be fulfilled within a few years, as we see in §2.7.2.

2.2.4 *Two Lagrangian algebras: functional equations and differential operators.* Lagrange did not invent either theory, but each one gained new levels of importance under the algebraic regime which he encouraged, and was to find a link with logic in De Morgan and Boole. On their histories, see respectively Dhombres *1986a* and Koppelman *1971a*; and for both Panteki *1992a*, chs. 2–5.

Functional equations can be explained by an example from Lagrange himself. To find the derived function of x^m for any real value of m he assumed it to be some unknown function $F(x)$ and showed from the assumed expansion

$$(1 + \omega)^m = 1 + \omega F(m) + \cdots \qquad (224.1)$$

that F satisfied the functional equation

$$F(m + n) = F(m) + F(n) = F(m + i) + F(n - i), \text{with } i \text{ real.} \quad (224.2)$$

By assuming the Taylor expansions of F about m and n respectively for the last two terms (and thus bringing the derived functions of F into the story) and then equating coefficients of i, he found that

$$F(m) = am + b, \text{with } a \text{ and } b \text{ constants;} \qquad (224.3)$$

and from the cases $m = 0$ and 1 it turned out that $b = 1$ and $a = 0$. Thus putting in (224.1)

$$\omega := i/x \text{ yielded } (x + i)^m = x^m + imx^{m-1} + \cdots, \qquad (224.4)$$

so that the derived function of x^m was shown to be mx^{m-1} by using only the Taylor expansion and algebraic means (Lagrange *1806a*, lecture 3: see also lectures 4–6).

Differential operators arise when the quotient dy/dx is interpreted not as the ratio $dy \div dx$ (§2.7.1) but as the operator (d/dx) upon y. The result of this operation was also written 'Dy' in order to emphasise the operational feature. In this reading, orders and powers of differentials were identified:

$$d^n y/dx^n = (dy/dx)^n. \qquad (224.5)$$

The most important application was to Taylor's series itself, which now took a form concisely relating D to the forward difference operator Δ:

$$\Delta f(x) := f(x + h) - f(x) = (e^{hD} - 1)f(x), \qquad (224.6)$$

where '1' denoted the identity operator. From results such as this, and summation interpreted as the (algebraically) inverse operator to differencing, Lagrange and others found a mass of general and special results, most of which could be verified (that is, reproved) by orthodox means.

2.2.5 *Autonomy for the new algebras.* However, some people regarded these methods as legitimate in themselves, not requiring foundations from elsewhere: it was permitted to remove the function from (224.6) and work with

$$\Delta = e^{hD} - 1. \qquad (225.1)$$

A prominent author was the mathematician François-Joseph Servois (1767–1847), who wrote an important paper *1814a* in Gergonne's *Annales* on the foundations of both these algebras. Seeking the primary properties that functions and operators did or did not obey, when used either on themselves or on each other, he proposed names for two properties which have remained in use until today. If a function f satisfied the property

$$f(x + y + \cdots) = f(x) + f(y) + \cdots, \qquad (225.2)$$

then f 'will be called *distributive*'; and if f and another function g satisfied the property

$$f(g(x)) = g(f(x)), \qquad (225.3)$$

then they 'will be called *commutative between each other*' (p. 98).

Had Servois been working with axioms—which in contrast to the late 19th century (§4.7), was not a normal procedure at the time—then he would have put forward two axioms for a general algebra. As it is, he knew the importance of the properties involved, and they gradually became diffused (by De Morgan and Boole among others, as we shall see in §2.4.7 and §2.5.2–3). These two algebras are important for reasons beyond their technical details; for they were among the first ones in which the objects studied were not numbers or geometrical magnitudes.[3] This feature was reflected in the practise of several authors to use the word 'characteristic' to refer to the letters of the algebra, not to the functions or operators to which they referred. Lacroix was such an author, and an example is given in his account of Servois's paper, where 'the characteristics [f and g] are subjected only to the sole condition to give the same result' in order to refer to '*commutative functions*' (Lacroix *1819a*, 728).

---

[3] On these and many related developments in post-Lagrangian algebras, see my *1990a*, chs. 3 and 4. Unfortunately, none of the histories of algebra has recognised the importance of these theories for the development of algebra in general.

Associativity had already been stressed by Legendre in connection with number theory, without name; this one is due to W. R. Hamilton.

2.3   SOME ENGLISH ALGEBRAISTS AND LOGICIANS

2.3.1 *A Cambridge revival: the 'Analytical Society', Lacroix, and the professing of algebras.*   While French mathematics was in a state of rapid development after the Revolution, most other countries slept pretty soundly. However, by the 1810s some movements were detectable, partly in reaction to the massive French achievements. Various reforms took place in the countries of the British Isles: we consider here the best known (although not the first of them), namely, the creation of the 'Analytical Society' by a group of undergraduates at Cambridge University in the early 1810s (Enros *1983a*). Its name exemplified the association of analysis with algebra mentioned in §2.2.3. While the Society ran only from 1812 to 1817, its enthusiasm for algebras was continued in the activities of its most prominent members. In particular, Charles Babbage, John Herschel and George Peacock published in 1816 their English translation of the second (1806) edition of Lacroix's textbook on the calculus (Lacroix *1816a*: the large treatise was cited above).

In order to clarify the philosophy of the new English mathematicians, a contrast with Lacroix would be in order here. As was noted in §2.2.3, he was under the strong influence of Condorcet, and thereby back to ency- clopaedistic philosophy. Following their advocacy of plurality of theories and even its classification, Lacroix himself had presented all the three main traditions of the calculus, especially in his large treatise but also in the shorter textbook version. Initially he had shown a strong adherence to Lagrange's position; but over the years he had moved gradually towards a preference for the theory of limits, while still presenting the other ap- proaches. By contrast, the young men at Cambridge voted *unequivocally* and uniquely in favour of Lagrange's approach, and in their editorial preface they even reproached Lacroix for his preference for limits over 'the correct and natural method of Lagrange' (Lacroix *1816a*, iii).

2.3.2 *The advocacy of algebras by Babbage, Herschel and Peacock*

> Since it leads to truth, it must have a logic.
> Robert Woodhouse on complex numbers
> (De Morgan *1866a*, 179; compare *1849b*, 47)

This love of algebra(s) was evident already in their senior (and presum- ably influential) Cambridge figure Robert Woodhouse (1773–1827), who even criticised Lagrange for not being algebraic enough; he wrote an essay *1801a* 'On the necessary truth' obtainable from complex numbers in exactly the spirit of the quotation above, which De Morgan seems to have recalled from his student days in the 1820s. The reliance upon algebra had prevailed with Babbage and Peacock in the 1810s, and continue in various forms in England throughout the century.[4] Indeed, Babbage and Herschel had already begun to produce such research while members of the Analytical Society, and they published several papers over a decade. Functional equations (then called 'the calculus of functions') was the main concern, together with related types such as difference equations: their formation and solution (partial and general), the determination of inverse functions, the calculation of coefficients in power-series expansions, applications to various branches of mathematics, and so on. The methods were algorithmic, rather wildly deployed with little concern over conditions for their legitimacy.[5]

The influence of French mathematics was quite clear, and various works, even earlier than Lagrange's writings, were cited. In return, Gergonne *1821a* wrote a summary of some of Babbage's results in his *Annales*. However, the philosophy of 'logique' did not enjoy the same influence: even in a paper 'On the influence of signs in mathematical reasoning' Babbage *1827a* only cited in passing (although in praise) one of the French semiotic texts, and otherwise set 'logique' aside. English logic was to gained inspiration from other sources, as we shall see in §2.4–5. First, however, another aspect of English algebra calls for attention.

While his friends were rapidly producing their research mathematics, Peacock was much occupied with the reform of mathematics teaching at Cambridge University. But in the early 1830s he produced a textbook *1830a* on the principles of algebra, which gave definitive expression to the philosophical position underlying the English ambitions for algebra. He recapitulated some of these ideas in a long report on mathematical analysis (*1834a*, 188–207).

A principal question was the status of negative numbers, and of the common algebra with which arithmetic was associated; complex numbers fell under a comparable spotlight (Nagel *1935b*). English mathematicians (and also some French ones) had long been concerned with questions such as the definability of $(a - b)$ when $a < b$. Peacock's solution was to distinguish between 'universal arithmetic' (otherwise known as 'arithmetical algebra') in which subtraction was defined only if $a > b$, and 'symbolical algebra', where no restrictions were imposed. The generalisation from the first to the second type of algebra was to be achieved via 'the principle of the permanence of equivalent forms', according to which '*Whatever form is Algebraically equivalent to another, when expressed in general symbols, must be true, whatever these symbols denote*' (Peacock *1830a*, 104; on p. 105 the

---

[4] On the algebras to be discussed here, see especially Nagel *1935b*, Joan Richards *1980a* and Pycior *1981a*. There were other interests in English mathematics, in which algebras were not necessarily marked: for example, the philosophy of geometry (Joan Richards *1988a*). On Cambridge mathematics in general in the early 19th century, see Becher *1980a* and my *1985a*.

[5] This algorithmic character is a common factor between Babbage's mathematics and his later work in computing (my *1992b*). On his work on algebra see Panteki *1992a*, ch. 2.

principle was mistakenly called *'algebraical forms'*). This hardly limpid language states that a form such as, say $a^n a^m = a^{n+m}$ in the first type of algebra for positive values n and m maintained its truth when interpreted in the broader canvas of the second type, which seems to be a version of the marks-on-paper algebra later known as 'formalism' often but mistakenly associated with Hilbert's proof theory (§4.7).

While the same laws applied in each algebra, this change in generality led to some change in emphasis: arithmetical algebra stressed the legitimacy of signs, while symbolical algebra gave precedence to the operations under which the elements of the algebra were combined. From this point of view Peacock was moving towards the modern conception of abstract algebras which were distinguished from their interpretations; but instead of adopting axioms he stressed the (supposed) *truths* of the theorems (validly) derivable in symbolical algebra.

In this way Peacock's philosophy of algebra involved an issue pertaining to logic, although its links to logic were developed by others. He did not take much interest in recent or contemporary developments in logic at his time. For example, he did not relate his symbolical algebra to Condillac's standpoint, where negative numbers were granted full status within the number realm on the grounds of an analogy with negation (*1798a*, 278–288), or his universal arithmetic to Lazare Carnot's opposition to negative numbers for their alleged non-interpretability in geometrical terms (*1803a*, 7–11). Neither did he react to the rather sudden revival of interest in logic in England in the 1820s, to which we now turn.

2.3.3 *An Oxford movement: Whately and the professing of logic.* While Cambridge began to come alive in mathematics during the 1810s and 1820s, Oxford executed a reform of the teaching of humanities students by introducing a course in logic. The study of logic in Britain was then in a peculiar state. The classical tradition, based upon inference in syllogistic logic, was still in place. But for a long time an alternative tradition had been developing, inspired by Locke and continued in some ways by the Scottish Common-Sense philosophers of the late 18th century. Critical of syllogistic logic, especially for the narrow concern with inference, its adherents sought a broader foundation for logic in the facultative capacity of reasoning in man, and included topics such as truth and induction which we might now assign to the philosophy of science. Showing more sympathy to the role of language in logic than had normally been advocated by the syllogists, they laid emphasis on signs as keys to logical knowledge (Buickerood *1985a*). At the cost of some simplification, this approach will be called 'the sign tradition'.

As has been noted, French 'logique' did not enjoy much British following; further, Kantian and Hegelian philosophies were only just starting to gain ground, and in any case logic as such was not very prominent in these traditions. Again, although the contributions of Leibniz had gained some attention in Germany (Peckhaus *1997a*, ch. 4), the news had not been received in Britain to any significant extent.

The leading figure in this reform was Richard Whately (1787–1863), who graduated at Oxford in 1808 in classics and mathematics and took a college Fellowship for a few years before receiving a rectorship in Suffolk. While there he wrote articles on logic and on rhetoric for the *Encyclopaedia metropolitana*, a grandiose survey of the humanities and the sciences conceived by the poet Samuel Taylor Coleridge. Several of the articles that appeared over the years until its completion in 1845 were of major importance; but none matched Whately's in popularity, especially the logic article, which first appeared in the encyclopaedia as his *1823a* and then, in a somewhat extended form, as a book in 1826 (Whately *Logic*$_1$). The year before he had moved back to Oxford; he left Oxford in 1831 to become Archbishop of Dublin, where he remained for the rest of his life.

The impact of the book both encouraged the Oxford reform and helped to stimulate it. Commentaries and discussions by other authors rapidly began to appear. Whately put out revised editions every year or so for the next decade (and also later ones), and many further ones appeared in Britain and the U.S.A. until the early 20th century. From its first edition of 1826 it carried the sub-title 'Comprising the substance of the article in the Encyclopaedia metropolitana: with additions, etc.'. Its first three Books comprised an introduction and five chapters, and a fourth Book presented a separate 'Dissertation on the province of reasoning', with its own five chapters. The 'additions etc.' mainly constituted an 'Appendix' of two items; and from the third edition of 1829 there was a third item and a new supplement to the chapter 'On the operations of the mind'. Later, the structure of the book was altered to four Books and the Appendix. Comparison of the first and the ninth (1848) editions shows that the changes of phrasing and small-scale structure throughout the work, and the additions, sometimes substantial, are far too numerous to record here. Instead I cite by page number the first edition, of which a photographic reprint appeared in 1988 under the editorial care of Paola Dessì. Further, I do not explore the influence upon Whately of the theologian Edward Copleston: according to Whately's dedication of the volume, it seems to have been quite considerable.

The great popularity of Whately's book is rather strange, as at first glance his treatment seems to be rather traditional: indeed, its original appearance in the *Encyclopaedia metropolitana* gained so little attention that even the date of its publication there became forgotten. He began the main text of the book by repeating the line about logic as 'the Science, and also as the Art, of Reasoning' (p. 1), and in the technical exegesis he stressed that logic should be reduced to its syllogistic forms. However, there were passages on religious questions which doubtless caused some of the attention (several of the extensive revisions mentioned above were also in these areas), and in other respects he put forward new views which were to be taken up by his successors (Van Evra *1984a*). For example, contrary to the normal tradition in England, he claimed that 'logic is *entirely conversant about language*' (p. 56: interestingly, in a footnote). His definition of a syllogism was formulated thus: 'since Logic is wholly concerned in the use of language, it follows that a Syllogism (which is an argument stated in a regular logical form) must be an "argument so expressed, that the conclusiveness of it is manifest from *the mere form of the expression*," *i.e.* without considering the *meaning of the terms*' (p. 88).[6] (The various French traditions sympathetic to this view were not mentioned in the historical sketch given in his introduction.) Again, in his analogy between logic and science he compared it with sciences such as chemistry and mechanics, and sought for it foundational principles and autonomy such as they enjoyed.

Among these sciences Whately claimed 'a striking analogy' between logic and arithmetic. Just as 'Numbers (which are the subject of arithmetic) must be numbers of *some things*', so 'Logic pronounce[s] on the validity of a regularly-constructed argument, equally well, though arbitrary symbols may have been substituted for the terms' (pp. 13, 14). However, he did not press the analogy with mathematics any further than this, and he did not introduce any mathematical techniques in his presentation (or indulge in any sophisticated assessment of sets or collections of things). Although 'Mathematical Discoveries [. . .] must *always* be of the description to which we have given the name of "Logical Discoveries" [. . .] It is not, however, meant to be implied, that *Mathematical* Discoveries are effected by pure Reasoning, and by that *singly*' (pp. 238–239). Similarly, in the reform at Oxford the logic course was offered as an *alternative* to one on Euclid; despite giving his book such a Euclidean title as 'Elements', he did not anticipate the insight to be made later that Euclid himself could be put under logical scrutiny (§2.4.3, §4.7.2).

One point of difference for Whately between logic and mathematics lay in the theory of truth. 'TRUTH, in the strict logical sense, applies to propositions, and to nothing else; and consists in the conformity of the declaration made to the actual state of the case' (p. 301); by contrast with this (correspondence) theory, 'Mathematical propositions are not properly true or false in the same sense as any proposition respecting real fact is so

---

[6] In this quotation I have put 'form' for Whately's word 'force', which seems to be a misprint although it appears in every edition that I have seen, including the original encyclopaedia appearance (*1823a*, 209). Boole made the same change when paraphrasing this passage in a manuscript of 1856 (*Manuscripts*, 109). De Morgan was to take the word 'force' to refer intensionally to a term (for example, *1858a*, 105–106, 129–130). See also footnote 21 on Jevons.

Whately was also well known in his lifetime for a wry and witty commentary on observation and testimony entitled *Historic doubts relative to Napoleon Buonaparte* (1819, and numerous later editions).

called; and hence the truth (such as it is) of such propositions is necessary and eternal' (the rather woolly p. 221).

We turn now to an important successor of Whately. However, he came to logic largely by other routes.

2.4 A LONDON PIONEER: DE MORGAN ON ALGEBRAS AND LOGIC

2.4.1 *Summary of his life.*   Born in 1806 in India, Augustus De Morgan studied at Cambridge University in the early 1820s, and was one of the first important undergraduates to profit from the renaissance of mathematics there. However, as a 'Christian unattached' (as he described himself) he could not take a position, and so in 1828 he became founder Professor of Mathematics at London University, then newly founded as a secular institution of higher learning (Rice *1997a*). Resigning in 1831, he resumed his chair in 1836, at which time the institution was renamed 'University College London' after the founding of King's College London in 1829, and the 'University of London' was created as the body for examining and conferring degrees. He resigned again in 1866, over the issue of religious freedom for staff, and died five years later.

De Morgan was prolific from his early twenties; his research interests lay mainly in algebras, logic and aspects of mathematical analysis, but he also wrote extensively on the history and philosophy of mathematics and on mathematical education. This section is devoted, in turn, to his views an algebra, his contributions to logic, and relationships between logic and mathematics.

2.4.2 *De Morgan's philosophies of algebra.*   De Morgan's views on the foundations of algebra vacillated over the years, and are hard to summarise.[7] In his first writings on the subject, including an early educational book *On the study and difficulties of mathematics*, he adopted a rather empirical position, in that algebraic theories were true and based upon clear principles; negative numbers were to be explained (away) by rephrasing the results in which they appeared or justified by the truths of the conclusions drawn from the reasonings in which they were employed (*1831a*, esp. ch. 9). But, like most English mathematicians of his time, he was influenced by Peacock's work on the foundations of algebra (§2.3.2). In a long review of Peacock's treatise he showed more sympathy than hitherto to the abstract and symbolic interpretation, allowing algebra to be 'a science of investigation without any rules except those under which we

---

[7] See Joan Richards *1980a* and Pycior *1983a*. I do not treat the influence upon De Morgan of the work of the Irish mathematician William Rowan Hamilton, or of the philosophical writings of Herschel and William Whewell.

may please to lay ourselves for the sake of attaining any desirable object' (De Morgan *1835a*, 99).

However, in the same passage and elsewhere in the review De Morgan referred to truth in the context of algebra, an imperative which informed all of his further thoughts on the subject. Thus he did not try to formulate the modern abstract position based upon axioms; for these axioms would have a status corresponding to hypotheses in science. With regard to Peacock, for example, while De Morgan also advocated the generality of algebra he did not wish to have recourse to the principle of the permanence of equivalent forms but relied upon truth and the interpretation of the symbols and of the theories of which they were components.

Instead, De Morgan used other language, which was also found in connection with logic: the distinction between 'algebra as an art', where it functioned merely as a symbolism, and 'algebra as a science', where the interpretation of the system was of prime concern. Interestingly, in the first of a series of articles 'on the foundations of algebra' he called the art a 'technical algebra' and the science a 'logical algebra, which investigates the method of giving meaning to the primary symbols, and of interpreting all subsequent symbolic results' (*1842b*, 173–174): although he soon confessed that 'logical' was a 'very bad' term (p. 177), there were certain links with logic which will be noted in §2.4.4. When in the next article he stated that x and $\div$ were 'distributive' over $+$ and $-$ (*1849a*, 288, with a reference to Servois), he did not grant these laws axiomatic status in a sense which we would recognise; and he did not even mention the instances of commutativity in the system.

In the first paper De Morgan noted some analogies which held between the common algebras and functional equations (*1842b*, 179). He could speak with authority, as a few years earlier he had written the first systematic account of this young algebra, as a long article *1836a* on 'the calculus of functions' published in the *Encyclopaedia metropolitana*. The presentation was technical more than philosophical, concerned with solutions to the equations (for one and for several independent variables), the inverse function, and so on; but this topic was to bear upon one of his main contributions to logic, as we shall see in §2.4.7.[8]

2.4.3 *De Morgan's logical career.* De Morgan was well aware of the changes taking place at Oxford: one of his early educational writings was a survey *1832a* of 'the state of the mathematical and physical sciences' there. He was partly inspired by Whately's book to take up logic, but his initial motivation was one which Whately had set aside: the logic involved in

---

[8] De Morgan's article is merely noted in Dhombres's extensive study *1986a* of the history of functional equations, because equations of functions in one variable are largely omitted; however, it has an extensive section on functions of two variables (De Morgan *1836a*, 372–391).

Euclidean geometry. The volume on 'studies and difficulties' contained a chapter 'On geometrical reasoning', in which he laid out the valid syllogistic forms, using 'O', '□' and '△' for the terms, and outlined the syllogistic form of Pythagoras's theorem (*1831a*, ch. 14). For background acquaintance with logic he cited there a passage from Whately's book, in its third edition of 1829, as 'a work which should be read by all mathematical students'.[9]

De Morgan again advocated studying the logic of geometry in another educational article *1833a*, and he took his own advice in a pamphlet *1839a* on the 'First notions of geometry (preparatory to the study of geometry)'. Here he laid out the logic which, as he stated in the preamble, 'he found, from experience, to be much wanted by students who are commencing with Euclid'; however, he did not then apply this logic to the ancient text. Most of the pamphlet was reprinted with little change as the first chapter of his main book on the subject, *Formal logic* (*1847a*). By then he had launched his principal researches, which appeared as a series of five papers 'On the syllogism' published between 1846 and 1862 in the *Proceedings of the Cambridge Philosophical Society*. There were some articles and book reviews elsewhere, especially a short book *1860a* proposing a 'syllabus' for logic; the total corpus is quite large.[10]

2.4.4 *De Morgan's contributions to the foundations of logic*

The law is good if one makes legal use of it.

De Morgan, motto (in Greek) on the
title page of *Formal logic*

De Morgan was not a clear-thinking philosopher, and his views are scattered in different places: also, they changed somewhat over time, although he did not always seem to be aware of the fact (different definitions of a term given in different places, for example). He worked largely within the syllogistic tradition, but he was much more aware than his contemporaries of its limitations, and extended both its range and scope: the preface of *Formal logic* began with the statement that 'The system given in this work extends beyond that commonly received, in

---

[9] De Morgan *1831a*, 212. Why, then, did he write on one of the front pages of his copy of this edition of Whately's book: 'This is all I had seen of Whately's logic up to Aug. 7, 1850'? Like his whole library, the copy is held in the University of London Library.

[10] Most of De Morgan's five papers on logic, together with the summary of an unpublished sixth paper and some other writings, are conveniently collected in De Morgan *Logic* (1966), edited with a good introduction by Peter Heath; its page numbers are used here. See also his correspondence with Boole, edited by G. C. Smith (Boole–De Morgan *Letters*); but note the cautions on the edition expressed in Corcoran *1986a*. Merrill *1990a* is a survey of his logic, especially that of relations; but for the connections with mathematics, largely missing (ch. 7), see Panteki *1992a*, ch. 6.

several directions'. We shall note some principal extensions in the next three sub-sections.

As with all logicians of his time and long after, De Morgan did not systematically distinguish logic from metalogic. The long chapter 'On fallacies' of his book made almost every other distinction but this one (see pp. 242–243 for some tantalising cases); and a particularly striking later example is his assertion that 'a syllogism *is* a proposition; for it affirms that a certain proposition is the necessary consequence of certain others' (*1860b*, 318).

De Morgan offered views on the character of logical knowledge in general; and we shall take his use of 'necessary' quoted just now to start with this theme. In the opening chapter of his book, on 'First notions', he stressed that logic was exclusively concerned with valid inference; truth was a secondary concept, dependent 'upon the *structure of the sentence*' (*1847a*, 1). The more formal treatment began in the second chapter with a specification of logic as 'the branch of inquiry (be it called science or not), in which the act of the mind in reasoning is considered, particularly with reference to the connection of thought and language' (p. 26).

Many of the forms of inference which De Morgan then investigated were dependent upon language; in particular, scientific induction, where he drew on probability theory to justify universal propositions rather than inference from particulars to particulars (chs. 9–11). The sub-title of his book is worth noting here: 'or, the calculus of inference, necessary and probable'—not the 'necessary and possible' of modern modal logic. There is also a link between logic and quantity, if the connection with probability is held to be that logic deals with the quantities 0 and 1. We shall meet the notion of quantity later in his work at §§2.4.6–7.

However, De Morgan did not wish to dwell upon 'the science of the mind, usually called *metaphysics*' (p. 27):

> I would not dissuade a student from a metaphysical inquiry; on the contrary, I would rather endeavour to promote the desire of entering upon such subjects; but I would warn him, when he tries to look down his own throat with a candle in his hand, to take care that he does not set his head on fire.

De Morgan's title 'Formal logic' may show influence from a recent *Outline of the laws of thought*, anonymously published by the Oxford scholar William Thomson (1819–1890); for he defined 'logic to be the science of the necessary laws of thinking, or, in more obscure phrase, a science of the *form* of thought' (*1842a*, 7) and then examined in detail the various forms that the notion of form could take. However, De Morgan did not handle too well the distinction between form and matter; Mansel *1851a* was to point this out in a thoughtful review of the book and of the second edition (1849) of Thomson (Merrill *1990a*, ch. 4). In his papers on the syllogism De Morgan somewhat changed his position on the nature of logic, or at least on his manner of expressing it. 'Logic inquires into the form of thought, as separable from and independent of the matter thought on', he opined in the third paper (*1858*, 75), in a manner reminiscent of his distinctions in algebra. However, later he claimed that mathematics has never 'wanted a palpable separation of form and matter' (p. 77); so now logic 'must be [...] an unexclusive reflex of thought, and not merely an arbitrary selection,—a series of elegant extracts,—out of the forms of thinking' (pp. 78–79). This is a kind of completeness assertion for logic: all aspects of thinking and inference should be brought out.

In revising the distinction between form and matter, De Morgan cast the copula in a very general form. In the second paper he recalled that in his book 'I followed the hint given by algebra, and separated the essential from the accidental characteristics of the copula' (*1850a*, 50, referring to *1847a*, ch. 3). The essentials led to the 'abstract copula', 'a formal mode joining two terms which carries no meaning' (p. 51). He laid down three laws that it should satisfy, giving them symbolic forms:

1) 'transitiveness' between terms X, Y and Z, 'symbolized in

$$X—Y—Z = X—Z', \qquad (244.1)$$

where '—' was 'the abstract copular symbol' and ' = ' was informally adopted as an equivalence relation between terms or propositions;

2) 'convertibility' between X and Y (which we would call 'commutativity': as we saw in §2.4.2, he did not use Servois's adjective); and

3) a completeness (meta)property of bivalent logic which he called 'contrariety: in $X—Y$ and [its negation] $X\text{--}Y$ it is supposed that one or the other must be' (p. 51). Since reflexivity ($X—X$) was taken for granted, he had in effect defined the abstract copula as an equivalence relation; but his sensitivity to relations and the state of algebra of his day did not allow him to take this step (that is, to see its significance). However, in effecting his abstraction and specifying the main pertaining properties he may well have recalled the abstraction applied to functions in forming functional equations.

2.4.5 *Beyond the syllogism*

These remarks [...] caution the reader against too ardent an admiration of the syllogistic mode of reasoning, as if it were fitted to render him a comprehensive and candid reader. The whole history of literature furnishes incontestable evidence of the insufficiency of the Aristotelian logic to produce, of itself, either acuteness of mind, or logical dexterity.

Blakey *1847a*, 162

In his book, which appeared in the same year as Blakey's caution (in an essay on logic), De Morgan pointed to some forms of inference which lay outside the syllogistic ambit. 'For example', a well-remembered one, ' "man is animal, therefore the head of a man is the head of an animal" is an inference, but not syllogistic. And it is not mere substitution of identity' (*1847a*, 114). To cover such cases he offered the additional rule 'For every term used universally *less* may be substituted, and for every term used particularly, *more*' (p. 115). While his treatment was not fully satisfactory, his modifications can be cast in a sound form (Sanchez Valeria *1997a*).

De Morgan also noted the case 'X)P + X)Q = X)PQ', which in his notation (p. 60) stated that if every X was both P and Q, then it was also 'the compound name' P and Q, and which 'is not a syllogism, nor even an inference, but only the assertion of our right to use at our pleasure either one of two ways of saying the same thing instead of the other' (*1847a*, 117). This remark occurred in a section in which he tried to formulate syllogistic logic in terms of 'names': that is, terms and the corresponding classes (the rather unclear pp. 115–126).

Partly in the context of this extension, De Morgan discussed at some length in ch. 7 limitations of the Aristotelian tradition. For example, on existence he noted the assumption that terms be non-empty, and criticised the medieval 'dictum de omni et nullo', where in universal affirmative propositions 'All Xs are Ys' all objects satisfying X must also satisfy Y, and in universal negative propositions 'No Xs are Ys' no object satisfying X may also satisfy Y.

These ideas show that De Morgan tried to push out the province of logic beyond syllogisms. In the next two sub-sections I note his two main extensions of its methods, and even of its province.

2.4.6 *Contretemps over 'the quantification of the predicate'.* This phrase referred to the cases in which the middle term of a syllogism was made susceptible to 'all' and 'some'. Thus, in addition to 'all Xs are Ys' and the other standard forms, there were admitted also the octet of new forms 'All/Some X is all/some/not any Y' (where 'some' must exclude the case 'all'), and the repertoire of valid and invalid syllogistic forms was greatly increased. The extended theory uses the word 'quantification' in the way to which we are now accustomed; and, while we shall see in §4.3.7 that that use has closer origins in Peirce's circle, the content here is similar.

The name was introduced by the Scottish philosopher William Hamilton (1788–1856). A student at Oxford University during the same period as Whately, Hamilton passed his career in his native country of Scotland, for many years at the University of Edinburgh. He seems to have introduced his new theory around 1840, and developed it in his teaching. De Morgan came across a similar form of the theory in 1846, in which he considered propositions of forms such as 'Most/Some of the Ys are Zs', and he described it in the first paper on the syllogism (*1846a*, 8–10). In an addition to this paper he discussed them in more detail, taking the collections associated with the quantified predicates to be of known sizes, as in 'Each one of 50 Xs is one or other of 70 Ys' (pp. 17–21). In his book he called these syllogisms 'numerically definite' and extended the notion further, in that he specified only numerical lower bounds of subjects possessing the predicated properties ('m or more Xs are Ys'). In his book he found the numbers associated with the predicates involved in the conclusion of valid syllogisms (*1847a*, ch. 8). His second paper contained a treatment of these forms of proposition different from Hamilton's in exhibiting an algebraist's concern for symmetries of structure between a form and its contrary forms (*1850a*, 38–42: see also the fifth paper *1862a*).

Hamilton responded to De Morgan's basic idea of quantification with accusations of plagiarism, and a row began which continued for the remaining decade of Hamilton's life (Laita *1979a*). De Morgan claimed, doubtless with justice, that his invention was independent of Hamilton (see especially the appendix to his book); and in fact priority for the innovation belongs to neither of the two contestants but to the botanist George Bentham (1800–1884), in a book on logic which was an extended commentary upon the first edition of Whately's book. Bentham had outlined his treatment of propositions, and then applied it to the analysis of some of the traditional forms of valid syllogism, stressing quite explicitly that his approach was superior to the normal classification (*1827a*, esp. pp. 130–136, 150–161). George was the nephew of Jeremy Bentham, and indeed acknowledged the influence of some manuscripts of his uncle; so maybe the idea goes back further![11]

Now in 1833 Hamilton harshly reviewed the third edition of Whately's *Logic* for the stress on language among other things; perhaps they had suffered poor relations at Oxford University. (He claimed that logic was better taught in Scotland than in England.) He also noted here several other books (Hamilton *1833a*, 199–200), and one of them was Bentham's. So he can be fairly accused at least of cryptomnesia (forgotten and maybe unnoticed access). Bentham's book sold very poorly, his publisher going bankrupt soon after its launch; he himself was presumably too deeply involved in botany to complain, and nobody noticed his work until 1850, when attention was drawn in *The Athenaeum* (Warlow *1850a*). Even such a bibliophile as De Morgan did not come across Bentham's book until his *1858a*, 140 where Warlow and an ensuing discussion were cited.[12]

These extensions of the syllogism need careful exposition (which Hamilton did not provide), for the relationship between the eight cases needs

---

[11] Previously George Bentham had published a short treatise *1823a* in French, exposing a classification of 'art-and-science' based upon Jeremy's philosophy of science. However, he explicitly set aside French *logique* in the preface of his *1827a*. For an advocacy of his originality, see Jevons *1873a*; and on predecessors to Bentham, see Venn *1881a*, 8–9.

[12] There is no copy of Bentham's book in De Morgan's personal library (on which see footnote 9). On them and Hamilton see Liard *1878a*, chs. 3–4.

careful examination since they are not all independent. In fact, there is little in the theory beyond the 'Gergonne relations', which Gergonne *1816a* had presented in a paper in the same volume of his *Annales* as his paper on 'analysis' cited in §2.2.5, in order to clarify the (intuitive) use of Euler diagrams (my *1977a*). The paper exercised little influence, the *Annales* gaining only a small circulation even in France:[13] De Morgan was one of the first to cite this paper, in his first reply to Hamilton cited above (*1847a*, 324), although he did not appreciate the significance of Gergonne's classification. However, he made other useful extensions to syllogistic logic in his *Syllabus* by adding to the list of categorical propositions forms such as 'Every X is Y', 'everything is either X or Y', 'some things are neither Xs nor Ys', and so on (most clearly in *1860b*, 190–199, with exotic names).

A related extension was presented in the appendix of De Morgan's fourth paper, which treated 'syllogisms of transposed quantity'. Here 'the whole quantity of one concluding term, or of its contrary, is applied in a premise to the other concluding term, or to its contrary', as in 'Some Xs are not Ys; *for every X* there is a Y which is Z: from which it follows, to those who can see it, that *some Zs* (the *some* of the first premise) are not Xs' (*1860a*, 242–246; he referred to his earlier (and briefer) mentions of this type of syllogism). The most interesting feature, which Peirce was to grasp (§4.3.6) but seemed to elude De Morgan himself, is that it is valid *only* for predicates satisfied by *finite* classes.

The episode of the quantification of the predicate may not seem now to be of great importance. However, at the time it brought publicity to logic; in particular, it stimulated Boole into print on the subject, as we shall see in §2.5.3.

2.4.7 *The logic of two-place relations, 1860.* (Merrill *1990a*, chs. 5–6) It is a curious feature of the history of philosophy that, while there had been awareness since Greek antiquity of roles for relations (Weinberg *1965a*, ch. 2), nobody seems to have taken seriously the fact that relational propositions, such as 'John is taller than Jeremy', cannot fall within the compass of syllogistic logic. De Morgan opened up this part of logic in arguably his most important contribution.

De Morgan touched upon relations from time to time. He contributed an article *1841a* on 'Relation (mathematical)' to the *Penny cyclopaedia*, restricting himself to cases in arithmetic and algebra though including the

---

[13] The extent of Gergonne's influence on mathematicians and logicians seems to have been far less than his philosophical writings merited. For another example, he published a perceptive article *1818a* on forms of definition which gained little recognition. However, the young J. S. Mill took a course with Gergonne in 1820 at the *Faculté des Sciences* of the *Université Royale de France* at Montpellier, and might have heard some of the same material.

operator form (225.1).[14] In his book he recalled the uses of the term in older writers on logic (*1847a*, 229). Within logic, we saw his abstract copula in §2.4.4, and will note his part-whole theory of class inclusion in §2.4.9, both of which embodied relations; and he even used the notion of a relation as a predicate in orthodox syllogistic logic, when pointing out that

> If I can see that
> > Every *X* has a relation to some *Y*
> > and Every *Y* has a relation to some *Z*,
> it follows that every *X* has a compound relation to some *Z*

(*1850a*, 55). Again, properties of the product of functions, akin to properties such as (224.3) in functional equations, were included in his discussion of the abstract copula in the context of relations and their compounding (p. 56):

> The algebraic equation $y = \phi x$ has the copula $=$, relatively to y and ϕx: but relatively to y and x the copula is $= \phi$. [...]. The deduction of $y = \phi \psi z$ from $y = \phi x$, $x = \psi z$ is the formation of the composite copula $= \phi \psi$. And thus may be seen the analogy by which the instrumental part of inference may be described as *the elimination of a term by composition of relations*.

He also commented on relations elsewhere; for example, whole and part, 'with its concomitants, I call *onymatic relations*' (*1858a*, 96). Indeed, relations were even granted priority over classes: 'When two objects, qualities, classes, or attributes, viewed together by the mind, are seen under some connexion, that connexion is called a *relation*' (p. 119).

However, not until the late 1850s, his own mid fifties, did De Morgan study the logic of relations, in his fourth paper *1860a* on the syllogism. Beginning by referring to the above two quotations as instances of the '*composition of relations*', he then treated relations (but only between two terms) in general. The paper is a ramble even by his standards, but there are two key passages.

'Just as in ordinary logic *existence* is implicitly predicated for all the terms' (p. 220), so relations were taken here to be likewise endowed; however, for some reason De Morgan did not mention appropriate universes of discourse. Symbolised by '*L*', '*M*' and '*N*', the corresponding lower case letters denoted the contraries; and periods were used to

---

[14] All the articles in this encyclopaedia were unsigned; but the British Library contains a copy with all the authors named in the margin, and De Morgan's name is given here. His widow's biography *1882a* includes a list of his (many) contributions, drawing also on his own copy; I have not traced it, but I share her doubts about the attribution to him of 'Syllogism' (in *23* (1842), 437–440). A more likely author is J. Long, the chief editor of the encyclopaedia; he wrote the general article on logic, which is interestingly entitled 'Organon' (*17* (1840), 2–11: De Morgan's pamphlet *1839a* is praised on p. 7 as a study of 'a purely formal logic').

distinguish a relation from its contrary; thus, for example,

$$X .. LY \text{ (or } X.lY \text{) and } X.LY \qquad (247.1)$$

respectively expressed that X was/was not 'some one of the objects of thought which stand to Y in the relation L' (p. 220). Compound relations were indicated by the concatenation 'LM', and quantification over relations by primes such that LM' 'signify an L of *every M*' and L, M 'an L of none but Ms'. The converse to L was written 'L^{-1}', or 'L^v' for 'Those who dislike the mathematical symbol';[15] further, '$L^{-1}X$ may be read "*L-verse of X*"' (p. 222). He also proved that 'if a compound relation be contained in another relation, [...] the same may be said when either component is converted, and the contrary of the other component and of the compound change places' (p. 224), a result of significance:

$$\text{if } LM))N, \text{ then } L^{-1}n))m, \text{ and } nM^{-1}l. \qquad (247.2)$$

Next De Morgan mentioned some main desirable properties of a relation, similar to those for the abstract copula (although he made no use of ' = ' in the paper). One was convertibility, 'when it is its own inverse', and where 'So far as I can see, every convertible relation can be reduced to the form LL^{-1}' (p. 225). With transitivity, 'when a relative of a relative is a relative of the same kind', 'L signifies *ancestor* and L^{-1} *descendant*', and he mentioned a 'chain of successive relatives, whether the relation be transitive or not', like the sequence of functional operations $\phi^n x$ for positive and negative integers n (p. 227).

De Morgan now applied this apparatus to syllogistic logic, with little concern for the extensions discussed in the previous sub-sections. All three propositions of a syllogism were cast in relational form and the various valid figures laid out (pp. 227–237). He mentioned in passing the syllogisms expressible in terms of onymatic relations, and did not (trouble to) present the pertaining numerically definite syllogisms; and his paper faded away in its final pages (the appendix dealt with the 'syllogisms of transposed quantity' noted in §2.4.6). But he noted in places the generality of his new concern: for example, that 'quantification itself only expresses a relation' between the quantified predicates (p. 234); or that 'The whole system of relations of quantity remains undisturbed if for the common copula "*is*" be substituted any other relation' (p. 235), so that some structure-similarity obtained between the calculi of relations and of classes.

---

[15] This type of notation for inverse functions had been introduced by Herschel in the 1820s, in connection with his work on functional equations (§2.3.1).

2.4.8 *Analogies between logic and mathematics*

But, as now we *invent algebras* by abstracting the forms and laws of operation, and fitting new meanings to them, so we have power to invent new meanings for all the forms of inference, in every way in which we have power to make meanings of *is* and *is not* which satisfy the above conditions. De Morgan *1847a*, 51

It is clear that De Morgan drew upon a number of similarities between logic and algebra: however, in one respect logic had to remain more fundamental. For even in the most abstract approach to algebra one is constrained by the need for the axioms to form a consistent system; but then a logical notion is underlying the algebra. He recognised this point in connection with the distinction between the 'form' and the 'matter' of an argument when he stated that 'logic deals with the pure form of thought, divested of every possible distinction of matter', including those pertaining to algebra and arithmetic (*1860c*, 248–249; see also *1858a*, 82).

However, De Morgan also pointed out many analogies between logic and algebra, and to a lesser extent with arithmetic. The quotation above belongs to the discussion of the abstract copula just described. Among other examples, he claimed (incorrectly) that elimination between algebraic equations functioned like inference in logic (*1850a*, 27). Similarities of property were sometimes reflected in the use of the same symbol. For example, he expressed the disjunction of propositions 'by writing $+$ between their letters' (*1847a*, 67: unexplained in *1846a*, 11). Again, for 'the convertible propositions' 'no P is Q' and 'some Ps are Qs' involving two terms P and Q he chose 'the symbols $P.Q$ and PQ', which the algebraist is accustomed to consider as identical with $Q.P$ and QP' (*1846a*, 4: no such point made at *1847a*, 60). Indeed, as we saw around (247.1–2), he used algebra-like notations deploying ' $=$ ', ' $-$ ' and/or brackets of various kinds to distinguish and classify types of proposition and valid forms of syllogistic inference (see, for example, *1850a*, 37–41). The procedures included rules for rewriting terms $P, \ldots$ in terms of their contraries $p, \ldots$; for example, 'All P are not q', symbolised '$P))q$', was convertible *salva veritate* to 'No P are Q', symbolised '$P).(Q$'. As a result no real distinction remained between subject and predicate from the symbolic point of view. The account in his *Syllabus* even included a 'zodiac' circle of 12 bracket-dot notations for valid syllogisms grouped in threes by logical opposition and placed at the corners of equilateral triangles (*1860b*, 163). His status in the history of semiotics should be raised.

Some of these collections of notations displayed duality properties, although De Morgan did not emphasise the feature. However, in using the symbol 'x' to represent the contrary term of a term X he deployed a symmetry of roles for X and x, and combinations of them using the dots and brackets of (247.1), which was rather akin to duality (see, for example, *1846a*, arts. 1–2 for the definition and an initial deployment).

Although De Morgan once opined that 'It is to *algebra* that we must look for the most habitual use of logical forms' (*1860a*, 241), he did not restrict himself to similarities with algebra and with arithmetic, but tried to encompass mathematics as a whole. Indeed, he introduced the expression 'mathematical logic' in his third paper on the syllogism, as 'a logic [which] will grow up among the mathematicians, distinguished from the logic of the logicians by having the mathematical element properly subordinated to the rest' (*1858*, 78). Of course he was referring to mathematical presence in general, not the specific doctrine of mathematical logic which will be the subject of several later chapters. However, he did use the word 'mathematical' in general contexts in his logic, often in connection with his discussion of collections, which we now note.

2.4.9 *De Morgan's theory of collections.*   If an algebra admits 'some' or 'all' into its brief, then stuff of some kind enters its concerns, be it of terms, individuals, properties or whatever; and it will form itself into collections, with associated properties of inclusion. Like all the logicians covered in this chapter, collections of things were handled by De Morgan part-whole (§1.2.2), *not* with the set theory to come from Georg Cantor (§3.2).

In his first paper on the syllogism, De Morgan soon stressed an important idea: 'Writers on logic, it is true, do not find elbow-room enough in anything less than the universe of possible conceptions: but the universe of a particular assertion or argument may be limited in any matter expressed or understood' (*1846a*, 2). Throughout these papers, and to a lesser extent in his book (*1847a*, 110, 149), he deployed the idea of a universe of discourse/objects/names with good effect. For example, he divided a universe U into (some) class A and its complement a, and for a pair of such '*contraries* or *contradictories* (I make no distinction between these words)', he noted that 'The contrary of an aggregate is the compound of the contraries of the aggregants; the contrary of a compound is the aggregate of the contraries of the components' (*1858a*, 119; compare *1860b*, 192). This is the form in which he gave the laws which are now known after his name.

Like most of his contemporaries, De Morgan did not *systematically* present all the properties that his collections satisfied; but here are a few cases. The earliest example occurs in his 1839 pamphlet, to be repeated in his book: if 'All the Xs make up part (and part only) of the Ys' and Ys similarly with Zs, then 'All the Xs make up *part of part* (only) of the Zs' (*1839a*, 26; *1847a*, 22). He associated the conclusion drawn with *a fortiori* reasoning.

Later in his book De Morgan specified identity as a property of objects: if X)Y and Y)X, then 'The names X and Y are then *identical*, not as names, but as subjects of application' (*1847a*, 66): unfortunately he immediately gave 'equilateral' and 'equiangular' in plane geometry as examples of identical names, having forgotten about figures such as rectangles. When the referent (not his term) of X was part of that of Y he described the terms X and Y respectively as 'subidentical' and 'superidentical' (p. 67).

Were these versions of identity to be interpreted intensionally or extensionally? In his third paper on the syllogism De Morgan gave his most detailed (though rather unclear) discussion. He distinguished between three senses of whole and part, 'giving rise into three *logical wholes*'. Firstly, 'arithmetical' was an extensional version with 'the class as an aggregate of individuals', where the aggregate was the extensional union of the parts of the class; or it was 'the attribute as an aggregate of qualities of individuals', where 'attribute' was a quality of the class as a whole. Secondly, 'mathematical' was used 'most frequently, of class aggregated of classes; less frequently, rarely in comparison, of class compounded of classes', where 'compound', in contrast to 'aggregate' and in some kinship with 'attribute', referred to a property adhering to every member of an aggregate. Finally, 'metaphysical' was 'almost always, of attribute compounded of attributes: sometimes, but very rarely, of attribute aggregated of attributes'.

To clarify this none too clear classification ('rarely'? 'frequently'?), De Morgan added: 'Extension, then, predominates in the mathematical whole; intension in the metaphysical' (*1858*, 120–121, with some help from pp. 96–100 and from *1860b*, 178–181). However, he did not pursue the major question of how much actual mathematics could be encompassed within the extensional realm; his use of the word 'mathematical' is rhetorical. A regrettable tradition was launched.

This issue exemplifies De Morgan's strengths and weakness as a logician. He had made major insights in this paper, and elsewhere in his writings he presented novelties to logic and suggested new connections, or at least analogies, with mathematics, especially algebra. However, he surrounded his fine passages with much discursive chatter, fun to read but inessential to any logical purpose. He did not gain the full credit that he deserves; but the reader has to turn prospector to find the nuggets. Much of his argument rested upon examples rather than general theorems or properties—which constitutes another similarity with his essay on functional equations. Furthermore, his contributions were to be somewhat eclipsed by the more radical innovations made by his younger contemporary and friend, George Boole.

2.5 A LINCOLN OUTSIDER: BOOLE ON LOGIC
AS APPLIED MATHEMATICS

2.5.1 *Summary of his career.* Boole must be among the most frequently mentioned mathematicians today, because of the bearing of his logic upon computing. In 1989 I saw the ultimate compliment, in Lima (Peru): a computer company displaying its name 'George Boole' in large letters on the side of its building. However, Boole himself did not relate his theory even to the computing of his day: on the contrary (*1847a*, 2),

> To supersede the employment of common reason, or to subject it to the rigour of technical forms, would be the last desire of one who knows the value of that intellectual toil and warfare which imparts to the mind an athletic vigour, and teaches it to contend with difficulties and to rely upon itself in emergencies.

In fact, many of the details of Boole's "famous" theory are not well known. While not a detailed account,[16] enough is given here to indicate later the differences between the tradition that he launched and the mathematical logic which was largely to supplant it.

Born in Lincoln in 1815, Boole passed the first 35 years of his life in and around that city. He had to maintain himself and even his family as a school-teacher, and was largely self-taught in mathematics; but nevertheless he began publishing research papers in 1841, in the recently founded *Cambridge mathematical journal*. His main interest lay in differential equations. His work in logic, our main concern, reached the public first as a short book entitled *The mathematical analysis of logic* (*1847a*, hereafter '*MAL*'), followed by a paper *1848a* in the *Journal*.

In the following year Boole moved to Cork in Ireland, as founder Professor of Mathematics in Queen's College, a constituent of the new Queen's University of Ireland. He stayed there for the remaining 15 years of his life, and wrote the definitive version of his logic, as the book *An investigation of the laws of thought* (*1854a*, hereafter '*LT*'). Reception of his ideas was rather slow; even his correspondence with De Morgan, while substantial (Boole–De Morgan *Letters*), did not focus strongly on the details of either man's system (Corcoran *1986a*).[17] In fact, as we shall see, their contributions to logic, while both mathematical and even algebraic in type, differed fundamentally in content. He seems to have had little contact even with William Rowan Hamilton in Ireland, although they had algebra and time as common matters of concern.

---

[16] Various rather trivial accounts of Boole's life and work, and some mistaken ones, can be found. MacHale *1985a* is the best biography, to be supplemented by two exceptional obituaries: the well-known Harley *1866a*, and the forgotten Neil *1865a*. Items concerned with specific aspects of his work will be cited *in situ*. Especially recommended is Panteki *1992a*, chs. 5 and 7; her *1993a* provides further little-known background. Jourdain *1910a* includes an important survey, using manuscripts which Harley had owned but which are now lost (see also footnote 21). Styazhkin *1969a* has a useful survey in ch. 5.

[17] De Morgan's obituary *1865a* of Boole shows the limitation of their relationship. Short, and as nearly concerned with his own work as with Boole's, it states that 'Of his early life we know nothing' and that he died 'at some age, we suppose, between fifty and sixty' (in fact, he was 49). The piece is anonymous, and I attribute it to De Morgan only because he is listed as a contributor to the volume of the journal (*Macmillan's magazine*) in which it appeared, and no other person named there could possibly have been the author.

The year after his second book was published, Boole married, and produced five daughters at regular two-year intervals. His wife Mary, a woman of considerable intelligence, helped him with the preparation of textbooks on differential and on difference equations, which appeared as Boole *1859a* and *1860a* respectively. He began work on the first one soon after publishing *LT*;[18] they made much more impact at the time than those on logic. During these years he also wrote extensively on the application of his logical system to probability theory. He also attempted a more popular account of that system which, however, was never completed; a selection of these and other manuscripts on logic has appeared recently as Boole *Manuscripts*.[19] There is no edition of his works, although all his four books have been reprinted. In addition, *MAL* appeared in 1952 in an edition of many of his writings and some manuscripts on logic and probability theory (Boole *Studies*).

2.5.2 *Boole's 'general method in analysis', 1844.* As was remarked briefly in §2.4.2, English mathematics became greatly concerned with operator methods of solving differential equations. One of the leading workers was D. F. Gregory (1813–1844), Scottish by birth but very English in his researches. In a monograph on these methods he laid out the basic laws of differential operators '*a*' and '*b*' operating on functions *u* and *v*. Citing Servois for terms (§2.3.2), he wrote (Gregory *1841a*, 233–234):

(1) $$ab(u) = ba(u)$$ [(252.1)]

(2) $$a(u + v) = a(u) + a(v)$$ [(252.2)]

(3) $$a^m . a^n u = a^{m+n} . u$$ [(252.3)]

The first of these laws is called the *commutative law* [...] The second law is called *distributive* [... The third] may conveniently be called the law of repetition [...].

By this time Gregory, the editor of the *Cambridge mathematical journal*, was encouraging new talent and taking Boole's first papers. In 1843 Boole had prepared enough material to write a large paper on this subject, which he submitted to the Royal Society, with Gregory's and De Morgan's encouragement. After difficulties with the referees, he had it accepted for the *Philosophical transactions*, where it appeared as *1844a* and later even won one of the Society's gold medals, the first occasion for a mathematical paper.

---

[18] See Boole's letters to MacMillan's of 30 August and 7 September 1855 in Reading University Archives, MacMillan's Papers, file 224/10.

[19] The Boole Papers have recently been put in some order, maintaining the original call-marks. Some years ago a smaller collection was acquired by Cork University; it includes an unpublished biography by his sister Mary Ann.

Boole's essay, entitled 'On a general method in analysis', treated 'symbols apart from their subjects'. Working out from the symbolic version (224.6) of Taylor's theorem, he produced a wide range of solutions of differential and difference equations and also summation of series and the use of generating functions. He started his account with the same three laws for differential operators to obey as were proposed by Gregory (whom he cited); he also used Servois's adjectives for the first two. However, he called the third 'the index law'; and he placed the laws at the head of the presentation, whereas Gregory's had appeared well into his book. After stating these laws, he noted at once that commutativity (252.3) applied only to differential equations with constant coefficients. Much of the paper was devoted to finding solutions to equations involving both commutative and non-commutative operators.

2.5.3 The mathematical analysis of logic, *1847: 'elective symbols' and laws.*   By the time of that paper De Morgan and William Hamilton were quarrelling over the quantification of the predicate (§2.4.6), prompting Boole to write up his own views about logic, in the short book *MAL* of 1847. While its content was substantially different from the subject matter of the two contestants—he ignored quantification of the predicate, in fact —some of their other issues were reflected (Laita *1979a*).

In a tradition of his time, Boole treated logic as a normative science of thought allied to psychology; indeed, it was fundamental to his operational theory (Hailperin *1984a*). In his introduction he spoke of 'mental operations' at some length (*1847a*, 5–7), and formulated his basic principles in the following way (pp. 15–16). Symbolising by '1' a 'Universe' which

> comprehend[s] every conceivable class of objects whether existing or not [...] Let us employ the letters X, Y, Z, to represent the individual members of classes. [...] The symbol x operating upon any subject comprehending individuals or classes, shall be supposed to select from that subject all the Xs that it contains. [...] the product xy will represent, in succession, the selection of the class Y, and the selection from the class Y of such individuals of the class X as are contained in it, the result being the class whose members are both Xs and Ys.

Although for some reason he did not mention his *1844a* or cite Gregory, Boole set down the basic 'laws with these mental acts [x] obeyed in a form closely similar with those for the differential operators (pp. 17–18). Given an 'undivided subject' $u + v$, with u and v 'the component parts of it', then the 'acts of election' x and y obeyed the laws

$$x(u + v) = xu + xv \qquad (1), \qquad [(253.1)]$$

$$xy = yx \qquad (2), \qquad [(253.2)]$$

$$x^n = x \text{ [integer } n \geqslant 2] \qquad (3), \qquad [(253.3)]$$

[...] From the first of these, it appears that elective symbols are *distributive*, from the second that they are *commutative*; properties which they possess in common with symbols of *quantity* [...]

The third law (3) we shall denominate the index law. It is peculiar to elective symbols, and will be found of great importance in enabling us to reduce our results to forms meet for interpretation.

The formulation of (253.3) in terms of x^n rather than x^2 is very striking; in a footnote he compared it with the law $+^n = +$, another consideration of Gregory (this time, *1839a* on 'algebraic symbols in geometry') which again he did not cite.

As normal for his time, Boole was not axiomatising a theory in any manner that we would practise today; rather he was laying down *laws* for his elective symbols to obey, in the algebraic tradition. He stated rather few of the laws and properties that his system required; (253.1) as the only distributivity law (over subjects $u, v, \ldots$), and no associativity laws (with consequent sloppiness over bracketing). He reserved the word 'axiom' for a property stated in the space occupied above by my second string of ellipsis dots: 'The one and sufficient axiom involved in this application is that equivalent operations performed upon equivalent subjects produce equivalent results'. We would regard this axiom as a metatheoretic principle.

Boole stressed interpretation. His introduction *began* with the statement that in 'Symbolical Algebra [...] the validity of the processes of analysis does not depend upon the interpretation of the symbols which are employed, but solely upon the laws of their combination' (p. 3), and we saw him mention interpretability at the head of this sub-section. However, some commentators were less familiar with this issue. For example, in December 1847 Arthur Cayley (1821–1895) wondered if it was true in this calculus that '$\frac{1}{2}x$ has any meaning', and Boole explained 'that this question is equivalent to whether $\sqrt{-1} \times \sqrt{-1} = -1$ in a system of pure quantity for although you may interpret $\sqrt{-1}$ in geometry you cannot in arithmetic'. In his reply Cayley disliked this analogy, but Boole insisted that $\sqrt{-1}$ should be treated 'as a symbol (i) which satisfies particular laws and especially this

$$i^2 = -1 \, [\ldots]'$$ (253.4)

(Boole *Manuscripts*, 191–195).

In Boole's algebra the cancellation law did not hold for multiplication: $zx = zy$ did not imply that $x = y$. Thus he needed the notion, novel for its time, of the 'indefinite symbol' v (or class), as it let him render as equations many relationships which otherwise would have had to appear as (some analogue of) inequalities. For example (p. 21), 'If some Xs are Ys, there are some terms common to the classes X and Y. Let those terms constitute a separate class V, to which there shall correspond a separate elective symbol v, then

$$v = xy\,[\,\ldots\,]'.\qquad\qquad(253.5)$$

However, he offered no laws which v should satisfy, and he did not distinguish between traditional forms of proposition and those involved in quantification of the predicate; for example, '$vx = vy$' could cover both 'Some Xs are Ys' and 'Some Xs are some Ys' (pp. 21–22).

Like De Morgan (§2.4.9), Boole's theory of classes was an extensional version of part-whole analyses of collections. Inclusion was the only relation, with proper or improper not always distinguished: 'The equation $y = z$ implies that the classes Y and Z are equivalent, member for member' (p. 19; see also p. 24). But again little information was given about ' − ' or ' + '; for example, he left rather implicit that ' + ' linked only *disjoint* classes.

2.5.4 *'Nothing' and the 'Universe'.* The symbol '0' first appeared on p. 21 when Boole rendered the categorical proposition 'All Xs are Ys' as

$$'xy = x,\ \text{or}\ x(1 - y) = 0'.\qquad\qquad(254.1)$$

Obviously '0' symbolised the mental act complementary to the elective symbol 1, but he gave it no formal definition nor stated its laws (of addition to any x, for example).

For the universe (p. 20),

> The class X and the class not-X together make the Universe. But the Universe is 1, and the class X is determined by the symbol x, therefore the class not-X will be determined by the symbol $1 - x$.

So '1' was serving double duty for elective symbols and for classes. The idea of an identity operator (or entity or whatever) in this world of expanding algebras was a novelty which took time to be understood, although it was already present in the Lagrangian (224.6): Cayley was to be another pioneer, in his paper *1854a* on matrix multiplication.

Another example of the conceptual difficulties arises later: 'To the symbols representative of Propositions [...] The hypothetical Universe, 1, shall comprehend all conceivable cases and conjunctures of circumstances', and x 'shall select all cases in which the Proposition X is true' (pp. 48–49). Boole offered no further explanation of this hypothetical Universe, which sounds the same as the "absolute" Universe presented before (253.1); but to have "everything" in that way is to have nothing at all, since non-partship of such a Universe is impossible. More importantly, within this Universe true propositions cannot be distinguished from tautological ones, or false propositions from self-contradictory ones (Prior *1949a*). Further, it led him to claim that a disjunction of particular propositions, but not a disjunction of universal ones, could be split into disjunctions; the (alleged) grounds were that the disjunction was hypothetical whereas the components were categorical (p. 59). On the role of universes he lagged behind De Morgan in insight.

Boole clearly thought that $0 \neq 1$, but the *status* of this proposition is not clear; since he had no symbol for 'not', it has to be an additional assumption. The closest that he came to the issue occurred when he mooted in *MAL* 'the nonexistence of a class: it may even happen that it may lead to a final result of the form

$$1 = 0, \qquad\qquad [(254.2)]$$

which would indicate the nonexistence of the logical Universe' (*1847a*, 65). But he did not extend his discussion to propositions such as

$$(1 = 0) = 0: \qquad\qquad (254.3)$$

we shall note at (445.1) that Schröder was to consider them. Naturally, Boole did not assert anything like '$x \neq 0$ (or 1) implies that $x = 1$ (or 0)', as the classial interpretation would have been lost. A. J. Ellis *1873a* made this point in contrasting Boole's treatment of propositions with that of classes; however, he formulated the contrast as being between algebra and propositions.

Boole also read '0' and '1' as two different states or situations. In the symbolisation of a proposition, '0' referred to 'no such cases in the hypothetical Universe' (p. 51). The paper *1848a* was still less clear; '0' took the stage, as an elective symbol, without cue, after the statement that 'There may be but one individual in a class, or there may be a thousand' but apparently not none (p. 127).

But Boole also interpreted '0' and '1' as *numerical* quantities. For example, on connections of logic with probability, after noting that 'every elective symbol [. . .] admits only of the values 0 and 1, which are the only quantitative forms of an elective symbol', he compared a manner of expressing hypothetical propositions with some unstated means using probability theory '(which is purely quantitative)', and added that 'the two systems of elective symbols and quantity osculate, if I may use the expression, in the points 0 and 1' (p. 82).

2.5.5 *Propositions, expansion theorems, and solutions.* Boole did not treat propositions X, Y, . . . as "atomic" entities, but presented his interpretation as propositions only when specifying the hypothetical type, 'defined to be *two or more categoricals united by a copula* (or conjunction)' (p. 48). But even now the constituent propositions did not stand alone but were encased in their truth-values: for example, '$(1 - x)y$' corresponded to

Some sleight of hand seems to be evident here. Since neither $0/0$ nor $1/0$ is 0 or 1, why should *only* the first be replaceable by a symbol v which obeys the index law; or alternatively, why should *only* the second demand that its term be set to zero? One can, of course see reasons for Boole's distinction in the consequences for logic, but what are they in the algebra? One can surely argue as good a case for the conversion of $1/0$ as for $0/0$, on the possible grounds that the index law was satisfied:

$$1/0 = \infty \text{ and } \infty^2 = \infty. \tag{255.10}$$

2.5.6 The laws of thought, *1854: modified principles and extended methods.* Boole later recorded that *MAL* had been written in haste, and that he regretted its publication (*1851a*, 252). He never specified the sources of his regret, but the points just discussed may have been among them. In addition, some casualness in presentation is evident: concerning 0, $+$, $-$ and division, for example. The paper *1848a* was not much clearer, and even introduced the new obscurity 'x 1 or $x =$ [*sic*] the class X' (p. 126). Further, in contrast to (253.1) but without comment, he now presented the distributivity law for elective symbols themselves rather than over their subjects; presumably this change was a slip, for in his system he really needed, and used, both laws.

Two copies of *MAL* contain extensive hand-written additions (see G. C. Smith *1983a*, and Boole *Studies*, 119–124), and a manuscript of around 1850 (pp. 141–166) constitutes more substantial a study than its title 'Sketch' suggests. Among the novelties Boole moved away from syllogistic logic towards the sign tradition inspired by Locke and others (§2.3.3). In a manuscript of the late 1840s he asserted that 'In general we reason by signs. Words are the signs most usually employed for this purpose' (*Manuscripts*, 14). These words may echo Whately: that signs are primary, and that 'language affords the *signs* by which these operations of the mind are expressed and communicated' (*1826a*, 55). They contrast with Boole's neutral remark in *MAL* that 'The theory of Logic is thus intimately connected with that of Language' (*1847a*, 5).

The next outcome was the second book, *1854a* on *The laws of thought*; Van Evra *1977a* contains a general survey of its logical contents. The title (but not the contents) closely follows that of Thomson *1842a* (§2.4.4). While basically the same algebra and expansion theorems as in *MAL* were presented and greater clarity was evident in general, various new results appeared, and also certain changes of emphasis and interest were manifest. The most substantial one is that over 150 pages were devoted to probability theory, which linked to logic via belief vis-à-vis certainty and the interpretation of compound events as logical combinations (in his sense) of simple ones (Hailperin *1986a*).

Among the main changes, the psychology was less prominent than before, although Boole began by repeating his intention of 'investigat[ing] the fundamental laws of those operations by which reasoning is performed' (p. 1) and '*x, y, &c., representing things as subjects of our conceptions*' (p. 27). Semiotics was much more to the fore, starting with ch. 2 on 'signs and their laws', where both words and symbols were so embraced. The story itself was worked largely as a theory of classes: 'If the name is "men," for instance, let *x* represent "all men," or the class "men"' (p. 8). The distributivity law took the 1848 form over symbols rather than (253.1) of *MAL* over subjects (p. 33).

The index law (253.3) was now framed as

$$x^2 = x \qquad (256.1)$$

rather than the previous $x^n = x$: Boole showed that $x^3 = x$ was not interpretable since factorisation included either the uninterpretable term $(1 + x)$ or the term $(-1 - x)$ of which the component -1 did not satisfy the corollary to (256.1)

$$x(1 - x) = 0 \qquad (256.2)$$

(p. 50: presumably similar points were to apply to $x^n = x$ for all n). Boole renamed the index law 'the law of duality', as a symmetric function of x and $(1 - x)$, and he used this important algebraic property at various later places in the book.

On connectives, 'Speaking generally, the symbol $+$ is the equivalent of the conjunctions "and," "or," and the symbol $-$, the equivalent of the preposition "except"' (p. 55). However, mutual exclusivity was still imposed upon '$+$' (p. 66), so that the union of 'things which are either x's or y's' was represented in terms of inclusive and exclusive forms (p. 56): respectively,

$$x + y(1 - x) \text{ and } x(1 - y) + y(1 - x). \qquad (256.3)$$

Similarly, for interpretation $(x - y)$ required that the class of ys was included within that of the xs (p. 77). The axiom of §2.5.3 concerning 'equivalent operations performed upon equivalent subjects' now became two 'axioms', which stated that when 'equal things' were added to or taken from equal things, the results were equals (p. 36).

1 was still the 'Universe', but it was specified within 'every discourse', where 'there is an assumed or expressed limit with which the subjects of its operation are confined' (p. 42): a recognition of the priority of De Morgan (§2.4.9) would not have been amiss. 0 was the class for which 'the class represented by $0y$ may be identical with the class represented by 0, whatever the class y may be. A little consideration will show that this condition is satisfied if the symbol 0 represent Nothing' (p. 47). Thus, while

he had clearly grasped the extensional aspect of class, he seemed not to wonder if the empty class was literally no thing (compare p. 88).

In between 0 and 1, the indefinite 'class' v again usually ranged from 0 to 1 inclusive (pp. 61–63); for example, when $0/0$ was converted (this time, by analogy) to the class v, '*all, some*, or *none* of the class to whose expression it is affixed must be taken' (pp. 89–90). However, when reducing 'Some Xs are Ys' to '$vX = vY$', v was 'the representative of *some*, which, although it may include in its meaning *all*, does not include *none*' (p. 124). Boole's unclarity is disappointing, for it had long been known in logic that certain inferences may fail if the antecedent or consequent involves empty predicates, especially with particular propositions (Hailperin *1986a*, 152–155).

Still no symbol was employed in *LT* for 'not', presumably for the symmetry inherent in the fact that 'we can employ the symbol y to represent either "All Y's" or "All not-Y's"' (p. 232). From the index law Boole claimed to prove 'the principle of contradiction', which took the symbolic form (256.2) (pp. 49–51); however, as his friend R. L. Ellis *m1863a* remarked, the absence of 'not' renders the law independent of the principle, which was interpreted by the law rather than expressed via it. But (256.2) was given other sorts of work to do, in particular, to distinguish the cases of $0/0$ and $1/0$; for the latter was 'the algebraic symbol of infinity' and 'the nearer any number approaches to infinity (allowing such an expression), the more does it depart' from (256.2) rather than from the index law which might admit (255.10) (p. 91).

Among results or remarks which made their *débuts* in *LT*, in the footnote in which he disproved the possibility of $x^3 = x$ after (256.1), Boole perceived—with no enthusiasm—the possibility of non-bivalent logics, in which 'the law of thought might have been different from what it is' (p. 50). Once again he ignored De Morgan, who had touched on the point in his own book (*1847a*, 149) and had raised it in a letter to him in 1849 (Boole–De Morgan *Letters*, 31).

The expansion theorem (255.4) duly appeared, proved by assumption of form and calculation of the coefficients, with MacLaurin's theorem now in a footnote (pp. 72–73). But it was supplemented by this important result for any equation $f(x) = 0$; that

$$f(1)f(0) = 0 \tag{256.4}$$

'*independently of the interpretation of x*' (p. 101), with analogues for several variables (p. 103). Several different proofs were given, usually drawing on (255.4) (pp. 101–104); as Harley *1871a* pointed out, it follows from the theory of roots of equations adapted to two-valued variables.

The importance of this result lay in its role in eliminating x from an equation $\phi(xyz\ldots) = 0$ containing x and other variables. This move greatly enriched his method of deduction, which was presented in chs. 7–8 with some nice short-cuts executed in ch. 9. One new extension was a theorem concerning '*any system of equations*' $V_r = 0$; that '*the combined interpretation of the system will be involved in the single equation,*

$$V_1^2 + V_2^2 + \&\, c. = 0\text{'} \tag{256.5}$$

(pp. 120–121). The purpose of squaring was to avoid the loss of terms by cancellation across the equations if simple addition were practised; the index law reduced the equation itself to linear form.

Boole solved class equations basically as in *MAL* (255.8–10), though now in a more general framework (pp. 90–98). Take as subject the class z from a given collection of given classes $u, v, \ldots$, form every combination uv, $u(1 - v)$, $(1 - u)v, \ldots$ of the remaining classes, express the logical premises as equations, and use the appropriate expansion and elimination theorems with z as subject to determine from the equations the coefficients c attached to each such combination m. If $c = 1$, then m was part of z; if $c = 0$, then not so; if $c = 0/0$, then any part vm of m was part of z; if c took any other value, then the proposition $m = 0$ imposed sufficient conditions for any class z to be found at all. Further, several equations could be reduced to a single linear combination of them (ch. 8).

One final feature, arising in Boole's treatment of probability theory, contrasts him with the philosophies of arithmetic of Frege and Russell. As we shall see in §4.5.3 and §6.5.4, they were to define cardinal numbers as sets of similar sets; for Boole, 'let the symbol n, prefixed to the expression of any class, represent the number of individuals contained in that class' and he treated 'n' as an operator, noting that it 'is distributive in its operation' over classes. He then read the frequentist interpretation of probability as the appropriate ratio $n(x)/n(1)$ for a class x (pp. 295–297). His further development of these ideas led him to some work on inequalities, in a context which we recognise today as linear programming (Hailperin *1986a*, 36–43, 338–350).

2.5.7 *Boole's new theory of propositions.* Boole proposed in *LT* a new distinction of propositions: instead of the categorical and hypothetical categories, he now worked with 'primary or concrete' and 'secondary or abstract' types. The second names clarified the distinction; the former type of proposition related 'to *things*' and the latter 'to *propositions*'. Primary propositions were categorical, but once again they were not treated as "atomic" entities; instead, they were components of secondary ones, which included hypothetical propositions such as 'If the sun shines the earth is warmed' (p. 53).

To us Boole's characterisation of this type of proposition places them in the metatheory, with his example wanting of interior quotation marks. However, lacking such a conception but desirous of giving the primary components objectual status, he replaced his hypothetical universe of the previous sub-section with *time*, as a location for propositions to consign their truth-value (and thereby become "things"). '*X*', '*Y*', and so on now denoted 'the elementary propositions concerning which we desire to make some assertion concerning their truth or falsehood', while *x* 'represent[s] an act of the mind by which we fix our regard upon that portion of time for which the proposition *X* is true' (pp. 164–165). Further, '1' now 'denotes the whole duration of time, and *x* that portion of it for which the proposition *X* is true', so that '1 − *x* will denote that portion of time for which the proposition *X* is false'; hence '*The propositions X is true/false*' were rendered respectively by '*x* = 1' and '*x* = 0' (pp. 168–169). The basic laws, and the means of combination, applied once more (with the usual restrictions on '+' and '−' in place again); the 'time indefinite' v seems to have been non-empty, although if v and *x* had no common period of truthhood, then $vx = 0$ (p. 171). This theory resembles a Boolean algebra of propositions, which however was to come only with Hugh MacColl (§2.6.4).

It is curious that Boole did not claim here to derive the law of excluded middle for propositions, which would take the form

$$x + (1 - x) = 1; \qquad\qquad (257.1)$$

for he did state the law of duality (256.2) in this context (p. 166). A "proof" is compromised by the absence of 'not' from his language; (257.1) is best seen as embodying a necessary assumption about truth and falsehood. Earlier in the book he had produced the rather similar result, that $(x + 1 - x)$ was the expansion of the function 1 via the expansion theorem (255.4) (p. 76).

While Boole noted that 'this notion of time (essential, as I believe it to be, to the above end [of explicating the theory of secondary propositions]) may practically be dispensed with' (p. 164), this was obviously not his personal wont. Indeed, he recalled his previous approach in *MAL* involving 'the Universe of "cases" or "conjunction of circumstances"', but found it far less clear than his new formulation (p. 176). This judgement followed a remarkable passage in which he speculated on the possibility of placing primary propositions in space but rejected it on the ground that 'the formal processes of reasoning in primary propositions do not require, as an essential condition, the manifestation in space of the things about which we reason', and that 'if attention were paid to the processes of solution [in certain stated problems in mechanics] alone, no reason could be discovered why space should not exist in four or in any greater number of dimensions' (pp. 174–175).

2.5.8 *The character of Boole's system*

All sciences communicate with each other [that which] they have in common. By common I mean that of which they make use in order to show

something; but not that to which their proof refers, nor the [final] outcome
of what they show.   Boole, motto on the title page of *MAL*
(in Greek, taken from Aristotle, *Posterior analytics*)

The logic which Boole offered was to be understood as a normative
theory of the products of mental processes (descriptive psychology was, of
course, not his brief). In *LT* it was grounded in the belief of 'the ability
inherent in our nature to appreciate Order' and thereby produce 'general
propositions' which 'are made manifest in all their generality from the
study of particular instances' (*1854a*, 403, 404). Thus in the underlying
philosophy of logic Boole stood at the opposite pole from the empiricism
of John Stuart Mill, for whom even the principles of logic, if true, were
formed by induction from experience (John Richards *1980a*).

Boole devoted ch. 5 of *LT* to 'the fundamental principles of symbolical
reasoning', where he began by arguing for his various principles from
instances, but the case was not overwhelmingly put. Even in his final
chapter on 'The constitution of the intellect' he rested his case on the
assertion that 'The laws of thought, in all its processes of conception and
of reasoning, in all those operations of which language is the expression or
the instrument, are of the same kind as are the laws of the acknowledged
processes of Mathematics' (*1854a*, 422).

Boole's system fulfilled some of the ambitions for a *characteristica
universalis* of Leibniz, who had even formulated the index law in the form
$xy = x$ for x and y as propositions (not his symbols). However, Boole
learnt of Leibniz's proposal (as then known) only in 1855 (Harley *1867a*),
after *LT* appeared; the informant was R. L. Ellis, who was helping to bring
Leibniz's logic and philosophy to British attention (Peckhaus *1997a*, ch. 5).

There was also an important religious connotation in Boole's position. A
Disssenter, he adhered to ecumenical Christianity, aloof from the xs and
$(1 - x)$s of the disputing Christian factions. He implicitly exhibited his
position in *LT* by devoting ch. 13 to sophisticated logical analyses of
propositions due to Samuel Clarke and Benedict Spinoza concerning the
necessary existence of '*Some one unchangeable and independent Being*' (p.
192). He also cited Greek authorities, and also Hegel, as sources for the
importance of unity among diversities (pp. 410–416). He alluded to his
stand a few lines from the end of the book by mentioning 'the Father of
Lights', and finished off with some enigmatic lines about the bearing of
religious belief upon his logic.

In his final years Boole enthused over the presentation of logic by the
French theologian A. J. A. Gratry *1855a*, in which claims such as God as
truth gained prominence, in addition to topics such as nullity versus unity,
universal laws of thought, and the exercise of the human mind. He held in
awe the theologian Frederick Denison Maurice, who advocated ecumenical Christianity and was therefore dismissed from his Professorship of

Divinity at King's College London, part of the established Trinitarian Church of England; Boole had a portrait of Maurice set by his bed as he lay dying (my *1982a*, 39–41).

MAL related heavily to syllogistic logic; the book contains passages on it set in smaller type, often taken from Whately's *Logic*, as Boole hinted on p. 20. But in *LT* he showed the revolutionary implications of his work; the details of syllogistic logic was demoted to the last of the 15 chapters strictly devoted to logic, and with quantification of the predicate again left out.

As logic, Boole's principal aim was consequences of premises rather than detailed deductions from them. The theory was always algebraic, with a strong kinship to differential operators: equations were the principal mode of working, facilitated in formation and manipulation by the indefinite symbol (or class) v. It was interpreted in terms of classes and propositions, and later to probability theory. (Non-)interpretability was a major feature, both of functions and of equations.

Differential operators were not the only link with mathematics. In *MAL* Boole noted 'the analogy which exists between the solutions of elective equations and those of the corresponding order of linear differential equations' (*1847a*, 72). In addition, he was enchanted by singular solutions of differential equations, since they had the character of unity from opposites required by the index law that he liked so much also in Maurice: as he put it, on the one hand, the '*positive* mark' of solving the equation, but also 'the negative mark' of lying outside the general solution (*1859a*, 140). However, as a result of ignoring quantification of the predicate, Boole's treatment of syllogisms was corrigible, partly for want of clarity over '$-$' and '$+$', but especially for failing to detect singular solutions (Corcoran and Wood *1980a*). For example, for the universal affirmative proposition 'All Xs are Ys', symbolised as

$$x(1 - y) = 0, \text{ he put forward } y = vx \qquad (258.1)$$

as 'the most general solution' (*1847a*, 25); but he should have noticed that $x = 0$ was missing from it, and also that it did not hold if $x = 0$ and v was a class such that $vy \neq 0$. Thus some solutions of his equations were not logical consequences of the premises under his system. The difference between necessary and sufficient conditions was again not under control; but this mathematician with a strong interest in singular solutions should have noticed analogous situations in his logic.

Collections were handled in normal part-whole style of the time; but Boole's reading of $(x + y)$ only for disjoint classes x and y (p. 66) was very take-it-or-leave-it. In a manuscript of 1856 he offered the following "deduction" of his definition:

$$x + y = (x + y)^2; \therefore xy + yx = 0; \therefore xy = 0 \qquad (258.2)$$

(*Manuscripts*, 91–92). It seems strange for a logician to have confused a sufficient condition with a necessary one: maybe he asserted this version of ' + ' to avoid having to read the intersection of the classes as a multiset, to which an object may belong more than once (such as the collection 3, 3 and 8 of roots of a cubic equation). If so, then he was in a strong tradition; for when multisets arrived in the 1870s and 1880s they gained little attention at first (§4.3.4, §4.2.8). But from a modern point of view, his theory can be read in terms of signed multisets, where an object can belong to a collection any finite number of times, positive or negative (Hailperin *1986a*, ch. 2).

Absent from Boole's theory were the quantifier as such (although *v* did some of the work); any logic of relations (even after the publication of De Morgan *1860a* he seems not to have taken them up); or any use of counter-arguments to establish results, or presentation of fallacies. In addition, his adherence to algebra prevented him from using pictorial representations such as Euler diagrams to depict an argument, although the untutored reader might thereby have been helped.

2.5.9 *Boole's search for mathematical roots*

De Morgan *develops* the old logic, Boole *converted* the forms of algebra
into exponents of the forms of thought in general.   Neil *1872a*, 15

In various manuscripts, especially from 1856 onwards, Boole sought a foundation for his logic in the philosophical framework used by other logicians of his time, such as Whately. More insistent than they on distinguishing between mental acts and their products, he proposed this scenario. The mind effects 'Conceptions' or 'Apprehensions' by the extensional processes of addition and subtraction of classes. The products were 'Concepts' which were subject to 'Judgement' as to their agreement or not. The products were 'Propositions', which were then subject to 'Reasoning' by inference among them to yield the 'Conclusion' (*Manuscripts*, ch. 5).

However, these procedures left Boole's philosophy incomplete. For, unlike other logicians apart from De Morgan, his logic used mathematics, so that a rich philosophy of mathematics was required—and this he never found, the relevant manuscripts being restricted to particular aspects such as axioms and definitions (*Manuscripts*, ch. 14). In particular, he did not break the following vicious circle, and may not have realised its existence. Whatever mathematical theory grounded his logic, it had to be consistent to fulfil its office; but consistency is *already* a concept taken from logic . . . . Thus did the mathematical roots of his logic elude him.

Boole's logic was *applied* algebra, the 'mathematical analysis *of* logic'. He remarked *en passant* but interestingly on this aspect when he followed an appraisal of a general treatment of logical equations with the comment in *LT* that 'The progress of applied mathematics has presented *other* and

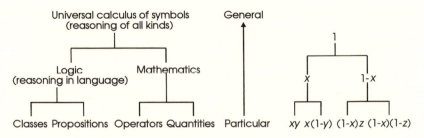

FIGURE 259.1. Schematic conjecture about Boole's system.

signal examples' of such unification of methods (*1854a*, 157, italics inserted). However, features such as the use of MacLaurin's theorem to prove the expansion theorem (255.4) suggest that an orthodox application of one theory to another may be too straightforward. Inspired by the proposal of Laita *1977a* that a universal calculus of symbols underlies both his mathematics and his logic, I offer in Figure 259.1 this representation of his system. It has a Boolean structure, as is indicated on the right hand side. This feature is important; for since Boole offered laws of thought, his system should apply to itself.

2.6 THE SEMI-FOLLOWERS OF BOOLE

2.6.1 *Some initial reactions to Boole's theory*

Mr. Boole began with a short account, which was read: he then published his larger work which is much less read, and would not have been read at all but for the shorter one.

> De Morgan to Jevons, letter of 15 September 1863
> (Jevons Papers, File JA6/2/114)

This information may surprise us, for whom *LT* is the main source on Boole. In fact it seems to begin to supersede *MAL* from around the time of this letter; in particular, as we shall see in the next sub-section, Jevons himself dealt with it alone. But even then the reception was modest; for example, for all his recent caution against syllogistic logic (§2.4.5), Blakey showed no understanding of Boole in his history of logic (*1851a*, 481–482).

Let us start with Boole's most fervent follower: his widow Mary (1832–1916), who prosecuted his ideas, mainly in philosophical and educational contexts and oriented around the alleged power of the mind, for the fifty years of her widowhood. While she became well known as an eccentric advocate, she had a good understanding of his ideas, and her testimony about him can be taken as basically reliable (Laita *1980a*). She also referred to his religious stimuli and to his praise of Gratry thus: 'Babbage,

Gratry and Boole [...] published their books. Then finding themselves confronted with dishonest folly, they left the world to come to its senses at its leisure' (M. E. Boole *1890a*, 424). However, by then these aspects of his system had been set aside completely by his successors, even though Victorian science in general was rather infatuated with connections with Christianity. One important link was to Spiritualism and related topics; several logicians were interested in psychical research.[20]

Although De Morgan's theory had little in common with Boole's, he was appreciative of his friend's achievements; 'by far the boldest and most original' generalisation of 'the forms of logic', he opined in an encyclopaedia article, making algebra 'appear like a sectional model of the whole form of thought' (*1860c*, 255). Interestingly, he misrepresented '+' as creating multisets, 'with all the common part, if any, counted twice' (pp. 255–256); and he did not use Boole's logical system in his own work.

Soon after Boole's death an interesting development occurred when the British chemist Benjamin Brodie (1817–1880) published in 1866 a Boole-like algebra for chemistry, as an alternative to the prevailing atomic theory. The main idea was that of chemical operations on a litre of substance-space, yielding a certain 'weight', such as x converting the litre into a litre of hydrogen, which had a certain weight. Succeeding operations x and y gave a 'compound weight' which was represented as xy and assumed to be commutative; joint operation was written '(xy)', equal to xy; collective operation was $(x + y)$ *and* separate operation was $x + y$. Since the result for a two-part compound was the same weight, the basic laws were

$$xy = x + y \text{ and } x = x + 1, \qquad (261.1)$$

where '1' denoted the litre with no weight in it.

Since it followed that $0 = 1$, Brodie's system did not enjoy a warm reception; but of historical interest is his correspondence with mathematicians (published in Brock *1967a*), partly inspired by a vigorous discussion at the 1867 meeting of the British Association for the Advancement of Science. De Morgan wrote several letters, stressing the functional aspects of operations and so criticising $(261.1)_1$ for equating 'symbols of aggr[egation] & comb[ination]', and noting that 'though Water = Oxygen × Hydrogen is certainly Oxygen + Hydrogen yet Oxygen + Hydrogen is not

---

[20] On this theme see my *1983a*. Mrs. De Morgan published in 1866 the first extended study of physical mediumship, to which he contributed a superb preface. When the Society for Psychical Research was founded in 1882, Mrs. Boole herself was a founder member of Council (although she resigned at once, feeling improperly placed as the only woman ...). Venn and Lewis Carroll were members; and Johnson sometimes helped his sister Alice, who was the first Research Officer, over mathematical matters.

necessarily Water'. (Compare him already on this sort of point in *1847a*, 48–49 and *1858a*, 120.) He preferred the alternative form

$$1 + xy = x + y \qquad (261.2)$$

to (261.1), for it 'is not only analytically perfect, but is also interpretable' (Brock *1967a*, 103, 109–110). Herschel doubted the utility of other notations that Brodie proposed (pp. 122–124).

2.6.2 *The reformulation by Jevons.*   Brodie's strongest critic was Stanley Jevons (1835–1882), who even wrote a piece for the *Philosophical magazine* but withdrew it after receiving criticism from the physicist W. F. Donkin (Brock *1967a*, 114–118). He was the first to work seriously on Boole's system, initially in a short book entitled *Pure logic* (Jevons *1864a*), to which the account below is largely confined. Then in his thirtieth year, he had recently been appointed tutor at Owens's College, Manchester.

Although he had taken courses in mathematics from De Morgan at University College London, Jevons concerned himself solely with Boole's system, and only as presented in *LT*. Subtitling his own book 'the logic of quality', Jevons followed Boole in detaching logic from the study of quantity, and gave several admiring references to Boole's work. However, he made some basic criticisms of Boole's system; while he presented them as his last chapter, it is best to take them now, as they obviously guided the construction of his alternative system.

Four '*Objections*' were made. Firstly, Boole's '*logic is not the logic of common thought*', even within its normative brief (art. 177). His reading of ' + ' was singled out for especial criticism, and entered into Jevons's second claim, that '*There are no such operations as addition and subtraction in pure logic*' (art. 184), and also the third, that the system '*is inconsistent with the self-evident law of thought*' that A or A is A (art. 193). Finally, '*the symbols* 1/1, 0/0, 0/1, 1/0, *establish for themselves no logical meaning*' (art. 197).

Jevons worked with 'terms', which covered '*name,* or any combination of names and words describing the qualities and circumstances of a thing' (art. 13). Without attribution, he used De Morgan's notations '*A*' and '*a*' for a term and its negation, and implicitly drew on the same symmetry of role between A and *a* which was noted in §2.4.8. A principal connective was ' + ', which stood 'for the conjunctions *and, either, or,* &c., of common language' and did not suffer the Boolean restriction to disjointness of its components (art. 16); however, he avoided the evident ambiguity of his explanation by using it only as 'or'. His account of ' + ' seemed to allow for both inclusive and exclusive disjunction of terms (arts. 64–72); but his examples in art. 179 used the inclusive sense, as in 'academic graduates are either bachelors, masters, or doctors'. He represented 'and' by 'combining' terms A and B in a Boolean manner to produce AB (art. 41). The other main connective was 'the copula *is*', symbolised ' = ', which registered 'the sameness of meaning of the terms on the two sides of a proposition' (art. 21).

All and nothing appeared in Jevons's system. He defined 'the term or mark 0' rather thoughtlessly as 'excluded from thought' (art. 94, where however he did state the basic laws 0.0 = 0 and 0 + 0 = 0); and he introduced a *'Universe of Thought'* specified like De Morgan's relative to a logical argument (art. 122, well into the text, and reflecting its subsidiary role in his system). But he also proposed the *'Law of infinity*, that *'Whatever quality we treat as present we may also treat as absent'*, so that 'There is no boundary to the universe of logic' U; in particular, its negation *'u* is not included in U'. Jevons was on the border of possible paradox here, but he made proposals in a footnote (to art. 159), which ended: 'this subject needs more consideration'.

Jevons also used 'U' to render 'some' as a term; but he denied U the property U = U and replaced it by an appropriate constituent terms in an argument: for example, 'A = UB, meaning that A is *some kind* of B [*sic*] is much better written as A = AB' (art. 144). However, this principle seems to infringe his 'Condition' that *'the same term have the same meaning throughout any one piece of reasoning'* (art. 14). (Boole had required his class v to satisfy normal properties (*1854a*, 96), but he used more than one such class when necessary.) Further, as with Boole, Jevons left unclear some questions of existential import of particular propositions.

Jevons was somewhat more conscientious than Boole in stating the basic laws of his system; but it is often less clear whether a proposition is a principle or a theorem, and, if the latter, how it was proved. For example, again like Boole he had no separate symbol for 'not' (not even in his ch. 7 on 'Negative propositions'). He gathered most of his principles together in art. 109; some were Boolean but others not, and the reference of the name 'Duality' was changed. His names and formulations, sometimes cryptic, are given here:

'Sameness':	'A = B = C; hence A = C',	(262.1)
'Simplicity':	'AA = A, BBB = B, and so on'.	(262.2)
'Same parts and wholes':	'A = B; hence AC = BC'.	(262.3)
'Unity':	'A + A = A, B + B + B = B, and so on'.	(262.4)
'Contradiction':	'Aa = 0, ABb = 0, and so on'.	(262.5)
'Duality':	'A = A(B + b) = AB + Ab[...] and so on'.	(262.6)

In addition, a 'law of difference' had been stated in art. 77 but omitted here, presumably for a lack of 'not'; making temporary use of '≠', it would read

$$A \neq B \text{ and } B = C; \text{ hence } A \neq C. \qquad (262.7)$$

Among his theorems, one on 'superfluous terms' (art. 70) became quite well known as 'absorption' for terms B and C:

$$\text{'B} + BC = B\text{'.} \qquad (262.8)$$

Jevons's method was to set up the premise(s) in equational form, to characterise logic itself as the 'science of science' (art. 37):

SCIENCE OF SCIENCE	$\{A = B = C\} = \{A = C\}$	REASONING
		[(262.9)]
SCIENCE	$A = B \qquad B = C$	JUDGMENT
		[(262.10)]
THINGS	$A \qquad B \qquad C$	APPREHENSION
		[(262.11)]

Then he used two modes of 'inference', both modelled on Boole's. In the 'direct' mode the premises were combined in suitable ways to cancel out middle terms; for example, the syllogism 'No A is B, Every A is C, $\therefore$ Some C is not B' came out as

$$A = Ab \text{ and } A = AC; \therefore AC = AbC = Ab; \therefore AC = Ab, \quad (262.12)$$

as required (art. 148). In the more general 'indirect inference' (ch. 11) all possible combinations of the terms in the premis(es) were listed, and combined with each of their terms, as a sum of products. Then each combination was appraised as an 'included subject' if it did not contradict either side of at least one of the premises, as a 'contradiction' if it contradicted one side of a premise, or as an 'excluded subject' if it contradicted both sides of every premise. The second type was to be deleted, leaving the other two as *'possible subjects'*, and their sum (in his sense of ' + ') as the consequences.[21] Thus the consequences pertaining to a (simple or compound) term t were found by equating it (in his sense of ' = ') to the sum of the consequences of which it was part; in other words, he found the term to which t was 'equal' given the premises. Various means of simplification and basic or derived laws such as (262.1–7) were found.

---

[21] Later Jevons developed this idea of contradiction with a proposition to form the notion of the 'logical force' of a proposition, the number of propositions which it negated (*1880a*, ch. 24). However, I do not think that this idea bears on the use of 'force' noted in footnote 6.

For example, from the premise A = BC, the three categories of consequence were ABC; ABc, AbC, Abc and Abc; and aBc, abC and abc. Selecting (say) b for the four possible subjects, two options arose. Thus (arts. 116–117)

$$b = ab\text{C} + abc = ab(\text{C} + c) = ab. \qquad (262.13)$$

Many of Jevons's examples were oriented around syllogisms, but in ch. 14 he reworked one of Boole's general cases and obtained the same consequences.

Jevons's procedures avoided Boole's expansion theorems, and dispensed with subtraction, division, 0 and 1, and most of the attendant methods; but his indirect mode of inference was rather tedious to operate, though more powerful. So in a paper *1866a* he announced his 'logical abacus', in which slips of paper containing between them all combinations of terms and their negations were prepared; the ones required for the given premises were selected and the consequences read off. He realised that the selection and appraisal could be better effected non-manually, and for the purpose he introduced in the paper *1870a* his 'logical machine', which produced the required inferences by mechanical means (Mays and Henry *1953a*). His procedure has some structural similarity with the truth-table method for determining the truth-values of propositions (§8.3.2).

Over and above these technicalities is the question of the relationship between mathematics and logic after these modifications. Jevons may not have fully considered it. In the introduction of *Pure logic* he stated that

> The forms of my system may, in fact, be reached by divesting his system of a mathematical dress, which, to say the least, is not essential to it [...] it may be inferred, not that Logic is a part of Mathematics, as is almost [*sic*] implied in Prof. Boole's writings, but that the Mathematics are rather derivatives of Logic.

(*1864a*, art. 6: compare his *1874a*, 191–192). This reads like a presage of Frege's or of Russell's logicisms, but is more of a preliminary speculation, and did not influence them.[22]

2.6.3 *Jevons versus Boole.* While his *Pure logic* was in press, Jevons sent Boole some proofs and corresponded with him; but the clash of position, especially concerning ' + ', was irreconcilable. For Jevons, (262.4) stated that any (finite) number of inclusive self-alternatives to A could be reduced to one instance without change of meaning: thus logical 'addition' differed from mathematical addition. For Boole,

> [...] it is not true that in Logic $x + x = x$, though it is true that $x + x = 0$ is equivalent to $x = 0$. You seem to me to employ your law of unity [(262.4)] in two

---

[22] Russell seems not to have drawn on Jevons at all; Frege's criticism of Jevons's definitions of numbers in terms of diversity is noted in §4.6.2.

different ways. In the one it is true, in the other it is not. If I do not write more it is not from any unwillingness to discuss the subject with you, but simply because if we differ on this fundamental point it is impossible that we should agree in others.[23]

The difference between the two men may be summarised as follows. Like many of the pioneers of new algebras in the 19th century, Boole was consciously extending the realm of algebras; but nevertheless he was still mindful of the properties of common algebra, which was formed as a generalisation of arithmetic. Thus he defined operations of addition, subtraction, multiplication and division, giving them these names because they satisfied laws (fairly) similar to those of the traditional versions. Jevons objected to this influence, and sought to reduce its measure in his version of Boole's system. However, he seemed to have confused the more general conception of algebra*s* (of which Boole was a practitioner) with the bearing of Boole's algebra upon quantity (which, as we saw at the end of §2.5.6, was very modest). This distinction can be related to that between universal arithmetic and symbolical algebra, and the use of the principle of the permanence of equivalent forms (§2.3.2).

2.6.4 *Followers of Boole and/or Jevons.*   Despite their differences, Jevons appreciated the novelties of Boole's system; in 1869 he opined to Macmillan, the publisher of all his books after *Pure logic*, that 'it must I am afraid be a long time before the old syllogism is driven out, and symbols of the nature of Boole's substituted in the ordinary course of instruction'.[24] Yet he did not encourage change: for example, his popular primer *1876a* on logic never mentioned Boole once, and his later books were largely restricted to syllogistic concerns. They were reprinted quite frequently, whereas of Boole's only *LT* received a reprint, in 1916, before recent times.

Some advocates of the new algebra of logic preferred Boole's version to Jevons's. For example, G. B. Halsted *1878b* wrote from the U.S.A. to *Mind* defending Boole's system, especially for its ability to express both the inclusive and exclusive kinds of disjunction via (256.3); he also rejected Jevons's association of Boole's mathematical approach with an algebra of quantity.

[23] The correspondence is published in my *1991b* (p. 30 here); parts of some of them are in Jourdain *1913d*, which was hitherto the only available source for some letters (p. 117 here). Otherwise on Jevons see, for example, Liard *1878a*, ch. 6.

The recent edition of Jevons's correspondence (1972–1981) extols his (important) contributions to economics uncluttered by his (important) contributions to logic. For example, his letters with Boole, Venn and De Morgan have been systematically omitted; a very few are included in his widow's edition (Jevons *Letters* (1886)).

[24] Jevons to Macmillan, 16 February 1869. The file of letters is held at the British Library (London), Add. Ms. 55173; this one is also excluded from the edition of Jevons mentioned in the previous footnote.

Boole's stoutest defender was John Venn, who concentrated on *LT* in his book *Symbolic logic* (the origin of this term) of 1881. For him Jevons's reforms meant 'that nearly everything which is most characteristic and attractive in [Boole's] system is thrown away' (Venn *1881c*, xxvii). He also defended Boole's definition of ' + ' on the grounds that both senses of 'or' could be expressed by means of (256.3).[25] 'I have done my best to make out in what relation [Jevons] himself considers that his exposition of the subject stands to that of Boole; but so far without success' (p. xxviii); but he was certain that Jevons's adherence to intensions led to various 'evils', such as 'the catastrophe' of not reading particular propositions extensionally (p. 36). However, in staying largely around the syllogistic tradition he was closer to *MAL* than to *LT*. In a lengthy review for *Mind* C. J. Monro *1881a* shared Venn's adhesion to Boole's principles, including over $(x + x)$ and the need for '0/0'.

Among other aspects of the book, Venn did not use De Morgan much, and on the 'Logic of Relatives' he commented: 'the reader must understand that I am here only making a few remarks upon a subject which [...] would need a separate work for its adequate discussion' (pp. 400–404), but which he did not then write. The best remembered feature of the book is the diagrammatic representation of logical relationships, now misnamed 'Venn diagrams'. The method so named is in fact usually Euler's procedure based upon the Gergonne relations (§2.4.6). His own way, which he published first in a paper *1880a* in the *Philosophical magazine*, was to draw closed convex curves in such a way as to exhibit all their possible intersections, and marking those which were empty in a given logical situation. It amounts to a pictorial representation of Jevons's method of taking the logical disjunction of all pertinent conjunctions.[26]

A significant newcomer was the Scotsman Hugh McColl (1837–1909), as he then called himself; I shall use his later version 'MacColl'. In a paper *1880a* on 'Symbolic reasoning' in *Mind*, and in related papers of the time, he offered himself as a 'peacemaker' (p. 47) between logic and mathematics. He divided the former field in the manner similar to Jevons: 'pure logic' covered 'the general science of reasoning understood in its most exact sense' (not Jevons's sense, as he noted), while 'applied logic' took this

---

[25] This point comes out especially clearly in Venn's correspondence with Jevons in March 1876 (letters in Jevons Papers, and Venn Papers, File C45).

[26] Later MacFarlane *1885a* outlined an alternative 'logical spectrum' based upon representing all of the candidate classes by a sequence of contiguous rectangles and half-rectangles. Convex curves cannot treat more than four classes; many modifications were proposed (for example, in Anderson and Cleaver *1965a*) before A. W. F. Edwards *1989a* found an indefinitely iterable algorithm. Venn *1881b* surveyed the history of logic diagrams in a piece for the Cambridge Philosophical Society, and in a companion survey *1881a* of notational systems he recorded over a score! Shin *1994a* analyses mostly Venn diagrams in terms of mathematical logic.

knowledge to *'special subjects'*, such as mathematics. For symbolism he offered (pp. 51–53):

$$\times \text{ (and)} \quad + \text{ (inclusive or)} \quad : \text{ (implies)} \quad = \text{ (equivalence)}$$
$$' \text{ (not)} \quad \div \text{ (not implies)} \quad 1 \text{ (truth)} \quad 0 \text{ (falsehood)}. \tag{264.1}$$

The latter two notions were unclearly indicated, ' = 1' and ' = 0' seeming to be the notions intended. While his treatment was oriented around syllogisms, he accepted the main lines of Boole's work; but in *1877a* he proposed that the propositional calculus be treated as a Boolean algebra, not done by either Boole or Jevons. Further, he read implication $A:B$ between propositions A and B as equivalent to $A = A \times B$ (Rahman and Christen *1997a*). He also subsumed quantification under this implication; for example, 'all X is Y' became 'an individual has attribute X: this individual has attribute Y'. MacColl's contributions of the 1900s to logic, better remembered, will be described in §7.3.6.

By contrast with these developments, De Morgan's contributions lay eclipsed, even his logic of relations.[27] Independently of De Morgan, R. L. Ellis *m1863a* had perceived the need for such a logic, but he did not fulfil it (Harley *1871a*). However, some effort was made by the young Scottish mathematician Alexander MacFarlane (1851–1913), in a three-part paper *1879–1881a* published by the Royal Society of Edinburgh, with a summary version *1881a* in *Philosophical magazine*. Using family relationships for his example, he wrote out the members related in equations such as

$$'sA = B + C + D' \tag{264.2}$$

for 'the sons of A are B, and C, and D', and developed quite an elaborate system for compounding relations and universes.

MacFarlane's paper followed a short book *1879a* on Boole's system. He kept most of it, including the expansion theorems, coefficients such as $0/0$, and the application to probability theory; but he used ' $-$ ' and ' $+$ ' without restrictions. He used separate symbols for nouns and adjectives, lamenting Boole's failure to do so. Although Venn *1879a* reviewed the book at some length in *Mind*, none of MacFarlane's work was influential; but it is of interest in treating both De Morgan and Boole.

Jevons's version of Boole's system gradually gained preference over Boole's own version. For example, when the Cambridge logician W. E. Johnson (1858–1931) wrote at length on 'The logical calculus' in *Mind*, he emulated Jevons in reducing the mathematical link; for example, with Johnson '1' and '0' became 'Truism' and 'Falsism' (*1892a*, 342–343). This

---

[27] For example, on 15 September 1863, in connection with Jevons's correspondence with Boole, De Morgan wrote to Jevons and offered to send him an offprint of his *1860a* on relations (Jevons Papers, Letter JA6/2/114); but Jevons appears not to have responded.

work was noted by Venn, in the second edition *1894a* of his book, for which Johnson read the proofs. While the basic purpose and design of the chapters was largely unchanged about 20% new material was added, raising the length to 540 pages. Some examples of the updating will be noted in §4.3.9 and §4.4.5.

The same change occurred abroad, especially regarding technical derivations, when the systems were studied by figures such as F. Kozloffsky and P. S. Poretsky in Russia, Ventura Reyes y Prósper in Spain (§4.4.4) and Hermann Ulrici in Germany (§4.4.1). We shall also see a rise in De Morgan's reputation when the fusion envisioned by MacFarlane was accomplished, and also Jevons's changes were adopted, by the two new major figures in algebraic logic: C. S. Peirce and Ernst Schröder (§4.3–4). For now, we turn to something entirely different.

2.7 CAUCHY, WEIERSTRASS AND THE RISE OF
    MATHEMATICAL ANALYSIS

2.7.1 *Different traditions in the calculus*

To the mathematician I assert that from the time when logical study was neglected by his class, the accuracy of mathematical reasoning declined. An inverse process seems likely to restore logic to its old place. The present school of mathematicians is far more rigorous in demonstration than that of the early part of the century: and it may be expected that this revival will be followed by a period of logical study, as the only sure preservative against relapse.   De Morgan *1860b*, 337

De Morgan concluded the main part of his last completed paper on the syllogism with this accurate prophecy. While he did not specify any branch of mathematics, undoubtedly mathematical analysis was one of the prime examples. The remainder of this chapter is devoted to a summary of the development of this discipline during the 19th century up to around 1870 (just before De Morgan's death, incidentally). While the main innovations took place in France and Germany, some notice was taken in Britain, and he was one of the first to encourage interest in his home country, as we shall note in §2.7.3.

Lagrange's approach to the calculus (§2.2.2), reducing it to algebraic principles, was the third and newest tradition (my *1987a*). It competed with theories stemming from Newton based in limits (but not pursued with the refinement that Cauchy was to deploy), and with the differential and integral calculus as established by Leibniz, the Bernoullis and Euler. Here the 'differential' of a variable x was an infinitesimal increment dx on x and of the same dimension as x, while $\int x$ was similarly an infinitely large variable of that dimension. The rate of change of y with respect to x, the slope of the tangent to the curve relating x and y, was written 'dy/dx', and was to be read literally as the ratio $dy \div dx$ of differentials, itself normally finite in value. The integral was written '$\int y \, dx$', which was to be understood again literally, as the sum ('$\int$' was a special forms of 's' adopted by Leibniz) of the product of y with dx: as the area between the curve and the x-axis, it was seen as the sum of infinitesimally narrow rectangles

This tradition was by far the most important one of the 18th century, which led the establishment of the calculus as a major branch of mathematics. Limits also gained some favour, although on the Continent they were presented without the kinematic elements present in Newton's 'fluxional' version in isolated Britain. So Lagrange had to meet stiff competition when selling his alternative approach; and it was mentioned in §2.2.2 that some of his contemporaries were not convinced of its legitimacy or practicability. We turn now to its most formidable opponent in the early part of the 19th century: Cauchy.

2.7.2 *Cauchy and the* Ecole Polytechnique. Born in 1789, Augustin-Louis Cauchy studied at this school in the mid 1800s (after Lagrange had finished teaching there), then entered the *Ecole des Ponts et Chaussées* in Paris and worked for a few years in the corresponding *Corps*. But his research interests developed strongly, and when Napoléon fell and the Bourbon Catholic monarchy was restored, Bourbon Catholic fanatic Cauchy was given in 1816 great and even artificial boosts to his career: appointment to the restored *Académie des Sciences* without election, and a chair in analysis and mechanics at his old school. During the Bourbon period (which ended with the revolution of 1830), he was in his element, and produced an amazing range and mass of top-class mathematics (my *1990a*, esp. chs. 10–11, 15). Our concern here is with his teaching of analysis at the *Ecole Polytechnique*, in which he set up many essential features of mathematical analysis as they have been understood ever since, especially the unification, in a quite new way, of the calculus, the theory of functions, and the convergence of infinite series. Most of the main ideas appeared in two textbooks—the *Cours d'analyse* (*1821a*) and a *Résumé* of the calculus lectures (*1823a*)—though some other results were published in research papers and later textbooks. A major inspiration and feature was his extension of the theory to complex variables; but I shall not need to treat it here, because it did not bear on the development of logic as such.[28]

The underlying link was provided by the theory of limits, in which the basic definitions and properties were presented to a measure of generality and degree of precision that had not been attempted before: 'When the values successively attributed to the same variable approach indefinitely a

---

[28] Among commentaries on Cauchy's analysis and its prehistory, see Bottazzini *1986a*, and my *1970a* (esp. chs. 2–4 and appendix) and *1990a* (esp. chs. 10 and 11). See also footnote 30.

fixed value, so as to differ from it as little as one might wish, this latter is called the *limit* of all the others' (Cauchy *1821a*, 19). He stressed, in a way then novel, that passage to this limit need be neither monotonic nor one-sided. He also represented orders of 'infinitely small' and 'infinitely large' by monotonic decrease of sequences of integers to zero. His choice of terms was unfortunate, as these infinitesimals did not at all correspond to the types such as dx mentioned in the previous sub-section. Nor did his infinities presage any Cantorian lore in this regard; on the contrary, elsewhere he explicitly denied the legitimacy of the completed infinite.

In terms of limits Cauchy cast many basic components of mathematical analysis, in the forms that have been broadly followed ever since. The convergence of the infinite series $\sum_j u_j$ was defined by the property that the remainder term r_n after n terms passed to (the limiting value) zero as n approached infinity; in this case the nth partial sum s_n of the series approached the sum s (pp. 115–120: he popularised the use of these notations). The exegesis following in ch. 6 included the first batch of tests for convergence of infinite series.

The continuity of a function $f(x)$ at a value x was defined in a sequential manner: that $f(x)$ *'will remain continuous with respect to the given limits, if, between these limits, an infinitely small increase of the variable always produces an infinitely small increase of the function itself'* (p. 43). Cauchy also re-expressed it for continuity 'in the vicinity of a particular value of the variable x', and proved in ch. 2 various theorems on continuous functions, of both one and several variables. Other material appearing in the *Cours*, ch. 5 included a study of functional equations, although his treatment was oriented more around conditions for the solution (especially for continuous functions) of simple equations, and the derivation of the binomial series, rather than Babbage-like manipulations (§2.4.2) to solve complicated ones. On functions in general, he insisted that they always be single-valued, so that even $\sqrt{x}$ $(x > 0)$ had to be split into its positive and negative parts. This restriction became standard in mathematical analysis, with fundamental consequences for Russell (§7.3.4).

The calculus appeared two years later in Cauchy's *Résumé* of 1823. There he defined the derivative and the integral of a function respectively as the limiting values (if they existed) of the difference quotient and of sequences of partition sums:

$$f'(x) := \lim[(f(x + h) - f(x))/h] \text{ as } h \to 0; \qquad (272.1)$$

$$\text{and } \int f(x)\,dx := \lim \sum_j \left[(x_j - x_{j-1})f(x_{j-1})\right], \; x_0 \leqslant x \leqslant X \qquad (272.2)$$

as the partition of chosen points $\{x_j\}$ within the finite interval $[x_0, X]$ became ever finer (lectures 3, 21). The great novelty of his approach lay not particularly in the forms of the defining expressions, for they had appeared before (usually in vague forms); it was the fact that the definitions were *independent* of each other, so that the 'fundamental theorem of the calculus', asserting that the differential and the integral calculi were inversedly related branches, could now really be a theorem, requiring sufficient conditions on the function for its truth, rather than the automatic switch from one to the other branch which had normally been the assumption made in the other versions of the calculus.[29] In his case his proof required the function to be continuous (in his sense) over the interval of definition of the integral (lecture 26).

In his exegesis of the calculus Cauchy proved versions of many of the standard results and procedures of the calculus: properties of derivatives and partial derivatives of all orders, differentials (though, as with infinitesimals he presented a new kind of definition bearing no resemblance to traditional versions) and total differentials, mean value theorems, term-by-term integration of infinite series, multiple integrals, differentiation and integration under the integral sign, integrals of simple functions, and so on. Four points need emphasis here, the first mathematical, the last two logical, and the second both at once.

Firstly, one of the main theorems was Taylor's, for which Cauchy provided forms for the remainder term and thereby imposed conditions for its convergence (lectures 36–37 and second addition). Lagrange's faith in the series, described in §2.2.2, was rejected; indeed, Cauchy went further, for in lecture 38 of the *Résumé*, and in more detail in a paper *1822a*, he refuted the assumption that a function can always be expanded in a series in the first place by providing counter-examples such as $\exp(-1/x^2)$ at $x = 0$.

Secondly, Cauchy's statements of the convergence of that series, and of the fundamental theorem, in terms of broad definitions of basic concepts and sufficient conditions for the (claimed) truth of the stated theorem, characterise the novel kind of rigour with which he invested his new doctrine; for he always presented theorems in terms of sufficient and/or necessary conditions laid upon functions, integrals, or whatever. Indeed, one must credit him for even thinking of stating conditions at all for the validity of several of the standard processes mentioned in the above exegesis.

Thirdly, Cauchy raised the status of *logic* precisely by stressing such conditions, and their weakening or strengthening when modifying theorems. However, he did not adopt any *theory* of logic known at the time—least of all the '*logique*' of Condillac (§2.2.2), with its associations with algebra which his new discipline was intended to supplant.

---

[29] Lagrange's allowance of exceptional values of x for the function (§2.2.2) was the best kind of awareness expressed hitherto. Between him and Cauchy, Ampère had essayed some ideas in this direction.

Finally, while Cauchy called his subject 'mathematical analysis', his proofs were almost always synthetic in the traditional sense of the term explained in §2.2.3; that is, he started from basic concepts and built up his proof with the theorem as its last line. This confusing use of the word 'analysis' flourishes throughout the rest of our story!

2.7.3 *The gradual adoption and adaptation of Cauchy's new tradition.* The reception was quite complicated in all countries, and is not well studied. The new approach was detested at the *Ecole Polytechnique* by both staff and students, as being far too refined for the students at an engineering school and remote from their concerns; the superior strength for heuristic purposes of the Leibniz-Euler tradition of differentials and integrals were preferred for applied mathematics, world-wide. However, when Cauchy left France in 1830 to follow the deposed Bourbon king into exile after the revolution in July of that year, many aspects of his doctrine were retained by his successors who taught the course in analysis at the school over the years (Navier, Sturm, Liouville and Duhamel), although in some cases it was diluted in precision and mixed in with elements of the other traditions.

In Britain De Morgan produced a large textbook on *The differential and integral calculus*. In a Cauchyan spirit he began with an outline of the theory of limits and gave versions of (272.1–2) as basic definitions; but he made no mention of Cauchy in these places (De Morgan *1842a*, 1–34, 47–58 (where he even used Euler's name 'differential coefficient' for the derivative!) and 99–105). He even devoted some later sections to topics consistent with his philosophy of algebra (§2.4.2) but which Cauchy did not tolerate, such as pp. 328–340 on Arbogast's calculus of 'derivations' (an extension of Lagrange's approach to the calculus which influenced Servois in §2.2.5), and ch. 19 on 'divergent developments' of infinite series. He did not even rehearse in this book the treatment of continuity of functions which he had given in 1835 in an algebra textbook: ' "let me make x as small as I please, and I can make $7 + x$ as near to 7 as you please" ' (*1835b*, 154–155). This is the first occurrence of the usual form of continuity which is used today and called the '(ε, δ)' form (G. C. Smith *1980a*). Cauchy had introduced these Greek letters into mathematical analysis; but they did not underlie his definitions of continuity, which we saw in the last sub-section to be sequential. De Morgan's definition appeared in a book on algebra because, in another difference of view from Cauchy's, he regarded the theory of limits as algebraic since it handled mathematical objects and properties such as $\sqrt{2}$ and $\sqrt{8} = 2\sqrt{2}$ (see, for example, *1836a*, 20).

At the research level two of Cauchy's most important first followers were young foreigners, who took up prominent problems in analysis and even refined his approach. N. H. Abel *1826a* studied the convergence and summation of the binomial series for both real and complex values of the arguments. J. P. G. Dirichlet *1829a* examined the sufficient conditions that a function should exhibit in order that its Fourier series could converge to it, and found that a finite number of discontinuities and turning values were required. At the end he threw off the characteristic function of the irrational numbers (as we now call it), as an example of a function which could not take an integral.

Abel was also one of the founders of elliptic functions in the 1820s, and his work and the independent contributions of Jacobi helped to spread Cauchy's approach in this important topic. Dirichlet's study was also influential, since Fourier series had become an important technique for applied mathematics, especially as a form of solution of differential equations (my *1990a*, esp. chs. 9, 15, 17–18). Further, he discussed some of the ensuing issues with the young Bernhard Riemann (1826–1866), who was inspired in 1854 to draft out a doctoral thesis at Göttingen University on these series. In fact a thesis on the foundations of geometry was chosen by examiner Gauss. Both texts appeared only posthumously, under the editorial care of Dedekind, apparently in 1867.

In his thesis *1867b* on geometry Riemann provided a philosophical study of space informed by mathematical insights (Ferreirós *1999a*, ch. 2). The chief idea was '*n*-fold extended magnitude' (space in general) upon which 'Mass-relationships' obtained; an important example was physical 'space' whose relations were studied in geometry, but discrete cases were also admitted (art. 1, para. 1). It is not necessary for us to pursue his line of thought, which is just as well given his cryptic style (Novak *1989a*); he admitted both continuous and discrete 'manifolds' ('*Mannigfalthigkeiten*') of objects falling under general concepts, with a part-whole relation implicitly adopted.

Riemann's thesis *1867a* on analysis contained a more direct use of set-theoretic notions (following Cauchy and Dirichlet), again formulated in cryptic but extraordinarily suggestive terms. Its appearance was a seminal event in the history of real-variable analysis: immediately several mathematicians started to explore and clarify various of its ideas. One part of the thesis tried to refine Cauchy's definition (273.2) of the integral by defining upper and lower bounds on the sequence-sums in terms of the maximal and minimal values of the function over each sub-interval defined by the partition: a clearer version of this idea using upper and lower sums is due to Gaston Darboux *1875a*. The main part dealt with various consequences of Dirichlet's conditions for convergence of Fourier series: we shall pick these up in §3.2.3, as they provided the origins of Cantor's creation of set theory.

2.7.4 The refinements of Weierstrass and his followers.   Riemann's paper was a wonderful source of *problems* for mathematicians; the main originator of *techniques* by means of which these and other problems in real-variable analysis could be tackled was Karl Weierstrass (1815–1897), who rose to great prominence in world mathematics from the late 1850s, especially with his lecture courses given at Berlin University. He accepted Cauchy's basic approach to real-variable analysis (and, like Cauchy himself, used limits and equivalent definitions of continuity and convergence also in complex-variable analysis[30]); but he came to see that in various ways its definitions and procedures did not match the aspirations for rigour which Cauchy had uttered.

Over the years Weierstrass and his disciples followed Cauchy's basic ideas on giving broad definitions and seeking sufficient and/or necessary conditions for theorems, working with limits, continuity, convergence, and so on, and producing detailed synthetic proofs; but they introduced several refinements. From the 1870s German figures dominated, such as Cantor, Paul du Bois Reymond, Hermann Hankel, Axel Harnack, Eduard Heine and Hermann Amandus Schwarz; but some other nationalities provided important contributors during the 1870s and 1880s, such as Darboux, Charles Hermite and Camille Jordan in France, Ulisse Dini and Giuseppe Peano in Italy, and Gösta Mittag-Leffler and Ivar Bendixson in Sweden. The most pertinent innovations are grouped below as five inter-related issues;[31] some will be described in more detail in §3.2 and §4.2.

Firstly, while Cauchy had a completely clear grasp of the basic definitions and use of limits, he was hazy on the distinction between what we now call the least upper bound and the upper limit of a sequence of values. For example, he used the latter notion in his *Cours d'analyse* when presenting the first batch of tests of convergence of infinite series, but he specified it with rather vague phrases such as 'the limit towards which the greatest values converge' (*1821a*, 129). The distinction had to be sorted out, and the different contexts for their respective use.

Secondly, theorems involving limits, and considerations of functions with infinitely many discontinuities and/or turning values and the definability of their integrals, focused attention on collections of points (or values) possessing certain properties. They were to be construed as sets, and were the main stimuli for the growth of point set topology, especially within Cantor's theory. Riemann's draft thesis was particularly fruitful in this context, for he constructed several examples of the type of function just mentioned and found their Fourier series; further, his definition of the integral worked in effect with sets of measure of zero without explicitly mentioning either set or measure.

Thirdly, and sometimes as examples of the last issue, the relationship between rational and irrational numbers needed closer examination. It was

---

[30] However, Weierstrass's founding of complex-variable analysis in power series was different from Cauchy's, and also from another approach due to Riemann (Bottazzini *1986a*, chs. 4, 6, 7).

[31] Among general secondary sources, see Pringsheim *1898a* and *1899a*, T. W. Hawkins *1970a*, my *1970a* (ch. 6 and appendix), and Dugac *1973a*.

well known that each type of number could be construed as the limit of a sequence of numbers of the other type; but it became clear that, especially in connection with theorems asserting the *existence* of some limit, the proof might require irrational numbers to be *defined* in terms of rational ones in order to avoid begging the question of existence involved in the theorem. Cauchy had faltered in his *Cours* when, for example, he drew on the real line structure when claiming to establish necessary and sufficient conditions for the convergence of an infinite series (*1821a*, 116: compare pp. 337, 341).

Fourthly, Cauchy and his successors tended to move fairly freely between properties of continuity and convergence defined *at* a point, in its neighbourhood, and over an interval of values. While the distinction between these different types of context was obvious, the consequences for mathematical analysis only began to be grasped in the Weierstrassian era. Then there were introduced *modes* of continuity and especially convergence: uniform, non-uniform, quasi-uniform. The need for these distinctions was increased when the '(ε, δ)' form of continuity came to be preferred over Cauchy's sequential form. The contexts included the convergence and term-by-term differentiation or integration of infinite series of functions, differentiation under the integral sign, double and multiple limits taken simultaneously or in sequence, and many aspects of handling functions of several variables. Quite a few variables could be present together: for example, in the series of functions $\sum_{j=1}^{n} u_j(x)$ not only were x and n at work but quite possibly also incremental variables on both of them ($x + h$ and $n + m$, say). Working out careful forms of definition and proof here, and keeping modes distinct from each other, required very meticulous scrutiny (Hardy *1918a*).

Finally, and notably in connection with the first and the fourth issues, the use of symbolism had to be increased in both considerable measure and a systematic manner. One type of case is of particular interest here: some nascent quantification theory, to express and indeed clarify the functional relationships between the different variables operating in a problem: in particular, to distinguish 'For all x there is a y such that ...' from 'There is a y such that for all x ...'.

2.8 JUDGEMENT AND SUPPLEMENT

2.8.1 *Mathematical analysis versus algebraic logic.*

The major place has been given to algebra and algebraic logic because during the period covered it emerged in this chapter as a group of (differing) uses of algebras to represent procedures in logic. By contrast, in the last section we saw no explicit logic, although ideas were born which will be taken up in the succeeding chapters on mathematical logic. No explicit clash between the two lines of work was in operation; however, some conflict in purpose and philosophy is evident.

In a famous and influential passage in the preface to *Cours d'analyse*, Cauchy wrote: 'as for methods, I have sought to give them all the rigour that one requires in geometry, so as never to have recourse to the reasons drawn from the generality of algebra' (*1821a*, ii). The key word is 'rigour', which he conceived in terms of the broad definitions and deductive lines of reasoning to prove *in detail* theorems which usually incorporated necessary and/or sufficient conditions upon the mathematical components involved. His allusion to geometry concerned the strict rigour which proofs of Euclidean geometry were then held to exhibit: exposure of Euclid's lacunae and flaws was not to occur for several decades (§4.7.1–3). But Cauchy was not appealing to intuition: on the contrary, as with Lagrange, no diagrams adorned his writings. Further, his disparagement of 'the generality of algebra' was directed especially against the Lagrangian tradition. However, it was precisely that tradition to which the English mathematicians adhered, from Babbage and Herschel to De Morgan and Boole; and the last two men found major sources of analogy and technique to guide and inspire their mathematicisations of logic. This clash will provide points of contrast during our examination of the further refinements and extensions of Weierstrass's version of mathematical analysis, which form a main theme of the rest of this book. First, however, we must briefly locate two "background" philosophers.

2.8.2 *The places of Kant and Bolzano.*

The thought of Immanuel Kant (1724–1804) bears somewhat upon our story regarding both logic and mathematics. He wrote little explicitly on logic, and the 1800 edition of his logic lectures is of somewhat doubtful authenticity (Boswell *1988a*). Regarding logic as providing 'the general rules for understanding relationships between bodies of knowledge', he largely followed the syllogistic tradition; in particular, he defined 'analytic' propositions as those cast in syllogistic form and in which the subject was contained within the predicate, and 'synthetic' simply as not analytic. His philosophy of mathematics was based upon the premise that mathematical propositions were 'synthetic a priori'; that is, they were neither analytic nor dependent upon our experience for their truthhood. Space and time were granted the same status; one consequence was the claim that Euclidean geometry was the only possible one (§4.7.4), a view which was to gain him a bad press by the late 19th century. In addition, his use of the traditional part-whole theory of collections, embodied in the containment property above, was to be challenged by Cantor's set theory. A survey of Kant's position and its relationship to some modern philosophies of mathematics is provided in Posy *1992a*.

The reception of Kant's epistemology among mathematicians and logicians was more mixed. The main assumption was the role of active 'thinking', which allowed the agent to use his power of 'intuition' to make 'judgements' about relationships between individual 'objects' and/or more general 'concepts'. Positivists usually dismissed such talk as mere 'idealism'; however, in all versions of logic some role was usually assigned to judgements.

Kant also discussed at length certain 'antinomies' of knowledge, such as the existence and also non-existence of a first moment in time. This dichotomy was heightened by Kant's semi-follower G. W. F. Hegel (1770–1831) into a methodology of 'thesis' and 'antithesis' resolved in a 'synthesis': it formed a central feature of the 'neo-Hegelian' movement in philosophy which became dominant at the end of the century, especially in the England of the young Russell (§6.1.3). However, it was not much used for solving the paradoxes which came to infect mathematical logic and Cantorian *Mengenlehre* in the 1900s.

This avoidance of Kantian principles was fairly typical of the reception of Kantian philosophy by Russell and followers; as a philosopher of both logic and mathematics he was found generally wanting (and Hegel even more so), especially for allowing synthetic *a priori* judgements, relying upon syllogistic proofs, and maintaining links between logic and psychology. Conversely, some thinkers of a generally Kantian persuasion were to criticise mathematical logics, disagreeing over the conception of analysis and wishing to see a greater role assigned to intuition, and in some cases doubting the legitimacy of the Cantorian actual infinite.

The situation was complicated by the rise to importance from the 1870s of various schools of 'neo-Kantian' philosophy, which remained active throughout our period. The most relevant change was to reject the Master's claim that space and time were *a priori* forms of perception or pure intuition and to treat them as constructions affected by pure thought in which logic played some role (Friedman *1996a*). Among the schools, the one associated with Marburg University is the most relevant, since they favoured thought and method(ology) over, say, sense-experience, or psychology. Among their members Ernst Cassirer considered logicism most closely (§7.5.2, §8.7.8). A further untidiness arises over the use of the word 'intuition': whether in some fairly strict Kantian sense (as often with philosophers) or in a looser sense of initial formulations of theories (as often with mathematicians).

These philosophical traditions were enormously influential during the 19th century, especially but not only in German speaking cultures; I have not attempted to do them justice here. By contrast, the work of the Bohemian philosopher and mathematician Bernard Bolzano (1781–1848) was then little-known. He achieved much in mathematics, logic and philosophy but gained little influence outside his immediate circle during his lifetime or afterwards; so he gains only this short review and a few mentions hereafter.

Bolzano's career falls into three phases. After training in philosophy, physics and mathematics at Prague University he joined the Theological Faculty but pursued mathematics as his main research interest. Two books and three booklets came out between 1804 and 1817; the most important item, *1817a*, contained a newly rigorous proof of the intermediate value theorem drawing upon formulations of the notions of limit, continuity of functions and convergence of infinite series strikingly similar to those found soon afterwards in Cauchy's *Cours* (1821) and thus making him a co-pioneer of mathematical analysis. He must have realised that this booklet was significant, for in 1818 he placed it also as a number in the *Abhandlungen* of the *Böhmische Gesellschaft der Wissenschaften*; thus, uniquely among these works, it is not rare.

For his living Bolzano taught religion at the Faculty, and drew from it, and from considerations of contemporary life, a Utopian socialist philosophy. This was Very Naughty, and as a result he was sacked in 1819. The second phase sees him living much in Southern Bohemia with a family called 'Hoffmann', where his major production was a four-volume epistemological work, *Wissenschaftslehre* (Bolzano *1837a*). Many of his ideas on logic were formulated here, of which two are particularly notable: his concern with 'deducibility', formulated in a manner general enough to individuate logical consequence (see esp. arts. 154–162); and his stress on objective truths as opposed to (though intimately linked with) judgements, as expressed in propositions (arts. 122–143, 198–218, 290–316). While not algebraic in the English sense, his logic used a relatively large amount of symbolism, and also the part-whole theory of collections (indeed, rather more substantially than with most contemporaries).

The political atmosphere seems to have calmed down enough for Bolzano to return in 1842 to Prague, where he continued to work until his death in 1848. During this phase mathematics was back on his agenda, producing especially a remarkable survey of the '*Paradoxien des Unendlichen*' which was published as a posthumous book *1851a*. The editor, Franz Přihonský, was one of a group of devoted followers who tried to sustain and spread his work, but with little success. There was also a huge collection of manuscripts to be edited, but little was done. Even a twelve-volume edition of his main philosophical and religious publications (including the *Wissenschaftslehre*), put out by a Vienna house in 1882 (Bolzano *Writings*), failed to capture the imagination.

But just around that time some of Bolzano's logic and mathematics began to receive attention. In an encyclopaedia article on the concept of limit Hankel *1871a* had mentioned the analysis booklet; maybe he had seen it listed in the entry for Bolzano in the first volume (1867) of the *Royal Society catalogue of scientific papers*. At all events, publicity now slowly increased: Otto Stolz wrote an article *1882a* in *Mathematische Annalen* on the analysis booklet and one of the early books. The book on paradoxes was reprinted in 1889, and the booklet in 1894 and 1905. During this century Bolzano's reputation has steadily risen, especially as more anticipations have come to light in the manuscripts; further editions and also translations have been made, dominated now by a *Gesamtausgabe* of both publications and (chosen versions of) manuscripts.[32] But in most contexts his successors have found only premonitions of now known notions and theories, albeit astonishing, rather than novelties directly to stimulate new work: the first figure of note to be significantly influenced was the philosopher Edmund Husserl (§4.6.1). More normally, Bolzano's meditations on the infinite(s) brought him to the edge of the results, already achieved, of an early admirer of the 1880s: Georg Cantor, whose own feats are chronicled in the next chapter.

[32] This edition, Bolzano *Works*, was launched with the splended biography E. Winter *1969a*, written by a leading Bolzano scholar; the manuscripts, which are held in the Vienna and Prague Academies, are being distributed among its various series and sub-series. The five early mathematical works were photo-reprinted in 1981 as Bolzano *Mathematics*. Several items have been translated into various languages: the trio mentioned in the text are available in English (only parts of the *Wissenschaftslehre*). His work has been subject to much commentary, variable in quality; of the general studies, Sebestik *1992a* is recommended. On the status of the principle of contradiction in Bolzano, Kant and many other figures, see Raspa *1999a*.

CHAPTER 3

Cantor: Mathematics as *Mengenlehre*

3.1 PREFACES

3.1.1 *Plan of the chapter.*   After summarising Cantor's life and career in the next section, the story is told of his creation of the branch of mathematics which we call 'set theory'; but when I wish to refer to his version of it I shall conserve even in translations the word '*Mengenlehre*' which he used especially in his final years of the mid 1890s and which became the most common name among German-writing authors thereafter (§4.2.1). First, §3.2–3 covers its founding between 1870 and 1885, and §3.4 treats the final papers. Important concurrent work of Dedekind is also included: on irrational numbers in §3.2.4, and on integers in §3.4.1–3. Then §3.5 presents a survey of some of the main unsolved mathematical problems and philosophical questions present in Cantor's work, followed by considerations of his philosophy of mathematics in §3.6.

The range and status of the *Mengenlehre* by the late 1890s is reviewed in §3.7. Throughout the emphasis falls rather more on the foundational and general features of the theory than on the mathematical aspects, which we now call 'point set topology'.

3.1.2 *Cantor's career.*   Georg Cantor was born in Saint Petersburg in 1845 as the first son of a cultured business man who was to exercise considerable influence on his formation; for example, instilling in him a strong religious spirit. According to a letter which Cantor sent to Paul Tannery in 1896, his father 'was born of *Isrealite parents*, who belonged to the *Portuguese Jewish community*' in Copenhagen but 'was *christened Lutheran*', while his mother was 'a born Saint Petersburger' of a Roman Catholic family (S. P. Tannery *1934a*, 306); so he was not a practising Jew, and was unrelated to the Jewish historian of mathematics Moritz Cantor. Three more children were born by 1849, and then in the mid 1850s the family moved to Heidelberg in Prussia, for sake of the father's health; nevertheless, he died there in 1863, leaving a considerable fortune.

Around that time Cantor went to Berlin University to study mathematics. One of his fellow students was Karl Hermann Amandus Schwarz; his principal professors were Karl Weierstrass, Leopold Kronecker and Ernst Kummer, and he followed the concerns of the latter two, writing both doctoral dissertations (the *Dissertation* and the *Habilitation*) on number theory in 1867 and 1869. To begin his academic career he replaced

PLATE 1. Portrait of Georg Cantor with his sister Sophie, with whom he was always close. He seems to be in his mid twenties, which would date the photograph around 1870, at the start of his career. First publication; made available to me by Cantor's descendants. Another portrait of the young pair was published, for the first time, in the *American mathematical monthly 102* (1995), 408, 426.

Schwarz as *Privatdozent* at Halle University, a second-ranking establishment in the German hierarchy, where Eduard Heine (1821–1881) was full *ordentlich* professor (his significance will be brought out in §3.2.1). To his disappointment, Cantor passed his entire academic career at Halle; but he was not rejected there, being promoted to *ausserordentlich* professor in 1872 and to a full chair in 1879, an additional post to Heine's.

Plate 1 is a photograph of Cantor from this period; published here for the first time, it shows him with his sister Sophie, who was very close to him. He married into a Jewish family in 1874, and had six children. His work developed steadily for the next decade; but in the summer of 1884 he seems to have had some sort of mental attack, possibly a mid-term crisis, which leaves the sufferer susceptible to depressive states. Although he seemed to recover and resume teaching duties, his research in *Mengenlehre* switched emphasis from the mathematical aspects to its philosophical and religious consequences. He also devoted much effort to attempting to prove that Francis Bacon wrote the plays of William Shakespeare; quite a popular topic in Germany at that time, for twenty years Cantor was to be a prominent figure, with support from Sophie (Ilgauds *1892a*).

Between 1891 and 1897 Cantor published two more papers on the *Mengenlehre*. However, his research activity was decreasing, and from the late 1880s he had been giving much time to professional affairs: the launching in 1890 of the *Deutsche Mathematiker-Vereinigung* (§3.4.5), and of International Congresses of Mathematicians from 1897 onwards.

Then, just when his external life began to flower with the general acceptance of his work, Cantor's internal life disintegrated. A serious concurrence of difficulties in the *Mengenlehre*, university politics, a dispute with the *Kultusministerium*, and the sudden death of his youngest son seems to have triggered a major collapse in the autumn of 1899, much more serious than the crisis in 1884. While he fulfilled his university duties for the majority of the following years until his retirement in 1913, he spent extended periods of the last twenty years of his life after 1899 in the University *Nervenklinik* and in sanitoria. To a modern view the surviving documentation suggests that he was manic depressive, and that his illness was endogenous, not basically caused by the controversies surrounding his work.[1]

After his death in 1918 Cantor studies were favoured in the 1930s by an excellent biographical article Fraenkel *1930a* for the *Deutsche Mathematiker-Vereinigung*, and a moderate edition Cantor *Papers* (1932) of his writings on *Mengenlehre* prepared by Ernst Zermelo,[2] followed five years

---

[1] See my *1971c* for much new information on Cantor's career, including evidence on his mental illness. I learnt from Bernard Burgoyne that some of the notes had been used by the Hungarian psychiatrist I. Hermann in a paper *1949a* on psychological aspects of set theory; he may have been drawn to Cantor by a collaboration with the logician Rosza Petèr (see her letters to P. Bernays during 1940 in Bernays Papers, 975: 3472–3474). Charraud *1994a* has also used the notes, in an interesting psychoanalytical study of Cantor which however suffers from shaky understanding of the mathematics.

An influential source of misinformation on Cantor's illness is Russell's autobiography, where he chose to publish two undeniably eccentric letters of 1905 from Cantor, and to preface them with the claim that Cantor 'spent a large part of his life in a lunatic asylum' (*1967a*, 217–220; the first also in Cantor *Letters*, 457). Russell heard this phrase uttered at an *Encyclopaedia Britannica* dinner in November 1902 (see his journal entry in *Papers 12*, 11).

[2] My lack of enthusiasm is stimulated by Zermelo's omission of some important footnotes from his papers (some will be mentioned in due course), frequent changing of notations (mostly from roman into italic format), and addition of his own technical terms (albeit in square brackets, but *omitted* here). For materials used in preparing the edition see Zermelo Papers, Box 6.

later by a careful edition of the main bulk of his correspondence with Richard Dedekind (cited as 'Cantor-Dedekind *Letters*'). This was prepared by Emmy Noether and the French philosopher and mathematician Jean Cavaillès,[3] who also wrote an excellent historical survey *1938a* of set theory (§9.6.5).

Proper research then largely languished for some decades; but in the mid 1960s Herbert Meschkowski traced some of the descendants and located the remaining *Nachlass*. It turned out that the materials had always been kept by the family, but that they had not been used by any of the scholars just mentioned, and that in the circumstances of the end of the Second World War the family house had been occupied and most of them had been destroyed or disappeared. In the late 1960s the remainder came into my hands, and I set them in some order: at my recommendation to his descendants, they were placed (and later recatalogued) in the University Library at Göttingen during the 1970s (Cantor Papers), to join those of contemporaries such as Dedekind, Felix Klein and David Hilbert.

This co-location carries more significance than is evident at first. Cantor was very lonely in his professional career, and compensated for his situation by carrying on intensive correspondences with a few colleagues, concerning both his *Mengenlehre* and his other work. The list of mathematical correspondents reflects his career in publishing on the former. From the beginnings in 1870 Schwarz was a major recipient for several years, until he turned against Cantor's work. Then it was the turn of Klein, who accepted for publication between 1878 and 1884 in *Mathematische Annalen* the longest sequence of papers. By then a new figure had arisen: the Swedish mathematician Gösta Mittag-Leffler (1846–1927), rich enough on his wife's money to launch in 1882 a new journal, *Acta mathematica*. Anxious to gain good copy for his first volumes and already appreciative of the quality of Cantor's work, he arranged for French translations to be prepared of most of the main papers that Cantor had already published. In the course of its preparation Mittag-Leffler received the usual torrent of letters, and also some original papers; but he was crossed off the list in 1885 when he recommended that a later paper be withdrawn for its lack of major new results (§3.3.2).

---

[3] Cantor-Dedekind *Letters* contains their correspondence up to 1882; according to a letter of 1932 from Zermelo to Cavaillès, the publisher of Cantor's *Papers*, J. Springer, rejected the edition (Dugac *1976a*, 276), so it appeared from Hermann in Paris. Some of these letters are now also in Cantor *Letters*, 30–60, and many are translated into English in Ewald *1996a*, 843–877. Noether took the manuscripts with her to the U.S.A. on her emigration in 1933, and after her death two years later they lay unknown with her lawyer, before being found and reported in Kimberling *1972a*; they are now held in the library of Dedekind's institution, the *Technische Hochschule* in Braunschweig. For the texts of the non-mathematical parts of the correspondence, see my *1974b*; for some reason Dugac's transcription (*1976a*, 223–262) lacks some folios. On the publication of the letters of 1899, see footnote 29.

Thereafter Cantor corresponded only spasmodically with mathematicians, turning more to theological figures and to Shakespearian experts; his last significant mathematical contact was with Hilbert, by letter, largely during the 1890s. During the 1900s he had an interesting exchange of letters with the English historian and mathematician Philip Jourdain, partly on the development of his work: Jourdain published short extracts from them in some of his own papers, and I have used the whole surviving exchange in my *1971a*. I cannot draw here in detail on all of Cantor's correspondence, but it casts valuable light on both the man and the mathematician; a good selection is available in Cantor *Letters*.[4]

3.2 THE LAUNCHING OF THE *MENGENLEHRE*, 1870–1883

3.2.1 *Riemann's thesis: the realm of discontinuous functions.* (Dauben *1979a*, ch. 1). The publication by Dedekind in 1867 of Riemann's thesis on trigonometric series (§2.7.3) launches the mathematics. Aware of the technical difficulty of extending Dirichlet's sufficient conditions for the convergence of Fourier series to handle functions $f(x)$ with an infinity of turning values and/or discontinuities (Riemann *1867a*, art. 6), he tried another approach and looked for necessary conditions, probably in the hope of finding some which were both necessary and sufficient. He worked with this series, which he called 'Ω':

$$\Sigma_{r=0}^{\infty} A_r, \text{ where } A_0 := b_0/2 \text{ and } A_r := a_r \sin x + b_r \cos x \ (r > 0) \quad (321.1)$$

and $-\pi \leqslant x \leqslant \pi$, with 'value' $f(x)$; the coefficients $\{a_r, b_r\}$ were *not* specified by the Fourier integral forms. He then defined a function $F(x)$ as the formal second term-by-term integration of Ω:

$$F(x) := C + C'x + A_0 x^2/2 - \Sigma_{r=1}^{\infty} A_r/r^2, \quad (321.2)$$

and sought various relationships between these two functions (arts. 8–9). Of main interest are the necessary and sufficient conditions on F for Ω to converge (or, as he put it, for its members to 'become eventually infinitely small for each value of x'); for he then wondered if the consequences of this property did not obtain 'for each value of the argument variable' x (art. 11). His own approach was to examine the series expansion of $(F(x + t) + F(x - t))$ (art. 12); his followers concentrated on the underlying perception that a trigonometric series might not be a Fourier series.

---

[4] Among the biographies of Cantor, the most valuable are Dauben *1979a* and Purkert and Ilgauds *1987a*; Kertesz *1983a* is not of this calibre, but it is well illustrated and contains more information on his Halle career, including on pp. 89–94 a full list of his lecture courses.

Of the various papers which followed, notable is the thesis *1870a* written at Tübingen University by Hermann Hankel on 'infinitely often oscillatory and discontinuous functions'. Not only an excellent mathematician but also an historian of mathematics, his early death in 1873 in his mid thirties was a serious loss. In his thesis he developed (321.1) to define functions which did or did not have a Riemann integral, and he went further than Riemann in exploring the topology of the real line. He also introduced the phrase 'condensation of singularities' to express the accumulation of points of oscillation and/or discontinuity of functions, and deployed Taylor's series (again like Riemann) to find expressions for them (art. 4). Among possibilities, he considered points which (in Cantor's later name of §3.2.6) were everywhere dense in a given interval (art. 7). Impressed by Dirichlet's mention of the characteristic function of the rationals (§2.7.3), he used Taylor's series to find in art. 9 (ex. 4) a complicated expression for this function, and in the second appendicial note he offered the definition of a similar function:

$$f(x) := \sum_{m=1}^{\infty} \omega^m (m \sin mx\pi)^{-2}, \ \omega > 0, \tag{321.3}$$

where $f(x) = \infty$ when x was rational and either tended towards ∞ or took finite values for irrational values of x. We shall see Peano unify these two approaches to find in (521.1) an expression for the Dirichlet function.

The quality of Hankel's study led to its reprint in 1882 as a paper in *Mathematische Annalen*, where it gained greater circulation; but before then a rare British contribution was made from Oxford by H. J. S. Smith *1875a*. Building upon both Riemann and Hankel to study 'the integration of discontinuous functions', he distinguished two kinds of 'system' of points along the line, as in 'loose' or 'close order' (arts. 9–10); the latter notion corresponded to Cantor's 'dense in an interval' (§3.2.9), and looseness was simply its negative. His most interesting exercise occurred in art. 15, where he defined a 'system' P_S of points with generic member specified by the series $\sum_{r=1}^{S} a_r^{-1}$ (a_r positive integers), and characterised the distribution of such systems in terms of partitioning a given interval into m equal sections, then all but the first again into m, and so on. The left hand points of these intervals would yield P_S after m iterations, and points of discontinuity could be allowed for a continuous function and still leave it Riemann-integrable. In a variant partition method (art. 16), he found a case where non-integrability would hold.

Several features related to Cantor's work to come are of interest here. Firstly, P_S looks like the ternary set (328.2) to be introduced by Cantor nearly a decade later; for example, it is of measure zero (as Smith proved, in the Riemannian language of §2.7.3). However, it is not as interesting an object, for Smith took an end of each of a collection of intervals and not Cantor's middle thirds, and all his systems were denumerable. Secondly, he introduced both a word *and a letter* for the collections; in fact, we shall see in §3.2.3 that Cantor had already taken the step three years earlier, but Smith did not cite him and can be regarded as an independent innovator.

Conversely, Cantor seems never to have read Smith, but he did propose a new method of condensing singularities in a paper *1882a* using the set-theoretic techniques which he had developed by then. For singularities had become an important part of point set topology and its application to the theory of functions and integration. A notable contributor was Paul du Bois Reymond (1831–1889), a close associate of Weierstrass (Mittag-Leffler *1923a*); he succeeded Hankel at Tübingen in 1874. He developed his techniques in response to various questions in the theory of functions and integration, including those raised by Riemann's thesis concerning the distribution of zeros, turning values and/or discontinuities of functions; he also used them to some extent in his 'calculus of infinitaries' ('*Infinitärkalkül*'), in which he studied the ways and rates at which functions $f(x)$ could go to $\pm\infty$ as x did. In the course of this work he constructed parts of point set topology for himself, although not without some errors of conception.[5] We shall note one example in §3.2.6.

As was mentioned in §3.1.1, we shall not explore this topic further. The rest of the section will treat another line of thought from Riemann's thesis, which was taken up by Cantor's senior colleague in Halle.

3.2.2 *Heine on trigonometric series and the real line, 1870–1872.* In a paper published in Crelle's journal Heine *1870a* concentrated on trigonometric series. He brought to the topic considerations of uniformity of convergence, using the alternative phrase 'convergent in the same degree', due to Weierstrass's teacher Christoph Gudermann; thereby he joined the concerns of Riemann with the (recently publicised) techniques of Weierstrass (§2.7.4). The bearing of these ideas was based upon the fact that each coefficient in $(321.1)_1$ was calculated by multiplying through the equation by the corresponding trigonometric function and integrating over $[-\pi, \pi]$ of x; but it was now known both that uniformity of convergence was required for the process to be safe, and that since the trigonometric functions were continuous, then under uniformity so was the sum-function. If uniformity did not obtain, then maybe more than one expansion of a function was possible, and the expandability of a discontinuous function was not clearly understood. These issues were accentuated by the considerations of Riemann's dissertation, especially the results just mentioned and the "ultra-Dirichletian" functions that he had found, to allow for the possibility that a function may not have any trigonometric expansion at all.

---

[5] On du Bois Reymond's successes and failure in set topology and integration, see T. Hawkins *1970a*, ch. 2. On his *Infinitärkalkül*, see especially Hausdorff *1909a* and Hardy *1924a*, and an historical survey in Fisher *1981a*.

Heine introduced the notion of 'convergence in general', by means of which he admitted the possibility of 'the exception that an *infinite number of points* might obtain' (*1870a*, 354). The intuitive idea was that, since the coefficients were defined as certain definite integrals over $[-\pi, \pi]$ of x, then their values would be insensitive to changes in the value of $f(x)$ for a finite number of values of x; the main purpose of the paper was to produce a proof of this result of Weierstrassian rigour. Another way of stating the result was in terms of uniqueness: if the series $(321.1)_1$ converged uniformly in general to zero, then each coefficient a_r and b_r must be zero (pp. 356–357). There were consequences also for understanding continuity, and he introduced here the definition of the '*uniform continuity*' of a function of two variables (p. 361, where he did use Weierstrass's adjective).

Heine cited Riemann's thesis twice: on p. 359 for one of the theorems on f and F, and in a footnote on p. 355 for discontinuous functions. For the latter point he thanked Dedekind, who seems to have sent him an offprint of the thesis; and in the sentence to which the footnote was attached he also acknowledged his junior colleague Cantor, 'to whom I made known my investigations', for remarks in this context and for the reference to an earlier paper introducing uniform convergence.

In a succeeding paper, Heine *1872a* discussed 'The elements of the theory of functions' within the tradition of Weierstrass, with whom he had discussed these matters (pp. 172 and 182). He began with a theory of irrational numbers, based upon 'number-series' of rationals $\{a_n\}$ in which 'for each given quantity η, so small yet different from zero, a value n exists which brings about that $a_n - a_{n+\nu}$ lies below η for all positive ν' and 'elementary series', 'in which the numbers a_n, with increasing index n, fall below any given quantity' (p. 174; he used only rationals, including zero). The '*more general number or number-sign*' was defined in terms of ordered sequences as

$$\text{'}[a_1, a_2, a_3, \ldots] = A\text{', with '}[a, a, a, \ldots] = a\text{'} \qquad (322.1)$$

for the special case of rationals $\{a\}$ (p. 176). The irrational numbers so formed constituted the 'first order'; those of 'arbitrary orders' were definable by iteration of these procedures (p. 180).

Some casualness was evident here. Heine tended to mix sign and its referent in a formalist way of the time seen already in De Morgan and Boole and awaiting us later also. Further, while he laid down the criterion for the equality of two numbers in terms of the sequence of their arithmetical differences forming an elementary one, he did not explicitly mention the case of the reordering of a sequence, and made no association of the numbers with the real line.

The rest of Heine's paper was taken up with an application of this theory to continuous functions, which he defined in its sequential form; for every 'number-series $\{x_n\}$ that the signs X possess, also $f(x_1)$, $f(x_2)$, etc. a number sequence with the number-sign $f(X)$' (p. 182, but with no allusion to Cauchy's similar formulation quoted in §2.7.2). He stressed that continuity at X was so defined; in a footnote he noted 'uniform' continuity, referring to his earlier paper, and at the end of the paper he proved (p. 188):

THEOREM 322.1 'A continuous function from $x = a$ to $x = b$ (for all individual values) is also uniformly continuous'.

The proof was effected by dividing up $[a, b]$ into a finite number of sub-intervals, expressing the continuity of the function over each one in the (ε, δ) way (§2.7.2), and taking a δ large enough to apply to all of them. The finitude of N, crucial to the proof, led Schönflies to give the unfortunate name 'Heine-Borel theorem' (*1900a*, 119) to a far more general theorem proved by Emile Borel in the 1890s (§4.2.2), in which a finite covering also occurs.

In the footnote Heine also thanked Cantor for a necessary and sufficient condition for the continuity of $f(X)$, and referred to a paper which his younger colleague had already published in this area of mathematics (p. 182). Cantor's acquaintance with such topics from Heine was to have a decisive effect on his own later career.

3.2.3 *Cantor's extension of Heine's findings, 1870–1872.* In an early paper 'On the simple number systems' Cantor *1869a* considered 'Systems of positive whole numbers' $\{b_r\}$, and in particular the question of whether expansions of the form

$$\sum_{r=1}^{\infty} [a_r / (\prod_{s=1}^{r} b_s)], \text{ where } 0 \leqslant a_r < b_r, \tag{323.1}$$

could generate irrational as well as rational numbers. So he was primed to respond to the issues raised by Heine, and between 1870 and 1872 he published five papers on them, switching for venue with the fourth one from Crelle's journal to the recently founded *Mathematische Annalen* (Dauben *1971a*).

In the first four papers Cantor handled expansions rather like (323.1), Riemann's theorems (and an extension due to Schwarz), uniform convergence, the uniqueness of expansion for continuous functions, and the possibility of a finite set of points excepted from the in-general convergence. While careful and intricate, they are only respectable footnotes to the work of his predecessors. But with the fifth paper, dated November 1871 and published as *1872a*, he opened a new era in these studies, with a proof that an *infinite* set of exceptional points (to use the modern term) could be allowed.

Drawing on his previous concern with irrational numbers, Cantor showed a fine grasp of the requirements that such a theorem would make upon him; for he gave a *definition* of them, basically following and developing that of Heine *1872a*, which had been completed a month before Cantor's paper. Taking the rational numbers (but excluding zero) as known, and forming a 'domain' ('*Gebiet*') A, he stated that if for any positive rational number ε and arbitrary positive integers n_1 and m,

$$\text{'}|a_{n+m} - a_n| < \varepsilon, \text{ if } n \geqslant n_1\text{'},\tag{323.2}$$

then the sequence $\{a_r\}$ '*has a specific limit b*' (p. 93). Numbers definable in this way formed a new domain, B, and (in)equality relations and arithmetical operations between them were specified in terms of the analogous properties of the absolute difference between members of the corresponding rationals: for example,

$$\begin{aligned}&b > b' \text{ if and only if there are an } \varepsilon \text{ and } n_0\\&\text{such that } |b_n - b'_n| > \varepsilon \text{ when } n \geqslant n_0.\end{aligned}\tag{323.3}$$

Cantor's way of expressing the property (323.2) seemed to assume the existence of the limiting value which was being defined into existence. Of course, he was quite aware of the point, and his theory was free of this criticism (which we shall see Russell make in §6.4.7); but his manner of expression was distinctly unfortunate. In addition, he did not properly treat the fact that the same irrational number could be produced by different sequences of rationals; if $b = b'$, which number was the irrational by definition? In a later paper he named such a sequence (323.2) a '*fundamental series*', and with great confidence stated that 'with the greatest simplicity' uniqueness of definition of an irrational was to be secured 'through the specialisation of the pertaining sets (a_ν)' of rationals, such as its decimal or continued-fraction expansion (*1883b*, 186, 185).

In addition, Cantor was to cause perplexity to commentators such as Dedekind (§3.2.4) when, after stating that 'The totality of number quantities' $\{b_r\}$ constituted a new domain B, he again followed Heine by allowing that 'now it generates in a similar way together with the domain A a new domain C', and rehearsed the procedures around (323.2–3) for 'number-quantities' $\{c_r\}$, and also onwards a finite number of times to obtain the domain L '*which is given as number-quantity, value or limit of the λth*' (*1872a*, 93–95). For him, since such procedures as specialisation and the construction of domains could be effected, the objects thereby defined definitely existed (see §3.6.1 on the philosophical background). He affirmed their status when he adjoined as an '*Axiom*' the assumption that to the 'distance from a fixed point o of the straight line with the $+$ or $-$ sign' corresponded the number-quantities of *each* domain (pp. 96–97). Note the implied omission of zero from the rational numbers; we shall appraise the status of zero in §3.5.3.

The second major feature of Cantor's paper was his handling of sets of points, which was developed out of this axiom. We have noted that in Cauchy, Dirichlet and Riemann collections of (mathematical) objects were being handled, albeit in a fairly informal way. By contrast, for Cantor 'I name for brevity a *value-set* a given finite or infinite number of number magnitudes' with the letter '*P*', 'and correspondingly a given finite or infinite number of points of a straight line a *point-set*' (pp. 97–98).

A fundamental concept was that of the '*limit-point*' ('*Grenzpunkt*') of a set P, defined as a point 'in such a situation that in each neighbourhood of it *infinitely* points of P are to be found' (p. 98). Then, corresponding to the limit-taking generation of higher domains, Cantor defined the 'derived point-sets' ('*abgeleitete Punktmengen*') of P, each one comprising the set of limit-points of its predecessor. Like the domains $A, B, \ldots, L$, the derived sets $P, P', \ldots, P^{(\nu)}$ were finite in number, the order of each producing a domain or point-set 'of the νth kind' (p. 98), and his extension of Heine's theorem on exceptional sets stated that they could in fact be of such a kind without infringing convergence in general within the interval $[0, 2\pi]$ (pp. 99–101). The proof drew on properties of Riemann's function $F(x)$, and Cantor concluded his paper by restating his result in function-theoretic terms: a discontinuous function which was non-zero or indeterminate over a 'point-set P' within $[0, 2\pi]$ of *every* kind 'cannot be developed by a trigonometric series'.

This theorem hints at Cantor's insight that there were sets for which $P^{(\nu)}$ was never empty for finite ν. In such a case an *infinitieth* derived set $P^{(\infty)}$ could exist, and so presumably possessed its own derived set $P^{(\infty+1)}$, and so on. But what were these sets? Cantor did not develop this line of thought at all in his paper, and only referred to it at the end of a later one (*1880a*, footnote to p. 355; sadly omitted from *Papers*, 148). Doubtless the theory was a bit *too* intuitive at that stage, and in any case did not (seem to) bear upon the theorem on trigonometric series. However, such considerations were soon to loom large in his thoughts; and he had the luck to gain a new confidant at exactly this time.

3.2.4 *Dedekind on irrational numbers, 1872.* We now consider the work of a major figure: Richard Dedekind (1831–1916), student of Gauss, editor of Riemann, and follower of Dirichlet. He passed the main part of his career at the *Technische Hochschule* at Braunschweig (a very respectable institution) despite receiving various possibilities for chairs at universities (including, we shall note in §3.2.6, at Halle).[6] Principally concerned with abstract algebras and number theory, he also took a strong interest in the foundations of mathematical analysis, partly in connection with teaching

---

[6] Some manuscripts from the Dedekind Papers pertinent to our theme were published in his *Works 3* (1932), and a substantial selection of manuscripts and correspondence from this and other sources is presented in Dugac *1976a*, pt. 2.

(of which he was unusually fond for a professional mathematician). As we saw in §2.7.3, he saw Riemann's two theses through the press in 1867, and was doubtless oriented by them to think about collections and continuity (Ferreirós *1999a*, ch. 3). In 1872 he put out a booklet on 'Continuity and irrational numbers' (Dedekind *1872a*). Its unusual manner of publication somewhat retarded its reception: for example, it escaped the attention of the reviewing journal *Jahrbuch über die Fortschritte der Mathematik*, and later Simon *1883a* referred to it as 'much too little known'. But it gradually gained attention, with reprints in 1892, 1905 and 1912, and has become a classic.

In his meticulous way, Dedekind recorded in the preface that he had come to his theory in the autumn of 1858 (when he was teaching at the *Technische Hochschule* in Zürich), with the key ideas being formulated on 24 November. In his draft he gave more details: discussion a week later with his colleague the analyst Heinrich Durège (who did not use the theory in his own work), and a lecture to a Braunschweig society in 1864.[7] However, he does not appear to have used much of the theory in his own lectures, and was motivated to write up his work for publication by receiving Heine *1872a* (on 14 March 1872, apparently). When writing his preface six days later he received an offprint of Cantor *1872a*, sent presumably at Heine's suggestion.

After reviewing the properties of the rational numbers and the corresponding relationships in the straight line (*1872a*, arts. 1–2), Dedekind turned to the 'continuity of the straight line' and 'the creation of the irrational numbers'. The distinction of category between line and number was essential to his assumption of a structure-isomorphism between them. (In a curious coincidence of notation with Cantor, he also wrote '*o*' for the origin-point of the line.) The 'completeness, gaplessness or continuity' of the line was assured by the 'Principle' that one and only one point on it could divide all its points 'in two classes' such that 'each point of the first class lies to the left of each point of the second class' (art. 3).

Similarly, numbers were divided into two classes A_1 and A_2 by a 'cut' ('*Schnitt*'), written '(A_1, A_2)'. They were of three kinds: I use round and square brackets to symbolise them. In the cases](and)[, the cut created a rational number; for the case)(, however, when 'no rational number be brought forth, we c r e a t e a new i r r a t i o n a l number α, which we regard as fully defined by this cut: we will say that the number α corresponds to this cut, or this cut brings it about' (art. 4). He also proved here the existence of irrational numbers by a lovely *reductio* argument that has never gained the attention that it deserves: assume that the equation

---

[7] The draft of Dedekind *1872a* is printed in Dugac *1976a*, app. 32; compare the recent edition in *1862a* of an earlier lecture course on the calculus. The source of all these exact dates was a diary (or collection of them), which has unfortunately been lost. On his work in analysis, see Sinaceur *1979a*, Zariski *1926a passim*, and (with caution) Dugac *1976a*, pt. 1.

$t^2 = Du^2$ in integers (D not a square) has a solution, and let u be the smallest integer involved; then exhibit a smaller integer also to satisfy the equation, a contradiction which establishes $\sqrt{D}$ as irrational. In a painstaking analysis (arts. 4–5), he also showed that the number-system thus defined satisfied the properties of ordering, continuity and combination required of the real numbers, and also that theorems on the passage to limits could be expressed (art. 6).

Dedekind stressed the distinction of category between cut and number in 1888; against the view of his friend Heinrich Weber that 'the irrational number is nothing other than the cut itself' he explained that 'as I prefer it, to create something *New* distinct from the cut, to which the cut corresponds [...] We have the right to grant ourselves such power of creation', and cuts corresponding to both rational and irrational numbers were examples (Dedekind *Works 3*, 489).

Dedekind also emphasised that 'one achieves by these means real proofs of theorems (such as e.g. $\sqrt{2}.\sqrt{3} = \sqrt{6}$)' (art. 6); and this claim excited the surprise of the analyst Rudolf Lipschitz, who wrote to Dedekind in 1876 that 'I hold that the definition proposed in Euclid [book 5, prop 5...] is just as satisfactory as your definition.' In reply Dedekind explained that the standard proofs were 'nothing than the crudest vicious circle', since not 'the slightest explanation of the product of two irrational numbers flows' from that for rational numbers. (In fact, Euclid there did not treat numbers at all, but geometrical magnitudes.) Dedekind's educational streak emerged in the added comment: 'Now is it really outrageous, the teaching of mathematics in schools rates as an especially excellent means of cultivating the mind, while in no other discipline (such as e.g. grammar) such great violations of logic would be tolerated only for a moment?' (*Works 3*, 469–471).

Thus the aim of Dedekind's study eluded even so distinguished a contemporary as Lipschitz. Another respect in which he was rather isolated from his colleagues concerned his philosophy of mind, and its bearing upon mathematics. His emphasis on the word 'creating' the new number exemplifies a philosophy which appeared also in his discussion of discontinuous space, clearly inspired by Riemann (*1872a*, end of art. 3):

> If space undoubtedly has a real existence, then it does n o t necessarily need to be continuous; numerous of its properties would remain the same, if it were discontinuous. And if we were to know for certain that space were discontinuous, nothing could hinder us, if we wished, from making it into a continuous [space] by filling out its gaps in thought into a continuous one; this filling out however, would consist in the creation of new point-individuals and would be executed in accordance with the above principle.

Even though Dedekind's philosophy was not fully appreciated, his definition of irrational numbers gradually came to be preferred over all others in textbooks and treatises.[8] The simplicity of his approach must have appealed: he structured the real line with his theory of cuts, and then associated real numbers with each cut, whereas other definitions took the real line for granted and obtained the irrational numbers via a process of construction (in Cantor's case the fundamental sequences (323.2)).

The booklet inaugurated for Dedekind a greater involvement in the foundations of analysis than he probably anticipated at the time, because he became amanuensis to Cantor's investigations into sets. He received an offprint of Cantor *1872a* while completing his booklet; and their friendship was established in April 1872, a few weeks after Dedekind had written the preface, when fate led them both to stay at the same hotel in Gersau in Switzerland. The correspondence was soon launched (Ferreirós *1993a*).

3.2.5 *Cantor on line and plane, 1874–1877.* Cantor's first paper devoted to set theory proper appeared in Crelle's journal as *1874a*. The title mentioned 'a property of the concept ['*Inbegriff*'] of all real algebraic numbers', namely, that they were denumerable. He did not use that word, but he stated the property in the standard way: that they could be laid out completely 'in the form' of an ordered sequence (art. 1). He also showed that the real numbers did not have this property (art. 2): taking any denumerable sequence S of them, he formed the sequence of nested open intervals $\{(\alpha_r, \beta_r)\}$ by taking the first one arbitrarily and defining the end-points of each one as the first two numbers which lay within the preceding interval. Monotonic sequences of numbers were thereby created: if these sequences were finite, then within the last interval at least one further member of S could be found; if infinite but converging to different values, then again a member of S was available; and if infinite and convergent to the same value η, then the property of nesting prohibited η from belonging to S. He ended by indicating consequences for rational functions. Dedekind had received versions of these and other details in Cantor's letters; indeed, according to his own note, he had contributed the proof of the case of η 'almost word for word' without acknowledgement, or use of his continuity principle either (Cantor-Dedekind *Letters*, 19).

Cantor now knew that the infinite came in different sizes. This conclusion was given a firmer form in the next paper, which appeared four years later, as *1878a*. Sets were now called 'manifolds' ('*Mannigfaltigkeiten*'), Riemann's word (§2.7.4) though with a different reference. When two of

---

[8] The history of Dedekind's "victory" has in fact not been traced, though much information on the English and American side is contained in Burn *1992a*. For an exhaustive account of foundational processes in analysis, including irrational numbers, see Pringsheim *1898a*. A significant figure in Germany early in this century was Oscar Perron: see, for example, his *1907a* and, much later, the preface to his book on irrational numbers *1939a* for his extraordinary sarcasm against Nazism in preferring Dedekind's theory over others' on the grounds of being good German mathematics.

them could be paired off by members, 'these manifolds have *the same power*, or also, that they are *equivalent*'. More significantly, he also stressed inequality of power, and the relations '*smaller*' and '*larger*'. Later in the paper he wrote of the 'scope' of a 'variable quantity'; and if two of these, *a* and *b*, could be paired off, then they too were equivalent, a relation expressed by the propositions '*a* ∼ *b* or *b* ∼ *a*' (art. 3).

Among the results proved was the equivalence of the irrational and the real numbers (art. 3), of the intervals $(0, 1)$ and $[0, 1]$ (art. 5), and of continuous manifolds of *n* and of one dimensions. The first proof of the last theorem drew on the uniqueness of the continued-fraction expansion of an irrational number (art. 1); the second was based upon decimal expansions, in the case of $n = 2$, that the point $(0 . x_1 x_2 \ldots, 0 . y_1 y_2 \ldots)$ in the unit square could be mapped onto the point $0 . x_1 y_1 x_2 y_2 \ldots$ of the unit line (art. 7). At the end of the paper Cantor concluded that the infinite came in only two sizes: namely 'functio ips. *ν* (where *ν* runs through all positive numbers)' and 'functio ips. *x* (where *x* can take all real values $\geqslant 0$ and $\leqslant 1$)'. He also characterised the latter case a few lines earlier as 'Equal to *Two*', which was his first statement of the conjecture known later as his 'continuum hypothesis'; it is discussed in §3.5.2.

In letters Dedekind had been bombarded with versions of every theorem, and indeed in June he had contributed the decimal-expansion proof himself, including the need to distinguish expansions such as 0.30000... from 0.2999... (Cantor-Dedekind *Letters*, 27–28). Cantor's reaction to this result was 'I see it, but I do not believe it' (p. 34, in French); and he transcribed most of the proof into his paper without acknowledgement . . . .

Allegedly Kronecker had held up publication of this paper in Crelle's journal. Cantor himself is the principal source of this story, though at the time he only told Dedekind in October that his paper had been with C. W. Borchardt (a co-editor of the journal) for three months (p. 40). In fact, if there was a delay, it cannot have been a long one (the date of submission of the paper, 11 July 1877, is not obviously out of line with others in the same volume); and, given the way in which Cantor had chosen to express himself, Kronecker deserves our sympathy. His philosophy of mathematics will be contrasted with Cantor's in §3.6.4.

3.2.6 *Infinite numbers and the topology of linear sets, 1878–1883.* The results of 1878 on the equivalence of sets of different dimensions led Cantor to consider in detail the question of correctly defining dimension. The success of his endeavours and those of some contemporaries was only partial (D. M. Johnson *1979a*, chs. 2–3), and the experience seemed to impel him to concentrate his studies largely on sets of points on the line. In his later work, *n*-dimensional sets were discussed sometimes, but the dimensional aspects themselves were not discussed. The main product was a suite of six papers with the common title 'On infinite linear point-manifolds', published between 1879 and 1884, the fifth part appearing also as a booklet. The venue was *Mathematische Annalen*, and Klein the relevant member of the editorial board; he became Cantor's chief correspondent for a while, receiving over 40 letters in 1882 and 1883. One reason followed from Heine's death in 1881; Cantor had asked Dedekind to put himself forward as successor (my *1974b*, 116–123) but Dedekind declined and so Cantor dropped him for many years (§3.5.3).[9]

Cantor broadly followed the order of interest of his earlier papers, beginning in *1879a* with an extended study of the derivation of 'point-sets', to quote the alternative name ('*Punktmengen*') to that of his title, which he introduced in the first paragraph. The exceptional sets 'of the λth kind', which were allowed in §3.2.2 under the rule of 'in-general convergence', were now grouped collectively as 'of the *first species*'; those with no empty derived set of finite order constituted 'the *second species*' (p. 140). As an important kind of example of the latter species he defined the '*everywhere dense set*' within the closed interval $[\alpha, \beta]$ (itself written '$(\alpha \ldots \beta)$'), by the property that its members could be found within every closed sub-interval of $[\alpha, \beta]$, however small; the property that it was contained within its first derived set was a theorem (pp. 140–141).

The rest of this paper was largely concerned with these two 'classes' of linear point-sets, each one defined by its common 'power' ('*Mächtigkeit*'): sets which were '*countable in the infinite*' including, he now knew, those of the first species; and those to which (interestingly) he gave no name but chose a '*continuous interval*' as the first 'representative', with the cardinality of the continuum (pp. 141–144). The (in)equality of cardinalities of two manifolds M and N was defined, as earlier, extensionally in terms of (no) isomorphism between their members (p. 141); and he began the second paper with the allied statement that 'the identity of two point-sets P and Q will be expressed by the formula $P \equiv Q$' (*1880a*, 145).

In this paper some basic machinery was presented (pp. 145–147). Disjoint sets were defined as '*without intersection*' (with no special symbol), and the union of 'pairwise' disjoint sets $\{P_r\}$ was written '$\{P_1, P_2, P_3, \ldots\}$'. For the inclusion of sets 'we say: P is *included* in Q or also that P be a divisor of Q, a multiplum of P'. The 'union' and 'intersection' of 'a finite or infinite number' of sets $\{P_r\}$ were written respectively as

$$\text{`M}\{P_1, P_2, P_3, \ldots\}\text{' and `D}\{P_1, P_2, P_3, \ldots\}\text{'.} \tag{326.1}$$

[9] Cantor's letters are in Klein Papers, 8:395–436; seven are transcribed in Cantor *Letters*. They are well used in Dauben *1979a*, chs. 4–5. In a letter of 15 November 1899 to Hilbert, Cantor claimed that Dedekind had stopped their correspondence around 1873 'aus *mir unbekannten Gründen*'! (Hilbert Papers, 54/14; quoted in Purkert and Ilgauds *1987a*, 154, and transcribed in Cantor *Letters*, 414).

where 'M' denoted 'multiplum'. Finally, for 'the absence of points [...] we choose the letter O; $P \equiv O$ thus indicates that the set P contains *not a single* point'. We note that Cantor was unclear over whether 'O' symbolised the/an empty set, or whether '$\equiv O$' denoted the property that a set were empty (compare Boole in §2.5.4).

Cantor's first use of these tools was to express certain properties of the sequence $P^{(\nu)}$ informally conceived in §3.2.2. Now the set '$P^{(\infty)}$' of a set P of the first species was explicitly introduced, as 'the derivative of P of order ∞', defined as the intersection of all its predecessors, and stated to be equal to the intersection of any infinite subset of them (p. 147). The idea of $P^{(\infty)}$ having its own derived set was now extended into prolonging the sequence to $(P^{(r)})$, where r was allowed to wander up through $(n_0\infty + n_1)$ to

$$2\infty, \ldots 3\infty, \ldots \infty^2, \ldots \sum_{r=0}^{\nu} n_r \infty^r, \ldots \infty^\infty, \ldots \infty^{\infty + n}, \ldots \infty^{n^\infty}, \ldots \infty^{\infty^\infty}, \ldots \text{ 'etc.'},$$

$$(326.2)$$

in a 'dialectical generation of concepts' (pp. 147–148).

It was at this point that Cantor added the footnote mentioned at the end of §3.2.2, concerning his possession of these ideas ten years earlier. Probably it was a retort to a claim of priority for the notion of the everywhere dense set recently made by du Bois Reymond (*1880a*, 127–128), whose own point-set topology was noted at the end of §3.2.1; he had named this type of set 'pantachic'.

The further refinement of the still intuitive formulation (326.2) was to be a major preoccupation for Cantor in later papers in his suite. In the third paper he reintroduced the concept of the '*limit-point*' of a set, but, in some contrast to §3.2.2, more like the form which we now distinguish as its accumulation point: 'in each neighbourhood of it, ever so small, points of the set P are to be found' (*1882b*, 149). He added that Weierstrass had proved that any bounded infinite set of points possessed at least one such point (the theorem now known as the 'Bolzano-Weierstrass').[10] He also attributed to Jacob Steiner's lectures *1867a* on projective geometry the name 'power' (p. 151), and ruminated on various properties of the cardinality and topology of sets of one and several dimensions. Most interesting

---

[10] Cantor had himself already stated this theorem in his major paper on trigonometric series (*1872a*, 98), without citing any mathematician. The name seems to be due principally to Cantor's friend Schwarz, in a paper on Laplace's equation published at the same time in Crelle's journal (*1872a*, 178). As a name it is unfortunate, as it associates Weierstrass's result with the very special case that a bounded set of values has an upper bound, proved much earlier in Bolzano *1817a*, art. 12. Paying tributes with inappropriate names both muddles together different levels of rigour (a matter of especial importance in this sort of mathematics) and also takes away from the quality of the earlier work: compare the 'Heine-Borel' Theorem 322.1.

was the view, echoing that of Dedekind in §3.2.3, that the axiom of §3.2.2 of the isomorphism between the real line and the real numbers extended to a hypothesis about the continuity of space, and that continuous motion was possible in a space made discontinuous by the removal of a denumerable set (p. 157): as we shall note in §3.3.3 and §3.3.5, early commentators were to pick up on this detail.

Cantor's fourth paper began with the notation

$$'P \equiv P_1 + P_2 + P_3 + \cdots \,', \tag{326.3}$$

to replace that for the union of pairwise disjoint sets quoted before (326.1) (*1883a*, 157): in both appearance and content this was now very like Boole's use of ' + ' (§2.5.3), but it is unlikely that he knew of Boole's work, at least in any detail. The principal new idea was of 'an *isolated* point-set' Q in n-dimensional space for which, in the notation of $(326.1)_1$,

$$'D(Q, Q') \equiv O\,'. \tag{326.4}$$

The importance of this type of set lay in the fact that one could be created for *any* set P, namely $(P - D(P, P'))$; and this insight led him to two 'important decomposition theorems':

$$'P' \equiv (P' - P'') + (P'' - P''') + \cdots + (P^{(\nu-1)} - P^{(\nu)}) + \cdots + P^{(\infty)} \,', \tag{326.5}$$

and its companion shorn of the last term $P^{(\infty)}$ for first-species sets. An isolated set was countable;[11] each component set in $(326.5)_1$ was isolated; if P' was denumerable, so was P; and first-species sets were denumerable, and so were those of second species when $P^{(\alpha)}$ was denumerable for any α of the '*infinity-symbol*' used after (326.1) (pp. 158–160). The use of the nervous word 'symbol' shows that the infinite was still somewhat out of his reach.

3.2.7 *The* Grundlagen, *1883: the construction of number-classes.*

That is a wonderful harmony, going into magnitudes, whose exact passage is the theme of the doctrine of transfinite numbers.

Cantor on the number-class, lecture of 1883 (*1887–1888a*, 396)

---

[11] Cantor proved this result by a measure-theoretic argument which was defective in as much as his definition of measure, formally introduced in *1884b*, art. 18, did not distinguish the measure of a set from that of its closure and so admitted of inadmissible additive properties (T. Hawkins *1970a*, 61–70).

In the fifth paper of the series, which comprised 47 pages (including 5 pages of endnotes), Cantor *1883b* reached new levels of both length and depth in developing his theory. He republished it at once as a pamphlet *1883c* with Teubner, the publisher of *Mathematische Annalen*, in a slightly revised printing and with a preface. It carried the new title 'Foundations of a general theory of manifolds', with the sub-title 'A mathematical-philo-sophical study in the doctrine of the infinite'. The account here will be confined to the foundational aspects and the construction of transfinite numbers (as he often now called them): the well-ordering principle and the continuum hypothesis are postponed to §3.5.1–2.

The word 'manifold' ('*Mannichfaltigkeit*') attached to this suite of papers was explained in the first endnote as 'each multiplicity, which may be thought of as a One, i.e. each embodiment ['*Inbegriff*'] of particular elements, which can be bound together, by a law into a whole'; he offered 'set' ('*Menge*') as a synonym (p. 204). The intensional form of this defini-tion will be noted in §3.4.6, on a later appearance. His choice of name was unfortunate, as it had been used already in a different context by Riemann (§2.7.3), Hermann von Helmholtz and others; we shall note Cantor's disapproval of their empiricist philosophy in §3.6.2.

The burden of the word 'general' was Cantor's attempt to ground his finite and transfinite arithmetic in a 'real whole number-concept', to quote from his first sentence.[12] He distinguished two kinds of reality: '*intra-sub-jective* or *immanent*', when numbers 'on the ground of definitions can take a quite specific place in our mind'; and '*trans-subjective* or also *transient*', when they 'should be regarded as an expression or an imagine of courses of events and relationships in the external world standing opposite the intellect' (p. 181). He accepted both kinds of reality, and saw the connec-tions between them to be established in 'the *unity* of the *all*, to which we ourselves belong' (p. 182).

Cantor distinguished two kinds of the infinite: 'proper' ('*eigentlich*'), which can be translated as 'real' or 'actual'; and 'improper' ('*uneigentlich*'), which was and is better known as the potential infinite (p. 165). He contrasted his current desire that the infinite numbers 'possess concrete numbers of real reference' (p. 166) with his previous use of 'infinity-sym-bol' (a footnote here, which Zermelo failed to include in his edition). So he replaced '∞', with its ambiguities of past use, with '*ω*', chosen as the last letter of the Greek alphabet and denoting the smallest transfinite ordinal (a footnote which Zermelo preserved on p. 195).

Cantor grounded ordinal numbers in sets in the following way. A '*well-ordered* set' was defined by the property that its elements exhibited 'a specific prescribed succession among them' with 'a *first* element' and a

---

[12] Cantor's word was 'real'; sometimes in this paper he also mentioned 'reellen Zahlen' in the mathematical sense, as contrasted to (hyper)complex numbers (see, for example, *1883b*, 165, 169).

specific successor for each one (apart from the last element of a finite set). The pre-eminence among types of order of well-ordering, with its alleged applicability to all 'well-defined' sets, was precisely the well-ordering principle, which will be discussed in §3.5.1.

Given this assumption, an ordinal was specified as 'the *number* of the elements of a *well-ordered* infinite manifold'; Cantor's use of 'Anzahl' for ordinals rather than cardinals contravened normal practice (p. 168). These ordinals were produced via two '*principles of generation*': that of 'the addition of a unit to an existing [and] already constituted number' (with 1 assumed as the first number), and thereby generated a succession of numbers with no greatest number (p. 195); and 'the logical function' (*sic*) of creating 'a new number' ω 'as *limit* of those numbers'. ω served as the new initial ordinal from which the renewed application of the first principle led to a fresh sequence $\{\omega + n\}$, after which was postulated the new limit-ordinal $2\omega, \ldots$.

The operation of these principles in tandem generated the sequence of ordinals (326.2), with the former 'dialectical generation of concepts' now better understood, and more properties of the sequences provided (pp. 196–203). One of them, stated for sets M with cardinality of the second number-class, that if a sub-subset M'' were isomorphic ('*gegenseitig eindeutig*') with M, then the intervening subset M' was isomorphic with both M'' and M (p. 201). The proof, only sketched, drew upon the well-ordering principle; the need for a general and sound proof became a major concern for Cantor and others from the mid 1890s, including Whitehead and Russell (§3.4.1, §4.2.5, (786.1)).

The 'number-classes' of these ordinals were introduced in a rather enigmatic way. The first class comprised 'the set of finite whole numbers'; 'from it follows' the second class 'existing from certain infinite whole numbers, following each other in specific succession'; then came to the 'third, then to the fourth, etc.' (p. 167). Details of only the second number-class were provided in the paper; but it became clear that one purpose of these classes was to serve as a means of defining transfinite cardinal numbers, or '*powers*'. The smallest such cardinal for an infinite set was defined by the property 'if it can be ordered isomorphically with the *first* number-class'. Cantor claimed that the cardinality of the class of ordinals possessing this property was not only not equal to that of the first class 'but that it actually is the *next higher* power', since 'The *smallest* power of infinite sets [...] will be ascribed to those sets which can be ordered isomorphically with the *first* number-class' (p. 167).

Cardinal numbers, both finite and infinite, were given epistemological priority over ordinals, in that they were defined independently of the orderings of which a set was susceptible. This was Cantor's position on the relationship between these two types of number, which will be a recurrent theme in this book; we shall note it again in §3.4.6.

3.2.8 *The* Grundlagen*: the definition of continuity.* In another important section Cantor studied the continuum of the real line and of continuous sets in general. He rehearsed his theory of irrational numbers of §3.2.3
in terms such as the definition

$$\text{`}\operatorname*{Lim}_{\nu=\infty} (a_{\nu+\mu} - a_\nu) = 0 \text{ (for arbitrarily composed } \mu)\text{'}} \qquad (328.1)$$

of a fundamental sequence (p. 186). Two features are worth noting: his
failure to specify the moduli of the differences; and the appearance of
'$\nu = \infty$' under the limit sign rather than the '$\nu \to \infty$' that would be
expected of a Weierstrassian, especially one who had defined ∞ as the
limit-ordinal ω a few pages earlier! He also defended his iterative definitions of higher-order numbers against Dedekind's criticism (§3.2.3), on the
interesting grounds that 'I had only the conceptually various forms of the
given in mind', not 'to introduce *new* numbers' (p. 188). But a retort to
Dedekind's own theory of cuts is harder to cope with: apparently 'the
numbers in analysis can *never* perform in the form of "cuts," in which they
first must be brought with great pomp and circumstance' (p. 185), to which
editor Zermelo understandably added '[?]'.

The real line is itself a set: how was it defined, and how many points did
it possess? On the first question Cantor made a definitive contribution in
requiring *two* 'necessary and *sufficient* characteristics of a point-con-
tinuum'. Firstly, he defined a set P to be '*perfect*' when it equalled each of
its derived sets $P^{(\gamma)}$ for all ordinals of the first two number-classes (p. 194:
by implication for ordinals of higher number-classes?); he distinguished
this property from the 'everywhere dense' of §3.2.6, and also from '*reducible*', where '$P^{(\gamma)} \equiv 0$' would occur for some γ of either the first or the
second number-classes (p. 193). Secondly, P had to be '*connected*': that is,
between any two members t and t' at least one finite collection of
fellow-members $\{t_r\}$ could be found such that 'the distances $[\{t_r t_{r-1}\}]$ are
collectively smaller than ε', an arbitrarily chosen (positive) number (p.
194). This second property was bought at the price of spatial, or at least of
metric, reflections; these were ensured under some measure-theoretic
considerations, which themselves included the questionable assertion that
'in my opinion, the involvement of the *concept of time* or of the *intuition of
space* [...] is not in order; *time* is in my view a representation that for its
clear explanation has for assumption the concept of continuity, which is
independent of it' (pp. 191-192). Later he was to replace connectivity with
a property free from metrical considerations (§3.4.7).

The need for both properties to define continuity was a major advance,
and characteristically Cantor crowed over the inadequacy of two of his
predecessors: Bolzano *1851a*, art. 38 for requiring only connectivity, and
Dedekind *1872a* for delivering only perfection. Further, given his own
(partly) set-theoretic definition of continuity and the construction of the

number-classes, Cantor was in a position to restate his continuum hypothesis, in a stronger form than the one noted in §3.2.4: 'that the sought power is none other than that of our *second number-class*' (the somewhat prematurely placed p. 192). 'I hope to be able very soon to be able to answer with a rigorous proof', he continued; the fate of these efforts will be recorded in §3.5.2.

In an endnote attached to his definition of continuity Cantor presented the remarkable 'ternary set', as it came to be known; its generic member z was defined by the expansion

$$z = \Sigma_{r=1}^{\infty} c_r/3^r, \text{ where } c_r = 0 \text{ (misprinted 'o'!) or 2,} \qquad (328.2)$$

and the totality of combinations of 0s and 2s produced the members (p. 207). He presented it as a countable and perfect set which was not everywhere dense in any interval. He did not himself use the set much, but its properties were to be of great interest to many of his mathematician successors.

3.2.9 *The successor to the* Grundlagen, *1884.* In a short special preface prepared for the pamphlet version of the *Grundlagen* and dated as of Christmas 1882 Cantor announced that his work 'goes very far'; but he also doubted that 'the last word to say was in place', and in the following April he sent to the *Annalen* a successor. This appeared as the 36-page paper *1884b*, in which the numbering of sections was continued but no endnotes were furnished. A further instalment was promised at the end of the paper, but it did not appear and maybe was not written.

In this instalment Cantor concentrated on topological properties of 'linear' sets in n-dimensional space, especially decomposition theorems, although he included many references to older sources. One of his main concerns was with the 'distributive properties' of sets, to use the name introduced later in connection with the Heine-Borel and such theorems: he did not himself have this result, but he stated a remarkably original theorem-schemum about points in a set possessing any '*property Y*' (p. 211). He also modified the definition of union to allow for overlapping sets (p. 226). His greater confidence over the status of transfinite ordinals was shown in the definition of '$P^{(\omega)}$' as the intersection of sets stated before (326.2). He also introduced an important new type of set: '*dense in itself*' ('*in sich dicht*'), for which 'a set P is a divisor of its derivative $P^{(1)}$' (p. 228). This completed a trio of definitions, that

$$P \text{ was closed/perfect/dense in itself if } P \subseteq / \equiv / \supseteq P^{(1)}. \qquad (329.1)$$

Cantor studied perfect sets in the most detail, partly in the hope of proving the continuum hypothesis (the closing statement on p. 244); his main result of this kind was that a perfect set was of the same cardinality as the closed interval $[0, 1]$ (p. 241), and thus of the continuum. But the result with more lasting consequences (pp. 222–223) stated:

THEOREM 329.1 A closed set of cardinality greater than the denumerable could be (uniquely) decomposed into a perfect set *P* and one *R* which was at most denumerable.

This theorem is now known as the 'Cantor-Bendixson', and he referred to correspondence with Ivar Bendixson (1861–1935) (p. 224). This exchange occurred partly in connection with his original formulation in *1883c*, 193, where *R* was held to be reducible (defined after (328.1)); Bendixson *1883a* corrected this stipulation. The contact was part of an important transfer of Cantor's circle of supporters, which we now recount.

3.3 CANTOR'S *ACTA MATHEMATICA* PHASE, 1883–1885

3.3.1 *Mittag-Leffler and the French translations, 1883.* Gösta Mittag-Leffler was the leading Swedish member of the coterie of mathematicians who fell under the spell of Weierstrass's tradition of mathematical analysis. Born in Stockholm in 1846 (the year after Cantor), he duly trekked to Berlin in the 1870s and soon was applying the new methods, with distinction, to elliptic functions and especially to complex-variable analysis. In 1881 he obtained a professorship at the newly founded university in his home town. But more germane to our story is that in the following year he married into a rich Finnish business family with whom he had become acquainted while holding a chair at the university there, and at once applied the financial windfall to the founding of a new mathematical journal.

From the start *Acta mathematica* was a major serial in its field, and Mittag-Leffler ran it until his death in 1927. He also built a magnificent house in a nice suburb of Stockholm, and assembled a superb library and a valuable archive not only of his own papers and correspondence but also of mathematicians in whom he was passionately interested. The two main heroes were Weierstrass and Cantor; for early on in his career he had read Cantor in *Mathematische Annalen* and appreciated the importance of *Mengenlehre*.[13]

The launch of his journal gave Mittag-Leffler a reason to develop his friendship with Cantor by asking for new papers, and also by suggesting

---

[13] Unfortunately there is no worthwhile obituary or biography of Mittag-Leffler, but some information and references are contained in Dauben *1980a* and Gårding *1998a*, chs. 7–8. In my *1971b* I announced the riches contained in his house, which has existed since 1919 as the Institut Mittag-Leffler. The archives include the manuscripts and some proof-sheets of various papers of Cantor to be discussed in this section, including the translations about to be described and the manuscript dealt with in §3.3.2–3; but apparently the Insititut has not employed an archivist to organise the holdings in the manner which their importance warrants, so that I cannot cite an item by callmark. A substantial selection of Mittag-Leffler's correspondence with Cantor is published in Cantor *Letters*.

that the bulk of the old ones be translated into French in order to be more accessible to the world mathematical community. Upon receiving Cantor's consent, Mittag-Leffler secured the assistance of Charles Hermite in Paris, and the translations were prepared there by one Darguet with revisions and corrections made by Cantor himself and Mittag-Leffler, and some of them by Hermite's younger colleagues such as Paul Appell and Henri Poincaré. Interestingly, 'Punktmenge' was rendered as the traditional French word 'système', and the pair '(un)eigentlich Unendliche' came out as 'l'infini (im)proprement dit'.

The ensemble, which I cite collectively as 'Cantor *1883a*', appeared as 104 pages of the last number of the second volume (1883) of the journal, in an order different from that of their original appearances: *1874a* on algebraic numbers; *1878a* on dimensions; two papers on trigonometric series, including *1872a*; the first four papers in the suite on linear point-sets; and finally the predominantly mathematical articles of the *Grundlagen*, but in a different (and rather more readable) order and with come cuts and revisions.[14] The historical and philosophical remainder, and the preface to the pamphlet version, were omitted, at Mittag-Leffler's request (and already with Hermite's prompting). The sixth paper in the suite had not yet been written, but Cantor contributed here his first original piece for Mittag-Leffler, also in French: a miscellany *1883e* of theorems on sets in an *n*-dimensional space, starting out from some in the *Grundlagen* and the decomposition (326.5).

Finally, Mittag-Leffler's student Bendixson contributed a melange *1883a* of his own decomposition theorems, especially Theorem 329.1. Upon seeing Bendixson's paper Cantor sent in his own paper *1884a* in French, devoted to 'the power of perfect sets of points' and to nesting sequences of closed intervals, and also publicising the ternary set (328.2). Mittag-Leffler explained the purpose of the paper in an explanatory note to the title, which Zermelo left out of his edition of Cantor's papers.

3.3.2 *Unpublished and published 'communications', 1884–1885.* Throughout 1883 and 1884 Cantor and Mittag-Leffler corresponded intensively about the developing *Mengenlehre*, and also non-friends such as Kronecker. Cantor dropped Klein and *Mathematische Annalen* in favour of his new contact, and by the autumn of 1884 he was promising four papers of various kinds, some successors to others; it started with a 'first communication' in German, on the 'Principles of a theory of order-types'.

During six weeks of the summer of 1884 Cantor, then in his 40th year, suffered his first mental crisis (Schönflies *1927a*). It started and ended suddenly after a few weeks, during which he displaced his research effort into other directions (the numbers of ways of expressing even integers

---

[14] The articles (some revised) of Cantor's *Grundlagen* were published, carrying their original numbers, in the order 1, 11, 12, 13, 2, 3, 14, 10; compare footnote 17. Of the endnotes, the mathematical trio 10–12, including the ternary set (328.2) were placed together as unreferenced 'Notes'.

as sums of two primes, and the belief that Bacon wrote the works of Shakespeare). These features strongly suggest that he had experienced a mid-term crisis; the effect will have affected the solidity of his psyche when he was struck by more serious attacks from 1899 onwards (§3.1.2).

Upon resuming work in August, he worked intensively on the continuum hypothesis (§3.5.2). Then in November 1884 he completed and sent off to Mittag-Leffler the first of his promised papers, and over the next four months he added to it two lengthy articles to the six already prepared. Mittag-Leffler designated the full paper for a place in volume 7 of the *Acta*; but when he reread the text in March 1885 upon receiving the first signature of proofs, he advised that 'It seems to me, that it would be better for you yourself not to publish these investigations before you can present new very positive results of new means of consideration' such as the continuum hypothesis; 'then your new theory would certainly have the greatest success among mathematicians'. As it was,

> I am convinced from it that the publication of your new work, before you can present new results, would hurt your repute among mathematicians very much [...] So the theories will be discovered again by somebody 100 years or more afterwards and indeed one finds out subsequently that you already had everything and then one gives you justice at last, but in this way you will have exercised no significant influence on the development of our science.

Mittag-Leffler's advice was well-meant (and his measure of the time-scale of Cantor's posterior recognition rather interesting); but it reflected his strong lack of enthusiasm for matters philosophical, and did not constitute a fair judgement of the paper. Cantor, already low in self-confidence, agreed at once, and in the following January (1885) he sent in a 'second communication', very mathematical in content, which appeared in the *Acta* as *1885b*. But this was his last paper to appear there: the frequency of his correspondence with Mittag-Leffler fell away quickly and virtually stopped by 1888, and in later years he was bitter in his recollection of the affair to correspondents.

Apart from such expressions, and a brief and largely unnoticed footnote in a later paper (*1887–1888a*, 411), the 'Principles' remained unknown until my astonished eyes saw it in the surviving *Nachlass* in 1969. I cite it as Cantor *m1885a*: I published it as part of my account *1970b* of the affair, where are to be found the quotations above, drawn from various other pertinent documents.[15]

---

[15] In my *1970b*, see pp. 101–103 for Mittag-Leffler's fateful letter, and pp. 104–105 for Cantor's reminiscences of the mid 1890s to F. Gerbaldi (where my editorial remark about Klein as another recipient is, I now think, mistaken) and Poincaré (of which the original has now been located in his *Nachlass*, still held by the family but denied to exist at the time of preparation of my paper). I have also since learnt that Mittag-Leffler's letters, which I had been told were copies, are in fact the originals, sent back to him after Cantor's death by his daughter Else. These and other letters of that period are published in Cantor *Letters*, 208–242.

3.3.3 *Order-types and partial derivatives in the 'communications'.*   One of Cantor's great achievements was to recognise the variety of orders in which the elements of (especially) an infinite set could be put. Examples of the resulting knowledge had appeared already, especially in *1884b*, 213–214; but in the 'Principles' he discussed the matter in detail and in some generality.

In the preamble Cantor explained his specific motivation to write the paper. The French mathematician and philosopher Jules Tannery (1848– 1910), whose elder brother Paul was mentioned in §3.1.2 as a correspon- dent of Cantor, had reviewed at length the first two volumes of *Acta mathematica* in his *Bulletin des sciences mathématiques*, devoting the last ten pages to the Cantor number (Tannery *1884a*, 162–171). He expressed reservations about some of Cantor's procedures and claims; for example, he preferred Dedekind's definition of irrational numbers, and cast doubt on the utility of Cantor's for science in view of the possibility (indicated just before (326.3)) of continuous motion in a discontinuous space. Cantor started his new paper by casting it as a reply to Tannery, to clarify his theory from philosophical and metaphysical points of view (*m1885a*, 82–83).

'The *real whole numbers* 1, 2, 3, . . . constitute a relatively *quite small species* of *thought-objects*, which I call *order-types* or also simply *types* (from δ τύπος)'. Further, those thought-objects 'which I call *transfinite* or *superfinite* numbers, [are] only special *kinds* of *order-type*'. Indeed, 'The general *type-theory*', his short name for 'Theory of *order-types*', 'constitutes an important and large part of pure *Mengenlehre* (Théorie des ensembles), also therefore, of *pure mathematics*, of which the latter is in my conception nothing other than pure *Mengenlehre*' (p. 84).

Thus an important theme of this book, mathematics as *Mengenlehre*, made its début, albeit in a text which did not reach the public. Cantor immediately stressed the close relationship 'to applied *Mengenlehre*' (which 'one takes care to call *natural philosophy* or *cosmology*') such as 'to *point set theory, function theory* and to *mathematical physics*'. He also associated his theory with chemistry (thereby continuing a link noted already in §2.2.1, §2.3.3 and §2.6.1) while distinguishing it from a specific 'theory of types' currently being pursued there. The chemical connections continued in his use of the word 'valency' as a synonym for 'power', a concept which he explained as 'the representation' or '*representatio generalis*' of a set M 'for all sets *of the same class as M*' (pp. 85–86: this term was already in the 1883 lecture).

After rehearsing these fundamental notions Cantor dwelt not on well- ordering but on '*simply ordered sets*' as a category to embrace all orderings; it was composed of members 'whether from *nature*, or through a *conven- tional lawful relationship*' and possessing a complete and transitive '*de- termined relation of rank*' (p. 86). Order-isomorphism between two such sets was specified as '*mutually similar*'. Each such set 'has now a determined *order-type* [...]; by it I understand that *general concept, under which fall collectively the given ord[ered] sets of similar ordered sets'*. For example, finite simple order was 'nothing other than the *finite whole numbers'*; the sequence of rational numbers $(1 - 1/\nu)$ was a type given the letter 'ω', the rationals within $(0, 1)$ were designated 'η', and the real numbers within $(0, 1)$ 'θ' (p. 87).

Much of the rest of Cantor's exegesis was taken up with related types; for example, that of the rationals within $[0, 1)$ was '$1 + \eta$', within $(0, 1]$ was '$\eta + 1$', and within $[0, 1]$ was '$1 + \eta + 1$'. More generally, for any type α there was the '*opposite type*' α_*, so that the following type-*equations* ensued:

$$\text{'}\alpha_{**} = \alpha\text{'; and examples such as '}(1 + \eta) = (\eta + 1)_*\text{' and '}\theta = \theta_*\text{'}$$

(333.1)

(pp. 87–88). He also took two simply ordered sets $\mathfrak{A}$ and $\mathfrak{B}$ with respective types α and β, and defined their sum $\alpha + \beta$ and product '$\alpha \cdot \beta$ or $\alpha\beta$' in terms respectively of their union and of a $\mathfrak{B}$-set of $\mathfrak{A}$s. In one of the articles added later he considered types for *n*-dimensional space, including 'α_*^ν' for the type in which the order of the νth dimension was reversed (p. 97: due to a printer's error this type was consistently misrendered as '$\alpha_{*\nu}$'). His treatment of well-order was rather brief, but he stressed its manifestations in finite and transfinite numbers (pp. 89–90).

The other extra article (pp. 92–95) and the published 'second communication' *1885b* dealt with this extended topology of respectively order-types and point-sets; but the first text was of course unknown and the second poorly organised. The best account was given in a long letter which Cantor sent to Mittag-Leffler in October 1884. The basic ideas were to write the operation of deriving a set P as '∂' (to produce the set ∂P), and to define five more operations on P:

'*Coherence*' $cP := P \cap P'$; '*Adherence*' $aP := P \cap P(P - P')$; (333.2)

'*Inherence*' $iP := c^\alpha P$; '*Supplement*' $sP := \partial P - P$; (333.3)

and the unnamed $seP := \bigcup(ac^\beta P)$, with $\beta < \alpha$, (333.4)

where α was an ordinal of the first or second number-class. 'The signs introduced in my new work are thus the *six*: a, c, ∂, i, r, s', Cantor told Mittag-Leffler in a sequel postcard, where he re-labelled (333.4) 'rP' and named it '*Remainder*';[16] properties of these sets were found, and new decomposition theorems presented, involving what I call 'partially derived' sets, such as $ac^\beta P$. While all the definitions were nominal and therefore

---

[16] Quoted from my *1970b*, 79; Cantor's long letter is on pp. 74–79 (also in his *Letters*, 208–214), with my explanation on pp. 70–72. I have used modern notation here; he deployed his symbols of (326.1, 3).

the defined terms eliminable, the aim was to help the topological analysis of sets with an enriched vocabulary. In addition, on the proof-sheets of the paper Cantor changed the name *'limit-point'* to *'chief-element'*, to reflect the extra conditions required to define this notion correctly for an ordered set (*m1885a*, 92–93).

It is a great pity that Cantor's new ideas, both philosophical and mathematical, came through at the time of his mid-term crisis, and the former kind did not meet with Mittag-Leffler's approval; apart from anything else, they were close to many of Whitehead's and Russell's later concerns. From this time on his contributions were made fitfully. We review them in the next section; this one ends with a short survey of the reactions of contemporaries to the work produced to date.

3.3.4 *Commentators on Cantor, 1883–1885.* Tannery's review of Cantor's papers in the *Acta* exemplified the growing interest in *Mengenlehre*. The *Jahrbuch über die Fortschritte der Mathematik* had been reviewing them, placing the reviews in the section 'Principles of geometry'. They were routine pieces, neither polemical nor missionary; several were written by the geometer Viktor Schlegel (1843–1905). Another reviewer was the historian and mathematician Max Simon (1844–1918), who also noticed the *Grundlagen* for the recently founded book review journal *Deutsche Literaturzeitung* (Simon *1883a*). In §3.2.3 we quoted from this review his lament that Dedekind's booklet on irrational numbers was 'much too little-known'; concerning Cantor, he appraised the notion of set, power (with the unhelpful explanation 'thus more or less, what one commonly calls set' (*'Menge'*), well-ordered sets and the transfinite ordinals and cardinals. As Tannery was also to note (§3.3.2), he remarked on the 'surprising theorem', actually in the third paper of the suite and stated just before (326.3), on the possibility of continuous motion in a discontinuous space. Interestingly, just as Cantor himself had recently proposed when the French translation of the *Grundlagen* was to be prepared, Simon recommended reading its articles in a fresh order.[17]

In 1885 two treatments of Cantor's work appeared, of quite different kinds. The Halle school-teacher Friedrich Meyer (1842–1898) published the second edition *1885a* of his textbook on algebra and arithmetic. Despite its elementary level, he emphasised the ideas of his distinguished townsfellow: in the second sentence of his introduction he mentioned 'the concept of set, especially the well-ordered set and the concept of power', soon followed by reference to 'a definition of number' (*'Anzahl'*). Cantor

---

[17] As was mentioned in footnote 14, Simon's order of sections was 1, 11–13, 2, 3, 14, 9, 10. His review of the *Grundlagen* appeared in May 1883, two months after Cantor had proposed 1–3, 11–14, 9, 10, *and then exactly Simon's order*, in letters of 15 and 18 (postmark) March to Mittag-Leffler (footnote 14). The final order is the second one but with 9 omitted. The similarity is strange; sadly, no correspondence between Simon and Cantor seems to survive.

was named several times afterwards, and at the end 'my friend Dr. Simon' was thanked for help. Further, the main text began at once with 'The concept set and quantity', and the first two chapters contained sprinklings of *Mengenlehre*; by the next page the novice reader was being confronted with 'various order-types' But the account of transfinite arithmetic was mercifully confined to two paragraphs of interjection into a routine account of the finite realm to indicate the transfinite ordinals and their basic properties (pp. 6, 10, 21).

The second treatment was written by Simon's philosophical colleague at the University of Strasbourg, Benno Kerry (1858–1889); he published in the *Vierteljahrsschrift für wissenschaftliche Philosophie* an excellent 40-page survey *1885a* of 'G. Cantor's investigations of manifolds'. Running through all the basic features of Cantor's current theory, he also picked up several interesting details. Early on he discussed the 'axiom' linking the line with the real numbers (pp. 192, also p. 217), and later on he discussed Cantor on continuity, and also Dedekind's definition of irrational numbers (pp. 202–204, 227). In a good summary of the sequence of derived sets he stressed the question of 'the reality of the concept' of the transfinite indices, especially when the 'kinds of index' extended to the 'Babylonian tower' $\omega^{\omega^{\omega^{\cdots}}}$ (p. 199); however, he rather underplayed the role of well-ordering (pp. 205–206). Recording the construction of ordinals in the second number-class, he noted that a third would follow, but was sceptical about the conception of this class and of its associated 'power' (pp. 211, 213, 230). By contrast, he saw a possibility of defining infinitesimals as inverses of transfinite ordinals (p. 220); and in connection with Cantor's definition of the measure of a set, he recalled the integral as 'alternatively a sum of infinitely-small spaces' (p. 229). His description of the definition 'powers' included a citation of Bolzano (pp. 206–208); this author was very well read indeed, as we shall see again in §4.5.4 when we note his reaction to the work of Frege.

3.4 THE EXTENSION OF THE *MENGENLEHRE*, 1886–1897

3.4.1 Dedekind's developing set theory, 1888. Cantor's former correspondent published a second booklet *1888a*; posing the question 'What are the numbers and what are they good for?', he gave a sophisticated and novel answer. Despite its rather unusual form of publication, it seems to have gained a quick reception (Hilbert *1931a*, 487). He reprinted it in 1893 and 1911, with new prefaces noting some recent developments. Like his other booklet, this one gained various translations and is still in print as a classic text; but it is deceptive in its clarity, for underneath lies a most sophisticated and also formal approach which actually makes it hard to understand. F. W. F. Meyer noted this aspect in an appreciative review *1891a* in the *Jahrbuch*.

As Dedekind mentioned in his preface (dated October 1887), his interest in the concept of number dates back to his *Habilitation* of 1854, when for his lecture he spoke before Gauss, Wilhelm Weber and his other examiners 'On the introduction of new functions in mathematics'. He had started out from 'elementary arithmetic', where 'the successive progress from a member of the series of absolute whole numbers to the next one, is the first and simplest operation' and led on 'in a similar way' to multiplication, exponentiation, and the other operations. 'Thus one obtains the negative, fractional, irrational and finally also the so-called imaginary numbers' (*m1854a*, 430–431). His later examples included the trigonometric functions (they related to the thesis itself, an unremarkable and unpublished essay on the transformation of coordinates), and elliptic functions and integrals.

Dedekind drafted his essay around the time of his previous booklet *1872a* on irrational numbers (§3.2.4); but he seems to have abandoned it in 1878 (the drafts are published in Dugac *1976a*, app. 56). He returned to it only on the occasion of the recent publication on the concept of number by Helmholtz and Kronecker, since he adhered to neither the empiricism of one nor the contructivism of the other (§3.6.2, 5) and wished to give his own approach some publicity.

In his preface Dedekind also referred to 'the laws of thought'. He did not intend Boole's view (§2.5.7), but the supposed power of the mind to create abstract objects which we saw also in §3.2.4. For sets he used the word 'System', probably taken from his reading of 'système' in French mathematics; it refers, in a naive way, to collections of mathematical objects (compare Hankel after (321.3)). It was specified in a way similar to that which Cantor had used (and will be quoted in §3.4.4); as 'various things $a, b, c \ldots$ comprehended from any cause under one point of view', where 'I understand by a thing any object of our thought' (arts. 2, 1). It seems that systems were different in category from things, but the matter was not clarified here. For example, he did not define systems of systems but instead the union and intersection of systems (arts. 8, 17), which he named respectively as 'collected together' and 'commonality' (*'Gemeinheit'*); for a collection of systems $A, B, C, \ldots$ they were notated

$$\text{'}\mathfrak{M}(A, B, C, \ldots)\text{' and '}\mathfrak{G}(A, B, C, \ldots)\text{'.} \qquad (341.1)$$

Dedekind also defined the relationship of 'part' between two systems A and S, and written '$A \ni S$' (art. 3); but this relation slid between membership and improper inclusion, a surprising slip to find in a careful reader of Cantor. (Proper inclusion was specified in art. 6 as 'proper part'.) These points were to be treated only in a manuscript conceived soon after publication of the booklet but written after 1899 and published in Sinaceur *1971a*. Entitling the piece 'Dangers of the theory of systems' he referred to art. 2 and noted that 'of the identification of a thing s with the system S standing from the single element s'; but even then he proposed to add only the strategy 'by which we want to indicate this system S again by a, thereby not distinguished from a, which will be permitted with some caution' (citing arts. 3, 8, 102 and 104 as among pertinent examples). He now drew on his philosophy of mind to claim 'the capability of the spirit to create a completely determined thing S from determined things a, b, c, . . . ' and thus to justify the difference between things and systems. To emphasise this point, he formally defined the 'null system' as the one 'for which there is no single thing'; he gave it the letter '0' and proved that it was unique. All of this was absent in 1888.

In tandem with his developing set theory Dedekind gave much attention to the transformation φ of a system S, in which any element s was mapped into s' (or $\varphi(s)$) and any part T of S went into part T' (*1888a*, art. 21). If $S' \ni S$, then φ was a transformation of S 'in itself' (art. 36). He laid particular stress on 'similar' ('*ähnliche*') transformations, under which different elements mapped into different elements. In this case $\varphi(S) = S$, and the (unique) '*inverse*' transformation $\bar{\varphi}$ returned s' back to s: $\varphi\bar{\varphi}$ 'is the identical transformation of S' (art. 26). Among consequences, 'One can thus divide up all systems in c l a s s e s' in which 'one takes up all and only the systems $Q, R, S, \ldots$ which are similar to a determined system R, the r e p r e s e n t a t i v e of the class' (art. 34)—an early example of the partition of a collection of objects by equivalence classes.

Dedekind proved that 'If $A \ni B$, and $B \ni A$, then $A = B$' (art. 5), as an obvious consequence of his definition of ' $\ni$ '. While preparing the text he had proved the much deeper equivalence theorem that if S' were transformed isomorphically into S and 'if further $S' \ni T \ni S$, so T is similar to S' (*m1887a*); but he left it out. This is surprising, for his formulation was more general than Cantor's (§3.2.8); we shall consider his proof in the context §4.2.5 of those published ten years later by Ernst Schröder and Felix Bernstein, after whom such theorems are often named.

3.4.2 *Dedekind's chains of integers.* Another main notion was that of a 'chain' ('*Kette*') relative to a transformation φ: a system K (which was part of a system S) was a chain if $K' \ni K$ (art. 37). After showing that S itself was a chain, as were unions and intersections of them (arts. 38–43), Dedekind took a part A of S and named the intersection of all chains containing A 'the c h a i n o f t h e s y s t e m A' (art. 44). He notated it 'A_0' —curiously in that, like Cantor (§3.5.3), he was chary of zero.

This was Dedekind's definition of a progression with an initial element in A. To distinguish 'the finite and infinite' progressions he offered the reflexive definition (art. 64) of infinitude which he had supplied to Cantor. (He seems not to have noticed C. S. Peirce's earlier discussion of the distinction between finite and infinite 'systems' in *1881b* (§4.3.4).) Then he allowed his philosophy of mind much reign with a "proof" that 'there are infinite systems' (art. 66); for he gave as evidence 'the totality S of all things, which may be objects of my thought', since as well as any of its elements s it contained also 'the thought s', that s can be the object of my thought', and so on infinitely. Some explanation from him would have helped; seemingly he was working within a Kantian framework.[18] As it was, this "proof" did not gain a good reception (§5.3.8, §7.5.2, §7.7.1).

So armed (as he thought) Dedekind characterised a system N as 'simply infinite' if 'there is such a similar transformation φ of N, that N appears as the chain of an element, which is not contained in $\varphi(N)$' and was called the 'b a s e - e l e m e n t' 1; thus one of the defining properties was '$N = 1_0$' (art. 71). This insight corresponded to Cantor's idea of well-ordering; and another similarity occurred when 'if we entirely ignore the special character of the elements' of N, then they became 'the n a t u r a l n u m b e r s or o r d i n a l n u m b e r s' of the 'n u m b e r - s e r i e s' (art. 73); also like Cantor, he ignored negative ordinals. Later he showed that all simply infinite systems were 'similar to the number-series' (art. 132). His theory resembles the Peano axioms for arithmetic, which we shall describe in §5.2.3 in connection with their named patron.

To found arithmetic suitably Dedekind treated mathematical induction with a new level of sophistication. Two main theorems (to us, metarules) were used, which he carefully distinguished in art. 130. Firstly, a 'theorem of complete induction (inference from n to n')' (arts. 59–60, 80) established a result for an initial value m and also from its assumed 'validity' for any n to that of its successor n'. In the ensuing treatment of arithmetical operations this theorem allowed him to prove results such as $m + n > n$ (art. 142). The second theorem was a deeper 'theorem of the definition of induction' (art. 126), which declared that, given any transformation θ of a system Ω into itself and an element ω in Ω, there existed a unique transformation ψ of the sequence N of numbers into a part of Ω which mapped 1 to ω and the ψ-transform of n' to the θ-transform of the ψ-transform of n. It legitimated inductive *proofs* in arithmetic by providing a justification for inductive *definitions*; for it guaranteed the existence of a means by which θ could successively locate the members of Ω, or any sub-system of them (art. 127). Despite all the concern with the foundations of arithmetic in the succeeding decades, the profundity of this section of Dedekind's booklet was not appreciated.[19]

---

[18] On a Kantian influence on Dedekind, see McCarty *1995a*; and on the acceptability of such proofs within phenomenology, see B. Smith *1994a*, 91–94. Bolzano had already offered this proof in his *1851a*, art. 20. Dedekind referred to it for the first time in the preface to the second (1893) edition of his booklet, and of Bolzano confessed that 'even the name was completely unknown' when he had completed this part of it. The draft version material mentioned in §3.4.1 (Dugac *1973a*, app. 58) does not contain this proof, and Cantor had sent him Bolzano's book in October 1882 (my *1974b*, 125).

[19] The classic location for an appreciation of Dedekind's treatment of inductive definitions is the textbook on foundations of analysis by Edmund Landau (*1930a*, preface and ch. 1), where he acknowledged an idea by Laszlo Kalmár. However, he made the point already in his obituary of Dedekind, in a version for which he thanked Zermelo (Landau *1917a*, 56–57).

Dedekind finally defined a finite cardinal as the number of members of any finite system 'Σ' (art. 161); provably unique to each Σ, it obtained also to any system similar to Σ, including finite number-series 'Z_n' from 1 to any n (arts. 159, 98). Thus, contrary to Cantor (§3.4.5), he gave ordinals priority over cardinals.

3.4.3 *Dedekind's philosophy of arithmetic.*

Of all resources which the human spirit [possesses] for the facilitation of his life, i.e. the work in which thinking exists, none is so momentous and so inseparable from his most inner nature as the concept of number. Arithmetic, whose only object is this concept, is now already a science of immeasurable extension[;] and it is not thrown into any doubt, that no barriers at all are set against this further development; just as immeasurable is the field of his application, because each thinking person even when he does not follow clearly, is a number-person, an arithmetician.

> Dedekind, undated manuscript (Dugac *1976a*, app. 58)

Three related aspects of Dedekind's treatment of arithmetic merit attention. Firstly, like Cantor and possibly following him, Dedekind specified ordinal numbers by abstracting the nature of element from a system and considering only their order; but, contrary to Cantor's position (§3.2.7), and also use of the word 'Anzahl', 'I look on the ordinal number, not the cardinal number [*Anzahl*] as the basic concept'. He did not really discuss the matter in the booklet (see art. 161), but it arose in contemporary correspondence with Heinrich Weber. Distancing himself from the usual use of 'ordinal', he specified his 'ordinal numbers' as 'the abstract elements of the simply infinite ordered system', and saw cardinals 'only for an *application* of ordinal numbers'.

Secondly, Dedekind did not define numbers from systems in the kind of way which Frege and Russell were to do; on the contrary, whether consciously or not, he followed Boole (end of §2.5.6) in *associating* numbers with systems, as the cardinal number of elements which a (finite) system contained. For example, in the letter (now lost), Weber seems to have suggested that the (finite) cardinals might be defined as classes of similar classes. Dedekind rejected this apparent anticipation of Frege (§4.5.3) and Russell (§6.5.2) and wished that cardinal to be 'something new (corresponding to this class) that the spirit creates. We are of divine species, and without any doubt possess creative strength, not merely in material things (railways, telegraph) but quite specially in intellectual things' (*Works 3*, 489).

Thirdly, Dedekind was aware of the model-theoretic (as we would now say) limitations of his formulations. His specification of equivalence relations between systems, and theorems about relationships (such as the cardinal number) between systems (arts. 162–165), showed that he saw some relativity in his formulations; and in a letter of 1890 to the school-teacher Hans Keferstein he made quite clear that the basic properties of numbers, such as the status of 1 and successorship to a number

> would hold for every system S that, besides the number sequence N, contain a system T, of arbitrary additional elements t, to which the mapping φ could always be extended while remaining similar and satisfying $\varphi(T) = T$. But such a system S is obviously something quite different from our number sequence N, and I could so choose it that scarcely a single theorem of arithmetic would be preserved in it. What, then, must we add to the facts above in order to cleanse our system S again of such alien intruders t as disturb every vestige of order and to restrict it to N? This was one of the most difficult points of my analysis and its mastery required lengthy reflection. If one presupposes knowledge of the sequence N of natural numbers and accordingly, allows himself the use of the language of arithmetic, then, of course, he has an easy time of it. He need only say: an element n belongs to the sequence N if and only if, starting with the element 1 and counting on and on steadfastly, that is, going through a finite number of iterations of the mapping φ (see the end of article 131 in my essay [on the theorem of definition by induction]), I actually reach the element n at some time; by this procedure, however, I shall never reach an element t outside of the sequence N. But this way of characterising the distinction between those elements t that are to be ejected from S and those elements n that alone are to remain is surely quite useless for our purpose; it would, after all, contain the most pernicious and obvious kind of vicious circle. [...] Thus how can I, without presupposing any arithmetic knowledge, give an unambiguous conceptual foundation to the distinction between the elements n and the elements t? Merely through consideration of the chains (articles 37 and 44 of my essay), and yet, by means of these, completely![20]

Dedekind realised that without the property of a chain his formulation could only specify a progression and therefore admit of elements additional to those intended; that is, he anticipated the notion of non-categoricity of an axiom system, when the various models are not in one-one correspondence with each other (§4.7.3). But he did not relate his insight to this definition in his booklet (arts. 1, 2):

> A thing a is the same as b [...] when all that can be thought concerning a can also be thought of b, and when all that can be thought concerning b can also be thought of a. [...] The system S is hence the same as the system T. In signs $S \equiv T$, when every element of S is also an element of T, and every element of T is also an element of S.

---

[20] Dedekind's letter is published in Sinaceur *1974a*, 270–278. An English translation of part of it was published in Wang *1957a* accompanied by an unreliable commentary, and in revised and complete form in van Heijenoort *1967a*, 98–103 (quoted here). For good discussions of Dedekind's theory of chains, see Cavaillès *1938a*, ch. 3; and Zariski *1926a* *passim*.

That is, equality between systems *was already defined categorically* by the specification of equality of two systems in terms of isomorphism between their members; thus the possibility of extra members was already eliminated—presumably by accident. Even this sophisticated mathematician-philosopher could miss a trick.

Finally, Dedekind's position may be called 'creative setism', covering the central places allotted both to mental powers and to sets and transformations between them. Although he referred in the preface of his booklet to 'the grounding of the simplest knowledge, namely that part of logic, which the doctrine of numbers handles', so that 'I name arithmetic (algebra, analysis) only a part of logic', he gave no formal presentation of his logic, and little indication of its content; in particular, his transformations are mathematical functions (or functors), not explicitly propositional functions and relations. Thus he was not a logicist in practice, though he may have had some vision of that kind.

3.4.4 *Cantor's philosophy of the infinite, 1886–1888.* While Dedekind was delivering the fruits of his thoughts on the foundations of arithmetic, Cantor turned away from the mathematical community after the debacle with Mittag-Leffler (§3.3.2) and towards the philosophical and theological aspects of his new doctrine. The products were two papers, a short one which appeared in two slightly different forms *1886a* and *1886b* in a variety of places,[21] and a long successor *1887–1888a* made up of versions of some recent 'communications' with colleagues and certain other material. While Cantor succeeded in gaining some attention from philosophers with these publications, they did not make much impact among mathematicians or logicians, and are not clear: the long paper is a repetitive wander around various points well discussed earlier, laced with long footnotes showing his remarkably detailed knowledge of the history of the infinite. However, in places he amplified certain points and published a few for the first time; references are confined to principal passages in the long paper.

On the general doctrine of the infinite, Cantor stressed strongly the difference between the absolute kind, as 'essentially inextendable' from the actual kind as '*yet extendable*' (*1887–1888a*, 385, 405). In response to

[21] This paper is unique in Cantor's corpus for its bibliographical complication. It began as a letter to the historian of mathematics and Mittag-Leffler's assistant Gustaf Eneström (draft of most of it in Cantor Papers, letter-book for 1884–1888, fols. 31–34), who placed it and some later additions with the Stockholm Academy as the paper Cantor *1886a*. Meanwhile, another version, containing also three letters for the long succeeding paper *1887–1888a*, was prepared (presumably by Cantor himself) and was published both as the paper Cantor *1886b* in the journal *Natur und Offenbarung* and as a short pamphlet. It also came out as a paper in the philosophical journal which also took the long paper, and the pair were reissued as the pamphlet *1890a*. This version was used in Cantor *Papers*, where a few words were printed twice on pp. 416–417. The titles and contents of these versions vary, but not substantially.

enquiries from theologians, he referred to the absolute as 'God and his attributes' and contrasted it with the actual infinities evident in nature such as 'the created individual entities in the universe' (pp. 399, 400). He also stressed again from the *Grundlagen* (§3.2.7) the difference between 'proper' and 'improper' infinities, claiming that their conflation 'contains in my opinion the reason for numerous errors', especially why 'one has not already discovered the *transfinite numbers* earlier' (p. 395).

3.4.5 *Cantor's new definitions of numbers.* Cantor also published here for the first time definitions of numbers from sets which became much better known from their re-run in the mid 1890s (§3.4.7). The opening sentence is the most pertinent (p. 387):

> Under *power* or *cardinal number* of a set M [...] I understand a general concept or species concept (universal) which one grasps by abstracting from the set the nature of its elements, as well as all relations which the elements have with each other or to other things, in particular the order which may govern the elements, and reflecting only on that common to all sets which are *equivalent* to M.

Dedekind also effected the first abstraction when defining ordinals from simply infinite systems (§3.4.2); we shall consider Cantor's use in §3.4.7.

Later in the paper Cantor introduced overbar notations: '$\overline{M}$' for the ordinal of M and '$\overline{\overline{M}}$' for its cardinal (p. 411–414), and (for example) '$\overline{\overline{5}}$' for the cardinal of the ordinal number 5 (p. 418). This part was described in a surly footnote as 'a short summary' of his withdrawn manuscript of 1884–1885 on order-types (§3.3.2); by and large the earlier version was clearer, and an unwelcome novelty here was the proposed synonym '*ideal numbers*' for '*order-types*' (p. 420), which did not endure. Each 'act of abstraction' (p. 379) was justified by the claim that 'a set and *the cardinal number belonging to it*' were 'quite different things', with the '*former* as Object' but 'the latter an abstract image of it *in our* mind' (p. 416).

3.4.6 *Cardinal exponentiation: Cantor's diagonal argument, 1891.* Cantor restored some of his contacts with mathematicians in the early 1890s when he played a major role in the founding of the *Deutsche Mathematiker-Vereinigung* (hereafter, '*DMV*') as a professional organisation separate from the hegemonies of Berlin and Göttingen, cities which were *not* used for its early annual meetings.[22] At its opening meeting, in Bremen, he treated the audience to a short but pregnant piece *1892a* presenting 'an elementary question of the theory of manifolds'—namely, a criterion for inequalities between cardinals by means of the 'diagonal argument', as it has become known. Two cases were taken, each of importance.

---

[22] On the early history of the *DMV*, see Gutzmer *1904a*. Its archives have recently been placed and sorted in the Archives of Freiburg University, too late for use here.

In the first part Cantor took the set M of all elements $\{E\}$ which could be expressed by the coordinates of a denumerably infinite coordinate space defined over (for examples) a binary pair of 'characters' m and w (presumably for 'Mann' and 'Weib'); a typical element would be, say,

$$E = (m, m, w, m, w, w, m, w, m, w, m, \ldots). \qquad (346.1)$$

'I now assert, that such a manifold M does not have the power of the series $1, 2, \ldots, v, \ldots$'; for this purpose he took a denumerable sequence S of elements whose collective coordinates formed a matrix-like array $\{a_{p,q}\}$. He then defined another element b_p of E by diagonalisation:

$$\text{if } a_{p,p} = m \text{ or } w, \text{ then } b_p = w \text{ or } m, \qquad (346.2)$$

which guaranteed it not to belong to S, as required.

A purpose of the form of expression (346.1) was to accommodate the decimal expansion of the real numbers, of which the non-denumerability was now proved, and by a method which avoided the cumbersome procedure of nesting intervals described in §3.2.5. The argument was direct (Gray *1994a*), not the *reductio* version in which it is sometimes construed today. Further, Cantor did not assume that S comprised the whole of M, although the result obviously held for M. His method went far beyond the selection of the diagonal members of an array of (say) functions $H(x, y)$ made by setting $x = y$. This procedure had been used before him by, for example, du Bois Reymond *1877a*, 156 in the context of his *Infinitärkalkül* (§3.2.1), and by Dedekind *1888a*, art. 125(m) in connection with definitions by induction (§3.4.2) and art. 159 concerning transformations between 'simply infinite' systems.

In the second part of the paper Cantor took as M the set of characteristic functions $\{f(x)\}$ (to use the modern name) of all subsets of the closed interval $L = [0, 1]$. Obviously M was not of lesser cardinality than L; to show that it was definitely greater he took the function $\phi(x, z)$ of two independent variables, where z was the member of L with which $f(x)$ was associated by the relation

$$f(x) = \phi(x, z) \text{ for } 0 \leqslant x \leqslant 1. \qquad (346.3)$$

He then considered the function

$$g(x) \neq \phi(x, z) \text{ for } 0 \leqslant x \leqslant 1: \qquad (346.4)$$

while an element of M, it took no value for z, thus establishing the greater cardinality of M. The argument assumed the well-ordering principle, as he noted at the end, together with the promise 'The further opening-up ['*Erschliessung*'] of this field is exercise for the future'.

3.4.7 *Transfinite cardinal arithmetic and simply ordered sets, 1895.* (Dauben *1979a*, chs. 8–9) During the 1890s Cantor worked at a new formulation of the principles of *Mengenlehre* (as he now called it, abandoning 'Mannigfaltigkeit' presumably for overuse). By mid decade he had work ready for the press, and he granted Klein again the honour of correspondent (Klein Papers, 8: 448–454), which he had broken a decade earlier (§3.3.2). A two-part paper appeared in the *Mathematische Annalen* as *1895b* and *1897a*. It became perhaps his best-known writing; Giulio Vivanti *1898a* and *1900a* described it in the *Jahrbuch*, and it was translated into French in 1899 and into English (by Jourdain) in 1915. Before that the first part quickly came out in Italian (§5.3.1). We shall note some of its features here; the second part is handled in the next sub-section.

'By a "set" I understand each gathering-together ['*Zusammenfassung*'] into a whole of determined well-distinguished objects *m* of our intuition or of our thought (which are called the "elements")'. This definition was similar to Dedekind's in §3.4.1, though we noted Cantor's priority in §3.4.5. It has often been quoted, usually without enthusiasm, on two counts. Firstly, poor old Cantor did not realise that this definition of a set admitted paradoxes; but in §3.5.3 I shall argue that it was so formulated precisely to *avoid* paradoxes. Secondly, its idealistic character, considered in §3.6.1, aroused philosophical reservations in various followers.

Cantor proceeded at once to his definitions of cardinals ('*Mächtigkeiten*') and their arithmetic, and in dealing with them first he showed perhaps more clearly than in the *Grundlagen* (§3.2.7) their epistemological priority over ordinals. He ran through once again the process of double abstraction of M to form its cardinal number $\overline{\overline{M}}$ rehearsed in his lecture of 1883, but he gave more details: one consequence was the need to restrict the definition of union to disjoint sets so that addition would avoid the difficulty of elements common to more than one set (art. 1). After stating the basic definitions of (in)equality between cardinals, and asserting an equivalence theorem without proof (art. 2, B), he proceeded to the arithmetical operations and properties such as the trichotomy law (that one of the relations ' < ', ' = ' or ' > ' always obtained between any two cardinals).

Cantor handled both the finite realm (art. 5, which suffers from comparison with Dedekind in §3.4.1–2 both for the lack of a definition of finitude and in its treatment of mathematical induction) and the transfinite range. In art. 6 he introduced the symbol '$\aleph_0$' and even also strung out the successors

$$\text{`}\aleph_0, \aleph_1, \aleph_2, \ldots \aleph_\nu, \ldots\text{'} \tag{347.1}$$

and their own successor '$\aleph_\omega$' without however entering into any details. Like Dedekind, he did not explain the use of the suffix '0'—and indeed left unclear the status of all the sufficial numbers (§3.5.3).[23]

The main novelty was inspired by the diagonal argument (art. 4). Cantor defined the '*covering for the set N with elements of the set M*', a single-valued function $f(n)$ from all elements n of N to the elements m of M; in a rather casual manner he introduced the set function $f(N)$ as 'the covering of N'. The definition let him proceed to that of the '*covering set from N to M*', written '$(N | M)$', comprising all possible coverings; for its cardinality gave a means of defining cardinal exponentiation:

$$\left(\overline{\overline{N}} \,|\, \overline{\overline{M}} \right) = \overline{\overline{M}}^{\,\overline{\overline{N}}}. \tag{347.2}$$

To stress the priority of cardinals over ordinals, Cantor introduced the 'simple' ordering of a set, where of any two members one always preceded the other, and then outlined their main arithmetical properties (art. 7: compare his stress in §3.3.3 on simple order and in arithmetic in the unpublished paper). As an example, he studied the rational numbers R over $(0, 1)$ (art. 9). Assigning the letter 'η' to their (simple) order-type, he showed that their cardinality $\overline{\eta}$ was $\aleph_0$ and noted various related properties. For example, $^*\eta = \eta$, where the pre-asterisk referred to the inverse of the order-type then indicated: in the manuscript of 1884–1885 he had used a suffixed asterisk, just before (333.1).

These results led to a survey of the continuum X of real numbers between 0 and 1 inclusive, with its order-type designated 'θ'. After noting that θ was not completely characterised by infinitude and perfection, Cantor replaced the metrical property of connectedness of 1883 (§3.2.8) by the requirement '*that between any two arbitrary elements x_0 and x_1 of X elements of R lie in rank*' (art. 11). He then claimed to be able to prove that '$\overline{M} = \theta$', as he put it right at the end of the part; the proof was based on demonstrating that a set M with the three properties just specified took the order-type '$\overline{X}$'. The issue of definition versus proof is at issue here; his use of irrational numbers as limits of sequences of rational numbers was clever, but one problem in the proof is that he drew upon the notions of the 'climbing' and 'falling fundamental series of first order' M and their 'limit-element'. These sequences had been defined in art. 10 as parts of a simply ordered infinite set, of order-types ω and $^*\omega$ respectively; the 'limit-element' of a 'climbing' *or* 'falling' simply ordered sequence (a_ν) of

---

[23] Cantor changed his numbering of alephs so as to start with 0 rather than 1 in July 1895, while the paper was in proof: see his letter (which contains no explanation) to Klein as editor of the journal in *Letters*, 356. He knew that 'aleph' also meant 'cattle' in Arabic, so that his cardinals were a cattle-herd ('*Rinderherde*': letter of 28 August 1899 in Hilbert Papers, 54/13).

members of M was the member which belonged to M and succeeded *or* preceded each a_ν, and was claimed to be unique. But the succession of notions is not very clear; indeed, the 'well-ordered sets' ω and $^*\omega$ had already been introduced in art. 7, with an appeal to finite cardinals to order the members, although only in the second part of this paper did he *formally* present well-ordered sets. Question-begging is in the air.

3.4.8 *Transfinite ordinal arithmetic and well-ordered sets, 1897.* Beginning his second part with a definition of the well-order-type (considered in §3.5.1) and properties of its segments (*1897a*, arts. 12–13), Cantor defined ordinal numbers as their order-types when the nature of the members was abstracted (art. 14). He then rehearsed their arithmetic, defining limit ordinals in terms of the idea of limit-element. He also presented sums of differences somewhat similar to the set-decomposition theorem (326.5) (art. 14, eq. (22)):

$$\text{'}\operatorname*{Lim}_{\nu} \alpha_\nu = \alpha_1 + (\alpha_2 - \alpha_1) + \cdots + (\alpha_{\nu+1} - \alpha_\nu) + \cdots\text{'};\quad (348.1)$$

the legitimacy of the procedure (or definition?) was taken for granted. He then treated only the second number-class; but he gave a much more detailed account than before of ordinal inequalities and of polynomials of the form

$$\sum_{r=0}^{\mu} \omega^r \nu_{\mu-r}, \text{ where } \mu \text{ and each } \nu_{\mu-r} \text{ were finite},\quad (348.2)$$

and their convertibility into transfinite products (art. 17). He showed that a number was *uniquely* expressible as (348.2), which he called its 'normal form' (art. 19). He concluded with a survey of 'The ε-numbers of the second number-class', the numbers which arose after a finite iteration of ordinal exponentiation as the roots of the equation $\omega^\xi = \xi$ (art. 20).

The second part stops rather than ends, for Cantor intended to proceed to at least one more part; but this pair was to be his last major publication on his subject. (Some unpublished results on simply ordered sets were to appear in F. Bernstein *1905a*, 134–138.) His plans for the third part are outlined in §3.5.3, among a survey of open and unresolved questions to which the next section is devoted.

3.5   OPEN AND HIDDEN QUESTIONS IN
      CANTOR'S *MENGENLEHRE*

3.5.1 *Well-ordering and the axioms of choice.* Here we take two issues, related but with the difference that Cantor was aware only of the first one. As we saw especially in §3.3.3, one of his major insights was to perceive the variety of orders into which an infinite set could be cast; and since one of his tasks was to provide a foundation for arithmetic, the order-type of finite and transfinite numbers had to be specified. This was the type which he called 'well-ordered'; and since arithmetic was a general theory, every set had to be orderable that way, even if it arose in some other order (for example, the rational numbers).

As we saw in §3.2.7, Cantor first addressed both definition and generality of well-ordering in detail in the early pages of the *Grundlagen* (*1883c*, 168, 169):

> By a *well-ordered* set is to understand any well-defined set, by which the elements are bound together by a determined succession, according to which there is a *first* element, and both for each individual element (if it not be the last in the succession) a determined successor follows and to each arbitrary finite or infinite set of elements belongs a determined element which is the *next following* element in the succession to all of them (if it be that there is none following them in the succession) [...]
>
> The concept of *well-ordered set* shows itself as fundamental for the entire theory of manifolds. That it is always possible to bring any *well-ordered* set into this law of thought, foundational and momentous so it seems [and] especially astonishing in its general validity, I shall come back in a later paper.

However, while he returned later to discuss his definition and give it alternative formulations, Cantor was not able to proceed beyond the optimism of the promised proof, which remained as an important task.[24] It became known in the 1900s as the 'well-ordering theorem'; Cantor gave it no name, and the preferable expression 'well-ordering principle' is of later origin. When a proof emerged in 1904, from Zermelo, it involved an axiom of choice (as he soon named it), which was concerned with the legitimacy of making an infinitude of independent selections of members from infinite sets, and that it turned out to be logically equivalent to the theorem itself (§7.2.5). Earlier Cantor and others on occasion made infinite selections without qualms—for example, different definitions of well-ordering itself, and in using set-decomposition theorems such as (326.5) that a union of denumerable sets was itself denumerable. He was also not aware of the bearing of these considerations on other results, such as the trichotomy law for transfinite ordinals mentioned before (347.1) (*1897a*, art. 14, B). Among similar stumbles, Dedekind noted in the preface *1893a* of the reprint of his booklet on integers that proving the reflexive and inductive definitions of infinity assumed that 'the series of natural numbers was already developed'.

---

[24] See, in particular, Cantor *1887–1888a*, 387–388; *1892a*; *1895b*, art. 6 (for the sequence of transfinite cardinals); and *1897a*, art. 12 (different definition). For commentary pertinent to this sub-section, see G. H. Moore *1982a*, ch. 1.

3.5.2 *What was Cantor's 'Cantor's continuum problem'?* One consequence of the well-ordering principle was that every transfinite cardinal
number was a 'power' (or, in his later notation, an aleph). For the
continuum, which aleph did it take? From the early 1900s this question
was called 'the continuum problem', and Cantor's answer 'the continuum
hypothesis'.[25]

This answer went through a variety of formulations as Cantor's theory
developed. The first version occurred at the end of *1878a*; as we saw in
§3.2.5, it took a rather unclear form, that sets came in only two sizes,
denumerable and '*Two*'. (This version is now sometimes called the 'weak'
form.) By the time of the *Grundlagen*, the definitions of the number-classes
(§3.2.7) allowed it to take the second form, 'that the sought power is none
other than that of our *second number-class*' (*1883c*, 192). Finally, cardinal
exponentiation (§3.4.7) gave him the *theorem* in his two-part paper that the
cardinality was $2^{\aleph_0}$ (*1895b*, art. 4); it was proved by using the binary
expansion of any real number, similar in form to the expansion (328.2)
used to define the ternary set. Then the hypothesis took the form

$$2^{\aleph_0} = \aleph_1 \qquad\qquad (352.1)$$

under which it is best known.

Curiously, Cantor did not explicitly state this form in the second part,
although $\aleph_1$ was discussed there (*1897a*, art. 16). He also did not treat any
version of the 'generalised continuum hypothesis' (as it became known)
that

$$2^{\aleph_r} = \aleph_{r+1}, \text{ where } r \text{ is any ordinal;} \qquad\qquad (352.2)$$

however, he may have perceived it in some intuitive form. For he knew
that the covering technique, which produced cardinal exponentiation as in
(347.2), was iterable, and it leads to the numbers

$$\aleph_0, 2^{\aleph_0}, 2^{2^{\aleph_0}}, \dots . \qquad\qquad (353.3)$$

If this sequence and (347.1) of alephs did not coincide arithmetically, then
transfinite cardinal arithmetic broke down; and since (352.1) claimed the
equality of the first member of each sequence, it would surely have been
natural to him to suppose that similarly (352.3) linked up the rest. Indeed,
in the footnote of the *Grundlagen* preceding the one which presented the
ternary set, he claimed a result amounting to (352.2) for $r = 2$ in terms of

---

[25] The title of this sub-section is an allusion to a famous and nice survey Gödel *1947a* of
this problem. For an excellent survey of its formulations by Cantor and reception by
contemporaries, see G. H. Moore *1989a*. The term 'continuum problem' was introduced in
the preface of the *Dissertation* F. Bernstein *1901a*, and in the revised version *1905a* (§4.2.5);
'continuum hypothesis' is due to Hausdorff *1908a*, 494.

the cardinalities of the sets of continuous and of continuous and discontinuous functions (*1883c*, 207), and in the second part of the paper introducing covering he had considered the cardinality of characteristic functions (346.3) of subsets of [0, 1].

As for proof of the hypothesis, is seems likely that Cantor hoped to use a decomposition theorem: show that the continuum C was the disjoint union of sets $\{P_j\}$ each one of known topological type, and appeal to lemmas on their respective cardinalities to add them up and obtain $\aleph_1$. The stumbling block, of course, was the continuum itself; the characterisation of C used metrical properties (§3.2.8), about which his lemmas rarely spoke. Evidence for this approach comes in a flurry of letters sent in June 1884 to Mittag-Leffler, shortly before his breakdown (§3.3.2); he was then producing many theorems of this kind (for example, Theorem 329.1 with Bendixson) and referred to some of them. He thought both that he had proved the hypothesis (which then stood in the second form described above) but then that the cardinal of C did not belong to (346.2) at all (Schönflies *1927a*, 9–11). In the end he was to get no further; the techniques associated with the third form doubtless seemed promising, but he was not able to profit, possibly because of the onset of deeper mental illness at the end of the century.

3.5.3 *"Paradoxes" and the absolute infinite.* (Jané *1995*) Cantor realised that his prolongation of the sequences of transfinite cardinal and ordinal numbers was unending, and that proposing a completion would lead to trouble. The key to his understanding of the point was mentioned in §3.4.4, where he distinguished between the actual and the absolute infinities. The former were the home of his doctrine of transfinite numbers; by contrast, as we saw, 'God as such is the infinite good and the absolute splendour' (*1887–1888a*, 386) with no place for humankind. Thus if man were to posit the existence of the largest ordinal β, then the process of its construction would entail that $\beta > \beta$ as well as $\beta = \beta$, thus infringing trichotomy. But for Cantor no paradox as such was involved: 'keep off the absolute infinite' was the conclusion, both β and for any analogous cardinal.

Although Cantor did not publish this analysis, he communicated it to various colleagues in correspondence. One of these was Jourdain, who received the above story in 1903 and was informed that Hilbert had been told around 1896 and Dedekind in 1899.[26] Hilbert had been the main contact, receiving several letters between 1897 and 1900: Cantor reported that he had developed his theory years earlier and even had ready the third paper of his new suite for the *Annalen*. He defined a set as 'ready' ('*fertig*') when it 'can be thought without contradiction as *collected together*

---

[26] Cantor's letter to Jourdain is published in my *1971a*, 115–116; Jourdain quoted it in *1904a*, 70. Hessenberg soon cited Felix Bernstein as a source for Cantor's priority (*1906a*, art. 98; on this work see §4.2.5).

and thus as *a thing for itself*'. Theorems included that a set of ready sets was ready, as was its power set; a particular case was the continuum. Excluded were sets containing β or its corresponding cardinal. He also hoped to winkle out a proof of the well-ordering principle.[27]

By late 1899 Cantor changed 'ready sets' to 'consistent multiplicities', with the other sets as 'inconsistent'. He had put Dedekind back on his visiting list in 1897, giving a lecture on this topic in Braunschweig;[28] in a suite of letters of July and August 1899 he sketched out a theory of consistent and inconsistent multiplicities, of which β was associated with the latter.[29]

However, a mathematical difficulty had to be faced; namely, the need to set up criteria for going up the sequence of ordinals as far as possible while avoiding β. (As we shall see in §7.4.4, this approach was to be called 'limitation of size' by Russell.) To this end Cantor returned to his definition of a set and assigned as 'inconsistent' (or, synonymously, '*absolutely infinite*') those multiplicities for which 'the assumption of a "being-together" ['*Zusammenseins*'] *of all* its elements leads to a contradiction'. He regarded the sequence (347.2) of alephs as inconsistent if "all" members were taken, and wondered if trichotomy could always obtain. But he did not avoid vicious circles of assumption and deduction, and never published his solution, which became known mainly through Jourdain quoting in papers short statements made to him by Cantor in letters.

A related curiosity is that while Cantor had zero in his theory of real numbers, its status as an integer was unclear. We noted that his sequence of ordinals began with 1 (§3.4.5), and that '0' was used without explanation in '$\aleph_0$' after (347.1); in his second 1899 letter to Dedekind he even consciously launched the series of ordinals with zero (*Letters*, 408). The hesitancy may have been caused by his abstractionist definitions of numbers from sets in §3.4.5, where again 1 was the first one so obtained; for if

[27] See Cantor's letters in Hilbert Papers, especially 54/3–9, 15–18; excerpts are transcribed in Purkert and Ilgauds *1987a*, 224–231 (with discussion on pp. 150–159, and in Purkert *1986a*) and in Cantor *Letters*, 390–400. Hilbert liked Cantor's approach but found 'ready' an unclear concept: see his report of a talk of 25 October 1898 to the *Göttinge Mathematische Gesellschaft* in the record book (Göttingen Mathematical Archive 49:1, fol. 43) and a note in his mathematical diary (Hilbert Papers 600/1, fol. 91). Note also Schönflies's letter of 12 July 1899 to some friends, kept in Klein Papers 11:735.

[28] Paul Stäckel took notes of the lecture (according to Fraenkel *1930a*, 265–266), but I have not found them. Schönflies (*1910a*, 251) stated that Cantor had defended the use of the law of the excluded middle against French criticisms (presumably by Borel and/or Poincaré; compare §4.2.2).

[29] These letters of 1899 were published in Cantor *Papers*, 443–451; but Zermelo fouled up Cavaillès's transcriptions, changing many spellings and even Cantor's mathematics in some places, and meshing the first two letters into one of the former date (28 July 1899, pp. 443–447: the break should be inserted on p. 443, at 'zukommt. Gehen'). On this vandalism, and the non-technical parts of these letters, see my *1974b*, 126–136; they are now reliably available in Cantor *Letters*, 405–411, and are translated into English, with some others and also letters to Hilbert, in Ewald *1996a*, 923–940.

a set were empty, how could one abstract from it to find its order-type or cardinal? Was it also well-ordered? His nervousness about the empty set '*O*', recorded after (326.1), could have a similar source. The issue is philosophically difficult, whiffing of paradox; and, as was shown in §3.4.1, even Dedekind was not lucid on the matter. The tri-distinction between zero, the empty set and literally no thing was to remain muddled until Frege and Russell, as we shall see in §4.5.3 and §6.5.3 respectively.

3.6 CANTOR'S PHILOSOPHY OF MATHEMATICS

3.6.1 *A mixed position*

The transfinite numbers are in a certain sense themselves *new irrationalities* [... they] *stand or fall* with the finite *rational numbers* [...]

Cantor *1887–1888*, 395

Although Cantor wrote extensively about the philosophical features of his *Mengenlehre* and was very well read in its history, he did not exhibit a very clear position. Some features will be exhibited in this section, partly to round off the story but also to prepare the later ground for negative as well as positive aspects of his influence. For a more detailed survey, see my *1980a* and Purkert *1989a*. The metaphysical and religious aspects have attracted attention recently; the background is surveyed in Bandmann *1992a*, pt. 1.

Cantor was a formalist in the sense that he felt that a consistent construction of a mathematical object guaranteed its existence. This seems to have been the motive behind the surprising construction after (323.3) of number-domains beyond those of the rationals and irrationals. It also underlay the quotation at the head of this sub-section; the construction of transfinite ordinals via the generating principles (§3.2.7) was also consistent, and so the constructed objects were on a par with the (mathematically respectable) irrational numbers.

This brand of formalism was Cantor's own. It is to be distinguished from the numbers-as-marks-on-paper type of formalism which we will find Frege attacking in §4.5.8–9. It differs also from the position which Hilbert was to promote from the late 1890s onwards (§4.7.3), in which questions such as consistency were examined in metamathematics; Cantor had no such category, so that with him consistency had only the status of a naive belief. There may well be a line of influence here, for Cantor's work had a strong effect on Hilbert.

In addition to this brand of formalism, Cantor exhibited traits of Platonism. The first part *1895b* of his final paper stated them explicitly in two of its three opening mottoes: a tag from Bacon (or, for him (§3.1.2), Bacon/Shakespeare), 'For we do not arbitrarily give laws to the intellect or to other things, but as faithful scribes we receive and copy them from the revealed voice of Nature'; and Newton's 'I feign no hypotheses'.

Finally, and with least enthusiasm among his contemporaries and followers, Cantor drew on idealist elements in granting a place to mental acts. The processes of forming a set by means of 'our intuition' and of abstracting from a set to form its order-type and cardinal (§3.4.7) are important and prominent examples, but not the only ones; others include associating an irrational number with a fundamental sequence of rationals and the 'specialisation' of one of them to guarantee uniqueness of definition (§3.2.3). This double use of idealism and Platonism relates to his acceptance of the '*immanent*' and '*transient*' realities of numbers noted in §3.2.7. In a way he linked them together in the first footnote of the *Grundlagen*, where he associated the formation of a set by abstraction 'with the Platonic ειδοσ, or ιδέα' from Plato's *Philebus* (*1883b*, 204). Maybe he saw abstraction as a generalisation of Socrates's teaching strategy of, say, associating five with the fingers and thumb of a hand. Some more explicit discussion from him on his philosophy would not have come amiss; in particular, he still seems to need order (Hallett *1984a*, ch. 3).

3.6.2 *(No) logic and metamathematics.*   Cantor encapsulated his philosophy in a phrase in the *Grundlagen*: 'The *essence* of *mathematics* lies precisely in its *freedom*' (*1883b*, 182).[30] We are free to construct objects, to draw upon mental processes, and to gain access to God—and later to have a professional association of mathematicians free from the dominating influences of Berlin and Göttingen (§3.4.5).

Since this freedom was presumably confined by the logical requirement of consistency, it is a curious irony that Cantor was cold to the developments of logic in his time. His only explicit point was to insist (often) that sets be 'well-defined', (for example, in the quotation in §3.5.1), and once he explained that this requirement entailed that 'on the ground of its definition and as a result of the law of the excluded third, it must be seen as *internally determined* as whether any object belonging to any same sphere of concept belongs to the considered manifold or not' (*1882b*, 150). Elsewhere he hoped to publish a version of his theory of ordinals 'developed forth with logical necessity' (*1887–1888a*, 380), but this language was only flourish. In §4.5.5 we shall consider his non-discussion in 1885 with Frege.

Frege was rather outside both the mathematical and logical communities. Much more prominent in Germany was the psychologist Wilhem Wundt, who published a treatise on logic in the 1880s. Like Boole (§2.5.3),

---

[30] This phrase is often quoted, but incorrectly, with the important emphasis 'precisely' ('*gerade*') omitted. The source is Schönflies *1900a*, 1 and repetitions later. Cantor was perhaps alluding to Hegel's *System der Philosophie* (1845): 'The essence of the spirit is hence formally freedom'.

he saw mathematics as applied logic, but he also involved the perception of time and space, counter to Cantor's assertion of the independence of arithmetic from such considerations; Cantor corresponded with him at some length on these and related issues (Kreiser *1979a*). Soon afterwards he repeated some of the same points in letters to the mathematician and historian Kurd Lasswitz, who was close to Wundt (Eccarius *1985a*).

Although Cantor was concerned with basic principles in his *Mengen-lehre*, he showed no interest in finding *axioms* for it; the word 'axiom' occurs very rarely (one case occurs after (323.3)). He also gave little welcome to a contemporary development in axiomatisation, which was called 'metamathematics'. This word did not carry the modern meaning noted in §3.6.1, in connection with Hilbert; it was then associated with the views of Riemann and von Helmholtz on the foundations of geometry, especially 'metageometry', as non-Euclidean geometries were then called, with the prefix 'meta' alluding to the readiness to admit hypotheses concerning the geometry of space (§6.2.1). As Helmholtz noted at the start of a paper *1878a*, 'the name has been given by opponents in irony, as suggesting "metaphysic"; but, as the founders of "non-Euclidean geome-try" have never maintained its objective truth, they can very well accept the name'. But for Cantor this reliance upon experience ran counter to his desire to abstract from it (as with Wundt), its acceptance of hypotheses infringed his (alleged) avoidance of them. As for its 'speculations with my works, they have not the slightest similarity and no proper contact' with his own theory of the infinite (*1887–1888a*, 391).

3.6.3 *The supposed impossibility of infinitesimals.* The formalist aspect of his philosophy seemed to have informed also his views on infinitesimals. As we saw in §2.7.4, Weierstrass inaugurated at Berlin a tradition of rigour in mathematical analysis in which infinitesimals were not used. For Cantor they were only a way of speaking about '*a mode* of *variability* of quantities' (*1882b*, 156), associated with the 'improper-infinite' (*1883b*, 172); however, more loyal than the king, he also claimed to be able to *prove* that their existence was impossible.

For heresy had been practised in the Holy Weierstrassian Empire: mathematicians such as du Bois Reymond, Otto Stolz, and the Italian Giuseppe Veronese attempted to develop theories of 'non-Archimedean quantities'. Cantor responded by presenting his proof of non-existence; but it was an unimpressive display. Defining 'linear number-quantities' as '*presentable as the image of bounded rectilinear continuous segments*' and satisfying Archimedes's law, he claimed that for any such quantity ζ for which $n\zeta < 1$ for any finite ordinal n, 'certain theorems of transfinite number theory' showed that $\nu\zeta$ 'is smaller than *any finite quantity ever so small*', contradicting the definition of a linear number (*1887–1888a*, 407–408). However, he never gave the details of these theorems; presumably they beg the question at hand, and maybe he realised this. In a letter

1895a to Vivanti, published at the time, he thought that the work of his opponents '*drifts in the air or much more is a* nonsense', and contains a '*vicious circle*'; unfortunately his own position was not dissimilar.

One does not have to be a formalist to think that demonstration of contradiction entailed non-existence; but this imperative seems to have underlined his opinion here. Thus Cantor was *not* to instigate the view that mathematical inverses of his consistently constructed transfinite numbers could lead to a consistent theory of infinitesimals.[31] He may also have thought there was no "room" in the continuum of real numbers as given by his 1872 definition (323.2) of real numbers (compare *1887–1888a*, 405–407); however, ironically it does *not* lead to Archimedes's principle

$$\text{if } 0 < a < b, \text{ then for some positive integer } n, \, ma > b \qquad (363.1)$$
$$\text{for all integers } m \geqslant n.$$

as a theorem, so that infinitesimals *can* be defined; only his 1895 definition (§3.4.7), or Dedekind's (§3.2.4), lead to Archimedean continua.[32]

3.6.4 *A contrast with Kronecker.*   We saw in §3.2.5 that Cantor thought that Kronecker had delayed publication of his paper (and also found reasons for sympathy with Kronecker, if in fact he did so act). No doubt there was conflict—indeed, for Kronecker himself it applied to the whole Weierstrassian approach—but his position is not easy to determine, since he never made an explicit statement of it. This point was made at the time; for example, by Dedekind in a frustrated footnote in his booklet on integers (*1888a*, art. 2; see also the manuscript published in H. Edwards and others *1982a*). However, one can gain some impression, especially from H. Edwards *1989a* and Jourdain *1913a*, 2–8.

It is well known that Kronecker was a constructivist for mathematics; but his famous motto 'The dear God has made the whole numbers, all else is man's work', made to a general audience at the Berlin meeting of the *Gesellschaft Deutscher Naturforscher und Ärtze* in 1886 (Weber *1893a*, 19), was only a hint. He seems to have objected to talk of 'any' or 'every' function, series or whatever in mathematics; he wished to know *how* it was to be put together. This comes out, for example in his work on algebraic

---

[31] This remark alludes to the inauguration in our time of non-standard analysis (Dauben *1995a*). The study of infinitesimals in Weierstrass's time would form an interesting history of heresy; for some information, see Dauben *1979a*, 128–132, 233–236.

[32] This point seems to have been made first in Stolz *1888a*; he mentioned his own definition of 'moment of systems of functions' and also du Bois Reymond's 'Nulls', both defined from considering ways in which $f(x) \searrow 0$ as $x \to \infty$, as defining infinitesimals by means *not* refuted by Cantor's claim. On Veronese's version of infinites and infinitesimals, see the extensive but unknown review *1895a* in the *Jahrbuch* by Ernst Kötter, interesting also as showing a mathematician from another area (in his case, principally geometry) studying *Mengenlehre*.

number fields, where the emphasis on constructivity and computational procedures using formulae makes a marked contrast with the competing approach of Dedekind and Weber.

In a paper 'On the number-concept' (to which Dedekind addressed his puzzled footnote) Kronecker *1887a* wrote of a 'bundle of objects' being counted in various ways, and of its cardinal ('*Anzahl*') as a property of the collection as a whole independent of any ordering (to which the ordinal was related: see esp. art. 2). He also replaced negative and rational numbers by talk of algebraic congruences (art. 5), so that, for example,

$$7 - 9 = 3 - 5 \text{ became } (7 + 9x) = (3 + 5x) \text{ (modulo } x + 1). \quad (364.1)$$

Focusing upon algebraic properties of numbers, his studies included finite sequences of integers, and the properties preserved when their order was changed by permutation.[33] But he saw no need to go beyond the rationals; for example, an elegant paper *1885a* on properties of the fractional remainder function of a rational quantity was called 'the absolute smallest remainder of real quantities'.

In his lectures on the integral calculus Kronecker *1894a* proceeded in much the same way with continuous functions as did the Weierstrassians; but all quantities (ε, and the like) were rational. He used limits, but only in contexts where the passage to the limiting case was determined, with known functions. He objected to Heine's Theorem 322.1 on uniform convergence on the grounds that the maximum of a collection of increments on the dependent variable was not defined, although its value could be approximated arbitrarily closely (pp. 342–345). Complex variables and integrals were handled as combinations of their real-variable analogues with $\sqrt{-1}$ in the right places (pp. 50–65).

Clearly there was plenty in Cantor for Kronecker to dislike; many features of the work on trigonometric series and irrational numbers even before wondering *how* one arrived at ω after $1, 2, 3, \ldots$ (*this* difficulty, not the infinite numbers as such). The extent of their personal differences will never be known, as Cantor seems to be our only source; apparently he effected a reconciliation in the summer of 1884 (Schönflies *1927a*, 9–12), when he was trying hard to prove the continuum hypothesis (§3.5.2), but there may not have been a great dispute on Kronecker's part anyway.

On the other hand, there were other differences between them. Kronecker did not attend the opening meeting of the *DMV* in 1891 for reasons of health (he was to die later that year), but he sent organiser

---

[33] In June 1898 Russell acquired, presumably by purchase, the offprint of *1887a* that Kronecker had sent 'Herrn Prof. Weierstrass mit freundschaftlichem Gruss' on 11 July 1887 (Russell Archives). The context is striking, for the two men had split severely in 1883 when Kronecker in effect accused Weierstrass of plagiarism in the theory of theta functions (Schubring *1998a*).

Cantor a pointed letter *(1891a,* 497):

> Therefore I do not even like the expression 'pupils' with us; we do not want or
> need any school [...] the mathematician must make himself at home intellectu-
> ally in his research sphere free from any prejudice, and freely look around and
> pursue discoveries [...]

3.7   CONCLUDING COMMENTS: THE CHARACTER OF
      CANTOR'S ACHIEVEMENTS

> As to infinites, I hold $1/0$ to be the infinite of infinites.
> For 0 marks the change from $+$ to $-$, which ∞ does not. [...]
> I am out of all fear about ∞^2. I believe in $\infty, \infty^2, \infty^3$, &c. &c.: and I intend to
> write a paper against the skim-milky, fast- and loosish mealy-mouthedness
> of the English mathematical world upon this point. My assertion is that the
> infinitely great and small have *subjective reality.* They have objective impos-
> sibility if you please; or not, just as you please [...]
> I therefore accept the concept *infinite* as a subjective reality of my con-
> sciousness of space and time, as real as my consciousness of either, because
> inseparable from my consciousness of either.
>
> De Morgan to Sir John Herschel, 29 April 1862
(S. De Morgan 1882a, 313)

De Morgan's statement, made before Cantor began to develop his
doctrine, exemplifies the point that Cantor was not the first mathematician
who thought that the infinite was not an indivisible entity but that orders
of the infinitely large could be distinguished: indeed, a variety of mathematicians and also eminent philosophers had put forward such views from
time to time, sometimes in connection with orders of the infinitesimally
small. De Morgan alluded to his ideas in a paper 'On infinity' written
shortly after his letter to Herschel, stating that 'The number of orders of
infinity is to be conceived as infinitely—as $\infty^{\infty^{\infty^{\cdots}}}$, indeed—exceeding the
unlimited multitude of values which a letter may take'.[34] However, not
even he developed his insight into a pukka theory, with (reasonably) clear
definitions and general criteria for cardinal and ordinal *in*equality and
arithmetic.

Such advances are due almost entirely to Cantor. Starting out from a
standard problem of the early 1870s in mathematical analysis concerning
trigonometric series, he gradually elaborated and also generalised his
Mengenlehre. Surpassing all predecessors in studying the topology of sets,

---

[34] De Morgan *1866a,* 168; this study may have been stimulated by his work on iterated
convergence tests (on which see my *1970a,* appendix), which also partly inspired du Bois
Reymond's *Infinitärkalkül* (§3.2.1). On the history of orders of the infinitely large, see
especially Schrecker *1946a* and Bunn *1977a;* the literature on infinitesimals touches on it
occasionally.

he clearly separated five distinct but related properties of sets:

1) *topology*: how is a set distributed along a line/plane/ ...?;
2) *dimension*: how is a linear set distinguished from a planar one, a planar set from ...?;
3) *measure*: how is its length/area/ ... to be measured?;
4) *size*: how many members does it possess?
5) *ordering*: in which different ways may its members be strung?

He also massively refined the notion of the infinite into theories of transfinite cardinal and ordinal arithmetic, and also introduced a range of order-types. The many links between the various sides were furnished basically by the infinitieth derived set $P^{(\infty)}$, and Cantor himself never abandoned his view that his *Mengenlehre* was an integrated theory. But most of his followers reacted otherwise.

CHAPTER 4

Parallel Processes in set Theory, Logics and Axiomatics, 1870s–1900s

4.1 PLANS FOR THE CHAPTER

In this chapter are collected six concurrent developments of great importance which, with one exception, ran alongside mathematical logic rather than within it. It is largely a German story, with some important American ingredients; among the main general sources is the reviewing *Jahrbuch über die Fortschritte der Mathematik*. Set theory is the main common thread, and §4.2 deals with the growth of interest in it, both as Cantorian *Mengenlehre*, and more generally.

Next, §4.3 describes the contributions to algebraic logic made by C. S. Peirce and some followers at Johns Hopkins University. The union of Boole's algebra with De Morgan's logic of relations led not only to the propositional calculus but also to the predicate calculus with quantifiers. In §4.4 some notice of the Grassmann brothers is followed by the contributions of Ernst Schröder, the main follower of Peirce. Working more systematically than his mentor, he articulated an elaborate algebra of logic, including relations, and developed a kind of logicism. The reactions of the Peirceans during the 1890s are also noted. By contrast, mathematical logic is introduced in §4.5, as practised by Gottlob Frege, now highly esteemed but then rather neglected; his work is taken from its start in 1879 to a major book in 1903.

Then §4.6 traces the early career of Edmund Husserl, trained under Weierstrass, developing with Cantor, and espousing phenomenological logic in important books of 1891 and 1900–1901. He then came into contact with the main subject of §4.7, David Hilbert, whose first phase of proof theory is described. It was stimulated by axiomatising geometry and arithmetic, but was also profoundly influenced by Cantor, and drew Ernst Zermelo into set theory, with spectacular consequences. Also included here is the allied emergence around 1900 of model theory (as it is now known), mostly in the U.S.A.

4.2 THE SPLITTING AND SELLING OF CANTOR'S *MENGENLEHRE*

4.2.1 *National and international support.* During the final years of the 19th century the importance of Cantor's *Mengenlehre* became generally recognised, but his own conception of it as an integrated topic was not often followed. Most mathematicians were primarily interested in the technical aspects; but the logicians and philosophers normally concentrated on the general and philosophical sides, including his vision of the *Mengenlehre* as a foundation for arithmetic and thereafter for "all" mathematics (§3.3.3). On the many developments of the 1900s, see especially Schönflies *1913a* (§8.7.6), T. Hawkins *1970a*, G. H. Moore *1982a* and Hallett *1984a*.

One type of occasion for publicity was the sequence of International Congresses of Mathematicians, which was launched at Zürich in 1897. Cantor had been a major figure in their founding, so it was meet that *Mengenlehre* should be featured. For example at Zürich, in the plenary address on analytic functions in the tradition of Weierstrass and his followers Adolf Hurwitz *1898a* included early on several pages of explanation of basic Cantorian concepts, including perfect and closed sets, the continuum, and the transfinite ordinals derived from the principles of generation (§3.2.7). However, as we shall see later (§4.2.7, §7.2.2), the treatment at these congresses was not always competent!

4.2.2 *French initiatives, especially from Borel.*   (Medvedev *1991a*) Courses in set theory began to be taught in a few centres, a practise which Cantor himself was never able to pursue at Halle. An important example of increased interest is provided by the three-volume *Cours d'analyse* by the Frenchman Camille Jordan (1838–1922). The first edition had concluded its last volume *1887a* with a collection of notes on set theory and related topics such as limits, continuity, irrational numbers and the integrability of functions; but six years later this material was moved and expanded to commence the second edition, on the grounds that such knowledge could not be presupposed among the students and was needed early (*1893a*, 1–54).

Jordan delivered his courses at the *Ecole Polytechnique*, traditionally the first choice of the mathematically talented in France. But over recent decades the *Ecole Normale Supérieure* had been rising in importance for mathematics. One of the key figures was Jules Tannery (1848–1910): placed first in 1866 to enter both schools as a student, he had chosen the latter and six years later was on the staff. We saw him in §3.3.3 as an early commentator on Cantor in a long review article *1884a*. Two years later he published a textbook *Introduction à l'étude des fonctions d'une variable* (Tannery *1886a*), which covered the *Mengenlehre* and related topics.

Among Tannery's students, one of the most notable was Emile Borel (1871–1956), who emulated him in 1889 as top student for both schools and also chose the *Normale*. Rapidly drawn into mathematical analysis by Tannery's lectures, he wrote a thesis *1894a* 'On some points in the theory of functions' while based at the University of Lille; it was quickly reprinted in the school's *Annales*, and was soon recognised as a significant contribution to point set topology. One of its results, rather casually presented,

became known as the 'Heine-Borel Theorem' (the origin of this unfortunate name was explained in §3.2.2); that if a bounded set of points on a
line can be covered by an infinitude of intervals, then a finite number will
do also. It was typical of his constructivist philosophy, which was similar to
Kronecker's (§3.6.4) in that he worked only with at most a denumerable
number of unions and complementations of given sets.

Appointed in 1897 to the staff of the school, Borel began with a lecture
course on functions which led to his first textbook, dedicated to Tannery
and presenting *Leçons sur la théorie des fonctions* (*1898a*). Its success led
his (and also Tannery's) publisher, Gauthier-Villars, to invite him to edit a
collection of volumes on this and related topics. A distinguished run was
launched, written mainly by members of Borel's circle (not only French); a
score of titles had appeared by 1920. Some aspects of set theory featured
in virtually all of them, often significantly. One of the most important
books was a volume *1904a* by *normalien* Henri Lebesgue (1875–1941),
building on his thesis *1902a* presented to the *Faculté des Sciences* of the
Université de Paris. He generalised the Riemann integral (§2.7.3) to a
theory of 'measure', with two major consequences (T. Hawkins *1970a*).
Firstly and more importantly, his theory greatly weakened the sufficient
conditions on theorems involving the processes of mathematical analysis
such as integrating or differentiating infinite series of functions, where
traditionally uniform continuity and/or convergence were required. Secondly, the exotic discontinuous or oscillatory functions which Riemann
himself had presented and Hermann Hankel and others had examined
(§3.2.1) were now integrable; for example, the characteristic function of
the rational numbers had no Riemann integral but Lebesgue measure
zero.

The following year another *normalien*, Réné Baire (1874–1932) (Dugac
1976b), built upon his *Faculté* thesis *1899a* to publish a volume *1905a* on
discontinuous functions. Extending Hankel's work on the classification of
functions, he took continuous functions $f_n(x)$ as the 'zeroth' class F_0 and
defined members of the first class F_1 as the (discontinuous) limiting
functions $\lim_{n \to \infty} f_n(x)$ of some sequence of functions from F_0. The
second class F_2 was defined similarly from F_0 and F_1, and so on. He
hoped that all functions could be expressed this way, but Lebesgue *1905a*
refuted him. Cantorian ideas of various kinds permeated all this work; for
example, Baire defined classes of functions up to F_α for any member α of
Cantor's second number-class, while Lebesgue drew upon both Cantor's
ternary set (328.2) and the diagonal argument (347.1) in constructing his
counter-example function.

Tannery's and Borel's remarkable entry performances were matched by
Jacques Hadamard (1865–1963), who also chose to be a *normalien*, in
1884. After graduation he too was based in the provinces for some years.
In 1897, when Borel began to teach at the school, Hadamard returned to
the capital with assistantships in both the *Faculté* and the *Collège de*

France. His main researches lay in mathematical analysis and its applications to other branches of pure mathematics such as number theory but also applications such as hydrodynamics (Maz'ya and Shaposhnikova *1998a*). While set theory did not feature in his work to a Borelian extent, it appeared enough to make him another focus, and a commentator on foundations.

4.2.3 *Couturat outlining the infinite, 1896.* (Couturat *1983a'*) These mathematicians formed much of the nucleus of the new generation in France for the new century; but the most important Frenchman for our story was an outsider. Once again a *normalien*, Louis Couturat (1868–1914) entered in 1887, specialising in philosophy. Much of his subsequent career was devoted to the interactions between mathematics, philosophy and logic. He also worked on their various histories, where his main figure was Leibniz, on whom he did important archival work in the early 1900s. Perhaps inspired by Leibniz's notion of a *characteristica universalis*, from then on he became passionately concerned with international languages. Much of his career was passed in the provinces, with occasional periods in Paris. His liking for logic seems to have condemned him to isolation from his mathematical compatriots: proud of their long Cartesian tradition of *raisonnement*, they despised the explicit analysis of reasoning. Poincaré's contempt for logic (and also ignorance of it) is unusual only in its explicitness (§6.2.3, §7.4.2, 5).

For one of his two doctoral theses, Couturat published as *1896a* his first and philosophically most important book: 660 pages on *De l'infini mathématique.* Impressed by Immanuel Kant, he began with a preface defending the place of metaphysics in philosophy, followed by in introduction seeking to distinguish the *a priori* and the *a posteriori* and considering the relationships between mathematics and physics. The compass of concern reduced still further in the text, which treated only number and quantity, although in great mathematical and philosophical generality. He was much influenced by Tannery's textbook, and also by an interesting study *1847a* of the relationships between algebra and geometry by the mathematician and economist Augustin Cournot (1801–1877), perhaps not by coincidence the second *normalien* (after the notorious Evariste Galois) of note in mathematics.

Part One of Couturat's book treated in 300 pages the 'generalisation of number'. Taking the integers for granted, he passed from the rationals through the irrationals (where on p. 60 he followed Tannery in adopting Dedekind's definition), transcendentals, negatives and imaginaries. The 'mathematical infinite' was handled in detail in the fourth and last *livre* of the Part; apparently influenced by Cournot, he presented various natural or intuitive encounters with the infinite in arithmetical or geometric contexts and resolved them, often by arguments in one of these branches but drawn from the other.

Part Two handled 'number and magnitude' (*'grandeur'*) again in four *livres*, this time in 280 pages. More philosophical in treatment, Couturat began by comparing 'empiricist' and 'rationalist' definitions of integers, largely Helmholtz versus Dedekind. Then he drew upon Kant's treatment of number, but including a brief début of Cantor's transfinite ordinals (p. 363). His sources on magnitudes included 'a magisterial lecture' by Tannery, apparently unpublished (p. 375); this time Helmholtz was contrasted with a largely Weierstrassian approach given in Stolz *1885a*. A very long discussion of the axioms of (in)equality (pp. 367–403) was followed by continuity; again Dedekind was the leading light but Cantor's definition was also noted (pp. 416–417).

The status and theory of the infinite was presented in the form of extensive dialogues between a 'finitist' and an 'infinitist' (pp. 443–503). Each speaker appealed to Great Men to support his position; Cantor was now more prominent, not only concerning ordinals but also his understanding of the isomorphism between the members of an infinite set and an infinite subset to counter the finitary tradition. Surprisingly, the alephs were not discussed.

The book was rather too long; in particular, the dialogues would need severe editing before being put on stage. In addition, in the final chapter on Kant's antinomies Couturat did not fully resolve the tension between his support for Kant and awareness of the limitations and even errors in the philosophy of mathematics (pp. 566–588). But overall he gave an excellent impression of both the range of mathematical situations in which the infinite was at issue and the philosophical questions which had to be tackled. In addition, much useful technical information was provided by a substantial appendix of notes (pp. 581–655) on hypercomplex numbers, Kronecker's theory of algebraic numbers (§3.6.4), the processes of limits in the theory of functions, and 40 pages on the *Mengenlehre* (but little on the alephs). A bibliography, well up to date, completed the book. Far beyond a typical doctoral thesis, it introduced or at least updated many readers to the new theories—including, as we shall see in §6.2.7, reviewer Russell.

4.2.4 *German initiatives from Klein.*   We saw in §3.2.6 and §3.4.7 that Cantor had published many of his main papers in *Mathematische Annalen*, thanks to the support of Felix Klein. This journal continued to take papers from Cantor's students and followers. Among the latter, the most noisy, though not the most competent, was Artur Schönflies (1853–1928). He came to the *Mengenlehre* relatively late after distinguished work in projective geometry and crystallography, but he took to it with a passion sustained for the rest of his life.[1] One of his first acts was initiated by Klein.

---

[1] Schönflies *Nachlass* was kept in the library of Frankfurt University, but it was destroyed by bombing in the Second World War. However, some interesting exchanges can be found in his letters in Klein Papers, Box 11, and in Hilbert Papers, 355.

In 1894 the *Deutsche Mathematiker-Vereinigung* (hereafter, '*DMV*') launched the *Encyklopädie der mathematischen Wissenschaften* as a vast detailed survey of all areas of mathematics at the time. Klein was the main instigator, and Teubner the publisher. French mathematicians soon began to prepare their own translation and elaboration of the project, as the *Encyclopédie des sciences mathématiques*, put out by Gauthier-Villars with Teubner. For the first of its six Parts, on arithmetic and algebra, Schönflies was invited to write a piece on *Mengenlehre*, which duly appeared as his *1898a*. It was divided about equally between the transfinite arithmetic and the point-set topology. While well referenced, and not only to Cantor's writings, it was pretty short, at 24 pages; he and Baire substantially reshaped and more than doubled its length in the French version Schönflies and Baire *1909a*, adding more than just the results found in the intervening decade.

Much more significant was the report on the *Mengenlehre* which Schönflies prepared for the *DMV*, in their annual series published in their *Jahresbericht* (Schönflies *1900a*): of book length, Teubner put it out also in this form. The order of material was hardly well, as Cantor might have said: generous to a fault were Vivanti's review *1902a* in the *Jahrbuch*, and Tannery's lengthy piece *1900a* in the *Bulletin des sciences mathématiques*. Starting by mis-quoting Cantor's statement that the essence of mathematics lay in its freedom (§3.6.2)—a mistake (in lacking 'precisely') which has been repeated infinitely ever since—the first section covered 'the general theory of infinite sets', taking cardinals first and proceeding to order-types, well-order and ordinals, and 'the higher number-classes'. Then followed a section on point set topology, including the sequence of derived sets (but not the motivation from trigonometric series). Perfect and closed sets dominated the account, followed by the content of sets (after the Riemann integral but before Lebesgue measure). Among 'point sets of a particular kind' Cantor's ternary set was included. The third section, on 'Applications to functions of real variables', took up nearly half of the report: Schönflies covered continuity, discontinuous and oscillatory functions of exotic kinds, the integral (nearly 30 pages, and intersecting with the earlier material on the content of sets), and the convergence of infinite series (ending with trigonometric series). Here he also named Borel's theorem on finite coverings 'the Heine-Borel theorem' because of its superficial similarity with Heine's Theorem 322.1 on the uniform continuity of functions (pp. 119, 51). A second part of the report appeared in 1908 (§4.2.7, §7.5.2).

Despite its drawbacks, the report also attracted new figures to the subject. Among the most significant were the English mathematicians Grace Chisholm Young (1868–1944) and her husband William Henry Young (1863–1942). She had taken a *Dissertation* under Klein in 1895 in a pioneering programme of higher education for women, and after her marriage the next year to this Cambridge University coach they went to the Continent to learn some genuine mathematics. The definitive choice of topic came when they visited Klein, who recommended them to try the

Mengenlehre as written up in Schönflies's report. The conversion decided their entire research career, the first of a married couple in mathematics, which lasted for 25 years (my *1972a*). With some financial independence provided by his earnings as coach, they lived in Göttingen until 1908, and came to know Cantor personally. Attracted to the topological aspects, William's first major achievement was 'a general theory of integration' constructed differently from Lebesgue's but more or less equivalent to it (Young *1905a*). His version was produced after Lebesgue; priority was readily acknowledged, and indeed the phrase 'Lebesgue integral' is Young's. They also published with Cambridge University Press a treatise on *The theory of sets of points* (Young and Young *1906a*), the first in English. (As Table 643.1 shows, Russell's *The principles* (1903) had concentrated more on the general aspects.) They also translated into English some of the *Encyklopädie* articles on mathematical analysis, to start an English edition; but they found only apathy from their compatriots on the island ('write textbooks', they were told). So they abandoned the project, and an edition was never prepared.

4.2.5 *German proofs of the Schröder-Bernstein theorem.* (Medvedev *1966a*) Unproven in the *Mengenlehre* was the equivalence of sets, as part of trichotomy; that is, that any cardinal was either equal, less than or greater than any other one. Cantor had proved equivalence, but only for sets of cardinality $\aleph_1$ (§3.2.7); the general result became a popular topic in the mid 1890s, with various proofs produced over the next decade. It was usually presented in two versions: I give both, with inclusions to be taken as proper. Firstly,

THEOREM 425.1 If set S is equivalent to its sub-subset R, then any subset U "between" S and R is equivalent to each.

As was noted in §3.4.1, Dedekind was the first prover, in his booklet on integers, but in a sketched manner (*1888a*, art. 63). For some reason he omitted a much clearer proof laid out the previous year in a manuscript *m1887a* which was to be published only in 1932, in his *Works*. By the assumption in the second version, a 'similar' (that is, one-one) mapping ϕ took S onto R. Defining the set $U := (S - T)$, he considered its chain U_0 under ϕ a new mapping ψ over S by the properties

$$\psi(s) = \phi(s) \text{ if } s\varepsilon U_0, \text{ and } \psi(s) = s \text{ if } s\varepsilon(S - U_0). \quad (425.1)$$

After proving that ψ was similar, he applied it to the two decompositions

$$S = (S - U_0) + U_0 \text{ and } T = (S - U_0) + (T - (S - U_0)) \quad (425.2)$$

related to the two clauses of the definition, where '+' indicated disjoint union of sets. Then he used the various relationships of inclusion between the sets to show that

$$\psi(S) \subseteq T \text{ and } T \subseteq \psi(S), \text{ so that } \psi(S) = T, \qquad (425.3)$$

from which the similarity of T and S was proved; that between R and T followed by imitation.

Dedekind seems to have communicated this jewel first only to Cantor, in 1899 (Cantor *Papers*, 449). Proofs of this type, found independently, were published only by Peano *1906a*, and Poincaré *1906b*, 314–315, the latter credited to a letter from Zermelo.[2] By then a quite different proof of this logically equivalent theorem had been in the literature for eight years:

THEOREM 425.2 If each of the sets M and N is equivalent to a proper subset N_1 and M_1 of the other one, then they are equivalent to each other (and so have the same cardinality).

For brevity I use ' $\sim$ ' to denote equivalence between sets. There must be a subset M_2 of M_1 for which $N_1 \sim M_2$; and so $M \sim M_2$. Hence the theorem reduces to the first version, that $M \sim M_1$. To prove it, define the disjoint sets

$$H_2 := M - M_1 \text{ and } K_2 := M_1 - M_2 \qquad (425.4)$$

and apply repeatedly to the trio of mutually disjoint sets M_2, H_2 and K_2 a similar mapping from M to M_1; this yields M_r, H_r and K_r respectively, each trio still disjoint. Let L be the intersection, maybe empty, of all the M_r after denumerably many applications. Then

$$M = L + \Sigma_{r=2}(H_r + K_r) \text{ and } M_1 = L + \Sigma_{r=2}(H_{r+1} + K_r). \quad (425.5)$$

Now map L and each K_r identically onto itself, and each H_r isomorphically onto its subset H_{r+1}; the equivalence between M and M_1 follows.

The theorem was all but named by Schönflies in his report, after the two independent creators of this proof (*1900a*, 16). The first was offered by Schröder, who (thought that he) had proved it in a long paper on finitude to be noted in §4.4.8 (*1898c*, 336–344); unfortunately, he had falsely assumed that the cardinality of each limiting set in the two sequences was equal to that of its predecessors. The slip was pointed out to him in a letter of May 1902 written by a school-teacher active in the foundations of mathematics, Alwin Korselt (1864–1947); Schröder replied that he had already noted it himself. This information was given in a short paper Korselt *1911a* in *Mathematische Annalen*: it contains also his own version of the first proof, which he stated he had submitted in 1902 to the journal that year but which for some reason had not then been published.

---

[2] Zermelo was to publish his proof himself in his paper on axiomatic set theory described in §4.7.6 (*1908b*, nos. 25–27). Poincaré's letter to him of June 1906 is published in Heinzmann *1986a*, 105.

No such slip in derivation tainted the version by the second figure, a young newcomer to the *Mengenlehre*: Felix Bernstein (1878–1956) (Frewer *1981a*). He spent the years 1896–1901 at various universities before writing his *Dissertation 1901a* under Hilbert's direction; a somewhat revised edition appeared in *Mathematische Annalen* as *1905a*. In both versions he mentioned this proof; but, like Poincaré with Zermelo later, it had already appeared with acknowledgement in Paris, in Borel's *Fonctions* (*1898a*, 104–107). He had presented it in the previous year to Cantor's own seminar at Halle University, where his father was professor of physiology.[3]

Further versions appeared in the fertile year of 1906, from Julius König *1906a*, and in Hessenberg *1906a*, arts. 34-37. Gerhard Hessenberg (1874–1925) belonged to a group of philosophers called 'the Fries school', after the neo-Kantian philosopher Jakob Fries (1773–1843). His proof was given within a long article on the 'Fundamental concepts of *Mengenlehre*', which was reprinted in book form. He paid much attention to equivalence, being especially impressed by the difficulty, evident since Cantor *1883b*, of proving such basic properties about sets. Like others of the time, he included his proof within a general discussion of trichotomy. Narrower in range but of greater philosophical weight than Couturat's book, he discussed in detail the more general aspects of the subject, such as order-types, transfinite ordinals, cardinal exponentiation, and definitions of integers. Some parts were unusual; for example, in ch. 22 on decidability he decomposed a set into the subset of members *known* to have a given property and the complementary subset. His views on the paradoxes, including one due to his colleague Kurt Grelling, are noted in §7.2.3.

By 1906 the role of the axioms of choice and the well-ordering principle were becoming evident, so that all proofs required not only examination but autopsy. In particular, Whitehead and Russell were to handle the Schröder-Bernstein theorem very carefully (§7.8.6).

4.2.6 *Publicity from Hilbert, 1900.* The leading German mathematician around 1900 was Klein's younger colleague at Göttingen, David Hilbert (1862–1943). His own work on foundational areas of mathematics (§4.7.1) had advanced sufficiently for him to be convinced of the basic correctness and importance of Cantor's *Mengenlehre* and of his own ideas on proof theory; and an occasion arose which allowed him to give both enterprises good publicity among mathematicians. A Universal Exhibition was held in Paris in 1900 to launch the new century (or, as the more mathematically minded might have noticed, to presage its commencement on 1 January

---

[3] Cantor mentioned Bernstein's achievement to Dedekind on 30 August 1899 after receiving Dedekind's own proof (Cantor *Papers*, 450). Compare Bernstein's own reminiscences to Emmy Noether following the text of *m1887a* (Dedekind *Works 3*, 449; translated in Ewald *1996a*, 836). For a nice comparison of these two proofs and of trichotomy, partly historical, see Fraenkel *1953a*, 99–104.

1901), and in this connection various disciplines held International Congresses in the city. The mathematicians met from 6 to 12 August for their 'Second' congress, succeeding the one held in 1897 in Zürich; it followed a corresponding jamboree for the philosophers (§5.5.1).[4]

Hilbert's general familiarity with mathematics gave him a fairly strong perception of its major open questions and research areas; so he chose to describe his view of the principal 'mathematical problems' awaiting attention in the century to come. The historian of mathematics Moritz Cantor was in the chair for the morning session of 8 August, when Hilbert spoke on 10 problems: the full version Hilbert *1900c*, which contained 23 problems, made history on its own, with two printings and translations into French (for the Congress Proceedings) and English.[5] Strikingly, and doubtless bearing order in mind, he placed Cantor's continuum problem (352.1) as the first problem (with the well-ordering principle as an associated question), and 'the consistency of the arithmetical axioms' as the second.

4.2.7 *Integral equations and functional analysis.* A significant application of set and measure theory to mathematical analysis was in integral equations. The task was to find which functions g, if any, satisfied an equation such as

$$f(x) = g(x) + \int_a^b h(x, y)g(y)\, dx, \tag{427.1}$$

with f and h known. The topic had arisen occasionally in the 19th century, usually in connection with differential equations or a physical application; but interest increased considerably from the 1890s. Hilbert became engaged from around 1905; for him (427.1) was a principal concern when f and h were continuous functions. A principal method of solution was to convert them into a denumerable number of linear equations with the corresponding number of unknowns.[6]

Finding sets of functions satisfying certain properties was a main method of solution, for such study of functions had also gained new interest in the 1890s; the name 'functional analysis' became attached to it later. They

---

[4] Other academic disciplines which held congresses in Paris in 1900 include geology, applied mechanics, physics, photography, medicine, ornithology, psychology and history. Two years earlier the International Council of Scientific Unions had been formed.

[5] A particularly useful work concerning Hilbert's lecture is the volume cited as Alexandrov *1971a*, where the text is followed by account up to the time of publication of the progress made on the problems which he posed. The range of problems suggested was rather limited; applications fared poorly, and probability and statistics even worse. Also absent is integral equations, upon which Hilbert was to concentrate for many years from 1904!

[6] Joseph Fourier pioneered some of these developments; for when reviving the trigonometric series in the 1800s he had stumbled into infinite matrix theory as a means of calculating the coefficients (Bernkopf *1968a*), and a decade later he developed his integral formula as a companion theory by finding the inverse transform of a given function from a double-integral equation (see, for example, my *1990a*, esp. ch. 9).

were conceived as objects belonging, in a set-like manner, to a 'space' by virtue of properties such as continuity, say, or differentiability (Siegmund-Schulze *1983a*). Publicity at the Zürich Congress came from Hadamard *1898a*, who outlined some of the basic ideas, including the use of set theory; however he displayed own limited knowledge of the *Mengenlehre* by misdefining the concept of well-ordering! Progress was leisurely, and explanation to outsiders essential; Maurice Fréchet (1878–1973) began his doctoral thesis *1906a* with several pages of *very* elementary explanation of the basic idea of functions being members of a space.[7]

Under this view, trigonometric series, which had drawn Cantor into sets in the first place (§3.2.3), were now construed as defining a space S of functions $f(x)$ expressible over some interval $[a, b]$ of values in a series (321.1) of sine and cosine functions which served as its basis. One of the most important theorems, proved in 1907 by Ernst Fischer (1875–1956) and Frigyes Riesz (1886–1969) and known after them, stated that if the sum of the squares of the coefficients were convergent, then there was indeed a function $f(x)$ belonging to S which was the sum of that series and for which

$$\int_a^b f(x)^2 \, dx \text{ was bounded,} \tag{427.2}$$

a property satisfied also by the component sine and cosine functions.

The integrals, and indeed the whole theory, were handled with a generality provided by Lebesgue theory of measure. But still greater generality was envisaged by a leading American mathematician, E. H. Moore (1862–1932). Impressed by the range of algebras and linear forms such as Fourier series in analysis and especially infinite matrices and integral equations, he sought a 'General analysis'. The governing principle of his theory was that '*The existence of analogies between central features of various theories implies the existence of a general theory which underlies the particular theories and unifies them with respect to those central features*', and drawing upon 'These theories of Cantor[, which] are permeating Modern Mathematics' (*1910a*, 2, 1). He told Fréchet in 1926 that he had chosen this name in imitation of the phrase 'general set theory';[8] in §6.6.3 we shall reveal his little-known role in the paradoxes.

Among his references, Moore cited the second part *1908a* of the report on the *Mengenlehre*, which Schönflies had recently published with the

---

[7] A curious feature of some Paris *Faculté* theses was their publication in Italy. Fréchet's came out in the *Rendiconti del Circolo Matematico di Palermo*, while Baire's and Lebesgue's had appeared in the *Annali di matematica pura ed applicata*.

[8] Moore to Fréchet, 16 February 1926 (Moore Papers, Box 3). His efforts to develop the theory, especially after 1911, are scattered through Boxes 5, 6 and 9–17. In 1908 he had been adapting Peano's logical symbolism for his purpose (letters to Veblen in Veblen Papers, Box 3). On Moore's theory see Siegmund-Schulze *1998a*; and on his great significance for American mathematics, Parshall and Rowe *1994a*, chs. 9–10.

DMV. We saw in §4.2.4 that in the first part, *1900a*, he had treated the basic features in his own way. Here he handled 'the geometrical applications', with a more detailed treatment of the topological aspects followed by the invariance of dimensions, continuous functions and curves, and elements of functional analysis. He was more in his special areas in this part, and its 331 pages (80 more than its predecessor) give a more confident and clearer impression. The first two and the final chapters updated and corrected the first part; in particular, the integral now included Lebesgue measure (pp. 318–325).

The most significant new theory for Schönflies was ordered sets, which he presented in his second chapter largely following an important pair of 60-page articles *1906a* and *1907a* by Felix Hausdorff (1868–1942). He greatly extended Cantor's treatment of non-well-ordered types, especially of non-denumerable sets, by using 'transfinite induction', as he christened it (*1906*, 127–128). He had come to the *Mengenlehre* around 1900 after a début in applied mathematics, and became one of its most distinguished practitioners (§8.7.6);[9] these articles were to influence Whitehead and Russell substantially (§7.9.5).

Thus the peculiar *Mengenlehre* of the late 19th century became the established set theory of the new century; further books appeared, as we see in §8.7.6. Yet the *Mengenlehre* had already been eclipsed by a still more general theory of collections which, however, gained little attention then or ever after.

4.2.8 *Kempe on 'mathematical form'*. (Vercelloni *1989a*, prologue) If Couturat was an outsider, Alfred Bray Kempe (1849–1922) lay almost out of sight. He was that characteristically British object, a highly talented mathematician who did not hold a professional appointment. He made his career as a lawyer, but his mathematical work earned him a Fellowship of the Royal Society in 1881—indeed, he was to be its treasurer from 1898 to 1919, and he was knighted in 1912 for those services (Giekie *1923a*).

Among his various mathematical interests, a remarkable achievement was contained in a long paper Kempe *1886a* on 'the theory of mathematical form', published in the Society's *Philosophical transactions*. I cite it by the number of the many short sections into which it is divided. Seeking 'the necessary matter of exact or mathematical thought from the accidental clothing—geometrical, algebraical, logical, &c' (sect. 1), he found it in 'collections of units', which 'come under consideration in a variety of garbs —as material objects, intervals or periods of time, processes of thought, points, lines, statements, relationships, arrangements, algebraical expressions, operators, operations, &c., &c., occupy various positions, and are otherwise variously circumstanced' (sect. 4). Individual units were written,

---

[9] On Hausdorff's work see especially Brieskorn *1996a*. A fine catalogue of his large *Nachlass* is provided in Purkert *1995a*.

say, '$a, b, \ldots$' separated by commas; but a pair 'ab' could be taken, and even 'may sometimes be distinguished from the pair ba though the units a and b are undistinguished', as in the sensed line ab from point a to point b (sect. 5). The same situation obtained for triads, $\ldots$ up to 'm-ads' for any positive finite integer m (sect. 7). Thus form, his key concept, was predicated of a collection 'due (1) to the number of its component units, and (2) to the way in which the distinguished and undistinguished units, pairs, triads, &c., are distributed through the collection' (sect. 9).

Kempe may have been inspired by the mathematical study of graphs launched by Arthur Cayley and J. J. Sylvester in the 1870s, for he had applied it in *1885a* to their theory of algebraic invariants. Indeed, they were the Society's referees for *1886a*;[10] while generally favourable, understandably they did not realise the extent of its novelty. His main advance over all predecessors was that *he allowed units to belong more than once to a collection*, unlike the single membership of set theory. We noted in §2.5.8 the example of the roots 3, 3 and 8 of a cubic equation; Kempe used cases such as the shape 'Y', construed as a collection containing one 'distinguished' central node together with three 'undistinguished' extremal ones (sect. 9). Sub-collections were 'components', and a disjoint pair was 'detached' (sects. 18–19); a collection of units in which every component was distinguished from each of its detached units was called a 'system' (sect. 25). This is curiously like Dedekind's phrasing in his booklet on integers published two years later (§3.4.1): 'various things $a, b, c, \ldots$ are comprehended from whatever motive under one point of view [$\ldots$] and one then says, that they form a s y s t e m' (Dedekind *1888a*, art. 2).

A very important kind of finite system of n units for Kempe was a 'heap'. It was 'discrete' when every component s-ad was distinguished from all others of the same number for all $s \leqslant n$; 'single' when every s-ad was undistinguished; and 'independent' in between, such as in the 'Y' (*1886a*, sects. 37–38, 44). A 'set' was defined as a collection of units such that any pair of undistinguished components could be extended by further units already in it. 'A system is obviously a set. A set is not necessarily a system' (the unclear sects. 130–131).

Special symbols were introduced in Kempe's theory of 'aspects' of a unit in a collection, which highlighted its location when mapped isomorphically across to a mate unit in another undistinguished collection (sect. 73); the notion corresponded in role to Dedekind again, and also to Cantor's

---

[10] Royal Society Archives, Files RR 9.287–288. Cayley suggested the title of Kempe's paper, while Sylvester stated that Kempe had thought of placing (a version of) it in the *American journal of mathematics* when he had been the editor. See also G. G. Stokes's letter on these changes in the Kempe Papers, Packet 19.

abstraction from (his kind of) set (§3.4.7). Among their 'elementary properties' (sects. 89–99), two *m*-ads being undistinguished was written

$$\text{'}abcd\ldots \succ\!\!\prec pqrs\ldots, \text{ when also, say, '}bc \succ\!\!\prec qr\text{'};\qquad (428.1)$$

but if distinguished, then

$$\text{'}abcd\ldots \leftrightarrow pqrs\ldots\text{', when also, say, '}abcd\ldots \succ\!\!\prec srqp\ldots\text{'};\qquad (428.2)$$

$$\therefore \text{'}a, b, c, d, \ldots \succ\!\!\prec p, q, r, s, \ldots\text{'}.\qquad (428.3)$$

In another strange anticipation of Dedekind's terms, he also considered 'chains' starting out with 'A succession of undistinguished pairs, *ab*, *bc*, *cd*, ...', which 'may be termed a *simple chain*' (sects. 211–221).

As in the case of 'Y', Kempe also used 'graphical representations of units' (sect. 39), usually graphs or grids of little lettered circles to represent particular cases. One of them was a mechanical linkage (sect. 82); maybe earlier work *1872a* on this topic had also helped to inspire him, for a linkage is a graph in wood or metal. His most extensive use of graphs provided a large classification of groups and quaternions (sects. 240–327).

Among other branches of mathematics, Kempe treated the geometry of the plane, especially concurrent and coplanar lines, and collinear and triads of points (sects. 350–359). But the last part, on 'logic' (sects. 360–391), was rather disappointing: an essentially unmodified review of the basic features of Boole's algebra of logic with Jevons's modifications (§2.6.2) interpreted in terms of 'classes', a term which Kempe did not explain.

4.2.9 *Kempe—who?* With one exception to be noted soon, the reception of the paper was silence; for some reason it was not even reviewed in the *Jahrbuch*. Perhaps this non-reception provoked him to seek more publicity at the end of the decade. A general paper *1890b* in *Nature* on 'The subject matter of exact thought' largely concentrated on the uses of the theory in geometry, with some emphasis on symmetric and asymmetric relationships (for example, as between the extremities of the unsensed and the sensed line). It came out soon after a more ample statement *1890a* placed with the London Mathematical Society, to which he later offered in his Presidential Address *1894a* a survey of his theory, ending with this definition of mathematics: '*the science by which we investigate those characteristics of any subject-matter of thought which are due to the conception that it consists of a number of differing and non-differing individuals and pluralities*'.

Mathematicians' ignorance of Kempe has always been great: his theory has been re-invented in recent years, under the name 'multisets', without knowledge of his priority (see, for example, Rado *1975a*). But he soon gained some surprising followers in two American philosophers: Josiah Royce in the early 1900s (§7.5.4), but quickly from C. S. Peirce. When the large paper appeared, Peirce wrote to Kempe about the theory of aspects,[11] with the result that Kempe sent to the Royal Society a short note *1887a* modifying some sections. But later the reaction was opposite; in retort to Peirce attributing to him the view that relationship was 'nothing but a complex of a bare connexion of pairs of objects' (Peirce *1897a*, 295: the context is described in §4.4.7), Kempe *1897a* replied that on the contrary, while often subsidiary, in general they lay among the basic units which he sought as 'the essential residue of the subject-matter of thought', and that lines in his diagrams served only to distinguish one arrangement of units from another one.

A more radical effect of Kempe occurred on 15 January 1889 (Peirce's own dating on the folios involved): presumably from looking at the various graphs in the original paper, Peirce suddenly conceived of a similar manner of representing the syntax of well-formed English sentences, in a theory which he came to call 'entative' and 'existential graphs'. For example (one of his), the 'Y', which was treated as a graph by Cayley and Sylvester and as a heap containing one distinguished and three undistinguished elements by Kempe, represented a 'triple relative' for Peirce. The development of this insight, quite foreign to Kempe's own purposes, became a major concern of Peirce for many years, and the recent recognition of its importance has made him a darling of the artificially intelligent.[12] Its consciously topological character signified a basic change from his severely algebraic approach hitherto to logic, a matter which dominates our next section.

4.3   AMERICAN ALGEBRAIC LOGIC: PEIRCE AND HIS FOLLOWERS

Much of my work never will be published. If I can, before I die, get so much accessible as others may have a difficulty in discovering, I shall feel that I can be excused from more. My aversion to publishing anything has not been due to want of interest in others but to the thought that after all a philosophy can only be passed from mouth to mouth where there is opportunity to object and cross-question; and that printing is not publishing unless the matter be pretty first class.

C. S. Peirce to Lady Welby, as transcribed by her in a letter to
Russell of 16 December 1904 (Russell Archives; Hardwick *1977a*, 44)

---

[11] Unfortunately this letter does not survive in the Kempe Papers; Packet 38 has three letters of 1905, where Peirce dwelt on recent interest from Maxime Bôcher and on existential graphs.

[12] The manuscripts involved are mentioned in Peirce *Writings 6*. On this theory see Roberts *1973a*; its modern significance is noted in the papers by Roberts and J. Sowa in Houser and others *1997a*.

4.3.1 *Peirce, published and unpublished.* Of all figures in this book Charles Sanders Santiago[13] Peirce (1839–1914) is the most extraordinary, many-gifted, frustrating and unfortunate. A son of Professor Benjamin Peirce (1809–1880) of Harvard University, his career was much oriented around that institution in positive and negative ways. After graduation from there, he worked for the Coast Survey as a mathematician and astronomer, achieving much scientifically and offending many personally.

However, by a variety of bad behaviours and social gaffes—among the latter, taking a Miss Juliette Pourtelai (or maybe 'Froissy') as mistress while married and, even worse, divorcing his wife Melusina in 1883 in order to marry her—he was left from the mid 1880s on to live on his own savings and earnings. Both were quite considerable, the respective proceeds of a good Survey salary and writings for American journals and dictionaries; but an excessive purchase of land in Pennsylvania combined with financial incompetence and bad luck in business left him heavily in debt. He lectured at Harvard occasionally, and corresponded widely, but he was on the academic fringe. He died Hollywood style without the music, on a cold April day without a stick of firewood in the box or scrap of food in the larder.

After that Harvard punished him further (Houser *1992a*). Juliette sold his manuscripts to the Department of Philosophy on condition that they be kept and an edition be made of them. A young graduate student, Victor Lenzen (1890–1975), was sent one winter's day with a horse and buggy to collect them (Lenzen *1965a*); but Juliette failed to tell him of the correspondence and financial papers stored in the attic, and they were destroyed by the farmer who bought the premises after her death in 1935.

At that time a rather sloppy six-volume edition of some manuscripts and publications had just been produced by the Department (Peirce *Papers*). Later, staff and students were allowed to take the original manuscripts as souvenirs until the Harvard librarians collected the rest and at least had them safely conserved even if unread. Juliette had also sold the library on the understanding that it would be kept together; but the books were widely scattered to the extent that some are thought to be now in other libraries.

While a thread of interest in Peirce's philosophy endured after his death, serious study dates only from the late 1950s, and came from outside Harvard. It included two more volumes of the edition (1958), properly

---

[13] Peirce seems to have added 'Santiago' to his given names sometime before 1890; for it is given, as 'S(antiago)', in the bibliography of Schröder *1890b*, 711, and one cannot imagine that Schröder invented it himself. Unfortunately the surviving correspondence between the two (Houser *1991a*) does not indicate the transmission of this name, which Peirce never published at that time; it is usually thought that he adopted it around 1903 (Brent *1993a*, 315 is too late with 1909), as 'Saint William' in honour of William James. His second name, 'Sanders', was for Charles Sanders, a granduncle by marriage.

done by Arthur Burks. A splendid biography was prepared as a doctoral dissertation for the University of California at Los Angeles in 1960 by Joseph Brent; but the Department refused him permission to publish any of the quoted manuscripts until the early 1990s, so that his achievement remained virtually unknown until a somewhat revised version was published as Brent *1993a*. Many of Peirce's manuscripts on mathematics and logic were edited by Carolyn Eisele and published in four volumes by the house of Mouton in 1976 (Peirce *Elements*). Then two years later a massive selected chronological edition of his writings in 30 volumes was launched at Indiana University under the leadership of Max Fisch, and is published by its Press (Peirce *Writings*). The main editorial task is to select material from the enormous mass of manuscript essays, draft letters (often pages long) and notes that Peirce left. There was much disorder, partly due to poverty: in his later years Peirce had to use the blank versos of essays written long before because writing paper was too expensive. Dating is thereby rendered difficult; handling of the texts by others has made the problem harder.[14]

Peirce's only academic phase was the years 1879–1884 at Johns Hopkins University in Baltimore, where he interacted with Sylvester, a highly volatile immigrant (Parshall *1998a*, 201–208). He built up a small but fine circle of students (§4.3.7) with a common interest in logic, which had been his infatuation since reading a copy of Whately's *Logic* around his 12th birthday.

4.3.2 *Influences on Peirce's logic: father's algebras.* (My *1997d*) Peirce is the next great contributor to algebraic logic after Boole and De Morgan; indeed, much of his work unified the two in developing a Boolean logic of relations. The influence of Boole himself was quite conscious: Peirce studied *The laws of thought* and adopted most of its aims and principles. He seems to have begun developing a theory of relations before reading De Morgan *1860a* on them (§2.4.7), but it confirmed the rightness of his approach. They met in 1870, early in his career and at the end of De Morgan's, when Peirce was in London en route with a Survey group to observe an eclipse in Sicily. Benjamin, the leader, gave him a charming letter of introduction for De Morgan (transcribed in my *1997a*), together

---

[14] On the history of the Peirce Papers, see Houser *1992a*; they are still kept at Harvard, and are available on microfilm. The edition is prepared at Indianapolis working out from photocopies and from many other sources, especially a vast collection of notes made by Fisch and his wife Ruth. The main single location for Peirce commentary is the *Transactions of the Charles S. Peirce Society*. A sesquicentennial conference held at Harvard in 1989 has produced a clutch of books with various publishers; the most relevant one here is Houser and others *1997a*, a large collection of essays. Alison Walsh is preparing a doctoral thesis under my direction on the links between algebras and logics in both Peirces. Among other literature, Murphey *1961a* is still a useful introduction to his philosophy in general.

with a copy of a new work of his own (B. Peirce *1870a*), which itself constituted the third formative influence on Charles.

Benjamin's own research interests lay largely in applied mathematics, including a strong enthusiasm for the quaternion algebra proposed by W. R. Hamilton in the 1840s. Here four independent basic units 1, i, j, and k, were taken, and the 'quaternion' q defined as a linear combination of them over a field of values $a, b, \ldots$:

$$q := a + ib + jc + kd, \qquad (432.1)$$

$$\text{where } i^2 = j^2 = k^2 = ijk = -1; \text{ and } ij = k \text{ and } ji = -k, \quad (432.2)$$

together with permutations among i, j, and k. Commutativity was lost, but associativity (Hamilton's word) preserved. Benjamin hit on the idea (Charles claimed credit for it ...) of generalising this case to take any finite number of units and enumerating the algebras with two means of combination which satisfied associativity and also other important properties. He noted commutativity and distributivity; and also these two, which he christened for ever:

'idempotent' when $i^m = i$, and 'nilpotent' when $i^n = 0$, integers $m, n \geqslant 2$.

$$(432.3)$$

Working with algebras with 1 up to 6 units, he found 163 algebras in all, with 6 subcases. He wrote the multiplication table for each case, where the product of each pair of units was displayed (a technique introduced in Cayley *1854a* in connection with substitution groups). One of the 'quadruple algebras' is shown in Table 432.1. The main task was taxonomy, not applications. Rather surprisingly, the catalogue excluded complex numbers, because he allowed them to appear in the coefficients of the units.

Peirce began with a philosophical declaration about mathematics that has surpassed the succeeding text in fame: 'Mathematics is the science that draws necessary conclusions'. Charles would quote it later with great approval, and even claim to have moved his father towards the position. But the slogan is enigmatic, since the sense of necessity is not explained. Maybe he was following a stress laid by George Peacock on necessary

TABLE 432.1.   A Quadruple Algebra in Peirce

	i	j	k	l
i	i	j	0	0
j	0	0	i	j
k	k	l	0	0
l	0	0	k	l

truths in symbolic algebra (§2.3.2), though enigma is there also. In drafts of the lithograph Peirce wrote 'draws inferences' and 'draws consequences', which seem preferable. Clear, however, is the active verb 'draws': mathematics is concerned with the *act* of so doing, not the *theory* of doing it, which belongs elsewhere such as in logic. Thus it was an anti-logicist stance, which Charles would always maintain.

As a sign of the financial poverty of American science in the 1860s, the Academy of Arts and Sciences (hereafter, 'AAAS'), recently founded as the prime such body in the country, could not afford to print the lengthy researches of one of its founding members. So in 1870 Benjamin's Survey staff came to the rescue, finding a lady in Washington with no mathematical training at all but a fine calligraphic hand who wrote out his scrawl with lithographic ink so that 12 pages could be printed together on a stone. The final product ran to 153 pages; he distributed the 100 copies produced to friends and colleagues, including (via Charles) to De Morgan, whose own work on double and triple algebras had been a valuable influence.

Charles was the first reader to stress the importance of the lithograph; in particular, while at Johns Hopkins in 1881, the year after Benjamin died, he had it printed in the usual way as a long paper in the *American journal of mathematics*, which Sylvester had founded in 1878. In a new headnote he hoped that his father's contribution would be recognised as 'a work which may almost be entitled to take rank as the *Principia* of the philosophical study of the laws of algebraical operation'. He also adjoined some 'notes and addenda' of his own. This version appeared in 1882 as a book from von Nostrand, with a short new preface by Charles. In its volume for 1881 the *Jahrbuch* promised a review; but, unusually and regrettably, none appeared. Nevertheless, it became sufficiently influential for the American mathematician J. B. Shaw to prepare a book-length survey *1907a* of the known results.

4.3.3 *Peirce's first phase: Boolean logic and the categories, 1867–1868.* (Merrill *1978a*) By 1882 Charles's own logical researches were well under way. His first public presentation had been given in 1865, his 26th year, in a series of 11 lectures 'On the logic of science' at Harvard (Peirce *m1865a*). Following the normal understanding at that time, he covered both inductive and deductive logic; in the latter part of the sixth lecture he treated Boole's contribution, while others outlined syllogistic principles. The following year he delivered the Lowell Lectures there, another eleven-some in the same area (Peirce *m1866a*), but with the balance more in favour of induction; it brought him to Boole the probabilist as much as to Boole the logician (pp. 404–405).

Peirce first published on deductive logic in two short papers accepted in 1867 by the AAAS. A short 'improvement' *1868a* was based upon dropping Boole's restriction of union to disjoint classes; later he recognised Jevons's priority (§2.6.2) for this move (*1870a*, 368–369). Then in *1868c* he reflected 'Upon the logic of mathematics', a recurring theme; in this début he stuck to syllogisms, with some symbols used for the basic connectives. In a footnote he mentioned De Morgan, and did not advance beyond him.

So far, so unremarkable: of far greater significance for Peirce's logic and especially philosophy was 'A new list of categories', presented to the AAAS in May 1867 between the other two papers and published as *1868b*. The Kantian in him put forward five categories based upon 'Being' and 'Substance', with the former divided into three 'accidents': the monadic 'Quality', referring to a 'ground', or general attribute; the dyadic 'Relation' referring to a correlate and a ground; and the triadic 'Representation', referring to ground, correlate and 'interpretant' or sign. The latter manifests an early concern with the theory of signs, or 'semiotics', to use the Lockean word (§2.3.3) which he was later to revive.

4.3.4 *Peirce's virtuoso theory of relatives, 1870.* The importance of this triad emerged in January 1870, when Peirce presented to the AAAS a 60-page paper on logic. They printed it in time for him to take it on his European trip in the summer and (for example) to give a copy to De Morgan, along with his father's lithograph; it appeared officially as a paper in the 1873 volume of the *Memoirs*, but I shall cite it as Peirce *1870a*. His main intention was made evident: he conjoined the modified Boole with De Morgan *1860a* (mentioned in the opening lines) in 'a notation for the logic of relatives', and the outcome was not merely a new collection of symbols but a substantial extension of the logics which Boole/Jevons and De Morgan had introduced.

The paper, 62 pages long in that printing, is notoriously difficult to follow, not least for frequent conflations of notions and symbols. The new theory of categories supplied his triad of 'logical terms', which were associated with classes; unfortunately he spoilt this care by characterising his trio as 'three grand classes' (*1870a*, 364), the noun being a technical term elsewhere. The first "class" was of '*absolute terms*', involving 'only the conception of quality' and so representing 'a thing simply as "a —"'. Then '*simple relative terms*' involved 'the conception of relation' such as 'lover of'. Finally, '*conjugative terms* [...]' involves the conception of bringing things into relation', such as 'giver of — to —' (p. 365). In this way he introduced a predicate calculus in symbolic logic, and with relations and not just classes; moreover, he went beyond De Morgan by bringing in three-place relations.

Peirce gave each kind of term its own kind of letter—roman t, italic *l*, cursive g —although sometimes he confused individuals with classes, and absolute and infinite terms (for example, around formulae (102.)–(108.)). Taking '∞' rather than the over-worked '1' to denote the universe, 'when the correlate is indeterminate' then 'l_∞' will denote a lover of something' (pp. 371–372): many of his examples involved lovers, including of servants, maybe revealing features of his private life. He used pairs of 'marks of reference' in compound relations to indicate the connections between components: for example, the wallpaper design '$g^+_{+\ddagger}1^{\parallel}_{\parallel}w^{\ddagger}h$' denoted 'giver of a horse to a lover of a woman' (p. 372).

Often these expressions and their verbal versions denoted classes, usually a 'relative'; that is, the domain satisfying a relation. This feature has often been misunderstood because Peirce's verbal account used relational words (Brink *1978a*). In symbolising the means of combining classes (including relatives), he maintained some analogies with arithmetical symbols. In particular, he continued to use Boole's '+' for the 'invertible' union of disjoint classes, but symbolised his preferred 'non-invertible' version with ' +,'; the corresponding subtractions were notated '−' and ' $\frac{}{,}$ ' (pp. 360–362). Similarly, intersection, or multiplication, was written '*x, y*' if commutative between the components, and '*xy*' if not; the corresponding divisions were notated '*x; y*' and '*x*∷*y*' respectively (p. 363).

Above all, instead of equality of classes as the primary relation Peirce took improper '*Inclusion in* or *being as small as*' (*sic*!), giving it the symbol '$\prec$'; proper inclusion was ' < '. *Thus implication took over from equivalence as a basic connective*: 'To say that $x = y$ is to say that $x \prec y$ and $y \prec x$' (p. 360).

Unlike Boole, Peirce worked with expressions like '$x + x$'; indeed, 'it is natural to write'

$$x + x = 2, x \text{ and } x, \infty + x, \infty = 2 \cdot x, \infty' \tag{434.1}$$

(p. 375), and he treated the denoted objects as multisets in the way which Kempe was to develop later (§4.2.8). One can understand his enthusiasm over Kempe's work, which must have come as an unexpected surprise.

Much of Peirce's exegesis was based on stating relationships between relatives and their 'elementary' components in linear expansions like quaternions (432.1), or more specifically after Boole's manner (255.5); sometimes the product form was used. The means of combination of classes were commutative multiplication and both types of addition. He also showed that the relationships between the 'elementary relatives' in a compound one could be expressed not only by an expansion but also as a multiplication table; one of his examples used nine units, and another was the quaternion case (432.4) in his father's lithograph (pp. 410–414). Later, in many short notes *1882a* which he added to his reprint of the lithograph, he restated an algebra in terms of its 'relative form', and he explained the general procedure in one of his addenda. In a short note *1875a* published by the AAAS he had shown the converse: that any of those tables could be given a 'relative form' as an expansion. These features show him contributing to matrix algebra (Lenzen *1973a*), then still a new topic.

Peirce's enthusiasm for algebraic symbols in *1870a* led him to use binomial and Taylor's series to produce his expansions. He used the symbols 'Σ' and 'Π' to abbreviate additions and multiplications, with superscript commas adjoined if the means of combination with subscript commas were used (first on p. 392); at this early stage the possible need for a horizontally infinitary logical language was not broached.

Peirce also used powers to symbolise 'involution' (p. 362), eventually explaining 'that x^y will denote everything for every individual which is an x for every individual of y. Thus l^w will be a lover of every woman' (p. 377). But in a surely unhappy move he also deployed powers to express negation: if x were a term, then its negative was 'n^x' (p. 380), and at once he stated the principles of contradiction and excluded middle respectively as

$$\text{'(25.) } x, n^x = 0\text{' and '(26.) } x +, n^x = 1\text{'.} \qquad (434.2)$$

Further, only a few lines *later* did he give 'the symbolic definition of zero', and none explicitly for 1; in a later summary they were given as

$$\text{'(34.) } x +, 0 = x\text{' and '(35.) } x +, 1 = 1\text{',} \qquad (434.3)$$

both credited to Jevons. The inverse operation, 'Evolution', was associated with taking logarithms (p. 363). One recalls Boole's use (255.4) of MacLaurin's theorem, and the consequences were no less wild, or at least difficult to follow.

Perhaps the hardest part of the paper is Peirce's theory of 'infinitesimal relatives'. They were "defined" 'as those relatives $[x]$ whose correlatives are individual' and number only one, so that x^2 can never relate two individuals; that is, like infinitesimals, $x^2 = 0$ (p. 391). The exegesis, successfully decoded in Walsh *1997a*, shows difference algebra in place 'by the usual formula,

$$\text{'(113.) } \Delta\varphi x = \varphi(x + \Delta x) - \varphi x, \qquad [(434.4)]$$

where Δx is an indefinite relative which never has a correlate in common with x' (p. 398). This curious clause is the clue to the theory, for he found an interpretation of higher-order differences under '$+$' and sought relationships between the pertaining relatives. However, it was not helpful to call such relatives 'infinitesimal' in this discrete theory, or to name as '*differentials*' (p. 398) the operation of differences corresponding formally to differentiation in the calculus. He applied his theory by, for example, forbidding anyone from both loving a person and being his servant, taking the class of lovers of servants of certain people, and forming the class of lovers of servants of some of them who love the others (pp. 400–408, my illustration).

4.3.5 *Peirce's second phase, 1880: the propositional calculus.* After this performance, innovative but confusing and probably confused, Peirce published very little on his algebraic logic for some years, although he worked hard on a book on it and published extensively in science and its 'logic' (to us, its philosophy: *Writings 4 passim*). But his five years at Johns Hopkins University, especially the interaction for the first time in his life with talented students, inspired him to major fresh developments.

One nice detail was that all the five basic logical connectives could be defined from 'not A and not B' of two propositions A and B. It is now abbreviated to 'nand'; Peirce gave it no name, but symbolised it 'AB'. Unfortunately, for some reason he never published his note *m1880b*; and it came to light only in 1928 when the Harvard edition of his *Papers* was being prepared.[15] By then the companion 'Sheffer stroke' for 'nor' (another Harvard product: §8.3.3) was well known.

Peirce's first Baltimore publication, possibly drawing upon a lecture course, was a complicated 43-page paper *1880a* 'On the algebra of logic' published in Sylvester's *American journal of mathematics* when he was 42 years old. As its title suggests, he presented his system in a more systematic manner; but it was less innovative than its chaotic predecessor in paying much more attention to syllogistic logic. He also went back to De Morgan's early papers on logic (§2.4.5) rather than the last one on relations.

The opening 'chapter', on 'Syllogistic', included an account of 'The algebra of the copula', which began by reviving the traditional word 'illation', the act '∴' of drawing a conclusion from a premise (p. 165). After stating the identity law as '$x \prec x$' for proposition x, Peirce stated one of his most important rules: conditional illation, with the inter-derivability of

$$x \quad y \quad \therefore z \quad \text{and} \quad x \quad \therefore y \prec z \qquad (435.1)$$

(p. 173: he displayed the inferences vertically). Negation was indicated by an overbar over the proposition letter or over '$\prec$', so that the *'principle of contradiction'* and of *'excluded middle'* were written on p. 177 respectively as

$$\text{'} x \prec \bar{\bar{x}} \ (17) \text{'} \quad \text{and} \quad \text{'} \bar{\bar{x}} \prec x \ (18) \text{'}. \qquad (435.2)$$

He presented many inferences, with syllogisms often used as examples, and also ran through his logic of relatives.

In the next chapter, on 'The logic of non-relative terms' (that is, purely classial ones), Peirce laid out many basic principles and properties of the propositional calculus, although their statuses as such were was not

---

[15] Peirce *m1880b* was found by editor Paul Weiss; see his letter of 19 November 1928 to Ladd-Franklin in her Papers, Box 73. In a later manuscript, of 1902, Peirce defined other connectives from 'nand', although only in passing (*Papers 4*, 215).

always clear. They included on p. 187 two 'formulae (probably given by De Morgan)' (§2.4.9) and 'of great importance:

$$\overline{a \times b} = \bar{a} + \bar{b} \qquad \overline{a + b} = \bar{a} \times \bar{b} \quad (30)'. \qquad (435.3)$$

Unfortunately he did not properly handle the 'cases of the *distributivity principle*'

$$'(a + b) \times c = (a \times c) + (b \times c)\ (a \times b) + c = (a + c) \times (b + c)\ (8)';$$

$$(435.4)$$

for he claimed them to be provable 'but the proof is tedious to give' (p. 184). There are *four* cases here, since the ' = ' in each proposition unites the '$\prec$' case and its converse; and it turned out that neither $(435.4)_1$ with '$\prec$', nor its dual, could be proved from the assumptions presented. This matter was one of Schröder's first contributions, in 1890 (§4.4.4); sorting it all out is quite complicated (Houser *1991b*). In addition, Peirce should have more clearly explained switches between terms and propositions and between lower- and upper-case letters.

In a final chapter on 'The logic of relatives' Peirce concentrated largely on the 'dual' kind '(A:B)' between individuals A and B, and its converse and their negatives. He showed that this quartet could be compounded with the corresponding quartet relating B and individual C in 64 different ways to deliver the quartet of relatives between A and C (pp. 201–204). The whole array could be read as the 64 truth-values for the 16 connectives between two propositions; but he did not offer this interpretation, putting forward instead other quartets of combination. He promised a continuation of the paper at the end, but only a short introduction on 'plural relatives' was drafted (*Writings 4*, 210–211).

4.3.6 *Peirce's second phase, 1881: finite and infinite.* (Dauben *1977a*) Peirce's next paper for Sylvester's journal, *1881a* 'On the logic of number', revealed his growing concern with the relationship between his logic and the foundations of arithmetic. He assumed 1 as 'the minimum number', and defined addition and multiplication of positive integers from 1 upwards, and then proved the basic properties (no trouble with distributivity this time). He also extended his definitions to cover zero and negative integers (pp. 304–306) by reversing mathematical induction via the lemma that

'If [for any positive integers] $x + y = x + z$, then $y = z$'. (436.1)

The contrast with the Peano/Dedekind axioms (§5.3.3) is striking; so is Peirce's concern with the distinction between finite and infinite, which came not from Cantorian considerations but De Morgan's syllogism of 'transposed quantity' (§2.4.6). Peirce gave as an example

> Every Texan kills a Texan,
>
> Nobody is killed but by one person,
>
> Hence, every Texan is killed by a Texan,

and realised that the form was valid *only* over predicates satisfied by *finite* classes (p. 309). Thus it was essential to define an infinite class, which he did inductively 'as one in which from the fact that a certain proposition, if true of any [whole] number, is true of the next greater, it may be inferred that that proposition if true of any number is true of every greater' (p. 301). He repeated this example of the syllogism several times in later writings (with 'Texan' replaced by 'Hottentot': perhaps some or all Texans had objected to this Unionist slur), and even contrasted the 'De Morgan inference' involved in it with the 'Fermatian inference' of mathematical induction.[16]

The reaction of mathematicians seems to have been indifferent or sceptical. For example, these papers were reviewed in the *Jahrbuch* (it missed Peirce *1870a* because it did not cover the Academy's journal). The author was C. T. Michaelis, a mathematician-philosopher of Kantian tendencies. Of Peirce's algebra *1880a* of logic, 'as in similar work of his predecessors and colleagues, much astuteness and careful diligence is shown; but whether logic gains overmuch through such refinement and intensification may be very doubtful' especially as 'the ties of syllogistic will be broken' (Michaelis *1882a*, 43), while Peirce's study *1881a* of number caused 'difficulties of comprehension, without raising the certainty of theorems' (Michaelis *1883a*). Such would be the common reaction of philosophers and mathematicians to all symbolic logics and logicisms!

4.3.7 *Peirce's students, 1883: duality, and 'Quantifying' a proposition.* The main fruits of Peirce's collaboration with graduate students at Johns Hopkins was a 200-page book of *Studies in logic* prepared under his editorship (Peirce *1883a*). The book seems to have been some time a-coming, due to financial difficulties which he helped to resolve.[17] In a ten-page review in *Mind*, Venn *1883a* generally welcomed the novelties of the book while regretting departures from Boole's principles. Indeed, the

---

[16] See, for example, Peirce *m1903b*, 338–340 for his Lowell Lectures at Harvard. The name 'Fermatian inference' does not appear in this particular passage; and it is not a happy name for orthodox mathematical induction, since it was inspired by Pierre de Fermat's method of 'infinite descent' in number theory where a sequence of successively *smaller* integers is taken until a proof by contradiction of the desired theorem is obtained.

[17] See the letters to Ladd of 8 August 1881 from Peirce on the need for $300, and of 1 October 1882 from co-author Allan Marquand wondering 'What has become of our logical efforts? Will they never see the light?' (Ladd-Franklin Papers, respectively Boxes 73 and 9). For a modern appraisal of the book on its centenary reprint, see Dipert *1983a*.

scope of the eight essays, by Peirce and four followers, was wide; for example, Peirce's own main piece *1883b* dealt with 'probable inference', and moreover in the direction of statistical distributions rather than the probability logic that had been studied by De Morgan, Boole and a few others. Three other contributions need notice here.

One algebraic benefit of Peirce's adoption of inclusive union had been that duality obtained between laws of union and of intersection; he had used it, though naively, in the distributivity laws (435.4). His student Christine Ladd (1847–1930) had already stressed duality in a paper *1880a* for Sylvester's journal extending De Morgan's work *1849b* with an operational algebra going from the arithmetical operations to logarithms and powers. She made great use of it in a long essay *1883a* here on 'the algebra of logic', in which she developed a term calculus and then used it to express the propositional calculus and solve particular exercises (Castrillo *1997a*). She used two copulas, a 'wedge' as a 'sign of exclusion' and an 'incomplete wedge' for 'incomplete exclusion': respectively, for propositions A and B,

$$\text{'A } \overline{\vee} \text{ B' for 'A is-not B' and 'A } \vee \text{ B' for 'A is in part B'.} \quad (437.1)$$

Following Peirce's use of '∞' for the universe of discourse, she expressed on p. 23 (non-)existence for a predicate x thus:

$$\text{'There is an } x\text{' as '}x \vee \infty\text{' and 'There is no } x\text{' as '}x \overline{\vee} \infty\text{'.} \quad (437.2)$$

She emphasised duality to the extent of presenting some of her definitions and theorems in such pairs; this feature was to stimulate Peirce himself later (§4.3.9).

Ladd's most striking innovation was based on the insight that the negation of the conclusion of a syllogism was incompatible with its major and minor premises. This situation could be expressed in the form 'ABC is false', where A, B and C were appropriate propositions; and the commutativity of conjunction led at once to the forms 'BCA' and 'CAB', so that two more syllogisms were handled (pp. 41–45). The trio came to be called 'the inconsistent triad' by Royce; the method was called 'antilogism' by Keynes (see Shen *1927a* in *Mind*, the most available presentation).

Peirce added a footnote to Ladd's (437.2) on the need for two copulas for existence and for non-existence, notions which he and his followers were now gradually transforming into quantification theory. The key figure was Oscar Mitchell (1851–1889), who handled adventurously 'A new algebra of logic' in his contribution *1883a* to Peirce's book. He stated that the extension of a term F comprised the universe not in Boole's manner '$F = 1$' but with a subscript as 'F_1'; if the extension was the class u, then

'F_u'; for vacuous terms, 'F_0'. Then, for example,

$$\text{'}F_1G_1 = (FG)_1\text{', and '}F_u + G_u = (F + G)_u\text{'.} \qquad (437.3)$$

(Like Ladd, he presented results in pairs.) More significantly, he allowed for *more than one universe*, such as '1' of time and '∞' for 'relation', or indeed any appropriate but prosaic universe; thus a term became a function of two of them. For example (both his), take the universe U of a village where the Brown family lives and V as some summer; then 'Some of the Browns were at the sea-shore some of the time' was written 'F_{uv}' for the classes u and v from these respective universes, while 'All of the Browns ...' was written 'F_{1v}'.

Mitchell saw such propositions as being of two 'dimensions', and realised that one could go further. 'The logic of such propositions is a "hyper" logic, somewhat analogous to the geometry of "hyper" space. In the same way the logic of a universe of relations of four or more dimensions could be considered' (pp. 95–96). These changes were not just notational: still more emphatically than Ladd, he stressed the *existence* of objects satisfying the term, which can easily be transferred to thinking existential quantifiers for u and the universal one for 1. The traditional opposition between affirmation and negation was being switched to that between existence and comprehension and from there towards quantification. While Peirce had more or less anticipated these ideas, Mitchell crystallised them clearly and with a compact symbolism which his master was to acknowledge and use with profit.

Mitchell also proposed a more efficient way of combining propositions, whether 'categorical hypothetical or disjunctive': draw inferences by forming their 'product' and erase the terms to be eliminated; no inference was possible if the middle term m was left (p. 99). To increase algebraic perspicuity, he used '$^{-1}$' instead of the overbar to denote negations, and so wrote, for example, the valid mood Barbara as

$$\text{'}(mp^{-1}) \times (sm^{-1}) \prec (sp^{-1})\text{'.} \qquad (437.4)$$

He also used display in converse pairs.

Regrettably, this paper was Mitchell's sole major contribution, although he published some papers on number theory in Sylvester's journal. After his time with Peirce he went back to college lecturing in his home town in Ohio and produced nothing more until before his early death (Dipert *1994a*).

To his book Peirce added a couple of 'Notes', of which the second, *1883c*, summarised 'the logic of relatives'. Distributivity was rather better handled (p. 455). Some advance in symbolism was evident, especially thanks to Mitchell, in the layout of collections of terms in matrix form, and in summation and product signs and subscripts. Thus pairs of the objects

A, B, … in the universe of discourse under a relative l ('lover' again) were aggregated in the linear expansion

$$\text{'}l = \Sigma_i \Sigma_j (l)_{ij} (I : J)\text{'}, \qquad (437.5)$$

where the coefficient was 1 or 0 Boole-style according as I loved J or not (p. 454). After symbolising syllogistic forms he brought in Mitchell's approach and presented propositions with multiple quantifiers.

4.3.8 *Peirce on 'icons' and the order of 'quantifiers', 1885.* The importance of symbols was emphasised in Peirce's next paper, the last in this sequence and one of his finest: 23 concentrated pages of Peirce *1885a* 'On the algebra of logic', offered as 'a contribution to the philosophy of notation'. The opening section presented one of his most durable innovations, developing *1868a* (§4.3.3) into 'three kinds of signs'. This new triad was motivated by the relationship between a sign, 'the thing denoted' and the mind. Normally the signs themselves, 'for the most part, conventional or arbitrary', were '*tokens*'. But should the triad 'degenerate' to 'the sign and its object', such as with 'all natural signs and physical symptoms', then the former is 'an *index*, a pointing finger being the type of the class'. Finally, when even this 'dual relation' degenerated to a 'mere resemblance' between the components, then the sign was an '*icon*' because 'it merely resembles' the corresponding object (pp. 162–164). He went on discuss their own relationships; in particular, the Euler diagrams for syllogistic reasoning were icons (of limited scope) supplemented by Venn's token-like use of shading (§2.6.4). Peirce was to become well remembered for this tri-distinction, mostly in later versions; the notion of an icon, treated here rather as the runt of the litter, has become especially notable.

In this paper Peirce markedly changed his treatment of the propositional calculus; for truth-values **f** and **v** now entered the algebra, in a manner implicit in Boole's law of contradiction (256.2). From

$$\text{'}(x - \mathbf{f})(\mathbf{v} - y) = 0\text{'} \text{ there followed } x = \mathbf{f} \text{ or } y = \mathbf{v}; \qquad (438.1)$$

that is, 'either x is false or y is true. This may be said to be the same as "if x is true, y is true"' (p. 166). Further, $(\mathbf{v} - \mathbf{f})$ was available to the algebra, including as a divisor since it 'cannot be 0' (p. 215). The status of this zero was not discussed apart from not being associated with falsehood itself: 'I prefer for the present not to assign determinate values to **f** or **v**, nor to identify the logical operations with any special arithmetical ones' (p. 168). He stated, as 'icons', five laws for the calculus, starting with identity but covering 'the principle of excluded middle and other propositions connected with it' with

$$\text{'}\{(x \prec y) \prec x\} \prec x\text{'}, \qquad (438.2)$$

a 'hardly axiomatical' proposition which is sometimes associated with him (p. 173).

In the third section, on 'first-intentional logic of relatives', Peirce acknowledged Mitchell in splitting a proposition into 'two parts, a pure Boolian expression referring to an individual and a Quantifying part saying what individual this is' (p. 177); a few pages later he called the latter the 'Quantifier' (p. 183). Then he gave a much more elaborate exhibition of multiple quantifiers in expressions, bringing out the importance of the *order* in which the quantifiers lay; but he did not individuate any formulae as icons.

In 'second-intentional logic', the name taken from the late medieval ages, Peirce defined the identity relation '1_{ij}' to state that indices i and j were identical, that is, that 'they denote one and the same thing' (p. 185). Four more icons were put forward to found its logic (pp. 186–187), starting with the principle that 'any individual may be considered as a class. This is written

$$\Pi_i \Sigma_j \Pi_k q_{ki} (\bar{q}_{kj} + 1_{ij})', \tag{438.3}$$

another example of mixedly quantified propositions in the paper. Finally, he rehearsed his views on the syllogism of transposed quantity of De Morgan, 'one of the best logicians that ever lived and unquestionably the father of the logic of relatives' (p. 188).

4.3.9 *The Peirceans in the 1890s.* Venn noted their contributions, with a score of references in the second edition *1894a* of his *Symbolic logic* (§2.6.4). He praised Mitchell the most, for the 'very ingenious symbolic method' (p. 193); but he did not highlight the logic of relatives, or even Ladd's antilogism. Let us turn to her later work.

Ladd's writings in the early 1880s launched a long and noteworthy career as a logician, the first of several female logicians from this time onwards. She combined it with other careers: colour physicist (another inspiration from Peirce); from September 1882 wife to and mother for the mathematician Fabian Franklin (1853–1939), then another member of the Johns Hopkins group and later a newspaper editor; teacher at Columbia University, New York; and proponent of feminist causes.[18] In a noteworthy stance which her husband supported, she always signed herself 'Christine Ladd(-)Franklin', not the normal submissive style 'Mrs. Fabian Franklin' of the time.

In a paper in *Mind* Ladd Franklin *1890a* presented her version of the algebraic propositional calculus, building upon her piece in Peirce's *Studies*. She showed first how many propositions as used in ordinary discourse

---

[18] The Ladd-Franklin Papers, mostly her material but some for her husband Fabian, forms a large and splendid source, but it needs much sorting. The failure to study her in detail in this age of feminist history escapes my male intuition.

are equivalent; for example, for terms x and y in the case 'All x is non-y', 'The combination xy does not exist' and 'There is no x which is y' (to quote three from her list of ten on p. 76). The 'entire lot of propositions to be named' was presented Benjamin-style in a $2^2 \times 2^2$-table, with the symbolism based upon Charles's '$x \prec y$' for 'All x is y'. Each row gave four equivalent propositions, including the second example above as

$$\text{case } `E \quad x \prec \bar{y} \quad y \prec \bar{x} \quad xy \prec 0 \quad \infty \prec \bar{x} + \bar{y}`, \tag{439.1}$$

with '∞' read as at (437.2). Each column presented four different propositions in the same form, for example (laid out in a row here):

$$\text{case } `0 \quad x\bar{y} \prec 0 \quad \bar{x}y \prec 0 \quad xy \prec 0 \quad \bar{x}\bar{y} \prec 0`, \tag{439.2}$$

of which the first stated 'No x is non-y' (pp. 79–80). Algebraic duality was very prominent, and later in the paper 'the eight copulas' were treated somewhat semiotically, with her wedge and its incomplete partner (437.1)$_2$ as one of the four pairs (pp. 84–86). The signs were chosen such that, as with her pair in (437.2), each universal *or* particular proposition used only logical connectives with an odd *or* even number of strokes.

Two years later Ladd published a review of Schröder in *Mind*, to be noted in §4.4.4. The same volume also contained another Baltimorean piece: Benjamin Ives Gilman[19] (1852–1933) presented some aspects of Cantor's theory of order-types in terms of relations. He used the symbol '$A \, r \, B$' to state 'The relation of anything A to anything B', with 'cr' for the converse relation (Gilman *1892a*, 518). While the paper is not remarkable —he had contributed to Peirce's *Studies* a modest item *1883a* on relations applied to probability theory—it was to attract the attention of Russell (§6.3.1).

Peirce himself was attempting to write mathematical textbooks, prepare a 12-volume outline of philosophy, develop his theory of existential graphs and so on and on; but none of these projects was ever finished (several logical ones are in *Papers 4*), and often not even his immense letters to colleagues and correspondents. In a long manuscript of around 1890 he argued that three-place relations could represent those of more places. He gave as example where a specific relation between A, B, C and D could be so reduced by bringing in an E related to A and B and also to C and D (*m1890b*, 187–188); but the generality was not established (for example, for all mathematical contexts). In a later piece *m1897b* on 'Multitude and number' he reviewed the principles of part-whole theory and then analysed inequalities arising in 'the superpostnumeral and larger collections' from

---

[19] Gilman passed his later career in arts education and aesthetics. He seems not to have been (closely) related to Benjamin Coit Gilman (1831–1908), who became President of Johns Hopkins University (and so was involved in Peirce's dismissal in 1884), and whose biography was written in 1910 by Fabian Franklin.

cardinal exponentiation; but he failed to handle them correctly (Murphey *1961a*, 253–274) and found no conclusive results such as Cantor's paradox.

During this period Peirce was also desperately trying to make money by publishing articles that were paid (Brent *1993a*, ch. 4). In an essay *1898a* 'On the logic of mathematics in relation to education' he affirmed his anti-logicist stance by stressing that (his kind of) logic was mathematical, and he cited De Morgan as a fellow traveller; he also quoted with enthusiasm his father's definition of mathematics as drawing necessary conclusions (§4.3.2).[20] Partly inspired by Ladd's paper *1883a* in his own *Studies* (§4.3.7), he worked across the turn of the century on symbolising the 16 logical connectives in four quartets of signs which imaged the relationships denoted; but he never published this fine extension of semiotics into shape-valued notations, and it has only recently been developed (Clark *1997a*). In a short note *1900a* to *Science*, edited by his former student J. M. Cattell, he asserted priority over Dedekind concerning the distinction between finite and infinite. We shall note this detail in the next section, for it had already interested Schröder, the main subject.

4.4 GERMAN ALGEBRAIC LOGIC: FROM THE GRASSMANNS TO SCHRÖDER

4.4.1 *The Grassmanns on duality.* Boole's logic was publicised in Germany especially by the philosopher and logician Hermann Ulrici (1806–1884), a colleague of Cantor at Halle University (Peckhaus *1995a*). A frequent reviewer in the *Zeitschrift für Philosophie und philosophische Kritik*, he produced there a long and prompt review *1855a* of *The laws of thought*. Treating in some detail Boole's index law and its consequences the laws of contradiction and excluded middle, he discussed 0 and 1 in connection with the latter. The former was to be understood as ' "Not-class or no class" ' whereas 'Nothing as class-sign thus contradicts the algebraic meaning of 0'; similarly, 1 was 'Alles' for a given context, not a 'Universum, totality, allness' ('*Allheit*': *1855a*, 98–100).

Boole might not have fully agreed, though we recall from §2.5.4 that all and nothing were tricky objects with him; but he would have been astonished by Ulrici's conclusion from a brief discussion of the expansion theorems (255.5–6) 'that mathematics is only an applied logic' (p. 102). Nevertheless, the review will have attracted Continental readers to this English author; at the beginning Ulrici stressed the contrast with the typical English empiricism of J. S. Mill's *Logic* (which he had reviewed earlier), and at the end he cited a passage from Boole's final chapter on

---

[20] Recently Haack *1993a* used some interesting texts in Peirce on mathematics and logic to argue that Peirce was sympathetic to some parts of a logicist thesis. For reasons such as this passage, I find welcome the rejection in Houser *1993a*.

the intellect to show that Boole 'stands much nearer to the spirit of *German* philosophy and its contemporary tendencies than most of his compatriots'. Much later Ulrici *1878a* guardedly reviewed Halsted *1878a* on Boole's system (§2.6.4) in a shorter piece.

The other main import into German algebraic logic was home-grown, although from another field (Schubring *1996a*). The Stettin school-teacher Herman Grassmann (1809–1877) had published in 1844 a book on *Die lineale Ausdehnungslehre*, a 'linear doctrine of extension' in which he worked out an algebra to handle all kinds of geometric objects and their manners of combination (H. Grassmann *1844a*). Two 'extensive magnitudes' a and b could be combined in a 'synthetic connection' to form '$(a \cap b)$', where the brackets indicated that a new object had been formed; he formulated novel rules on their removal. Conversely, an 'analytic connection' decomposed '$(a \cup b)$' such that

$$\text{given } b \text{ and } c = a \cap b, \text{ then } a = c \cup b. \qquad (441.1)$$

He examined the basic laws of '$\cap$' and '$\cup$', especially 'exchangeability' (commutativity) and distributivity; and properties such as linear combination and the expansion of a magnitude relative to a basis (to us, by implicit use of a vector space). Also a philologist, he may have chosen the unusual word 'lineale' for his title to connote 'Linie-alle'—all linear.

Grassmann was influenced philosophically by the *Dialektik* (1839) of the neo-Kantian Friedrich Schleiermacher (1768–1834), whose lectures he had heard while a student in Berlin (A. C. Lewis *1977a*). In particular, he drew upon pairs of opposites, of which (441.1) is one of the principal cases. Known in German philosophy as 'Polarität', it covered many other features of his theory: pure mathematics (or mathematics of forms) and its applications, discrete and continuous, space and time, and analysis and synthesis.

As he well knew, Grassmann's theory enjoyed a remarkable range of applications, which indeed are still sought and developed; the recent English and French translations of the *Ausdehnungslehre* were not prepared just for historical homage. Indeed, the uses went beyond geometry and physics which he had had in mind, including to arithmetic (as we see in the next sub-section) and to new algebras and thereby into logic. This last inquiry was effected by his brother Robert (1815–1901), a philosopher and logician by training, and a teacher and publisher by profession: they also ran a local newspaper together.

Robert's best-known publication was to be a group of five little books under the collective title *Die Formenlehre oder Mathematik* (R. Grassmann *1872a*). In this visionary compendium he went beyond even Hermann in generality. To start, *Formenlehre* laid out the laws of 'strong scientific thought' of 'Grösen' (the word for 'shine') which denoted any 'object of thought'; each of them could be composed as a sum of basic 'pegs' ('*Stifte*') 'e' (set roman, not italic). Like Hermann, he stipulated two means of

integers only later, both positive and negative thus (art. 55):

$$`\ldots | -3 | -2 | -1 | 0 | 1 | 2 | 3 | \ldots'. \qquad (442.3)$$

Multiplication was defined in arts. 56–58 for integers a and b from

$$`a.(b+1) = ab + a\text{' and '}a.(-b) = -ab'$$

$$\text{with } a \neq 1 \text{ and } b > 0, \text{ and '}a.0 = 0'. \quad (442.4)$$

Many of the proofs were based upon mathematical induction, used to a degree perhaps new in a textbook. In this and the definitions of integers Grassmann's later influence upon Peano is evident (§5.2.2); the effect upon Schröder came first, in his own textbook *1873a* 'on arithmetic and algebra', which started a long association with the house of Teubner. In the subtitle he mentioned 'the seven algebraic operations': addition and subtraction at the 'first level', multiplication and division at the second, and exponentiation, roots and logarithms at the third (a trio which spoilt the polarity!). In a variant upon the Grassmanns, he put forward mathematics as 'the doctrine of numbers', rather than of magnitudes; and he stressed the algebraic bent by seeking an 'absolute algebra' of which common algebra was an example. Another one was algebraic logic, as he noted when reporting his late discovery of Robert Grassmann (pp. 145–147).

Schröder developed his system somewhat in an essay *1874a* written for the school in Baden-Baden where he taught; probably nobody read it, but he had now read Boole. He presented his theory quite systematically in a 40-page pamphlet *1877a* from Teubner on *Der Operationskreis des Logikkalküls*. The second noun made its début in symbolic logic here, I believe; the first one showed the main influence from the Grassmanns, especially Robert's *Formenlehre*. After the usual nod towards Leibniz's 'ideal of a logic calculus' (p. iii), he presented two pairs of 'grand operations' on classes: 'determination' (conjunction) and 'collection' (disjunction), and 'division' (abstraction) and 'exception' (complementation, in Boole's way (255.2)) (pp. 2–3). He emphasised duality by laying out definitions and theorems in double columns, with quirky numberings, all features to endure in his logical writings. He reworked Boole's theory of solving logical equations, presenting as his 'main theorem' that for classes a, x and y

$$xa + ya_1 = 0 \text{ was equivalent to } xy = 0 \text{ with } a = ux_1 + y, \quad (442.5)$$

with class u arbitrary, where a_1 was the class complementing a relative to a universe 1 (p. 20, thm. '20°'). He solved a particular problem from Boole's *The laws of thought* (pp. 25–28); like Boole (§2.5.5), he did not seek singular solutions. He also did not cite Jevons.

The booklet enjoyed some success. Robert Adamson *1878a* gave it a warm welcome in *Mind*, and Venn was complimentary in his textbook (*1881a*, 383–390). Peirce used it in his teaching at Johns Hopkins University, and Ladd was influenced by it to highlight duality in her paper *1883a* in his *Studies* (§4.3.7). Above all, it led Schröder to a huge exegesis which was to dominate his career—untaught 'lectures on the algebra of logic'.

4.4.3 *Schröder's Peircean 'lectures' on logic.*

(Dipert *1978a*) The main product of Schröder's career was a vast series of *Vorlesungen über die Algebra der Logik* (*exacte Logik*) which he published with Teubner in three volumes. They appeared at his own expense; as a bachelor, he may not have found this too onerous, but apparently only 400 copies were printed. At his death in 1902 the second volume was incomplete; three years later the rest of it appeared (making a total of nearly 2,000 pages), including a reprint of an obituary *1903a* written by Schröder's friend Jacob Lüroth (1844–1910) for the *DMV*. The editor was Lüroth's former student the school-teacher Eugen Müller (1865–1932), who also put out as Schröder *1909a* and *1910a* a two-volume *Abriss* of Schröder's logic, edited out of the *Nachlass*. This posthumous material will be described in more detail in §4.4.9, but the contents of the entire run is summarised in Table 443.1.[21] The first two volumes contained excellent bibliographies and name indexes, but sadly none for subjects; the third volume had no apparatus at all. Each (part) volume is given its own dating code, and cited by Lecture or article number if possible. Of the many topics indicated in the Table, the account below concentrates upon algebraic aspects, duality and the part-whole theory of collections. Some main features are described in my *1975b* and *passim* in Mehrtens *1979a*; on the general background influence of Leibniz, see Peckhaus *1997a*, ch. 6.

4.4.4 *Schröder's first volume, 1890.*

Schröder used largely unchanged the main technical terms and symbols from his earlier writings. In this first volume of over 700 pages, published in his 50th year, he introduced the basic properties of 'domains' ('*Gebiete*') across a given 'manifold' ('*Mannigfaltigkeit*') with subsumption ('*Subsumtion*' or '*Einordnung*') as the basic relation, symbolised ' $\in\!\!\!\!\!\!-$ '. Both duality and polarity were stressed in the frequent use of pairs of definitions, theorems or even discussions printed as double columns on the page. Schröder probably took over this nice

---

[21] In 1975 Schröder's volumes were reprinted in a slightly rearranged form. His corrigenda and addenda in vols. 1 and 2 were incorporated into the text (as is stated on the copyright page), or moved to the end of vol. 1; the obituary Lüroth *1903a* was transferred to the head of vol. 2; and the *Abriss* was included, and repaginated to run on after vol. 3. Paul Bernays *1975a* reviewed this version from a modern point of view.

TABLE 443.1. Summary of the contents of Schröder's *Vorlesungen über die Algebra der Logik* (1890–1905) and *Abriss der Algebra der Logik* (1909–1910). The book is divided thus: vol. 1 (1890), Lectures 1–14 and Appendices 1–6; vol. 2, pt. 1 (1981), Lectures 15–23; vol. 2, pt. 2 (1905), Lectures 24–27 and Appendices 6–8. vol. 3, pt. 1 and only (1895) had its own numbering of Lectures, each one titled. In the Data column, 'a/b–c; n' indicates Lecture a, articles b–c, n pages. The order of topics largely follows that of text. An asterisk by a word or symbol marks a (purported) definition. My comments are in square brackets. The two-part *Abriss* was divided into unnumbered sections, which guide the division below; and also into short articles, which are indicated by number, followed by the number of pages.

Data	Description
Section A; 37	**Volume 1, Introduction** Philosophy; induction, deduction; contradiction. 'Presentations' and 'things'.
Section B; 42	Nouns and adjectives. Names; general, individual, species. Classes and individuals.
Section C; 46	Concepts. Pasigraphy. Intension and extension. Judgement, deduction and inference. Purpose of the algebra of logic.
1/1–3; 42	**Volume 1, Lectures** *Subsumption and *judgement. Euler diagrams. *'Identical calculus' of *'domains' of a *'manifold'.
2/4; 23	First two principles of subsumption; properties. *Equality, *0 and *1.
3/5–7; 26	Identical *'addition' and *'multiplication'; Peirce. *'Consistent manifolds'.
4/8–9; 37	Calculus of classes, including the *null class; their *'addition' and *'multiplication'. *'Pure manifolds'.
5/10–11; 28	Propositions lacking negation; multiplication and addition. Propositions '0' and '1'.
6/12; 17	Non-provability of the law of distributivity (§4.4.2).
7/13–15; 43	*Negation; its laws. Duality principle. Negative judgements.
8/16–17; 23	Complementary classes. Laws of contradictions and of excluded middle. Double negation. Dual theorems of subsumption.
9/18; 31	Applications to logical deductions, impreciseness; examples from Peirce, Jevons.
10/19; 38	Expansions of logical functions [mainly following Boole].
11/20–22; 44	Synthetic and analytic propositions. 'Pure theory of manifolds'. Simultaneous solutions and elimination, for one and for several unknowns.
12/23–24; 43	Subtraction and division as inverse operations. Negation as a special case. General symmetric solutions.
13/25; 38	Examples taken from Boole, Venn, Jevons, MacColl, Ladd Franklin and others.
14/26–27; 33	Other methods of solution: Lotze, Venn, MacColl, Peirce.
Apps. 1–3; 22	To arts. 6 and 10. Duality; other properties of multiplication and addition. Brackets.

TABLE 443.1. *Continued*

Data	Description
Apps. 4–5; 30	To art. 12. Group theory and functional equations; 'algorithms and calculations'.
App. 6; 53	To arts. 11, 19 and 24. 'Group theory of identical calculus'; combinatorics.
15/28–30; 48	**Volume 2, Part 1** Propositional calculus, sums and products of domains. Duality.
16/31–32; 36	Basic theorems of propositional calculus. Consistency; truth- and duration-values.
17/33–35; 33	Categorical judgements; Gergonne relations. Basic relationships of domains.
18/36–39; 61	Logical equations and inequalities. Sums and products of basic relationships. Negative domains. Propositions for n classes, including De Morgan's.
19/40–41; 38	Solved and unsolved problems. Mitchell; dimensions. Uses of elimination.
20/42–44; 39	Traditional views of syllogistic logic. Ladd Franklin's treatment. Correction of old errors. 'Subalternation and conversion'.
21/45–46; 52	Propositional and domain calculi. *Modus ponens/tollens.* Applications to examples of De Morgan, Mitchell, Peirce.
22/47; 32	*Individual and *point; basic theorems.
23/48–49; 51	'Extended syllogistic' [quantification of the predicate: §2.4.6]. *'Clauses' (products of propositions); basic properties.
24/50–51; 36	**Volume 2, Part 2** Additions to Vol. 1, esp. art 24 on general symmetric solutions.
25/52; 27	Review of recent literature: MacFarlane, Mitchell, Poretsky, Ladd Franklin, Peano.
26/53–54; 29	Controversy over Ladd Franklin *1890a*. Particular judgements. 'Negative' characteristics of concepts.
27/55–56; 18	'Formal properties in the identical calculus'. Modality of judgements.
App. 7; 49	McColl's propositional calculus, with the use of integrals (§2.6.4).
App. 8; 29	Kempe in the context of the 'geometry of place' (§4.2.8).
1/1–2; 16	**Volume 3, Part 1 [and only]** Plan. *Binary relatives. 'Thought regions of orders and their individuals'.
2/3–5; 59	Basic assumptions. Expansion of a relative; matrix and geometrical representation.
3/6–7; 39	General properties of binary relatives. Duality, conjunction. Propositional calculus.
4/8–10; 33	Algebra of binary relatives; product expansion. Basic 'correlation of modules' with identity. *Null relatives.
5/11–14; 51	Basic laws of compounding of propositions. Types of solution. including by iteration of functions. Simple examples.

TABLE 443.1. *Continued*

Data	Description
6/15–16; 40	Development of a general relative in 2^8 rows or columns.
7/17–20; 52	Elementary 'inversion problems'.
8/21–22; 53	Types of solution for problems in two or three letters.
9/23–24; 59	Dedekind's theory of chains (§3.4.2); complete induction.
10/25–27; 63	'Individuals in the first and second thought-regions'; ordered pair. 'Systems' as unitary relatives; connections with 'absolute modules'.
11/28–29; 85	Elimination, mostly following Peirce *1883c* (§4.3.7); methods of solution.
12/30–31; 96	15 kinds of mapping; uniqueness. Dedekind similarity and equipollence (§3.4.2).
1–32; 26	**Abriss, Part 1** Main assumptions, including propositional calculus and domains.
33–75; 23	Deduction. 0 and 1. Multiplication and addition of domains; negation.
76–107; 24	**Abriss, Part 2** Domains for propositions and 'relations'. 'The propositional calculus as a theory of judgements'.
108–121; 18	Theory of logical functions; normal forms.
122–150; 34	Elimination and methods of solution.
151–165; 22	'Inequalities'; normal forms, elimination, Boole's approach.

practice from the projective geometers: J. V. Poncelet and J. D. Gergonne had introduced it in the 1820s (with a French-style priority row, of course) when stating dual theorems about point/lines/planes and planes/lines/points (Nagel *1939a*). The Grassmanns were present in the use of analogies between algebra and logic, including the same names and symbols in the calculi of domains and classes (and in the later volumes, in propositions and relatives), and also in the organisation and removal of brackets in symbolic expressions. But Peirce was the main source, as Schröder made clear in his foreword. However, the enthusiasm was not uniform; in his bibliography he recommended especially those items marked with an asterisk, and of Peirce's strictly logical papers only the opening trio of 1868 (§4.3.3) and the final piece *1885a* (§4.3.8) were so honoured.

The calculus was grounded in these 'principles' of subsumption (Schröder *1890a*, 168, 170):

$$\text{I. } a \not\in a. \quad \text{II. if } a \not\in b \text{ and } b \not\in c, \text{ then } a \not\in c. \tag{444.1}$$

He called the first *'Theorem of identity'*, but did not really furnish *proofs* of either one. However, he was aware of the chaos about laws (435.4) of distributivity as left by Peirce, devoting art. 12 to the clean-up by assuming (p. 243) a new

'Principle $III_\times$': if $bc = 0$, then $a(b + c) \nleq ab + ac$. $\qquad$ (444.2)

Although the book carried the subtitle 'exact logic', some imprecisions are evident. One concerns definitions; although Schröder used 'Def.' sometimes and admitted only nominal definitions (p. 86), it is not always clear whether the overworked ' = ' symbolised identity, equality, or equality by definition. For example, he explained in his first Lecture that, as its symbol ' $\nleq$ ' suggested, subsumption between domains covered both the cases of inclusion and equality; yet he merged the latter with 'the *complete* agreement, sameness or identity between the meanings of the same connected names, signs or expressions' (pp. 127–128), and he called his theory the 'identical calculus' (pp. 157–167). He even named on p. 184 the following definition '*identical equality* (*identity*)' for domains a and b:

'(1)' If $a \nleq b$ and $b \nleq a$, then $a = b$ $\qquad$ (444.3)

'(read *a equals b*)' (p. 184): Husserl will spot the slip in §4.6.2.

A list of Schröder's basic notions included not only domains but also '*classes* or species of individuals, especially also concepts considered in terms of their range' (p. 160), which reinforces the extensionalist character of the theory and thereby makes the difference between identity and equality more moot. Both manifolds and classes contained 'individuals' as 'elements', named by 'proper names' ('*Eigennamen*': pp. 62–63). The intensionalist aspect was associated in this list with '*concepts* considered in terms of their content, especially also ideas'; he even distinguished a horse, the idea of a horse, the idea of the idea of a horse, ... (p. 35: compare §4.5.4) and dwelt a little on the concept of a concept (p. 96). However, he only skated around philosophical issues—a little disappointing after a thorough survey of the zoo of terms used in naming collections (pp. 68–75). He also found a paradox.

Schröder defined a 'pure' manifold as composed '*of unifiable elements*', presumably by some governing property or intension. Classes of such individuals were elements of a 'derived' ('*abgeleitete*') manifold, 'and so on' finitely up (p. 248). This is a kind of type theory; but it would be foolhardy to follow Church *1939a* and see this construction as a theory of types anything like that which Russell was to create, for Schröder worked only with one type of manifold at a time. But this led him into trouble further on, when he solved for domains) x the following dual pair of equations:

$$x + b = a; \qquad\qquad x \cdot b = a; \qquad (444.4)$$

$$\therefore x = ab_1 + uab =: a \div b, \quad \therefore x = ab + ua_1b_1 =: a :: b, \quad (444.5)$$

where u was an arbitrary domain. Now elementhood to these solution domains 'should be interpreted as relating to the *derived* manifold, and not to the original one' for x 'be contained *as an individual* in a class of domains' in the solution (p. 482). But if the class $(a \div b)$

itself comprises only one domain, the sign for subsumption would be open to misunderstanding, in that it seems to allow subsumption (as part) where, as mentioned, only equality can hold. To avoid such drawbacks, one must strictly speaking make use of *two kinds of sign of subsumption*, one for the *original* and one for the derived manifold.

But Schröder did not pursue his strict speech, which would have led him to some kind of set theory instead of the part-whole theory to which he was always to adhere. To his description of subsumption he added a footnote, that in Cantor's 'famous' *Mengenlehre* and Dedekind's 'epoch-making' work on number theory and algebraic functions 'subsumption plays an essential role' (p. 138)—not incorrect, but off the point in either case. Russell was to seize on its use of part-whole theory as one of his criticisms of algebraic logic (§8.2.7).

One major playground for analogy was the domains 0 and 1 for a given manifold. Schröder "defined" them on p. 188 in a dual manner down even to the numberings:

'$(2_\times)$' *'identical Null'*:	'(2_+)' *'identical One'*:	
$0 \nmid a$ for all domains a.	$a \nmid 1$ for all domains a.	(444.6)

He then argued that each of these domains was unique, and by implication that '1' was the manifold itself (p. 190: see also p. 251). On considering '*the class* [M] *of those manifolds, which are equal to* 1', he reasoned that necessarily $0 \nmid M$, so that $0 = 1$, which could hold only for 'a *completely empty* manifold 1' (pp. 245–246). $0 \neq 1$ was still more assumed than proved. Classes also had a 0 and a 1, understood respectively as 'Nothing' and 'All' (esp. p. 243). The empty domain of a derived manifold was written 'O', a 'large Null' (p. 250).

In this connection Schröder also defined on p. 212, in dual manner, 'consistent' manifolds, rather akin to pure ones:

'$((1_\times))$ Negative p o s t u l a t e'	'$((1_+))$ (Positive) p o s t u l a t e'
No domain has the property $(2_\times)$;	Elements are 'mutually agreeable,
all mutually disjoint within the	so that we are able *to think of the*
manifold.	*manifold as a whole*'.

(444.7)

Cantor may come to our mind, over both the positive property (§3.4.7) and the adjective in the context of paradoxes (§3.5.3); but Schröder has priority, and the two theories seem to be independent (my *1971a*, 116–117).

This volume received several reviews; those by Husserl and Peano will be considered in §4.6.2 and §5.3.2 respectively, when the work of the reviewers is discussed. Among the others, the most unexpected piece came from Spain. In 1891 Zoel Garcio de Galdeano (1842–1924) at the University of Zaragoza started a mathematical journal, *El progreso matemático*, and its opening trio of volumes contained several pieces on algebraic logic. His reviews *1891a* and *1892a* of the first two volumes totalled 22 pages; he made no particular criticisms but reasonably covered features, including the use of double columns. His colleague Ventura Reyes y Prósper (1863–1922) chipped in with seven short articles on logic (del Val *1973a*): a short article *1892a* on Schröder was followed by *1892b* on 'Charles Santiago Peirce y Oscar Honward [*sic*] Mitchell' and *1892c* on the classification of logical symbolisms.

Reyes y Prósper's first article, *1891a*, dealt with Ladd-Franklin, on the occasion of a visit by her to Europe (when she met Schröder[22]). She reviewed Schröder's first volume in *Mind*, stressing the influence of Peirce, concentrating on properties of subsumption, and finding unclear the treatment of negation (Ladd-Franklin *1892a*). By contrast, in the *Jahrbuch* Viktor Schlegel *1893a* found Boole and Robert Grassmann to be the main sources, and never mentioned Peirce! At six pages, his review was very long for that journal; a similar exception was made for Korselt by the editors of a journal in mathematics education, for they took from him a two-part review of 36 pages, in view of the 'high significance of the work'. Korselt *1896–1897a* provided a rather good summary of the basic mathematical features and methods, and noted difficulties such as the laws of distributivity; but he did not analyse foundations or principles very deeply.

4.4.5 *Part of the second volume, 1891.* In 400 pages Schröder *1891a* dealt mainly with propositions and quantification (again not his word), rather mixed together; for example, the outlines of both calculi were given in the opening Lecture 15. In the analogies the arithmetical signs were given (too?) much rein, to mark the logical connectives; disjunction ('+') was inclusive (p. 20), to match the union of domains. But the symbols most affected by multi-use were '0' and '$\dot{1}$' (as he now wrote it, to indicate that a different kind of manifold was involved: unconvincingly, he rejected on p. 5 the need for '$\dot{0}$'). These symbols now not only denoted respectively contradiction and tautology but also, when prefaced by ' = ' and read as one

---

[22]According to Schröder *1905a*, 464; see also his letters to Ladd-Franklin around that time in her Papers, Box 3, which also has letters of 1895–1896 from Reyes y Prósper. There are no relevant materials in Galdeano's *Nachlass* at the University of Zaragoza (information from Elena Ausejo).

compound symbol, symbolised truth-values; thus, for example, the arithmetical example '$(2 \times 2 = 5) = 0$' (p. 10) is rather disconcerting to read! Indeed, the two categories were intimately linked in this '*specific principle of the propositional calculus*' for a proposition A:

$$\text{`}(A = \dot{1}) = A\text{'};\tag{445.1}$$

that is, a proposition was equated/identified with its truth (p. 52). All kinds of multiply interpretable corollaries followed; for example from many, on p. 65

$$\text{`}(\dot{1} = \dot{1}) = \dot{1}\text{' and `}(0 = \dot{1}) = 0\text{'. It was assumed that } 0 \neq \dot{1}\tag{445.2}$$

to avoid triviality; he claimed it to be provable (p. 64). Among other cases, the 'theorems' of contradiction and of excluded middle were respectively

$$\text{`}AA_1 = 0\text{' rather than } (A = \dot{1})(A = 0) = 0\tag{445.3}$$

$$\text{and `}A + A_1 = \dot{1}\text{' rather than } (A = \dot{1}) + (A = 0) = \dot{1}\tag{445.4}$$

as one might expect (p. 60): compare Russell at (783.5).

To us Schröder has meshed logic with its metalogic; at that time logic would have been linked with the assertion of a proposition (compare §4.5.2 with Frege) or with a judgement of its truth-value, and indeed he called '0' and '1' 'values' (p. 256). But he also followed Boole's temporal interpretation of these symbols (§2.5.7) in terms of the 'duration of validity' of the truth of a proposition between never and always true (p. 5). One motive was to claim that categorical and hypothetical propositions were basically different; for example, for him only the former could take the values 0 and $\dot{1}$.

Subsumption now denoted this sort of implication between propositions A and B: 'If A is valid, then B is valid' (p. 13). The basic notions and principles were broadly modelled upon (445.1–4). The layout was *very* messy, between a rehearsal of the calculus of domains $a, b, c, \ldots$ on pp. 28–32 and its re-reading for propositions $A, B, C, \ldots$ both there and, with re-numberings, on pp. 49–57:

$$\text{`Principle I of identity'. `}(A \nleqslant A)\text{'};\tag{445.5}$$

$$\text{`Subsumption inference': }\overline{\text{II}}. \, (A \nleqslant B)(B \nleqslant C) \nleqslant (A \nleqslant C)\text{'};$$
$$\tag{445.6}$$

$$\text{`*}\overline{\text{III}}. \, (A + B = \dot{1}) = (A = \dot{1}) + (B = \dot{1})\text{'},\tag{445.7}$$

this latter read in terms of propositional validity; but not

$$\text{'Equality } (\bar{1}). \text{ Def. '} (A \nleftarrow B)(B \nleftarrow A) = (A = A)' \qquad (445.8)$$

because of the 'vicious circle' allegedly involved in the two ' = 's. The 'i d e n t i c a l N u l l and O n e' propositions were defined for domains on p. 29 and numbered on p. 52:

$$\text{`}^{co}\!\left(\bar{2}_\times\right)\text{'} \quad \text{`}0 \nleftarrow A\text{'.} \ \Big| \ \text{`}^{co}\!\left(\bar{2}_+\right)\text{'} \quad \text{`}A \nleftarrow \dot{1}\text{'.} \qquad (445.9)$$

Propositional equivalence did *not* use analogy (p. 71); for reasons concerning period of validity, instead of

$$A = B = (A \nleftarrow B)(B \nleftarrow A) \text{ he offered '} A = B = AB + A_1 B_1\text{'.} \quad (445.10)$$

Quantification theory was based upon Peirce *1885a* (§4.3.8), with a strong emphasis on the 'duality' between the union 'Σ' and disjunction 'Π' of domains (art. 30); the algebra made the text look like an essay on series and products. Multiple additions or multiplications were used, but not mixtures ('$\Pi\Sigma$' or '$\Sigma\Pi$') involving quantifier order; in the account of 'clauses' (art. 49) each term in the products was written out. Presumably the truth-values of propositions should have been defined in a manner analogous to (444.6) for empty and universal domains (p. 29):

$$\left[\left(\bar{2}_\times\right)?\right]\text{Def. } \prod_A (X \nleftarrow A) \ \Bigg| \ \left[\left(\bar{2}_+\right)?\right] \text{ Def. } \prod_A (A \nleftarrow X)$$
$$= (X = 0)' \qquad \qquad = (X = 1)'. \qquad (445.11)$$

Much of this second volume was concerned with syllogistic logic. The 'incorrect syllogism of the old times' was replaced by a modern version (art. 44), including the 'extended' quantification of the predicate (art. 48), extensions of De Morgan's propositional laws (435.3) (art. 39), and Ladd-Franklin's inconsistent triad (§4.3.7: pp. 61, 228) and copulas (§4.3.9: art. 43).

One of the most interesting Lectures, 21, dealt with 'individuals' and 'points', the ultimate parts of any manifold (or class). Schröder recorded on p. 326 Peirce's definition of an individual (*1880a*, 194), that any part of an individual must be empty. But his own definition (p. 321) used the (impredicative) property as a non-empty domain i which could never be a part of both any domain and its complement:

$$\text{'} (i \neq 0) \prod_x \{(ix \neq 0)(ix_1 \neq 0)\} = \dot{1}\text{'.} \qquad (445.12)$$

He gave various other versions of this property, including on p. 325 that it be non-empty and a part either of any domain or of its complement. Oddly, this version appeared again twenty pages later (p. 344) as a seemingly independent definition of the property 'J^a' that a was a point:

$$\text{'}J^a = (a \neq 0) \prod_x \{((a \nleftarrow x) + (a \nleftarrow x_1)\}\text{'}. \tag{445.13}$$

He then defined the cardinality of a class a (*sic*!), 'num . a', thus:

$$\text{'}(\text{num} . a = 0) = (a = 0)\text{'}, \text{'}(\text{num} . a = 1) = J^a\text{'}, \tag{445.14}$$

$$\text{'}(\text{num} . a = 2) = \sum_{x,y} J^x J^y \ (x \neq y) \ (a = x + y)\text{'}. \tag{445.15}$$

and so on finitely; note ' = ' hard at work again. This sequence does not anticipate Russell's logicist definitions of cardinals (§6.5.2), or try to; it belongs to a tradition of *associating* numbers with collections.

After the extensive reaction to the first volume, this one was poorly noted; for example, neither the *Jahrbuch* nor *Mind* reviewed it. But Galdeano *1892a* devoted several pages of *El progreso matemático* to a reasonable survey of the principal definitions and some of the applications, especially those of algebraic interest. He also reported on p. 355 that his colleague Reyes y Prósper was translating the book into Spanish; but nothing was published.

In England, Venn praised Schröder's work to date in the second edition of his *Symbolic logic*, giving a score of references, mostly to the lectures. But they always concerned particular details, such as symbols of individual problems; no *connected* statement was made about his 'admirably full and accurate discussion of the whole range of our subject' (Venn *1894a*, viii).

4.4.6 Schröder's third volume, 1895: the 'logic of relatives'. In his mid fifties Schröder published as *1895a* his third volume, the first part of it and in the end the only one. The topic, 'the algebra and logic of relatives', is arguably his most important contribution, greatly developing Peirce's theory. The Lectures were numbered afresh, 1–12, over 650 pages. No bibliography was given, presumably because nothing new was to be cited; in his opening paragraphs he recalled De Morgan's and especially Peirce's contributions.

If a 'thought-domain' was comprised of individuals $A, B, C, D, \ldots$, then it was 'first-order', and

$$\text{'}1^1 = A + B + C + D + \cdots\text{'}; \tag{446.1}$$

its 'second-order' companion was similarly composed of a collection of 'binary relatives' (to us, ordered pairs)

$$\text{'}1^2 = (A:A) + (A:B) + (A:C) + (A:D) + \cdots\text{'} = \text{'}\Sigma_{ij}(i:j)\text{'}. \tag{446.2}$$

This was Schröder's introduction to his theory (pp. 5–10): the first expansion of '1^2' used a Peircean (§4.3.2) matrix-style expansion in rows, which was discussed in painstaking detail in art. 4; the second version gave a generic form which he used more often. The theory of individuals itself was worked out in detail in Lecture 10, where classes were also recast as 'unitary relatives'.

In this part Schröder concentrated on binary relatives; presumably the ternary, quaternary, ... ones would have been treated in its second part had he lived to write it (compare p. 15). He did not follow Peirce in handling the domains (using the word in our sense) of relatives, but construed a 'binary relative a' (regrettably, the same letter again) extensionally as a class of ordered pairs, expressible in terms of its 'element-pairs' as

$$\text{'}a = \Sigma_{ij}a_{ij}(i:j)\text{'} \tag{446.3}$$

(pp. 22–24). The relative coefficient' of each pair was '$a_{ij} = (i$ *is an* a *of* $j)$', a proposition which gave the values 1 *or* 0 to the coefficient when it was true *or* false (p. 27). Logical combinations or functions ('$*$', say) of relations could be defined as an expansion in the manner of (446.3) as

$$a * b = \Sigma_{ij}(a * b)_{ij}(i:j)\text{'} \tag{446.4}$$

(p. 29), where '$*$' took values such as '$^-$' for negation, '$^\smile$' for the converse relative, '$+$' for disjunction, '$\cdot$' for conjunction, and the cases about to be described.

As usual '0' and '1' were busy, used not only for the 'null' and universal relations respectively but also identity ('1'') and diversity ('0''): following (446.3),

$$1 = \Sigma_{ij}1_{ij}(i:j) = \Sigma_{ij}i:j \quad \Big| \quad 0 = \Sigma_{ij}0_{ij}(i:j) = \tag{446.5}$$

$$1' = \Sigma_{ij}1'_{ij}(i:j) = \Sigma_i(i:i) \quad \Big| \quad 0' = \Sigma_{ij}0'_{ij}(i:j) = \Sigma_{ij}(i \neq j)(i:j) \tag{446.6}$$

(pp. 24–26). The empty space in $(446.5)_2$ follows Schröder on p. 26, with a reading of the relative as (another) '*nothing*'; but he did not resolve the issues raised of empty names. '$1'_{ij}$' is in effect the Kronecker delta, recently introduced in Kronecker's lectures in Berlin (§3.6.4); Schröder seemed not to know of this, but he presented his coefficient in the same way on p. 405.

Duality was again prominent, the topic of much of art. 6 and elsewhere with the use of dual columns. For example, Schröder defined this pair:

'the *relative product*', '*a of b*'	'the *relative sum*', '*a then b*'	
'$a; b$' $= \Sigma_h a_{ih} b_{hj}$'.	'$a \dagger b$' $= \Pi_h(a_{ih} + b_{hj})$'.	(446.7)

(pp. 29–30: their own duals, '*Transmultiplication*' with 'Π' '*Transaddition*' with 'Σ', were introduced on p. 278). Quantification was also well to the fore, with explicit use of mixed types, especially on p. 41 this important case on reversion of order:

$$ `\sum_u \prod_v A_{u,v} \;\not\in\; \prod_v \sum_u A_{u,v}`. \qquad (446.8) $$

Among other examples, he devised a classification of many kinds of relatives by five-string characters, each one with its dual or serving as self-dual (arts. 15-16).

In his opening paragraphs Schröder also promised to take note of Dedekind's booklet on integers. He devoted a very appreciative Lecture 9 to the theory of chains reworked in terms of relations and their subsumption. This may seem a misunderstanding, but we recall from §3.4.2 that Dedekind himself had worked mostly with parts and wholes and in fact had not individuated membership. The treatment of mathematical induction omitted Dedekind's deep theorem on definability but included a reworking of parts of the theory in terms of iterated (mathematical) functions and functional equations (one of Schröder's other interests). Later, Lecture 12 on transformations began with a general classification and presentation (art. 30) before focusing upon Dedekind's kind of 'similar' isomorphisms between 'systems', called on p. 587 'one-one' ('*eineindeutig*'). In the preface to the second (1893) edition of the booklet, Dedekind had praised the first two volumes of Schröder's book, and made notes on them (Papers, File III, 30); he then acknowledged priority in *1897a*, 112 in the context of the law of distributivity (compare *1900a*, art. 4). The overlaps lay mainly in collections and in lattice theory, especially in Schröder's fourth and sixth appendices; Dedekind does not seem to have responded to Schröder's theory of relatives.

As with the second volume, the reaction was slight, although once again Schlegel *1898a* took six pages in the *Jahrbuch* to give a warm and rather nice survey of the main notions and methods, and the reworking of Dedekind. More penetrating, but also much more rambling, were a pair of papers by Peirce.

4.4.7 *Peirce on and against Schröder in* The monist, *1896–1897*.   Peirce's venue was a journal launched by the zinc millionaire Edward Hegeler, a

German immigrant who had founded the Open Court Publishing Company initially to publish translations of books in and on German philosophy and scholarship. He also started the journal *The open court* in 1887 partly to sustain this aim; the translation *1892a* (§4.4.2) appeared there. *The monist* was launched three years later, with a rather broader remit, and it became recognised internationally; for example, in the 1910s it was to be an important venue for Russell (§8.2.6). The editor was a fellow immigrant, the philosopher and historian Paul Carus (1852–1919), a former student of Grassmann and later a son-in-law of Hegeler.[23] At this time he published both an article by Schröder (§5.4.5) and, after some difficult correspondence, two pieces on Schröder by Peirce.

Although footnoted as reviews of the third volume, Peirce's papers, his first on logic since *1885a*, were commentaries on Schröder and Peirce, together with various other things of current interest. The first one, *1896a*, carried the optimistic title 'The regenerated logic'; while Schröder's volumes were a main source, he criticised them on various points. Concerning the propositional calculus, the main one was to reject Schröder's distinction between categorical and hypothetical propositions, since all propositions could be cast in the latter form (p. 279). He also discarded Schröder's assignment of a time-period of validity to hypothetical propositions, since '*Every* proposition is either true or false' and ' "this proposition is false" is meaningless' (p. 281). But his main preoccupation was with the 'quantifier' (p. 283); he disliked Schröder's use of quantification of the predicate, because it stressed equations rather than 'illation' (or inference: p. 284).

Similarly, in his second commentary Peirce *1897a* queried Schröder's keenness to find equational solutions of logical premises, and the merit of finding algebraically general solutions rather than considering their bearing upon logic itself, because solution and premiss could equally be reversed (pp. 321–322). He appraised as Schröder's 'greatest success in the logic of relatives' (p. 327) the classification by five-string characters; the patronising tone is easy to detect. Among other topics, he touched upon his existential graphs, commented upon Kempe (whose reaction was quoted in §4.2.9), and ended with some unoriginal remarks on Cantor's diagonal argument.

---

[23] The Open Court Papers form a vast and outstanding source for the development of American philosophical and cultural life from the 1880s onwards; Carus's own correspondence (with Peirce and Schröder among many) is especially important. So far three collections of manuscripts have been moved at different times from the company house (when still in La Salle), and are numbered 27, 32 and 32A; I shall cite by Box number, such as 32/19. They overlap and collectively are not complete; neither is my search of them, regrettably. I have not used the manuscripts, proofs and letters for *The monist* and *Open court*, for the file for each issue was tied up like a cylinder and kept in a huge wicker basket; thus they require special processing before consultation, and many are not yet available. The Company has published its own bibliography in McCoy *1987a*, and a biography of Carus in Henderson *1993a*.

Despite nearly two more decades of intensive work to come, these commentaries were Peirce's last papers on logic, a subject which he defined rather surprisingly as 'the stable establishment of beliefs' (*1896a*, 271); apart from illustrating his existential graphs, they are far from his best. They also show differences between the two algebraic logicians, Schröder driving the algebra hard while Peirce preferred the logic. In 1893 Schröder had told Carus how difficult it was proving to prepare this third volume, with the first two 'pure children's games' by comparison[24]; he must have been disappointed by his mentor's reaction.

4.4.8 *Schröder on Cantorian themes, 1898.*   Following traditional logic and Peirce in particular, Schröder always used the part-whole theory of classes in his logic; but outside it he studied aspects of *Mengenlehre* closely. In particular, he considered Cantor's and Peirce's definitions of infinitude in a long paper *1898c* expressed in his logical symbols. One section treated simply ordered sets, largely following Burali-Forti *1894a* (§5.3.8); another treated equivalent sets, with his proof of the theorem named after him and Bernstein which we saw in §4.2.5 was faulty. He ended with a hope for a general recognition that 'algebraic logic is an important instrument of mathematical research itself'.

In a shorter successor Schröder *1898d* restated from his book the concept of the cardinality of a finite manifold; for example, for (445.14),

$$\text{`(Num}\,.\,a = 2) = (0'a\,.\,a \not\in 0'; a0')\text{'}. \tag{448.1}$$

Further thoughts on relations at this time led him to rethink his views on the relationship between mathematics and logic; we shall record the outcome in §5.4.5, along with Peano's reaction, in connection with Peano's review of his book.

Although these papers were published by the Leopoldina Academy in Halle, Cantor's town, their relationship was not warm. Both men had also placed papers recently in the same volume of *Mathematische Annalen*— Schröder *1895b* on relations applied to Dedekind's theory of transformations, then Cantor *1895b* as the first part of his last paper on *Mengenlehre* (§3.4.7)—and both corresponded soon afterwards with editors. Schröder told Klein in March 1896 of 'Mr. G. Cantor, from whose geniality I am far distant; to want to place my modest talent in comparison, he has occupied himself with his own researches, although a deepening of them always

---

[24] Schröder to Carus, 30 September 1893; 'mein erster und zweiter Bände waren resp. werden sein das reine Kinderspiel dagegen' (Open Court Papers, Box 32/3: the same file covers also the translation of his discourse *1890a* described in §4.4.2). He was then still working on the second part of the second volume.

hovers for me as a desideratum'.[25] Exactly a year later Cantor told Lazarus Fuchs, the editor of the *Journal für die reine und angewandte Mathematik*, that 'in my opinion the sign language of the logic calculus is superfluous to mathematics. I will not regret it, if you do not publish the relevant papers in your [Crelle's] journal'.[26]

4.4.9 *The reception and publication of Schröder in the 1900s.* One of Cantor's firm admirers was Couturat, who enthused over the definition of continuity, and of well and simple order in a piece *1900b* in the French philosophical journal the *Revue de métaphysique et de morale* (his favourite watering-hole, as we shall see in §6 and §7). But earlier in the same volume he was sceptical about Schröder's handling of integers both in the book and in the recent papers. Schröder's definition (445.11) of an individual as incapable of being part of two disjoint classes surely 'is prior to the definition that one gives' of 1 in (446.5), so that a vicious circle arose (*1900a*, 33). He also doubted that a nominal definition of integers were possible, and wondered if the use of notions such as isomorphism in Dedekind's theory of chains, which had inspired Schröder, really was logical.

At the same time but in different mood, Couturat presented a warm and extensive two-part review *1900c* of Schröder's volumes in 40 pages of the *Bulletin des sciences mathématiques*. Mostly he just described the main features, since they would not have been familiar to most readers. He concluded the first part by praising the definition of the individual, and stressing that an 'algorithmic calculus' of deduction was now available. Presumably his doubts noted above arose between preparing the two parts, for he cited them at the end of the second part. By 1905 he found great fault in Schröder's conflation of membership and inclusion, describing it in a letter to Ladd-Franklin as a 'colossal error'.[27]

---

[25] Schröder to Klein, 16 March 1896: 'Herr G. Cantor, mit dessen Genialität ich weit entfernt bin; meine bescheidnen Anlage im Vergleich stellen zu wollen, hat sich mit seiner Forschungen beschäftigt, obwohl einer Vertiefung in diese mir stets also Desideratum vorgeschwebt' (Klein Papers, 11: 766). From this and a previous letter of 11 March it emerges that Schröder sent Klein the manuscripts of these two essays, and also an essay on sign-language to be described in §5.4.5, for *Mathematische Annalen*, but that Klein rejected them.

[26] Cantor to Fuchs, 16 March 1897: 'Die Zeichensprache des Logik-kalkuls ist m.E. für die Mathematik entbehrlich. Ich werde es nicht bedauern, wenn Sie die betreffenden Schröder-schen Abhandlungen in Ihrem Journal nicht abdrücken' (Dirichlet *Nachlass* (for some reason), Berlin-Brandenburg Academy Archives, Anhang II, no. 74). Schröder never published there at this time; maybe Cantor knew about the two papers relating to his own work (see the previous footnote).

[27] Schröder's third volume 'contient une erreur colossal sur le symbolisme de Peano' (Couturat to Ladd-Franklin, 12 December 1905, in her Papers, Box 3). On the context of this letter, see §7.4.2.

As was mentioned in §4.4.1, after Schröder's death Eugen Müller edited the second part of the second volume in 1905, and prepared the *Abriss* in two parts (Schröder *1909a, 1910a*). This travail was effected on behalf of a commission set up by the *DMV* to handle Schröder's *Nachlass*. According to his forewords, Müller seems only to have had to edit the first part but to write much of the second. He ran through most of the main ideas of the first two original volumes in welcomely crisp style, with the newer ones rather more evident in the second part. One was 'normal form' (*'Normalformen'*), products of sums for functions of domains and of logical expansions (arts. 110–111, 153–154); this term may have come from its use in the theory of determinants, and/or maybe from Hilbert (§4.7.5). He also twice cited Löwenheim *1908a* on somewhat similar forms of solution (arts. 117, 127), an early piece written by one of Schröder's few admirers outside his circle (§8.7.5). Presumably the announced third part would have covered relatives; but it never appeared, maybe because Müller's teacher Lüroth, another member of the commission, nicely summarised the theory in a long essay *1904a* in the *Jahresbericht*, soon after his obituary *1903a* there of Schröder.

The posthumous part of Schröder's second volume began with a reprint of this obituary, and then contained three Lectures appraising events 'since the appearance of the first $1\frac{1}{2}$ volumes' (*1905a*, 401). The main topic was a disagreement with Ladd Franklin's criticism of him on negative judgements in her review of the first volume; his reply constituted a rather ponderous wallow through negated propositions of various kinds (art. 53).

Despite all this effort, Schröder's logic made little impact outside the commission members, and the *Abriss* was much of a tombstone. Further, all of his *Nachlass* seems to have been destroyed during the Second World War: the part that Müller had held was lost in a bombing campaign of Frankfurt am Main in 1943 that also eliminated Schönflies's, and the rest was destroyed with Frege's (§4.5.1) two years later at Münster.

Apart from this loss, it is not easy to assess the longer-term influence of Schröder's book. It was the only compendium on algebraic logic, Peirce's contributions being scattered among several papers and some difficult to follow anyway; and the theory interested algebraists as well as logicians. But direction and strategy is often hard to determine; and the length and expense cannot have encouraged sales anyway. Maybe it was a pity that he paid for publication himself; had Teubner picked up the bill, they might have asked for a much tighter text. In 1912 J. N. Keynes opined to Ladd-Franklin with typical Cambridge snobbery that it 'is rather full of German stupidities, but the core is sound' (her Papers, Box 73). At all events, the algebraic tradition of logic of which Schröder and Peirce were the chief representatives was largely to be eclipsed in the new century by the mathematical logic of Peano, Whitehead and Russell—and of Frege, whose contributions are reviewed in the next section.

4.5   FREGE: ARITHMETIC AS LOGIC

> The aim of scientific work is *truth*. While we internally *recognise* something *as
> true*, we *judge*, and while we utter judgements, we assert.
>
>   Frege, after 1879 (Frege *Manuscripts*, 2)

4.5.1 *Frege and Frege'*. The position of Frege in this story is rather
strange, and often misrepresented; so, unusually, we have to begin after his
end. Much commentary is available on an analytic philosopher of language
writing in English about meaning and its meaning(s), and putting forward
some attendant philosophy of mathematics. The historical record, however,
reveals a different figure: Gottlob Frege (1848–1925), a mathematician
who wrote in German, in a markedly Platonic spirit, principally on the
foundations of arithmetic and on a formal calculus in which it could be
expressed. Some features (for example, on definitions and axioms) were
applicable to all mathematics, and indeed to well-formed languages in
general; but even the titles of two of his books make clear that he
developed a logicistic philosophy of *only* arithmetic, with an (unclear)
measure of extension to mathematical analysis. His views on geometry
were explicitly different (§4.7.4), and he did not attempt the philosophies
of (say) probability theory, algebra or mechanics. Further, his highly
Platonic concern with objective 'thoughts' ('*Gedanken*') and centrally pre-
occupied with the (possible) 'reference' ('*Bedeutung*') of well-formed
phrases or propositions, especially with naming abstract objects such as
truth, rules him out as a founder of the Anglo-Saxon tradition of analytic
philosophy of this century.

During his lifetime the reaction to Frege's work was modest though, as
we shall see, not as minute as is routinely asserted: Russell's claim to be
his first reader after publicising him in 1903 (§6.7.8) is ridiculous. However,
after that exposure the audience was not notably greater or more sympathetic, seemingly because his calculus had been shown by Russell to be
inconsistent and because he chose then to pursue childish polemics (§4.5.9).
Only in his last years and soon afterwards were his merits publicised; but
usually they fell upon the consequences of his contributions to formal logic
and to language (§8.7–§9 *passim*). Hence was born that philosopher of
language and founder of the Anglo-Saxon analytic tradition; most of the
massive Frege industry, especially in English, is devoted to him and his
development.[28] To distinguish him from the logician rather neglected in

---

[28] It seems that Frege' moved further away from his parent over time. His version of Frege
1892a rendered 'Bedeutung' reasonably as 'reference' in the first (1952) and second (1960)
editions of his papers; but in the third (1980) it had become 'meaning', which marks an
important change of philosophy. Other similar changes include 'identity', a relation applicable to many items of the Frege' industry itself. For an authoritative survey of Frege', with
insights also on Frege, see, for example, Dummett *1991a*.

Frege's lifetime, I shall name him as 'Frege'', with the prime used in the spirit of the derived function '$f'(x)$' in Lagrange's version of the calculus (§2.2.2). This book is concerned with Frege.

As a more welcome consequence of the creation of Frege', all of Frege's books have been reprinted, and an edition prepared of most of his papers and pamphlets (Frege *Writings*: it is cited by page number below when necessary). The surviving manuscript sources have also been published. He corresponded quite extensively, and in 1919 prepared quite a lot of the letters received to give to the chemist and bibliographer of chemistry Ludwig Darmstaedter (1846–1927), who was building up a massive collection of contemporary and historical manuscripts. (Frege's covering description *m1919a* is a nice draft summary of much of his work, which the recipient would not have understood!) After Frege's death in 1925 his *Nachlass* was inherited by his recently adopted son Alfred, who sent those letters to Darmstaedter and retained all the rest until he gave it in 1935 to the logician and historian of logic Heinrich Scholz (1884–1956) at Münster University (Bernays Papers, 975: 247). With his assistant Hans Hermes, Scholz transcribed many (but not all) documents before the War, and luckily had a transcript at home when the originals were destroyed by bombing of the University on 25 March 1945. But the editions were not completed until the mid 1970s by Scholz's successors (Frege *Letters* and *Manuscripts*, the latter cited from the second edition of 1983). Readers of Frege' have available much inferior partial editions, not used here.

Let us review Frege's career, such as it was (Kreiser and Grosche *1983a*, Gabriel and Kienzler *1997a*). After training in mathematics in Jena in Saxony, Frege prepared his *Dissertation* at Göttingen in 1873 on complex numbers in geometry. The next year he wrote his *Habilitation* back in Jena, allowing him to work there as *Privatdozent*. To his intense disappointment he stayed at this second-ranking university for his entire career, rising to *ausserordentlicher Professor* in mathematics in 1879 through the support of the physics Professor Ernst Abbe. In that year Johannes Thomae (1840–1921), an analyst and function theorist (and also a former colleague and close friend of Cantor), was appointed *ordentlicher Professor*. Frege's relations with him declined later (§4.5.9), perhaps because he himself became only *Honorarprofessor*, a level between *ordentlicher* and *ausserordentlicher Professor*, in 1896. He retired in 1918.

Frege published quite steadily: four books and a few pamphlets, about 20 papers and some reviews (including lengthy ones). At first the papers and reviews appeared with local Jena organisations, and probably found audiences to match; but from the mid 1880s he used nationally recognised philosophical journals, and in the 1900s the *Jahresbericht* of the *DMV*, which he joined in 1897 and served (with fellow arithmetician Thomae!) as accounts auditor between 1899 and 1901. The treasurer, and editor of the *Jahresbericht*, was August Gutzmer (1860–1924); he came to Jena from

Halle as a second *ausserordentlicher Professor* in 1899 and was promoted the next year, but moved back to Halle in 1905.

Given Frege's sadly modest place in our history, the account in this section is restricted. In some atonement, further features will be described in connection with his exchanges with Husserl (§4.6.3), Hilbert (§4.7.4) and Peano (§5.4.4), and his late writings and revised position of the early 1920s are noted in §8.7.3. Among surveys of his work (as opposed to Frege''s), the collections Demopoulos *1995a* and Schirn *1996a* are recommended. Unless otherwise stated, the translations from Frege are mine; I quote many of his original technical terms, the word 'notion' being as usual my umbrella word for any of them.

4.5.2 *The 'concept-script' calculus of Frege's 'pure thought', 1879.* (Demopoulos *1995a*, pt. 2)

> The number of means of inference will be reduced as much as possible and these will be put forward as rules of this new language. This is the fundamental thought of my concept script.   Frege *1896a*, 222

In his *Habilitation* Frege *1874a* described a variety of 'methods of calculation' to help 'an extension of the concept of quantity'; they included functional equations (with an application to Fibonacci series, called the 'Schimper sequence') and integration techniques using determinants for functions of several variables. No references were given and little seems to be original; so the bearing upon the generality of quantity is not evident. But it shows the early tendency of his interests, which were to flower in his first book, published in Halle in 1879, his 31st year (Frege *1879a*).[29]

In just under 100 pages Frege outlined his 'concept-script' ('*Begriffsschrift*') for 'pure thought'. That is, he sought an *objective* basis of 'thoughts' independent of mental acts, belief structures, or psychological assumptions: this imperative was always to govern his work. But the rest of this title, 'modelled upon arithmetic', was unfortunate, for it suggests analogies, and in various places he emphasised extending normal theories of magnitudes; and the last section had a marked mathematical tinge. However, analogies were explicitly *avoided*, precisely because he wished to build up a symbolic calculus from basic notions; indeed, very few symbols show kinship with either arithmetic or algebra.

After stating his aims in a preface and making the customary nod of the time towards Leibniz's '*calculus ratiocinator*', Frege laid out his principal

---

[29] Two reliable English translations exist of the *Begriffsschrift*: one by S. Bauer-Mengelberg in van Heijenoort *1967a*, 1–82; the other in Frege *1972a*, 101–203, by T. W. Bynum, who also translates some related papers of that time and not quite all of the reviews (on them, see Vilkko *1998a*), and supplies a comprehensive though rather biased survey of Frege's life and work.

notions in the first of the three sections. A 'proposition' (*'Satz'*) was regarded as a unified whole if prefaced by the 'content sign' '—', and its affirmation or negation judged if the sign '❙' was placed contiguously to the left (arts. 1–4). Truth-values played no role: an affirmed 'judgement' (*'Urtheil'*) meant that the content 'occurred', referring to a 'fact'. The notion bears some similarity to our highlighting of meta-theory as against object theory, but Frege himself was not thinking in such a framework; his signs expressing the content act more like tokens than like names.

Like Peirce (§4.4.9), 'The distinction between hypothetical, categorical and disjunctive propositions appears to me to have only grammatical significance' (art. 4). The conditional judgement between antecedent proposition B and consequent A was displayed in a simple tree layout '❙┬A', where the vertical line was 'the conditional stroke'; but Frege's ⌊B account of the various pertinent combinations of affirming or denying A or B was rather ponderous (art. 5). Negation of A was marked by a small vertical line placed such that in '⊤A' it divided the application of the content sign into A to its right and not-A to its left (art. 7). These were the two primitive logical connectives, chosen 'because deduction seems to me to be expressed more simply that way' than with other selections (art. 7). Among various rules of inference available he chose for convenience *modus ponens* (not so named), symbolised by a thick horizontal line between premises and consequent (art. 6). 'Identity of content' was presented as the property that two symbols 'A' and 'B', not their referents, had the same content '$(A \equiv B)$' (art. 8); this view was not to endure.

Next Frege decomposed a proposition into an *'indeterminate function of the argument A'* (this symbol yet again!), written '$\Phi(A)$'; if two arguments were involved, '$\Psi(A, B)$' (arts. 9–10). He could have added that this dissection replaced the tradition of subject and predicate. It was a pity that he used the word 'function' without adjectival qualification; for, as he emphasised at the end of art. 10, this type of function was quite different from those used in mathematical analysis. The *'judgement that the function is a fact whatever we may take as its argument* α*'* was called 'generality' (*'Allgemeinheit'*) and symbolised '❙—$\overset{\alpha}{\smile}$—$\Phi(\alpha)$' (art. 11): he stressed the independence of this calculus from the propositional by introducing German letters such as 'α' over the 'cavity' (*'Höhlung'*). This brought in universal quantification; the existential case was defined from it as 'not for all not ...' by placing negation signs to left and right of the cavity (art. 12).

Frege's presentation was usually quite clear; for example, while not axiomatic, he made clear his assumptions. However, he was curiously reticent about his choice of them: (apparent) self-evidence seems to have been a factor. In the second section of his book he gave various examples of well-formed (and numbered) formulae in the two calculi (arts. 13–22); again the account is clear and easy to follow, with a sequence of nesting trees of steadily greater complication. The symbolism uses up a lot of space, but it is easy to read and reduces the need for brackets. If Frege were left-handed, then it might have been natural for him to write that way.

While not explicitly stating the rule, Frege substituted symbols quite carefully, warning about not doubling the use of letters in a formula or swapping German and Latin letters. (His treatment of quantification seems to be substitutional rather than objectual, although probably he did not then recognise the distinction.) To make explicit details of a derivation, he often placed to its left a scheme of the form '(n): $a \mid b$', which informed that 'b' had been substituted for 'a' (either or both possibly a tree) in a previous formula (n) (art. 15).

But Frege opened his third section with a mysterious design; I present it schematically as follows:

$$\mathbf{I} \vdash [[\text{Expression}] \equiv \text{Greek}] \tag{452.1}$$

(art. 24). Apart from the two words all the symbols are his, and several were explained only afterwards. The double bar indicated that it was both a judgement and a nominal definition; the array of Greek letters abbreviated the Expression, which came from the predicate calculus with quantification. The Greek letters had '*no independent content*' but served as place markers in which referring letters (in this case, German ones) could be sited—another substitution technique, in fact, and of an original kind. The verbal counterpart of the Expression read: '*if from the proposition that* $\mathfrak{d}$ *has the property F, whatever* $\mathfrak{d}$ *may be, it can always be inferred that each result of an application of the procedure f to* $\mathfrak{d}$ *has the property F*' (end of art. 24). The use of 'procedure' to describe the function $f(\mathfrak{d}, \mathfrak{a})$ which permitted the inference of $F(\mathfrak{a})$ $F(\mathfrak{d})$ for all $\mathfrak{a}$ and $\mathfrak{d}$ was hardly helpful, but clearly 'hereditary' (his word) situations were at hand in this section on 'the general theory of sequences', whether in ordinary talk such as the son of a human being human, or in mathematical induction. The latter type of case was his main concern, and he presented three kinds: the version of the above form (formula 81); the second-order kind involving also quantified F, as it now had to be written (91); and that case where the sequence started with the initial member (100).

Later the names 'first-order' and 'second-order' would become attached to the kinds, without or with functional quantification, and the relation be known as '(proper) ancestral' according as it did (not) include the first member. Curiously, Frege omitted the first-order proper ancestral; further, the presence of function f of two variables did not inspire him to develop a *general* logic of relations, either here or later.

Three cases of priority arise. Firstly, MacColl *1877b* had anticipated Frege with the propositional calculus, using a broadly Boolean framework (§2.6.4); but Frege seems not to have read him. Secondly and conversely, he preceded by four years Peirce's group over the predicate calculus and quantification (§4.3.7). Now Ladd's paper ended with a literature list, including the *Begriffsschrift* (*1883a*, 70−71); but she cited Schröder's review with it and seems to have known of it only that way, so they had been working independently. Finally, Frege's theory of heredity contains the essentials of Dedekind's theory of chains in his booklet on integers, already drafted (§3.4.1) but unknown to any one else; Dedekind stated in his preface *1893a* to the second edition that he had read Frege only in 1888. Thus none of these similarities suggests influence.

Frege published his first book in the year 1879 of his promotion, and its existence in manuscript had been a factor; but after it appeared his colleagues were apparently disappointed by his preoccupation with a topic of seemingly marginal significance for mathematics. To make his aim clearer, he published a short paper *1879b* with the local Jena scientific society immediately after the book was completed, symbolising two mathematical theorems: that three points are collinear, and that any positive integer can be expressed as the sum of four squares. But nobody got excited; in particular, none of the several reviewers.

For example, in the *Jahrbuch* Michaelis *1881a* noted the generality of Frege's theory but judged that 'it seems doubtful, that mathematicians would much use of Frege's concept-script'. In a longer review in a philosophical journal he expressed scepticism over the record of mathematics interacting with philosophy and saw no revolution here, since the 'concept-script has only a limited scope' (Michaelis *1880a*, 213). He also doubted that the theory of 'ordering-in-a-sequence' could be reduced to logic because it was 'dependent upon the concept of time' (a true Kantian speaking, as in §4.3.6!), while number was 'primarily mathematical' (p. 217). But he admired the calculus itself, and gave a good prosodic description of it.

A long review in a mathematical journal came from Schröder. Like Frege, he paid for his main books and rarely taught their content; but there was little intersection between their logics. In the bibliography of the first volume of his lectures (§4.4.3) Schröder was to mark Frege's book with an asterisk, indicating special importance; but in his review he was critical of the tree symbolism, pointing out as an example how clumsily inclusive disjunction read: four branches and three negations (Frege's art. 7), as opposed to his own Boolean '($ab + a_1b_1$)' (Schröder *1880a*, 227). He also found the use of various letters 'only detracts from the perspicuity and rather offends good taste' (p. 226). The first point relates to utility, but the second is a matter of logic and bears more upon the reviewer than the author.

Behind these and some other criticisms lies the role of analogy: strong in Schröder, absent in Frege. In a paper on the 'purpose of the concept-script' written soon afterwards as a reply to Schröder and published by the local scientific society in Jena, Frege *1882a* stressed that judgements rather than concepts were his prime category. He also introduced without definition 'the extension of the concept' (*'der Umfang des Begriffes'*), which seems to be his version of the set of objects satisfying it (p. 2). He also pointed out 'the falling of an individual under a concept, which is quite different from the subordination of one concept to another' (pp. 2–3), a distinction corresponding to that for Cantor between membership and proper inclusion for sets; he criticised Boole for conflating this distinction, a point to be repeated many times by mathematical logicians against their algebraic competitors. Reviewing some of Boole's procedures (and also citing Mac-Coll), he rejected as confusing the multiple uses of signs such as ' + '; as for his space-consuming version of disjunction, he retorted that formulae in algebraic logic could be very long.

This paper drew on a long manuscript in which Frege *m1880a* had compared his calculus with Boole, especially the versions of a propositional calculus (§2.5.6). After a survey of Leibniz's contributions (as then known), he then described his own calculus in detail, symbolising several examples of implications in arithmetic, including mathematical induction. But he revealed little knowledge of Boole's system, not even discussing the merits of their quite different aims (for example, Boole "burying" the proofs, Frege wanting to expose them in full detail); so not surprisingly his paper was rejected, by three editors. Klein was one of them, for the *Mathematische Annalen*; in his letter of August 1881 he pointed out Frege's ignorance of the Grassmanns (Frege *Letters*, 134–135). A succeeding essay *m1882b*, refused by a fourth journal, is better in being much shorter. The reputation of young Frege among mathematicians must have been mixed.

4.5.3 *Frege's arguments for logicising arithmetic, 1884.* Frege's next book *1884b*, published in his 36th year, devoted its 130 pages to 'the foundations of arithmetic' (*'Die Grundlagen der Arithmetik'*). The contrast with the *Begriffsschrift* was marked. Instead of producing symbolic wall-paper, he wrote almost entirely in prose, possibly following an encouraging suggestion made in September 1882 by the psychologist Carl Stumpf (1848–1936) (Frege *Letters*, 256–257). Instead of ignoring others' views, he discussed them extensively, often critically. Instead of treating sequences in terms of heredity with no particular numbers used, he put forward his logicist philosophy, that arithmetic could be obtained from his logic alone.[30]

In his introduction Frege announced his three guiding principles: 1) to 'keep apart the psychological from the logical, and the subjective from the objective'; 2) 'the reference of words must be asked in the context of a

---

[30] An English translation, Frege *1953a*, is available, though it is in part a translation'; thus I have not always followed it. In particular, I do not render 'gleich' as 'identical', or 'zukommen' as 'to belong to' because of the close association of that verb in this book with set theory. The original German is printed opposite in this edition, and moreover with the original pagination preserved—a nice touch. The centenary edition Frege *1986a* prepared by Christian Thiel contains also a valuable editorial introduction, the reviews and some other commentaries; it inspired an excellent review (Schirn *1988a*).

proposition, not in its isolation'; and 3) to distinguish concept from object. (The second assumption is now called his 'context principle' or—very unhappily—his 'holism'; given its wide remit, his presentation was rather offhand.) He began his main text by urging the need in the introduction to definite numbers in the new age of mathematical rigour (art. 1); he must have had the Weierstrassians in mind as one example, although he never attended a course and commented later on the difficulty in procuring copies of the lecture notes (*1903a*, 149). After some preliminaries, the rest of the book divided into two equal halves.

In the first half Frege reviewed a wide range of philosophers of number taken from British or German authors, and found them all wanting (Bolzano was unknown to him). For example, Mill's empirical approach (§2.5.8) could not distinguish the arithmetic involved in two pairs of boots from that for one pair of them (art. 25), and confused arithmetic with its applications (art. 17: Mill might not have accepted the distinction). Among Frege's compatriots and perhaps with certain recent events in mind, Schröder's textbook *1873a* on arithmetic (§4.2.2) was a favourite target. The main failure was to take numbers as composed of repetition of units (Frege *1884b*, arts. 29 and 34), which was no better than taking 'colour and shape' as basic 'properties of things' (art. 21); in consequence numbers were muddled with numerals (art. 43 and 83). He also objected to Schröder's use of isomorphism between collections, on the grounds that this technique was used elsewhere in mathematics (art. 63). Idealism was attacked for requiring 'my two, your two, a two, all twos'; in one of his best one-liners, 'it would be wonderful, if the most exact of all the sciences had to be supported by psychology, which is still groping uncertainly' (art. 27). Dependency upon space and time was also thrown out (art. 40), and just distinguishing objects would not do (art. 41)—a striking opinion in view of Kempe's contemporary meditations on multisets, for a different purpose (§4.2.8).

In a profound discussion of 'one' Frege criticised predecessors of all ilks for confusing the number with the indefinite article (art. 29–33), although some of his points rested on word-plays with 'ein' and 'Einheit' (English is better served by 'one and 'a'). This was the first lesson that Russell was to learn from him (§6.7.7).

After these failures Frege presented his own theory 'of the concept of number'. The epistemological election lay between the synthetic *a priori* and the analytic. The first choice was the Kantian one, and therefore subject to criticism: facile invocations of intuition (of 100,000, for example), and dependence upon physical situations which should not bear upon arithmetic (art. 12). So the vote went to Leibniz: analyticity with logic, both construed objectively (art. 15).

One motive for Frege's choice was again generality (art. 14):

Does not the ground of arithmetic lie deeper than that of all empirical knowledge, deeper even than that of geometry? The arithmetical truths govern the domain of the numerable. This is the widest; for not only the actual and the intuitive but also all that is thinkable belong to it. Should not the laws of numbers have the most intimate connection with those of thought?

Another piece of common ground lay in equality ('*Gleichheit*': also identity?), which was taken in Leibniz's form: 'things are the same as each other, of which one can be substituted for the other without loss of truth' ('*salva veritate*': art. 65).

The definitions of numbers within logic seem to have been inspired by the following insight. A decent theory should cover both 0 and 1 and not accept the tradition since antiquity (for example, in Euclid) of ignoring the former and treating the latter as something special; for Frege 0 is not nothing, but it has to do with non-existence in some sense; existence had long been recognised as a predicate of an unusual kind; so let *all* numbers be of that kind.

In this way Frege's logicism for arithmetic was born; numbers 'attach' ('*zukommen*') to concepts F via nominal definitions by 'falling under' ('*fallen unter*') them in the way that existence does, as a second-order notion. But an important distinction was presented, rather briefly, in arts. 52–53: between 'properties' ('*Eigenschaften*') of a concept and its 'marks' ('*Merkmale*'), which were properties of objects which fell under it. Thus in the expression 'four thoroughbred horses' the adjective was a mark of the concept and a property of each horse, while 'four' was the number attached to it: in Cantorian language, properties of a set were marks of its members. This fruitful passage ended with the situation where 'a concept falls under a higher concept, so to say [one] of second order' (art. 53), a repeat from *1882a* on subordination.

With these notions in place, Frege proceeded to his own theory of Numbers ('*Anzahlen*') with a heuristic argument in art. 55, followed later by formal definitions (for which I use ' := '):

0) *the starter*: 0 to concept F if the proposition 'a does not fall under F' was true for all objects ('*Gegenstände*') a; thus 0 := attached to the concept 'not equal to itself' (art. 74);

1) *the unit*: 1 to F if the true propositions 'a does not fall under F' and 'b does not fall under F' required that a and b had to be the same object; thus 1 := attached to the concept 'equal to 0' (art. 77);

n) *the sequence move*: $(n + 1)$ to F if there were an object a falling under F and n was attached to the concept 'falling under F, and not [the same as] a'; thus $(n + 1)$:= attached to the concept 'n belongs to the sequence of natural Numbers beginning with 0' (art. 83, after a detailed account of mathematical induction).

Arithmetic was based upon (Leibnizian) equality between Numbers. After a lengthy discussion, with examples taken from various parts of mathematics, Frege described more amply than before the 'extension of the concept' ('*Umfang des Begriffes*'), a special kind of object comprising the collection of objects which fell under the concept (Parsons *1976a*). Then he defined the 'Number' attached to *F* as the extension of the concept 'equinumerous [*gleichzahlig*'] with *F*'. Thus the proposition asserting the equality of the extensions of concepts *F* and *G* was logically equivalent to that stating that the same Number attached to each concept (arts. 68–69).

Two important notions have crept in. Firstly, Frege invoked the truth-values of propositions, first in the definition of equinumerousness just quoted; but he did not discuss his change from the reliance on facts in the *Begriffsschrift*, nor did he present any definition of truth. Secondly, in a footnote to art. 68 'I believe that for "extension of the concept" we could simply write "concept"'; and while he pointed to objections, he did not seem to realise what a mess the move would cause (Schirn *1983a*). The end of the footnote is his limpest sentence anywhere: 'I assume that one knows what the extension of the concept is'. Russell's paradox was to show that he did not know it sufficiently well himself, but the notion is already enigmatic. It amounts to a Cantorian set, containing members rather than parts: Frege seems to have invented this set theory for himself, although he had read at least Cantor's *Grundlagen* of the previous year (§3.2.7) and even praised the theory of transfinite numbers (arts. 85–86), while criticising the use of isomorphisms (art. 63). Further, how can the truth-values of propositions using equinumerousness be assessed if one or both of the concepts are not explicitly numerical? While he touched on this point (art. 56, for example), he did not resolve it: a vicious circle seems present, and his complaint about Schröder and Cantor using isomorphism rings hollow.

For some unknown reason Frege's book provoked very few reviews; it did not even receive one in the *Jahrbuch*, although his Breslau publisher was known there for other books. Part of the small attention paid was a short review by Cantor. He approved of the general aim and the avoidance of space, time and psychology (this from him!); but he criticised details, regarding 'extension of the concept' as 'in general something completely indeterminate', disagreeing that his own notion of 'power' (cardinality) was the same as Frege's Number, and briefly rehearsing his theory of cardinals and ordinals (Cantor *1885c*). His second point was a mistake, perhaps caused by the fact that for him 'Anzahl' referred to an ordinal (§3.2.7), a difference which Frege *had* observed in his remarks on Cantor. In a brief reply Frege *1885b* explained the blunder, noting that Cantor had misunderstood Number as related to a concept *F* instead of to the concept of equinumerousness to it. He was polite; but resentments may have been excited, and an opportunity for their release was provided several years later (§4.5.5).

One might have expected Cantor and Frege to be close; but this is true only geographically, Halle and Jena being 40 miles apart. There is no evidence that they even met, although this presumably happened at some annual gatherings of the *DMV*.

4.5.4 *Kerry's conception of Fregean concepts in the mid 1880s.* In a short paper 'On formal theories of arithmetic' Frege *1885a* contrasted two kinds: a nice one based upon grounding arithmetic in logic, and a boring one based upon viewing arithmetic as composed of 'empty signs', leading to 'no truth, no science' such as knowing that $\frac{1}{2} = \frac{3}{6}$. This paper and the book, together with the *Begriffsschrift*, inspired a substantial and rather negative reaction from Benno Kerry. We met him in §3.3.4 as an acute commentator on Cantor in his *1885a*; his comments on Frege occurred within an eight-part suite of articles 'On intuition and its psychic propagation', which appeared in the same journal, *Vierteljahrsschrift für wissenschaftliche Philosophie*, from 1885 until posthumously in 1891. Based upon his *Habilitation* at Strasbourg University (Peckhaus *1994a*), he included Frege in a wide survey of the literature: he had even read Bolzano. Most of his remarks on Frege are contained in the second and especially the fourth parts (Picardi *1994a*); the examples below are taken from the latter.

Kerry had studied with the philosopher and psychologist Franz Brentano (1838–1917) for a time, and so was well aware of subtle psychological issues in philosophy. He rehearsed various concerns of 'psychic works' on 'inner perceptions', and so on (Kerry *1887a*, 305–307), matters which Frege wished to avoid considering. More pertinently, Kerry wished to rescue arithmetic for the synthetic *a priori* from 'the F[regean] logification ['*Logificirung*'] of the general concept of Number' (p. 275), and included a nine-page footnote on affirming or denying analytic and synthetic judgements (pp. 251–260). Some of his criticisms of Frege were based upon his own misunderstandings: for example, the senses of 'one' beyond the arithmetical (pp. 276–278), and the (apparent) impossibility of setting up an isomorphism between empty extensions, thus blocking Frege's definition of 0 (pp. 270–273). But he enquired carefully into Frege's enigmatic notion of extension of the concept, and the status of that notion (p. 274):

> [...] that the judgement 'the concept "horse"' is a simply graspable concept' of the concept 'horse' is also an object, and indeed one of the objects which falls under the concept 'simply graspable concept'.

He did not claim this situation to be paradoxical, but it was distant from Kantian territory.

4.5.5 *Important new distinctions in the early 1890s.* Kerry was the first serious student of Frege's theory. A reply did come, though tardily: perhaps discouraged by the continuing non-impact, Frege published nothing for some years, although he seems to have developed his logicism and symbolism. Early in the new decade he put out two papers (one inspired by Kerry) and a pamphlet; each work carried a title of the form 'X and Y' and explained the distinction between the pair of notions involved. The trio seems to have been written or at least thought out together, in an intensive refinement of his theory. I start with the paper which contained the most far-reaching distinction.

Frege began the paper 'On sense and reference' (*1892a*) by stating that now 'Gleichheit' carried 'the sense of identity', thus marking a change of previous normal practice, or at least indicating a new precision. Claiming that in the *Begriffsschrift* he had taken identity as a relation between names, he announced a second change by introducing the distinction for 'signs' ('*Zeichen*'), be they single letters, or one or more words: between their 'sense' ('*Sinn*') and their 'reference' ('*Bedeutung*') to some object. He gave examples from mathematics, science and ordinary life of signs with different senses but the same referent, such as 'the point of intersection of [lines] *a* and *b*' and 'the point of intersection of [lines] *b* and *c*' for three coincident lines; and of signs with no referent at all, such as 'the least rapidly convergent series' (pp. 143–145), and presumably 'Odysseus' (p. 148). 'A proper name (word, sign, combination of signs, expression) expresses ['*drückt aus*'] its sense, denotes or designates ['*bedeutet oder bezeichnet*'] its reference' (p. 147). Conversely, an object had these signs as its 'designation' (p. 144). Distinct from both notions was the subjective 'connected idea' ('*verknüpfte Vorstellung*') of the referent pertaining to a thinker (p. 145).

Such distinctions had long been recognised by philosophers and logicians, with names such as 'signification' and 'application' (to quote the very recent example Jones *1890a*); Frege's novelty lay in the range of use. For example, he re-oriented his view of propositions by placing centre stage truth-values, two only: 'There are no further truth-values. I call the one the True, the other the False' ('*das Wahre, das Falsche*': p. 149). This latter pair of notions served as the sole reference of true or of false propositions, as Leibniz's definition of identity taught (p. 150). In particular, all arithmetical propositions became names of the True—hence his frequent use of noun clauses rather than propositional forms (for example from §4.5.2, 'the falling of an individual under a concept' not 'the individual falls under the concept'). He then described the way in which the reference of a compound proposition was to be determined via its connectives (pp. 152–157)—not unlike testing by truth-tables but perhaps closer to using a valuation functor.

This paper was one of Frege's most influential contributions, not least upon its author (Thiel *1965a*); in his later writings he was much more systematic in deploying or avoiding quotation marks, and in distinguishing a word from its reference. He used its proposals in the pamphlet, which contained a lecture given to the local scientific society (but *not* published in their journal, unlike his *1879a* or *1882a*). This time he dealt with the distinction between 'Function and concept' (Frege *1891a*). He regarded a function as 'unsaturated ('*ungesättigt*'), which became 'saturated' when a value for the variable was inserted (pp. 127–129). Perhaps he chose this surprising analogy from chemistry to suit his audience: it would have helped them if he had stated explicitly that he was replacing the traditional distinction between subject and predicate. He also stressed more clearly than before that *all possible* values of the argument were admitted, so that values which might have been better construed as inadmissible sent the resulting proposition to the False. Presumably his context principle (§4.5.3) inspired this strategy.

Frege defined a new object relative to a function $F(x)$, corresponding to the curve specified by $y = f(x)$: its 'value-range' (*'Wertverlauf'*), the set of ordered pairs of values of its arguments x and of its 'values' (*sic*) $F(x)$. For symbols he invoked Greek letters and drew upon the diacritical apostrophe to write '$\acute{\varepsilon}F(\varepsilon)$' (pp. 129–131). In the important special case of the concept, a function which took only truth-values for its values, its value-range was named 'extension of the concept' (p. 133). He introduced this notion casually, and did not mention his earlier use of the phrase (§4.5.3) where it seemed to name a set of objects rather than ordered pairs of them. Indeed, this author of a paper *1884a* on 'the point-pair in the plane' did not mention ordered pairs at all here. He could also have clarified the relationship between the two types of function; that (for example) the zeroes of the mathematical function $f(x)$, x variable, give the values of x when the propositional function (or concept) $(f(x) = 0)$ refers to the True (and otherwise to the False).

Frege reworked the basic notions of the concept-script in terms of truth-values of asserted contents (*1891a*, 136–141). He finished with an explanation of functions more marked by brevity than clarity of functions of the second 'level (*'Stufe'*); either functions of functions, or functions of two variables like '*f*' involved in (452.1) (pp. 141–142). A short review appeared in the *Jahrbuch*: Michaelis *1894a* judged that 'As with all Frege's work, the reviewer also has the impression that it gets lost in subtleties'.

In his pamphlet Frege deployed sense and reference in all sorts of contexts, such as

$$\text{'``}\acute{\varepsilon}(\varepsilon^2 - 4\varepsilon) = \acute{\alpha}(\alpha.(\alpha - 4))\text{''' and '``}2^4 = 4.4\text{''' }\qquad (455.1)$$

(*1891a*, 130, 132); he also identified (as it were) mathematical equality such as here with identity, and maintained this position in later writings. He also introduced the technical term 'thought' (*'Gedanke'*) when stating that the propositions '$2^4 = 4^2$' and '$4.4 = 4$' express different ones; but its role was explained only in the other paper, *1892a*. Published in the journal that had taken Kerry's suite, it served partly as a reply to Kerry, whose comments had motivated several parts of the draft version (Frege *Manuscripts*, 96–127).

Frege's main concern was to tackle the distinction between 'Concept and object'. He accepted Kerry's puzzled reading as correct: 'the concept "horse"' was indeed no concept but designated an object (Frege *1892b*,

170–171). But the reply is glib; some major questions of a paradoxical kind arise concerning the different ways in which a horse is named by 'horse' and by 'the concept "horse"'' (de Rouilhan *1988a*, ch. 4).

Frege addressed more completely other of Kerry's concerns; for example, the senses of 'is' beyond that of the copula (*1892b*, 168–169). Of his own theory he confessed that 'I did not want to define, but only give hints while I appealed besides to the general sense of language' (p. 170)—a phrase which suggests that he saw his aim, especially with his concept-script, of capturing Leibniz's *characteristica universalis* as an *ideal language*.

Frege repeated his criticism of the failure, this time by Schröder, to distinguish an object 'falling under' a concept from a concept subordinated to another one (p. 168). He also applied to propositions his distinction of sense from reference, which 'I now designate with the words "thought" and "truth-value"'' (p. 172). Even here he was cryptic; the clearest and most detailed presentation of these distinctions was given in a letter of May 1891 to Husserl, rendered here as Figure 455.1 (Frege *Letters*, 96: the context is explained in §4.6.3). In contrast to subjective 'ideas' ('*Vorstellungen*'), 'thought' was intended in an objective sense, rather like state of affairs, sharable among thinkers and indeed independent of anyone thinking them. Presumably but regrettably, he came to this schema only after his two papers and pamphlet had been accepted for publication. In a later manuscript he noted that a proposition need not contain any proper names (*m1906c*, 208).

In another paper from this period Frege reversed previous roles with Cantor when he reviewed Cantor's pamphlet *1890a* reprinting recent articles on the philosophy of the actual infinite (§3.4.4). Perhaps in unhappy memory of last time, his barbs were sharp. After again praising his enterprise, 'Mr. Cantor is less lucky where he defines' (Frege *1892c*, 163); but he chose Cantor's use of 'variable finite' to definite finitude, which could have been better conveyed in terms of indefiniteness rather than variability but was hardly a failure. Again, 'If Mr. Cantor had not only reviewed my "Grundlagen der Arithmetik" but also had read it with reflection, then he would have avoided many mistakes', such as 'impossible abstractions' (p. 164). He also recalled Cantor's error over 'extension of

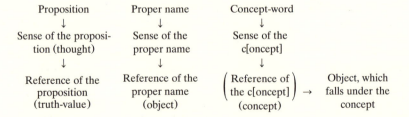

FIGURE 455.1. Frege's schema of sense and reference.

the concept', and attacked his epistemological dependence upon abstraction in definitions of cardinal and ordinal numbers (p. 165). In a draft version of the review (*Manuscripts*, 76–80) Frege was even more sour, especially on this last matter (Dauben *1979a*, 220–226). Cantor did not reply to the published version.

4.5.6 *The 'fundamental laws' of logicised arithmetic, 1893.* (Demopoulos *1995a*, pt. 3)

Frege has the merit of [...] finding a third assertion by recognising the world of logic, which is neither mental nor physical.   Russell *1914c*, 206

Armed with his new distinctions, Frege could now work out in detail 'the fundamental laws of arithmetic' ('*Grundgesetze der Arithmetik*') in his calculus. The first volume, containing 285 pages, appeared, apparently at his own expense, from a Jena house as Frege *1893a*,[31] when Frege was in his mid forties. In a long foreword he began by stating his mathematical aims and scope, and lamenting the silence over the *Grundlagen*: then mathematicians, 'who give up false routes of philosophy unwillingly' (p. xiv), were allowed to leave the classroom while he waxed philosophical. Criticising at length the empiricist version of logic *1892a* recently published by Benno Erdmann (1851–1921), Frege stressed that 'I recognise a domain of what is objective, non-real ['*Nichtwirklichen*'], while the psychological logicians [such as Erdmann] take the non-real without further ado as subjective' (p. xviii)—the third realm which Russell was to spot.

The first part of the volume was devoted to the 'Development of the concept-script'. In the opening articles Frege crisply laid out his basic notions and signs: function (including of two variables) and concept, (un)saturation, thought and truth-values, sense and reference, course-of-values, generality, negation and the connectives, identity ('*Gleichheit*'), and the three types of letters. The content-sign '—' of the *Begriffsschrift*, now named 'the horizontal' (art. 5), was presented as a special function-name which mapped true propositions to the True and anything else (for example (his), 2) to the False. When combined with the vertical judgement sign '|' it became the judgement sign '⊢', which denoted the 'assertion' of a proposition (arts. 5-6). There was a newcomer: 'the function \ ξ' which 'replace[d] the definite article' by taking as value the object falling under the concept represented by 'ξ' if unique (such as the positive square root of 2 for the concept 'positive square root of 2') and otherwise the extension of that concept (art. 11). This notion grounded his theory of

---

[31] There has been only a reprint edition of the *Grundgesetze*, in 1962. Parts of this first volume were sensitively translated into English by Montgomery Furth, with a perceptive introduction (Frege *1964a*).

definite descriptions—which was motivated, as with Russell after him (§7.3.4), by the need for mathematical functions to be single-valued.

This time Frege presented three rules of inference: *modus ponens*, transitivity of implication, and a complicated one for compound propositions with some parts in common; he included various 'transition signs' ('*Zwischenzeichen*'), mostly horizontal lines, which showed how a formula below it was derived from those above (arts. 14–16). Rules of various kinds were summarised in art. 48, immediately after a listing of the eight 'basic laws', with three for the propositional calculus (including negation), three for universal quantification over functions, and one for the extension of the concept. The other rule, introduced in art. 20, replaced equinumerousness in the *Grundlagen* by the assumption that the equality/identity of two value-ranges was logically equivalent to the equivalence of the quantified corresponding functions:

$$\text{'}\vdash (\dot{\varepsilon}f(\varepsilon) = \dot{\alpha}g(\alpha)) = (\overset{\alpha}{\smile} f(\alpha) = g(\alpha)) \text{ (V'.} \qquad (456.1)$$

This is Law V, which Russell was to find to be susceptible to paradox (§6.7.7). Although a principle for extensionality, it is now called his 'comprehension principle'. He used no names for any of his laws; and once again he was silent on their choice, seeming to use self-evidence as a criterion.

After presenting the double-bar sign (452.1) (art. 27), Frege gave much attention to forms of definition. Perhaps by reflecting upon the dubious definition of equinumerousness in the *Grundlagen*, he favoured only nominal ones (art. 33). One of them, concerned functions of functions '$X(\Phi(\xi))$'; since only objects could be arguments for functions, 'Φ' would have to be replaced by its value-range (art. 21). To improve upon (455.1) he used a new function, '$\xi \cap \zeta$' (I follow his unhelpful choice of Greek letters) which replaced the value '$\Phi(\Delta)$' of the function for argument 'Δ' by the combination '$\Delta \cap \dot{\varepsilon}\Phi(\varepsilon)$'; as usual, he extended the definition to cover all kinds of arguments (arts. 34–35). He used this function frequently in later exegesis: the chief property for a mathematical function was

$$\text{'}\vdash f(a) = a \cap \dot{\varepsilon}f(\varepsilon)\text{'} \text{ for argument } a \qquad (456.2)$$

(arts. 54–55, 91). He also stratified functions into 'levels' ('*Stufen*') by the kinds of quantification, if any; for example the function in (456.1) was of second level, and quantification of f was third level, and so on (art. 31).

Self-membership being excluded, a theory of types was embodied. However, the logic of relations remained rudimentary, especially when compared with Peirce's, which Frege seems not to have known. For example, in defining a 'double value-range' of a function of two variables, and the associated 'extension of the Relationship' ('*Beziehung*') when it took only truth-values, he did not stress the role of ordered pairs of objects (art. 36, with '$\xi + \zeta$' used as illustration). He also defined the extension of the converse of a Relationship (art. 39). Later he dealt with compounding Relationships 'p' and 'q', and for once a schematic representation of the process (art. 54, formula (B):

$$\text{symbol '}p \sqcup q\text{', \quad picture '}w \twoheadrightarrow u \twoheadrightarrow v\text{'.} \qquad (456.3)$$
$${}_{p} \quad {}_{q}$$

These last notions were introduced in the opening of the second part of the volume, in which Frege worked out the 'Proofs of the fundamental laws of the Number' in great symbolic detail; Frege' and even Frege scholarship is usually silent about it, but see Heck *1993a*. The spatial symbolism works very nicely, but Frege chose some ghastly symbols for his various notions, presumably wishing to avoid analogy but often losing both sense and reference for the reader. For example, almost all the numerals refer to pages, articles or theorems! The perplexity could have been reduced by an index of symbols, though several appear in those for laws and definitions at the end of the volume. The text switched regularly between articles talking about the plan in 'analysis' ('*Zerlegung*') and those effecting the 'construction' ('*Aufbau*'); correspondingly, quotation marks around formulae were alternately present or absent.

In the first part Frege had sketched out the theory of defining 'Numbers' as the sequence stated verbally in the *Grundlagen*, launched with '$\emptyset$' (using '$\top \varepsilon = \varepsilon$' to refer to the True), then 'λ' (via '$\varepsilon = \emptyset$') and the relation 'f' of 'successor of' ('*Folge*': arts. 41–43). The detailed exegesis included properties such as the uniqueness of the successor and (its converse) of the predecessor of a Number (arts. 66–77, 88–91) and basic features of '$\emptyset$' and 'λ' (arts. 96–109). Then attention switched to many properties of 'endless' ('*Endlos*') sequences of Numbers with no final member (Cantor's well-order, not mentioned), including a definition of the concept 'Indefinite' which corresponded to Cantor's $\aleph_0$ (art. 122, Cantor not mentioned). He also treated 'finite' ('*endlich*') sequences which did stop (arts. 108–121). In art. 144 he at last formally defined an ordered pair:

$$\text{'}\mathbf{I}\hspace{-0.3em}\vdash\dot{\varepsilon}(o \cap (a \cap \varepsilon)) = o\,;a\text{',} \qquad (456.4)$$

where 'the semi-colon herewith is [a] two-sided function-sign'. His theorems included versions, stated in terms of indefinite sequences, of Dedekind's validation (§3.4.2) of mathematical induction and the isomorphism of such sequences (art. 157, Dedekind not mentioned though noted in the introduction to the volume).

Despite much acute precision, some unclarities remain. A significant one concerns the balance between intensional and extensional notions, and even the specification of some of them. Names such as 'extension of the concept' suggest that concept itself is an intensional notion of some kind, as indeed is corroborated in various places. In particular, in a letter probably written around this time (Frege *Letters*, 177) he opined to Peano that

> one may freely regard as that which constitutes the class not the objects (individui, enti) which belong to it; for then these objects would be annulled with the class which exists out of them. However, one must regard the marks that are the properties which an object must have, as that which constitutes the class in order to belong to it.

He wrote in similar vein when discussing Husserl (Frege *1894a*, 455: the contexts are explained in §5.4.5 and §4.6.3 respectively). In addition, one can hardly conceive extensionally of an empty course of values, so close to the important Number 0. On the other hand, he required that a function-name be *always* saturated when completed by a proper name, which carried an extensional ring (Furth *1964a*, xxvii–xliv). Maybe he had not fully thought out this distinction across his calculus.

The volume received very few reviews. Peano's, the most important, will be noted in §5.4.5. Michaelis *1896a* wrote one paragraph in the *Jahrbuch*, mentioning as new notions the diacritical apostrophe (he had obviously forgotten reading about it in 'Function and concept') and the description functor. After a brief hint of Frege's logicist programme, he referred to the summary of results at the end of the book, 'which in its peculiar form may put off many readers'.

4.5.7 *Frege's reactions to others in the later 1890s.*   After publication of this volume Frege continued work on its successor(s). Various manuscripts show new considerations, such as the sense and reference of concept-words (*Manuscripts*, 130–136). They seem to relate to critiques of two contemporary logicians which he published in philosophical journals in the mid 1890s. His views on Husserl will be aired in §4.6.3; we note here his 'critical elucidation' *1895a* of Schröder's first volume.

One major issue was Schröder's subsumption relation, which conflated Frege's 'falling under' and 'falling within' (membership and improper inclusion). Frege proposed to distinguish them as '*subter*' and '*sub*' respectively, and to solve Schröder's paradox of 0 and 1 (§4.4.2) by invoking the intransitivity of the former relation (pp. 198–199). But he showed again his poor knowledge of Boole in claiming Boole's universe of discourse was 'all-embracing' ('*allumfassend*': p. 197), which is true only for the first book (§2.5.4). He also made play with Schröder's various uses of '0' and '1', and of mixing concepts with objects. There was no reply in the posthumous part of Schröder's second volume. Given their fundamental differences, it is amusing to see that *each* man had seen himself as fulfilling the vision of a '*calculus ratiocinator*' made by that necessarily Good Thing, Leibniz (for example, Frege in the preface to the *Begriffsschrift*, and Schröder in the introduction of his first volume)!

Comments of a similar kind were inspired by an article *1894a* on integers in the *Revue de métaphysique et de morale* by the school-teacher Eugene Ballue (1863–1938). Frege's reply *1895b*, his only publication in a foreign language, criticised Ballue's focus on numerals rather than on numbers, or at least mixing the notions, and also for defining one as a 'unit' (*'unité'*) and larger numbers as 'pluralities' of it. Ballue did not reply in print, but he corresponded with Frege for a couple of years thereafter (Frege *Letters*, 2–8), admitting some 'lack of precision in the [technical] terms' of his article. He also reported correspondence with Peano (even transcribing one letter), and noted that Peano had not yet treated Frege's work; this lacuna was soon to be filled (§5.4.5).

A more sarcastic version of the same line was inspired in Frege by the opening article in the *Encyklopädie der mathematischen Wissenschaften* (§4.2.4), a survey of 'the foundations of arithmetic' by Hermann Schubert (1848–1911). Largely concerned with the historical and cultural aspects, Schubert *1898a* did not launch this great project well; starting with counting processes, he advanced little further in a routine survey of arithmetical laws and operations, and some algebraic aspects such as the principle of permanence of forms (§2.3.2). Frege's theory was not discussed, although the *Grundlagen* was listed in a footnote (p. 3). Doubtless Frege had thought of a more suitable author for the article, and he replied to this product with witty savagery: for example, 'the numbers as product of counting. Really! Is not the weight of a body the outcome of the weighing as well?', and would a collection of peas lose its peaness after being abstracted? (Frege *1899a*, 241, 244). He also doubted the legitimacy of the principle of permanence (§2.3.2) as a source for proofs (p. 255). Schubert did not reply to this attack, and may not have seen it; for it appeared only as a pamphlet, from Frege's Jena publisher.

4.5.8 *More 'fundamental laws' of arithmetic, 1903.* For four years Frege did not publish again, until the second volume of the *Grundgesetze* appeared near the end of 1903, when he was in his mid fifties. A small delay was caused by the need to respond to Russell's paradox in an appendix; we note that in §6.7.7 and treat here the volume as originally conceived. Exactly the same length as its predecessor, it contained the same mixture of symbolic wallpaper (hardly read) and prosodic discussion (overly read).

Without explanation for the pause of a decade since the first volume, Frege continued the second part of the book on the 'construction of sequences' by dealing with topics such as the isomorphic comparison of sequences (arts. 1–5) and the summation of numbers (arts. 33–36). He ended by using again the concept 'endless' (*'Endlos'*) to distinguish indefinite from finite Numbers (arts. 53–54).

The rest of the volume contained (not all of) the third part of the book, dealing with real numbers. After rehearsing again his stipulations of well-formed nominal definitions of concepts and functions (arts. 55–65), Frege attacked various theories of real numbers recently proposed by contemporaries. Cantor was taken to task on various matters, such as (indeed) sloppily associating the existence of a real number with a fundamental sequence (323.2) of rational numbers (arts. 68–69); however, the basic strategy, similar to definition by equivalence classes, was hardly as hopeless as Frege wished to convey. Among other authors, while praising aspects of Dedekind's theory of cuts, Frege noted that it contained no investigation of the possibility of constructing the irrational number from the cut (arts. 138–140). Indeed, Dedekind had put the move forward as an axiom (§3.2.4).

Frege's other main target was the opening pages of Thomae *1898a*, the second edition of a textbook on complex-variable analysis. The situation in the *Mathematischer Seminar* at Jena around that time had deteriorated to such an extent that the course in logic was given by *Professor* Thomae while *Honorarprofessor* Frege handled topics such as remedial geometry. Perhaps in revenge, Frege obsessively denounced his senior colleague for talking about numerals instead of numbers, muddling symbols with their referents, allowing ' = ' to cover both arithmetical and definitional equality without explanation, and regarding zero as a 'purely formal structure' (arts. 88–103). Thomae had also compared arithmetic with chess as games; Frege pointed out that chess included moves as well as rules (that is, different categories), and so did arithmetic (arts. 88; 96, 107–123).

The whole passage in the volume, over 80 pages long, is the main source of 'Frege against the formalists', as it is now often called. However, while Thomae's presentation is sloppy,[32] it is doubtful any formalist intended to hold so absurd a position as that which Frege criticised. And when Frege denounced another favourite butt, Heine *1872a*, for saying that a sequence of numbers continued to infinity by sarcastically claiming that 'In order to produce it, we would need however an infinitely long blackboard, infinitely much chalk and an infinitely long time' (art. 124), the stupidity lies with Frege.

The last part of the volume is by far the most important, for it contained Frege's own theory of real numbers (Simons *1987a*). He conceived these objects to be ratios of magnitudes of any kind, from which itself the theory should be independent. To set up the required machinery he drew upon the concept of Relationship and on its extension, which was now also

---

[32] Curiously, the passages from Thomae's second edition cited by Frege were rewritten from the first edition *1880a*, which in general was less reprehensible though of the same philosophical ilk. Thomae also shared with Frege of the *Begriffsschrift* the same Jena publisher.

called 'Relation' ('*Relation*'), without mention of Schröder's logic of relations; 'extension of the concept' became 'class' ('*Klasse*': arts. 161–162). Then real numbers formed a class of Relationships, and each one was defined as a Relationship of Relationships. To specify these he drew upon the bicimal expansion of a real number a, of *any* kind:

$$a = r + \sum_{k=0}^{\infty} 2^{-n_k}, \text{ with } r \text{ the proper part;} \qquad (458.1)$$

then a could be captured by taking the sequence $\{r, n_1, n_2, \ldots\}$ (art. 164: he ignored the ambiguity of expansions ending with non-stop 1s, but it can be dealt with). This sequence could be infinite, finite—or empty in the case of integers, which were notated '1' in contrast to the Number '1'. The negative of any number was defined from the converse of its Relationship, and '0' by compounding any Relationship with its converse since (458.1) was not available (art. 162).

In the rest of the volume Frege established the properties required of Relationships to allow the constructions to be effected, drawing heavily upon functions of functions and compounding. After proving commutativity and associativity (arts. 165–172), he defined the 'positival class' of magnitudes from which, among other things, the least upper and greatest lower bounds of a collection of real numbers could be defined (arts. 173–186); the special case of the 'positive class' comprised members which satisfied Archimedes's axiom and thereby avoided infinitesimals (art. 197). He ended by promising more details about this class (art. 245), maybe on using the proper ancestral to generate the sequences of numbers specified by (458.1) and passing to further properties such as upper and lower limits. Presumably he also intended to exhibit the basic arithmetical operations, properties and relations in terms of the sequences defined from (458.1) or from notions derived from them, and proceed to related topic such as upper and lower limits. However, before he could start saving up to publish the third volume Russell's paradox arrived (§6.7.7).

Why did Frege take a decade to publish this volume? The second part was presumably completed by 1893; and apart from the passage on Thomae most of the rest could have been ready then also. Had he needed several years to pay for it? If so, the return on investment was small. As last time, there were very few reviews; in particular, in the *Jahrbuch* the school-teacher Carl Färber *1905a* wrote one paragraph, solely on the prosodic middle, and found 'many replies of Frege as pedantic or nit-picking'—harshly phrased, but not unfair. However, a more considered reaction was also published in that year.

4.5.9 *Frege, Korselt and Thomae on the foundations of arithmetic.* Korselt placed in the *Jahresbericht* of the *DMV* a commentary *1905a* on Frege's second volume, in the form of an exchange between 'F.' and 'K.'.

First K. appealed to Bolzano[33] to argue that F.'s rules for 'sharp definitions' were too strong and indeed not achievable in principle; he doubted whether the 'inner nature' of, for example, 'point' could be captured in the way that F. sought for 'number' (p. 372). Cantor's theory of real numbers seemed to be such a case; while suggesting improvements in presentation, he wondered if doubter F. had 'either not understood Cantor's definition or it goes with him like an absent-minded rider, who looked for his horse and sat upon it' (p. 376). Again, while Thomae's enterprise was 'to be considered as failed' K. wondered 'how should one otherwise know, that one has come across the "essence" of an object?' (pp. 379–380). He also defended the practice of abstraction in mathematics since it was executed only on 'certain conditions' (but F. rightly wanted to know which ones, and why?). Again, K. (naively) queried the merits of worrying about definitions since only 'one indicates an uncomfortably long expression or an arbitrary figure of known conditions with a short name, which itself is a figure of the theory?' (p. 381). Dedekind's creation (323.2) of irrational numbers seemed reasonable to K., since 'cannot also thoughts, concepts and theorems be created?' (p. 386).

Overall K. gave an intelligent appraisal of all current theories. For whatever reason F. did not answer (their swords had already crossed over geometry in §4.7.4), but instead went for Thomae's reply *1906a* to him in the same journal: a 'holiday chatter' on 'thoughtless thinkers' such as, apparently, the chess player. Thomae had concluded from his alleged attachment to numerals that 'for instance one might let the number three grow in the following figures

$$_3 \; 3 \; 3..3 \; 3.., \qquad\qquad [(459.1)]$$

but then there are the doom-laden little dots', which under Frege's characterisation denoted 'four more threes' (p. 437). In other words, he rightly rejected the kind of formalism attributed to him by Frege, as treating mathematics as *instances* of signs, sizes included, instead of the ideographical *form* of each sign. Thomae's ironic conclusion was that

Mathematics is the most unclear of all sciences.
Written in the dog-days of the year 1906.

[33] Later in the paper Korselt urged that

One should study Bolzano, not only his 'Paradoxes of the infinite' [*1851a*] or the 'three problems' [*1817a*] but above all the 'Wissenschaftslehre' [*1837a*]. If a mathematician and a publisher could yet be found for the voluminous manuscripts that the Vienna Academy of Sciences possesses! That would be a task for the *Deutsche Mathematiker-Vereinigung*!

(*1905a*, 380): evidently he was not aware of the Bolzano holdings in Prague (§2.8.2). Korselt himself is little known; his writings would certainly repay careful study. His surviving letters to Frege, dating largely from 1903 (Frege *Letters*, 140–143), dealt with Schröder's mistaken proof of the equivalence Theorem 425.1 and his own solution to Russell's paradox (§7.5.2).

In reply Frege *1906b* felt sure that he 'had destroyed Thomae's formal arithmetic for ever' and the recent chatter 'only strengthened [...] this conviction'. Thomae *1906b* began his ironic and witty answer; '22 years ago Mr. Frege let me know unequivocally in conversation, that he held me as incapable of understanding his deeper deductions. Now he pronounces the same *urbi et orbi*'. The *Honorarprofessor* replied with a new account *1908a* of the 'impossibility' of Thomae's approach. The editor of the *Jahresbericht*, Gutzmer, then back at Halle, may have felt pressure from his contact with these two fomer colleagues at Jena (§4.5.1) to accept all this stuff.

This last scratch at Thomae's eyes was Frege's final publication before his retirement in 1918, although he continued to lecture on his theory and may have been writing a textbook on it (*Manuscripts*, 189–190). We note his last period in §8.7.3.

It is not surprising that Frege had a poor reception in general. Intemperate polemics, partly based upon silly criticisms, are not the only reasons; unattractive are seemingly excessive fussing about names, the use of normal words like 'function' in unfamiliar ways, highly forgettable symbols in the technical accounts (although not, I hope, the nice if impractical spatial layout), and, after 1903, the presence of Russell's paradox in his system. Indeed, his logic remains rather mysterious; the logicism is easier to grasp. His failure to acknowledge sources does not help either (and helps the Frege'ers to know that he thought up everything for himself). In particular, Kreiser *1995a* has shown recently that Frege's father Karl-Alexander (b. 1809) published a grammar-book *1862a* for schoolchildren which just happens to emphasise a context principle on the primacy of propositions, the role in them of logical connectives, their expression of 'thoughts', the distinction between objects, propositions and names, and the designation of an object of a concept by adding 'the', and even a spatial layout of symbols (but without lines) to symbolise the subordination of adverbs to verbs. Well, fancy that.

Two serious concerns of Frege have not yet been noted. One was his lack of respect for Hilbert's way with the foundations of geometry, due in §4.7.4; the other is his response to Husserl, to whom we now turn.

4.6 HUSSERL: LOGIC AS PHENOMENOLOGY

4.6.1 *A follower of Weierstrass and Cantor.* (Schuhmann *1977a*) An unusual member of the Weierstrass school (§2.7.4) was Edmund Husserl (1859–1938), who took courses with The Master in 1878 and 1879 (when Klein, Max Planck, Otto Hölder and Aurel Voss were also around). His special interest was in the calculus of variations, and his version of the course given in 1879 was so good that it was used in the Weierstrass edition (see the editorial remarks in Weierstrass *Works 7* (1927)). Husserl then wrote a *Dissertation m1882a* on the subject at Vienna University under the supervision of Weierstrass's follower Leo Königsberger (1837–1921) (Biermann *1969a*).

But thereafter Husserl devoted his career to philosophy, hoping to achieve there standards of rigour comparable to those in mathematics exhibited by Weierstrass's lectures, and by similar means of exposing clearly the basic principles and building up the exegesis in a rational manner. While in Vienna he had also studied with Brentano, from whom he learnt that the act of perception was directed towards (more than) one object (in the general sense of that word), which therefore inhered with the act itself, and that psychology was to be understood primarily as the analysis of acts of consciousness (Gilson *1955a*). Husserl was to call this brand of philosophy 'phenomenology', the philosophical analysis of reasoning with especially reference to consciousness. Brentano was more an inspirer than practitioner of it, partly because he did not focus upon philosophical issues beyond supporting positivism whenever possible.

Husserl was also perhaps the first philosopher outside Bohemia to be influenced significantly by Bolzano; he discovered him first through the article Stolz *1882a* (§2.8.2), and then especially via the enthusiasm of Brentano. One point of attraction was the notion of presentations in themselves beyond any particular instances of them; another was pure, objective logic itself, which grew in importance in his philosophy. Thus he was no simple idealist: on the contrary, he sought objective contents independent of any thinker's (ap)perception of them. Rigour and rationality coupled to perception and inherence: the elaboration of these insights was to dominate his philosophical endeavours life-long.[34]

The opportunity to launch them came in 1886, when Husserl moved to Halle University as a *Privatdozent* and wrote his *Habilitation 1887a* 'On the concept of number. Psychological analyses'.[35] The main supervisors were Erdmann (§4.5.6) and Stumpf (a former student of Brentano); but he also came in contact with mathematicians, especially Cantor (who also told him about Bolzano) and Hermann Grassmann's son, also Hermann. He expanded the work into his first book, *Philosophie der Arithmetik* (*1891a*).

Husserl's next book was two volumes of *Logische Untersuchungen* (*1900a*, *1901a*). Partly because of it, he was promoted in 1901 to *ausserordentlicher Professor*, and moreover at the more prestigious Göttingen University,

---

[34] For a general history of phenomenology, including chapters on Husserl, Brentano and Stumpf, see Spiegelberg *1982a*; Husserl's own brand is surveyed in Smith and Woodruff Smith *1995a*.

[35] The file on Husserl's *Habilitation* examination in June 1887 is held at Halle University Archives, *Philosophische Fakultät* II, *Reportorium* 21, no 139: Stumpf chaired the jury, to whom Cantor expressed satisfaction over the mathematical aspects of the examining. The documents are transcribed in Gerlach and Sepp *1994a*, 161–194, a useful book on Husserl's Halle period and his thesis. On the influence of Cantor on Stumpf's psychology of consciousness, see B. Smith *1994a*, 86–96.

where Hilbert was one of his new colleagues. Five years later he received a personal full chair. In 1916 he obtained a full chair at Freiburg im Breisgau; he retired in 1928, two years after Zermelo joined the faculty. He wrote incessantly throughout his life, and also corresponded extensively (Husserl *Works, Letters*); but much of his philosophy has no specific mathematical concern, and he never attempted a logicism. Thus the treatment of his work here will be brief, and confined almost entirely to the main publications of his Halle period. Most of his other publications then were long reviews of books in German on non-symbolic logic; he also wrote many manuscripts on arithmetic and on geometry (*Works 12* and *1994a*). Some later work and followers appear in §8.7.8.

4.6.2 *The phenomenological 'philosophy of arithmetic', 1891.* (Willard *1984a*, chs. 2–3) Although Cantor was mentioned only twice in Husserl's *Habilitation*, his influence seems to be quite marked: the choice of the number concept as his topic (Weierstrass may also be detected), and the distinction of cardinal and ordinal by 'Zahl' and 'Anzahl' (§3.2.7). Focusing on 'o u r g r a s p o f t h e c o n c e p t o f n u m b e r', not the number as such, he highlighted the intentional act of 'abstraction' from maybe disparate or heterogeneous somethings to form 'embodiments' ('*Inbegriffe*': pp. 318–322). His phenomenology refined Cantor's naive idealism, and indeed may have been a motivation for it (Hill *1997a*). For example (an important one), he applied 'specialisation' (Cantor's word, after (323.3) and in §3.6.1) to the counting process to specify numbers out of sequences as successions of ones from 'something' ('*Etwas*': *1887a*, 336). Two bases furnished 'the psychological foundation of the number-concept': '1) t h e c o n c e p t o f c o l l e c t i v e u n i f i c a t i o n; 2) t h e c o n c e p t o f S o m e t h i n g' (pp. 337–338).

Husserl soon expanded his *Habilitation* of 64 pages into a book of five times the length; but it appeared after delay (or hesitation?) as *Philosophie der Arithmetik. Logische und psychologische Untersuchungen* (*1891a*). It was dedicated to Brentano, despite his friendly protests, and a lack of interest which took him 13 years to spot the dedication![36] Husserl followed the line of his *Habilitation*, to near repetition of text in the first three chapters; they comprised about half of the first part, which was devoted to 'the concepts of multiplicity, unity and Number' ('*Anzahl*'). Much of the second part, on the symbolisation of Number and its logical roots' was new in text though not in context. Husserl began by claiming that 'numbers are no abstracta' and distinguished, say, '3' from 'the concept 3': 'the arithmetician does not operate with the number concepts as such at all, but with the generally presented objects of this concept' (p. 181); again, 'Is it not clear,

---

[36] See Brentano's letters to Husserl of May 1891 acknowledging receipt of the book, and of October 1904 upon discovering the dedication (Husserl *Letters 1*, 6–7, 19–20; note also Husserl's recollection in *1919a*, 312).

that "number" and the "presentation of counting" is not the same?' (p. 33). Similarly, on 'Presentations of multiplicities' ('*Vielheitsvorstellungen*'), 'We enter a room full of people; an instant suffices, and we judge: a set of people', though he stressed that 'an instant' was an over-simple phrase in 'the explanation of the momental conception of sets' (pp. 196–197). More generally, he noted 'figural moments', acts of perception which create out of a collection 'e.g. a row of soldiers, a heap of apples, a road of trees, a line of chickens, a flock of birds, a line of geese etc.' (p. 203). But he did not contrast Cantor's *Mengenlehre* with the part-whole tradition (a brief waffle about 'infinite sets' occurs on pp. 218–222), and he seems not to have known Kempe's recent theory of multisets (§4.2.8).

This concern with perception bore centrally upon Husserl's philosophy of arithmetic, in which he saw Numbers as 'multiplicities' ('*Vielheiten*') of units; in rather sloppy disregard of the tradition of distinguishing extensions from intensions, he used 'Menge' and 'Inbegriff' as synonyms. Since his philosophy also drew upon counting members of multiplicities, the grasp of numbers involved numeral systems, which he discussed at length in ch. 12. He developed X-ary arithmetic for any integer X in a rather ponderous imitation of Cantor's principles (326.2) of generation of ordinals: '$1, 2, \ldots, X$', with successors '$X + 1, X + 1 + 1$' through multiples to polynomials in X (pp. 226–233). X was always finite; he was not following Cantor into the transfinite ordinals, maybe because of their dubious perceptibility. Further, the central place of counting in his philosophy of arithmetic casts doubt upon the primacy of cardinals stated in the preface (p. 10).

Husserl's number system was prominent in his final chapter, which treated 'The logical sources of arithmetic' (not 'foundations', note); for again 'the method of sensed ['*sinnliche*'] signs is thus the logical method' (p. 257). Thus, despite the mention of 'logic' in the sub-title of his book, its role was linked only to relationships between numbers, not the numbers themselves: 'from the development of a g e n e r a l a r i t h m e t i c in the sense of a general theory of operations', as he put it in his final words (p. 283). The status of 0 and 1 was also not clear: 'One and None—they are the only ['*beiden*'] possible n e g a t i v e answers to the How many. [. . .] But logical this is not' (p. 131), in a passage where unit and unity were rather mixed together.

This attitude makes a great contrast with Frege, whose *Grundlagen* Husserl had read since completing his *Habilitation*. The difference is beautifully captured by their reactions to exactly the same passage from Jevons: 'Number is but another name for diversity. Exact identity is unity, and with difference arises plurality' (*1883a*, 156). For Husserl in both *Habilitation* and book this procedure was satisfactory, although Jevons's following remarks on abstraction were psychologically naive (*1887a*, 319–321; *1891a*, 50–53). By contrast, in the *Grundlagen* Frege had found the whole approach to be indefensible, in its use of successions and especially in assumptions about units (*1884b*, art. 36).

Husserl was also critical of Frege, partly for avoiding psychological issues which for him were central (Husserl *1891a*, 118–119) but also on other matters. The most important was the equivalence of extensions of concepts: 'I cannot see, that this method marks an enrichment of logic' since it worked with 'ranges' (*'Umfänge'*: p. 122). In particular, he did not find convincing Frege's Leibnizian definition of 'equality' (§4.5.3) because 'it defines identity instead of equality', reversing the correct relationship because 'Each same characteristic grounds the same judgements, but to ground the same judgements does not ground the same characteristics' (p. 97; compare p. 144). Given the paradox that Russell was to find in Frege's comprehension law (456.1), Husserl's intuition was very sharp; Frege's own modification of his calculus was to involve modifying identity (§6.7.7). Less clear is Husserl's claim that 'More difficult [than counting] is it, correctly to characterise psychologically the role which the r e l a t i o n s   o f   e q u a l -
i t y are assigned by the number-presentations' (p. 142).

Husserl completed his book in April 1891 by writing a short preface; in the same month he prepared a long review of Schröder's first volume *1890a*, which appeared later in the year as Husserl *1891b*. It shows further moves towards objectivity, perhaps inspired in part by reaction against Schröder. For example, having appraised Schröder's calculus as an 'algorithmic logic of extensions' (*'Umfangslogik'*: p. 7), he stressed more strongly than in the book the 'ideal content of concepts', which 'no person possesses' as Schröder seemed to assume (p. 17). Schröder's failure to handle this distinction led to 'all confusion', and Husserl expended upon various examples and consequences. One of these was Schröder's paradox of 0 and 1 after (444.6), where for once Husserl noted the merits of the membership relation in the *Mengenlehre* (pp. 35–36). He also disliked some technical features; for example, since subsumption incorporated equality as well as inclusion, the definition (444.3) of 'identical equality' using it was 'an obvious circle-definition' (p. 30). But he seemed to misunderstand Schröder's use of 'Principle' to denote an axiom when criticising Schröder's new one for distributivity (pp. 37–38). Another change was bibliographical: for the first time in print Husserl mentioned the Peirceans (p. 3).

Schröder referred little to Husserl in the posthumous part of his lectures, and only once to this review (*1905a*, 484). Meanwhile, others had reacted to Husserl's book.

4.6.3 *Reviews by Frege and others.*   One of the first reviews of Husserl's *Arithmetik* came from Jules Tannery; although writing in the *Bulletin des sciences mathématiques*, he concentrated on the philosophy. Warming to the book in general and Husserl's doubts over Leibnizian identity, he declared that *'axioms are conditions imposed upon definitions'* (*1892a*, 240), a kind of conventionalism which his younger compatriot Henri Poincaré was to expound later (§6.2.3, §7.4.3, 5). In the *Jahrbuch* Michaelis *1894a* was still more positive; perhaps recalling Frege (§4.5.2), he concluded that Husserl's book 'may be considered by far the best that has been written on the foundations of arithmetic for a long time'. However, neither reviewer much penetrated the philosophy or the psychology: for that a sterner piece, in a philosophical journal, came from Frege.

Frege mainly just contrasted his philosophy with Husserl's. For example, he attacked the mixture of logic and psychology (Frege *1894a*, 181), which for Husserl was intentional. Maybe deliberately, he misunderstood as a 'naive opinion' Husserl's remarks on heaps and swarms in connection with numbers, diagnosing as cause 'because he seeks in words and combinations of words specific presentations as their references' (pp. 186–187) without allowing that heapness or swarmhood could be part of that reference. Indeed, he seems not have realised that for Husserl 'presentation' had an objective ring, maybe following Bolzano, not his own subjective connotation. But he also rightly detected some confusion between multiplicities and Numbers (p. 179), and he could have been more critical than pp. 188–189 on the handling of 0 and 1.

Doubtless Frege's review nudged Husserl further along the path towards objectivity; but the extent of its impact needs careful appraisal (Hill *1994a*). The Frege' industry routinely informs us that the review quite transformed poor Husserl's philosophy; but elementary attention to chronology and sources (Hill *1991a*, pt. 1) shows that this claim refers far more to the False than to the True. We noted Husserl's use of 'ideal content of concepts' in his review of Schröder, so that he was already shifting his position even while his book was in press; later (*1900a*, 179) he retracted only a few pages of censure of Frege (including the comments on equivalence, which were worth retaining!), and left intact his basic approach and other reservations of Frege's theory (on identity, for example).

One of these concerned sense and reference: instead of Frege's distinction for proper names, recently introduced (§4.5.5), Husserl worked in *Arithmetik* with 'a two-fold reference' of an 'abstract name', both 'as name for the abstract concept as such' and 'as name for any object falling under this concept' (*1891a*, 136). In recognition of this difference, Frege explained his own position in the beautiful schema given in §4.5.5. His letter was a response to Husserl sending both his book and the review of Schröder; when reviewing the former, Frege seems not to have noted the changes evident in the latter. In reply Husserl politely pointed out several similarities between them; for example, observing the distinction between a logic as such and its calculus (Frege *Letters*, 100).

4.6.4 Husserl's 'logical investigations', 1900–1901. During the 1890s logic moved to centre stage in his phenomenological concerns as he sought his version of the objective. Bolzano's work made its full impact during this period. The principal outcome was one of his major publications, the *Logische Untersuchungen*, published in two volumes and dedicated to Stumpf (Husserl *1900a, 1901a*). A lightly revised second edition appeared in 1913 and 1921; I use it here, as it is much more accessible (but not always the English translation *1970a*). Here a few features of his view of logic and its relationship to mathematics are noted.

The first volume contained Husserl's 'Prolegomena to pure logic', a long essay on psychologism, where, perhaps unhappily, he enjoined both German idealism and the sociological reductionism of Mill (*1900a*, art. 13). Of his various criticisms, one concerned the unavoidably limited horizons of human experience, which surely prevented delivery of the generality required by a philosophy of mathematics. A relatively well-known passage used a mathematical example (art. 46):

> All products of arithmetical operations go back to certain psychic acts of arithmetical operating. [...] Quite other is arithmetic. Its domain of research is known, it is completely and exhaustively determined by the familiar series of ideal species 1, 2, 3, ... [...] The number Five is not my own or anyone else's counting of five, it is also not my presentation or anyone else's presentation of five. It is in the latter regard a possible *object* of acts of presentation [...].

The passage was inspired by one in Cantor using 'five' (*1887–1888a*, 418–419).

The 'pure logic' which Husserl sought was a normative science of objective contents, requiring 'the fixing of the pure categories of meaning' ('*Bedeutung*'), objects, their relationships and laws, and 'the possible forms of theories or the pure theory of manifolds' (*1900a*, arts. 67–70). The application of mathematics to logic recalls Boole, though the details were quite different; for under 'pure manifold', whose laws determine '*the theory's form*', he included Riemann's theory, Hermann Grassmann's calculus and Cantor's *Mengenlehre* (art. 70). Finally, he drew in probability theory, though without clear intent (art. 72).

The larger second volume contained six investigations of the title. Husserl discussed others' work in some detail; after Brentano the author most cited was Bolzano, mainly for his *Wissenschaftslehre*. His own exegesis sought to articulate the pure logic from the main notions of his descriptive psychology: expression, meaning, attention, objects, experiences, contents, and so on. Of greatest mathematical interest is the third investigation, where he extended the part-whole theory in *Arithmetik* into an elaborate classification of kinds of part, such as (not) spatio-temporal and (in)dependent, and their relationships to aspects such as redness of objects (Smith and Mulligan *1982a*). The discussion shows that phenomenology deserves a *much* better place among the philosophies of mathematics than it normally gains. But Husserl's pure logic itself seems to be rather fugitive (with 'pure' being passed from one notion to another!); for example, he did not discuss logical connectives or quantification theory, which surely should come into a logic influenced by mathematics. His silence over Peirce and Schröder is loud.

4.6.5 Husserl's early talks in Göttingen, 1901. The next stage of Husserl's development is rather surprising. At the end of his *Habilitation* one of his six 'theses' stated that irrational numbers needed 'logical justification' (*1887a*, 339). Perhaps in fulfilment, he had announced in the preface of *Arithmetik* a second volume to deal with negative, rational, irrational and complex numbers; indeed, apparently it was 'largely ready' (*1891a*, 8, 7). But his philosophical uncertainties prevented the volume from being completed (the surviving manuscripts are published in *Works 12*, 340–429).

However, when he moved to Göttingen in 1901, the year of publication of the second volume of the *Untersuchungen*, Husserl gave two lectures to the *Göttinge Mathematische Gesellschaft* in November and December on 'the imaginary in mathematics'. The word 'imaginary' covered all these types of number (*m1901b*, 432–433); but instead of trying (and failing) to grasp them by phenomenological means, the 'way through' was now provided by specifying a consistent axiom system and the manifold or domain ('*Gebiet*') of objects determined by it. One of the main properties was defined thus (p. 443):

> A formal axiom system, which contains no inessentially included axiom, is called definite, when each theorem which decidedly has a sense through the axiom system, thereby falls under the axiom system, be it as consequent, be it as contradiction, and that will apply overall, where it can be shown on the basis of the axioms that each object of the domain is reduced to the group of numerical objects, for which each relationship fulfils the true identically and every other is therefore false.

Thus Husserl's notion of definiteness was oriented around arithmetic ('group' above carries no technical meaning), but was related to propositions which were not derivable from any axiom system. To us it sounds very close to Hilbert on axiomatics: so it did to Husserl, who distinguished between definiteness 'relative' to a particular domain and the unrestricted absolute version which ' = complete in t h e H i l b e r t i a n sense' (p. 440); the Club minutes of the lectures use 'vollständig' for the first sense and 'definit' for the second.[37] Now Hilbert had recently spoken to the Club on axiomatics (§4.7.3), with Husserl present; but Husserl seems to have formulated his own approach independently before arriving in Göttingen,

---

[37] See Göttingen Mathematical Archive, 49:2, fol. 93 for Husserl's two lectures, which took place on 26 November and 10 December 1901. Both minutes contain the phrase 'Durchgang durch die Unmögliche', but this seems to be a mishearing or -reading of Husserl's phrase 'Durchgang durch das Imaginäre' (*m1901b*, 440) by the Club secretary, Hilbert's doctoral student Sophus Marxsen. On 12 November Husserl had spoken about the work of De Morgan and the German philosopher J. B. Stallo (1823–1900) (fol. [92]). The lectures at the Club were listed routinely in the *Jahresbericht* of the *DMV*.

in connection with his treatment of manifolds in the *Untersuchungen* (Hill *1995a*, Majer *1997a*). The converse is also true; Hilbert had found his own way to axiomatics during the 1890s, as we shall now see.

4.7 HILBERT: EARLY PROOF AND MODEL THEORY, 1899–1905

4.7.1 Hilbert's growing concern with axiomatics. Husserl's use of axioms was a sign of the mathematical times, for their role grew quite noticeably during the last 30 years of the 19th century. Two branches of mathematics were largely responsible (Cavaillès *1938b*): abstract algebras, mostly group theory (Wussing *1984a*, pt. 3) but also other structures (some traces were seen in §4.4 with Dedekind and Schröder); and geometries, now various with the acceptance of the non-Euclidean versions. As a mathematician, Hilbert was an algebraist; his earliest work dealt with invariants and algebraic number theory. The latter also brought him to axiomatics; but his first detailed exercise was in geometry.

Hilbert was concerned with geometries throughout the 1890s (Toepell *1986a*). While still at Königsberg he gave a course on projective geometry in 1891, followed three years later by one on foundational questions such as the independence of axioms and particular ones such as connection and continuity (called 'Archimedes's axiom'). Some wider publicity came in a short note *1894a* in *Mathematische Annalen* on defining from certain axioms 'the straight line as the shortest connection between two points'.

After his move to Göttingen in 1895 Hilbert continued working on the projective side, becoming especially interested in the proof in Isaac Schur *1898a* of Pascal's famous theorem on the collinearity of the three points of intersection of the opposite sides of a hexagon inscribed in a conic, which did not use continuity. He treated this theorem in a special short course at Easter 1898 'On the concept of the infinite', which dealt with geometrical spaces and continuity rather than Cantor's *Mengenlehre*. This brought him to a course in the winter semester of 1898–1899 on 'the foundations of Euclidean geometry', of which several dozen copies were made; it led to one of his most famous publications. He was then in his late thirties.

As part of the growing interest in axiomatics, it had become clear that Euclid had not specified all the assumptions that he needed; so some of the gaps were filled (Contro *1976a*), especially by Moritz Pasch (1843–1930) with an emphasis on the ordering of points, and then by Peano with a treatment also using lines and planes (§5.2.4). Hilbert decided to fill all the remaining gaps. An unusual occasion for publicity arose in June 1899, when a statue was unveiled to celebrate the work of Gauss and the physicist Wilhelm Weber. Klein thought that some accounts of scientific work related to their interests should be prepared, and so two booklets were written. Physics professor Emil Wiechert described electrodynamics, in honour of the heroes' creation of the *Magnetische Verein*; and Hilbert drew on his lecture course to present the 'Foundations of geometry', with especial reference to the Euclidean version (*Geometry*₁ (1899)). The essays were published together as a book by Teubner, Hilbert receiving 235 Reichsmarks for his part (Hilbert Papers, 403/6).

Over the decades Hilbert's essay expanded from its original 92 pages to over 320 pages in the seventh edition (1930). Some of this extra material arose from additions or changes to the text, even to the axiom system; but most of it was reprints of articles on geometry or the foundations of mathematics written in the interim (Cassina *1948–1949a*), for the book inspired him to a general study of the foundations of geometry and also arithmetic. The words 'formalism' and 'metamathematics' became attached to his philosophy and techniques during his second phase, which ran from the late 1910s to the early 1930s (§8.7.4); he gave it no special name during the first one, which ran until 1905, but 'axiomatics with proof and model theory' is a reasonable characterisation.

4.7.2 *Hilbert's different axiom systems for Euclidean geometry, 1899–1902.* In his first edition Hilbert presented 20 axioms: I 1–7 on 'Connection' ('*Verknüpfung*'), II 1–5 on 'Ordering' ('*Anordnung*'), III for the parallel axiom (in a Euclidean version rather than one of the equivalents found since), IV 1–6 on 'Congruence', and V on 'Continuity'. Then he proved various elementary properties of points, lines and planes; angle was defined from IV 3 as the 'system' of two intersecting half-lines. The second chapter dealt with the independence of the axioms, which he demonstrated by working with a corresponding co-ordinate geometry and assuming the consistency of the real numbers which it used. While the independence of each group of axioms seems well shown, that within a group was not fully handled, and some redundancy was soon found (§4.7.3). Then he handled planar areas, and proved Pascal's theorem and a similar one due to Desargues. In the final chapter he made some straight-edge constructions in the plane, assuming congruence; they led to remarkable links to number theory which may have been a little out of place and contrasted sharply with the regular use of diagrams in the earlier chapters. In the *Jahrbuch* Friedrich Engel *1901a* summarised the book in some detail and judged that 'it gives a satisfactory, nay definitive answer to many pertinent questions for the first time'.

Hilbert's stress on the consistency and independence of axioms, and on the axioms (not) needed to prove particular theorems, characterised his philosophy of mathematics at this time. Concerning a lecture by Hermann Wiener *1892a* to the *DMV* on proving Pascal's and Desargues's theorems, he had stated that 'one must be able to say "tables, chairs, beer-mugs" each time in place of "points, lines, planes"' (Blumenthal *1935a*, 402–403); but this famous remark is normally misunderstood and Hilbert may not have thought it through at the time.

Firstly, Hilbert was advocating model theory for the axioms (intuitively at this early stage), not the mere use of words nor the marks-on-paper formalism that Frege detected in the symbol-loving arithmeticians (§4.5.8); intuitive knowledge of Euclidean geometry motivated the axiomatising enterprise in the first place. Unfortunately he did not make this point in the book, although the lecture course had contained consideration of 'intuition' ('*Anschauung*': Toepell *1986a*, 144–147): one obvious consequence is that intuitive knowledge of beer-mugs is different. Secondly, he treated concepts such as 'point' as implicitly defined via axioms. Thirdly, the same versatility could not be demanded of the logical connectives used to form and connect his propositions; for example, 'and' cannot become 'wine-glass'.

Typically of mathematicians' casual attitude, Hilbert took logic for granted in his book; but he soon began to attend to it (§4.7.4). The Paris lecture of 1900 on mathematical problems (§4.2.6) showed him already to be aware of the consistency of an axiom system. He continued to develop his approach to geometries, in papers and also before the *Göttinge Mathematische Gesellschaft*. The talk on 18 February 1902 may have surprised the audience, for he presented a quite different axiomatic treatment using groups of continuous motions, the latter defined in terms of mappings;[38] the details appeared in a long paper *1902b* in *Mathematische Annalen*, curiously given the same title as the book. He cited Lie for the algebra and Riemann and Helmholtz for the geometry, but not his colleague Klein for either. This axiom system was much simpler than its predecessor, so that the proof-theoretic task was reduced. The number system was used to definite a 'number plane' of co-ordinates, and set theory was prominently used. These features led the young American mathematician E. B. Wilson (1879–1964) to conclude an acute and sceptical commentary *1903a* on the paper that a better title for it would be 'Geometric analogues of ensembles', on the grounds that its reliance upon numbers and sets could not capture geometry itself.

Despite the difference of approach in this paper, Hilbert reprinted it in later editions of his book. The second edition (1903) won for him in the following year the third Lobachevsky prize, awarded by Kazan University for contributions to geometrical knowledge; as a member of the jury, Poincaré wrote a long and admiring report *1904a*. The changes in this edition included not only misprints; for the text was also altered, at one point in a very important way.

4.7.3 *From German completeness to American model theory.* Hilbert soon imitated for arithmetic his success with geometry in a short paper *1900a* for the *DMV* 'On the concept of number'. This time axioms I 1–6 covered 'connection' by addition and multiplication, II 1–6 for 'calculation'

---

[38] Göttingen Mathematical Archive, 49:2, fol. [96].

via the equality relation, III 1–4 for 'ordering' by inequalities, and IV 1–2 for 'continuity'. The first axiom of this last group was Archimedes's, as usual; it guaranteed the existence of the real numbers and thereby the real line, hence underpinning geometry. The other axiom was a significant innovation:

> IV 2. (A x i o m o f c o m p l e t e n e s s). It is not possible to add another system of things, so that the system of numbers resulting from the composition of the axioms I, II, III, IV 1 will be thoroughly filled; or briefly: the numbers form a system of things, which due to the maintenance of the collective axioms is not capable of further extension.

This use of 'completeness' is not that to which we have become accustomed from Hilbert's second phase, but follows Dedekind, perhaps consciously. To us it is a kind of meta-axiom about sets or manifolds, like Husserl's absolute definiteness (§4.6.5) but independent of it: the other axioms are assumed to have captured all the objects required by the theory. He ended the paper by reporting Cantor's conclusion (§3.5.3) that the set of all alephs was 'inconsistent (unready)'. In similar vein, he told members of the *Göttinge Mathematische Gesellschaft* (including Husserl) on 29 October 1901 of one consequence: admit the completeness axiom but omit Archimedes's, and the system is contradictory.[39]

Clearly this kind of assumption was not confined to arithmetic, so Hilbert added the corresponding axiom in the second (1903) edition of his book; in a review for the *Jahrbuch* Max Dehn *1905a* described it as a ' "between" axiom', a 'second continuity axiom' equivalent to Dedekind's cut principle. Hilbert must have found it very soon after publishing the first edition, for it was directly added to the French translation *1900a*; it also entered the English one *1902a*. That translation, published by Open Court, was reviewed in the Company journal *The monist* by Veblen *1903a*, who mentioned a recent American contribution to the axiomatisation. It had been made by E. H. Moore, who had received an honorary doctorate from Göttingen University at the statue ceremony in 1899, at the relatively early age of 38 years. (Hadamard, three years younger still, was also given one.) He soon published a paper *1902a* in his *Transactions of the American Mathematical Society* showing the redundancy of Hilbert's axioms II 4 in the group on ordering (a claim concerning I 4 on connections was soon withdrawn).

This was the reference made in his review by Oswald Veblen (1880–1960), then a student of Moore at the University of Chicago, writing a thesis on the axioms of Euclidean geometry but based upon Pasch's use of 'order' between 'points' (the only two primitives). He published his doctorate as a paper *1904a* in the *Transactions*, presenting a system of 12 axioms

---

[39] Göttingen Mathematical Archive, 49:2, fol. 91. Husserl's notes 'from memory', carrying the date of 5 November, are reproduced in *Works 12*, 444–447.

for the purpose. Inspired by Hilbert's notion of completeness, he defined an axiom system to be 'categorical' if 'there is *essentially only one class*' of objects satisfying the axiom system, so that any two classes would be isomorphic; otherwise, it was 'disjunctive' if further axioms could be added. These terms were not his own; he acknowledged them (p. 346) as due to the Professor of Philosophy at Chicago, John Dewey, no less (but no mathematician). The second term has not endured, but the first, and its attendant noun 'categoricity', became standard; probably Dewey thought of it in rough mathematical analogy with 'categorical' in logic (his reaction to symbolic logic will be noted in §7.5.4 and §8.5.5). The notion may have stimulated Moore to devise the similar methodology of analogous theories in his general analysis (§4.2.7).

Veblen's thesis helped launch 'postulate theory', important in the rise of American mathematics (Scanlan *1991a*): the name 'model theory' has become more common. The notion of modelling was not new: non-Euclidean geometries had used it, and Boole's reading of his algebra of logic as in terms of elective symbols or of classes (§2.5.3) is another example. But the theory was treated much more systematically from now on, not least for the recognition of (non-)categoricity. Benjamin Peirce's work on linear associative algebras (§4.3.2) also played a role in showing how various systems of postulates could be handled; indeed, Moore's Chicago colleague L. E. Dickson (1874–1954) exposed closer links in a paper *1903a* in the *Transactions*.

In 1907 Veblen himself co-authored a textbook in mathematical analysis. Although categoricity was not represented, it was a pioneer work in the (dubious) educational practise of using axioms from the start (Veblen and Lennes *1907a*, ch. 1). The authors took the axioms from the other major American postulationalist, E. V. Huntington (1874–1952). He passed his entire career at Harvard University, except to write a *Dissertation 1901a* on the geometrical interpretation of real numbers and vector algebra at Strasbourg University under the direction of Dedekind's friend Heinrich Weber. On his return he studied the former kind of quantities by means of postulate theory; indeed, he started a little before Veblen and was more prolific. His main mathematical interest was finding axiom systems for various mathematical theories and studying their consistency, independence, completeness and 'equivalence' (his word for categoricity). He published most of his studies in the *Transactions*: the early cases included *1902a* for 'absolute continuous magnitudes', *1902b* for positive integers and rational numbers, and *1903a* for real numbers; a later long study of the continuum (in another journal) will be noted in §7.5.6.

In addition, Huntington *1904a* examined Schröder's algebraic calculus, and so brought model theory to logic. By such means the central place of interpretation in Hilbert's conception of axiomatics flowered naturally into model theory in American hands.

4.7.4 *Frege, Hilbert and Korselt on the foundations of geometries.* (Boos *1985a*) Frege saw both Hilbert's lecture course and the book in the winter of 1899–1900 and sent objections by letter (*Letters*, 147–152, 60–79). His logicism was not an issue, since it did not include geometry: the main point concerned the use of axioms rather than definitions to determine or specify objects (Demopoulos *1994a*). Typically for Frege, in Hilbert's groups I and III of axioms 'the referents of the words "point", "straight line", "between" have not been given, but will be assumed as known' (*Letters*, 61): typically in Hilbert's reply, 'The complete definition of the concept point is given first by the finished construction of the system of axioms. [...] point in the Euclidean[,] non-Euclidean, Archimedean, non-Arch[imedean] geometry is something different each time' (pp. 68–69).

Frege left the matter for a time, presumably while he finished the second volume of the *Grundgesetze*. But then he sent a short two-part paper *1903b* to the *DMV* in which he rehearsed again his view that axioms rather than definitions gave precision. He repeated his doubts about axioms by comparing some of Hilbert's second group, on connection, with a group of his own for congruence in arithmetic (pp. 267–268); but since his first axiom could be false, the point was poorly made.

Among other parleys, Frege rejected Hilbert's assumption that consistency guaranteed existence, since the latter rested for him on criteria of reference. For example (p. 269), Hilbert's axiom I 7, that 'On each straight line there exists at least two points', was no better than considering

'Explanation. We think to ourselves of objects, which we name gods.

Axiom 1. Each god is all-powerful.

Axiom 2. There exists at least one god.'

Since Hilbert allowed geometrical axioms to be interpreted in terms of beer-mugs, he might not have objected to this satire. He did not reply to the paper, but Korselt responded with his first printed comments on Frege. Among other matters, he treated ironically Frege's own account of reference (Korselt *1903a*, 402):

Should one not finally be able to agree over the 'meaning' of an expression, then this is only an indication ['*Zeichen*'] that one or more [disagreeing persons] must make more sentences about this sign ['*Zeichen*'] or with this sign. 'The sign has no meaning' will thus name: 'No sentences are known to us, which rule the use of these signs in general or in a given domain'.

Rather pointedly, he recommended that mathematicians read Bolzano's *Wissenschaftslehre* in order to avoid falling into contradiction (p. 405). In the *Jahrbuch* Dehn *1905b* wrote a brief review of this exchange, judging that Korselt's reply showed in an 'enlightening way, the objections [of Frege] as untenable'.

Frege replied at great length to both Korselt and Hilbert in a three-part paper. Some of his comments (*1906a*, 282–284) sunk to the level of polemic then being directed at Thomae (§4.5.9). Otherwise, he ran again through his preference for definitions. He also launched a long attack on axiomatics based on proposing in nonsense language that 'Each Anej bazes ['*bazet*'] at least two Ellah', and wondered what it might mean (p. 285). Covertly it seems to say that Frege at Jena is at least equal to Cantor and Gutzmer at Halle; in any case it reduces the issue to the choice of words, not to definitions versus axioms.

More importantly, perhaps confronted by the novelty of several geometries rather than the one and only arithmetic, Frege seems to have confused sense and reference himself several times. In particular, he corroborated Korselt's irony, for he construed it as denying that the parallel axiom 'might have the same or similar wording in all geometries, as if nothing were to depend on the sense' ('*Sinn*': p. 293), whereas surely reference is involved.

Frege's performance is variable in quality; in any case, by 1906 his authority in this area had been compromised by the failure of his law (456.1) of comprehension. This time Dehn *1909a* was even briefer in the *Jahrbuch*, merely recording Frege's main points without comment. Korselt's reply *1908a* was oblique, in that he presented his own view of logic, based as usual on Bolzano (for example, the important account of logical consequence in Bolzano *1837a*, art. 155). A proposition-in-itself, Korselt's 'proposition', corresponded to Frege's 'thought'; it contained 'presentations' ('*Vorstellungen*') as parts, either with classes or individuals. Also needed were relations of various types and numbers of places, rather like some of Cantor's order-types (not cited); he used them in a partly symbolic listing of the axioms of geometry from Hilbert's second edition.

Hilbert was perhaps too enchanted with representing geometrical axioms by beer-mugs to think through the consequences for concepts and for logic, and did not sufficiently stress the place of intuitive theory prior to its axiomatisation; but Frege was equally blind to model theory, in any form (Hintikka *1988a*). In a manuscript of the time (*Manuscripts*, 183) although mercifully not in print, he even declared that

> Nobody can serve two masters. One cannot serve truth and untruth. If Euclidean geometry is true, then non-Euclidean geometry is false, and if non-Euclidean geometry is true, then Euclidean geometry is false.

A main issue behind the non-discussion with Hilbert, and also Korselt's contribution, was Frege's adherence to the correspondence theory of truth, in contrast to Hilbert's preference for consistency of axiom systems. Another theme was definitional equivalence between systems, a topic crossing the boundaries of mathematics, logic and philosophy which was only being born at this time (Corcoran *1980a*), precisely because of Hilbert's work on axiomatics and the American launch of model theory. Meanwhile Hilbert was forging links between logic and proof theory.

4.7.5 *Hilbert's logic and proof theory, 1904–1905.* (Peckhaus *1990a*, chs. 2–3, 5) Hilbert chose to talk about 'the foundations of arithmetic' at the next International Congress of Mathematicians, which took place at Heidelberg in August 1904. After surveying various positions of the topic, including those of Kronecker, Frege and Dedekind, he outlined his own approach, 'unfortunately too short, because of the limited time accorded to each communication' (Fehr *1904a*, 386). Cantor and Jules König took part in the discussion; the broader context will be explained in §7.2.2.

The published version, Hilbert *1905a*, carried the interestingly different title 'On the foundations of logic und arithmetic', although the basic content may have been the same. He began his own treatment by positing the existence of two 'thought-objects', '1 (one)' and ' = (equals)' and forming combinations of them by concatenation: for example,

$$\text{`(1)( = 1)( = = = )' and `(11) = (1)(1)'} \tag{475.1}$$

(the brackets seem to be primitive also). The formulae which we call 'well-formed' belonged to the 'class of beings ['*Seienden*'], with its associated 'correct proposition' ('*richtige Aussage*') a; the rest went to the complementary 'class of non-beings', with $\bar{a}$. The other logical connectives were '$u.$' for conjunction, '$o.$' for disjunction, '$|$' for implication, and the rather clumsy symbols '$A(x^{(o)})$' and '$A(x^{(u)})$' for first-order existential and universal quantification over proposition A containing the 'arbitrary' x. The axioms were

$$\text{`1. } x = x. \quad 2. \ \{x = y \text{ u. } w(x)\} \,|\, w(y)\text{'} \tag{475.2}$$

for some (unexplained) propositional function w.

To this machinery Hilbert added 'three further thought-objects u (infinite set, infinity), f (successor), f' (accompanying operation)', and developed arithmetic based upon the axioms of Dedekind and Peano (not cited):

$$\text{`3. f(u}x) = \text{u(f}'x) \quad 4. \text{ f(u}x) = \text{f(u}y) \,|\, \text{u}x = \text{u}y \quad 5. \overline{\text{f(u}x) = \text{u1}}\text{'}; \tag{475.3}$$

he did not make clear the need for quantifiers, and introduced rules of inference only later. But he argued for consistency of the system in a novel way: propositions provable from them have the same number of thought-objects on either side of the equality sign, whereas candidate contradictories do not.

The paper is very suggestive though not too clear; arithmetic and logic are somewhat intertwined, with logic primarily used to make proofs more explicit, not for a deeper purpose such as Frege intended. Hilbert reads somewhat like the formalists whom Frege attacked, although the use of 'thought-object' showed that he was working with the referents of his symbols. This aspect came out more clearly in a superb lecture course *m1905a* on the 'Logical principles of mathematical thought' given at

Göttingen in the summer semester of 1905.[40] For he began by contrasting three ways of presenting arithmetic. In the 'geometrical' way appeal was made to diagrams (fols. 3–9). The 'genetic' way was somewhat more formal, in which rational numbers were treated as ordered pairs and irrational numbers treated from their decimal expansions; he cited as examples the textbooks Pasch *1882a* (§6.4.7) and Frege's favourite (§4.5.8) Thomae *1898a*. Finally came the 'axiomatic' way, his preference; the first two chapters of the first part contained axioms for arithmetic and geometry. On the whole he followed respectively his paper and the second edition of his book, but the treatment of consistency and independence was rather more elaborate. A long third chapter gave axioms for 'science': specifically, mechanics, probability theory and physics.

More original was the second part (fols. 122–188), on 'the logical principles'. In the first chapter Hilbert ran though many aspects of set theory, especially (non)denumerability and power-sets; interestingly, he did not attempt an axiomatisation. Then followed the 'logical calculus', whose symbols for connectives were ' ≡ ' for identity, '|' again for implication, and in a reverse from normal, ' + ' for conjunction and ' · ' for disjunction; they linked 'beings' ('*Seienden*'), not necessarily propositions, therefore. An axiom system was given for them and for the special beings '0' and '1' (fols. 143–152); the consequences included two 'normal forms' for logical expansions (fols. 160–163). In this and other details he seems to have drawn upon Schröder, who was not named.

The existence of these beings was guaranteed by a remarkable 'axiom of thought' or 'of the existence of an intelligence', no less: 'I have the capacity to think of *things*, and to indicate them by simple signs $(a, b, \ldots, x, y, \ldots)$ in such a perfectly characteristic way, that I can always recognise them unequivocally' (fol. 143).[41] Again Hilbert cited no sources, but he was aware of the Fries circle of neo-Kantian philosophers mentioned in §4.2.5; indeed, he held in high esteem its young member Leonard

---

[40] Two texts for Hilbert's course *m1905a* survive: one by E. Hellinger with some notes by Hilbert, kept in the Mathematics Faculty Library; the other by 'cand. math.' Max Born (no less), kept in the University Library and cited here. There are no substantial differences between the two versions; an edition is planned. Hilbert lectured on the systems for geometry and arithmetic to the *Göttinge Mathematische Gesellschaft* on 3 November 1903 and 25 October 1904 (Göttingen Mathematical Archive, 49:2, fols. 105 and [108]). Hermann Fleischer, then a Göttingen student though not under Hilbert, spoke about Peano on 19 January and 23 February 1904.

[41] The original text reads: 'Ich habe die Fähigkeit, *Dinge* zu denken und sie durch einfache Zeichen $(a, b, \ldots, x, y, \ldots)$ derart in vollkommen charackteristischer Weise zu bezeichnen, dass ich sie daran stets eindeutig wiedererkennen kann'. Lower down are translated these passages: 'sehr interessanter Hilfsmittel einer Begriffsschrift' (fol. 138); and 'Weierstrasschen Strenge' and 'der Beweis, dass in der Mathematik kein „Ignorabimus" geben kann, muss das letzte Ziel bleiben' (fol. 168). On the thread of set theory throughout Hilbert's work on foundations, see Dreben and Kanamori *1997a*, where however Cantor's letters of the late 1890s (§3.5.3) are not noted.

Nelson (1882–1927), who sought the *a priori* by analysing a theory into its components, a procedure quite congenial with axiomatics. One recalls also the power of the mind as advocated by Dedekind (§3.4.3), whom Hilbert mentioned as a pioneer logical arithmetician. Frege was also cited, for the 'very interesting resource of a concept-script' (fol. 138); but he was no source for an axiom of this kind.

Hilbert ended his course with his own philosophical considerations. After acknowledging Cantor and Dedekind and referring to 'Weierstrassian rigour' in proofs, he urged that 'the proof, that there can be no "Ignorabimus" in mathematics, must remain the ultimate aim' (fol. 168), echoing an optimism put forward in his Paris lecture *1900c* (§4.2.6). His fervour was stimulated by awareness of paradoxes in set theory which he had mentioned earlier in his course. One of them arose from the class of all power-classes, which we recognise as a version of the paradox of the greatest cardinal (fol. 136). We shall see in §6.6.1 that it was to lead Russell in 1901 to discover his own paradox, that the class of all classes not belonging to themselves belongs to itself if and only if it does not. Remarkable, then, is Hilbert's other main example—*this paradox itself*, apparently already known to his younger colleague Zermelo (fol. 137). How had he come to set theory?

4.7.6 *Zermelo's logic and set theory, 1904–1909.*   (Peckhaus *1990a*, ch. 4)

Like Husserl, Ernst Zermelo (1871–1953) began his mathematical career with a *Dissertation* on the Weierstrassian calculus of variations, at Berlin in 1894. He became *Privatdozent* at Göttingen in 1899, staying until accepting a chair at Zürich in 1910. Leaving in 1916,[42] he lived privately until becoming in 1926 *Honorarprofessor* at Freiburg am Breisgau, where Husserl was soon to be a colleague.

Soon after arriving in Göttingen, Zermelo's main interest switched from applied mathematics to set theory, and remained so for the rest of his career; Hilbert was probably the main influence.[43] The discovery of the paradox seems to have been one of his earliest findings, but largely unknown because for some reason he did not publish it, and only mentioned it in print once (*1908a*, 116–117). Hilbert made no special fuss either; when Frege told him in 1903 of Russell's discovery, he merely replied that Zermelo had priority of three or four years by then (Frege *Letters*, 80). Although Hilbert's lecture course had always been available in Göttingen, specific knowledge of Zermelo's priority came to light only in

---

[42] The usual reason given for Husserl's departure from Zürich is poor health. Another reason states the he took a holiday in Germany one summer and wrote 'Gottseidank kein Schweizer' in the registration book of the hotel, where the Swiss Education Minister stayed a few days later … (Fraenkel *1968a*, 149). The truth-value of this story is not certain.

[43] The Zermelo Papers contains rather few early manuscripts (see mainly Box 2), and seemingly none about the paradox; however, the collection of letters in Box 1 is quite good. His own letters to Hilbert are kept in the Hilbert Papers, 447.

the 1970s, in connection with Husserl. While preparing a volume of Husserl's *Works*, the editors found at the page of his own copy of the review *1891b* of Schröder discussing the paradox (444.6) of 0 and 1 a note recording a communication of April 1902 from Zermelo (maybe a letter now lost), laying out the paradox in the form 'the set of all sets which do not contain themselves as elements [...] does not contain itself as element'.[44] Husserl did not add that the paradox is (presumably) constructible within his own theory of manifolds, of which *Mengenlehre* was a case (§4.6.4), and he seems never to have pursued the matter.

Zermelo did publish on *Mengenlehre* at this time, especially in *Mathematische Annalen*. One paper introduced the axiom of choice in 1904; discussion is postponed until §7.2.6 when we note Russell's independent detection at around the same time. Another was the full-scale axiomatisation of set theory in *1908b*. Primarily intended to block out the paradoxes, his axioms included extensionality, the basic construction of sets, power-set, union, infinity and choice. Several of them captured the concerns of Cantor and Dedekind, especially their exchange of 1899, although Zermelo seems not to have been privy to it (G. H. Moore *1978a*); maybe Hilbert had told him about the letters that he had received from Cantor at that time, which give some hints (§3.5.3). Like Cantor, he followed an approach which Russell had recently called (§7.4.4) 'limitation of size' (Hallett *1984a*, ch. 7). He did not attempt to define the notion of set; maybe he followed Hilbert's penchant for intuition of some kind.

Zermelo also left the logic implicit; and this decision disfigured his system, in that another axiom, of separation, declared that a set could be formed of the objects satisfying any propositional function which was 'definite' for some overall set. Russell noticed this defect at once (see his letter of 8 March 1908 to Jourdain in my *1977b*, 109); Weyl *1910a* had to make clear that this vague adjective meant that the function was constructed by only a finite number of logical connectives and quantifiers and set-theoretic operations. Zermelo's paper contained a rather odd recipe of Peano's symbols mixed with Fregean notions such as assertion and the use of truth-values, with a side-salad of Schröder for first-order quantification (Peckhaus *1994b*). These features are rather surprising, because he was paid to teach a logic course in 1906 and 1907 (though poor health delayed him until the summers of 1908 and 1909), and in 1907 he had been appointed *Honorarprofessor* for mathematical logic. This was the first such post in Germany (Peckhaus *1992a*); Frege's title a decade earlier (§4.5.1) had been in mathematics.

In his paper Zermelo proved various basic theorems; his proof of Schröder-Bernstein was cited in §4.2.5. At the same time he wrote another paper, which appeared as *1909a* in *Acta mathematica*, on the related

---

[44] See Rang and Thomas *1981a*; Husserl's note is published in *Works 22* (1978), 399, and in English translation in *1994a*, 442.

theme of the role of mathematical induction in handling finite sets. He used Dedekind's notion of the chain (§3.4.2), but he defined infinitude inductively instead of reflexively, and so was able to avoid using an axiom of infinity.

This concern to show that an axiom may *not* be needed in a given situation is typical of Hilbertian proof theory, as we have seen above. The influence of Hilbert on Zermelo extended not only to consider set theory but also to treat it axiomatically; and the *latter* aspect makes them both thoroughly modern mathematicians (Mehrtens *1990a*, ch. 2). Zermelo's approach contrasts starkly with that of Schönflies, whom we saw practise *Mengenlehre* in Cantor's non-axiomatic way early on in this chapter on parallel processes, in which set theory has been the main linking thread.

CHAPTER 5

Peano: the Formulary of Mathematics

5.1 PREFACES

5.1.1 *Plan of the chapter.* Giuseppe Peano was an important contributor to mathematical analysis and a principal founder of mathematical logic, as well as the leader of a school of followers in Italy. Our concern here is with their work until around 1900, when Russell met Peano; their later contributions will be noted in subsequent chapters.

The account focuses upon logic and the foundations of arithmetic and analysis, including set theory. Peano's own writings are the main concern; they seem to have gained the main reaction at the time, not only with Russell. §5.2 traces his initial contributions to mathematical analysis and acquaintance with logic between 1884 and 1890. Then §5.3–4 surveys the developments made in the 1890s by Peano and his followers (who are introduced in §5.3.1). In 1895 he started to publish *Formulaire mathématique*, a primer of the results which they were finding; the title of this chapter alludes to it. Finally, their work around 1900 is described in §5.5, especially their contributions to the International Congress of Philosophy held in Paris in August 1900, which Russell heard (§6.4.1). Conclusions are drawn in §5.6 about the achievements which Peano had made and inspired.

5.1.2 *Peano's career.* Born in 1858 the second son of a farmer in the town of Cuneo to the north of Turin, Peano's ability emerged early, and his lawyer uncle Michele in Turin took care of his education. He enrolled as a student there in 1876, and was to pass his entire career in the University: over the years he received the usual promotions, becoming Extraordinary Professor in 1890 and obtaining a full chair five years later (which he held almost until his death in 1932), and in between these appointments he was elected to the Turin Academy of Sciences in 1891. He also held a post at the Military Academy in the town from 1886 to 1901. Married in 1887, he had no children. He was active in both University and Academy affairs, and in some other societies and journals in Italy and abroad.

Soon after graduating in 1880 Peano started publishing. His research interests lay within mathematical analysis (including Cantorian set theory, to which he became an important early adherent) and the foundations of geometry; in both contexts he came across the logic of this day and became a major contributor to it, applying it to various mathematical issues, especially the foundations of arithmetic. He contributed also to other areas of mathematics, especially geometry and mechanics, and was much interested in history and education in mathematics.

By the 1890s Peano was not only making important contributions of his own but also inspiring a distinguished school of compatriots. Their publications comprised papers and books in the usual way (and, in his case, also booklets). In addition, in 1891 he launched a journal, entitled *Rivista di matematica*, and from mid decade he also edited the *Formulaire* mentioned above; both publications continued until the mid 1900s. His principal publisher for this pair, and also his books and booklets, was the Turin house of Bocca; we note his début with them in §5.2.1.

The last title of Peano's primer, *Formulario mathematico*, was written in uninflected Latin, which became a principal interest during the last 30 years of his life. His mathematical researches (which up to then had almost always been written in French or Italian) declined considerably from this time, although he continued a strong interest in mathematical education. Plate 2, published here for the first time, shows him possibly in the 1910s, when he was in his fifties.

Although Peano published some substantial textbooks, research monographs and long papers, the majority of his 230 titles refer to short papers.

PLATE 2. Sketch of Giuseppe Peano in perhaps the 1910s. First publication; made available to me by Peano's grandson Agosto Peano.

He was an opportunist mathematician, finding a new result, say, or an ambiguity in an established proof; further, his attitudes to foundational questions (in particular, definitions) was unusually developed for a mathematician of his time. But he did not have the mentality of a Weierstrass or a Russell to pursue the consequences of these insights to their (logico-)mathematical conclusions.

As a result, while in his lifetime Peano gained and preserved a world-wide reputation as mathematician, logician and international linguist (§9.6.8), he soon became rather forgotten. Since his death in 1932 he has gained less attention than any other major figure discussed in this book. But two scholars have studied him notably: Ugo Cassina (1897–1964), with a selected edition of his *Works* (1957–1959) and an ensemble of articles which were gathered together into two books *1961a* and *1961b*; and H. C. Kennedy, who translated some of Peano's writings into English in the edition Peano *Selection*; (1973) and also produced a biography Kennedy *1980a*. In addition, two events have led to commemorative volumes: Terracini *1955a*, on the occasion of the opening of a new school in Cuneo on the centenary of his birth; and Peano *1986a'*, the proceedings of a meeting organised by the University of Turin and held in the Academy in 1982 to celebrate the 50th anniversary of his death. Among other noteworthy literature is a volume produced by three Italian scholars (Borga and others *1985a*), and also Rodriguez-Consuegra *1988b* and *1991a*, ch. 3.

The total writings of Peano and his followers run into thousands of pages; further, Peano himself repeated certain theories with evident enthusiasm which however becomes tiresome for readers. So the account given in this chapter is *very* selective; many changes of notation and presentation are not rehearsed, and some features are described from an important text or context which however may not mark their début. In line with his usage, their word 'class' is normally adopted whatever kind of collection is involved, although the phrase 'set theory' is retained. Some of his notations used square brackets; his and mine are distinguished by context.

Prior to Peano's *entrée* into logic the subject had received a wide range of studies in Italy throughout the 19th century (Mangione *1990a*). However, they were almost entirely non-mathematical in character, and Peano himself seemed not to know much of them; and the importance of his achievements and their consequences has largely obliterated them from memory.

5.2 FORMALISING MATHEMATICAL ANALYSIS

5.2.1 *Improving Genocchi, 1884.* One of Peano's teachers at Turin was Angelo Genocchi (1817–1889). Trained and practising as a lawyer, he began to study mathematics seriously only when he was in his mid thirties,

around 1850. He held chairs in algebra and then analysis at the University of Turin, and in this latter capacity he gave an excellent lecture course in mathematical analysis. The publisher Bocca wanted to have a written version, as there were very few books of that level in the subject in Italian; Genocchi was not minded to produce the text, but he agreed to Peano's offer to do so. However, when the volume on 'the differential calculus and principles of the infinitesimal calculus' appeared as Genocchi *1884a*, it carried the explanation 'published with additions by Dr Giuseppe Peano' on the title page, which did not please the senior author at all; so he placed in the journal literature a disclaimer of responsibility for the book. Peano maintained, however, that he had been authorised to prepare the book (see, for example, the preface of his *1887a*, a successor study of 'geometrical applications of the infinitesimal calculus'), and documentary evidence has borne him out (Cassina *1952a*). He was then in his mid twenties.

Comparison between the summary lecture notes of Genocchi and the book, made in Bottazzini *1991a*, shows that the text basically followed Genocchi's intentions and content as taken down by Peano and others, even though the words were in Peano's hand, and that his own contributions were confined to the 'Annotations' (Peano *1884a*). Rather naively, he placed them at the head of the book; they constituted a fine contribution. Later the book gained the honour of a German translation (Genocchi *1898–1899a*), which included also some of Peano's later writings on logic and analysis (§5.3.8); Genocchi had then been dead for a decade.

Genocchi's text looks like a fairly standard analysis textbook of that time; Cauchyan in its basic cast but with Weierstrassian input in various important respects. The normality included no historical remarks (not even to explain the point just made) or references, and very few diagrams. He covered all the basic theory of differential and integral real-variable calculus, together with a limited treatment of the theory of functions and of infinite series; many elementary special functions were worked through as exercises or examples. Unusual was ch. 6 on basic complex-variable analysis, although he eschewed contour integration; on occasion he also drew on determinants, especially Jacobians and Hessians.

In a few respects the book was perhaps a little below par; for example, in its treatment of limits and upper limits. Peano's own annotations show that he was already aware both of difficulties in Weierstrassian analysis and of several other current developments in mathematics; he also displayed his knowledge of historical writings. Three features stand out. Firstly, in the opening annotation on numbers and quantities (Peano *1884a*, vii), his citations included Cantor's paper *1872a* on trigonometric series and Dedekind's booklet *1872a* on irrational numbers; a little later (p. xi), in connection with upper limits, he mentioned Cantor again, and also Heine *1872a* (§3.2.2–4). Secondly, as a contribution to the ever-expanding world of functions he gave on p. xii the first symbolic representation of the characteristic function of the irrational numbers, which Dirichlet had proposed as a pathological case in 1829 (§2.7.3):

$$\lim_{n \to \infty} [\phi(\sin n!\pi x)], \text{ where } \phi(x) := \lim_{t \to 0} \left(\frac{x^2}{x^2 + t^2} \right). \qquad (521.1)$$

Thirdly, among a suite of remarks on functions of several variables (where the provability of theorems was deepening the level of rigour in analysis), he stressed on p. xxv Genocchi's example of one where mixed derivatives were unequal (that is, $f_{xy} \neq f_{yx}$), followed by the example

$$f(x, y) := xy/\sqrt{x^2 + y^2}, \text{ with } f(0,0) = 0, \qquad (521.2)$$

which took discontinuous first-order derivatives at (0,0) and so lost its Taylor expansion (p. 174). The young man touched upon a good range of problems in analysis, and showed his awareness of current researches.

5.2.2 *Developing Grassmann's 'geometrical calculus', 1888.* Peano's contacts with logic and current algebras were publicised in his next book of the following year, a study *1888a* of Hermann Grassmann's *Ausdehnungslehre*. As we saw in §4.4.1, this theory was a novel algebra in which means were given of generating lines, planes and volumes, and types of combination of them. Peano's version, while not free from unclarities (for example, in the interpretation of combination of letters), helped to continue the spread of these ideas, and their embodiment in vector algebra and analysis and in linear vector functions.[1] In a review in the *Jahrbuch* his compatriot Gino Loria *1891a* welcomed its use of logic (and set theory) as a contribution to this 'so interesting branch of the exact sciences'.

After presenting the basic 'geometric formations' (ch. 1), Peano described the three 'species of formation': lines AB (in that order) and their multiples, lines BC generating planes (especially triangles ABC as P moves along BC), and planes similarly generating tetrahedral volumes ABCD (chs. 2–4). Each formation was signed, and thereby became vectorial, with left- and right-hand conventions imposed to define positives and negatives. The rest of the account developed various aspects of the definitions (chs. 5–7), ending with related parts of the calculus such as vector derivatives and integrals, and vector spaces.

Thus Peano did not follow the treatment of vector algebra and analysis that has become standard fare since; indeed, he did not even present the vector product (which in Grassmann was called 'outer multiplication'). He also followed a more axiomatic style than Grassmann himself had used, a

---

[1] On Peano's version of the theory, see Bottazzini *1985a* and Freguglia *1985a*, 177–182. On the general background of vector algebra, see Crowe *1967a*, ch. 3.

feature which was to grow in importance in his work from this book onwards. For after mentioning Grassmann in his title he referred to 'the operations of deductive logic', and in his preface he stated that his reading had included also logicians such as Boole, Jevons, Schröder and Peirce (not MacColl, however); and in an introductory chapter he outlined those operations.

While Peano's basic ideas drew largely upon the algebraic tradition, he also used and indeed popularised some notations of Grassmann. He worked in art. 1 with 'classes' ('*classi*'), including the universal 'all' ⊘ and the empty 'null' ○ and combined them by the (Grassmannian) symbols '∩' and '∪', denoting operations 'called in logic *conjunction* [. . . and] *disjunction*'; the latter operation was also represented by concatenation. Inclusion was denoted by ' < ', so that 'A < B' stood for 'every A is a B', with the converse writing 'B > A' also available. 'The signs < and > can also be read *less than* and *greater than*'; in later writing he would abandon this analogy. Complementary classes were denoted by the negation sign or by an overbar: ' − X' or '$\overline{X}$'.

In a style recalling Schröder *1877a* in its use of duality (§4.4.2), Peano laid out in art. 2 the basic 'identities', such as

$$A \cap - A = O \text{ and } A \cup - A = \oslash , \text{ and } AB = BA \text{ and } A \cup B = B \cup A.$$

$$(522.1)$$

Like Grassmann, he stressed commutativity, distributivity and associativity where applicable. Among major theorems in art. 3 he produced Boole's expansion formula (255.5) for a function of two variables:

$$f(X, Y) = f(\oslash , \oslash)XY \cup f(\oslash , O)X\overline{Y} \cup f(O, \oslash)\overline{X}Y \cup f(O, O)\overline{\overline{X}\overline{Y}}.$$

$$(522.2)$$

For the formulation of classes from 'numerical functions' $f(x, y, \ldots)$ Peano used notations such as '$x: [f(x) = 0]$' for the class of zeroes of f, where 'the sign : may be read *such that*' (art. 4). He did not specify whether these classes were formed in the part-whole or the Cantorian sense, with which he was beginning to become familiar and which would come soon to take a central place in his work.

Peano laid out propositions as equations, that is involving equality of classes and/or equivalence between propositions; and in art. 6 he stated duality principles such as that '*every logical equation transforms itself into another equal one, where are changed the two members and the signs* = , < , > *which join them into* = , > , < '. The balance between equivalence and implication was to change over the years. In his list of propositions he did not offer any axioms.

Peano followed Boole in associating ' = ○' and ' = ⊘' with falsehood and truth of propositions: explicitly though not too clearly, '○ expresses an *absurd* condition. ⊘ expresses the condition of *identity*'; thus, for example, 'Some A are B' was symbolised '−(AB = ○)' (art. 8). Not much of this machinery was used in the main text of the book, but he imitated Grassmann's use of ' = 0' to write, for example, 'the point A lies in the plane α' as 'Aα = ○' (ch. 2, art. 3).

After publishing the book Peano continued to collect information on its various topics, not only logic but also vector mathematics; he annotated his own copy of the book with references and commentators, including Mac-Coll and Frege among logicians (Bottazzini *1985a*). Symbolic logic and set theory grew in importance rapidly for him in the ensuing years.

5.2.3 *The logistic of arithmetic, 1889*

Mathematics has a place between logic and the experimental sciences. It is pure logic; all its propositions are of the form: 'If one supposes *A* true, then *B* is true'. Peano *1923a*

Peano's next publication in this area, a short booklet of xvi and 20 pages in Latin, has become one of his best-known works; however, it may not have been well known at the time, for it is now difficult to find. In *Arithmetices principia novo methodo exposita* (Peano *1889a*) he increased the role of logic in mathematics (or, as Loria *1892a* put it in the *Jahrbuch*, he wished to show how logic could help mathematics) with a more extended survey of logical notions in (the xvi pages of) the preface.

In various ways Peano's presentation of logic followed that of *1888a*: citation of the same literature (and also MacColl), and a list of some basic 'propositions of logic'. But there was no emphasis on duality of theorems, and some notions were introduced which moved his account away from the algebraic tradition. Above all he worked with the set theory of 'cl.^{mus} Cantor' so that, while he still used the word 'class', it now referred to objects with not part-whole but Cantorian composition; they contained 'individuals' in the sense of '$a \varepsilon b$ is read *a is b*' and also admitted the possibility that '$a \supset b$ means *the class a is contained in the class b*' (p. 28, no. 50). He seems to have had proper inclusion in mind, but his definition covers also the improper kind.

Peano defined the empty class Λ within the class K of classes as 'the class which contains no individuals'; but he reduced non-membership to a false proposition, which was also symbolised by 'Λ'! Thus, sadly (no. 49),

$$a \varepsilon \text{K} . \supset \therefore a = \Lambda \mathrel{\vcentcolon=} x \varepsilon a \mathrel{\vcentcolon=}_x \Lambda. \tag{523.1}$$

He used here universal quantification: 'If the propositions a, b contain the determinate quantities $x, y, \ldots$ [$\ldots$] then $\supset_{x, y, \ldots} b$ means: whatever be the $x, y, \ldots$, from proposition a one deduces b' (p. 25). However, he seems not to have noticed the anticipation by Peirce *1883a* (§4.3.7), and he did *not* develop here an *explicit* predicate calculus, nor introduce the existential quantifier. Further, his understanding of Cantor was not always secure: allegedly, if a class s contained as sub-class the class k, then if k was a unit class, it was also 'an individual' (that is, a member) of s (p. 28, no. 56).

At (523.1) Peano also mentioned the dual universal class, V; but he promised to make no use of it, in contrast to the ready deployment of the Grassmannian symbol '⊘'. He maintained this position in all of his writings on logic, thus laying himself open to the difficulties of an unrestricted universe which we noted in §2.5.4 concerning Boole, and also some paradoxes of set theory.

To a greater extent than in his previous book, Peano stressed implications rather than equivalences between propositions; in particular, '*Theorema* (Theor. or Th)' took the form $\alpha \supset \beta$, where '⊃' denoted 'one deduces' from the '*Hypothesis* (Hyp or more briefly "Hp")' α to '*Thesis* (Thes. or Th.)' β, where α and β were propositions (p. 33); these terms were used also in his later writings, sometimes (including here) with other abbreviations.

As these quotations show, Peano was quite liberal in using the same symbols to denote classes or propositions. Parallels in connectives were also utilised, as we saw with '⊃'; among others, '∩' did double duty as the conjunction of propositions and as the intersection (not his word) of classes, with '∪' similarly doubling as inclusive disjunction and as class union. '−' was 'not' in all contexts (pp. 24, 27). The worst sufferer was '=': it covered 1) equality between classes, defined by the property that each one was contained within the other, with the consequent property that they contained the same members (p. 28, at no. 51); 2) equivalence between propositions, that each one implied the other (p. 25, no. 3); and 3) equality by definition, with the abbreviation '*Def.*' promised (p. 33) but not always delivered, so that definitions were not always clearly individuated. Quite often theorems were stated in terms of a class not being empty, a property expressed at the end of a symbolic line by '$- = \Lambda$'.

Peano used duality in a different way here: to take pairs of symbols in horizontal or vertical mirror image which represented in some way converse notions. He even introduced a functor '[ ]' called '*sign of the inverse*' (p. 28); for example, '∍' 'is read *the entities such that*', and served as dual to ε:

'Thus $\ni \alpha\, y . =: [x\, \varepsilon] . x\, \alpha\, y\, [\ldots]$ We deduce that $x\, \varepsilon \ni \alpha\, y = x\, \alpha\, y$'

$$(523.2)$$

(p. 29). His use of '∍' was striking; for example, '$\ni < u$' denoted the class of all real numbers less than u (p. 29). Curiously, he was not to use '∍' much again until *1900a* (§5.4.7).

This passage also exemplifies two other striking features of Peano's system: the well-remembered convention of dots to replace the use of brackets (p. 24), systematising the practise of predecessors such as Lagrange; and the undeservedly forgotten use of connectival variables, in which 'Let $x \alpha y$ be a relation between indeterminates x and y (e.g., in logic, the relations $x = y$, $x - = y$, $x \supset y$ [...]'. He also used the square brackets in mathematical contexts; for example, '[sin]' was the inverse sine function (p. 31).

After these preliminaries Peano presented his axioms for the class N of integers on p. 34 as follows:

1. $1 \varepsilon N.$ [(523.3)]

2. $a \varepsilon N . \supset . a = a.$ [(523.4)]

3. $a, b \varepsilon N . \supset : a = b . = . b = a.$ [(523.5)]

4. $a, b, c \varepsilon N . \supset \therefore a = b . = . b = c : \supset . a = c.$ [(523.6)]

5. $a = b . b \varepsilon N : \supset . a \varepsilon N.$ [(523.7)]

6. $a \varepsilon N . \supset a + 1 \varepsilon N.$ [(523.8)]

7. $a, b \varepsilon N . \supset : a = b . = . a + 1 = b + 1.$ [(523.9)]

8. $a \varepsilon N . \supset . a + 1 - = 1.$ [(523.10)]

9. $k \varepsilon K \therefore 1 \varepsilon k \therefore x \varepsilon N . x \varepsilon k : \supset_x . x + 1 \varepsilon k :: \supset . N \supset k.$ [(523.11)]

Some comments are in order. Firstly, in the preface of the booklet Peano devoted *separate* columns to logical and arithmetical signs, placing 'K' (for classes) in the former category. However, these axioms were a mixture in that (523.4–7) dealt with equality, 'which must be considered as a new sign, although it has the appearance of a sign of logic' (p. 34; compare p. 30); in later presentations he removed this quartet, declaring categorically that 'they belong to Logic' (Jourdain *1912a*, 281) and thus maintaining his distinction between the two kinds of theory. Secondly, the induction axiom (523.11) was stated in first-order form, with no quantification over K; and the universal quantification over x characterises it as of the strong form (in modern terminology) in involving all integers preceding x. The high status of induction recalls the textbook Grassmann *1861a* on arithmetic (§4.4.2), which indeed Peano cited at the head of his booklet. Neither he nor his immediate followers were to enter into such issues, nor the demonstrability of the existence or the uniqueness of the defined objects. Thirdly, he did not discuss the difference between the informal numbers used to enumerate the axioms (and many other contents of the booklet) and the "proper" numbers defined therein: in this regard the opening of (523.3), '1. 1', is striking. Finally, in stating that repeated succession always produces a novelty, axiom (523.10) amounts to an axiom of infinity.

These axioms also show another contrast with Peano's preface, on logic; main properties were listed, with *no* attempt made to axiomatise the calculus (for example, *modus ponens* was absent). It is not surprising that they were not sufficient to justify all the deductions made in his proofs. Apart from that, however, the treatment was impressively concise, passing though the basic arithmetical operations, the specification of rational and irrational numbers (although he did not attempt to rehearse any of the definitions discussed in §3.2.3–4) and elements of point-set topology, centred on the interior of a class. While he did not introduce propositional functions, in art. 6 he proposed 'the sign ϕ' as 'a *presign of a function on the class s*' to allow statement of ϕx of the members x of s, together with the '*postsign $x\phi$*' (for example, respectively $x + a$ and $a + x$). He gave a flavour of functional equations (§2.2.4) in noting the use of $\phi\phi$, $\psi\phi$, and so on.

In stressing the property

$$\phi x = \phi y . \supset . x = y \tag{523.12}$$

Peano mentioned Dedekind's 'similar transformation' (§3.5.2); so a comparison between these two works and their authors needs to be made. He had seen Dedekind's booklet by the time of writing his *Arithmetices*, for he referred to it in his preface. However, later he claimed that he found his axiom system independently, while also noting the 'substantial coincidence with the definition of Dedekind' (*1897a*, 243).[2] Their mathematical and philosophical aims are indeed similar, even down to the definition of a simply infinite system; but three differences are worth stressing.

Firstly, as the title of Dedekind *1888a* shows (§3.4.1), he sought to individuate numbers, whereas Peano took number as one of his primitive concepts and sought to present its main properties. Soon afterwards he made this point himself, concluding that 'the two things coincide' (*1891c*, 87–88), although his attached demonstration of the independence of his axioms by presenting a variety of interpretations of them (§5.3.3) did not lead him to notice that the system as a whole only defined progressions. Secondly, Dedekind transformed the principle of mathematical induction into other forms and examined its foundations with theorems on transformation; once again, Peano set it as primitive at (523.11). Thirdly, Dedekind claimed that arithmetic was part of logic, although he did not characterise logic in any detailed way; by contrast, Peano stressed the distinction

---

[2] This passage is constantly overlooked by scholars who assert that Peano acknowledged that his axioms came from Dedekind. Reporters include Bachmann *1934a*, 38, and later van Heijenoort *1967a*, 83 (citing Peano *1891c*, 93), and Wang *1957a*, 145 (citing Jourdain *1912a*, 273); but neither original source provides the evidence. For a detailed examination of Peano's treatment of arithmetic and also analysis, see Palladino *1985a*.

between arithmetical and logical notions and described both categories in detail (although his use of Cantorian set theory made the distinction less clear than he seemed to realise), and went further into mathematics by outlining some point-set topology. Overall, while his booklet had neither the depth (nor length) of Dedekind's of the previous year, it showed better sweep.

5.2.4 *The logistic of geometry, 1889.* For the next two decades, Peano's work was to be dominated by the development and application both of Cantorian point-set topology, and of set theory within his own logic. To a lesser extent he also treated the foundations of geometry; in the same year as the *Arithmetices* he 'logically expounded' upon it in another booklet, *1889b*.

This study was a valuable contribution to the clarifying of Euclidean geometry which was to lead to a large body of work by himself and by some of his followers (§5.5.4–5) and to culminate in Hilbert's famous essay of 1899 (§4.7.2); it was also to serve as a major influence upon studies of geometry among certain of his followers (Freguglia *1985a*). Little logic as such was presented, but much emphasis was laid on definitions, especially of geometric entities. ' = ' was given the usual hard work, including between both classes and propositions, 'identity' of points, and by definition (pp. 59–62). The set theory looked very Cantorian, with 'ε' used extensively; however, only Boole was cited for the 'principal operations of Logic' (p. 57), and the classes have to be understood in the part-whole sense.

Although Peano stressed at the beginning that he was dealing only with 'the fundamentals of the Geometry of position' (p. 57), at the end he indicated that study of the motion of a rigid body required 'the concept of *correspondence* or of *function*', which 'regards it [as] belonging to Logic' (p. 91). The distinctions between propositional functions, mathematical ones, and general mappings were to become important issues in the later development of log(ic)istic thinking, especially for Russell.

The axioms were laid out in a fully symbolic manner; even more than the *Arithmetices*, this was wallpaper mathematics. A long series of prosodic notes afforded explanation. A particularly interesting axiom (p. 64) asserted that ' "The class [**1**] *points* is not empty" ' ('*nulla*'), of which 'we shall not have occasion to make use' (p. 83); so a profound point on existential assumptions was seized but dropped. The machinery of construction was based upon taking lines as classes of points, based upon these initial definitions of lines relative to points a and b (p. 61):

$$a'b =: \mathbf{1} . [x \, \varepsilon] . (b \, \varepsilon \, ax) \text{ and } ab' =: \mathbf{1} . [x \, \varepsilon] . (a \, \varepsilon \, xb); \quad (524.1)$$

that is, the points x such that respectively b belonged to the line ax or a to xb, thus defining $a'b$ as the prolongation of the line ab beyond b to the right and *ab'* beyond *a* to the left (end points excluded). Proceeding to further definitions of this kind for points and classes, and classes and classes, he found properties for lines, planes (the class **2**) and spatial figures (**3**), in a spirit close enough to Grassmann to render surprising the absence of his name from the booklet.

In the following year Peano published in *Mathematische Annalen* in French a short note *1890a*, containing another of his most durable contributions: the space-filling curve, which, as he noted, underlined the importance of Cantor's discovery of the equinumerousness of the unit line and the unit square (§3.2.5). Cantor's influence is present in the construction also: like him, Peano used expansions of the coordinate of each point, and indeed in the ternary form which Cantor was later to use to define the ternary set (328.2) in 1883. The geometrical zig-zag representations of Peano's curve were produced later in Hilbert *1891b* (who also reviewed Peano's paper for the *Jahrbuch* in *1893a*) and E. H. Moore *1900a*.

5.2.5 The logistic of analysis, 1890. Of other papers by Peano of this time which bore upon mathematical analysis, the most noticeable was a long paper *1890b*, also published in French in *Mathematische Annalen*, on existence theorems for ordinary differential equations. The most striking feature of this paper was the strongly symbolic rendering of many of its results and proofs, like the geometry booklet. He began with 20-page 'first part' outlining his logical and set-theoretic machinery, and the dot convention. The elements of logic were presented in less detail than in the previous works (for example, no results of the propositional calculus were given at all), and again were not always clear; for example '$a = b$' stated of classes that they were 'identical' (although of propositions that they were 'equivalent'). As usual, no universal class was mentioned, and the disjunction of propositions was taken to define the truth of one or some of them (pp. 120–121).

On the other hand, the new sign 'ι (initial of ἴσος)' was introduced to represent 'equal to', and thereby relieve some of the strain on ' = '; 'thus instead of $a = b$ one can write $a \,\varepsilon\, \iota b$' (p. 130). More importantly, in consequence an individual was distinguished from its unit class: 'In order to indicate the class constituted of the individuals *a* and *b* one writes sometimes $a \cup b$ (or $a + b$, following the more usual notation). But it is more correct to write $\iota a \cup \iota b$' (p. 131). This was an important refinement to Cantorian set theory: while Cantor used both membership and inclusion, his sets were usually large ones, so that he did not emphasise this distinction.

Syllogistic logic was also affected: the forms

$$`a \supset b . b \supset c : \supset . a \supset c` \text{ and } `a \,\varepsilon\, b . b \supset c : \supset . a \,\varepsilon\, c` \qquad (525.1)$$

(of which the former would be the rule of *modus ponens* for him) 'are exact; but from premises $a \, \varepsilon \, b \, . \, b \, \varepsilon \, c$ one cannot draw consequences. One sees also more clearly that one must distinguish well the two signs ε and $\supset$' (p. 131).

Peano also uttered some tentative remarks on the relationship between mathematics and his logic. After introducing his main vocabulary he claimed that 'All propositions of any science can be expressed by means of these notations, and of words which represent the entities of this science. They alone suffice to express the propositions of pure Logic' (p. 123). However, categories may be conflated here; and an example soon occurs, in his discussion of the notion of function: 'The idea of function (correspondence, operation) is primitive; one can consider it as belonging to Logic. As an example taken from common language, let us put h = "homme", p = "le père de"'; but he went straight on to give the logarithmic and the sine functions as examples from analysis (p. 128). Not for a decade did he again assign logical status to functions in his writings on logic (after (548.1)).

The rest of the first part of the paper was largely devoted to point-set topology, including fully symbolic statements of some of Cantor's theorems. The main context for using of pairs of signs was for a function f and its inverse(s) $\bar{f}$, and class membership ε and abstraction $\overline{x \, \varepsilon}$; and Peano also proposed the notation 'b/a' for a function which set a correspondence between a class a and a class b (pp. 125–126). An interesting but unhappily notated distinction lay between the definitions of the classes

$$`\bar{f}y = \overline{x \, \varepsilon} \, (y = fx)` \text{ and } `f'y = \overline{x \, \varepsilon} \, (y \, \varepsilon \, fx)`; \qquad (525.2)$$

the second specified the domain of f and the first the subclass of xs mapping to a given y of its range (p. 130). He mentioned that the second class was already used as at (524.1) in his booklet on geometry, and that the first could be stated in terms of it; but the use of the prime clashed with its role for derived functions in mathematical analysis, and he did not deploy this distinction later.

The second part of the paper ran through an existence proof for continuous functions. At one point Peano had to define a function $f(t)$ over $[0, 1]$ by assigning to it values chosen arbitrarily from certain classes of numbers when t belonged to a certain sub-class of rational numbers and to its complement. 'But as one cannot apply an infinity of times an *arbitrary* law with which to a class a one makes correspond an individual of this class, one has formed here a *determinate* law with which one makes correspond to each class a, an individual of this class under convenient hypotheses' and then he gave some rules (pp. 149–151). However, neither here nor later did he develop this clear insight into the problems of selection of members from an infinity of classes; so once again this opportunist mathematician missed a lovely opportunity—in this case, to be the father of the axioms of choice (§7.2.5–6).

5.2.6 Bettazzi on magnitudes, 1890. Concurrently with Peano's logical début, in Pisa Rodolfo Bettazzi (1861–1941) became interested in Grassmann's algebras, especially when a prize problem on them was proposed by the *Accademia dei Lincei* in Rome in 1888. He responded with a long essay which was crowned the following year and published as *1890a*, comprising two parts and an appendix.

Like his inspirer, Bettazzi started none too intuitively, with an operation S applied to members of an ensemble of objects $A, \ldots, L$ to produce the object M. For two initial objects he proposed an operation and its converse D, called 'divergence', and like Grassmann shortly before (522.1):

$$\text{if } S(A, B) = C, \text{ then } D(C, A) = B; \qquad (526.1)$$

he examined properties such as commutativity and ordering (pp. 3–24). He then applied his algebra to a wide range of cases: classes of magnitudes in one and several dimensions, including those containing infinitesimals. Then in a second part, on 'Number and measure', he treated in detail integers, rationals, irrationals (Cantor's on p. 88 but Dedekind's in the appendix on pp. 175–176) and worked through one and several dimensions to hypercomplex numbers. As Giulio Vivanti *1891a* pointed out in a long review in the *Bulletin des sciences mathématiques*, the procedures drew upon formal definitions of objects from given properties rather than real definitions in the opposite direction; but then other objects might possess the same properties.

Bettazzi made use of parts of Cantorian set theory, but he explicitly avoided the 'ultra-infinite numbers' (p. 150). However, in a nearby discussion of defining infinitesimals he spoke of 'conveniently limiting the arbitrariness of the selection of the primary magnitudes' (p. 147), which brought him close to the axioms of choice, like Peano above. Some years later Bettazzi *1895a* considered an axiom for infinite selections but rejected it, maybe because in 1892 he had moved to the Military Academy in Turin and so became personally involved with Peano and his followers. To that school we now turn.

5.3 THE *RIVISTA*: PEANO AND HIS SCHOOL, 1890–1895

5.3.1 The 'society of mathematicians'. We come now to the time when Peano launched his journal in 1891. Initially called *Rivista di matematica*, it was also known as *Revue des mathématiques* from its fifth volume (1895); it concluded three volumes later in 1906, with the last one entitled '*Revista de mathematica*', in uninflected Latin. I shall always refer to it as '*Rivista*'.

He was the editor, and operated without a named editorial board; as usual, the publisher was Bocca. Starting with 272 pages, the volumes decreased to around 180 pages each; they appeared in monthly signatures. Most articles were less than 15 printed pages, and normally were written in Italian, although the later volumes carried quite a proportion in French. To help produce them he bought a printing press, and even operated it himself on occasions. In *1916a* he discussed mathematical typesetting, especially the practise of placing all symbols along the line, a feature of his own notations which aided their widespread acceptance.

Considerable attention was paid to set theory and mathematical logic, including some papers commenting upon the *Formulaire* of the main logico-mathematical results which Peano also edited (we note its début in §5.4.1). Some translations were made, including two of Cantor's most important papers: *1892a* on the diagonal argument, and (only) the first part *1895b* of his final paper (§3.4.6–7), which appeared as Cantor *1892c* and *1895c* respectively. The journal also took papers in mathematical analysis, geometries and algebras, and some material on history and education; there were frequent book reviews, and occasional sets of problems.

Peano did not work alone: at this time he also led a school of talented mathematicians and philosophers, whom he was to describe later as a 'society of mathematicians' (§5.3.5), to develop his programme, especially its set-theoretic and mathematical aspects. They contributed much material to these publications as well as books of their own and papers elsewhere (Kennedy *1980a*, ch. 12). All born between the late 1850s (as was Peano himself) and the early 1870s, several encountered him initially as undergraduates at Turin, and most served at some time as his assistant. They were sometimes humorously called 'the Peanists'. Let us mention the principal ones.

Cesari Burali-Forti (1861–1931) is now famous for his paradox (not how he construed it, as we shall see in §6.6.3); in addition, he was author of several other valuable contributions (§5.3.7, §5.5.3) and, also following Peano, he worked on Grassmann's theory. Mario Pieri (1860–1913) and Alessandro Padoa (1868–1937) specialised in set theory and geometry, and extended their master's sensitivity to definitions (§5.5.4–5). Bettazzi, just noted, contributed to set theory, with an especial interest in finite classes; so did Vivanti (1859–1949), more independent and critical a contributor than the others, who in addition took a great interest in history. Giovanni Vailati (1863–1909) and Giovanni Vacca (1872–1953) are also notable for their historical knowledge. With Vacca, this included mathematical induction, and his second career as a sinologist led him to maintain contact with Peano in later years over languages in general (see Peano's *Letters* to him). Vailati concerned himself with philosophy (including Peirce's) and education, and for him there are editions of both *Works* and *Letters*.

5.3.2 *'Mathematical logic', 1891.* Peano launched his *Rivista* with two papers on the subject to which he gave the name that it still carries. The first one, *1891a*, outlined the 'Principles of mathematical logic' in the first ten pages of the journal; its successor, *1891b*, presented a suite of 'Formulae' in a later issue of the opening volume. The first paper was quickly translated in *El progreso matemático* as Peano *1892b*, as part of the Spanish interest in logic noted in §4.4.4; Reyes y Prósper *1893a* reported there on the new 'symbolic logic in Italy', in contrast with 'the Baltimorean logic' of the Peirce school.

Neither paper by Peano contained fundamental novelties of notion or notation, but both exhibit interesting details. The 'Principles' gave a special emphasis to Boole's index law (253.3), stressing that 'This identity does not have analogy in algebra' (*1891a*, 93); but he did not mention Boole at this point, and in his historical note at the end (p. 100) he referred to Jevons's name (262.2) 'law of simplicity'. In other notes he reported that *The laws of thought* was 'rare in Italy' (p. 101); and for the first time he cited Frege there, in connection with symbols for implication (compare §5.4.4). Cantor was named only in another note, but 'ε' clearly denoted Cantorian membership, especially in art. 3 on the 'Applications' of logic to arithmetic.

At its start Peano presented the 'Formulae' as a catalogue of the 'identities of Logic' and stressed the place of definitions; in the latter context he not only introduced but also used the notation '[Def.]' at the end of a symbolic line to mate up with the ' = ' in the middle somewhere and so relieve the strain on that hapless symbol (*1891b*, 103). He also introduced '$(\begin{smallmatrix}a\\x\end{smallmatrix})p$' to denote the proposition or formula obtained by replacing a constituent x of the proposition or formula p with a (p. 104, with extension to several simultaneous substitutions); this was useful especially in explaining steps in proofs. Further, '[Pp.]' was located like '[Def.]' to indicate the status of a 'primitive proposition' of a logical system, although in this category he embraced also rules of inference; he also did not give any indication of the means used to determine the primitive status. But he increased the axiomatic flavour of his approach, a feature strengthened by a very systematic numbering of every symbolic line and the use of these numberings to enclose within square brackets the lines of derivation of a given theorem. All these features were to be adopted by Whitehead and Russell.

This time the treatment of classes was brief, and placed at the end; but the Cantorian sense was clearly indicated by Peano's definition of 'ε': 'we shall write $x \varepsilon s$ to indicate that x is an individual of the class s'. In a reversal of roles from the *Arithmetices* (§5.2.3), equality between two classes was defined by the property of possessing the same members, with the consequent property that each one was contained within the other (p. 110). Rather casually, he used without explanation his subscript notation to indicate universal quantification in some succeeding results; he did not return to the matter in three pages of 'additions and corrections' made to the paper later (pp. 111–113).

However, another detail of note slipped in here; the exclusive disjunction of propositions a and b (p. 113):

$$\text{`}a \circ b = a - b \cup b - a\text{'} \quad \text{[Def.].} \tag{532.1}$$

Peano acknowledged it from Schröder, and later that year he published in the *Rivista* a review *1891d* of the first volume and the first part of the second volume of Schröder's lectures. He concentrated on the mathematical features, 'I being incompetent' on the philosophical side (p. 115); but even then his treatment was somewhat preliminary, for on Schröder's 22nd Lecture of individuals 'I do not intend now to dwell' (*'fermarmi'*) (p. 121). Again, Schröder worked with the part-whole relation (444.1) 'subsumption' between collections rather than with Cantorian set theory: for Peano 'I indicate the *same* relation with $a \supset b$', noting that by contrast Boole 'retained as fundamental the concept of equality' (p. 115, italics inserted) but not discussing the more refined machinery which Cantor had provided. He also merely recorded Schröder's (and Peirce's) algebraic way of handling universal and existential quantification (he did not use these terms). And while pointing out the ambiguity in the algebraic tradition that '(Root of a given equation) = 0' could indicate either that the equation had the sole root 0 or that it had no roots (p. 116), he did not clarify the status of nothing-like "things" in the tradition which he had adopted. 'The Algebra of Logic is now in the course of formation' (p. 121); but so was his own mathematical logic. Schröder's views on their differences are aired in §5.4.5.

5.3.3 *Developing arithmetic, 1891.* In the *Rivista* Peano and his colleagues gave treatments of various branches of mathematics. Mostly he deployed Cantorian set theory, although some features of logic and of definitions were also brought out.

In a pair of lengthy 'Notes' on integers Peano *1891c* used a functorial device '$a \backslash b$' with 'α' as operator, which mapped any member x of class a to some member $x \alpha$ of class b. This notion replaced (and clarified) the class-relation 'a/b' (§5.2.4). He presented his axiomatisation of integers in a more symbolic form (also removing axioms (523.4–7) as not specifically arithmetical), so that they read:

$$\text{`}1 \, \varepsilon \, N, \, + \, \varepsilon \, N \backslash N, \quad a, b \, \varepsilon \, N \, . \, a + \, = b : \supset . \, a = b, 1 - \, \varepsilon \, N\text{'} \tag{533.1}$$

$$\text{and `}s \, \varepsilon \, K \, . \, 1 \, \varepsilon \, s \, . \, s + \supset s \, . \, N \supset s\text{'}. \tag{533.2}$$

He proved the independence of the last three axioms (concerning the first two, 'There can be no doubt' . . .); for example, the class of all integers, positive and negative and including 0, satisfied the first three axioms but not the fourth one (pp. 93–94).

Zero was allowed into the story because Peano had also proposed these definitions of 0 and 1:

$$\text{'}s\,\varepsilon\,K\,.\,\alpha\,\varepsilon\,s\backslash s\,.\,\alpha\,\varepsilon\,s:\mathbf{C}\,.\,a\,\alpha\,0 = a\text{' and ditto '}a\,\alpha\,1 = a1\text{'}\qquad (533.3)$$

(pp. 91, 88); '$a\,\alpha\,0$' was to be read '$a(\alpha\,0)$' (p. 89), so that they stated that $a + 0 = a$ and $a + 1 = a + $. However, the lack of quantification over a and α rendered them rather unclear; and in any case they involved a vicious circle, especially relative to the assumed 1 in (533.1). He touched on some other aspects of arithmetic, including a further meshing of the distinction between logic and arithmetic with an inductive sequence of definitions of numbers of members of a class u, associating 0 with the empty class and defining 'num a' of a class a as one up from that of $(a - \iota x)$, where x was one of its members (p. 100): maybe he was following Boole (§2.5.6). He also treated real numbers, where he symbolised Dedekind's definition (p. 105). He also introduced on p. 101 a valuable definition: the 'sum' ('*somma*') $\cup$ 'k of a class u of classes:

$$\text{'}u\,\varepsilon\,\mathrm{KK}\,.\,\mathbf{C}\,.\cup\text{'}u = \overline{x\,\varepsilon}\,(x\,\varepsilon\,y\,.\,y\,\varepsilon\,u\,.\,\blacksquare\, = \Lambda)\,\text{Def.'}.\qquad (533.4)$$

By 1892 the mass of symbolised theories had grown sufficiently critical for a 20-page supplement to be published with the April issue of the *Rivista*, cataloguing, in order, 'Algebraic operations', 'Whole numbers', 'Classes of numbers', 'Functions' and 'Limits'. Several notes in that and the succeeding volume discussed the various sections.

5.3.4 *Infinitesimals and limits, 1892–1895.* Around this time several *Rivista* authors considered the legitimacy of infinitesimals. In a short note Peano *1892a* developed Cantor's cryptic argument for the impossibility of these worrying objects (§3.6.3). Starting out from the notion of 'segment' (including end points) u, he defined u to be infinitesimal relative to another segment v if $Nu < v$ for any finite integer N; then he deduced from

$$(\infty + 1)u = \infty u \text{ and } 2\infty u = \infty u \qquad (534.1)$$

that ∞u could not be 'terminated', contradicting the definition of a segment. This does not get us much further; and in surveying the literature he referred only to recent discussions in the *Rivista* and not to German material mentioned in §3.6.3. Further, as Vivanti *1893a* pointed out in a *Jahrbuch* review, since Peano defined his $(\infty + 1)u$ as the limit of $(nu + u)$, it equalled ∞u rather than secured a truly Cantorian ωu, and so the argument failed.[3] However, the notion of segment was put to excellent use in a further symbolic rendering Peano *1894c* of the foundations of geometry, where again definitions and set theory dominated over logic as such.

Two other publications of that year, written in French, advanced Peano's concerns more considerably. One was a lengthy study *1894a*, published in the *American journal of mathematics*, of 'the limit of a function'; here he examined the significance of distinguishing limits from upper limits and least upper bounds, a feature of Weierstrassian analysis then gaining considerable attention (Pringsheim *1898a*). Once again he laid out his symbolic repertoire, indeed, he subtitled the paper 'Exercise in mathematical logic'. While the proofs were not fully formalised, he laid out the basic logical connectives and properties of classes, and explicitly described some steps in derivations (for example, *1894a*, 231–235). For fuller details he referred the reader to the *Formulaire* and to a recent book by Burali-Forti (p. 229). Before we consider these works, however, we note their precursor, which was another publication of 1894.

5.3.5 *Notations and their range, 1894.*   This item was again a booklet (curiously, with no publisher named on the title page): Peano *1894b*, entitled *Notations de logique mathématique*. The work was subtitled 'Introduction to *Formulaire de mathématique* published by the *Rivista di matematica*' (art. 1).

Peano began by stating that 'Leibniz announced two centuries ago the project of creating a universal script', an anticipation which he and some of his followers were to become fond of recalling in the opening sentences of their general writings on logic. (Note the contrast with Schröder and Frege, who invoked Leibniz's vision of a *calculus* (§4.4.2, §4.5.2).) In 52 pages he covered much of the ground already considered above, starting with 'classes' and their 'relations and operations' (arts. 2–7): (half) open and (half) closed intervals between the values a and b were distinguished by the nice notations

$$\text{`}a - b\text{', `}a \vdash b\text{', `}a \dashv b\text{' and `}a \dashv\vdash b\text{'.} \qquad (535.1)$$

Then followed 'Properties of the operations of logic', in a presentation which recalled his treatment in *1888a* of Grassmann in that the properties were laid out in dual pairs (§5.2.2); but this time it was classes *of the Cantorian kind* which possessed them (art. 8). Only then there followed a brief statement of the analogous forms for propositions, in which the deduction '$a \supset b$' was again interpreted in terms of proposition b being a

---

[3] Cantor corresponded against infinitesimals with Vivanti at this time, and with Peano two years later, in letters which appeared in the *Rivista* as Cantor *1895a*; for private letters to Peano on this topic, and also on the Italian translation mentioned in §5.3.1, see Cantor *Letters*, 359–370.

consequent of proposition *a* (art. 9). The properties, called 'identities' rather than 'axioms' (art. 8), included Boole's law (253.2) and its dual for union, 'called by Jevons "the law of simplicity"' (§2.6.3); and 'interesting properties of negation' due to De Morgan, whose verbal formulation (§2.4.9) he cited and also gave (not for the first time) the symbolic formulation

$$\text{'6. } -(ab) = (-a) \cup (-b) \quad 6'. \; -(a \cup b) = (-a)(-b).\text{'.} \quad (535.2)$$

For the sake of duality he introduced the universal class, V, as $-\Lambda$, and also the dual pair of relations; but once again he made no use of them. The dot convention was explained in detail, followed by many examples from arithmetic and the propositional calculus (arts. 10–12).

Next Peano devoted a part to '*Variable letters*', giving examples of bound ones (he used no term) such as x in $(fx)_{x=a}$, and then rehearsing again universal quantification (arts. 13–14). He also recalled how properties between classes could be expressed in terms of quantified propositions about their members: for example, the definition (523.2) of the empty class again, and improper inclusion between classes *a* and *b*

$$\text{'}a \supset b . = \; : x \, \varepsilon \, a . \supset_x . x \, \varepsilon \, b\text{'}. \quad (535.3)$$

He noted again the shortcoming (525.1) of syllogistic logic in not distinguishing ε from $\supset$; regarding the (over-worked!) latter symbol he now read it between propositions as producing a hypothetical proposition (arts. 15–16).

Class abstraction was still denoted by '$\overline{x\,\varepsilon}$', as in *1890b* (§5.2.5) rather than '$\ni$' of *1889a* (§5.2.3); but Peano introduced the symbol 'p_x' for 'a proposition containing a variable letter x'. Again, as in §5.2.3, he did not initiate explicitly a calculus of propositional functions, for p might contain other free and bound variables. When extending the abstraction to two variables in a proposition p he only stressed the difference between '$\overline{x\,\varepsilon p}_{x,y}$', '$\overline{y\,\varepsilon p}_{x,y}$' and '$x,y\,\varepsilon p_{x,y}$', where the latter case denoted a class of ordered pairs (art. 17).

The part on '*Functions*' was dominated by mathematical ones; Peano preferred 'fx' to '$f(x)$', dismissing fears of misinterpreting it as a product by appealing to mathematician predecessors such as Lagrange for this usage (art. 23). In that tradition brackets were used sometimes but not always (they were deployed in §2.2.2–3).

A notation analogous to the functorial device '$a \backslash b$' of Peano *1891c* between classes *a* and *b* at (533.1) was introduced: '$b \, f \, a$' mapped members of *a* to members of *b*, so that, for example' sin ε q f q' stated that the sine function went from real numbers to real numbers (art. 23). Inverse functions were now given inverse notations such as f for 'f'; however, perhaps with this typesetter in mind, 'we shall make little use of it'. He also repeated here from *1891c* his treatment of the number num u of members of a class u (art. 19); 'num' was a function, with an inverse 'num a [which] signifies "class of objects in number of a"', which rather muddled the class a with its members (art. 27).

In the part on 'Relations' Peano stressed that any relationship between 'two objects' constituted a relation (art. 30), and he followed his symbolic treatment (533.1–2) of the axioms for integers in using 'α' as his symbol for a general relation, with '$\alpha\,|$' as its *'inverse'* and '$-\alpha$' as its *'negative'*. He reduced α to a function by decomposing it into ε ('is') and a function ϕ; the special case of equality was rendered as $\varepsilon\,\iota$, where '$\iota[\ldots]$' signifies *equal'* (art. 31). Then he rehearsed his notion (525.2)$_2$ '$f'y$' of the inverse of a function f with respect to a member y of its range, but now formulated in terms of a relation α and with a new notation free from primes:

$$\text{'}y\,\alpha\,|x = x\,\alpha\,y\text{'} \quad \text{Def.?.} \tag{535.4}$$

In art. 33 he even (and in this order!) defined, and specified the existence of, the range and domain u and v of α by introducing two new kinds of functor: for example,

$$\text{'}x\,\alpha\uparrow v. = \,:y\,\varepsilon\,v\,.\mathfrak{I}_y\,.x\,\alpha\,y\text{'} \quad \text{and} \quad \text{'}u\downarrow\alpha\,y. = \,:x\,\varepsilon\,u\,.x\,\alpha\,y.- =_x \Lambda\text{'}.$$

$$\tag{535.5}$$

Rather sloppily, he verbalised them as 'each v' and 'some u' respectively.

5.3.6 *Peano on definition by equivalence classes.* Somewhat tardily, Peano considered symbolising mathematical theories in general (arts. 34–35), and then definitions, having avoided '[Def.]' hitherto even in contexts such as (535.4) where it seems to be in play (arts. 36–42). But his remarks were important, for he gave his version of what became known as his theory of 'definition by abstraction'. The phrase may have come to him under the influence of Cantor, for in art. 39 he gave Cantor's definition of transfinite cardinal and ordinal numbers by abstraction (§3.4.7) as examples. At all events, his theory appeared in art. 38, after some examples of definition under hypothesis:

> There are ideas which one obtains by abstraction, and [...] that one cannot define in the announced form. Let u be an object; by abstraction one deduces a new object ϕu; one cannot form an equality
>
> $$\phi u = \text{known expression}, \qquad [(536.1)]$$
>
> for ϕ is an object of a different nature from all those which one has considered hitherto. Thus one defines equality, and one puts
>
> $$h_{u,v}\,.\mathfrak{I}:\phi u = \phi v\,. = \,.p_{u,v} \quad \text{Def.} \qquad [(536.2)]$$

where $h_{u,v}$ is the hypothesis on the objects u and v; $\phi u = \phi v$ is the equality that one defines; it signifies the same thing as $p_{u,v}$, which is a condition, or relation, between u and v, having a well known meaning.

Peano then specified the three required conditions upon this equality, which he called '*reflexive*', '*symmetric*' and '*transitive*', and symbolised with respect to p respectively as

$$\text{'} p_{u,u} \text{ is true', '} p_{u,v} \supset p_{v,u} \text{' and '} p_{u,v} \cdot p_{v,w} \cdot \supset \cdot p_{u,w} \text{'.} \qquad (536.3)$$

This theory is a form of definition known now as definition by equivalence classes across the collection of objects $\{\phi u\}$. Peano was quite clear that it was not nominal in form: using as 'the new object' the example of the upper limit $l'a$ (introduced in art. 19) of a class a of rational numbers, 'we will not say that it is $l'a$', but he proceeded to define '$l'a = l'b$' (art. 39). He also affirmed a nominal interpretation of identity: 'The equality $a = b$ always has the same meaning: a and b are identical, where a and b are two names given to the same thing' (art. 40). In some respects his procedures resemble those of Bettazzi *1890a* (§5.2.6), but he did not cite it.

After further discussion of definitions and remarks on '*Demonstrations*' (arts. 41–44), Peano concluded that 'the problem proposed by Leibniz is thus resolved', and introduced the *Formulaire* as a depot for 'the collections of propositions on the different subjects of mathematics that we will receive, and all the corrections and complements that will be indicated to us'. The first 'volume' (or edition, really) appeared in 1895 (§5.4.1); before that, however, one of his followers popularised mathematical logic for the Italian public.

5.3.7 *Burali-Forti's textbook, 1894.*   The first textbook in the new subject was prepared by Burali-Forti, Peano's assistant at the time, and at 33 two years his junior. Based on a lecture course in the University of Turin, it was published in the well-known series 'Manuali Hoepli', and contained 158 small pages. Burali-Forti *1894b* naturally followed the master, at times down to small details or examples—in particular, he must have seen a version of Peano's *Notations*, which seems to have been published later in the year—but it seems most unlikely that Peano had plagiarised from him. And in any case he had some points of his own to make.

The book comprised four chapters. 'General notions' treated rather more of set theory and mathematical examples than logic itself. Then 'Reasoning' largely handled propositions, including the notion of a '*chain of deductions*':

$$\text{'} a \supset b \,.\, b \supset c \,.\, c \supset d \,.\, d \supset e \,.\, e \supset f \text{', or '} a \supset b \supset c \supset d \supset e \supset f \text{'} \qquad (537.1)$$

as an abbreviated form (p. 18); similar chains for equivalences (' $=$ ') were proposed on p. 29. This idea followed Peano (first at *1891b*, 105, no. 11, up to '*d*' of (537.1)). In places he showed awareness of Schröder (who was cited in the preface) by laying out theorems in dual pairs. The next chapter, 'Classes', however, definitely involved collections of the Cantorian kind, and included Peanist emphases such as distinguishing an individual *a* from its unit class '*ιa*' (p. 94).

On definitions, Burali-Forti proposed '*x* $=_{Def}$ *α*', to be understood as '*x is identical to α*' (p. 26); the notation, though not the interpretation, has become well known. Sometimes he also marked definitions similarly to Peano at (532.1), writing '(Def)' on, and at the end of, the line. One case was his definition '$\Lambda = a - a$ (Def)' of the 'the absurd', which however was defective in leaving '*a*' free (p. 49); Peano had treated '$a - a = \Lambda$' as a 'Pp.' when *a* was a proposition (see, for example, *1891b*, 109). However, Burali-Forti defined the empty class in the manner of Peano's (523.1) of *Arithmetices*, with the double use of 'Λ' (p. 82). Peano himself was to tidy up such matters when he spoke before Russell and others in 1900 (§5.5.2).

In his main discussion of definitions (pp. 120–149) Burali-Forti distinguished between four 'species': nominal, nominal under hypothesis, of 'any definition of an entity in itself', and by abstraction. He stressed the difference between the first two species (as definitions of names) from the third one, of which his least unclear examples were those of the positive integers. These definitions were based upon the primitives 'one', 'successive' ('suc') and 'number', in a Peanist fashion; indeed, he also stated the Peano axioms from *Arithmetices*, although presumably by oversight he omitted (523.10), which stated that suc *a* ■ = 1 (pp. 136–138). He noted that the axioms delivered 'the property of the product [...] without introducing the concept of *number of* the *individuals of a class*' (p. 137). On definition by abstraction, he imitated closely Peano's line (536.1–2) in the *Notations*.

5.3.8 *Burali-Forti's research, 1896–1897.* Burali-Forti took these thoughts further soon afterwards with a paper *1896a* on 'The finite classes' published by the Turin Academy. His principal aim was to rework in Peanese Cantor's presentation the previous year of finite cardinals (§3.5.6), and using Dedekind's reflexive definition of infinitude (§3.5.2). Where Cantor had written of an equivalence between sets, Burali-Forti took as his basic notions class and function, the latter called '*correspondence*' and given the Cantorian symbol ' ~ ', and sought to define integers independently of size, measure and order (p. 34). The Peano axioms were stated, all five this time (p. 41), and then applied to define the number (num *u*) of a class *u*, 'an *abstract* entity function of *u* and that *u has in common* with all the equivalent classes *v*' (p. 39, italics added). To establish mathematical induction, he defined (non-unique) '*normal* classes *formed with u*'; they were nesting chains of non-empty sub-classes of *u* possibly containing 'seq *u*', a class obtained 'from *u* by adjoining an element *y* not belonging to u' (p. 43) and certainly including at least one unit class (p. 47: for the class of those classes he proposed on p. 43 the name 'Un').

Burali-Forti ended by defining integers inductively:

$$\text{'1} = \iota(\text{N'Un}) \text{ Def', 'N'}u = \text{N'}v . = . u \sim v\text{' and}$$
$$\text{'}v \, \varepsilon \, \text{seq } u . \supset . \text{N'}v = \text{N'}u + 1\text{'}. \tag{538.1}$$

Some interesting features attended either end of this number sequence. He allowed that 'num $u = 0$, when the class u does not contain elements' without either demur or further use (p. 41); in addition, he rejected Dedekind's idealist construction of an infinite class (§3.5.2) and asserted of 'there exist infinite classes' that 'we place it in an explicit manner among the hypotheses every time that it is necessary' (p. 38).

In his next paper, Burali-Forti *1897a* on transfinite ordinals, he showed similar tendencies. It is famous today for the origins of the paradox now known after his name; we shall see his own, different, interpretation in §6.6.3, but here we note his use of definitions. He gave a certain 'order of members of a class' u the letter 'h', so that u so ordered was notated '(u, h)' (art. 2). Then, in another distinction of category, following Cantor on order-types *of* sets (§3.4.7), he wrote 'T'(u, h) for "order type of the us ordered by criterion h"', and defined this *'abstract object'* by an equivalence relation (art. 6). In these papers Burali-Forti was working at the edges of the vision of both his master Peano and father-figure Cantor; we shall note his next steps in §5.5.3.

5.4 THE *FORMULAIRE* AND THE *RIVISTA*, 1895–1900

5.4.1 *The first edition of the Formulaire, 1895*

The *Formulario di Matematiche* has for aim to publish all the propositions, demonstrations and theories, gradually that they be expressed with the ideographic symbols of mathematical Logic; as also the relative historical inductions.   Peano in the *Rivista* (*1897a*, 247)

We have now reached the time mentioned in §5.3.1, when Peano began to publish under his editorship a primer of logico-mathematical results compiled with the help of his 'society of mathematicians'. It was called *Formulaire de mathématiques* in the first edition of 1895 (which contained nearly 150 pages) and in succeeding editions of 1897–1899 and 1901 (published by the Paris house of Carré and Naud and publicised there by Louis Couturat *1901a*), and *Formulaire mathématique* in 1902–1903; but the fifth edition of 1905–1908, which contained over 500 pages, was named '*Formulario mathematico*' in his now favoured uninflected Latin. Apart from the 1901 edition, they were published by Bocca. I shall refer to it always as '*Formulaire*'; the basic details are listed under 'Peano *Formulary*' in the bibliography.

Peano organised each edition like the factory manager, assigning the various parts to his operatives while also contributing himself. As well as areas of wallpaper symbolism, they also contained elaborate numberings of propositions and definitions, and valuable historical notes and extensive references to the original literature. A few supplementary papers to some editions appeared in the *Rivista*. The various editions are surveyed and compared in Cassina *1955a* and *1956a*; on notations see Cajori *1929a*, 298–302.

As the first edition was being readied, Peano sought some publicity from Klein and told him on 19 September 1894 that printing was slow and publication envisaged 'in a very limited number of copies':

> the aim of Mathematical logic is to analyse the ideas and reasonings which feature especially in the mathematical sciences. The analysis of ideas permits the finding of the fundamental ideas with which all the other ideas are expressed; and to find the relations between the various ideas, that is the logical identities, which are such forms of reasoning. The analysis of ideas leads even to indicate most simply[,] by means of the conventional signs, of which convenient combinations of signs then represent the compound ideas. Thus is born symbolism or symbolic script, which represent propositions with the smallest number of signs.[4]

The main text covered, in order, mathematical logic, algebraic operations, arithmetic, 'Theory of magnitudes', 'Classes of numbers', set theory, limits, series, and aspects of algebraic numbers. The ten-page introduction to mathematical logic, Peano *1895b*, began with a terse catalogue of properties of propositions, with a strong emphasis on logical equivalence and on the properties of 'Λ' and its inverse 'V'; one curious feature was that he used the propositional analogue of De Morgan's law $(535.2)_2$ (whom he cited on p. 186) as the definition of inclusive disjunction (p. 180, no. 7). Then followed a shorter list of main definitions and features of classes (including some use of universal quantification but omitting (535.2)), and details of how functions mapped from range to domain (to which he

---

[4] Klein Papers, Box 11, Letter 190A. The translated passages read: 'in un numero limitatissimo di esemplari', and

> Lo scopo della Logica matematica è di analizzare le idee e i ragionamenti che figurano specialmente nelle scienze mathematiche. L'analisi delle idee permette di trovare le idee fondamentali, colle quali tutte le altre idee si esprimono; e di trovare le relazioni fra le varie idee, ossia le identicà logiche, che sono tante forme di ragionamento. L'analisi delle idee conduce anche ad indicare le più semplici mediante segni convenzionali, coi quali segni convenientemente combinati si rappresentano poi le idee composte. Cosi nasce il simbolismo o scrittura simbolica, che rappresenta le proposizioni col più piccolo numero di segni.

did not give names). The symbols were supplemented near the end of the book by much-needed explanatory notes, together with historical references, especially to writings in algebraic logic.

The Peanists published much more during the rest of the decade, not only in the *Formulaire* and the *Rivista* but also elsewhere. Peano gave a summary bibliography in *1900a*, 306–309.

In a paper Peano *1897c* on 'Studies' in the new field, delivered to his colleagues in the Turin Academy of Sciences, the coverage was normal, but a few differences are worth recording. He laid greater stress on 'primitive ideas' in the system (pp. 204–207), and in general emphasised definitions. He now used the adjective 'apparent' to characterise variables which we now customarily call 'bound' (p. 206); he had made the point before (for example, in *Notations*: *1894b*, arts. 13–15) but without assigning a name, and he did not offer one for free variables.

Peano had a few new points here to make about classes. One was a somewhat closer approach to a predicate calculus with

$$\text{`}x \, \varepsilon \, a \, . = . \, p_x \text{'} \tag{541.1}$$

(*1897c*, 209); but he did not offer it as a definition of a predicate, for, as was explained after (535.3), p_x was not necessarily a propositional function and so did not entail any particular classhood. Another novelty was a fresh definition of equality to the empty class; instead of (523.1) on the absurdity of membership to it, the defining property was that being contained in every class b (p. 211):

$$\text{`}a \, \varepsilon \, K \, . \, \supset \, \therefore \, a = \Lambda := : b \, \varepsilon \, K \, . \, \supset_b . \, a \, \supset \, b \quad \text{Def.'}; \tag{541.2}$$

he assumed that a class, and only one, was thereby defined.

Peano also introduced two new notations (pp. 214–215). One was the symbol '$x \, ; \, y$' instead of '$x \, , \, y$' for the ordered pair (as usual 'considered a new object') to avoid confusion with '$x \, , \, y \, \varepsilon \, \alpha$', which stated that the two individuals belonged to a class. The other was the symbols '$\exists$', which indicated that a class was not empty;

$$\text{`}a \, \varepsilon \, K \, . \, \supset \, : \exists a \, . = . \, a \sim = \Lambda \quad \text{Def.'}, \tag{541.3}$$

where ' $\sim$ ' was his current symbol for negation. Despite his frequent stress on analogies between classes and propositions, he did not extend this term to define existential quantification, but he explained elsewhere that 'the notation $a - = \Lambda$ has been recognised by many collaborators as long, and too different from ordinary language' (*1897b*, 266).

5.4.2 Towards the second edition of the Formulaire, 1897. This last statement appeared in Peano's survey *1897b* of 'Mathematical logic', which launched the first part of the second edition of the *Formulaire* (the further two parts are noted in §5.4.6). He brought it out in time to present it at the First International Congress of Mathematicians at Zürich in August 1897 (§4.2.1).

At 64 pages this survey far surpassed its predecessor or the Turin paper. But there were fewer novelties; even the old notation for the ordered pair was used (p. 256). Peano gave prominence and even priority to classes instead of to propositions: possibilities of vicious circles arise, of course, and maybe were not fully appreciated; for example, his first definition of two individuals belonging to a class naturally used the properties that the first did *and* so did the second (p. 221, no. 11), before any explanation of 'and' had been given.

In places the changes were not necessarily desirable. In one case, after defining the intersection *ab* of two classes *a* and *b* as usual as the ensemble of their common members (p. 222, nos. 14–14′), Peano did not repeat the brother definition of class union but instead deployed De Morgan's law $(535.2)_2$ (p. 226, no. 201), as he had done for propositions last time. Equality of x and y was defined in terms of their belonging to the same classes (p. 225, no. 80); the short explanation on p. 258 quoted Leibniz on the identity of indiscernibles (as usual), and so did not help.

In line with his desire to print symbols along the line, Peano's two-row way of indicating substitutions of letters (§5.3.2) was replaced by overbars; for the two-letter case, '$a(p, q)\overline{(x, y)}$' denoted 'that which becomes the proposition a when for the letters x and y one *substitutes* the letters, or the values, or the expressions indicated by the letters p and q' (p. 220: in an alternative notation he allowed 'a' to come at the end). Overbars continued to serve for converse notations, such as '$\overline{x\,\varepsilon}$' for class abstraction; and he extended it to functional abstraction when 'we indicate by $a\bar{x}$ the sign of function f, such that $fx = a$. Thus one has $(fx)\bar{x} = f$' (p. 277). He also now introduced both the terms 'real' and 'apparent' for variables (p. 243).

Another important pair of converse notions drew upon the inversion of a symbol rather than the overbar of (525.2): the function 'Ⅎ' which took each member x of class a to a unique member 'xu' of class b under a 'correspondence' u between the classes, and its inverse function 'f':

$$\text{`}a, b\,\varepsilon\,K\,.\,\supset\,\therefore\,u\,\varepsilon\,a\,\text{Ⅎ}\,b\,.\,=\,:\,x\,\varepsilon\,\alpha\,.\,\supset_x\,.\,xu\,\varepsilon\,b\,.\quad\text{Df.'},\quad (542.1)$$

$$\text{`}\qquad »\qquad u\,\varepsilon\,a\,\text{f}\,b\,.\,=\,:\,x\,\varepsilon\,\alpha\,.\,\supset_x\,.\,ux\,\varepsilon\,b\,.\quad\text{Df'}.\quad (542.2)$$

(p. 236, nos. 500–501). This strategy was to influence both Russell and Whitehead, the latter negatively (§6.8.2).

One feature of this account was the stress laid upon 'primitive propositions' ('Pp.', as usual) with other propositions derived from them; it showed his concern with issues connected with axiomatics. To this category of primitives Peano assigned 'The simplest forms, by the combination of which one can compose the others', although he admitted at once that 'The choice of primitive propositions is also in part arbitrary' (p. 247).

5.4.3 *Peano on the eliminability of 'the'*. (Zaitsev *1989a*) Peano was similarly attentive to definitions (marked by 'Df.', as in (542.1–2), from now on the usual abbreviation), and to the definability of other notions from them; and in a related context occurred a striking passage in the survey, concerning 'the' (pp. 234–235, with commentary on pp. 268–270). Firstly, he introduced the symbol '$\bar{\iota}$', as the converse to 'ι' for forming the unit class, so that

$$\text{'}\bar{\iota}\iota x = x\text{'} \text{ and '}x = \iota a \,.=\,.\, a = \iota x\text{'}. \tag{543.1}$$

Next he replaced (541.2) with a definition of the empty class itself, using the same property:

$$\text{'}\Lambda = \text{'}\bar{\iota}K \cap \overline{x\,\varepsilon}\,[a\,\varepsilon\,K\,.\,\mathfrak{I}_a\,.\,x\,\mathfrak{I}\,a]\,[\text{Df.}]\text{'}, \tag{543.2}$$

adding that '$\Lambda\,\varepsilon\,K$' (nos. 434–436: on this claim, see §5.4.7). But then he elaborated upon a brief remark in the paper for the Turin Academy (*1897c*, 215) to explain that 'the' was *eliminable*. With b as a second class, he transformed

$$\text{'}\bar{\iota}a\,\varepsilon\,b\text{'} \text{ into '}\exists\,\overline{x\,\varepsilon}\,[a = \iota x\,.\,x\,\varepsilon\,b]\text{'}, \tag{543.3}$$

'another [proposition] where there is no more the sign $\bar{\iota}$' (*1897b*, 269). At last existential quantification had arrived.

Now, taking the maximum of a class of real numbers (K'q) as an example, Peano showed that the maximum need not exist: in

$$u\,\varepsilon\,K\text{'}q\,.\,\mathfrak{I}\,.\,\max u = \bar{\iota}\{u \cap \overline{x\,\varepsilon}\,(u \cap \cap(x + Q) = \Lambda)\}\,\text{Df.'} \tag{543.4}$$

(where Q was a positive number), 'we do not affirm that the class [in (543.4)] exists effectively; that is to say we do not affirm the existence of the maximum' (p. 269). By in effect taking the unit class of either side of (543.4) under the hypothesis, he obtained

$$u\,\varepsilon\,K\text{'}q\,.\,\mathfrak{I}\,.\,\iota\,\max u = u \cap \overline{x\,\varepsilon}\,(u \cap (x + Q) = \Lambda)\text{'}, \tag{543.5}$$

from which the existence (in his sense (537.3) of the non-emptiness '$\exists$' of a class) could (not) be asserted (p. 270). Perhaps he was inspired to this example by Burali-Forti's use of it (*1894b*, 122), where the existence was taken for granted; at all events, in a mathematical context he had formulated a theory of definite descriptions with sufficient conditions corresponding to those which Russell was to propose in 1905 for natural languages (§7.3.4).

One other novelty deserves attention. In keeping with his separation of logical and arithmetical notions, Peano modelled the relations of classes into number theory: with *a* and *b* now denoting integers, he read inclusion of *a* within *b* as *a* being a divisor of *b*, intersection and union as their greatest common divisor and least common multiple respectively, and the empty class as 1 (pp. 262–263). In a manuscript Padoa *m1897a* extended this approach by associating equivalence of propositions with equality of integers, membership of *a* with being a prime divisor of *a*, class abstraction with the product of all such divisors, and non-membership of *a* with the product of all primes which were not divisors. In connection with the latter he also proposed this more precise definition of the complement ∼ *a* of a class *a* relative to V:

$$a \, \varepsilon \, K . \supset . \sim a = \overline{x \, \varepsilon} \, (b \, \varepsilon \, K . a \cup b = V . \supset_b . x \, \varepsilon \, b) \quad \text{Df.} \quad (543.6)$$

However, for some reason he kept these ideas to himself.[5]

5.4.4 *Frege versus Peano on logic and definitions.* During these years Peano also reviewed in the *Rivista* recent publications by two other leading logicians, which we treat in this and the next sub-section. Greatly different from each other, Peano stressed the differences of each from himself.

Peano had mentioned or cited Frege occasionally in his writings from 1891;[6] more detailed contact arose with his review *1895a* of the first volume (1893) of Frege's *Grundgesetze* (§4.5.6). In emphasising similarities between his own 'mathematical logic' and Frege's 'ideography' he came unintentionally close to a position nearer to Frege's logicism than his own non-logicism when claiming of his own programme that 'Mathematics is now in possession of an instrument ready ['*atto*'] to *represent* all its propositions, and to analyse the various forms of reasoning' (p. 190, my curious italics). He took Frege's primitives to be assertion, truthhood, negation, implication, and universal quantification, and compared them with his own trio comprising the last three (though he identified the latter with his own quantification (535.2) involving 'p_x' rather than with predicates); thus his own system 'corresponds to a more profound analysis' (p. 192). He found 'inconvenient' Frege's use of Greek, Latin and German letters (which, we recall from §4.5.2, represented respectively free and bound variables, and gap-holders); he also judged Frege to 'occupy himself

---

[5] So did Leibniz, in that in 1679 he wrote a manuscript on the arithmetical interpretation of predicates. Peano and Pieri seem not to have known of it, as it was published only in Couturat *1903a*, 42–43; however, Vacca had consulted Leibniz manuscripts in Hannover and contributed historical remarks to the *Formulaire* from 1899.

[6] Peano's first reference to Frege was given in §5.3.2: on the others see Nidditch *1963a*, which otherwise is not historically reliable.

scatteredly with the rules of reasoning' (p. 194). At the end he took Frege's definitions of integers and the use of succession as 'identical in substance' with those in the *Formulaire* (p. 195).

Overall Peano's review was not very penetrating, and Frege responded in 1896 with two pieces: a lecture to the *Deutsche Mathematiker-Vereinigung* at Lübeck in September 1895,[7] laid before the Leipzig Academy in the following July and published there as *1896a*; and a shorter letter *1896b* sent to Peano in September which Peano published in the *Rivista* with his brief reply *1896a*. This pair formed part of a private correspondence, which began in January 1894 and included exchanges of publications (Frege *Letters*, 176–198). I take them first together, and then the lecture.

The first question for Frege *1896b* was the question of primitives. He doubted that Peano had captured everything in his trio, and lamented the absence of the concept of assertion in Peano's system (p. 294). He also included equality, which, we recall from §4.5.5, he placed under identity (pp. 288–290). Peano answered that his threesome covered only basic operations and relations between propositions, not logic *in toto*, and that ' = Df' was really just one sign; in a private letter of 14 October 1896 he admitted that he should have included membership.

Frege further claimed that Peano's use of ' = ' for equivalence and for equality was illegitimate; but Peano replied that the former really constituted a single symbol, and pointed to Burali-Forti's ' =$_{\text{Def}}$' (§5.3.7) as an alternative of this "contiguous" kind. On definitions in general Frege repeated his insistence that they be formulated in a 'complete' manner (§4.5.3), rather than in Peano's way under hypothesis (p. 292); but Peano defended his use of hypotheses, and surely with some justice, for they correspond in role to imposing a universe of discourse, which we saw in §2.5.4 to be essential to Boole for avoiding paradoxes. Frege also disliked Peano's use of letters denoting functions alone (such as '*f*' instead of '*fx*' on p. 292), but only noted his own use of Latin, Gothic and Greek letters briefly at the end.

Frege took up this and some other points in more detail in his lecture, where he drew principally upon Peano's *Notations* (§5.3.5), which Peano had sent him. Near the end he defended his use of different letters, relating it in part to Peano's talk of 'apparent' letters (*1896a*, 233). He began the essay by stressing the difference between the truth of propositions and the conclusions ('*Schlüsse*') of arguments and his and Peano's notational systems: in the course of the latter occurred his well-remembered remark that 'The comfort of the typesetter is not yet highest of possessions' (p. 222). In connection with identity, he now also lamented the absence in Peano of the distinction between the sense and the reference of a proposition (p. 226). On Peano's all-purpose conception of deduction he

---

[7] See *Jahresbericht der DMV 4* (1894–1895: publ. 1897), 8, 129 (title only): these references supplement the editorial information in Frege *Letters*, 180.

noted the three different kinds which were recorded around (535.2–3); he regarded as correct only the third one, where deduction was interpreted in terms of the truth-values of antecedent and consequent propositions (pp. 228–229: compare §4.5.2).

By and large Frege showed himself to be the sharper logician and philosopher; but on mathematical matters he was less strong, for he puzzled over Peano's use of classes in a way which revealed his own misunderstandings. He did not recognise that class abstraction, as after (353.3), was effected in the manner of Cantor, whom he did not mention once (p. 235); indeed, he judged Peano's 'concept script [as] a descendant of Boole's calculating logic' (p. 227), which is out of date for Peano by four years.

Curiously, the issue of logicism itself was not addressed in this exchange. The same is true of Peano's next encounter, where it also arose, along with veteran topics from this campaign such as primitives and definitions.

5.4.5 *Schröder's steamships versus Peano's sailing boats.* Peano's second review was of Schröder, who went through an intellectual conversion under the influence of its logic. In §4.4 we saw that his *Vorlesungen* were broadly Boolean, in that mathematics was used to analyse logic; but after completing his volume on the logic of relatives he reversed these roles. He publicised his change in a paper delivered to the International Congress of Mathematicians of 1897 (§4.2.1):[8] 'I may incidentally say, that pure mathematics seems to me merely a branch of general logic' (*1898a*, 149). Dedekind's cryptic claim that arithmetic was part of logic (§3.4.1) was one inspiration; Peirce's logic of relatives was adduced as a more specific one, since it provided means of expressing all the basic 'categories' for mathematics, such as 'multiplicity, number, finitude, limit-value, function, mapping, sum'.

Schröder took as his five basic categories identity, intersection, negation, conversion of a relation, and relation in general, and showed how the other 18 required notions could be defined from them (for example, the null manifold and universal relation, and subsumption). He proposed an 'absolute algebra', in which algorithmic methods would be applied to the algebra of logic to turn out all possible combinations of relation, connective and proposition, together with the laws appropriate to each case; then the entirety of mathematics was to be *cumulatively* delivered, case after

[8] In a letter of 15 December 1897 to Paul Carus, editor of *The monist* (and author there of a waffly review *1892a* of Schröder's first volume), Schröder had planned to deliver his lecture in English, as a 'neutraler Boden zwischen Deutsch und Französisch'; but he added that he had spoken in German because of the tiny proportion (10 out of about 230) of native English speakers at the Congress (Open Court Papers, Box 27/1). His version appeared in the journal as Schröder *1898b*; Carus put virtually the above phrase at its head. The letter is translated in Peckhaus *1991a*, 194–197, an article where similarities with Frege's and Russell's logicisms are stressed; here I emphasize the differences.

case. This cataloguing approach to logicism rather resembled his extensional conception of classes (§4.4.4); it contrasted greatly with the organic construction of logicism from mathematical logic, already tried by Frege and soon to be adopted by Russell under influence from Peano (§6.5).

Among the algebraic laws, those of associativity were especially important for Schröder, maybe recalling Benjamin Peirce on listing all linear associative algebras (§4.3.2). He also emphasised the newly emerging algebra of group theory, but some of his methods were of a lattice-theoretic character, in both form and intended generality. This feature marks a point of similarity with some work of Dedekind in abstract algebra (Mehrtens *1979a*, chs. 1–2).

Another purpose of Schröder's paper was to contrast his own approach to the 'pasigraphy' (that is, universal writing) with that of Peano and his followers. He quoted but rejected from *Notations* Peano's claim that Leibniz's vision of a universal language was resolved (§5.3.6), stressing in particular the absence of the latter from Peano's programme. However, he associated his subsumption relation with both membership and improper inclusion in Peano, thus reducing set theory to part-whole theory. In addition, he claimed a kind of squatters' rights for the pasigraphic task at hand: the Peanists were 'still making use of sailing boats, while the steamships are already invented' (*1898a*, 161).

Peano chaired this lecture, and when it was published he reviewed it in the *Rivista* as his *1898a*. Surprisingly and as with Frege, he ignored the issue of logicism, although it ran counter to his own division of mathematical from logical notions; he also failed to address the issue of a logic of relatives. Instead he treated two other matters. The first was the comparison between his and Schröder's symbolisms; he emphasised the difference between his own use of Cantorian set theory and Schröder's part-whole methods, in which membership and improper inclusion were conflated. More penetratingly than in his review *1891d* of Schröder's lectures (§5.3.2), he rehearsed examples of the differences, such the non-transitivity of membership after (525.1) (pp. 298–300), and effectively cast doubt on the impression given by Schröder that Cantor's set theory was on board the steamship.

Secondly and 'more important', Peano assessed as illegitimate Schröder's identifications among his primitives of equality with his identity relation '1'' (446.6) and of conjunction with universal quantification; the first point is perhaps a quibble but the second carries substance, showing the difference between an algebraist's and an analyst's understanding of the relationship between 'and' and 'for all'. He also objected that Schröder's five categories did not cover all primitives (like Frege on Peano himself two years earlier!): he added in conventions over use of brackets, 'variable letters' (presumably the concept of the variable), and the notion of definition, which 'evidently cannot be defined', in contrast to his own deployment of ' = Df.' (p. 301).

To stress the last point Peano turned again to zero, which Schröder "defined" as '$0 = a - a$ "we say nothing [is] the a not a"' (p. 303). Building upon such insights as (543.3−5) in the *Formulaire*, he had already warned 'not to confuse, in our formulae, logical symbols with algebraic ones' (*1897b*, 265), in the context of Schröder's uses of 'zero'; now he replaced that empty class with

$$\text{`}\Lambda = \imath x \ni (a \, \varepsilon \, \text{Cls}. \supset_a . a - a = x)\text{'} \qquad (545.1)$$

(*1898b*, 303). The expression was not graced by 'Df.' (ironic, in view of his above objection), and no mention was made of the recent definition (543.2). Further, there were three unexplained débuts in notation; the inverted iota '$\imath$' for 'the', '$\supset$' to replace '$\mathfrak{O}$', and 'Cls' to replace 'K'. But more important is the *form* of (545.1): *the* class was to be defined *such that* ..., which Schröder's language could not permit him to say. These points were to be of major importance for Russell, when he heard Peano and Schröder discuss the definition of classes at Paris in 1900 (§6.4.1); and shortly before that Peano was to dwell on them with profit himself.

5.4.6 *New presentations of arithmetic, 1898.* So far most of the interest in Peano's contributions had been developed in Italy, but a more international public was being secured. Three examples are worth noting.

Firstly, the Paris *Revue de métaphysique et de morale* gave Peanism an airing, doubtless on the initiative of Couturat (§4.2.3). Vailati produced a rather scrappy piece *1899a*, on the historical background in Leibniz and on some of Peano's writings;[9] perhaps to compensate Couturat himself wrote a much more substantial essay *1899a*, to which Vailati contributed two historical errors by crediting Servois with the name 'associative' and W. R. Hamilton with 'distributive' instead of the other way around (p. 618: compare (225.2−3)). Couturat greatly welcomed Peano's initiative, and also various details; for example, he found Peano's derivation of some of the laws of substitution for the propositional calculus superior to Schröder's since he deployed 'not' explicitly (pp. 214−218). However, he also had some pertinent reservations, especially concerning the limited role given to duality, in contrast to the prominence evident in Schröder. For example, on p. 643 he lamented Peano's strategy (*1897b*, 226−228) of taking De Morgan's law (535.2)$_2$ as the definition of class union, on the grounds that duality was thereby impaired. He also noted various changes which had appeared in the various versions; one of these is outlined in §5.5.2.

---

[9] In the previous volume of the *Revue* Vailati placed his *1898a*, seemingly his own French translation of a pamphlet *1898a* published at Turin as the introduction to a lecture course on the history of mechanics. He covered induction in science and compared it with deduction and syllogisms, but he did not mention Peano's programme.

Secondly, Padoa gave a course of 11 lectures on mathematical logic in October and November 1898 at the University of Brussels, to 'students of philosophy and of mathematics' (*1898a*, 3). Based upon two recent issues of the second edition of the *Formulaire* in order, content and current notations, he did introduce a few variants: for example, 'Ks' for 'is a class' and 'ε' for 'is a' (pp. 13, 17), and '$x \equiv y$' to stand for '$-(x = y)$' between propositions (p. 32) to introduce some refinement in the handling of negations. His coverage included all Peanist logic and set theory (including Euler diagrams on pp. 26–31) and the basic properties of integers based upon the Peano axioms (pp. 50–51). The account began with the customary (§5.3.5) recollection of Leibniz's aspirations for a universal language, and ended with a hope that enough had been said to convey 'the importance and the usage' of the subject (p. 80). However, Belgian contributions came only 25 years later (§8.6.3).

Thirdly, there appeared the German translation of Peano's edition of Genocchi's textbook on mathematical analysis (§5.2.1); and the second volume (1900) contained translations of five papers, including *1890a* on the space-filling curve (§5.2.4) and two of his papers related to logic: the Turin Academy 'Studies' *1897c* (§5.4.1), and parts of a recent *Formulaire* treatment of arithmetic. The original of this latter study was the 60-page second part (on arithmetic) Peano *1898b* of the second edition of the *Formulaire*, where instead of the 1 of (523.1) he started off the axioms with '0 = "zero"', just like that (and redolent of Burali-Forti *1896a* at the end of §5.3.8): he modified his definition of 'N_0' to incorporate the new member. He now stated the induction axiom (523.11) in the form

$$\text{'} s \,\varepsilon\, \text{Cls} . 0 \,\varepsilon\, s . x \,\varepsilon\, s . \supset_x . x + \varepsilon\, s : \supset . N_0 \supset s \text{ Pp'} \qquad (546.1)$$

(*1898a*, 217). The surrounding commentary shows that, presumably unintentionally, only a sequence $\{s, s + , \ldots\}$ was being defined: the set theory as such was rather incidental. The rest of the part included rational numbers and the Euclidean algorithm, and touched upon Cantor's theory of transfinite ordinals.

In the current sixth volume of the *Rivista* Peano had recently placed a commentary *1897a* on this *Formulaire* presentation. We have heard from it in §5.2.3, where he stated that he had found his axioms independently of Dedekind; and at the head of §5.4.1, for the list of rules by which the *Formulaire* was compiled. The treatment itself was pretty similar to the *Formulaire* account, including starting the axioms for integers with 0. But in a successor paper, *1899a* on irrational numbers, he went further on this topic than hitherto, for he surveyed all the current versions in set-theoretical notation. Citing Burali-Forti's textbook (§5.3.7), he also discussed in this context various forms of definition, a theme of growing interest in his school: while not making a choice among the definitions, on p. 267 he obviously liked Cantor's (323.2). By contrast, he also noted on p. 264 the definition using segments given in Pasch *1882a*, 1–3; however, a few pages earlier he had already stated that 'It is possible, always speaking of segments, to construct a complete theory of the irrationals', but opined that the resulting formulae were of 'a form rather different from those in use today in Algebra' (p. 259). He did not elaborate on the point; as we shall see in §6.4.7, Russell did not follow it.

Also in 1899 there appeared the third part of the second edition of the *Formulaire*. Unlike the other two parts it was written by Peano alone; and it began surprisingly, for the first 90 of its 199 pages contained versions of the earlier treatment of logic and arithmetic before passing on to limits, complex numbers, vectors, and elements of the differential and integral calculus. No major changes in policy were made, but some notations were updated, and more developed indexes and bibliographies furnished. We note the next two editions (1901–1903) in §5.6.1.

5.4.7 *Padoa on classhood, 1899.*   In the *Rivista* Padoa supplied a valuable series of 'Notes' *1899b* to Peano's first part *1897a* of the second edition of the *Formulaire* (§5.3.7); he also covered Peano's recent treatment *1898a* of arithmetic. His main point was to consider the class 'Cls' of classes, and the role of classes of classes in general. With Peano '$x \,\varepsilon\, a$' indicates the proposition "x is an a"' (*1897b*, 221); Padoa proposed rewriting it as '$x \,\grave{e}\, a$', and to 'insert the P[roposition]'

$$\text{'}x \,\grave{e}\, a \,.\, \supset\, .\, a \,\grave{e}\, \text{Cls'} \tag{547.1}$$

(*1899a*, 106; as a 'Pp' on p. 108, no. ·4).

Padoa did not work out or even explain the philosophical consequences of this change, which brought him to the territory of higher-order classes and (maybe) of pertaining predicates. For example, in (547.1) itself he proposed the *same* connective, '$\grave{e}$', on either side of the implication. In addition, he imitated Peano (§5.4.2) in defining the equality of x and y in terms of being in the same classes (p. 108, no. 9), a definition which apparently 'do[es] not require comments' (p. 109). But he explored some of the set-theoretic repercussions. These included the need to prove that Cls itself was one (that is, 'Cls $\grave{e}$ Cls': p. 107, no. ·22), and to re-define the power-class 'Cls $'u$' of class u (p. 114, no. 450). The unit class was now notated '(ιx)', with round brackets to distinguish it from the previous symbol (no. ·52); for their own class he preferred the name 'Elm' (for 'Elemento') to Burali-Forti's 'Un' (538.1), with membership in the sense of 'ε', of course), for its lack of arithmetical connotation (p. 117).

The empty class gained Padoa's attention; indeed, its existence must have prevented him from asserting the implication converse to his basic principle (547.1). He claimed that '$\Lambda \,\varepsilon\,$ Cls […] 'is not demonstrable', but his own treatment was not too clear, for he proved that '$\Lambda \,\grave{e}\,$ Cls' via showing that

$$`(\iota x) - (\iota x) = \Lambda` \qquad (547.2)$$

(p. 110). He adjoined a definition of being equal to Λ in terms of containment within any class:

$$`a = \Lambda . =: a \,è\, Cls : b \,è\, Cls . \supset_b . a \supset b \quad Df`, \qquad (547.3)$$

the property which Peano had used (with 'ε') at (541.2) for his definition of Λ.

Padoa used the pairing of symbols in various contexts: class membership and class abstraction, the latter notated by '$\ni$' (p. 112); and the theory of the couple, where he replaced Peano's '$x ; y$' of individuals with 'the couple of a and of b' of classes a and b, written '$(a \vdots b)$' and defining è-membership to them as

$$`(x ; y) \,è\, (a \vdots b) . = . x \,è\, a . y \,è\, b \ Df` \qquad (547.4)$$

(p. 120, no. 68). This reworking was made on the apparent belief that the order of members would not now have to be specified, although it seems only to be transferred from the original couple; but his preference for classes rather than individuals was typical of his approach which, presented only as these series of notes, did not gain the attention that it deserved.

One cause of the lack of response may have been the fact that Padoa did not clearly work out the outcomes for arithmetic. However, he envisaged the need to grant classhood to the ensemble N_0 of positive integers, together with the accolade '$0 \,è\, N_0$' following Peano's recent commencement of the axioms of arithmetic with it at (546.1) (p. 107).

5.4.8 *Peano's new logical summary, 1900.* Peano showed signs of reaction to Padoa when he started the seventh volume of *Rivista* with another long catalogue *1900a* of 'Formulae of mathematical logic'. Apart from joining up the passages of symbols with the prosodic discussions, the general style and content was not changed; but some interesting additions were made, partly because of Padoa. The symbol '$\ni$' for class abstraction reappeared after its *entrée* and exit at (523.2) over a decade earlier (p. 314). The "official" introductions were made of the inverted iota 'ι' for 'the', and of 'Cls' (pp. 351, 313), which we saw appear three years earlier (§5.4.3); and in this connection he made further progress on 'the' and the empty class. The form (543.2) of definition of Λ was repeated, but with a question mark placed after 'Df' (as with many other definitions in this paper) to indicate 'possible definition', on the grounds that the defining term involved notions—in this case, 'ι' and '$\ni$' themselves—absent from the defined terms; then he gave another example, similar to (543.3), of the eliminability of 'the' (pp. 351–352). Curiously, earlier in the paper Peano had given another definition of Λ, akin to Padoa's (547.3) and to his (541.2) in requiring Λ to be a *member* of every class:

$$\text{`}\Lambda = x \ni (a \, \varepsilon \, \text{Cls} \, . \, \supset_a \, . \, x \, \varepsilon \, a) \, \text{Df'};\qquad\qquad (548.1)$$

and in contrast to the former treatments, he now recognised that its status as a class was a 'Pp' (p. 338). In his exegesis of the class calculus, he stated Boole's expansion theorem (255.6) for 'a logical function of two classes x and y', in terms of the extreme values V and Λ (p. 345).

The final sections of the catalogue (pp. 358–361) were taken up with functions, formally granted logical status (p. 311). In addition to rehearsing again possible domains and ranges of a function, and properties such as single-valuedness, he added two new notions. The first (p. 356) introduced '|' as 'the sign of inversion', in which if u were 'a sign of function' and ux were 'an expression containing the variable letter x', then '$ux \,|\, x$' was this expression 'considered as function of x' (p. 356). To us this is functional abstraction; but Peano stressed that 'By the sign | one can indicate substitution' (p. 358), which was how Padoa had used it: '$x \,|\, y$' was his instruction to replace y by x (Padoa *1899a*, 107). The second notion was definite functions' F (as opposed to the usual 'f'), in which not only such a function but also its domain of values were fixed (Peano *1900a*, 359–361, including its own inverse function 'F^{-1}'). Mathematical issues were always close to this logician's attention. But now an opportunity came to address the philosophers.

5.5 PEANISTS IN PARIS, AUGUST 1900

5.5.1 *An Italian Friday morning.* Peano published this number of the *Rivista* a few days before going to Paris for the International Congress of Philosophy. Originally conceived to run from 2 to 7 August, the congress actually took place from 1 to 5 August (launching a series of such gatherings for philosophy), so as not to overlap with the Second International Congress of Mathematicians which ran from 6 to 12 August and where Hilbert presented his list of major unsolved mathematical problems (§4.2.6).[10] It was organised by the editorial committee of the *Revue de métaphysique et de morale*, with Couturat largely responsible for the section on 'Logic and history of sciences'; we saw him publicising Peanism in §5.4.6. In another section international languages were discussed, to the

---

[10] Contrast Congresses in 'Philosophy' and of 'Mathematicians'. Their original dates are given in, for example, the *Rivista 6* (1896–1899), 187–188. The plan of the Logic section is shown with Couturat's letters to Pieri in Pieri *Letters*, 44–45. For accounts of the Peanists around this time, see Vuillemin *1968a*, 169–194; Borga *1985a*, 41–75; and Rodriguez-Consuegra *1991a*, 127–134.

interest of Couturat himself and of Peano; a few days later the mathematicians also took up this topic.

Peano went to Paris with Padoa; in addition, papers by Pieri and Burali-Forti were presented by Couturat. Other participants included also adherents to other traditions, such as Schröder (§6.3.2), P. S. Poretsky, MacColl and Johnson (whose paper was presented in his absence by Russell). Non-speaking participants included Vacca and Whitehead, who stayed on for the mathematicians' show, as did Peano, Padoa and Schröder.

A huge report on the Congress of nearly 200 pages rapidly appeared in the September issue of the *Revue*. No overall editor was named, but it seems safe to cite the 50 pages of it recording the five sessions of Couturat's section as Couturat *1900e*. He also published a shorter account under his name in the newly founded Swiss journal *L'enseignement mathématique*: the Italians featured prominently there, as did his regret that Cantor had not been able to fulfil a promise to attend and speak on 'Transfinite numbers and the theory of sets' (Couturat *1900f*, 398, 401–404). In addition, the American mathematician Edgar Lovett *1900a* wrote at length on the mathematical aspects for the American Mathematical Society. The proceedings of the Congress were published in the following year, with this section covered in the third of the four volumes (Congress *1901a*). Let us turn now to the Italian quartet, who occupied most of the morning session on Friday 3 August.

5.5.2 *Peano on definitions.* The published version of Peano's talk *1901a* corresponded largely to parts of the discussion in *Notations* (§5.3.5). He stressed that definitions should be in the form of equations, and he criticised the formulation of geometry in Euclid's *Elements* for presenting, for example, 'the point has no extension' as a definition of point. Even if an equational form was adopted, difficulties could arise; in Euclid Book 7, '(unity) = (quality of that which is one)' faced objection to the co-presence of 'unity' and 'one' across the equality (pp. 362–364).

Peano's other main point was to explain the 'law of homogeneity', with the example $0 = a - a$. We noted it in §5.3.7 and §5.4.5 with Peano and Burali-Forti when a was a proposition or a class; here no specification was made, for Peano noted that 'it is not a complete proposition; one has not said which value we attribute to the letter a'. Even when specified to numbers, it was complete but not homogeneous since a was still free; but 'The proposition

$$0 = \text{(the constant value of the expression } a - a,$$

whatever be the number $a)$', (552.1)

with 'the' explicit, was 'a homogeneous equality' and so 'a possible definition' (pp. 365–366). His caution was well placed; for the definition assumed that a constant value obtained in the first place.

The discussion of this lecture included objections by Schröder. We postpone the details to §6.4.1, when we record the momentous effect that they made upon the young Russell.

5.5.3 *Burali-Forti on definitions of numbers.*

Analysis is absolutely independent of postulates.
Attributed to Weierstrass in Burali-Forti *1903a*, 193

Peano's paper was rather light, lacking the discussion of other forms of definition such as induction and abstraction that we saw in §5.3.7. Maybe he chose to leave the matter to Burali-Forti, who in his contribution *1901a* contrasted those types of definition of integers. First, some preceding work must be sketched.

Burali-Forti was much involved in mathematical education at this time. He wrote several textbooks on arithmetic and algebra for the Turin publisher Petrini, most of them with the school-teacher and former Peano student Angelo Ramorino. One of these books treated 'rational arithmetic' in both senses of the adjective; they even referred the reader to Burali-Forti's paper *1896a* on finite classes (§5.3.8) and to the Peano axioms, where (523.10) was omitted for some reason (Burali-Forti and Ramorino *1898a*, 5–7: compare Cantor's follower Friedrich Meyer in §3.3.4).

Burali-Forti took these connections further in the first volume of *L'enseignement mathématique*, where he placed an essay *1899b* summarising his ideas on equality and 'derivative elements in the science'. After running through the properties of equivalence relations and explicitly linking 'the equality (or identity)' of two objects, he followed the *Formulaire* in defining identity by the property of belonging to the same classes (p. 248). Then, after surveying the usual relationships between equal classes, and also between 'correspondences' (functions) from one class to another one (pp. 248–253), he applied the machinery to define rational numbers, "$(\frac{m}{n})_g$' for integers m and n, as 'equal to the *unique correspondence* f_g *among the* g *and the* g *such that, whatever be the element* a *of* g, $n(f_{[g]}a) = ma$'. He then defined irrational numbers essentially via Dedekind's 'principle of continuity' (pp. 255–257).

The point was to stress that these nominal definitions depended upon, and had to be distinguished from, definitions via functions; thus, for example, '*cardinal number* indicates a class, *cardinal number of* indicates a correspondence between the *classes* and the simple elements which are the elements of the class *cardinal number*' (pp. 257–258). Feeling that '*to make abstraction* appears here like a logical operation', Burali-Forti converted Cantor's definition of cardinal numbers of sets by double abstraction (§3.5.4) into one of 'CARDINAL NUMBER OF' as '*one of the correspondences* f *between classes and simple elements*'. As in his *1896a*, he rejected the possibility that 'the class of correspondences *f* may contain a single element', essentially on the grounds that members under correspondence could change for a given function (p. 259). Setting aside the class which he associated with cardinal number itself, he did not take Russell's later step of defining it as the class of such functions (§6.5.2).

Burali-Forti discussed these matters in much more detail in the *Rivista* in a paper *1899a*, called 'book' and very long for the journal at 37 pages. The first chapter dealt with 'magnitudes', his speciality (he usually prepared that part of the *Formulaire*). He outlined a general theory of classes, to which name he regarded 'homogeneous magnitude' as a synonym; their totality, '∪'Cls', 'represents the total class' (the somewhat puzzling pp. 145–146). In his presentation of integers in the second chapter he defined two zeros: '0_+' relative to additions, as the ('1') magnitude x such that $y + x = y$ for all y (p. 150); and its mate '0' for multiplication, the x for which $xy = 0_+$ for all y (p. 156). He also gave a nominal definition of N_0; the other integers arose as the successors of 0_+ arose under the operation '$+$' (p. 155). He then ran through rational and irrational numbers (chs. 3 and 4) in broadly the same way as before, and then powers (ch. 5). He deployed an impressive array of notations for functions, including a right half-arrow '$a \upharpoonright m$' for the power-function a^m (p. 172, unexplained at $*72 \cdot 2$); but his aim of the 'immediate application, in higher secondary schools' of his theory (p. 141) presages the lunatic aims of the '"new" mathematics' of our times (§10.2.4).

Types of definition of numbers was the theme which Burali-Forti *1901a* presented in Paris, read out for him by Couturat. After rehearsing the various differences between them he took integers as an example; he defined N_0 as a class of 'similar' (that is, reflexive, symmetric and transitive) operations such as '$+$' over a class of 'homogeneous magnitudes' $\{x\}$ which when applied to x produced the class $N_0 x$ of objects with these three properties: that if x and y were members, so was $x + y$; that x itself and zero (for which $y + 0 = y$ for all y) were members; and that all other members took the form $y + x$, where member y was not equal to 0 (pp. 297–298). The last requirement raised the question of the status of equality, but he strengthened his position by arguing that N_0 was defined, and moreover unique. He also defined the rational number m/n as the result of the operation such that

$$n[(m/n)x] = mx, \qquad (553.1)$$

so that, for example, '$\frac{1}{5}x$ = the magnitude which multiplied by 5 gives x' (p. 305). The irrational number mx was defined like Peano in §5.4.6 (but not cited), via upper limits as $l'(ax)$, where a was a 'limited class of rational numbers' (p. 306). He concluded by asserting of Dedekind's definition of irrational numbers (§3.2.4) that it proceeded by abstraction, but was 'perfectly logical'!

5.5.4 *Padoa on definability and independence.* Padoa had been specialising in the logic and modelling of 'deductive theories' (a phrase which he often used), building upon Peano's concern with the independence of axioms (§5.3.3 and elsewhere) and the definability of concepts (§5.4.6). Unfortunately he kept some of his work to himself: we noted at (543.6) the extension made in the manuscript *m1897a* of Peano's arithmetical model of set theory, and around the same time he made an interesting study Padoa *m1896a?* of the propositional calculus which, had it been published, would have raised his status among pioneers of model theory (Rodriguez-Consuegra *1997a*).

In this manuscript Padoa divided the symbols of 'any abstract deductive science' into the class of those (such as logical connectives, perhaps) fixed 'by knowledge acquired in advance', and the class X of those whose referents were (presumed to be) indeterminate. When each member of X received a referent, 'one obtains a system of referents ['*significati*'], which I call [an] *interpretation* of X'; he also specified 'an untrue interpretation' A, although he did not specify any theory of truth and muddled the class with any associated propositions $a, b, \ldots$ (p. 325). He ran through a range of properties of propositions under interpretation(s); arithmetical analogy again played a role, such as in 'a being divisible by b' when '$(a\bar{b}) = A$' but not so for a, b or $(\bar{a}b)$ (p. 327, def. X). He also distinguished 'absolute' from 'ordered' independence of postulates, the latter defined for a postulate relative to its predecessors in an assigned order; and he defined the 'indecomposability' of a relative to X when either a or $\bar{a}$ was true under an interpretation of X and a was not divisible by any other proposition (p. 328, def. XIII).

In Paris Padoa *1901a* used some of these ideas when he outlined at length procedures for determining both the independence of axioms and the definability of concepts. On the former he briefly rehearsed the method of modelling which Peano had already deployed, where the target axiom was false but the others true (pp. 321–323); but on definitions he was similar in thought but more original and expansive. Without naming him, he criticised Peano's affirmation of the simpler propositions (§5.4.2) by doubting that we could 'imagine a *rule* to choose infallibly *the simpler among two ideas*' (pp. 316–317), and instead advocated the same kind of modelling strategy. For him 'the undefined symbols' were subject to 'several (and even infinitely many) *interpretations*' relative to which they 'can be regarded as the *abstraction*' from the pertaining theories (pp. 319–320, his italics). After taking one interpretation that '*verifies* the system of *unproved*' propositions, all of them 'continue to be verified if we suitably change the meaning of the undefined symbol *x only*'; thus 'it is not possible to deduce a relation of the form $x = a$, where a is a sequence of other defined symbols, from the unproved propositions', and 'the system of undefined symbols is **irreducible** with respect to the system of unproved' propositions (pp. 320–321).

Padoa did not prove his (to us, meta)theorem, presumably regarding it as an obvious cousin to Peano's procedures for establishing the independence of propositions. Thus, for example, he did not distinguish logical from non-logical symbols for x from among the sequence a: given the rather fluid lines of distinction between the two categories in the Peanist canon, this is not surprising. But in the rest of the paper he applied the method (and that for propositions) to the arithmetic of integers (positive, negative and zero) in a clear way. He set up his vocabulary of undefined symbols—'ent' (integer), 'suc' (successor of) and 'sym' (the symmetric of; for example, -7 of 7)—and his '*unproved* propositions', and after working out a detailed list of basic properties (pp. 325–356) he finished off with a 'commentary' in which he demonstrated irreducibility by providing suitable interpretations (pp. 356–365). Soon afterwards he gave 'the ideographic transcription' of the theory in a paper *1901b* in the *Rivista*.

Some time in 1900 Padoa also gave a lecture course on 'algebra and geometry as deductive theories' at the University of Rome. The treatment of logic and set theory was broadly similar to that of the Brussels course two years earlier (§5.4.6), although some advances and changes were made and the emphasis on mathematics was stronger. He also took an explicitly model-theoretic view of his system, with the symbols of the 'formal aspect' taken as 'deprived of meaning' (Padoa *1900a*, 17). Among notations, he now preferred ' " $\sim$ " (the stenographic n of Gabelsberger)' for negation (p. 13). After showing the 'Absolute independence of the Pp', he briefly described his views on 'irreducibility of the primitive ideas' (pp. 17–20). Symbolising the successor of a number x by ' $\rightarrow x$ ', he gave the Peano axioms in the order (523.8, 9 (with the two sides reversed), 3, and 10–11 combined) (p. 22); he did not specify the initial integer, but later he defined 0, as the number with no successor (p. 27). He also used ' $\leftarrow x$ ' for the predecessor of x (p. 29) and ' $\downarrow x$ ' for 'the contrary of x ' (that is, its negative); this led to the semiotically elegant recursive definition (p. 30):

$$\text{`}\downarrow 0 = 0\text{' and `}\downarrow \rightarrow x = \leftarrow \downarrow x \text{ Df Rcr'.} \tag{554.1}$$

The coverage of arithmetic advanced to rational numbers and powers. In a paper *1902a* on the integers in the *Rivista* he dropped the Peano axiom (523.3), which asserted the numberhood of (in this version) 0; only 'N' and 'suc' were used, and 0 was defined as $\iota(N - N_1)$, where N_1 was the class of integers possessing successors (p. 47).

When the mathematicians convened in Paris straight after the philosophers, Padoa gave them a short summary *1902b* of *1901b*; and also *1902c* on definitions in Euclidean geometry, where the main aim was the reduction of undefined notions to 'point' and 'is superimposable upon'. In a footnote he noted that the same conclusion had been reached independently by his colleague Pieri, who indeed had concentrated upon geometry in his own work and had had a paper read out to the philosophers.

5.5.5 *Pieri on the logic of geometry.* (Marchisotto *1995a*) Prior to the Congress of Philosophy Pieri had published steadily on geometry (Cassina *1940a*); in particular, two long papers were published by the Turin Academy. In the first, Pieri *1898a* on 'the geometry of position' as a 'logical deductive system', he laid out his postulates ('P') concerning projective points, lines and planes (which he denoted by notations '[0], [1], [2]' of the type which Peano had introduced at (524.1) in his booklet *1889b* on geometry). Some properties concerned existence: specifically, of one point, of another one, and of at least one third one (P 2, 6, 13). One striking passage concerned the nominal and set-theoretic definitions of order ('verso', and 'new abstract entity') and of the sense of a direction; having defined 'natural ordering' for points along a line, he saw how to define 'order of' by abstraction but preferred a 'true and proper [nominal] definition of the name', as the 'class of all the natural orderings of a line' (p. 37: compare Burali-Forti *1899b* on integers in §5.5.3). At the end of the paper he used Peano's method to test his postulates for compatibility.

In the second paper, Pieri *1899a* similarly treated 'elementary geometry as a hypothetical deductive system', a title which doubtless Padoa was to note. The subtitle was 'Monograph of point and of motion', referring to his undefined notions. He stressed at once (p. 175) that 'motion' had nothing to do with mechanics; 'function' and 'transformation' were synonyms (and preferable ones, Burali-Forti might well have thought), and he used Greek letters to denote them and their compounds and inverses (μ, ν, $\mu\nu$, $\bar{\mu}$, and so on: p. 175, P5). After postulating the existence of one and of two points (P2, P3) a and b, he also assumed that there was a 'motion' to get from one to the other (pp. 181–182). Then he launched a sequence of definitions: of a 'conjunction' of a and b as a class of all points collinear with a and b, of the 'stretch' ab as the class of all, and of the 'segment' $|ab|$ as the 'stretch terminated by the points a and b' inclusive (pp. 182–184, 208); later he remarked that definition by abstraction of the addition of segments could be replaced by a nominal definition of the class of (previously defined) congruent segments, but normally we call '"sum" any segment of the said class' (p. 216). Loria *1901a* praised the paper highly in the *Jahrbuch*, translating all the axioms. Maybe this exposure helped stimulate Hilbert, no reader of Italian, to his second treatment *1902b* of geometries, also using motions as transformations (§4.7.2); however, Pieri was not cited there.

In Paris Pieri *1901a* communicated a paper with a title developing that of the last one; now geometry was 'envisaged as a purely logical system'. After recalling recent history since Pasch, he meditated in general upon 'primitive ideas' and their irreducibility. For geometry he again put forward 'point' and 'motion' for this office, and he claimed that from them and 'from the more general logical categories of *individual*, of *class*, of *membership*, of *inclusion*, of *representation*, of *negation* and some others' he could 'give a nominal definition of all the other concepts' and thereby 'one obtains a geometrical system' (pp. 383–384). In some ways the logic of his system is superior to Hilbert's; for instance, he used only these two primitive notions, and avoided adopting 'line' as one of them. His master was to make such a claim for him in a report Peano *1904a* written for the Lobachevsky Prize which nevertheless Hilbert won (§4.7.2).

In his Paris communication Pieri turned to philosophical questions. Doubting that one could find '*luminous evidence*' for either premisses or primitive notions, he pursued the idea they were invariant with respect to '*a maximal group of transformations*' (*1901a*, 389), and argued for 'point' and 'motion' under that criterion. At one point he mentioned superimposability of geometrical figures (p. 392), and at some stage soon after completing this paper in May 1900 he must have come to the recognition which was to dawn upon Padoa also; that the number of primitive notions could be reduced still further. Thus the Peanists were coming to like mind; for Burali-Forti also referred to superimposability as an example of definition by abstraction in his own Paris text (*1901a*, 292). Russell must have been all ears on the morning of 3 August.

5.6 CONCLUDING COMMENTS: THE CHARACTER OF PEANO'S ACHIEVEMENTS

5.6.1 *Peano's little dictionary, 1901*

The classification of the various modes of syllogisms, when they are exact, has little importance in mathematics.
In the mathematical sciences are found numerous forms of reasoning irreducible to syllogisms.   Peano *1901b*, 379

Peano provides a text suitable to conclude this chapter: the part for 'Mathematical logic' of a 'Dictionary of mathematics', published in the *Rivista*. The quotation above comes from it, and shows his recognition of the advances in logic that were imperative for mathematical needs; but other entries reveal the partial nature of his successes. Primitive propositions did not have an entry, although they were mentioned in 'Axiom', 'Postulate' and 'Lemma'. 'Deduction' covered all forms of implication or inference between propositions; and under 'Proposition' he now stated categorically (as it were) that 'Mathematical logic operates solely on conditional propositions' (*1901b*, 381). 'Definition' was a disappointing entry, with only nominal equational forms discussed, although the form by

'Abstraction' had its own entry.[11] 'Equals' was a sign, apparently, which 'one indicates with the symbol = '; but 'Identity' held between 'objects'. This rendered somewhat unclear his definition (p. 376) of a unit class a (in which he used Padoa's 'Elemento' of §5.4.7):

$$\text{'(The class } a \text{ è an element)} = [\exists a : x, y \, \varepsilon \, a . \supset_{x,y} . x = y].'. \quad (561.1)$$

'Class' was a 'primitive idea', and as synonyms to 'Classe' he listed 'Insieme', 'Sistema' and 'Gruppo'. 'To belong' and 'To contain' were both explained. The empty class was 'Null'; the propositional analogue 'Absurd' was not given a symbol. Although universal quantification was used in (561.1) and one other place, it was not explained. However; 'propositions containing variables' (themselves not given an entry) 'p_x' were introduced, under 'Condition'; quantification also crept in unclearly under 'All' for classes, where

$$\text{'(all } a \text{ è } b) = (a \supset b)'. \quad (561.2)$$

'Relation' was explained as $p_{x,y}$, resulting in 'The class of the couple $(x ; y)$' cross-reference was made to 'Function' (a logical notion again, after the claim of *1890a* in §5.2.5), which held between classes (as its range and domain). He laid out his basic symbols under 'Ideography, in German "Begriffsschrift" ', doubtless with Frege in mind. The primitives were now these:

$$\text{'= Cls } \varepsilon \ni \supset \cap \cup - \exists \iota \, \text{ɿ}'. \quad (561.3)$$

Also in 1901 Peano published the third edition of the *Formulaire*, this time in Paris. While its overall coverage of mathematics was more or less the same as in its 1899 predecessor (§5.4.6), at 239 pages it was 30 pages longer, mainly due to the addition of material already published by colleagues in the *Rivista*. The opening part on mathematical logic was closely based on his own recent presentation *1900a* in that journal. He dated his preface as of 1 January 1901, doubtless intentional symbolism of another kind.

For the next edition, of 1902–1903, Peano was back with Bocca, and at 423 pages substantially longer again. The main reasons were the introduction of two new parts on the calculus (pp. 145–200) and on differential geometry (pp. 287–311), and much more extensive treatments of real numbers (pp. 59–121, including definitions of irrational numbers and properties of derived sets) and of elementary functions (pp. 225–249). The

---

[11] In letters of 1901 to Vacca, Vailati expressed his dissatisfaction with Peano's dictionary *1901a*, and opined that the types of definition stressed by Peano elsewhere are common in mathematics, using Euclid's definition of proportion (Vailati *Letters*, 188, 195). In the first letter he also mentioned speaking at a teachers' conference addressed also by Padoa (§5.6.2).

end matter now included a name index. A large collection of 'Additions' was made during printing (pp. 313–366); they were contributed by 21 hands (p. viii), including some foreigners such as Couturat and Korselt, and the American W. W. Beman (b. 1850), who had just published his English translations of Dedekind's booklets on numbers (Dedekind *1901a*). The programme was consuming more and more mathematics, and was surely wanting to eat it all up; but none of the Peanists took this step.

5.6.2 *Partly grasped opportunities*

I have given two names to the sign ⊃ '*one deduces*' and '*is contained*', one reads it still in various other ways. This does not signify that the sign ⊃ has several meanings. I represent better my idea in saying that the sign ⊃ has a single meaning; but in ordinary language one represents this meaning by several different words, according to circumstances. Analogously with the sign Λ.   Peano to Frege, 14 October 1896 (Frege *Letters*, 189)

Peano continued to develop his programme to publish in Peanese in the 1900s; for example, his proof *1906a* of the Schröder-Bernstein Theorem 425.1 was so clothed. But we can now sum up his career, and the main features of him and his followers.

In §5.1.2 I characterised Peano as an opportunist mathematician; similarly he was an opportunist logician. In both disciplines he made not only excellent presentations of known work but also valuable contributions of his own to foundational questions: clarifying and developing Grassmann's theory, the space-filling curve, the distinctions between membership and inclusion and between an individual and its unit class, universal quantification over individuals, a compact and printer-friendly library of notations and interesting principles of notational pairs, pioneering sensitivity to considerations of definitions in formalised theories, the importance of 'the', and so on. The extent to which symbolism could be effected was very impressive: it extended to the running-heads, especially in the *Formulaire*, which frequently read 'Σ', say, or '+'.

With regard to arithmetic and mathematical analysis Peano can be seen as a link between Weierstrass and Russell (my *1986b*); similarly, on the foundations of geometries, he connects Pasch with Pieri and Hilbert. In contrast to Frege (already around) and Russell (to come), one might say that Peano *presented* arithmetic in a *symbolic* language which *contained* logical techniques rather than *grounded* it in an *ideal* language which *expressed* such features.

But Peano's stance is hard to characterise precisely, and the quotation at the head of this sub-section from a letter to Frege hints towards some reasons. There are some incoherences within his philosophy of logic, such as his all-embracing 'deduction' with its long-suffering 'sign ⊃'. More importantly is his insistence that logic and mathematics were distinct subjects. His special concern with arithmetic, analysis and geometries probably reinforced his position concerning each case. However, many of his basic notions, including some of those mentioned above, involved collections (initially of the part-whole kind but then of the Cantorian version); and the place of this subject within the intersection of logic and mathematics made the line of division rather hard to discern even when he listed logical and mathematical signs in separate columns. For example, the distinction (if there were one) between equality and identity raised demarcation disputes which he did not resolve.

Peano normally conceived of classes intensionally, but he left these classes to do the work and sought no alternative basis for them; thus the predicate calculus attendant upon (and derivative from) them was often incoherent, both technically and with regard to its philosophical implications. However, in connection with Peano Quine *1986a* argues that it would be beneficial to the needs of set theory to return to a remuddling of an individual with its unit class.

An interesting example of this situation occurs in a remark on Frege: Peano quoted a formula involving universal quantification over individuals, but his purpose was to object that the 'thesis' (the consequent of the deduction) might not be present on the 'hypothesis', not the form of the proposition itself (*1897c*, 207). So we see another limitation to his programme: that quantification was implicitly restricted to ranges of individuals, in contrast to Frege's use of functional quantification.

In addition, the structure of the symbolic language varied widely in the various versions; only partly committed to axiomatisation, Peano changed forms and status of several key propositions. As an example (from several) of this feature take the following equivalence, involving three propositions:

$$a \supset . b \supset c := : ab \supset c. \tag{562.1}$$

It first appeared in the *Arithmetices* (§5.2.3) as one of the list of unproved propositions (*1889a*, 26, no. 42). In the 1891 review (§5.3.2) each implication was proved separately, without commentary (*1891b*, 106, nos. 19–20). In the *Notations* (§5.3.5) it appeared in the above form (*1894b*, art. 12, no. 13), not proved but included in an excellent list of proved propositions. He added names here to each implication: 'We shall call *to import* the hypothesis *a*, the passage from the first to the second member, and *to export* the hypothesis *a*, the inverse passage'. He assigned credit for (562.1) to Peirce *1880a*, art. 4, and made the point that 'this formula transforms a proposition containing two deductions into one which contains a sole sign $\supset$, and reciprocally.' However, neither this formula (a rewrite in terms of implications of Peirce's (435.1)) nor any other one possessed the property which Peano had described, for in Peanese it reads

$$a \supset . b \supset c := : b \supset . a \supset c. \tag{562.2}$$

The proposition then appeared in the *Formulaire*, but with varying statuses and not always with the new names. In the first edition (§5.3.7) each implication was proved separately: then the equivalence was deduced, and its reference number was assigned an asterisk to indicate its importance (*1895b*, 179, nos. 37–39): no names were attached, but Peirce's paper was cited at the end (p. 186). The second edition (§5.4.2) followed the same three-proposition plan, but with each one assigned an asterisk: they were stated with *a*, *b*, and *c* as classes and with universal quantification applied over their members; the implication from left to right was set as a 'Pp.', with the names given (*1897b*, 225, nos. 72–74); and the subsequent discussion was quite lengthy, but without Peirce (pp. 257–259). The paper associated with the third edition (§5.6.1) used the class form and the names, but gave only the equivalence; Peirce was cited, *in situ* (*1900a*, 337). All logic, but perhaps not too logical.

Broadly the same remarks can be made regarding Peano's followers. While they examined several branches of mathematics more deeply than he did (geometries are an obvious example), they did so very much under his approach. For an important example, we find serious thoughts about nominal definitions in Pieri and Padoa, but worked out in branches of mathematics rather than in a general way. The programme was awaiting a fresh pair of eyes, perhaps from outside the 'society of mathematicians'.

The ambitions of this society extended beyond formal treatments of logic and mathematics. We noted in §5.5.3 that Burali-Forti applied parts of the programme to school mathematics: Peano himself acted similarly with a book *1902a* of 144 pages on 'General arithmetic and elementary algebra'. Although logic was not mentioned in the title, it occupied the opening pages, after a reprint of the school curriculum and prior to a treatment of integers, rational and real numbers (with irrational numbers defined as the upper limits of certain classes of rationals), and some applications. At one point he wondered if, since $+5 = 5$, then $3 + 5 = 35$ (p. 52). Shades of the 'New mathematics' of 60 years later, unfortunately; one can be glad that *this* speculation was not pursued.

Nevertheless, the poor Italian schoolteachers were also treated to Padoa *1902a* on 'mathematical logic and elementary mathematics' at a teachers' congress held at Livorno the previous August. Stressing the merits of 'logical ideography' for understanding known languages and of proving propositions in mathematics (pp. 6, 9), he also discussed primitive propositions and symbols, and even their respective independence and irreducibility (pp. 11–14). One can easily imagine that more was imparted than absorbed.

5.6.3 *Logic without relations.* One surprising lacuna in Peano's programme is his failure to produce a general logic of relations, despite his occasional use of them (§5.3.5). His follower Edmondo De Amicis *1892a* ponderously recounted properties 'between entities of a same system' in the *Rivista*, while Vailati *1892a* presented some relation(ships) between propositions; but no comprehensive theory was thereby produced. The surprise increases when one recalls Peano as a careful reader and citer of Peirce and Schröder, where such a logic (conceived within their own tradition) fills many pages. Causes of the lacuna need to be found.

One reason was that Peano conceived of a relation extensionally as an ordered pair, so that no special treatment seemed necessary (compare him in 1904 in §7.5.1). Again, he would have had to formulate a theory in terms of propositional functions of several variables, whereas he failed explicitly to individuate such functions of one variable: as was emphasised after (535.3), his symbol 'p_x' denoted a proposition p containing the free variable x without reference to the "internal" logical structure of p, so that propositional functions were not exhibited even if only one such function was involved, as in (541.1).

The status of relations typifies the strengths and weaknesses of Peano's contributions to logic.[12] He may only have half-grasped certain of his opportunities, but he opened up many of them in the first place; and his followers, especially Burali-Forti, Padoa and Pieri, developed them in ways and to an extent which have never been fully used since Russell, to whom we now turn.

---

[12] As this book completes its production process, I learn that Peano's descendants have recently placed his *Nachlass* in the *Biblioteca Comunale* of his home town of Cuneo. Apparently it includes thousands of letters (information from Livia Giacardi).

CHAPTER 6

Russell's Way In: From Certainty to Paradoxes, 1895–1903

> I hoped sooner or later to arrive at a perfected mathematics which should
> leave no room for doubts, and bit by bit to extend the sphere of certainty
> from mathematics to other sciences. Russell *1959a*, 36

6.1 PREFACES

6.1.1 *Plans for two chapters.* This chapter and its successor treat Russell's career in logic from 1897 to 1913. The point of division lies in 1903, when he published the book *The principles of mathematics*, where he expounded in detail the first version of his logicist thesis. This chapter traces the origins of that enterprise in his student ambitions at Cambridge University from 1890 to 1894 followed by six years of research under a Prize Fellowship at Trinity College and then a lectureship there; *The principles* was the principal product.

In addition to the birth of logicism, we shall record the growing positive role of Cantor's *Mengenlehre*, the influence on Russell of Whitehead from 1898, and especially their discovery of the Peano school two years later. But we also find his paradox of set theory (1901), which compromised the logic of the new foundations. Convinced of the seriousness of the result, he then collected all paradoxes that he could find (§7.2.1–2), in the hopes of diagnosing the underlying common illness. The next chapter covers the years of collaboration with Whitehead which was to lead to the revised version of logicism presented in the three volumes of *Principia mathematica* (1910–1913) (hereafter, '*PM*').

The use of some technical terms needs to be explained. Throughout the period Russell and Whitehead referred to 'class(es)', both in their early phase when they were using the part-whole theory and also when referring to Cantorian sets after converting to Cantor and the Peanists: 'set' was then a neutral word, referring to a collection, such as 'sets of entities' in *The principles* (*1903a*, 114–115). I have followed the *same* practise, which of course is converse to modern parlance: however, I use 'set theory' to refer to the theory in general, reserving '*Mengenlehre*' for cases where Cantor's own conception is involved. Technical distinctions between classes and sets date only from later developments of axiomatic set theory.

While I quote Russell and Whitehead writing of 'contradictions', I have preferred to use the word 'paradoxes'. Both words have a variety of different meanings (Quine *1962a*), and it was sloppy of Russell not to make any distinction between results such as his own paradox of set theory and correct but surprising theorems or constructions that turn up in mathematics. I shall also write about 'the propositional and predicate calculi' and of 'quantification', where he spoke Peanese about 'free' and 'apparent variables'. Finally, I am using 'logicism' to describe his philosophical position, though this now common word was introduced with this sense only in the late 1920s (§8.7.6, §8.9.2).

Finally, Russell used 'analysis' in two different ways which he did not clearly distinguish and which therefore have misled many commentators. In its narrow sense it means breaking down a theory or body of knowledge into its basic units; its more general sense includes this one together with the companion synthetic process of construction of complexes from these units (Hager *1994a*, ch. 4). 'The business of philosophy, as I see it', he wrote later but seems to have thought from early on, 'is essentially that of logical analysis, followed by logical synthesis' (*1924a*, 176). Cauchy's 'mathematical analysis' has this general sense in its unfortunate name (§2.7.2), and Russell seems to have been following this tradition.

6.1.2 *Principal sources.* In addition to the original texts and historical surveys, some general sources are available. Of Russell's own reminiscences, his *My philosophical development* (*1959a*) and the first volume of his autobiography (*1967a*) are the most significant; but they are not reliable. He had a strong memory, and so relied on it more than was warranted (as often happens with people so gifted). Sometimes the errors are just of dating (not trivial in history, of course), but others are more serious. For example, a main theme of this chapter is that he massively over-simplified the story of writing *The principles*. Again, he claimed that one day he dictated to a secretary 'in a completely orderly sequence' the ideas which became the book *Our knowledge of the external world* (*1967a*, 210), and the tale has been much cited as evidence of the human capacity for mental preparation; however, letters of the time show that he struggled hard with the manuscript for months (§8.3.2).

In addition, the collections of letters that were added to the chapters of Russell's autobiography were not always well chosen or explained. For example, it includes a long string of letters to one Lucy Donnelly, who was never mentioned in the text (pp. 163–184); she was an American friend of a cousin of Russell's first wife Alys and later a patron of his last wife Edith, and met him when he lectured at her college, Bryn Mawr, in 1896 (§6.2.1).[1]

---

[1] In addition to these general reservations about Russell's autobiography, further doubts surround the third volume. A few parts are stated to be written by others, but even sections of it under his name totally lack his style. The most striking case for me is a lecture against

Other victims of silence include G. H. Hardy, with whom Russell actually enjoyed a long and varied friendship (my *1992a*).

Russell published his life story in the 1960s to gain funds for the various world-significant enterprises that were then operating under his name. For the same reason he sold his *Nachlass* to McMaster University in Hamilton in Ontario, Canada where it forms the basis of the splendid Russell Archives (hereafter, 'RA'). After his death in 1970 the rest of his unpublished materials went there, and when Edith died eight years later his library of books, offprints and journals was transferred also. When his second wife Dora died in 1986, some more of his manuscripts were found and transferred to the Archives; they are cited as '(RA, Dora Russell Papers)'.

Manuscripts until 1903 cause tricky problems of dating; for up to and including *The principles* of that year Russell often transferred folios of a rejected draft to its successors, so that several of them, including that book, are chronologically mixed. In addition, he used the new public facility of typing bureaux from time to time,[2] so that some items exist in both holograph and typed forms, often differently incomplete. Many of these manuscripts are now appearing alongside his published papers, essays and book reviews in an edition of his *Collected papers* edited by a team based at McMaster. Cited as 'Russell *Papers*', it is planned in 30 volumes: his logic and philosophy will occupy volumes 2–11, following the initial volume published in 1983, which covered his years at Cambridge to 1899. The mass of manuscripts on and around logic are surveyed in my *1985b*.

The edition excludes Russell's books, and most of his notes on others' writings on logic and mathematics, which survive in two large notebooks and several files of loose sheets. It also deliberately omits the masses of unpublished correspondence, though letters are used in the editorial matter. Further, the first volume of a selection of his letters has appeared (Russell *Letters 1*). His most important correspondents during his years as a logician were Couturat (Schmid *1983a*), Frege, Hardy, Whitehead; and Philip Jourdain, who took a lecture course with him in 1901–1902 at

---

American policy delivered at the London School of Economics on 15 February 1965 (*1969a*, 205–215). I heard it, in the Old Theatre, and still remember vividly the puzzled and even shocked reception by the audience, many of whom were acolytes; for he seemed not to be in contact with its contents (my *1998a*, 25–27: on p. 26, line 8, read 'script, but in so'). Luckily Russell's last years are not the subject of this book (a touch in §10.1.1); he certainly did not go senile, but his judgement in a number of important areas seems to have become faulty.

Russell's prepared this autobiography at various times from 1931 up to publication. He had written one in the early 1910s; unfortunately it seems to be lost.

[2] I hope your Dissertation is growing with all speed, and that you will have it typed by my people', wrote Russell to G. E. Moore on 20 July 1898, mentioning 'The Columbia Literary Agency, 9 Mill Str. Conduit Str. [London] W.' (Moore Papers, 8R/33/7; copy in RA).

Cambridge (§6.8.2) and later wrote to him at length on logic, *Mengenlehre* and their histories (my *1977b*).

PM is normally cited by theorem number, as '(*PM*, ∗41·351)', with the modern six-pointed star; following Peano, the original text used the eight-pointed '∗'. When volume and page numbers are needed, they come from the second printing of the 1920s (§8.4.4).

Russell's massive bibliography is magisterially catalogued in Blackwell and Ruja *1994a*. Any item, published or manuscript, will be cited by article or formula number if possible, but page numbers in the edition will be used when necessary, and also references to other material there. Russell published several papers on logic in French; they appear in the edition also, but I cite, and normally quote, the English translations that are provided. Finally, manuscripts are usually cited by page number in the edition; several are cited as, say, '*m1904c*', in which case I recall that *no* published work is named '*1904c*'.

Finally, since 1971 the Archives has published a journal entitled '*Russell*', which is the single principal source of information of Russell studies in general. This activity has grown enormously especially from the 1980s, with his philosophy prominently featured and logic appearing from time to time (my *1990b* surveys work in these areas). But the mathematical background and indeed foreground of logic is often not well treated: the commentary on Russell's philosophy usually lacks serious attention to Peano, Cantor, or the last 1,600 of the 1,800 pages of *PM*.

In these two chapters we see one of the two lives that Russell lived at that time, that of a philosopher-scholar working quietly and often on his own in the country; from 1905 to 1911 he lived at Bagley Wood near Oxford (Plate 3), in a house then the only one in the area, designed for him by a college friend (my *1974a*). For the rest of his time he was The Honourable Bertrand Russell, in London and other Important Places, knowing everybody and throwing himself into major social issues of the time such as Free Trade in the mid 1900s (*Papers 12*, 181–235). As a young member of the British aristocracy, he felt deeply the responsibility of his class at that time; Inheriting The Earth and so obligated to tend it carefully. Indeed, in this respect the philosopher Russell was of the same cast; this late Victorian (as he thought of himself) set up a logicist empire of mathematics and philosophy, and devoted much energy to its meticulous construction, especially after discovering its infection by paradoxes. Let us consider now the origins.

6.1.3 *Russell as a Cambridge undergraduate, 1891–1894.* (Griffin and Lewis *1990a*) Russell's parents died when he was an infant, and he passed a lonely childhood educated by tutors. His interest in mathematics developed quite quickly, especially for Euclidean geometry. By his teens he was proving things himself, for in 1890 he sent in to the *Educational times* a

PLATE 3. Russell outside his house at Bagley Wood with his friend
Goldsworthy Lowes Dickinson, maybe in 1905 (RA). The picture appeared in
my *1977b*, at which time the other figure had not been identified. It features
also as the frontispiece of Russell *Papers 4* (1994), with identification.

solution of a non-trivial problem set there about a property of a parabola
touching all sides of a triangle (my *1991a*).

Going up to Trinity College Cambridge as a minor scholar in mathematics in October 1890, Russell took the Part 1 Mathematics Tripos after
three academic years. Immediately he sold his mathematical books—an
action which suggests little enthusiasm for the experience. In his reminiscences he gave very few details; by contrast, his predecessor by one year,
Grace Chisholm (1868–1944) at Girton College, was eloquent on the
matter: 'At Cambridge the pursuit of pure learning was impossible. There
was no mathematician—or more properly no mathematical thinker—in
the place' (my *1972a*, 131). But this does not conform with the situation in
applied mathematics, with figures of the calibre of J. J. Thomson, J. J.
Larmor and Lord Rayleigh in and around town; indeed, the enrolment of
Trinity in the year after Russell's included E. T. Whittaker, who soon

became one of their distinguished successors. In addition, her judgement was harsh on Whitehead, although he had published very little by that time (his early thirties); Russell liked him as a teacher.

The disillusion seems to be more justified in pure mathematics, and at the undergraduate level, where, in Chisholm's view, Arthur Cayley 'sat, like a figure of Buddha on its pedestal, dead-weight on the mathematical school of Cambridge' (p. 115). Russell himself recalled that he never heard of Weierstrass while a student (*1926a*, 242); yet Cambridge gave Weierstrass an honorary doctorate in 1893. The initiative may have been taken by E. W. Hobson and A. R. Forsyth, the analysts at the University at the time; both knew Weierstrass's work, especially Forsyth *1893a* on complex analysis.[3]

The main defect with the Part 1 Tripos seems to have been the system of crammer-training; it reduced education to rehearsing techniques for answering Tripos questions, and replaced academic nourishment by aspiration for a high place on the list of Wranglers (the curious name for the mathematics graduates). 'Everything pointed to examinations, everything was judged by examination standards, progress stopped at the Tripos', recalled Chisholm, 'There was no interchange of ideas, there was no encouragement, there was no generosity' (p. 115). She left Cambridge to discover real mathematics at Göttingen where she wrote a *Dissertation* under the direction of Felix Klein in 1895, and after returning to England married one of the coaches, W. H. Young (1863–1942) (§4.2.4). Russell also travelled away, but in the mind.

6.1.4 *Cambridge philosophy in the 1890s.* (Griffin *1991a*, chs. 2–3) After passing the Part 1 Mathematical Tripos as joint 7th Wrangler, Russell turned to philosophy for Part 2. After some resistance, around the time of the examination he fell in with the dominating doctrines of Kantian and especially neo-Hegelian philosophy. The most prominent representative at Cambridge was J. M. E. McTaggart (1866–1925), but the leading British figure was F. H. Bradley (1846–1924) at Oxford. Since Russell practised this philosophy with some enthusiasm for the rest of the decade and held Bradley in high regard, some main features need to be noted, with Bradley's *The principles of logic* as the main source (*1883a*, cited from the second edition of 1922, which is almost unaltered and much more accessible; the 'Additional notes' to many chapters and new material at the end are not used).

Both kinds of philosophy stressed the importance of mental constructions and the objects thereby produced; in the neo-Hegelian form, they

---

[3] Much later Forsyth *1935a* lamented in the *Mathematical gazette* upon the quality of Tripos life in his time at Cambridge; but in reply Karl Pearson *1936a* gave it a warmer accolade. The differences may lie in the perceptions of a pure and of an applied mathematician.

were the *only* items for analysis, with facts treated on a par with propositions. Bradley emphasised judgement of the existence, content and meaning of ideas. Logic was an important handmaiden, for it distinguished categorical from hypothetical propositions and supplied basic notions like negation and principles such as identity, contradiction, excluded middle and double negation (Book 1, ch. 5). But his attention to matters symbolical was restricted to a short chapter on Jevons's system (§2.6.2), where he lamented its limitation to syllogistic logic and also showed himself not only resolutely but also triumphantly unmathematical (pp. 386–387).

Proof by contradiction was used frequently to produce sceptical conclusions from the given premises. In particular, taking a 'thesis' and its conflicting 'antithesis', a resolution was effected in the form of a 'synthesis' in some higher level of theorising (Book 3, pt. 2, chs. 4–6). In *Appearance and reality*, which appeared just before young Russell joined the faith, the ultimate goal was 'the Absolute', the realm of everything including itself (Bradley *1893a*, esp. chs. 14 and 26).

Bradley concluded that a relation was internal to the objects related (quite opposite to Peirce, whose work he did not seem to know): 'Relations, such as those of space and time, presuppose a common character in the things that they conjoin' (*1883a*, 253). Continuity and the continuum of space and time were fruitful source of contradictions, such as the same body in different places (p. 293). Among arithmetical examples, one and one only made two if they were manipulated in some way; otherwise they remained as one and one (p. 401). These kinds of cases were to attract Russell strongly, as we shall soon see.

6.2 THREE PHILOSOPHICAL PHASES IN THE FOUNDATION OF MATHEMATICS, 1895–1899

> [...] I don't know how other people philosophize, but what happens with me is, first, a logical instinct that the truth must lie in a certain region, and then an attempt to find its exact whereabouts in that region. I trust the instinct absolutely, tho' it is blind and dumb; but I know no words vague enough to express it. If I do not hit the exact point in the region, contradictions and difficulties still beset me; but tho' I know I must be more or less wrong, I don't think I am in the wrong region. The only thing I should ever, in my inmost thoughts, claim for any view of mine, would be that it is in a direction along which one can reach truth—never that it is truth.   Russell to Bradley, 30 January 1914 (RA)

Russell effected a sort of synthesis out of his education, in that he applied this philosophy to study foundational aspects of mathematics over the rest of the decade. He started out with some issues in dynamics (*Papers 2*, 29–34); they drew him to geometry, upon which he then concentrated. He also gradually took more interest in arithmetic and *Mengenlehre*. The selected survey in this section follows the order of these main concerns: the choice is partly guided by his later interests, which tended to focus upon arithmetic, *Mengenlehre*, continuity, infinity and geometries.

6.2.1 *Russell's idealist axiomatic geometries*. (Griffin *1991a*, ch. 4) In 1895 Russell won a Fellowship at Trinity with a study of geometry; this success led him to a career as a philosopher rather than as an economist or politician (*1948a*). His dissertation was examined by Whitehead and the philosopher James Ward: the manuscript has disappeared, but a chapter appeared in *Mind* as *1896a*; by oversight he left the word 'chapter' on p. 23 (regrettably changed to 'paper' in *Papers 2*, 285). Later that year he lectured on the topic in the U.S.A., at Bryn Mawr College and Johns Hopkins University, and after his return he published with Cambridge University Press a revised version of the dissertation as *An essay on the foundations of geometry* (*1897c*). Appearing in June in a run of 750 copies, it contained a few diagrams in its 200 pages; dedicated to McTaggart, effusive thanks were offered to Whitehead in the preface. He also wrote some other papers and manuscripts, now all gathered together in *Papers 2*. From the first essay up to 1899 Russell's position was basically unchanged; I shall usually quote from the book, and concentrate on his attention to axioms.

After a 50-page 'Short history of metageometry', using the word to cover all non-Euclidean geometries (§3.6.2), various philosophies of geometry were analysed. Two of them were found especially wanting. Firstly, Riemann's theory of manifolds (§2.7.3) was criticised for failing to stipulate the space in which they were to be found (Russell *1897c*, 64–65); to readers of Riemann who understood him better, his ability to formulate all properties of the manifold without recourse to any embedding space ('intrinsically', we now say) is precisely one of his virtues. Secondly, the recently deceased Hermann von Helmholtz, who for Russell 'was more of a philosopher than a mathematician' (p. xii), had moved too much the other way in advocating a totally empiricist philosophy of geometry, especially the claim that it could be deduced from mechanics (Helmholtz *1878a* (§3.6.2) was one of Russell's main sources). He concluded a long discussion thus with this typically Victorian flourish of capital letters (p. 81):

> But to make Geometry await the perfection of Physics, is to make Physics, which depends throughout on Geometry, forever impossible. As well might we leave the formation of numbers until we had counted the houses in Piccadilly.

His views on these German predecessors were held still more strongly by the German neo-Kantian philosopher Paul Natorp (§8.7.1) in a commentary on his book (Natorp *1901a*, art 3).

Russell's own position was guided by the neo-Hegelian philosophy that he had imbued. Instead of distinguishing between Euclidean and non-Euclidean geometries, he divided geometry into its 'projective' and 'metrical' branches by the criterion that the former involved only order but the latter also 'introduces the new idea of motion' (*1897c*, xvii) in order to effect measurement. These geometries were human constructions given space and time as an 'externality', and in this sense they were applied mathematics; however, synthetic *a priori* knowledge was present, and the main aim was to locate its place and role—central for projective geometry but only in parts of the metrical branch.

Russell did not present his position very clearly. For example, he found three *a priori* axioms for projective geometry, but presented them twice in somewhat imprecise and different ways, even in different orders (pp. 52, 132). The second account assumed that

P1) The 'parts of space' are distinguished only by lying 'outside one another', although they are all 'qualitatively similar';

P2) 'Space is continuous and infinitely divisible', finally arriving at a point, 'the zero of extension';

P3) 'Any two points determine a unique figure, called a straight line', three points a plane, four a solid, and so on finitely many times.

For the 'very different' (p. 146) metrical geometry Russell also proposed three *a priori* axioms, which he correlated with the 'equivalents' in the projective trio (p. 52). Nevertheless his order was different again; I shall mimic the one above:

M1) '*The Axiom of Dimensions*', that '*Space must have a finite integral number of Dimensions*' (p. 161);

M2) '*The Axiom of Free Mobility*', that '*Spatial magnitudes can be moved from place to place without distortion*' (p. 150), thus permitting the possible congruence between two figures to be examined;

M3) '*The Axiom of Distance*', that 'two points must determine a unique spatial quantity, distance' (p. 164), which was zero only when the two points coincided.

6.2.2 *The importance of axioms and relations.* In his book Russell tried to grant axioms M1)–M3) *a priori* status also. Concerning M1), the fact that we live in a world of three dimensions was 'wholly the work of experience' although 'not liable to the inaccuracy and uncertainty which usually belong to empirical knowledge' (pp. 162, 163). Together with Euclid's parallel postulate and straight line axiom (that two straight lines cannot enclose a space), they were 'empirical laws, obtained' by investigating 'experienced space' (pp. 175–176). M2) and M3) were *a priori* in the double sense of being 'presupposed in all spatial measurement' and 'a necessary property of any form of externality' (p. 161; see also pp. 173–174).

Most ambitious, however, was Russell's claim that M1)–M3) were also sufficient for metrical geometry. The reason was that the 'metageometers' have constructed other 'metrical systems, logically as unassailable as Euclid's [...] without the help of any other axioms' (p. 175). However, he overlooked the possibility that some other geometry might be constructed in which other axioms were needed: this was exactly the option allowed for by Riemann's approach, under which all geometries, Euclidean or non-Euclidean, were placed on the same epistemological level.

In his final chapter Russell considered the consequences of his position with regard to Kantian and Bradlean understanding of space and time. While the treatment is uncertain, with philosophical vicious circles spinning, it is striking that he relied upon axioms: we have seen already that mathematical theories were very often not axiomatised, Peano and Hilbert standing out among the exceptions of the time (and giving a very different treatment of geometries from Russell's: §5.2.4, §4.7.2). As part of his childhood interest in mathematics Russell had been profoundly puzzled when learning that in his *Elements* Euclid was forced to assume something to prove his theorems with such impressive rigour (*1967a*, 36). He must also have been drawn to axioms by his exposure from schooldays to that work, which was given great emphasis in English education; indeed, national controversies had raged over its manner of teaching in the Association for the Improvement of Geometrical Teaching (from 1897, the Mathematical Association: Price *1994a*). Although he saw the limitations of Euclid's rigour (as he related in an essay *1902b* in the Association's *Mathematical gazette*), the place of axioms remained important with him, and grew during his logicist phase. Certainly he did not learn it from neo-Hegelian philosophy where, as Bradley put it in the opening sentence of his *Logic*, 'It is impossible, before we have studied Logic, to know at what point our study should begin. And, after we have studied it, our uncertainty may remain'.

Among the axioms, M2) had two philosophical consequences. Russell also formulated it as '*Shapes do not in any way depend upon absolute position in space*' (*1897c*, 150), which not only imposed the opinion that space was relative but also focused his attention upon relations in general. 'All elements—points, lines, planes—have to be regarded as relations between other elements', he is reported to have told his Bryn Mawr audience in November 1896, in connection with projective geometry, 'thus space is simply an aggregate of relations' between points, lines and planes (*Papers 1*, 342). In the book, '*Position is not an intrinsic, but a purely relative, property of things in space*', indeed, even externality itself was 'an essentially relative conception' (*1897c*, 160); hence 'points are wholly constituted by relations, and have no intrinsic nature of their own' (p. 166). Thus, the relation of distance between two points was unique: by the Axiom of Distance, 'A straight line, then, is not the *shortest* distance, but is simply *the* distance between two points' (p. 168). In a manuscript 'Note on order'

he considered in detail relation(ships) between points for projective geometry and even laid out collections of axioms specific to them (*m1898c*,
345–347).

Russell's interests in axioms and in relations, which remained strong
throughout his mathematical career, took part of their common origin
from geometry, especially its projective part. As a neo-Hegelian he saw
relations themselves as internal to the objects related, but surely they
could not share all the same properties; for example, the property 'being a
factor of' between integers is not itself a factor. Such difficulties may have
led him to reject the position held in a paper in *Mind* 'On the relations of
number and quantity', where he treated them as different categories
(Michell *1997a*); 'number' was 'applied' to produce 'measure', while 'quantity', some 'portion' of the 'continuum of matter', yielded a 'magnitude'
(*1897b*, 72–73). Soon afterwards he also dropped his advocacy of the
relativism of space (§6.3.1).

6.2.3 *A pair of* pas de deux *with Paris: Couturat and Poincaré on
geometries.* (Sanzo *1976a*) Russell's book does not seem to have excited
the mathematicians; for example, Arthur Schönflies told the *Göttinge
Mathematische Gesellschaft* on 18 June 1897 that 'the author is a philosopher, the picture thus philosophically presented. For mathematicians it
offers no interest'.[4] However, it inspired a letter to Russell from Couturat,
then entering his thirties, which initiated an extensive correspondence.
Couturat also reviewed the book in two papers in the *Revue de métaphysique et de morale*, as part of his survey of foundational studies in mathematics mentioned in §4.2.3; Russell wrote a reply. This trio seems to have
provoked Henri Poincaré (1854–1912) into print, with similar results in the
same venue. The list of six papers is

> Couturat *1898a* (May), Couturat *1898b* (July), Russell *1898d*
> (November);
> Poincaré *1899a* (May), Russell *1899c* (November), Poincaré *1900a*
> (January).

Couturat also soon became involved in a French translation of Russell's
book, to which both he and Russell made some revisions and additions; it
appeared from Gauthier-Villars as Russell *1901f*.[5]

---

[4] Göttingen Mathematical Archive, 49; 1, fol. 20; 'Der Verfasser ist Philosoph, das Bild
philosophisch gehalten. Für Mathematiker bietet es kein Interesse'.

[5] The French edition of Russell's book, and also his exchange of papers, were discussed in
many letters with Couturat and some with the translator, Albert Cadenat. In addition, RA
holds a set of manuscript notes on revisions for the translation which Russell prepared in the
winter of 1898–1899 and kept in his own copy of it when it appeared. On 9 October 1900 the
house of Teubner asked Hilbert about the idea of translating into German both this book and
Whitehead's *Universal algebra* (Hilbert Papers, 403/15): Hilbert's reply is not known, but
neither translation has ever been effected.

While remembered as a critic, Couturat was not a very critical one. The main features of his long review *1898a* included some Kantian antinomies of space (not discussed above) and the empirical status of Euclid's axioms; in the follow-up *1898b* he treated the concepts of magnitude and quantity, where he wandered off into some group theory. Russell's rather unimpressive response *1898d* concentrated on axioms. He proposed that if a penny were rolled exactly one revolution on a horizontal surface and the length of this line compared with that of the radius, then the closeness of the ratio to π would give information on the 'space-constant' of empirical space (p. 326). But he also used his relativism to argue for 'the à priori character of Euclidean space' on the grounds that no absolute magnitude existed (pp. 327–328); but this does not easily fit with his empiricism. At the end he even went for a conventionalist view, that Euclid's axioms 'constitute the simplest hypothesis for explaining the facts' (p. 338).

This last argument, quite uncharacteristic of Russell, was the preference of Poincaré. Then 45 years old, he was drawn into action by Russell's rejection in the reply to Couturat of his own view that axioms were conventions, so that their truth-value need not be considered (p. 325; compare *1897c*, 30–38). Finding Couturat's review to be a 'very banal eloge', Poincaré *1899a* really was critical, on several issues. He rightly savaged Russell's sloppy formulation of projective geometry by P1)–P3), pointing out that P1) should have said that a straight line was determined by two points rather than the other way round; that the plane was specified as containing all three lines determined by pairs of three given points; and that a plane and a line always meet, possibly at the point of infinity, an important concept which Russell had not discussed at all (pp. 252–253). Among several other issues, he felt that Russell had exaggerated the similarities between projective and metrical geometries; in particular, the former was *not* necessary for experience (pp. 263–269), and qualitative aspects of geometry lay largely in topology (which he was then developing in a remarkable way: Bollinger *1972a*).

In this and his second piece Poincaré criticised Russell's talk of externality, and especially of the truth or falsehood of axioms on empirical grounds involving space; for him axioms were only convenient conventions, as was the fact that we live in three dimensions (*1900a*, 72–73). Further, Russell's proposed experiments to (dis)prove Euclid's axioms were actually exercises in mechanics or optics, whatever the geometry (pp. 78, 83–85).

In his reply to Poincaré's first paper, Russell *1899c* bowed suitably low over the failure of his axiomatisation, and gave a strikingly detailed and symbolised formulation of a new system of axioms.[6] But he stood firmer

---

[6] I report Russell's *intention* here; understandably but regrettably Couturat omitted a long symbolic completeness proof of his axiom system, based upon showing that a procedure due to Karl von Staudt produced a unique quadrilateral. It was first published in *Papers 2*, 404–408.

against conventionalism. As in his reply to Couturat, he saw essential and indeed welcome aspects of empiricism in metrical Euclidean geometry; he found no difficulty in seeking the truth-value of propositions such as 'There exist bodies (e.g. the earth) whose volume exceeds one cubic millimetre' (p. 398), and held out for isolating the notion of distance. But he may not have realised that Poincaré's conventionalism involved a sharp distinction between physical space and material bodies existing and moving in that space, and thus between a body and the portion of space which it occupies (O'Gorman *1977a*). The choice of geometry for space, and properties such as its continuity and congruence between figures, were conventional, and so had no causal effect on bodies. Russell's statement about the size of the earth concerns a body (including a convention about the unit of measurement), not geometry as such.

The exchange did not seem to leave any major mark on the positions of the opponents. In any case, Russell's interest in geometries decreased thereafter; his last major essay was an article, apparently written early in 1900 and entitled 'Geometry, non-Euclidean' when it appeared as *1902c* in the tenth edition of the *Encyclopaedia Britannica*. The first part was historical, largely following his geometry book with the division into three periods and his three axioms; he provided more details about non-Euclidean metrics. But the short philosophical part was naturally a long way from 1897, especially on the *a priori* nature of Euclidean geometry; in his final paragraphs on the '*Philosophical value of non-Euclidean Geometry*' he now concluded that 'There is thus a complete divorce between Geometry and the study of actual space' (p. 503). But he also emphasised here the merit of considering 'different sets of axioms, and the resulting logical analysis of geometrical results', and this philosophy was to remain durable.

6.2.4 *The emergence of Whitehead, 1898.*   In the exchanges with his Paris *confrères* Russell admitted to changes of expression and even mind on several aspects since the publication of his book; for example, to Couturat he corrected his remark on Helmholtz quoted in §6.2.1 to 'the *possibility* of Geometry cannot depend upon Physics' (*1898b*, 327). The exchange with Poincaré had also reinforced some growing doubts about the relativity of space, but he went public on the issue only in 1900, especially at the Paris Congress of Philosophy (§6.3.2).

Russell also kept a set of critical annotations about the *Essay* in his own copy of it (RA). Against a remark on p. 120 about points specifiable only by means of properties such as the straight line between them, he judged that 'This is a mistake. Pts., like str. lines, must be supposed to differ qual[itative]ly'. And of a quotation on p. 171 from William James's *The principles of psychology* (1890) that 'relations are facts of the same order with the facts they relate', such as 'the sensation of the line that joins the two points together', he confessed to himself that 'I have nowhere in my book grasped the meaning of this remark. Wh. gives a truer view than mine; everything spatial is <u>both</u> a relation and an object'.

Russell was alluding to a large volume *1898a* of more than 600 pages by Whitehead, entitled *A treatise on universal algebra with applications* and published by Cambridge University Press. Then in his 38th year, Whitehead had been a Fellow of Trinity College since 1884 thanks to a (lost) dissertation on Clark Maxwell's theory of electromagnetism; he had taken about seven years to write this his first book. This time it was Whitehead to thank Russell in the preface, for help over non-Euclidean geometry, a topic on which he himself was about to publish a paper (to add to two on hydrodynamics).

The title suggests a marked change of interest from those two papers, but it was ill-chosen, apparently at a very late stage. Whitehead had taken it from a paper by J. J. Sylvester *1884a* which only treated matrices, an algebra which he did not treat extensively in his book though he often deployed determinants. His book contained no all-embracing algebra, but instead a collection of newish algebras with applications mainly to geometries and a few aspects of mechanics. The chief inspiration came from the *Ausdehnungslehre* of Hermann Grassmann, whose work was beginning to gain general attention at last (§4.4.1). The British had taken little interest so far, however, so Whitehead ended with a bibliography of Grassmann's main writings as one of the historical notes appended to some chapters.

In the preface Whitehead saw symbolic logic both as pure 'systems of symbolism' valuable for 'the light thereby thrown on the general theory of symbolic reasoning' and in application as 'engines for the investigation of the possibilities of thought and reasoning connected with the abstract general idea of space' (p. v). Further, perhaps under the influence of Benjamin Peirce (§4.3.2), he defined mathematics 'in its widest significa-tion' as 'the development of all types of formal, necessary, deductive reasoning', so that 'the sole concern of mathematics is the inference of proposition from proposition' (p. vi). He did not furnish any extended philosophical discussions; but he had obviously not washed in neo-Hege-lianism, and Russell's note to himself above shows the superiority of interpreting points as members of a manifold.

In a short Book 1 on 'Principles of algebraic symbolism', Whitehead mentioned both Grassmann's and Riemann's theories of manifolds, cited Boole on (un)interpretability (§2.5.3), and emphasised general algebraic operations and their various laws. Then followed a Book on 'The algebra of symbolic logic', based upon Boole but using ' + ' without restriction and acknowledging MacColl (§2.6.4) for the propositional calculus. Like most contemporaries of all nationalities, he seems not to have read Hermann's brother Robert, who had explored the links between that calculus and algebraic logic more explicitly (§4.4.1). And in any case this Book played no essential role in the remaining five. Reviews of the book concentrated on this algebra and logic: a 40-page description Couturat *1900d* in the *Revue*; a rather discursive survey, also of Grassmann's work, in Natorp *1901a*, arts. 1–2; and a feeble notice MacColl *1899a* only of Book 2 in *Mind*.

Book 3, on 'Positional manifolds', ran through principal features of projective geometry in n dimensions, but in a largely algebraic manner expressing a point as a linear combination $\Sigma_r \alpha_r e_r$ of some basis $\{e_r\}$; Whitehead covered (hyper)planes and quadrics. Grassmann came to the fore in the 100-page Book 4 on the 'Calculus of extension', where Whitehead's coverage included not only the basic means of combination but also some aspects of matrix theory. Applications arrived in Book 5 on 'Extensive manifolds of three dimensions'; the main one was to systems of 'forces', but these were treated kinematically and the theory was virtually vector algebra. Measurement was introduced in the longest and most interesting Book, 6 on 'The theory of metrics' (156 pages), where he followed Cayley for axioms and worked through the theory in some detail for elliptic and hyperbolic geometries, adding some applications to mechanics and kinematics.

In the final Book 7, 'Applications of the calculus of extension to geometry', Whitehead treated vector algebra and analysis, including the vector and scalar products ('Vector area' and 'Flux' on pp. 509 and 527 respectively). He also handled some standard partial differential equations and elements of potential theory.

Overall the volume gives an unclear impression, resoundingly belying its title; Whitehead had mixed logic, algebra and geometry together, but the fusion had eluded him. While it marked an important stage in the development of his philosophy in general (Lowe *1962a*, ch. 6), he seems not to have seen ahead clearly. His next major mathematical foray was a long paper of 1899 on aspects of group theory, which he submitted to the Royal Society but then withdrew after finding many of his results in recent work by the German mathematician Georg Frobenius (my *1986a*). He intended to write a successor to this 'Volume 1', treating quaternions, matrices and Peirce's algebras (*1898a*, v); but he never fulfilled it, for the contact with Russell was gradually to develop into a formal collaboration (§6.8.2) which was to embody his philosophical aspirations, or at least several of them (§8.1.1–2). It has enjoyed no substantial influence (or detailed historical appraisal), although Russell reported on 16 August 1900 after the Paris Congress that 'Whitehead has a great reputation; all the foreigners who knew Mathematics had read and admired his book' (*Letters 1*, 202).

6.2.5 The impact of G. E. Moore, 1899. Russell communicated this news about Whitehead to his special philosophical friend G. E. Moore (1873–1958). 'It was towards the end of 1898 that Moore and I rebelled against both Kant and Hegel' (Russell *1959a*, 54). A year junior to Russell at

Trinity and trained in the same neo-Hegelian philosophy, Moore came the more rapidly to regard it as dangerous to mental health. He announced his revolt mainly in a paper *1899a* in *Mind* on 'The nature of judgement', where he proposed in anti-idealist vein:

1) facts are independent of our experience of them;
2) judgements (or propositions) deal primarily with concepts and relationships between them rather than with mental acts;
3) existence is a concept in its own right; so that
4) truth is specifiable relative to these various existents.

Instead of an all-embracing monism of the Absolute, he advocated pluralities, and moreover Out There rather than in the mind: for example, the truth or falsehood of existential judgements such as 'the chimera has three heads' was determined by relationship between the concepts chimera, three, head and existence. If 'the judgement is false, that is not because my *ideas* do not correspond to reality, but because such a conjunction of concepts is not to be found among existents' (p. 179). His new position also held no sympathy for phenomenology, where the act of perception of an object inhered with the object perceived (§4.6).

In the same vein, and volume of *Mind*, Moore *1899b* also published a review of Russell's *Essay* on geometry. While unable to tackle its mathematical side, he attacked the use of psychology and psychologism to identify the genesis of knowledge with knowledge in general, for example for failing to show that time was necessary for diversity of content (p. 401). He also queried the status of 'ideal motion' to move a figure onto another one as allowed by the Axiom of Free Mobility, for it surely assumed the congruence to be appraised (p. 403). Russell had already moved on from the position in his book towards a Moorean stance; this review, especially concerning the theory of judgement, nudged him along further.

6.2.6 *Three attempted books, 1898–1899.* Fitted out with Whitehead's geometric and logical algebras and Moore's external reality, Russell drafted a monograph on the foundation of mathematics during the summer of 1898. Its long title shows evidence of the new influences: 'An analysis of Mathematical Reasoning Being an Enquiry into the Subject-Matter, the Fundamental Conceptions, and the Necessary Postulates of Mathematics'. Much of it is lost or was transferred into later drafts; the surviving holograph and typescript *m1898a* show that a substantial text had been prepared (Griffin *1991a*, ch. 7).

The philosophical ground was still traditional: judgements prominent, subject-predicate logic and the part-whole theory of collections. But the place of Whitehead is evident in the title of the first of Russell's four Books: 'The Manifold'. He took the word as synonymous with 'Class', denoting a collection of 'terms' construed intensionally under some predicate; an extensional collection formed an 'assemblage' (pp. 179–180). The last chapter of the Book treated 'the branch of Mathematics called the Logical Calculus' (p. 190), exhibiting a "mathematicism" in tune with Boole, say, but converse to the logicism soon to come. It encompassed a fragment of Whitehead's Boole/Grassmann way of treating predicates $a, b, \ldots$ and their complements $\bar{a}, \bar{b}, \ldots$ relative to the Whiteheadian universe i. In 'Book II Number' the formula

$$\text{`}a = ab + a\bar{b}\text{', where } a \cdot \bar{a} = 0 \text{ and `}a + \bar{a} = i\text{'} \qquad (626.1)$$

was used to interpret judgements of adding the integers associated with a and b (pp. 201, 193). Development of the algebra convinced him in his holograph that 'the relation of whole and part underlies addition, and hence all Mathematics' (p. 205).

Whitehead's approach was evident also the discussion of number in Book 1. Cardinal integers were extensional manifolds (p. 196), but the connection between the two notions remained obscure. In the chapter on 'Ratio' Russell mooted the strategy of taking it as primitive and treating an integer as a special case of ratio: '20 would mean that the thing of which it is predicated has to the unit the relation 20:1'; similarly, 1/20 was construed as 1:20 (p. 207).

Among the fragments of 'Book III Quantity', the chapter 'On the Distinction of Sign' is notable for the immediate emphasis on the 'connection with order, and the two senses in which a series may be ordered' (p. 216); the link pervaded the chapter, and spilled into others. In particular, in connection with 'position in space or time' Russell indicated converse relationships such as 'A's adjective of being east of B, and B's adjective of being west of A' leading to the contradiction of space mentioned in §6.1.4 (p. 225). The consequences were fundamental for this young idealist: 'relations of this type pervade almost the whole of Mathematics, since they are involved in number, in order, in quantity, and in space and time' (p. 226).

The role of Moore came through in the greater place now accorded by Russell to concepts, and to more detailed examination of kinds of judgement. He referred quite frequently to 'existents', terms possessing the (primitive) property of existence. His examination of predication drew much on the pertaining classes. Partly in connection with such needs, numbers no longer had their old idealist home: 'Anything of which a cardinal integer can be asserted must be the extension of some concept—must be, in fact, a manifold' (p. 196).

Soon after setting aside this book, Russell tried another one, 'On the Principles of Arithmetic'. Two chapters survive, on cardinal and ordinal integers, largely following the predecessor as taking manifolds as basic and emphasising relations between terms; he thought that cardinals were epistemologically prior to ordinals (*m1898b*, 251).

The next book-to-be, on 'The Fundamental Ideas and Axioms of Mathematics' followed broadly the same approach but returned to the previous scale of ambition. An apparently complete 'Synoptic Table of Contents' shows that eight Parts were involved: on 'Number', 'Whole and Part', 'Order', 'Quantity', 'Extensive Continuity', 'Space and Time', 'Matter and Motion' and 'Motion and Causality' (Russell *m1899b*, 265–271). 'I find Order & Series a most fruitful & important topic', he told Moore on 18 July 1899, 'which philosophers have almost entirely neglected' (G. E. Moore Papers, 8R/33/14). An intensional approach is evident in the course of cogitating about 1, *one* and allied terms: 'A *class* may be defined as all terms having a given relation to a given term' (*m1899b*, 276). A lengthy discussion of inference stressed the logical order of propositions involved (pp. 291–294).

These forays show Russell in an enthusiastic state of mind but with neither the basic notions nor the mathematical range fully under control. For example, in the last text he simply added truth 'to the list of predicates' with the hiccup that 'To define truth is impossible, since the definition must be true' (p. 285).

Noticeably absent from all these drafts was Cantor's *Mengenlehre*; but Russell had come across it in 1896 and found it steadily more interesting. Let us now examine this parallel process.

6.2.7 *Russell's progress with Cantor's* Mengenlehre, *1896–1899.* The initial contact came not from Cambridge mathematics or the principal existing commentaries, but from Ward, who gave him perhaps in 1895 the pamphlet version *1883c* of Cantor's *Grundlagen* (Russell *1967a*, 68: he also received Frege's *Begriffsschrift*, but could make nothing of it). He may have started to read Cantor then; but the principal initiation came later that year when he was asked to review for *Mind* a book on atomism by the French philosopher Artur Hannequin. It contained over 20 pages on Cantor, drawing on the French translations *1883d* (*1895a*, 48–69), and on this evidence Russell found against Cantor in the review, mainly on the simple grounds that since the first number-class 'has no upper limit, it is hard to see how the second class is ever to begin' (Russell *1896b*, 37).

Later in 1896 a more substantial French volume on this new subject came Russell's way, and for the same reason: *Mind* asked him for a piece on Couturat's *De l'infini mathématique* (*1896a*). As we saw in §4.2.3, this book played a notable role in diffusing *Mengenlehre* and the foundations of arithmetic to a wider public than the mathematicians; Russell's review *1897a* must have helped to inform English-speaking philosophers, for he both surveyed many of Couturat's themes and showed their philosophical richness. But he was still not convinced by the rehearsal of Cantor's arguments for the existence of actually infinite numbers; and he also demurred against 'the axiom of continuity', as expressed by the Dedekind cut principle (p. 64, not so named). He also found 'mathematical zero' to be 'grossly contradictory', although 'quantitative zero is a limit necessarily arising out of the infinite divisibility of extensive quantities' (p. 64). At least he had advanced beyond the idealism run riot in an incomprehensible manuscript inference of the previous year that 'In *reality*, 0 sheep means so many cows' (*Papers 2*, 17); but much rethinking was still needed.

Russell applied himself to *Mengenlehre* with ardour in the winter of 1896–1897. A notebook called 'What shall I read?', kept between 1891 and 1902, shows that the French translations of Cantor were on the menu then, together with Dedekind's booklet on continuity (*Papers 1*, 357, 358).[7] He started to transcribe and comment upon Cantor's work at length in the right hand pages of a large black notebook (transcribed in *Papers 2*, 463–481); the opposite pages were left blank for later comments, such as bewilderment at the laws of combination of transfinite ordinals. But *Mengenlehre* seems to have dropped away while he had his Whitehead and Moore experiences; it came back to his reading list only in April 1898 with Dedekind's booklet on integers (which he had just bought), and in July 1899 with Cantor's main suite of papers (§3.2.6), including the full *Grundlagen* received from Ward some years earlier (*Papers 1*, 360, 362). He started a second large notebook on some of these works, which was to run into the 1900s and cover many other writings past and present (RA, Dora Russell Papers).

Russell's progress with Cantor's *Mengenlehre* grew with his understanding. In 1896 he had found it mistaken for failing to obey the normal rules (about infinities, for example). When he passed from (competent) commentators to the Master, he found it to be a rich source of both mathematical ideas and a solution to some idealist contradictions. By 1899 it was moving more centre stage, especially for the bearing of Cantor's theory of different order-types upon relations and order. But the full impact was still to come, and amidst other changes of philosophical import.

6.3 FROM NEO-HEGELIANISM TOWARDS 'PRINCIPLES', 1899–1901

6.3.1 *Changing relations.*   One central tenet issue of idealism was that relations were internal to the terms related. Russell had already found against this view, especially in concerning asymmetrical relations such as 'greater than' between terms; for example, it was surely necessary to pick out the ' > ' in 'A > B' to distinguish it from 'B > A' (*Papers 2*, 121, of 1898). Such doubts were reinforced during the winter of 1898–1899 when he prepared a course of lectures on the philosophy of Leibniz, delivering

---

[7] It is worth noting that, apart from their own latest writings, some of the figures who were to influence Russell had only read each other fairly recently: Peano on Dedekind in 1889 (§5.2.3), and on Frege by 1891 and again in review in 1895 (§5.4.4); and Frege on Dedekind in *1893a*, vii.

TABLE 631.1.   Russell's Classification of Relations

Symmetrical (631.1)	$ArB \supset BrA$ $(ArB$ and $BrC) \supset ArC$	equality, simultaneity, 'identity of content generally'
Reciprocal (631.2)	$ArB \supset BrA$ $(ArB$ and $BrC)$ not $\supset BrC$	inequality, separation in space or time, 'diversity of content generally'
Transitive (631.3)	$(ArB$ and $BrC) \supset BrC$ $ArB \supset$ not BrA	whole, part, before, after, greater, less, cause, effect
One-sided	None of the above	predication, occupancy of space or time

them in the following Lent (that is, spring) Term at Cambridge.[8] When he wrote up his lectures in book form he opposed Leibniz's internalist opinion that a proposition necessarily contained a subject and a predicate; he cited ones containing 'mathematical ideas' such as 'There are three men', which could not be construed as a sum of subject-predicate propositions, 'since the number only results from the singleness of the proposition' (*1900b*, 12); we may also sense here a philosopher primed to find the quantifier. He concluded, against Leibniz, that '*relation* is something distinct from and independent of subject and accident' (p. 13), a change that must have dented his idealism considerably.

During this period of preparation Russell considered in detail 'The classification of relations' in an essay *m1899a* read in January to the Cambridge Moral Sciences Club in which he described three main kinds together with a residual category. He used the names and examples as in Table 631.1; he seems to have adopted from Gilman *1892a* (§4.3.9) the notation '*ArB*' for a relation *r* between terms *A* and *B*.[9] To save space I use ' $\supset$ ' for Russell's 'if ... then'.

From various examples, Russell decided that 'diversity is a relation, and the precondition of all other relations' (p. 142). Curiously, he did not consider the converse of a relation in general, although they arose in two of his kinds: he was still a long way behind Peirce and Schröder. But he maintained his move towards the externalist interpretation with this poser at the end of his manuscript: 'When two terms have a relation, is the relation related to each?'.

---

[8] G. E. Moore attended Russell's lectures on Leibniz, and later helped him with the Latin texts and other aspects of the book (Moore Papers, 10/4/1–2 (notes) and 8R/33 *passim* (correspondence)). For a valuable survey of the book and its effect on immediate Leibniz scholarship, see O'Briant *1984a*.

The publication Couturat *1903a* of many hitherto unknown Leibniz manuscripts changed understanding on several matters, as Russell readily acknowledged in his review *1903b* in *Mind*. Couturat's interest had been stimulated by conversations with Vacca at the International Congress of Philosophy at Paris in 1900 (Lalande *1914a*, 653).

[9] Russell did not mention Gilman here, but he cited him in the paper on order discussed in §6.4.2 (*1901a*, 292, with the relation letter '*R*').

6.3.2 *Space and time, absolutely.*   Closely connected to the status of relations was the relativity of space, time and motion. Russell's unease about relativism became in the spring of 1899 a switch to the absolutist positions, first for space and then also for time—two more *voltes faces* of the retiring idealist. Of his various writings of this period (*Papers 3*, 215–282) I take his lecture *1901g* to the International Congress of Philosophy in Paris. This was the first occasion that he addressed an international audience of this calibre.

Russell wrote with the polemical conviction of a convert, citing Moore for his current philosophical line (pp. 252, 257). 'Since the arguments against absolute position have convinced almost the entire philosophical world', which included himself until rather recently, 'it would perhaps be well to respond to them one by one' (p. 249). He then mentioned Leibniz as one culprit, but he chose as standard target the *Metaphysik* (1879) by the German phenomenologist Hermann Lotze (1817–1881), who apparently was 'full of confusions' over senses of being (p. 253). Indeed, some criticisms seemed to be directed more against idealism as such rather than relativism; for example, that the view that a proposition in geometry had to be linked to time (pp. 248, 253), and that propositions had to have a subject-predicate form (p. 251).

Russell's main argument for absolutism was that each event then had its own location, so that for example, the simultaneity of two different events can be appraised (pp. 241–243). He started his exegesis with new definitions of symmetrical and transitive relations (631.1, 3), but stipulating only the respective first conditions of his classification manuscript (p. 241). 'Following Schröder' with the notation '$\breve{R}$' (after (446.4)) for the converse of R, he noted that the terms of such a relation lay in a series and doubted that when R denoted posterity relativism could properly express such a series of events in time, since the relation itself was supposed to be "absorbed" in the events (p. 242). He also noted similarities between his theory and that presented by Schröder *1901a* at the Congress on 'an extension of the idea of order' beyond Cantor's range to cases where several members of a collection could take the same rank; this was the only time that the theories of the two men converged.

6.3.3 *'Principles of Mathematics', 1899–1900.*   Russell added to the proofs of his book on Leibniz a footnote approving Leibniz's opinion that 'infinite aggregates have no number' as 'perhaps one of the best ways of escaping from the antinomy of infinite number' (*1900b*, 117). So even in 1900 Cantor's theory was not accepted. Indeed, given his idealist concern with continuity, Russell's reaction to Cantor's formulation of it had been surprisingly slight.

But after reading Cantor's *Grundlagen* in July 1899 (§6.2.7), *Mengenlehre* featured more prominently in Russell's next attempt to write a book on the foundations of mathematics. As its quoted title above shows, he was moving towards a more definitive conception; indeed, in contrast to its predecessors with their many discarded or transferred parts, this manuscript is pretty complete, about 170 pages in print in *Papers 2* (*m1899–1900a*). Further, its division into Books was to be followed fairly closely in *The principles* of 1903 (Table 643.1 below); indeed, in his habit of transferring manuscripts in well-ordered series, he took into it several portions of 'Analysis of Mathematical Reasoning', including much of Part 3 on quantity and the chapter on distinction of sign.

Cantor featured mainly in Part 5, 'Continuity and Infinity', where Russell discussed his formulation of continuity (pp. 110–115) and the generation of transfinite numbers, chiefly ordinals (pp. 116–125); however, in an earlier chapter on 'Infinite collections' he rehearsed again his doubts from the footnote in the Leibniz book (pp. 33–34). He still did not appreciate Cantor's general theory of order-types (§3.3.3), for it did not feature as much as it deserved in his Part 4 on 'Order'. However, he emphasised strongly the underlying importance of order: the logical order of propositions in inference, whole and part itself, ordinal numbers, and space and time. He gave a comparable status to the various kinds of relations and the series which they generated.

Elsewhere the Whitehead approach was again strong. In particular, manifolds were now collections, with the part-whole relation given Part 2 to itself. A chapter on 'Totality' concentrated upon 'all' or 'any' members of a whole which might share a predicate; but the quantifiers, already well known to Peirce and Schröder, were still absent. Integers remained difficult to define from collections and might have to be indefinable, though Russell mooted again from the 'Reasoning' manuscript the idea of defining integers as special cases of ratios (§6.2.6). Part 3 on 'Quantity', taken over from that manuscript, attempted a very general theory; some of his difficulties with infinity (and also with zero) arose from efforts to make them quantities. In places his ideas resembled those of Bettazzi's monograph *1890a* (§5.2.6), of which however he was still unaware.

Russell also wrote at some length on the calculus, seemingly using De Morgan's old textbook (§2.4.2), on which he had made notes in 1896 (*Papers 2*, 519–520); he even adopted the antiquated name 'differential coefficient' as the title of a chapter (*m1899–1900a*, 131). It opened with such a lamentable summary of Leibniz's approach that one must conclude that his recent reading of that philosopher had omitted the calculus entirely. His account of limits concluded that 'dy/dx is the limit of a ratio, not a ratio of limits' (p. 135), which is Cauchy's approach (272.1), which he did not mention at all. So his principles of mathematics were still somewhat scattered, and also scrappy; no Part was devoted to geometries, although various aspects arose in the discussions of space and time, and of mechanics.

Russell seems to have worked on this manuscript until June 1900. In that month he also completed a draft of his Congress offering on the absoluteness of space and time and sent it off to Couturat. A month later, in the company of Alys and the Whiteheads, he followed it to Paris.

6.4 THE FIRST IMPACT OF PEANO

> I am obliged to you that you gave me the sad announcement of the death
> of Peano. He indeed is the man whom I much admired, from the moment
> when I came to know him, for the first time, in 1900, at a Philosophical
> Congress, which he dominated on account of the exactness of his mind.
>
> Russell *1932a*, to Sylvia Pankhurst

6.4.1 *The Paris Congress of Philosophy, August 1900: Schröder versus Peano on 'the'.* As was described in §5.5.1, this event, unprecedented in scale, generated considerable interest; the products included four volumes of proceedings, and three lengthy reports on the logical and mathematical sessions, two from organiser Couturat *1900e* and *1900f*, and Lovett *1900a* for the U.S.A. In addition to presenting his own paper *1901g* on absolute order in space and time—which received a tepid discussion—Russell also read for W. E. Johnson an abstract on 'logical equations'. Presuming that they attended, he and Whitehead will have heard, among others, Poincaré (on mechanics), MacColl and Schröder, and abstracts read from MacFarlane and Poretsky. But the magic time was the morning of Friday 3 August, when the Peanists gave their concert (§5.5); Peano and Padoa in person, Burali-Forti and Pieri in summaries read out by Couturat.

As we recall, Peano had spoken on definitions in mathematics. An ensuing discussion, presumably around 10 o'clock, first stimulated Russell's excitement in him. Peano *1901a* rejected definitions such as

$$0 = a - a \qquad (641.1)$$

on the grounds that '*a*' could not be allowed to float free. Schröder objected to this ban, citing as an example his own specification (445.3) of the contradiction 0 as (*a* and not-*a*) for any proposition *a* (Lovett *1900a*, 169–170). But Peano stood his ground; as Russell recalled to Norbert Wiener in 1913, Schröder's proposed definition $(446.1)_1$ of his empty domain 0 was ill-formed (my *1975b*, 110):

> There is need of a notation for 'the'. What is alleged does not enable you to put
> '0 = etc. Df.'. It was a discussion on this very point between Schröder and Peano
> in 1900 at Paris that first led me to think Peano superior.

This *personal contact* with Peano was the crucial factor for Russell (and Whitehead); first the 'the' question, and then reflection about the Peanists in general from Friday lunch-time onwards. Russell had received one offprint from them, Pieri *1898a* on geometry, in 1898 (§5.5.5); and he had seen the paper Couturat *1899a* in the *Revue* on Peano, for he mentioned it in a letter to its author on 9 October 1899 (copy in RA). But these texts had not been enough.

After the Congress Russell stayed abroad for a few days before returning to England. Later he stated that he received and read Peano's works at the Congress (*1959a*, 65); but in fact Peano had with him for sale only the current issue (volume 7, number 1) of his *Rivista di matematica* (my *1977b*, 133), and Russell had to wait until the end of August before the earlier numbers and other material came in the post. He was busy enough, however, since the proofs of the Leibniz book had been around since June. Moore read these, and Russell also told him on 16 August of an 'admirable' gathering, with 'much first-rate discussion of mathematical philosophy. I am persuaded that Peano and his school are the best people of the present time in that line' (Russell *Letters 1*, 202).

Russell received from Peano the first two editions of the compilation *Formulaire des mathématiques*, the first six volumes of the *Rivista* (now trading under the title '*Revue de mathématiques*'), and the short book *1889b* on geometry. He read again Cantor's *Grundlagen*, and by November he had also consumed Pieri's offprint, Dedekind's booklet on integers again, Bolzano's book on paradoxes, Bettazzi's monograph on quantities, Pasch's lectures on geometry, and at last Cantor's final pair of papers *1895b* and *1897a* on general sets (the first in the Italian translation *1895c* in the *Rivista*); next Febuary's reading included Hilbert's book on the foundations of geometry (*Papers 1*, 363–364). Among other works, he bought in September a set of Schröder's lectures then published (RA). Late in 1900 he also looked at the first (and then only) volume of Frege's *Grundgesetze*, but made little of it (letter to Jourdain in my *1977a*, 133); this was a pity, for he had told Moore in August that the meaning of 'any' had been of special interest in Paris, and here Frege was perceptive.

6.4.2 *Annotating and popularising in the autumn.* One of Russell's first reactions to the Peanist experience was to add comments and references to the several folios of the current manuscript on 'Principles of Mathematics' (§6.3.3). The most striking addition, dated October, filled most of the space surrounding his titling of Part 2, on 'Meaning of whole and part'.[10] To the left he put:

> I have been wrong in regarding the Logical Calculus as having specially to do with whole and part. *Whole* is distinct from *Class*, and occurs nowhere in the Logical Calculus, which depends on these notions: (1) implication (2) and (3) negation.

[10] This important folio is reproduced in Russell *Papers 3*, plate 2; the top part is also on the front cover of Rodriguez-Consuegra *1991a*.

His word 'class' referred to Cantorian sets, centre stage in Peano's logic; he cited Bettazzi's book for whole-part theory. To the right he resolved that 'I must preface Arithmetic, as Peano does, by the true Logical Calculus, to be called Book I, The Individual'.

Also in the autumn Russell wrote two papers for *Mind*. The first, *1901a* 'On the notion of order', drew much on Part 4 of the current 'Principles' manuscript: definitions of (in)transitive and (a)symmetrical relations, series generated by them, and a lengthy discussion of logical order before turning to examples in integers, space and time. But in the text he cited Peanists several times (including Pieri's offprint), Bolzano's book, and De Morgan's paper *1860a* on relations. The second paper was based upon the English draft of his Paris talk on absolute order, and so was largely pre-Peanist in content; but he cited the new master twice on matters of geometry (*1901e*, 265, 269), and for the first time in print he mentioned Frege, on the objectivity of cardinals (p. 278).

Russell also proposed to editor G. F. Stout a more popular essay for *Mind* on the Peanists, an idea which Stout welcomed. He produced in the autumn an excellent survey, starting out with Weierstrass's emphasis on rigour and not only emphasising distinctions such as between a term and its unit class but also indicating the Peanists' mathematical range, of which 'the theory of Arithmetic [...] is] I think Peano's masterpiece' (*m1900c*, 358). He also took note of some German work, such as Schröder and writings (not Hilbert's) on the foundations of geometry. Unfortunately the essay was not published; maybe Stout changed his mind, but then the *Mathematical gazette* would have been a suitable venue. At all events, British readers never saw a most timely and competent piece of enlightenment.

6.4.3 *Dating the origins of Russell's logicism.* In his reminiscences Russell tells us that during the rest of 1900 he wrote yet another book manuscript at great speed, which formed the substance of *The principles of mathematics* (Russell *1903a*); in the intervening period some revision was carried out, especially on the two opening Parts and the last one (*1959a*, 72–73; *1967a*, 145). But he told Jourdain a different story in April 1910 (my *1977b*, 133):

> During September 1900 I invented my Logic of Relations; early in October I wrote the article that appeared in RdM VII 2–3 [Russell *1901b* in Peano's *Rivista* (§6.5.2)]; during the rest of the year I wrote Parts III–VI of my *Principles* (Part VII is largely earlier, Parts I and II wholly later, May 1902) [...]

Russell received back the manuscript from the Press after publication, and kept it in his files. Like this letter, it suggests a different story from the well-known recollection (my *1997b*); a very heterogeneous text, not only because of transferral of folios from 'Principles' but especially for the chronology of the writing, which follows the order of Parts 3-4-5-6-1-2-1again-7. Further, Parts 1 and 2 were referred to only in general ways in the later ones; in particular, a mention in Part 5 that 'irrationals could not be treated in Part II' (p. 278) refers to 'I or II' in the manuscript, and in a similar remark four pages later 'I' was altered to 'II' for publication. In addition, unlike the other three Parts, in the manuscript of Parts 3–6 the chapters are numbered from 1 onwards in each Part instead of the consecutive system that was printed (numbers 19–52); the texts are not divided into the numbered articles printed (149–436); and there are no printers' markings. It seems likely that another version of them was prepared (probably a typescript), which he and the printer used.

These elements of evidence suggest two surprises: *that Parts 1 and 2 did not exist at all in 1900*, at least not beyond sketch form; and *that the book conceived in 1900 did not advocate logicism.* These hypotheses, and study of the manuscript of the book and pertinent letters and diaries, suggest this scenario:

1) In the autumn of 1900 Russell was sure that Peano's programme was important for him, with its logic and the central role given to Cantor's set theory, and so could provide Parts 1 and 2 with the grounding that he had been seeking; however, a logic of relations had to be introduced. He also followed the Peanists in maintaining some distinction between mathematics and logic, although he was not sure what or where it was, especially regarding set theory. So he re-wrote Parts 3–6 of 'Principles': Part 5, on infinity and continuity, was especially pertinent.

2) In the new year (and century), Russell decided that the distinction did not exist: instead, pure mathematics was *contained* in Peanist logic. (His special sense of 'pure' will be explained in §6.5.1.) However, he did not yet have a detailed conception of this logic, apart from the need for relations, which he quickly sketched out; still awaiting clarity were the constants and indefinables, and the status of set theory.

3) In January, and definitively in May, he rethought a discussion in Part 5 of Cantor's diagonal argument, and thereby found his paradox.

4) Around the same time Russell thought out more clearly the basic notions of his logic, and thereby refined logicism. The notion of variable was now crucial, for Part 1 carried 'The Variable' as its new title (§6.7.1). However, propositional functions and quantification still remained rather in the shadows. Part 2 on 'Number' was also written, including the definition of cardinal integers as classes of similar classes, basic for arithmetic and therefore for logicism.

5) By the spring of 1902 Part 1 could be developed further; the prominence of the variable was tempered by deeper consideration of propositional functions, so that the Part was now called 'The indefinables of mathematics'. Despite the presence of the paradox, logicism could still be stated, in more detail, and the book readied for publication by further referencing and changes and two new appendices.

This proposed chronology, outlined in more detail in Table 643.1, guides the design of the rest of this chapter. After a preface and an elaborate

TABLE 643.1. Russell's Progress with *The principles*, August 1900–February 1903.
ProM ='Principles of mathematics' *m1899–1900a*. *Pr = The principles
of mathematics*. *Papers* entry gives the first page(s) of the text(s).

Month(s)	Papers 3	Activity	Here
August 00		Hears Peanists; likes their logic and use of set theory	§6.4.1
September 00		Learns Peanese: invents logic of relations	§6.4.1–2
October 00	590	Drafts paper *m1900c* on relations	§6.4.4
October–Dec 00	351	Writes manuscript *m1900d* on Peanists	§6.4.3
November 00		Writes Parts 3–5 of *Pr* in Peanist spirit, using ProM	§6.4.5–7
December 00		Writes Parts 3–5 of *Pr* in Peanist spirit, using ProM	§6.4.8
January 01?		Envisions logicism: 'pure mathematics' in his logic	§6.5.1
January 01	363	Writes popular essay *1901d* on mathematics	§6.5.1
Jan–May 01?	385	Approaches his and Burali-Forti's paradoxes	§6.6.1
February 01	310/613	Completes paper *1901b* on relations: sent to Peano	§6.5.2
March–April 01	630	Drafts paper *1902a* on well-ordered series	§6.5.4
?–May 01		Refines logicism: clarifies logical indefinables and constants	§6.7.1
Apr–May 01?		Finds his paradox of set theory	§6.6.2
May 01	181	Drafts Part 1 *m1901c* for *Pr*; includes his paradox	§6.7.1
June 01?	423	Writes *1902d* for Whitehead: definition of cardinals	§6.5.3
June 01		Writes Part 2 of *Pr*, using ProM	§6.7.2
August 01	384/661	Completes paper *1902a* on series: sent to Peano	§6.5.4
April–May 02	208	Writes Part 1 of *Pr*	§6.7.3–4
May 02		Writes Part 7 of *Pr* (much from ProM)	§6.7.5
May 02		Readies manuscript of *Pr*	§6.7.6
June 02–Feb 03		Handles proofs: adds many footnotes, rewrites passages	§6.7.6–7
July?–Nov 02		Writes appendix A on Frege's work	§6.7.8
November 02		Completes appendix B on the theory of types	§6.7.9
December 02		Writes preface of *Pr*	§6.8.1
February 03		Indexes *Pr*	§6.7.6
May/June 03		*Pr* published in Britain/in U.S.A.	§6.8.1

TABLE 643.2. Summary by Parts of Russell's 'Principles of mathematics' (1899–1900) ('ProM') and *The principles of mathematics* (1903) ('*Pr*'). The Summaries of *Pr* use many chapter titles but do not always follow the order of chapters.

ProM; chs.	*Pr; chs., pp.*	*Summary of main contents of* The principles
1: 'Number'; 6	1: 'The indefinables of mathematics'; 10, 105	'Definition of pure mathematics'; 'Symbolic logic', 'Implication and formal implication'; 'Proper names, adjectives and verbs', 'Denoting'; 'Classes', 'Propositional functions', 'The variable', 'Relations'; 'The contradiction'
2: 'Whole and part'; 5	2: 'Number'; 8, 43	Cardinals, definition and operations; 'Finite and infinite'; Peano axioms; Numbers as classes; 'Whole and part', 'Infinite wholes'; 'Ratios and fractions'
3: 'Quantity'; 4	3: 'Quantity'; 5, 40	'The meaning of magnitude'; 'The range of quantity', numbers and measurement; 'Zero'; 'Infinite, the infintesimal, and continuity'
4: 'Order'; 6	4: 'Order'; 8, 58	Series, open and closed; 'Meaning of order', Asymmetrical relations', 'Difference of sense and of sign'; 'Progressions and ordinal numbers', 'Dedekind's theory of number'; 'Distance'
5: 'Continuity and Infinity'; 9	5: 'Infinity and continuity'; 12, 110	'Correlation of series'; real and irrational numbers, limits; continuity, Cantor's and ordinal; transfinite cardinals and ordinals; calculus; infinitesimals, infinite and the continuum
6: 'Space and Time'; 4	6: 'Space'; 9, 91	'Complex numbers'; geometries, projective, descriptive, metrical; Definitions of spaces; continuity, Kant; Philosophy of points
7: 'Matter and motion'; 7	7: 'Matter and motion'; 7, 34	'Matter'; 'Motion', definition, absolute and relative, Newton's laws; 'Causality', 'Definition of dynamical world', 'Hertz's dynamics'
	Appendix A: 23 pages	Frege on logic and arithmetic
	Appendix B: 6 pages	'The doctrine of types'

analytical table of contents, the main text of the latter was divided into seven Parts with 59 chapters and 474 numbered articles, 498 pages in all. By intention, the text was largely prosodic, with a modest use of symbols and rather few formulae or diagrams; the formal version was planned for a sequel volume (p. xvi).

The length and range of both book and its own manuscript could generate an historical analysis of comparable length, 'with an appendix of leading passages' (to quote the sub-title of his book on Leibniz). Quite a few folios came from 'Principles' (Table 643.2 compares the book with this manuscript), and some even earlier (King *m1984a*). A few were discarded but kept—for example and not only, the folio heralding his paradox (§6.6.2). Later, many changes and additions were made in proof.

Despite its fame, a book never out of print since its re-issue in 1937 (§9.5.4), no comprehensive survey of its contents seems to have been written (Vuillemin *1968a* is one of the best studies); indeed, many commentators seem unable to get much beyond Parts 1–2 and the two appendices. Both published and written versions are noted here, along with several associated manuscripts and published papers which are now gathered together in *Papers 3*.

6.4.4 *Drafting the logic of relations, October 1900.*   Russell was bowled over by reading the Peanists; mathematical range combined with logical power, especially the use of predicates and quantification, and especially the overthrow of subject-predicate logic with the distinction between membership and inclusion. But he soon found fault with them; in particular, they had failed to develop a logic of relations. Thinking out many of the required details in September, he wrote out a draft manuscript *m1900c* of a paper for Peano the next month, in which he affirmed his belief in the central importance of relations for logic and mathematics. I note here some main features, reserving some details for the final version in §6.5.2.

Russell wrote fully in Peanese, with all the notations, 'Pp' for both axioms and rules of inference, ' = Df', the numbering of propositions, wallpaper look, the lot. His opening flourish criticised Schröder and Peirce; like the Peanists, he did not appreciate their achievements, or De Morgan's before them. Again he used, with acknowledgement, only Schröder's '$\breve{R}$' for the converse of relation R, and also '1' and '0' in (446.6) for identity and diversity respectively. As he told Jourdain in April 1910, 'I read Schröder on Relations in September 1900, and found his methods hopeless, but Peano gave just what I wanted' (my *1977b*, 134). Thus much of the logic which he developed repeated details of the structure which the algebraists had already furnished. (This was the major issue between Russell and Wiener (§8.2.7), which stimulated the reminiscence quoted in §6.4.1.) But the differences were substantial: in particular, he construed relations as intensions defined by some property external to the objects related.[11] He denoted the 'domain' and 'converse domain' of a relation by using whenever possible the corresponding lower case Greek letter, such as 'ρ' and '$\breve{\rho}$' for relation R (*m1900c*, 590).

Among other preliminaries Russell distinguished between the compound '$R_1 R_2$' of relations R_1 and R_2, and the class '$R_1 \cap R_2$' of ordered pairs in common between them (p. 591). Padoa's symbol 'Elm' for the class of unit classes (§5.4.6) was frequently used, for it was easier in Peanese to handle unit classes than their individual members. In an interesting paragraph he floated the idea that diversity might replace identity as a 'logical indefinable' (pp. 593–594).

A striking pair of symbol-strings occurred within a few lines on pp. 591–592:

$$\text{'}\varepsilon\ \varepsilon\ \text{Rel' and '}x\ \varepsilon^2\ y\text{'}. \qquad (644.1)$$

The second formula simply used Peanese to say that x belongs to the class of classes y (because there exists a class z belonging to y and containing x as member), while the first stated that membership was itself a relation and so belonged to the class of them. However, has the symbolism slipped into a formalism? Is not the first 'ε' a noun while the second is a verb? This conflation of use and mention is an early case of many to be found in Russell's logic.

Between these two lines occurs a hiccup when Russell defined the class of individuals by the property of belonging to a class. However, since a class can belong to a class of classes, then this definition or that of Cls (not given here) needs refinement. The status of individuals was to remain a considerable difficulty in Russell's logic (§7.8.3).

The mathematical exercises concentrated on arithmetic. After defining the similarity of two classes by the existence of a one-one relation taking one class for its domain and the other for converse domain, Russell defined the class of cardinal numbers as the converse domain of the one-one compound '$S\breve{S}$' of any many-one relation S, so that two similar classes had the same cardinal (pp. 595–596)—and Peano's principle of abstraction (536.1) now became a theorem. He also rehearsed various basic definitions and properties of ordinal numbers, including transfinite ones, where he introduced the name 'progression' for an infinite well-ordered series (p. 597). I leave the details to the more ample presentation in the published version (§6.5.2).

---

[11] The differences between the distinctions of intensions and extensions, and between internality and externality, often confuse students of relations. Russell could have been more explicit in *PM 1*, 26, and in recollection in *1959a*, 54–62 (externality) and 87–89 (extensionality).

In addition to material which would become very familiar in Russell's later logical writings, the draft included articles on 'groups' with applications to 'distance' and 'angles' (pp. 594–595, 609–612). The mathematics came from Cayley and Klein, and his interest in it harked back to his geometry book (*1897c*, 28–38). 'Group' for Russell was basically a permutation group composed of a class of one-one relations which contained the converse of each member (the identity relation '1" was assumed present) and the compound of any two members which had equal domains): in Peanese (*m1900c*, 594),

$$\text{`Group} = G = \text{Cls'}1 \rightarrow 1 \cap K \ni \left\{ P \, \varepsilon \, K . \supset_P . \, \breve{P} \, \varepsilon \, K : \right.$$

$$\left. P, R \, \varepsilon \, K . \supset_{P,R} . \, PR \, \varepsilon \, K . \, \pi = \rho \right\} \text{Df'}$$

(644.2)

(with 'Df' serving double duty). The motivation came from Whitehead, as we shall see in §6.5.3. Unfortunately, although Russell's contact with Whitehead had increased by the time of his final version, he left out these articles—an early sign of his narrowing conception of mathematics, which unfortunately was to continue through the decade.

After this exercise Russell then rewrote Parts 3–5 of his previous 'Principles' into a new book during November, and Part 6 the month afterwards. The concluding article of each Part was a general summary; we shall use it for guidance, together with the table of contents.

6.4.5 *Part 3 of* The principles, *November 1900: quantity and magnitude.* (Manuscript, Byrd *1996a*; summary, art. 186) This Part, 'Quantity', largely followed the previous version (§6.3.3): 'Magnitudes are more abstract than quantities: when two quantities are equal, they have the *same* magnitude' (p. 159) as defined via transitive and symmetrical relations (p. 163). He stressed that both notions were general, dependent upon order (to be analysed in the next Part), but not upon divisibility and so not necessarily restricted to continuous or discrete ranges. Among derived notions, an important one in connection with measurement was 'the terms intermediate between any two' a_0 and a_n in a series, which 'may be called the *stretch* from a_0 and a_n'. A 'whole composed of these terms is a quantity, and has a divisibility measured by the number of terms, provided their number is finite' (p. 181).

In places this Part was less pithy than its predecessor, but one main extension lay in ch. 22 on the 'quantitative zero'; it had 'a certain connection both with the number 0 and the null-class in Logic, but it is not (I think) definable in terms of either' (p. 184). He considered but rejected various other possible definitions; for example, identity would not do because 'zero distance is not actually the same concept as identity' (p. 186).

He finally plumped for defining a zero for each kind of magnitude rather than some "universal" type, and used as defining expression 'the denial of the defining concept' of its kind, such as *'no pleasure'* for the zero magnitude of pleasure. It was a special relation, holding 'between *no pleasure* and *pleasure*, or between *no distance* and *distance*', say; it was 'not obtained by the logical denial of pleasure, and is not the same as the logical notion of *not pleasure*' (p. 187).

6.4.6 *Part 4, November 1900: order and ordinals.* (Manuscript, Byrd *1996a*; summary, art. 248) This Part, 'Order', elaborated in prose many of the ideas of the draft paper on relations together with applications to measurement; but continuity and infinity were postponed as much as possible until the next Part. Russell explained at length the generation of series as domains and/or converse domains of relations of various kinds (chs. 34–35), with special emphasis on transitive asymmetrical relations (hereafter, 'TAR') in ch. 26 for producing order in general (ch. 36) and progressions in particular (ch. 38). This led him to define ordinal numbers, with negative ones generated by the converse of a relation (p. 244). He then gave in ch. 30 an account of Dedekind's definition of integers (§3.4.1), with the transformation theory reworked in terms of relations. Although he appreciated the significance of the theorem which guaranteed mathematical induction, nevertheless he concluded that Dedekind produced 'not the numbers, but any progression' (p. 249).

One assumption in Dedekind's treatment was that ordinals were prior to cardinals. Russell noted this also, but sided with Cantor's opposite viewpoint (pp. 241–242). The issue was to recur as logicism developed.

As in his draft for Peano, Russell's account of TAR included a criticism of Peano's principle of abstraction which he restated as defining a concept such as the number of a class by such relations. He noted its assumption of an entity satisfying this relation, which he cast as an axiom, 'my principle of abstraction' (p. 220). This principle was to play quite an important role in Russell's drive for nominal definitions in mathematics, although he tended to replace it by talk of the corresponding classes of classes or relations in order to avoid the charge of ambiguous definition (Rodriguez-Consuegra *1991a*, 189–205).

Another major use of TAR was in defining the concept of distance, which Russell handled in a rather peculiar ch. 31 to end the Part. Seeking a definition more general than that used in mathematics itself, he presented it as a one-one TAR between two terms of the generated series; thus the sum of two distances came from compounding the corresponding relations. If the second distance was the reverse of the first, then zero distance was produced: symmetry was abandoned, and the relations were mutually converse (*1903a*, 253). While somewhat obscure, the treatment played a significant role in the Part following.

6.4.7 *Part 5, November 1900: the transfinite and the continuous.* (Manuscript, Byrd *1994a*; summary, art. 350) At 112 pages 'Infinity and Continuity' was the longest Part of the book. Deeply influenced by set theory—not just order but also the theory of transfinite numbers and the real number system—it was the first extended account in English of this material. Russell showed a good mastery of the current situation; on the transfinite cardinals, for example, he emphasised the Schröder-Bernstein Theorem 425.1 (cited from Borel *1898a*, 108–109 and Zermelo *1901a*) and doubted that trichotomy could be proved (p. 306).

Russell began his own way of handling the theorems with one of his major innovations: 'The correlation of series' (ch. 32) by means of order-isomorphic relations, so that 'when one series is given' as generated by relation P, 'others may be generated' by applying a one-one 'generating relation' R and its converse $\breve{R}$ so as to product $\breve{R}PR$; it was a TAR if P was also (p. 261). In the next March he was to develop the method further in his second paper for Peano, and sometime to add three important new articles to the book (§6.5.4).

One main innovation in this Part was Russell's definition of irrational numbers. He took the class of rational numbers, an 'everywhere dense set' for Cantor (§3.2.3), as a 'compact series' (not the sense of this adjective which has endured), and defined an irrational number as a 'segment', that is, the class of rational numbers less than some given one. 'My contention is, that a segment of rationals *is* a real number' (p. 272), or more precisely after much discussion of limits, 'a segment of rationals which does not have a limit' (p. 286). His approach resembled Dedekind's cut (§3.2.4), which he had analysed in the interim (pp. 278–284); but instead of positing a number corresponding to a cut Russell gave another nominal definition in terms of classes of classes.

For Russell the advantage of his procedure was that the existence of these numbers was guaranteed only this way, and he criticised at some length the theories of Dedekind, Cantor and Weierstrass on this issue (pp. 280–285). However, his Peanist enthusiasm had rather led him astray. He used the word 'existence' here in Peano's sense (541.3), defined of a class that it be non-empty. But for mathematical purposes existence has to be understood far more generally, and none of his victims can be accused of error in principle merely for not defining non-empty classes even if one may criticise them on other grounds. A further irony is that he had been anticipated in this definition by Moritz Pasch—not in the textbook *1882b* on geometry which had much pleased him in earlier work and also in Part 6 to come here, but in the contemporary one *1882a* on the calculus; he learnt of his predecessor only in 1910 from Jourdain (my *1977b*, 139).

The 'infinitesimal calculus' received in this Part a rare, and short, exposition from Russell (ch. 39). Although he cited Leibniz several times, his appraisal of Leibniz's methods as 'extremely crude' (p. 325) shows that he had still not carefully read the mathematics. His account naturally followed the dictates of Weierstrass, whose approach he had learnt from the textbook literature, especially Stolz *1885a*, Dini *1892a* and C. Jordan *1893a* (pp. 328–329); in particular, '*dy/dx* [...] is not a fraction, and *dx* and *dy* are nothing but typographical parts of one symbol' (p. 342), quite opposed to Leibniz's own reading of it as the ratio *dy ÷ dx* of differentials (§2.7.1).

Russell's main concern was with the status of infinitesimals and the continuum, to which he devoted three chapters (40–42). Again in line with the Cantorian *Diktat* (§3.6.3), he concluded that 'infinitesimals as explaining continuity must be regarded as unnecessary, erroneous, and self-contradictory' (p. 345). On this last calumny he considered Zeno's famous paradoxes, giving the usual but irrelevant solution of Achilles and the tortoise in terms of limits (pp. 350, 358-360).[12] When he came to write Part 1, this paradox would play a different role (§6.7.4); for now, as November 1900 drew to an end, he reverted to his old subject.

6.4.8 *Part 6, December 1900: geometries in space.* (Manuscript, Byrd *1999a*; summary, art. 436) For mathematical directions in this Part, on 'Space', Russell drew upon Pasch and Whitehead. Hilbert's book was cited only for three details (pp. 384, 405, 415); a foundational approach allowing interpretations as beer-mugs would have seemed alien to Russell. He organised this Part along the lines of his book on geometry, specifying separately the projective, descriptive and metrical branches in terms of appropriate relations between points and then examining the relationships between them (chs. 45–47, 48). Point itself was an indefinable class-concept for each geometry (p. 382, with Pieri *1898a* highly praised); its existence was defended against the relativistic critics by appeal to absolute space (ch. 51).

Russell's affirmation of absolutism was not the only philosophical change since the days of the geometry book: Moorean empiricism having replaced idealism, principles had to be changed. In particular, instead of assuming space to be an *a priori* externality (§6.2.1), he provided 'Definitions of various spaces' (ch. 49); each one was a class of terms (or entities) endowed with relations between them appropriate to the axioms of the associated geometry, and its continuity could be formulated entirely by Cantorian means (ch. 50). The number *n* of dimensions of a space was defined from the series of series of ... (*n* − 1 times) of terms, each one generated by a TAR (pp. 374–376). Cantor's proof of the equi-cardinality of line and plane (§3.2.5), and its extension to more dimensions, showed

---

[12] The irrelevance of solving the Achilles-tortoise paradox in terms of limits lies in the fact that, in the primary sources such as Aristotle, *the argument is valid*: 'Achilles is still running', for it is *not* stated that either contestant is moving with uniform velocity, so that each one could be slowing down all the time (my *1974c*). See also footnote 18.

for Russell that his definition could be extended to ω dimensions, thus making clear that the numbers themselves were ordinals (p. 376).

Russell did not confine himself to the various axioms but also considered appropriate metrics, drawing on discussions of distance and measurement in earlier Parts, and related notions. For example, in metrical geometry 'An angle is a stretch of rays [from its vertex], not a class of points' (p. xxxv in the table of contents, summarising p. 416).

So Russell finished the old century with a fine reworking of geometry. However, the basis of the philosophy was still not fully thought out; Parts 1 and 2, on basic logical notions and definitions of cardinals, were still not down on paper. As always throughout his years of studying the foundations of mathematics, the weak part was the foundations themselves; the mathematical roots lay tangled in the ground. Let us leave him at the end of the century with this productive four-Part draft, and catch up with the activities since Paris of his friend Whitehead, work which was already beginning to intersect with his own.

6.4.9 *Whitehead on 'the algebra of symbolic logic', 1900.* Whitehead's reaction to the Paris Congresses was different from Russell's, for he continued in a largely algebraic style with an examination of Cantor's theory of finite and infinite cardinals, spiced up with the new Peano/Russell logic. Maybe the combination of the Peanists with the philosophers and Hilbert announcing Cantor's continuum hypothesis as the first of his problems to the mathematicians sparked this interest. At all events, he produced three papers and an addendum within two years, publishing them all in the *American journal of mathematics,* slightly over 100 pages in total length. The venue may seem surprising; but, in addition to its reputation it was edited by Frank Morley (1860–1937), who had been a fellow student with Whitehead in the mid 1880s but had emigrated to the U.S.A. and was then professor at Johns Hopkins University.[13] In patriotic U.S.A. style, Whitehead's Peanist eight-point asterisks used to number the theorems were printed as five-pointed stars '★'!

Whitehead's first paper *1901a,* written in 1900 and proof-read the following February, dealt with 'the algebra of symbolic logic'. Finding algebraic logic to be 'like argon in relation to the other chemical elements, inert and without intrinsic activities', he sought to inject it with the juices of the theory of equations: factorising Boolean expansions into 'prime' linear terms in its predicate variables $x, y, \ldots$ and their complements $\bar{x}, \bar{y}, \ldots$ relative to some universe i, with coefficients $a, b, \ldots \bar{a}, \bar{b}, \ldots$; finding a necessary and sufficient condition that such an expansion admits

---

[13] Some letters from Whitehead survive in the *Nachlass* of Morley (Humanities Research Center, University of Texas at Austin). On 5 November 1902 he opined that Schröder's symbolism for the logic of relations 'is entirely useless for *mathematical research*' and that only Peano's programme would do.

a unique solution; extending theory to cases where these quantities were *denumerably infinite* in number; forming symmetric functions of these quantities; examining the groups of substitutions and of transformations of the variables (hence Russell's awareness of groups in (644.2)); and seeking invariants under these transformations. In the third paper, finished in July 1901 according to a February footnote added to its predecessor, he calculated the 'order' (meaning the cardinality) of some of the classes of groups that he had found; he did not confine himself to finite group theory, for he interpreted answers such as 24^n as 'equal to the power of the continuum at least' if the size n of the group was infinite (Whitehead *1903a*, 171). In such ways he brought algebra into transfinite arithmetic; his treatment of the finite cardinals is described in §6.5.3.

6.5   CONVOLUTING TOWARDS LOGICISM, 1900–1901

6.5.1 *Logicism as generalised metageometry, January 1901.*   Unlike his book on geometry, Russell did not use the word 'metageometry' in Part 6 of *The principles*. But the plurality of geometries was prominent: the theorems of a geometry depend hypothetically upon the axioms and other assumptions required (Nagel *1939a*). One passage in the manuscript is especially striking: 'In this way, Geometry has become (what it was formerly mistakenly called) a branch of pure mathematics, that is to say, a subject in which the assertions are that such and such consequences follow from such and such premises, not that entities such as the premises describe actually exist' (also printed thus, at p. 373). This passage, or at least the thoughts in it, may well have solved for Russell his demarcation problem between logic and mathematics; generalising this conception of metageometry, he envisioned logicism as the philosophy which defined *all* pure mathematics as hypothetical, and that the Peanist line between mathematics and logic did not exist. All mathematics, or at least those branches handled in this book, could be obtained from mathematical logic as an all-embracing implication, for this new category of 'pure mathematics'; the propositional and predicate calculi (including relations) with quantification provided the means of deduction, while the set theory furnished the "stuff": terms or individuals, and classes or relations (of classes or relations ...) of them. Maybe a trace memory of this origin of logicism came to him when he introduced the reprinting of the book many years later: 'I was originally led to emphasis this [implicational] form by the consideration of Geometry' (*1937a*, vii).

Later on in Part 6 occurs a similar passage: 'And when it is realized that all mathematical ideas, except those of Logic, can be defined, it is seen that there are no primitive propositions in mathematics except those of Logic' (*1903a*, 430). Unfortunately, unlike the passage from page 373 just cited, this one belongs to a sector of the manuscript which is lost,[14] so we cannot tell if it was so written in December 1900; my guess is that it does contain some rewriting. But it also shows other features of the developing logicism, such as an emphatic discussion of the need for nominal definitions (p. 429):

> [...] a definition is no part of mathematics at all, and does not make any statement concerning the entities dealt with by mathematics, but is simply and solely a statement of a symbolic abbreviation; it is a proposition concerning symbols, not concerning what is symbolized. I do not mean, of course, to affirm that the word *definition* has no other meaning, but only that this is its true mathematical meaning.

The details of the logicistic vision were still not clear, but Russell made his first public statement of it in a popular essay 'On recent work on the principles of mathematics'. Written in January 1901 (my *1977b*, 133), it was published in the July number of a 'most contemptible' (Russell in *Papers 3*, 363) American journal called *International monthly*, nevertheless selling over 6,000 copies.[15] This essay has become one of his best known works of this genre, largely because he included it in the anthology volume *Mysticism and logic* in 1918. His announcement took the form of an aphorism which soon became very well known: 'mathematics may be defined as the subject in which we never know what we are talking about, nor whether what we are saying is true' (*1901d*, 365). Its kernel is the hypothetical character given to mathematics; but its import *as* the birth announcement of logicism has understandably escaped readers. One early example was a ramble Vailati *1904a* around this 'most recent definition of mathematics' by one of the Peanists, who taught Russell's logicism at the University of Turin then (see his letter of 26 July to Giovanni Vacca in *Letters*, 235). Another hint came a little later with the unfortunate identity thesis that 'formal logic, which has thus at last shown itself to be identical to mathematics' (Russell *1901d*, 367), although it lacks elaboration to clarify its seemingly whimsical tone: 'those who wish to know the nature of these things need only read the works of such men as Peano and Georg Cantor' (p. 369). We return to this matter in §6.7.3.

Also regrettable is Russell's equally famous aphorism in the essay: the howler, often quoted by fellow misinterpreters, that 'Pure mathematics was discovered by Boole, in a work which he called the *Laws of thought* (1854)'. It launched a whole paragraph of incomprehension (p. 365), typical of the way in which one kind of symbolic logician did not at all understand the

---

[14] No manuscript survives between fols. 81a and 169 of Part 6. The corresponding published version starts around the middle of p. 413 (in a much rewritten passage) and ends at p. 453, line 15 ↑ with '(4) succession;'.

[15] See the letter from the *International monthly* to the Open Court Publishing Company, 9 September 1900 (Open Court Papers, Box 27/27).

work of the other tradition. Presumably he used here the adjective 'pure' in its traditional sense rather than his new one, in which case earlier examples can be found (for example, Diophantos in the +4th century); but either way he misunderstood Boole's use of uninterpretable formulae (§2.5.3), although Whitehead had clearly explained it in *Universal algebra* (*1898a*, 10–12 and elsewhere). Again, he did not convey the central fact that Boole always saw his work as mathematics *applied* to logic (§2.5.8). Furthermore, like the algebraic logicians to follow him, Boole's logic was *only qualitative*: Russell's vision was of *a logic both qualitative and quantitative*, with constructions of real numbers, continuity and geometry to be made.

This remark on Boole is not the only hiccup in the essay: in the 1918 reprint Russell had to add some footnotes to correct mistaken opinions made about Cantor (pp. 374, 375); the second one will be handled in §6.6.2. But logicism was emerging, as Russell absorbed the logic that he had recently learnt and then developed it in two important papers for Peano's journal.

6.5.2 *The first paper for Peano, February 1901: relations and numbers.* The final version of Russell's paper on relations, prepared by February in French, was sent to Peano in March. Peano's letter of thanks and acceptance contained the extensionalist statement that 'The classes of couples correspond to relations' (Kennedy *1975a*, 214), explaining the absence of such a theory from his programme but doubtless also confirming to Russell the need for his paper! (compare §7.4.1). It appeared later in the year in two consecutive issues of the *Rivista* as Russell *1901b*.

In the first article of the paper Russell ran through in somewhat more detail than in his draft (§6.4.4) the criticism of Schröder and Peirce, notations such as 'R' and '$\breve{R}$', and basic properties such as compounds and converses. Among many other passages repeated here were the ungrammatical proposition '$\varepsilon \; \varepsilon \; \text{Rel}$' (644.1); but the idea of diversity as indefinable was surprisingly dropped, as were the articles on groups, distance and angles.

The next article, elaborating art. 3 of the draft, treated 'Cardinal numbers' on the basis of similarity between two classes u and v:

$$\text{`*1·1} \quad u, v \; \varepsilon \; \text{Cls} . \supset \; : u \; \text{sim} \; v . = . \; \exists \, 1 \to 1 \cap R \ni (u \supset \rho . \breve{\rho} u = v) \; \text{Df'}.$$

$$(652.1)$$

He added nervously, that 'If we wish to define a cardinal number by abstraction, we can only define it as a class of classes, of which each has a one-one correspondence with the class "cardinal number" and to which belong every class that has such a correspondence' (p. 321). These nominal definitions took him that one crucial step beyond the Peanists, but he wrote hypothetically, and also only in words. Perhaps as a concession to non-logicist Peano or maybe in shared doubt, his surrounding theorems dealt with 'the number of a class' in the Peanist tradition (§5.3.3) rather than numbers themselves. But the passage reads oddly for another reason: art. 1 had ended with the categorical statement that 'the cardinal number of a class u will be the class of classes similar to u' (p. 320)! However, that remark was added only on the proofs, as the end product of a tortuous analysis, through the draft and final versions to these proofs, of the similarity relation and its domain and converse domain (Rodriguez-Consuegra *1987a*, 143–150).

In art. 3 Russell reworked from the draft 'Progressions', his name for denumerable series, as well-ordered classes. His first presentation of a symbolic definition of a number seemed again nervous: 'ω, or rather, if one wishes, a definition of the class of denumerable series. The ordinal numbers are, in effect, classes of series' (p. 325: the original French, 'si l'on veut', is perhaps less weighty). The uncertainty was justified, however, albeit unintentionally; for (already in the draft) he forgot to use u in series form and so had actually defined $\aleph_0$. He acknowledged and corrected this error in the second paper for Peano (*1902a*, 391). Here he went on to prove results which corresponded to the basic operations of arithmetic; for example (*1901b*, 328):

$$a + b = c \text{ to } aR^b c \text{ and } x + ab = y \text{ to } x(R^a)^b y. \qquad (652.2)$$

Then multiplication could be defined by associating a relation B with each muliplicand (p. 331):

$$\text{'}aBc . = . ab = c\text{'}, \text{ with } a, b \text{ and } c \text{ numbers.} \qquad (652.3)$$

The notion of ratio lost more status, for division was effected by using the corresponding relations: b/c was linked to $B\check{C}$ (p. 332).

Perhaps the most important novelty relative to the draft was art. 4, on 'finite and infinite'. Perhaps building upon Cantor and especially Dedekind (§3.4.2), Russell defined a finite and an infinite class by the respective properties of not being, or being, similar to the class created by removing one member. In an interesting following paragraph he noted the alternative definition of finite numbers by mathematical induction, and confessed himself unable to deduce either definition from the other one (p. 335). He had come across an issue which was to help him to recognise an axiom of choice three years later (§7.1.6), and grudgingly to adopt an axiom of infinity two years after that (§7.6.1).

The last two articles of the paper were largely rewritings in Peanese of Cantor's theories of everywhere dense sets (now 'compact series'), and of progressions and their inverse order-type 'regressions' within them. Russell reworked in detail many results in Cantor's recent paper *1895b*, including the construction of compact sub-classes and properties of limit members. His treatment relied on the existence for each series u of a one-one 'generating relation' R whose domain contained u and converse domain included all relata of members of u (forming themselves a sub-class of u) together with further properties to ensure well-order (p. 341).

6.5.3 *Cardinal arithmetic with Whitehead and Russell, June 1901.* Despite some slips and unclarities, this paper was a brilliant début in Peanese, one of the best in the *Rivista* to date. But absent from it were Russell's nominal definitions of cardinals; they were given in the middle of Whitehead's second paper, which was completed by June and published in the *American journal of mathematics* as *1902a*. Whitehead began by introducing Peano's symbolism and Russell's logic of relations; on Cls (also written 'cls'), 'a class whose extension is formed by all classes', he included the proposition

$$\text{'Cls } \varepsilon \text{ Cls . Cls} = \text{Cls'} \tag{653.1}$$

(p. 372), which was to raise Jourdain's eyebrows in the *Rivista*, for one (*1906a*, 134–135).

Whitehead also introduced the important notion 'cls^2 excl', the class of mutually exclusive (or disjoint) class, whose cardinality was the product of those of the given classes. The 'multiplicative class' '$d^{\times}$' of a class d was defined on p.383 as

$$\text{'}d \text{ } \varepsilon \text{ cls}^2 \text{ excl} . \supset \therefore$$

$$d^{\times} = \text{cls} \cap m \ni \{p \text{ } \varepsilon \text{ } d . \supset_p . p \cap m \text{ } \varepsilon \text{ } 1 : m \supset \cup \text{'}d\} . \text{Df'}; \tag{653.2}$$

it was to be crucial in the devclopment of logicist arithmetic. The applications centred on proving in as general manner as possible theorems on the addition and multiplication of cardinals, especially $\aleph_0$, and extending the binomial theorem to infinite indices. In a small addendum paper Whitehead *1904a* proved that strict inequality of cardinals was preserved under addition.

Russell was credited by Whitehead with adapting into Peanese Cantor's proof by diagonal argument (§3.4.6) that the cardinality of a class was less than that of its power-class (*1902a*, 392–394). But his main contribution was art. 3, listed separately as Russell *1902d*, in which he gave the nominal definition of cardinals as classes of classes. 0 was the class '$\iota\Lambda$' of the empty class Λ; 1 the class of all unit classes, thus allowing him to replace Padoa's 'Elm' by '1'; and so on, with the defining expressions formulated to avoid the vicious circle implicit in my chatty formulation above. After defining the class 'Nc' of cardinal numbers as the class of classes of classes

z for which there exists a class having z as its cardinal, Russell unfortunately slipped up in his definitions from 1 upwards (p. 435); he pointed out the mistake to Frege December 1902 (Frege *Letters*, 251) and the following November to Couturat (RA), but not in *The principles*. I give both definitions, in chronological order:

$$\text{'}1 = \text{cls} \cap u \ni (x \, \varepsilon \, u \, . \, \supset \, . \, v \sim \iota x \, \varepsilon \, 0) \text{ Df'}. \tag{653.3}$$

$$\text{'}1 = \text{cls} \cap u \ni \{ \exists \, u \cap x \ni (v \sim \iota x \, \varepsilon \, 0) \} \text{ Df'}. \tag{653.4}$$

The first definition fails because it allows the class $0 \, \varepsilon \, 1$.

After defining the class 'Nc fin' of finite cardinal numbers by the property of mathematical induction—which he took as known rather than primitive—Russell worked through the basic operations and (in)equations of finite arithmetic. However, his construction was incomplete, in that he used multiplication but did not define it. After defining the class of infinite cardinals as the complement of Nc fin relative to Nc, he also proved various results of its different arithmetic, finishing off with the Schröder-Bernstein Theorem 425.1 to establish trichotomy (p. 430, citing Borel *1898a*, as in §6.4.7).

6.5.4 *The second paper for Peano, March–August 1901: set theory with series.* A draft of Whitehead's paper seems to have inspired Russell to treat ordinals with comparable detail. A second paper for Peano was apparently drafted in the spring of 1901, and revised into French during July and August (*Papers 3*, 630–673); like the first, it also appeared the following year in two consecutive issues, as Russell *1902a*. He provided a 'General theory of well-ordered series', largely as presented in Cantor's recent two-part paper (§3.4.7). It enriched Russell's budding logicism by expressing in Peanese his conviction of the importance of order for logic and mathematics.

Although much material was reorganised between draft and published version, there were no major changes of content and far less hesitancy than in the first paper. Presumably for reasons of diplomacy, Russell discarded the draft remark that the Peanists' 'endeavour to dispense with *relation* altogether as a fundamental logical notion' (*Papers 3*, 632). The major theme of this paper was the class 'of relations generating well-ordered series', defined from the appropriate kind of transitive relation P, with domain π (*1902a*, 390). The class λP of relations order-isomorphic to P launched a lasting concern with 'the relation *likeness* [L] between two relations' P and P' with common domain π under the generating relation (§6.4.7) S with domain σ:

$$\text{'*}2{\cdot}1 \quad (P)L(P') . = . \, P, P' \, \varepsilon \, \text{Rel} .$$

$$\exists \, 1 \rightarrow 1 \cap S \ni (\sigma = \pi \cup \breve{\pi} . P' = \breve{S}PS) \quad \text{Df'}, \tag{654.1}$$

together with the class λP of relations order-isomorphic to P (p. 392). 'As the properties of likeness are important, I shall develop some of them', especially concerning order-isomorphism and terms in the generated series (pp. 393–395, 407–409). Likeness led via the class Ω 'of relations generating well-ordered series' (p. 390) to

$$'[*]2\cdot12 \quad N_0 = \text{Cls} \cap x \ni \{\exists\, \Omega \cap P \ni (x = \lambda P)\}' \text{ Df}, \quad (654.2)$$

where this time without doubt 'N_0 is the class of ordinal numbers. An ordinal number is a class of well-ordered similar relations', although the status of (654.2) as definition was not mentioned (p. 393, 'N_0' misprinted). After a previous, and this time correct, definition of ω (p. 391), he advanced up to ω_1 (p. 416), the starter of Cantor's third number-class; he even proved results such as $\omega_1 = \omega^{\omega_1}$ (p. 420). He also established Cantor's theorem that the series of ordinals less than any given one was well-ordered, but he confessed that 'We do not know how to demonstrate that the class of all ordinals forms a well-ordered series' (p. 405), and later that 'There is no reason, so far as I know, to believe that every class can be well-ordered' (p. 410).

Russell worked out many features of ordinal arithmetic. In contrast to his definition of mathematical induction in the first paper, he now had in *6·1 'a generalized form of complete induction. In a well-ordered series, if s be a class to which belong the first term of the series and the successor of any part of the series contained in s, then the whole series is contained in s' thanks to the transitive and asymmetric relation which generated the terms of s as its field (pp. 404–405). Maybe there was influence from Burali-Forti (§5.3.8) as well as Peano in this reduction of assumptions.

While trying later in this article to show that the ordinals were associative under multiplication (*6·47), Russell confessed that 'I do not know how to extend the method of Prop 6·47 to a product of an infinite number of ordinals' (p. 408). Further thoughts on precisely this technique would make him a pioneer of the axioms of choice three years later (§7.1.6).

In addition, Russell considered the transfinite cardinals, starting out from a definition of $\aleph_0$:

$$*7\cdot32\ '\alpha_0 = \text{Cls} \cap u \ni \{\exists\, \omega \cap P \ni (u = p)\}\ \text{Df}', \quad (654.3)$$

where 'I have replaced Cantor's aleph by α, since this letter is more convenient' (p. 410); 'p' was the 'range' of P, that is, the union of its domain and converse domain (p. 390). In a short manuscript on 'continuous series' produced probably later in 1901 (*Papers 3*, 431–436), he went further by trying to produce Peanist definitions of rational and real numbers in imitation of Cantor's way of producing continuous order-types (§3.4.6).

In a 'Note' within the paper Russell extended the definitions to cases where *P* and *P'* might not be well-ordered, thus creating 'relation-arithmetic' instead of Cantorian ordinals (*1902a*, 407–408). He developed the theory during the winter in various notes (*Papers 3*, 437–451). Perhaps around this time and certainly under the influence of the content of this paper, he added three important articles (299–301) to Part 5 of *The principles*. In the first he outlined the basic principles of likeness between relations, and relation-arithmetic, presaging his most original contribution to *PM* (§7.8.5). In the second he recalled Cantor's two principles of generating finite and then transfinite ordinals (§3.2.6); but he felt that Cantor's well-ordering principle 'seems to be unwarranted', giving the absence of a proof of the well-ordering of the continuum as an example. The third one took the special case of 'the type of the whole series of all ordinal numbers'—where talk of the greatest one led to a 'contradiction' of Burali-Forti (p. 323). Logicism had seemed to be going wonderfully well; however, just at this sunny time the roof fell in. As he recalled later (*1959a*, 73), 'after an intellectual honeymoon such as I have never experienced before or since' in writing *The principles* in the autumn of 1900, 'early in the following year intellectual sorrow descended upon me in full measure'.

6.6   FROM 'FALLACY' TO 'CONTRADICTION', 1900–1901

> A *paradox* is properly something which is contrary to general opinion: but it is frequently used to signify something self-contradictory [...] *Paralogism*, by its etymology, is best fitted to signify an offence against the formal rules of inference.   De Morgan (*1847a*, 238, 239)

6.6.1 *Russell on Cantor's 'fallacy', November 1900.*   In the essay in the *International monthly* Russell described as a 'very subtle fallacy' Cantor's belief (§3.5.3) that there is no greatest cardinal (*1901a*, 375). He was following a passage in the manuscript of *The principles* written in November 1900 on 'The philosophy of the infinite' where, with his usual enthusiasm for faulting Cantor before reading him carefully, he found two supposed errors:

1) there *was* such a number, namely that of the class 'Cls' of all classes, so that
2) the diagonal argument (346.1) could not be applied to it to create a class of still greater cardinality.

Applying that argument to Cls by setting up a mapping to its power-class in which each class of classes was related to itself and every other class to its own power-class, he thought that 'Cantor's method has not given a new term, and has therefore failed to give the requisite proof that there are numbers greater than that of classes' (Coffa *1979a*, 35).

But at some time in the ensuing months Russell thought over this argument, and diagnosed a different illness. Maybe he had read Cantor's paper *1890a* introducing the diagonal argument, perhaps in the Italian translation *1892b* in Peano's *Rivista*, which he now possessed. Or maybe he had told his result to Whitehead, who was then working on Cantor's theory of cardinals (§6.5.3) and who would doubtless have found unbelievable the deduction above. In a letter to Jourdain now lost, Russell reported that 'In January he had only found that there must be *something* wrong' concerning this argument (Jourdain *1913e*, 146). He now found fault neither in the idea of no greatest cardinal nor in the diagonal argument but in the new class thrown by up the mapping—and the new news was very serious.

6.6.2 *Russell's switch to a 'contradiction'*.   We recall from §3.4.6 that Cantor's diagonal argument showed that the cardinality of any class α was less than that of its power-class $P(\alpha)$ by attempting to set up an isomorphism between the classes but finding a member of $P(\alpha)$—that is, a class —to which there was no corresponding member of α. After noting that some classes belonged to themselves while the rest did not do so, Russell now took his deduction to show that the class of all classes which did not belong to themselves belonged to itself if and only if did not do so—and, by a repetition of the argument, *vice versa also*. This is his paradox.

The passage in *The principles* was withdrawn (but Russell kept the folio), and in May 1901 the revised argument was expressed in terms of predicates in the chapter on 'Classes and Relations' of an attempted 'Book 1 The Variable' (§6.7.1) (*m1901c*, 195):

> We saw that some predicates [for example, 'unity'] can be predicated of themselves. Consider now those (and they are the vast majority) of which this is not the case. [. . .] But there is no predicate which attaches to all of them and to no other terms. For this predicate will either be predicable or not predicable of itself. If it is predicable of itself, it is one of those referents by relation to which it was defined, and therefore, in virtue of their definition, it is not predicable of itself. Conversely, if it is not predicable of itself, then again it is one of the said referents, of all of which (by hypotheses) it is predicable, and therefore again it is predicable of itself. This is a contradiction, which shows that all the referents considered have no common predicate, and therefore do not form a class.

He was to summarise this deduction on 24 June 1902 in his second letter to Frege (Frege *Letters*, 215–217). In May 1905 he outlined it to Jourdain (my *1977b*, 52), and a few weeks later he gave a much more detailed account to Hardy (my *1978a*).

This result was a true paradox, a *double contradiction*, not another neo-Hegelian puzzle to be resolved by synthesis. Although Russell always called it simply 'The Contradiction', he surely realised its significance quickly; if he ever thought otherwise, then Whitehead, Hardy or Jourdain would have soon set him straight. But no doubt his neo-Hegelian habit of seeking contradictions helped him to find it, and early on in his Peanist phase. There is a striking contrast here with Frege, who had been in the same area of work for over 20 years but had not found it. We recall from §4.7.6 that Zermelo had found the paradox in 1899; but he seems to have told nobody outside the Göttingen circle, so that it was new to both Frege and Russell.

The change of interpretation, occurring at some time over a period of intense work, may explain Russell's uncertainty over its date. In his autobiography he gave May 1901 (*1967a*, 147); but in earlier reminiscence he stated Spring (*1959a*, 75–76), and twice—surely wrongly—June (*1944a*, 13; *1956a*, 26). Even nearer the time he was no better, giving Jourdain the June date in the recollection of 1910 noted in §6.6.1 but Spring in 1915 (my *1977b*, 133, 144).

Whenever the change occurred, something had gone wrong. Where was the error: in the set theory, or the logic, or both? Maybe somewhere else? And could the paradox be Solved, or only avoided?

6.6.3 *Other paradoxes: three too large numbers.* (Garciadiego *1992a*, ch. 4) There should have been further consequences of the new interpretation of Cantor's argument; for Cantor's claim that there was no greatest cardinal was perhaps another paradox—or maybe two of them. Russell's deduction had drawn upon the diagonal argument, which itself leads to a paradox concerning the exponentiation of cardinals. Writing again his 'Cls' for the class of all classes and using Cantor's overbar notation, it takes forms such as

$$\overline{\overline{\text{Cls}}} < 2^{\overline{\overline{\text{Cls}}}} \text{ and } \overline{\overline{\text{Cls}}} \geqslant 2^{\overline{\overline{\text{Cls}}}}; \tag{663.1}$$

the first property follows from the power-class argument while the second relies upon the definitions of Cls. On the track of this paradox, Russell was diverted from it by his switch of thinking, and it has rarely been mentioned in the discussion of paradoxes (my *1981a*). Instead, the usual paradox of the greatest cardinal is a different one based just on the sequence of cardinal without exponentiation; in the above notation, it could read

$$\overline{\overline{\text{Cls}}} < \overline{\overline{\text{Cls}}} \text{ and } \overline{\overline{\text{Cls}}} = \overline{\overline{\text{Cls}}}. \tag{663.2}$$

Russell never mentioned (663.2) at all, and (663.1) only in one of his papers (*1906a*, 31) and in his popular book on logicism (*1919b*, 135–136), naming it after Cantor. (By contrast, more modern accounts of the paradoxes usually present (663.2) and ignore (663.1).) The most relevant passage in *The principles* occurs in a discussion of the diagonal argument

(p. 362), fairly heavily rewritten at some stage and occurring shortly before the passage on Cantor's 'fallacy' which he was to replace.

In his lists of paradoxes Russell stressed much more strongly that of the greatest ordinal number, which takes forms such as (663.2) with one overbar instead of two and inequality read in ordinal terms. Presumably he gave it greater publicity because of its intimate connection with order and thereby with relations, two staples of his philosophy. The two paradoxes are closely linked; for if the Cantorian cardinal $\aleph_\beta$ generates a paradox, then ordinal β must be pretty large also. He learned of trouble with ordinals in January 1901 in correspondence with Couturat (*Papers 3*, 385), from whom he borrowed the paper Burali-Forti *1897a*. After defining a 'perfectly ordered class', explicitly different from Cantorian well-order, Burali-Forti had shown that the trichotomy law did not apply to its members; thus its order-type Ω could satisfy the order-inequalities

$$\Omega + 1 > \Omega \text{ and } \Omega + 1 \leqslant \Omega \qquad (663.3)$$

without logical qualms. But he had also confused the situation by repeating, from an earlier paper *1894a* on simply ordered classes, a mistaken definition of well-order; he corrected himself only in an addendum *1897a*. Thus the possibility of paradox was mixed in with different kinds of order and with mistakes (as the Youngs *1929a* were to point out rather heavy-handedly, quoting Cantor's sarcasm from a letter to them).

Russell's reaction to Burali-Forti's deduction was to apply it to Cantor's well-order-type, obtain the result analogous to (663.2), and award it also the status of paradox. He named it after Burali-Forti first in a note added at the end of his second paper in the *Rivista*, where he wished to deny the property of well-ordering to the inequality relation (*1902a*, 421); and later in *The principles* in the new article 301 noted in §6.5.4 (G. H. Moore and Garciadiego *1981a*).

Cantor had known this result already, and followed his policy of avoiding the absolute infinite, as with the paradox of the greatest cardinal (§3.5.3). So did E. H. Moore, who found it a little later than Burali-Forti and took it to be really paradoxical; but, despite his strong interest in *Mengenlehre* (§4.2.7), he only wrote about it in a letter of September 1898 to Cantor.[16] So Russell was unaware of that predecessor.

Whatever the historical situation about these three strange results, Russell did see them *as paradoxes*. They were very much his creations, including the names.

---

[16] I found Moore's letter to Cantor in the Institut Mittag-Leffler, near Stockholm, in 1970; it was published in Garciadiego *1992a*, 205–206, with the provenance indicated on p. xx. Cantor's reply has not survived, but in 1912 he recalled to Hilbert corresponding with Moore (Cantor *Letters*, 460). Moore presented the paradox on 11 March 1898 to the 'Mathematical Society of the University of Chicago', in one of several talks on the *Mengenlehre* (University Archives, Society Records, Box 1, Folder 6, fols. 62v–67).

6.6.4 *Three passions and three calamities, 1901–1902.* Russell's intellectual honeymoon was truly over: the construction of logicism would be far trickier than he had imagined. But this paradox was only the second of three great difficulties which struck him during the first year of the new century.

The first in chronological order occurred during March and April of 1901, when Russell and Alys stayed for six weeks together with the Whiteheads at Downing College, Cambridge. The pleasure of the time was spoilt by continuing pains suffered by Mrs. Whitehead, with whom he may have been covertly in love: one day 'we found Mrs. Whitehead undergoing an unusually severe bout of pain. She seemed cut off from everyone and everything by walls of agony, and the sense of the solitude of each human soul suddenly overwhelmed me' (*1967a*,146). The effect of this mystical experience inspired his pacifism, his urge to tackle social problems, and his anguish over the loneliness of life.

The third calamity was the collapse of Russell's marriage in the spring of 1902 (p. 147),

> when we were living with the Whiteheads at the Mill House in Grantchester [near Cambridge ...]; suddenly, as I was riding along a country road, I realised that I no longer loved Alys. I had had no idea until this moment that my love for her was even lessening. The problem presented by this discovery was very grave.

Perhaps as a personal confessional, Russell started keeping an occasional journal in November 1902, and maintained it until April 1905. On his birthday, 18 May 1903, about the time when *The principles* appeared, he reminisced of events one year earlier (*Papers 12*, 22–23):

> This day last year I was [...] finishing my book. The day, I remember, stood out as one of not utter misery. At the time, I was inspired; my energy was ten times what it usually is, I had a swift insight and sympathy, the sense of new and wonderful wisdom intoxicated me. But I was writing cruel letters to Alys, in the deliberate hope of destroying her affection; I was cruel still, and ruthless where I saw no self-denial practised. [...]
>
> As regards the achievements of the year, I finished the book at the Mill House on May 23. [... On one day in June] came Alys's return, the direct question, and the answer that love was dead; and then, in the bedroom, her loud, heart-rending sobs, while I worked at my desk next door.

'The problem presented by this discovery was very grave': Russell could have said this about any of these three setbacks. Each of them was sudden or at least unexpected; each shattered previous expectations and beliefs; each destroyed a foundation of hope and optimism based on successful personal achievement. The personal anguish over a woman with whom he was secretly in love must have stood like a paradox against his coldness towards the woman who was his wife.[17] The combined effect was decisive on his work and personality, and left in his writings a streak of cynicism and perhaps facile pessimism which has made him in the last decades so much a man of his time.

Russell's autobiography reveals the extent of the impact perhaps more than he intended. For in one section he describes *in neighbouring paragraphs* the discovery of his paradox and the loss of love for Alys, and a few pages later he follows a frank description of his unhappy married life over the following years immediately with an account of his failures to solve the contradiction (*1967a*, 144–149). In addition, this trio of calamities corresponds like an isomorphism with the trio of hopes which he stated at the beginning of his autobiography: 'Three passions, simple but overwhelmingly strong, have governed my life: the longing for love, the search for knowledge, and unbearable pity for the suffering of mankind' (p. 13). These striking juxtapositions and stark contrasts may not have been made intentionally, but they cannot be coincidental.

6.7 REFINING LOGICISM, 1901–1902

My present view of the relation of mathematics and logic is unchanged. I think that logic is the infancy of mathematics, or, conversely, that mathematics is the maturity of logic.   Russell to J. Ulrich, 23 May 1957 (RA)

6.7.1 *Attempting Part 1 of* The principles, *May 1901.*   Russell's detailed forays into Cantor's and Peano's territories must have helped Russell to understand which undefined notions and logical constants (whether undefined or not) were needed for logicism. In May 1901 he outlined short summaries of eight chapters to make up 'Part I Variable' of *The principles* (§6.6.2). The first chapter treated the 'Definition of Pure Mathematics' (Russell *m1901c*, 185):

Pure mathematics is the class of all propositions of the form '*a* implies *b*', where *a* and *b* are propositions each containing at least one variable, and containing no constants except constants or such as can be defined in terms of logical constants. And logical constants are classes or relations whose extension either includes everything or at least has as many terms as if it included everything.

With these sentences, followed by one ponderously describing the relation between a collection and its members, Russell encapsulated his logical career: logical constants, their relationship to each other and the choice of indefinable ones; the variable, its character and role; the machinery of classes and relations, based upon set theory especially as utilised by the

---

[17] Many years later Alys wrote her own recollections of the collapse of their marriage (my *1996a*).

Peanists; the range and content of the class of pure mathematics so developable; explanation of the sense of 'pure', quite different from normal; and the details of the logical inference required to deduce these desired mathematical propositions from chosen logical axioms. The vision was clearly stated (p. 187):

> [...] the connection of mathematics with logic, according to the above account, is exceedingly close. The fact that all mathematical constants are logical constants, and that all the premises of mathematics are concerned with these, gives, I believe, the precise statement of what philosophers have means in asserting that mathematics is *à priori*.

One of the main tools was the theories of classes and especially of relations that Russell had just developed. He outlined the main features in a separate chapter, following with a discussion of the variable in which any temporal connotation was condemned. The manuscript is incomplete, but remaining is part of a survey of 'Peano's symbolic logic' (pp. 203–208).

6.7.2 Part 2, June 1901: cardinals and classes. (Manuscript, Byrd *1987a*; summary, art. 148) After these essential preliminaries, Russell could now write the Part on 'Number', by laying out cardinal arithmetic within this logic: his own nominal definition of cardinals as classes of similar classes, the pertaining arithmetical operations and their arithmetic, and the definition of the infinite class of finite cardinals without reference to numbers themselves but by generation from transitive and asymmetrical relations. Thanks to his own insights and Whitehead's exegesis, finite and infinite could be nicely distinguished, and mathematical induction did not have to be taken as primitive. He also showed that the Peano postulates (523.3, 8–11) for ordinal arithmetic came out as theorems, thus making clear by this example the deeper level of foundation which he could attain (pp. 127–128).

Comparisons with the corresponding Part of 'Principles' show how Russell's priorities had changed with his conversion. That one had been entitled 'Whole and Part' (§6.3.3): in *The principles* the topic received just ch. 16, of six pages. Logicism was shaping up nicely; but the paradox, surely important, lacked Solution. Part 1 needed reworking.

6.7.3 Part 1 again, April–May 1902: the implicational logicism. For several months after June 1901 Russell seems not to have much modified his book; in August he completed his second paper for Peano, and during the winter he gave a lecture course at Trinity College (§6.8.2). If a typescript of Parts 3–6 was prepared, as was mooted in §6.4.3, then perhaps it was done during this period.

Two major concerns were Solving the paradoxes, and choosing the indefinable notions of his logic. In April 1902 Russell planned Part 1 of *The principles* in 11 chapters (*Papers 3*, 209–212), including 'Denoting',

'Assertions' and as a finale 'The Contradiction'. He followed the scheme closely in the final writing.

The Part began with a 'Definition of Pure Mathematics' (*1903a*, 3–4), which elaborated upon the version a year earlier:

1. Pure Mathematics is the class of all propositions of the form '*p* implies *q*', where *p* and *q* are propositions each containing at least one or more variables, the same in the two propositions, and neither *p* nor *q* contains any constants except logical constants. And logical constants are all notions definable in terms of the following: Implication, the relation of a term to a class of which it is a member, the notion of *such that*, the notion of relation, and such further notions as may be involved in the general notion of propositions of the above form. In addition to these, mathematics *uses* a notion which is not a constituent of the propositions which it considers, namely the notion of truth.

2. The above definition of pure mathematics is, no doubt, somewhat unusual. Its various parts, nevertheless, appear to be capable of exact justification—a justification which it will be the object of the present work to provide.

Curiously, Russell's list did not include the notion of variable, which he soon emphasised as 'one of the most difficult with which Logic has to deal, and in the present work a satisfactory theory [. . .] will hardly be found' (pp. 5–6); but specifying ranges of values for variables in a given context formed part of the premises *p* (pp. 36–37). Hence (p. 8),

9. Thus pure mathematics must contain no indefinables except logical constants, and consequently no premisses, or indemonstrable propositions, but such as are concerned exclusively with logical constants and with variables. It is precisely this that distinguishes pure from applied mathematics. In applied mathematics, results which have been shown by pure mathematics to follow from some hypothesis as to the variable are actually asserted of some constant satisfying the hypothesis in question. [. . .]

10. The connection of mathematics with logic, according to the above account, is exceedingly close.

Russell clearly stated logicism here, and as an inclusion thesis; pure mathematics is *part* of this logic. However, at the end of the Part he declared that his thesis 'brought Mathematics into very close relation to Logic, and made it practically [*sic*] identical with Symbolic Logic' (p. 106). As already in the popular essay (§6.5.1), he was to state logicism as an identity thesis on two later occasions (§8.3.7, §9.5.4), the latter in the reprint of this book! But this position is indefensible; logic can be used in many contexts where mathematics is absent (for example, 'I am hungry', and 'if I am hungry, then I will eat'; hence 'I will eat'). The point is not at all trivial; apart from the question of whether or not mathematics is running, say, syllogistic logic or the law courts, there is the possibility that *only some* of the principles of logic are required for grounding (pure) mathematics. In *The principles*, however, he assumed that all of them were needed.

Russell's logicism seemed to require that both logic and pure mathematics were analytic, at least in the sense that logic and definitions alone would deliver the content. He was curiously silent on this matter; and in a passage in Part 6, written in 1900, he had claimed without explanation (or reflection, it seems) that 'logic is just as synthetic as all other kinds of truth' (p. 457), with a footnote reference back to his presentation of the propositional calculus, then not yet composed (Coffa *1980a*)! To that account we now turn.

6.7.4 *Part 1: discussing the indefinables.* (Manuscript, Blackwell *1985a*; summary, art. 106) The rest of this Part, which was given the title 'The Indefinables of Mathematics', went through the required basic components of Russell's logic. They were adopted precisely and only as the epistemological starting points of logicism, *not* as self-evident entities, which is the position frequently mis-attributed to him; as he was to warn clearly in the preface, 'the indefinables are obtained primarily as the necessary residue in a process of analysis', so that 'it is often easier to know that there must be such entities than actually to perceive them' (p. xv, 'analysis' used in the narrower sense explained in §6.1.1).

The 'indefinable logical constants' (p. 3) were implication, membership, 'such that', relation, 'propositional function, class, denoting, and *any* and *every term*' (p. 106). They made up the 'eight or nine' (*sic*) indefinables promised on p. 11.

Implication was important cement in building the house of logicism. Russell divided it into two kinds (p. 14): 'material' between propositions $p \supset q$, where p had to be false or q be true in order for it to hold; and a 'formal' version using universal quantification of individuals over propositional functions and requiring the last four or five indefinables above. He may have taken these adjectives from De Morgan *1860c*, 248–249, where they arose in a discussion of consequence connected with his distinction between form and matter (§2.4.8). For Russell the latter kind of implication was 'not a relation but the assertion' of a proposition, which I render in symbols as

$$\varphi(x) . \supset_x . \chi(x). \tag{674.1}$$

The notion of assertion played the role of inference between propositions, conveying 'the notion of *therefore*, which is quite different from the notion of *implies*, and holds between different entities' (*1903a*, 35). In a footnote he mentioned that Frege had a 'special symbol ['⊢'] for assertion', in order to make it explicit in symbolic work; he was to soon to adopt it, at (721.1)

Russell explained the role of assertion by solving a clever puzzle about 'What Achilles said to the tortoise' published a few years earlier in *Mind* by Lewis Carroll *1895a*. The tortoise asked Achilles to note down these premises in his notebook:

A. Things that are equal to the same are equal to each other.

B. The two sides of this triangle are things that are equal to the same.

But in attempting to deduce the conclusion

Z. The two sides of this triangle are equal to each other,

the tortoise showed that there were unexpected difficulties. For, as Achilles admitted, the logical principle

C. If A and B are true, then Z must be true

was undeniably relevant and therefore had to be entered in the notebook. But this fact had to be written down also:

D. If A and B and C are true, then Z must be true.

Thus an infinite intermediate sequence of propositions C, D, ... was set up, implying that Z could never be deduced from A and B. But we make deductions like this constantly.[18]

To us this puzzle calls for the distinction between logic and metalogic, with the *modus ponens* rule of inference distinguished from propositions *in* logic. However, at that time this approach was absent. For Russell 'The principles of inference which we accepted lead to the proposition that, if p and q be propositions, then p together with "p implies q" implies q'—that is within his understanding of implication, and assertion doing the rest, such as resolving Carroll's paradox (*1903a*, 35). Similarly, concerning the paradox of implication given by $p \supset q$ and $p \supset \sim q$, proposed in Carroll *1894a*, was solved by the principle that 'false proposition imply all propositions'; however, it is just restated (p. 18).

Upon this somewhat shaky basis Russell presented a system of ten axioms for propositions (pp. 16–17). It was one of the very few axiomatisations of a theory in the book.

Regarding general notions, Russell took '*term*' as 'the widest word in the philosophical vocabulary' (p. 43), with 'the words unit, individual, and entity' as synonyms. He also divided terms into '*things* and *concepts*'; examples of the latter category included 'Points, instants, bits of matter, [...] the points in a non-Euclidean space and the pseudo-existents of a

---

[18] The failure ever to deduce proposition Z from A and B could also be taken as an interpretation of the validity of Zeno's supposedly paradoxical argument (§6.4.7). Carroll himself wrote a short manuscript of 1874 recording 'An inconceivable conversation between S[ocrates] and D[odgson] on the indivisibility of time and space' (Library of Christ Church, Oxford) which shows that he may have had this insight, and if so before he wrote or at least published his paper *1895a* (my *1974c*, 16).

novel' (p. 45). However, he also allowed himself to use '*object* in a wider sense than *term*, to cover both singular and plural, and also cases of ambiguity such as "a man"', which indeed 'raises grave logical problems' (p. 55). Presumably he had in mind his stricture that 'every term is one' (p. 43) so that, for example, classes and relations as many might be objects but were not terms. Thus, while still in philosophical tune with Moore, he moved away from Moore's emphasis on concepts (§6.2.5), partly because of reservations about the universality of universals (pp. 51–52). But this did not bring him towards Aristotle, who indeed was never mentioned in the book, although there were a few unenthusiastic remarks about syllogistic logic.

These considerations bore upon 'Denoting' (ch. 5), which covered far more than definite descriptions using 'the'; for 'characteristic of mathematics' are the six words '*all, every, any, a, some* and   *the*' (p. 55). Russell could not handle any of them to his own satisfaction, but 'the' fared the best: doubtless recalling a morning in Paris, he noted that it had been emphasised by Peano, but 'here it needs to be discussed philosophically' (p. 62). He noted that a definite description did not have to denote a term, since in cases such as 'the present King of France' no denotation was available (Griffin *1996a*); however, while he brought out well its importance for theories of identity, he could not find a workable criterion for its legitimate occurrence. Denoting was soon to gain a central place in his further analysis of logic (§7.2.4).

The notions 'all' and 'any' appeared again in ch. 8, 'The variable', where Russell discussed different ranges which it might cover. His treatment went far wider than that conceived by mathematicians, to the full realm of objects which logic might treat. He also discussed quantification here, though not much; a chapter on its own would have been more appropriate. But a related notion received a belated chapter: propositional functions (ch. 7), which had been rather passed over in the earlier drafts. One of their main roles was to determine classes, via the indefinable '*Such that*' (p. 83); he also wondered about functions ϕ predicated of themselves to produce '$\phi(\phi)$', but his paradox made such matters uncertain (p. 88).

This use of propositional functions may suggest that Russell gave classes an intensional reading; but in his account of them in ch. 6, which amplified and in some ways modified a short exegesis in ch. 2, he preferred the extensional view. One of his reasons was that mathematicians take this view of classes when they deal with them (p. 67), but he gave no evidence to support this contention, which seems an unlikely generalisation: if mathematicians think about the issue at all, they are (and were) naive intensionalists. Cantor was a major example, with his intensional conception of a set by abstraction (§3.4.7) and other procedures of this cast.

Throughout the book and in Russsell's later logical writings, words such as 'proposition', 'propositional function', 'variable', 'term', 'entity' and 'concept' denoted *extra*-linguistic notions; pieces of language indicating

TABLE 674.1. Russell's Distinctions Concerning Denoting

Name	Example	Denotes?
predicate	*human*	does not denote
class-concept	*man*	does not denote
concept of the class	*men* or *all men*	class of all men as many
class as many	*men*	object denoted by *men*
class as one	*human race*	class of all men as one

them included 'sentence', 'symbol', 'letter' and 'proper name'; that is, a word 'indicated' a concept which (might) 'denote' a term (John Richards *1980a*). He is normally misread by commentators because they render him in Frege's quite different scheme (§4.5.5).

The wide scope of denoting played a role in Russell's complicated array of distinctions, with all six little words above in place. For a class *u*, ' "all *us*" is not validly analyzable into *all* and *u*, and that language, in this case as in some others, is a misleading guide. The same remark will apply to *every, any, some, a*, and *the*' (pp. 72–73). But he did not bring out here a use of '*some*', where 'some *a*' meant that 'some one particular *a* must be taken' (p. 59)—a sense of existential quantification different from his usual one, where any one would do.

In a difficult account Russell made various distinctions of which the Table 674.1 is, I hope, a fair rendering. One consequence was that he felt that much talk of class was actually of (intensional) class-concepts; in particular, Peano was held to identify the two (p. 68). But for Russell, taking an example of great importance (p. 75),

> *Nothing* is a denoting concept, which denotes nothing. The concept which denotes it is of course not nothing, *i.e.* it is not denoted by itself. [...] *Nothing*, the denoting concept, is not nothing, *i.e.* is not what itself denotes. But it by no means follows from this that there is an actual null-class; only the null class-concept and the null concept of a class are to be admitted.

A related achievement was clarifying the tri-distinction between nothing, the empty class, and the cardinal and ordinal zeroes (pp. 128, 244); his importance here, and that of his anticipator Frege (§4.5.3), are far too little recognised.

In his account of relations (ch. 9) Russell briefly stated the main technical terms and notations. He gave examples of self-relation, such as 'class-concept is a class-concept' (p. 96); however, he did not rehearse the proposition (644.1) concerning the self-membership relation. He found it 'more correct to take an intensional view of relations, and to identify them rather with class-concepts than with classes' (p. 99); but this stance sits uneasily with that on classes noted above.

In final ch. 10 of this Part, Russell discussed 'The Contradiction'; he had briefly presented it in terms of classes and of relations in their respective chapters (pp. 80, 97), and now it appeared for impredicable class-concepts (pp. 101–102). Analysis of his intractable class led him to examine in further detail classes as one and classes as many; however, he was not able to solve the paradox. A meticulous dissection of membership was also unsuccessful, despite proposed restrictions on the use of propositional functions (p. 104, with much rewriting on proof). No solution satisfied him when he finalised his manuscript; 'Fortunately', he added in the last sentence of the Part before the summarising article, 'no other similar difficulty, so far as I know, occurs in any other portion of the Principles of Mathematics' (p. 105).

6.7.5 *Part 7, June 1902: dynamics without statics; and within logic?*   This Part, 'Matter and Motion', was put together largely by importation from 'Principles' (§6.3.3); one folio is dated June 1900. He treated some aspects of dynamics, following studies from around 1898 (*Papers 2*, 83–110); for some reason he ignored statics. In addition to containing much of the oldest text in the book, it is the weakest Part as well as the shortest (34 pages): he seemed to be unaware of a rich field of work in the foundations of mechanics at that time, especially in Germany (Voss *1901a*, Stäckel *1905a*). One suspects an understandable desire to get this big and tiresome book finished as soon as possible.

Russell's basic strategy was to treat 'rational Dynamics' as 'a branch of pure mathematics, which introduces its subject-matter by definition, not by observation of the actual world', so that 'non-Newtonian Dynamics, like non-Euclidean Geometry, must be as interesting to us as the orthodox system' of Newton (*1903a*, 467). He then used the continuity of space, as established in Part 6 by Cantorian means, to establish realms within which motion could take place (ch. 54).

In the next chapter Russell sought to establish causal chains as implications; unfortunately he made the obviously mistaken assumption that 'from a sufficient [finite] number of events at a sufficient number of moments, one or more events at one or more moments can be inferred' (p. 478). Maybe he drew upon analogies from logic, such as the members of a finite class (p. 59) or from finite stretches (§6.4.5); but it was an elementary gaffe (§6.8.1).

Apart from this, the enterprise undertaken in this Part sounds too good to be true, or more especially to be logicistic; how, or why, should logic care about rotation? Are the propositions of this Part really expressed only in terms of logical constants and indefinables? It is worth noting that *PM* was to contain no treatment of dynamics (although unfortunately also no explanation of its absence); by then Russell had thought out better this aspect of logicism, and must have seen that Part 7 belonged more to its origins in the 1890s than to the new position of 1903. His definition of logicism, as quoted in §6.7.1 and §6.7.3, is unclear in that he did not lay down any restriction over the *kinds of values* over which variables could range; thus intruders such as terms from dynamics could be admitted.

6.7.6 *Sort-of finishing the book.* The last article of the Part, 474, was received by the Press on 27 January 1903 (according to their date stamp on the first folio). Here Russell reviewed the entire book. After an analysis in Part 1 of 'the nature of deduction, and of the logical concepts involved in it', among which

> the most puzzling is the notion of *class* [. . .] it was shown that existing pure mathematics (including Geometry and Rational Dynamics) can be derived wholly from the indefinables and indemonstrables of Part I. In this process, two points are specially important: the definitions and the existence theorems,

the latter being 'almost all obtained from Arithmetic'. The known types of number and of order-type apparently provided the stuff of space and of geometries, which could be correlated with continuous series to 'prove the existence of the class of dynamical worlds'; thus it followed that 'the chain of definitions and existence-theorems is complete, and the purely logical nature of mathematics is established throughout'. With these words he finished his first presentation of logicism, including dynamics but excluding not only statics and mathematical physics which sit so akin to it but also abstract algebras, probability and statistics, ...

It seems that Russell completed the manuscript rather suddenly; as well as lifting most of Part 7 from the previous version, he found that the Parts 3–6 needed much less revision than he expected. This swift wish-fulfilling sort-of-finish of the book must be understood against his difficult personal circumstances, especially his non-relationship with Alys (§6.6.4). At all events, in May 1902 he finally stopped rewriting his book (or thought he did, anyway), and sorted out the numberings of chapters and articles (more or less). In June he signed a contract with Cambridge University Press, and shipped off the manuscript to them.

But the manuscript shows that the fiddling was not over. While handling the proofs (which have not survived) between June and the following February he added a lot of footnotes, especially many of the references to pertinent literature which he now read at greater leisure: for example, a nice summary on pp. 310–311 of the state of play over Cantor's continuum hypothesis. He also entirely rewrote a few articles and added two appendices, and maybe prepared the lengthy analytical table of contents (its manuscript is also lost). During early February 1903 he prepared the index (*Papers 12*, 18), and at last it was over.

6.7.7 *The first impact of Frege, 1902.* In his preface Russell acknowledged his two principal inspirations thus (p. xviii):

> In Mathematics my chief obligations, as is indeed evident, are to Georg Cantor and Professor Peano. If I should have become acquainted sooner with the work

of Professor Frege, I should have owed a great deal to him, but as it is I arrived independently at many results which he had already established.

Russell read some of Frege's work *in detail* only in June 1902; he told Couturat of his previous ignorance in letters of 25 June and 2 July (copies in RA). One early reaction was to add three remarks and five footnote references to his text, all but one to Parts 1 and 2. He also altered art. 128 and most of art. 132 of Part 2 from doubts about treating a 'number as a single logical subject' to a stress that the '*one* involved in *one term* or *a class*' should not be confused with the cardinal number one defined earlier, citing the *Grundlagen* (*1903a*, 132–136). The rewriting in proof of p. 104 mentioned at the end of §6.7.4 was partly inspired by Frege, but finally he omitted the most explicitly dependent passage (Blackwell *1985a*, 288). Later he added an appendix on Frege, which we shall consider in the next sub-section.

In addition, in his preface Russell stated that Frege's work had corrected him on 'the denial of the null-class, and the identification of a term with the class whose only term it is' (*1903a*, xvi). However, he had learnt these features of set theory from Cantor and especially Peano, and had used them in his text (for example, pp. 23 and 106)! Perhaps he was recalling his criticism of Peano on these points on pp. 32 and 68: however, while not systematic in his philosophising, Peano seems unfairly charged since at (541.2) he had defined the null class by one of the properties accepted by Russell himself.

Russell wrote his first letter to Frege, in German, on 16 June 1902, two weeks *after* sending in his book manuscript to the Press.[19] To fit in with Frege's notion of value-range using ordered pairs (§4.5.5), he stated his paradox in terms of the predicate that cannot be predicated of itself. He also mentioned that he had written to Peano about it; but he never seems

---

[19] This detail of chronology leads me to demur from the interpretation given by some historians, such as G. H. Moore and Garciadiego *1981a* that Russell appreciated the significance of his paradox only after hearing Frege's reaction in June 1902 (compare §6.6.1).

The survival of the Russell-Frege correspondence is a minor miracle. As was described in §4.5.1, Frege included his correspondence with Russell in his planned donation to the bibliographer Ludwig Darmstaedter, and after his death his adopted son Alfred made the transfer. Darmstaedter's collection later came into the *Stiftung Preussischer Kulturbesitz* in Berlin, was conserved in a mine during the Second World War, and survived afterwards in West Germany while the card catalogue was kept in the *Staatsbibliothek* in East Berlin. (Ignorant of this split, I had a surprised and surprising response to my request to see them in 1969 when I came across the catalogue by accident.) The collection and two catalogues are now together again, in the new building housing the *Stiftung*.

Meanwhile in 1935, while planning his edition of Frege's correspondence (§4.5.1) Heinrich Scholz was allowed to borrow Frege's letters from Russell and make photostats and typescripts of them (compare §9.6.3). He placed the originals in Frege's *Nachlass* at Münster and sent Russell the photostats. After the destruction of the *Nachlass* in the War Hans Hermes borrowed the photostats from Russell in 1963 to check against the typescripts. They were safely returned and now survive in the Russell Archives, as do Hermes's pertaining letters.

to have received an answer. By contrast, in a prompt reply written six days later, Frege related the paradox to the system in *Grundgesetze der Arithmetik*. His stratification of functions (§4.5.6) prevented a first-level one saturating itself as an object, but if ' "A concept will be predicated of its own range" ', then trouble followed; 'the ground, upon which I thought to construct arithmetic, would fall into tottering' (*'in's Wanken geräth'*: Frege *Letters*, 213).

Frege diagnosed the illness as lying in his Law 5 (456.1), which associated the equivalence of two propositional functions (which, we recall, were for him concepts taking truth-values) with the equality of their value-ranges (hereafter, 'VR'); for the function corresponding to Russell's class belonged to its own VR if and only if it did not do so. This association infringed his otherwise strict demarcation of objects from concepts; for, as he put it to Russell on 20 October 1902 (p. 233),

> Accordingly a concept can have the same range as another, even though this range falls under it but not under the other one. It is only necessary, that all other objects part from the concept-range itself, which fall under a concept, also fall under the other one and vice versa.

The second volume of the *Grundgesteze* being in press, Frege quickly added a 13-page appendix, admitting that Law 5 'is *not so evident*, as the others' (*1903a*, 253, my interested italics), and outlining the above solution; however, as he realised, it led to a very complicated stratification of levels. Further, it forbids the assumption of more than one individual, which is an unacceptably tight restriction on arithmetic.[20] Indeed, even the system of his earlier *Begriffsschrift* admits paradox via its rules of substitution and detachment, even though it does not have VRs (Thiel *1982a*, 768–770). He was not to publish again on logic for fifteen years (§8.7.3).

Russell's system seems unredeemably prone to paradox; but some ways out for Frege's have been proposed. For example, concepts such as 'does not belong to itself', when quoted like this, do not bear content and so render meaningless propositions such as 'the concept "does not belong to itself" belongs to the concept "does not belong to itself" ', from which the paradox follows (Sternfeld *1966a*, 131–136, difficult to follow since 'function' is used in both the ordinary and Frege's technical senses). But then concepts such as 'not identical with itself' would also be forbidden, thus removing Frege's definition of 0 (§4.5.3). Again, Frege might have argued that since classes were logical objects (otherwise arithmetic lost its *a priori* character), their names were not subject to the distinction between sense and reference, unlike objects such as VRs which could be the reference of

---

[20] For analyses of Frege's solution of Russell's paradox, see Sobocinski *1949–1950a*, art. 4 (by Leśniewski) and Quine *1955a*. A considerable literature has developed on the consistency of the first-order part of Frege's system given that his law of comprehension is second-order: like Frege at the time, I shall not explore this interesting feature.

names with different senses; hence paradoxical propositions could not be constructed. But Frege did not develop a philosophy of VRs of a sophistication comparable to Russell's theory of classes (Angelelli *1967a*, ch. 8).

6.7.8 *Appendix A on Frege*. With his first letter to Russell Frege enclosed offprints of five papers, to supplement the books which Russell already possessed. Russell used all these sources to add an appendix to *The principles* on Frege's work, which was virtually unknown in Britain (though, as we have seen, not so on the Continent). He sent the text to the Press in mid November 1902, by when he and Frege had exchanged 14 letters; when printed it occupied pp. 501–522, at slightly smaller font size.

With his title, 'The logical and arithmetical doctrines of Frege', Russell precisely captured the scope of Frege's logicism. He noted the development of Frege's ideas, especially new notions, and treated seven areas where they overlapped with his own, approving or dissenting as seemed appropriate (p. 501). I shall note a few points from each.

On '*Meaning and indication*' Russell referred to Frege's 'Sinn' and 'Bedeutung'; the mistranslation of the latter as 'meaning' is a modern innovation (§4.5.1). He allied them respectively with his own 'concept as such and what the concept denotes' (p. 502), and rightly rejected Frege's claim that proper names had meaning as well as indication. He might have emphasised that Frege identified a word with its indication, whereas he separated them (§6.7.3).

'*Truth-values and Judgement*' was for Russell the same as his distinction between asserted and unasserted propositions; the first term seems to be an early use by anybody, while the latter one was his rather unhappy rendering of Frege's 'Gedanke'. But he understandably doubted Frege's Platonic proposal (§4.5.5) that any true proposition such as 'the assumption "$2^2 = 4$" indicates the true, we are told, just as "2^2" indicates 4' (p. 503).

In '*Begriff and Gegenstand*' Frege's first technical term meant 'nearly the same thing as *propositional function*', including relations for more than one variable (p. 507: two pages earlier he associated it with his version of assertion given on p. 39). The second term was allied to '*thing*'. Russell had trouble here, and also in his notes made later in 1902, with Frege's notion of function as a place-holder, which he rendered here as '$2(\ )^3 + (\ )$' from the case $2x^3 + x$ (p. 505).

Russell stressed that Frege's 'very difficult' theory of '*Classes*' (p. 510) dealt with membership as symbolised by Peano's 'ε' and not the traditional part-whole approach; he rendered 'Werthverlauf' as 'range', and associated this notion with his class as one (p. 511). Impressed by Frege's distinction between an individual and its unit class, he went again over various intensional and extensional formulations of classes, and felt even more drawn to the latter reading (pp. 515–518). His discussion complemented his new text for arts. 128 and 132 by emphasising the various different senses of '*one*' (pp. 516–517).

Concerning '*Implication and symbolic logic*' Russell merely noted that Frege's implication relation did not require the antecedent to be a proposition. He again reported that Frege had a special sign for assertion.

Over '*Arithmetic*' Russell acknowledged that 'Frege gives exactly the same definition of cardinal numbers as I have given, at least if we identify his *range* with my *class*', and noted that Frege's theory of hereditary relations generated series as a means of handling mathematical induction (pp. 519–520): he did not comment upon the absence of a more general theory of relations. He surveyed Benno Kerry's criticisms of Frege (§4.5.4), awarding most marks to the latter (pp. 520–522).

Russell did not attempt a complete survey of Frege's system; and in his contrasts he did not convey the great philosophical gulf between his own positivistic and reductionist spirit and Frege's Platonic world. But his survey greatly helped to give Frege a less tiny audience, although §4.5.2 reveals as absurd his later claim to have been the first reader of the *Begriffsschrift* (*1919b*, 25; *1956a*, 25).

6.7.9 *Appendix B: Russell's first attempt to solve the paradoxes.* Russell added a note in press to Appendix A on Frege's solution to the paradox, opining that 'it seems very likely that this is the true solution' (p. 522); at the time he worked on it in detail, but without success (*Papers 4*, 607–619). He was trying various stratagies to solve the paradox himself, looking again at Cantor's diagonal argument and trying to restrict in some ways the membership of classes and their formulation from propositional functions (*Papers 3*, 560–565); but to no avail. 'It is the distinction of logical types that is the key to the whole mystery', he judged in *The principles* at the end of a paragraph in the chapter on the paradox added in proof (*1903a*, 105), and late in 1902 he proposed a solution in Appendix B (pp. 523–528).

A significant change is evident at once: propositional functions $\phi(x)$ were raised to a still higher level of importance, for the 'ranges of significance form *types*, the class of x's for which $\phi(x)$ was a proposition; Russell also considered the hierarchy of classes of classes, classes of classes of classes, and so on finitely (p. 524: compare already p. 517 in Appendix A). But the theory was more primitive than the mature version to come in *PM* (§7.8.1–2). Since 'A *term* or *individual* is any object which is not a range' (p. 523), then 'predicates are individuals' (p. 526), which is not only peculiar but also confused the relationship between propositional functions and classes. He was worried about the legitimacy of apparently 'mixed classes', such as 'Heine and the French' (p. 524). He let the class of cardinal integers be a type of its own, which endangered the definition of 0 since each type could have its own range of zero members (the curious p. 525: surely any other integer would also be so endangered).

Further, propositions also formed a type; but then Russell's paradox could be constructed by applying Cantor's diagonal argument to the proposition M given by 'every member of a class m of propositions is true'. M corresponds one-one with m, and may or may not belong to it; and if one forms the class w of non-belonging propositions M, then 'every member of class w of propositions is true' belongs to w if and only if it does not (p. 527). He seems to have forgotten this paradox, for he never noticed that it was constructible in some later type theories (§7.3.7, §7.8.1). He concluded here that the possibility of a hierarchy of propositions 'seems harsh and highly artificial' (p. 528), although in *PM* he was to present one (§7.8.1).

Frege did not comment to Russell on this theory after receiving a copy of *The principles* in May 1903 (*Letters*, 239–241); but he would not have liked it, for he had already objected to types in logic when disputing the efficacy of Schröder's logical system (§4.5.7). Russell was not satisfied himself, and knew that he had a big task on his hands.

6.8 THE ROOTS OF PURE MATHEMATICS? PUBLISHING *THE PRINCIPLES* AT LAST, 1903

6.8.1 *Appearance and appraisal*

We should say that Mr. Russell has an inherited place in literature or statesmanship waiting for him if he will condescend to come down to common day.

> Anonymous review of *The principles* (*The spectator 91* (1903), 491)

According to Russell's journal, the preface was written on 2 December 1902 (*Papers 12*, 14). In it he outlined the scope and also limits (especially regarding dynamics) of the book, and indicated the 'more specially philosophical' portions, which included the whole of Part 3, much of Parts 1, 2 and 7, and the appendices, but rather little on Part 5 (p. xvi). The book was sub-titled 'VOL I.' on the title page; the logicist thesis 'will be established by strict symbolic reasoning in Volume II' (p. xv). He thanked Whitehead for reading proofs, Johnson (the Press's reader) for comments, and Moore for philosophical background.

The principles of mathematics appeared in May 1903, around Russell's 31st birthday—about the age when Frege had published his *Begriffsschrift*, and Peano his *Arithmetices*. It was his fourth book, the third with Cambridge University Press. The print-run was of 1,000 copies at 12/6d each, or $3.50 when it went on sale across the water in June. Among compatriots he gave copies to Whitehead, Johnson, Moore, Bradley, G. F. Stout, Jourdain and (I think) Hardy; copies went abroad to (at least) Couturat, Frege, Peano, Vailati and Pieri. The book seemed to sell steadily; in June 1909 the Press told him that the last 50 copies were at the binders (RA).

The audience for the book comprised mainly the sector of the philosophical and mathematical communities interested in each others' concerns, especially the audience for set theory which had been growing rapidly for around a decade. Indeed, the book played an important role in awakening the British to some parts of Cantor's theory, and to mathematical logic. Various Peanist terms came into English or at least became better known, such as 'propositional function', 'material' and 'formal implication', and 'indefinable' (Hall *1972a*); however, 'mathematical logic' in this context still had to wait (§7.6.3).

But Russell knew that the book as published was rather a shambles. Within days of issue he wrote to Frege on 24 May 1903 that in Parts 1 and 2 'there are many things which are not thoroughly handled, and several opinions which do not seem correct to me' (Frege *Letters*, 242); and two months later he told his friend the French historian Elie Halévy that 'I am very dissatisfied with it' (Russell *Letters 1*, 267). The previous 28 December he even confessed to his friend Gilbert Murray that 'this volume disgusts me on the whole' (RA).

The unsolved paradox was doubtless one main reason; but Russell must have recognised that the presentation was somewhat disordered and even contradictory across and even within some chapters. His apparent decision not to write Parts 1 and 2 until he had tested out Peano's programme in 3–6 was very sensible, since he had a good idea of what they would contain; books are often written out of order of reading (this one is an example). But he did not bring the later Parts in line with positions and assumptions finally laid out in the openers, nor did he tidy up the overlaps (for example, on infinity and continuity between Parts 2 and 5). The manuscript had needed an overhaul, and he knew it.

Some reviews appeared, with varying degrees of understanding of its content. (The anonymous quotation at the head of this sub-section scores high marks for prophecy!) The first one appeared anonymously, in the *Times literary supplement* in September. Hardy *1903b* concluded there that Russell 'seems to have proved his point' about logicism, and was glad to learn of Frege, 'of whom we must confess we had never heard' (p. 851); but he found the book 'a good deal more difficult than was absolutely necessary' by being 'much too short' and condensed, given its unfamiliar doctrines. Specific criticisms included the incompetent handling of causality in Part 7 (§6.7.5); for let a particle be 'projected from the ground, and take the second time to be that at which it reaches the ground again. How can we tell that it has not been at rest?' (p. 854). On the logical aspects, he stressed the unintuitive character of implication, that 'every false proposition implies every other proposition, true or false' (p. 852).

This last feature was also mentioned in a review by a mathematician in a German philosophical journal. Felix Hausdorff *1905a* was rather sarcastic about the book for giving 'the impression of pointless intellectual athletics' in its 'orgy of subtleties'; his summary estimate 'with two words' required the five words 'sharp and yet not clear' (p. 119). However, he also gave a good survey of the contents, similar in some ways to Hardy's, stressing the importance of propositional functions (which he rather unhappily translated as 'Urteilsschema'), describing the paradox, sceptical about Part 7 but happy with Parts 4 and 5. His attitude as a mathematician exemplifies well the remark of Friedrich Engel in a generally positive notice *1905a* of the book for the reviewing *Jahrbuch*:

> [...] the most productive mathematicians do not at all have much inclination, to devote themselves to such philosophical speculations about the ultimate foundations of their science; just as little as the practising musician has the need to concern himself with the calculating science on which musical logic, which his ear teaches him, properly touches.

The two longest reviews, each close to 20 pages, appeared in 1904. In the May issue of *Bulletin des sciences mathématiques* Couturat *1904a* concentrated on Cantor from the mathematical background, having written elsewhere on the Peanists (§5.4.6). He noted the three kinds of geometry, and also the dynamics, which he accepted into logicism without qualms. He also wrote then a long series of articles on 'The principles of mathematics' inspired by Russell's book, which we note in §7.3.1.

Six months after Couturat's review, the American mathematician E. B. Wilson *1904a* covered both the book and Russell's *Essay* on geometry for the American Mathematical Society. After citing Couturat's review and stating his to be supplementary, he dwelt on the Peanists, whose work 'is very little known and still less appreciated' in the U.S.A. (p. 76), referring to their four lectures at Paris in August 1900 and ending his piece with a list of some of their main works. He contrasted the treatments of geometries in the two books, and also noted the dynamics in *The principles*; its presence led him understandably to speculate 'why not thermodynamics, electro-dynamics, biodynamics, anything we please?' (p. 88). Neither reviewer paid much attention to the paradox. In letters to Couturat of 5 April and 12 May 1905 Russell liked Wilson's review but found Hausdorff's 'disappointing' ('*désespérant*'); Couturat disagreed on the latter opinion on 28 June (RA and copies).

The shortest review was Peirce *1903a* in the general American periodical *The nation* (B. Hawkins *1997a*). Although he prepared 15 folios of notes, he published only a few lines, clearly showing the gulf between algebraic and mathematical logics. However, they included the remarkably accurate prediction that 'the matter of the second volume will probably consist, at least nine-tenths of it, of rows of symbols'.[21]

---

[21] In his review Peirce *1903a* turned, with warmth, to the recent book *What is meaning?* by Lady Welby (1837–1912), his correspondent at the head of §4.3 and main British follower in semiotics. Her manuscripts are held at York University, Toronto; I have not used them, but I have profited from her heavily annotated sets of *Mind* and *The monist* in the University of London Library.

At least this was more than *The monist*, where no review was published. Nothing appeared in Peano's *Rivista*, either, for Vacca failed to deliver. Moore drafted a long and dull one for the *Archiv für systematische Philosophie*, which he had the good sense to set aside (Papers, File 15/2). Stout, who wrote to Russell on 3 June 1903 of the book that he was 'immensely impressed by it, but all the same believe it to be fundamentally wrong' (RA),[22] asked Johnson to review it for *Mind*. As usual, nothing arrived, and eventually the London logician A. T. Shearman (1866–1937) produced 12 pages. Welcoming the book as the most important one on logic since Boole's *Laws of thought*, he concentrated upon logicism and the paradox. His solution of the latter was based on the proposal that in 'not predicable of itself is not predicable of itself' the first occurrence of the clause was a quality which could not become a subject, as in the forbidden 'happy is happy' (*1907a*, 262); Russell had already been through such considerations. Shearman also welcomed the account of Frege's work without the 'extreme cumbrousness' of the original notation (p. 265).

Like the book itself, the general reception was mixed.

6.8.2 *A gradual collaboration with Whitehead.* (Lowe *1985a*, ch. 10) Russell's contributions to Whitehead's second paper (§6.5.3) constituted the first public piece of collaboration between the two men; he told Jourdain in 1910 that it occurred in January 1901 (my *1977b*, 134). Their teaching also converged: Russell, his six-year Prize Fellowship over in 1901, gave the first course in mathematical logic in Britain at Trinity College in the winter of 1901–1902 to a small audience which however included colleague Whitehead and student Jourdain. (During this time he experienced the two calamities while with the Whiteheads described in §6.6.4.) The small amount of surviving material suggests that in addition to the basic Peanist logic he seems to have covered quite a bit of set theory and some aspects of geometry, and apparently outlined a plan of the joint book which would become *PM* (*Papers 3*, 380–383). On 2 October 1902 he described it to Couturat as 'a book "On the logic of relations, with applications to arithmetic, to the theory of groups, and to functions and to equations of the logical Calculus"',[23] and the next 7 January he reported that the contents of the course would be in it.

---

[22] In an undated note sent to Russell maybe at this time (RA), Stout distinguished between 1) 'a class taken simpliciter', as conveyed distributively by 'every man' or 'all men', and the only kind of class allowed to belong to other classes; 2) 'class quâ class', without any specifying predicate; and 3) a 'class as many' as conveyed collectively by 'a man' or 'any man'. The paradox was avoided by membership restrictions; but various pieces of mathematics would also disappear, a consequence which he did not examine.

[23] The quotation reads: 'un livre "Sur la logique des relations, avec des applications à l'arithmétique, à la théorie des groupes, et aux fonctions et aux équations du Calcul logique"' (copy in RA).

They also tried to spread their new doctrine. In 1902 Russell tried but failed to have Peanist logic and *Mengenlehre* introduced into a new philosophy course at the University of London proposed as part of its reorganisation after an Act of Parliament in 1898 (*Papers 3*, 680–685: the description of the University at the start of the headnote is mistaken). That autumn Whitehead taught 'applications of logic to set theory' at Cambridge, with young Trinity Fellow Hardy present (Hardy *1903a*, 434), and perhaps Jourdain also. Whitehead had been elected Fellow of the Royal Society in June after nomination by Forsyth (Society Archives).

Yet there were notable differences of interest and emphasis between the two men. An early and striking example is provided by a letter which Whitehead wrote to Russell on 16 November 1900 (Garciadiego *1992a*, 185–186). He found that Peano's treatment of arithmetic in the second edition (1899) of the *Formulaire* (§5.4.6) unfortunately led him 'to have prematurely identified his symbols with those of ordinary mathematics. The result is that he is led into some inconsistencies'. His example involved Peano's transformation (542.1) of a member x of a class a into member '*xu*' of class b under the 'correspondence' functor u. Whitehead showed that this concatenation of symbols could be confused with the way of writing the multiplication of numbers to the extent that further theorems of Peano led to this nonsensical property of members of the class N_0 of finite integers:

$$\text{`}a, b \, \varepsilon \, N_0 \, [\ldots] . \supset . a + b = ab = a \times b!! \text{'}. \qquad (682.1)$$

(Compare Peano at this time on $3 + 5 = 35$ in §5.6.2.) Russell quoted this "result" in the opening of his first paper in Peano's *Rivista*; but he judged that 'the definition of *function* is not possible except though knowing a new primitive idea, that of *relation*' (*1901b*, 314). His reaction is quite different from Whitehead's algebra-based criticism; such contrasts would permeate their whole partnership.

Nevertheless, the process of building a logico-mathematical system with the cancerous paradox still unSolved drew Russell and Whitehead together during 1901 and 1902. Both men were engaged on similar studies of foundational questions, and Whitehead must have seen his wanderings in algebras and cardinals after his Volume 1 of 1898 as less clearly focused than Russell's way ahead singing Peanist melodies about logicism following his own 'VOL I'. Thus it was a reasonable decision for them to pool resources entirely, with Russell largely determining philosophical policy.

CHAPTER 7

Russell and Whitehead Seek the
Principia Mathematica, 1903–1913

7.1 PLAN OF THE CHAPTER

This chapter covers the period during which Whitehead and Russell collaborated to work out their logicistic programme in detail. Mostly they prepared *Principia mathematica* at their respective homes at Grantchester near Cambridge and Bagley Wood near Oxford; thus much discussion was executed in letters, of which several survive at Russell's end.

This chapter divides into two halves around 1906 and 1907 because of their change of strategy. After accumulating more paradoxes and axioms, and much work on denoting (§7.2–§7.4.5), Russell developed intensively a logical system which he called 'the substitutional theory' (§7.4.6–8); but then he abandoned it and switched to the one which was to appear in *PM* (§7.7-9). At the division point are noted some of the reactions of others to logicism and related topics, especially set theory, and the independent activities of Whitehead (§7.5–6).

Another difference between the two halves concerns access to Russell's writings: the first one is comprehensively covered in Russell's *Papers 4*, but the succeeding volume will not be ready for the second half until after the completion of this book. In some compensation, two compilations of papers by Russell and others are available: Russell *Analysis* (1973) and Heinzmann *1986a*.

7.2 PARADOXES AND AXIOMS IN SET THEORY, 1903–1906

7.2.1 *Uniting the paradoxes of sets and numbers.* The task was to find a logical system of propositions and propositional functions, with quantification over them and also over individuals, using set theory as fuel, in which as much mathematics as possible could be expressed but the paradoxes avoided and indeed Solved. 'Four days ago I solved the Contradiction', Russell had told himself in his journal on 23 May 1903, while finishing *The principles*, 'the relief of this is unspeakable' (*Papers 12*, 24). But, like stopping smoking (which Russell himself never attempted), it was easy to do, lots of times. 'Heartiest congratulations Aristotles [*sic*] secundus', wrote Whitehead in a telegram the following 12 October after another solution; however, Russell wrote on it later: 'But the solution was wrong' (RA, reproduced in Garciadiego *1992a*, 187).

TABLE 721.1. Three Paradoxes

Paradox	ϕ	f	
Russell's	not belonging to itself	identity	(721.1)
Burali-Forti's	is an ordinal	ordinal of w	(721.2)
Cantor's	is a cardinal	cardinal of the power-class of w	(721.3)

Before publishing *The principles*, Russell had found that his own paradox could be expressed in terms of relations. He showed Frege on 8 August 1902 that if relations 'R and S are identical, a[nd] the relation R does not hold between R a[nd] S. One sets this equal to $(R)T(S)$, where T should be a relation. With $R = T$ one then obtains a contradiction' (Frege *Letters*, 226–227). He published this version in a paper of 1906, where he also generalised it to cover all three paradoxes of classes in terms of any relation f between classes u (*1906a*, 35):

> Given a property ϕ and a function f, such that, if ϕ belongs to all the members of u, $f'u$ always exists, has the property ϕ, and is not a member of u; then the supposition that there is a class w of all terms having the property ϕ and that $f'w$ exists leads to the conclusion that $f'w$ both has and has not the property ϕ.

The paradoxes arise as shown in Table 721.1.

7.2.2 *New paradoxes, mostly of naming.*

(Garciadiego *1992a*, ch. 5) New paradoxes soon arrived, mostly concerning naming and definability. At the International Congress of Mathematicians at Heidelberg in August 1904, the Hungarian mathematician Julius König (1849–1913) proposed a refutation *1905a* of Cantor's continuum hypothesis by claiming to show that the continuum was not well-ordered. Cantor was present, and after the lecture took part in a discussion with Hilbert and Schönflies: 'only simple remarks were made and not objections to the proof', according to a report (Fehr *1904a*, 385), 'However, Mr. Cantor reserved for himself the right to a more considered examination of the problem', and indeed emotionally urged colleagues to locate the mistake. Hausdorff *1904a* quickly found it, in the use of a theorem about cardinal exponentiation under invalid conditions (Kowalevski *1950a*, 198–203, wrongly credits Zermelo). The discussion continued at a gathering of mathematicians at Wengen in the Bernese Overland: König's failure 'is a great triumph for Cantor who spoke of nothing else in Wengen', Hilbert told Klein in a letter.[1]

However, the following year König came back with a different refutation *1905b*; like the first, it appeared in *Mathematische Annalen*. The non-denumerable cardinality of the continuum entailed that not all real numbers

---

[1] Frei *1985a*, 132. See also König's admission of error to Hilbert on 7 September 1904, and Hausdorff's analysis three weeks later (Hilbert Papers, 184/3 and 136/2).

could be defined in a finite verbal expression; hence there must be a smallest such number—but this property itself was stated *in* a finite verbal expression. He concluded that Cantor's well-ordering principle was false (this claim related to the controversy over the axioms of choice to be described in §7.2.5), and by consequence the continuum hypothesis also.

Around the same time, the French school-teacher Jules Richard (1862–1956) published a similar paradox, in a note *1905a* in a general science journal; it was reprinted the following year in *Acta mathematica*. He also considered the set F of finitely definable numbers and wondered about a number definable by applying Cantor's diagonal argument to the members of F; a different number was produced, but the argument itself furnished it by a finite definition. His version was independent of König's; it grew partly out of some consideration of his own recent book on the philosophy of mathematics (Richard *1903a*, 107–113), and he had been led to write his note by reading of the Heidelberg contretemps.

Russell heard of these paradoxes from the publications or from correspondents such as Couturat and Jourdain. But another version arrived directly, from a resident in Summertown in north Oxford, not far from him at Bagley Wood: G. G. Berry (1867–1928), a junior librarian at the Bodleian Library and in Russell's opinion 'a man of very considerable ability in mathematical logic'. Stimulated by reading *The principles*, he wrote ten letters to Russell between 1904 and 1910 (published in Garciadiego *1992a*, 166–184): his only known writings on logic, which also reveal an impressive familiarity with Cantor's theory of transfinite numbers. The paradox which he gave to Russell came in the first letter of 21 December 1904, in which he also considered the class of ordinals not finitely definable: 'This least member of the class is then the least ordinal which is not definable in a finite number of words. But this is absurd, for I have just defined it in thirteen words' (p. 168). Russell's reply is unfortunately lost, but he published the paradox (first in *1906h*, 645).

According to a note left by Russell, Berry also brought him a paradox to his front door (Garciadiego *1992a*, 166):

> The first time he came to see me at Bagley Wood he was bearing, as if it were a visiting card, a piece of paper on which I perceived the words: 'the statement on the other side of this paper is false'. I turned it over & found the words: 'the statement on the other side of this paper is false'. We then proceeded to polite conversation.

This recollection seems incorrect: doubtless Berry had written 'false' on one side of the paper and 'true' on the other one. But it is the origin of 'the visiting card paradox', as it has become known.[2] Surprisingly, Russell never included it in his lists of paradoxes; publicity was to be brought by

---

[2] Russell also mis-stated the visiting card paradox in his autobiography (*1967a*, 147) but corrected himself soon afterwards in a letter *1967b* to a newspaper.

Jourdain, especially in a paper *1913f*. Presumably Russell regarded it only as a variant on the classical Greek paradox of the liar, which arises from pondering upon the truth-value of the proposition 'This proposition is false'.[3] He did list that paradox, first in *1906h*, 632; he might have been encouraged to do so by this nice version sent to him in January 1905 by his friend Oliver Strachey (RA):

> David said, (in a moment of wrath):—'All my remarks between the hours of 2 and 3, are, have been, or will be lies'. All his other observations in this period were palpable falsehoods, and this one was made at 2.30. Was it true or false? In my present state of mind this seems to be another edition of the damnable Contradiction,—probably soluble by what you call the harsh and highly artificial suggestion of various types of propositions. [. . .] Any sign of lunacy in this letter please put down to your Appendix B and blame yourself (like suicides or the heroes of Smiles' Self-Help).

Down the side of the letter he questioned: 'What I mean is Can a proposition be its own subject?'. Russell wrote on it: 'Answered more or less' (but now lost), and he repeated this version to Jourdain the following April (my *1977b*, 44, 50). Berry found it independently, and told Russell in November 1906 (Garciadiego *1992a*, 179–180).

7.2.3 The paradox that got away: heterology. One paradox which never seems to have come to Russell's attention is that named after the German philosopher Kurt Grelling (1886–1942), who published it in a paper written with his friend Leonard Nelson (1882–1927) (*1908a*, art. 4). Some words can be predicated of themselves: in English, 'word' is a word, 'noun' a noun, and so on. This property is called 'autological', and is obviously itself autological. Others English words are not autological; 'German', say, or 'verb'. They are called 'heterological'—but this word is heterological if and only if it is not so.

Grelling and Nelson both belonged to the Fries school of philosophers (§4.2.5); their paper appeared in its *Abhandlungen*. Close to David Hilbert in Göttingen, they came to paradoxes partly through his interest in them (Peckhaus *1995b*). Their own is an interesting one, not least for endangering the law of excluded middle (which kind of word is 'gracious', say?), as they noted in a survey of candidate 'solutions' and 'corrections' (arts. 6–7),

---

[3] In a letter of 4 May 1995 to Jourdain, John Venn responded to a request for a reaction to Russell's paradox by recalling the liar paradox and this variant: 'Some young man learnt rhetoric of a sophist on the engagement that he should pay his teacher 1000 talents if he won his first course, otherwise nothing. He refused to pay, when the sophist thoughtfully prosecuted him. Each side claimed a verdict on the terms of their agreement. The judges, puzzled, dismissed the case' (Jourdain Papers, Notebook 1, fols. 256–257).

Among other variants is one attributed to the Astronomer Royal G. B. Airy (1801–1892). His meticulousness at the Observatory at Greenwich is rumoured to have gone as far as writing 'empty box' on a piece of paper and putting it inside the object—thereby, however . . . .

in which they sought to avoid 'circular definitions' and propositions (arts. 13–15). It suggests extensions to phrases and propositions, such as 'I kan spel verry acurratly'. Strangely, Russell never mentioned it, although he corresponded with Grelling in 1909 on a possible translation of *The principles*, and in the following year on type theory (RA). The translation was not done, but later Grelling translated some of Russell's philosophical books; so the contact must have continued.[4] In his *Dissertation* written at Göttingen under Hilbert's direction, Grelling *1910a* reviewed the development of finite arithmetic without using an axiom of infinity, relying much on Zermelo's recent investigation *1909a* (§4.7.1) and working out from Russell's definition of cardinal numbers as classes of similar classes.

Another member of the Fries school was the mathematician and philosopher Gerhard Hessenberg (1874–1925). As we saw in §4.2.5, he published in the *Abhandlungen* a long article *1906a* on 'Basic concepts of *Mengenlehre*', which was also sold in book form. As a Kantian, Hessenberg was prone to see paradoxes everywhere and came only to the naming paradoxes in the 23rd of his 30 chapters—a rather dull treatment which however may have inspired Grelling and Nelson two years later. The next chapter treated the 'ultrafinite paradoxes' such as Russell's, which he appraised as 'not especially mathematical, from that also comprehensible to lay people, but at the same time not dangerous for mathematicians, who have nothing to do with the class of all things'. His solution followed from assuming that 'a class is to be distinguished from each one of its elements' (art. 97, together with praise for Poincaré *1906b* (§7.4.5)). He took Burali-Forti's paradox more seriously, since it related to his extensive discussion of order-types; in fact, he found it to be 'completely unsolved', and tried to sketch a solution in terms of theorems claiming it impossible to adjoin elements to the class of all ordinals (arts. 98–99).

7.2.4 *Russell as cataloguer of the paradoxes.* Apart from Grelling's, Russell collected or created all the 'paradoxes of logic', as he came to call them (the title of his paper *1906h*, for example). His motive was not only a practical one, the doctor seeking out all occurrences of the virus in the body logic, whether in classes, names, propositional functions, propositions or truth-values; he also wanted to Solve them in some philosophical sense,

---

[4] Landini *1996a*, 312 suggests that Russell ignored Grelling's paradox because it permitted denoting to be a self-referring predicate, which was not allowed. This may constitute Russell's solution, but it can hardly account for his silence; surely, he did not identify it with the impredicative versions of his own paradox.

A younger friend of Grelling and Nelson was Alexander Ruestow (1885–1963), who wrote a thesis at Erlangen University in 1908 on the history and possible solutions to the paradoxes. He published a version as the book *Der Lügner* (1910); although the publisher (and then his employer) was Teubner, it seems not to have gained the attention that it deserved. I have not found it; some data are given in Peckhaus *1995c*. Ruestow later become well-known as an economist.

and not merely avoid them. This desire for a "global" remedy may have prevented him from wondering if the paradoxes differed in any basic way. Thus he made no reaction to another of Peano's flash (but typically undeveloped) insights at the end of a miscellany paper *1906b* in his *Rivista*: after formalising Richard's definition of the troublesome number *N*, Peano noted that it was partly symbolic and partly verbal, so that 'the example of Richard does not pertain to Mathematics, but to Linguistics; an element, fundamental in the definition of *N*, cannot be defined in an exact manner (according to the rules of Mathematics)'. This line of attack, and the distinction of paradoxes which it excited, was not to be taken up until the 1920s (§8.4.6); Russell was not one of the instigators, although Peano mentioned the remark to him in a letter of September 1906 (Kennedy *1975a*, 218). Cantor was close to it also; in a letter of 8 August 1906 to Hilbert he pointed out the difference between definitions as such, always finite, and the concepts which they define (Hilbert Papers, 54/29).

Russell also did not much consider the logical forms of the paradoxes. In Cantor's and Burali-Forti's results, given the premise *p that* there exists a greatest cardinal or ordinal respectively, opposing conclusions (*c* and $\sim c$) are deduced about it:

$$p \supset c \text{ and } p \supset \sim c; \therefore \sim p. \tag{724.1}$$

Reductio ad absurdum proofs can have this logical structure, sometimes in the condensed form given by $c = p$:

$$p \supset \sim p; \therefore \sim p. \tag{724.2}$$

(This is the version called 'reductio' in *PM*, $*2 \cdot 01$, although without distinction of '$\supset$' from '$\therefore$'—or of *reductio* from the method of indirect proof, which is effected by deducing contradictory consequences from $\sim p$.) But with Russell's paradox, from the premise *r* that his class exists, we deduce the following about the proposition *b* that it belongs to itself:

$$r \supset . b \supset \sim b \text{ and } r \supset . \sim b \supset b; \therefore r \supset . b \equiv \sim b. \tag{724.3}$$

The differences may be reconciled via *reductio*, so no basic issue arises; in its terms, the paradoxes of the greatest numbers and of naming exemplify the first form while Russell's, the liar and Grelling's take the second.

Questions of form should be distinguished from those concerning the existence assumptions that have to be abandoned in each case (my *1998b*). For example, there is no barber who shaves those and only those who do not shave themselves, thus there is no barber (seemingly Russell's reaction in *1918–1919a*, 261); by contrast, eliminating Russell's paradoxical class affects set theory and logic quite fundamentally, as he was to find for several years to come.

7.2.5 *Controversies over axioms of choice, 1904.* In other letters of 1906 Peano expressed to Russell his opinions on the current discussion of the axioms of choice (Kennedy *1975a*, 216–217). It bore strongly upon logicism; indeed, Russell deserves greater credit for its recognition than is generally realised. Luckily, it has been well examined by historians, especially in the early 1980s.[5] Thus the summary here is brief; some general points now, and Russell's role in the next sub-section.

As part of his investigations of set theory (§4.7.6), Ernst Zermelo published at the end of 1904, his 34th year, a short paper *1904a* in *Mathematische Annalen* in which he proved Cantor's well-ordering principle (§3.5.1). His proof rested upon admitting a new axiom proposed with his friend Erhard Schmidt (1876–1959); in his axiomatisation of set theory Zermelo *1908b* called it 'the axiom of choice', which soon became the standard name. Given a non-empty class *M*, assume the existence of a functional 'covering' to associate with each non-empty subclass *M'* one of its members as 'distinguished'; then the collection of them may be treated as a class on a par with its parent classes. The proof followed by associating any member of any well-ordered subclass of *M* with its complementary subclass.

The assumption of infinite selections had been made before by many authors, including Weierstrass, Dedekind and Cantor (§3.5.2), but its status had rarely been questioned. But Peano was one who did: in the above letter to Russell he recalled that he had stated it *as* a principle in his paper on differential equations described in §5.2.5 (*1890b*, 150). However and typically, he had not developed his insight, and serious attention dates only from Zermelo's paper.

A sharp controversy quickly developed over the legitimacy of the axiom. Soon after publishing *1904a*, Zermelo surveyed it himself before the *Göttinge Mathematische Gesellschaft* on 15 November 1904.[6] Running for several years, it was an exceptionally rich source of controversy among mathematicians and philosophers (though Hessenberg was rather agnostic in *1906a*, ch. 25). The following questions attracted especial attention:

1) Different forms of the axiom were found; for this reason I use the plural 'axioms'. Were they in fact logically equivalent, and were some philosophically more acceptable than others? One important issue was

---

[5] On the history, see especially G. H. Moore *1982a* for an account in English, Medvedev *1982a* for one in Russian, and Cassinet and Guillemot *1983a* for a French version enriched with a large collection of translations of original texts written in English, German and Italian. My book *1977b* based upon the correspondence between Russell and Jourdain may also be used, as the axioms of choice became an obsession with Jourdain. After the dust of the time had largely settled, Sierpinski *1918a* gave a magisterial presentation of the places in mathematical analysis and set theory where the axioms were needed, and also where they could be avoided.

[6] Göttingen Mathematical Archives, 49:2, fol. 109; sadly lacking in details.

whether the choices were made simultaneously or successively; another concerned denumerably versus non-denumerably many choices.

2) Which theorems were logically equivalent to the axioms, and so became candidate axioms themselves? In a sequel paper Zermelo *1908a* showed that the well-ordering principle was one; over the decades more and more results were found to be either equivalent to, or necessary or sufficient for them (G. H. Moore *1982a*, app. 2).

3) Was the infinitude of independent choices a legitimate mathematical procedure? If so, was the choice class defined, or merely constructed? The French school of analysts became much exercised with this matter, especially in mutual correspondence which I cite as 'Letters *1905a*'. Emile Borel, subscribing to a constructivist view of mathematics similar to Kronecker's (§3.6.4), worked only with classes put together by at most a denumerable number of unions and complementations of classes (§4.2.3); thus to him all forms of the axiom were unacceptable. At first Henri Lebesgue was chary of its use, for example in his analysis *1905a* of René Baire's classification of functions (§4.2.2) he strove hard (but unsuccessfully) to avoid it; but he let it slip into his later work (G. H. Moore *1983a*).

4) If acceptable at all, where in mathematics were the axioms needed? Very many places were found in set theory and mathematical analysis, and also in other branches of mathematics; the gold-digging went on for quite a time.

5) Could proofs using the axioms be reworked without them? Sometimes this was clearly so, often apparently not—and on occasion the matter was unclear. An interesting example is the Bolzano-Weierstrass theorem (§3.2.6), where both answers seemed to apply. Eventually it was realised that four different theorems lay under this title, depending upon the way in which definition of infinitude was used and what kind of point was claimed to exist. The following version of the theorem does need an axiom: an inductively infinite bounded class of points possesses at least one limit point (Sierpinski *1918a*, 122).

6) Do they lead to paradoxes of their own? After all, Zermelo's paper came out shortly before the appearance of the naming paradoxes, which could be applied to ordered classes. Certain consequences took the same logical form. In particular, the Italian mathematician Giuseppe Vitali (1875–1932) published a little pamphlet *1905a* related to the new theories of integration and measure of the time (§4.2); assuming available the required properties of a measure such as infinite additivity, he used an axiom to construct a class whose measure was both zero and greater than 1 (an example of (724.1)), and conclude that it was not measurable.

7.2.6 Uncovering Russell's 'multiplicative axiom', 1904. Among works published before 1904 which showed implicit use of infinite selections, Russell's *The principles* is a prominent example. Various passages on, for example, definitions of the infinite, the well-ordering principle, and the multiplication of cardinals, show him on the brink of its discovery. In May

1905, in his second letter to Russell (Garciadiego *1992a*, 170), Berry pointed to this last context (a passage on p. 118); presumably in his reply he learnt that Russell had in fact found the axiom exactly there the previous summer. As Russell was to recall to Jourdain in March 1906 (my *1977b*, 80),

> As for the multiplicative axiom, I came on it so to speak by chance. Whitehead and I make alternate recensions of the various parts of our book, each correcting the last recension made by the other. In going over one of his recensions, which contained a proof of the multiplicative axiom, I found that the previous proposition used in the proof had surreptitiously assumed the axiom. This happened in the summer of 1904. At first I thought probably a proof could easily be found; but gradually I saw that, if there is a proof, it must be very recondite.

Note that his dating of the discovery places it slightly before the conception and writing of Zermelo's paper, so reversing the priority over finding Russell's paradox (§4.7.6).

We saw that Whitehead had stressed the importance of defining infinite multiplication logicistically via the definition (653.2) of the multiplicative class $d^\times$ of a class d: pursuing this line, Russell found this surprising need, using the name 'multiplicative axiom' because of this context. In contrast to Zermelo's assumption of a covering functional, he accepted the need for the class involved. At first he thought that he assumed less than Zermelo had because of its restriction to mutually disjoint classes (my *1977b*, 63), but in March 1906 Jourdain gave him a rather loose argument for their equivalence (pp. 81–83); a tighter proof appeared in *PM*, *258·32–37.

Like his contemporaries, Russell realised that the proofs of many theorems were now puzzling or maybe defective; but logicism faced a special extra difficulty. Whatever form his logic were to take, it had to be finite, both horizontally in the lengths of expression and formulae and vertically in the lengths of proofs. But the first constraint was now threatened: how could he find in his logic a propositional function to express an *infinitude* of independent selections of members from classes?

Spotting this quandary in January 1906, Jourdain wondered if the axioms might be stated in terms of the possibility of splitting a many-many relation into an infinite class of many-one relations in some appropriate way. Russell agreed, proposing an alternative version about the existence of a class which had only one member in common with each class in a class of mutually disjoint classes (my *1977b*, 67–69); but, quite apart from the state of logicism at the time, especially the relationship between classes and propositional functions in general, the place of the axiom was not clear. He corresponded with Jourdain, and also Hardy, on examples of its use; and he worked on it from time to time, most notably in a manuscript *m1906h* of 31 folios, in which he compared a variety of forms expressed in terms of classes, relations or substitutions (the latter his logical system at that time, as we shall see in §7.4.6). He was to remain very sceptical about this axiom, and looked forward to a proof of it from standard procedures and notions of set theory (§7.8.7); but he never dealt with the quandary spotted by Jourdain.

7.2.7 *Keyser versus Russell over infinite classes, 1903–1905.*   Another axiom for set theory confronted Russell, this time sent from the U.S.A. In the early 1900s the mathematician and philosopher Cassius Keyser (1862–1947) became interested in mathematical induction (hereafter, 'MI'), especially in connection with definitions of infinity. After a survey paper *1901a* on the finite and the reflexive infinite published by the American Mathematical Society, he presented his main conclusion in a lecture to them; that the existence of an infinite class was assumed as an 'axiom of infinity'. He considered Poincaré's view argued in *1894a*, that much of 'the nature of mathematical reasoning' rested upon MI as a primitive procedure, and saw the axiom buried in Poincaré's metaphysical belief that '*the affirmation of the power of the mind which knows itself capable of conceiving of the indefinite repetition of the same act as soon as it is possible once*' (Keyser *1903a*, 427, translating Poincaré). Similarly, Dedekind's theory of chains to generate the cardinals (§3.4.2) included the 'theorem on the definition by induction' which assumed the axiom in the ψ transformation, and again in Dedekind's (and also Bolzano's) claimed proof of the existence of an infinite class by taking an object, the thought of it, the thought of the thought of it, and so on—but *how*? (pp. 429–434).

A committed Christian, Keyser rehearsed his views in an aroma of theology in a paper *1904a* in the April issue of *The Hibbert journal*, a 'quarterly review of religion, theology and philosophy' recently founded from funds established 50 years earlier by the Victorian philanthropist Robert Hibbert (1770–1849). Among texts criticised by Keyser was Russell's *The principles*, which had appeared in the interim; so Russell quickly wrote a short reply *1904f* in the next issue. Following his line in his book (*1903a*, 357–358), he asserted that Dedekind's theory implied the actual infinite but did not presuppose it, so that the axiom was not needed. However, in a retort Keyser did not budge, on the grounds that 'trying to prove that proof is a possible thing' involved an unavoidable vicious circle (*1905a*, 382). At this time Russell still regarded propositions as objects, which bolstered his defence; but in 1906 he changed to Keyser's position (without the theology) after further work on his logical system (§7.7.2).

7.3 THE PERPLEXITIES OF DENOTING, 1903–1906

> Frege and Russell are not ordinary language philosophers, but ideal language philosophers.   Jan Dejnožka (*1996a*, 222)

7.3.1 *First attempts at a general system, 1903–1905.*   (Russell *Papers 4*, Parts 1–2) Russell wrote some thousands of folios in and around logic between 1903 and 1907. Among those that he kept, many form fairly connected essays on specific topics; but some long ones read like logical experiments, in which he started out from one possibility or issue but then wandered around a whole range of options, producing partly developed and mutually inconsistent forays. Numerous nice features adorn the corpus; the coverage in this chapter is perforce limited.

Some manuscripts were versions of 'Volume II'. In one group, cited collectively as Russell *m1903c*, the treatment was systematic enough to use Peano's numbering of propositions by asterisked digit strings and method of dots for brackets, as well as various of his notations; Whitehead was especially fertile in inventing new ones. They used '$\phi | x$' for the 'value' of a propositional function ϕ of one argument, with '$|$' specified as an indefinable, before switching for a time to '$\phi'(x)$'. For abstraction they moved away from Peano's symbol ' $\ni$ ' for 'such that' either to existential quantification or to the right apostrophe as in '$\dot{x}(X)$' for the 'form' of the well-formed formulae (or 'expression') X in which x was a constituent. The form X was not necessarily a propositional function ϕ; but in such a case, the two notions were assumed to be inverse (p. 53):

$$'\vdash \; : \dot{x}(X) = \phi . \supset . \phi | x = X \quad \text{Pp.'}. \tag{731.1}$$

Here, and throughout these manuscripts, Russell used Frege's assertion sign (§4.5.6), and seemingly in a similar way. In March 1908 he explained the need for this sign to the writer Horace M. Kallen thus (copy in RA):

In common language, the effect aimed at is produced by the use of a verb instead of a verbal noun, e.g. 'Caesar died' instead of 'Caesar's dying'. But when you are using symbols, this is impossible, and therefore a special symbol is required. Thus I should say, in words,

'The proposition "$x = x$" is true' or "is the law of identity" or etc.

But if I wish to assert '$x = x$', I write

$$\vdash \, . \, x = x.$$

Russell also tried out Frege's modification to Law 5 to solve his paradox (§6.7.7), and a variant idea involving two kinds of membership to a class; but without success in both cases (pp. 3–15; see also *Papers 4*, 611–619). He essayed a few adventurous definitions, such as this one for the empty class (p. 30):

$$'*16 \cdot 1 \; \lambda = x \ni \{(\phi) . \phi x\} \; \text{Df'}. \tag{731.2}$$

He also gave outlines of the theories of propositional functions and relations, again trying modifications to avoid the paradoxes without being close to any Solution (pp. 38–72).

A long and important manuscript is a collection of sheets marked 'FN' in the top left-hand corners, and dating from the autumn of 1904 (Russell m *1904d*). The first page has not survived, but we may surmise that the title was something like 'Fundamental Notions'. The pagination goes up to 888, but he seems to have left gaps for possible later insertions. At all events, the 304 surviving folios show that he stopped around every 50 or so and wrote out the main assumptions or axioms tried since the last pause. In one particularly good folio, reproduced as Plate 4, he went over the batch of principles proposed to allow the proof of Cantor's power-class theorem and then assessed them in pencil under the four-valued calculus 'True', 'False', 'Probably true' and 'Doubtful'.

By June 1904 Russell was able to answer a request for information from Couturat (the context is explained in §7.4.1) with a collection of notes *m1904a* on the 'Outlines of symbolic logic' for these topics. Assertion was included among the indefinables; so was abstraction from ϕ, which was written '$\phi'(\hat{x})$' (and remained so far apart from the latter omission of the apostrophe while the associated class abstraction was symbolised '$\hat{x}(\phi'\hat{x})$'. A passage in 'FN' written soon afterwards contains a plan for 'Part I. Symbolic Logic' in five Sections and articles up to $*28$ (*m1904a*, 149–150).

7.3.2 *Propositional functions, reducible and identical.*

In addition, some durable notions were taking shape. One of these, explained to Russell by Whitehead in a letter of 23 April 1904, was the 'reducibility' of a function '$\phi!x$' (with exclamation mark), propositional or not, meaning that 'an equivalent expression is $x \in u$' for some class u (*Papers 4*, xxiv). While they had not yet attempted to impose any type theory on the predicate calculus, the aim of associating a function with a "simpler" one was similar. Russell wrote out axioms for 'Reduc' and some allied notions in his notes for Couturat (Russell *m1904a*, 84), and in an essay for himself at the time he included the sufficient condition that the truth-value of a reducible function was not affected if it were contained among its own values, and likewise for its negation function (pp. 89–90). In FN, which included a section on reducibility in the plan for the Part (p. 149), he referred to this condition as 'the "vicious-circle" principle' (*m1904d*, 138). This marks the début of another important idea in *PM* (§7.4.5), although here not adjoined to a type theory.

Russell also thought of the converse situation: a function ϕ was 'irreducible' if it were satisfied by some but not all members of any class u and also of its complementary class not-u. Moving the other way, 'Irreducible sets [*sic*] are got by a zigzag' which found terms satisfying some ϕ and then not-ϕ (written '-ϕ': pp. 120–121). This procedure, which may have its origins in one of his late revisions to *The principles*, 103–105, gave birth to a 'zigzag' theory of solving the paradoxes by discarding from logicism such "complicated" propositional functions and their classes; the name may

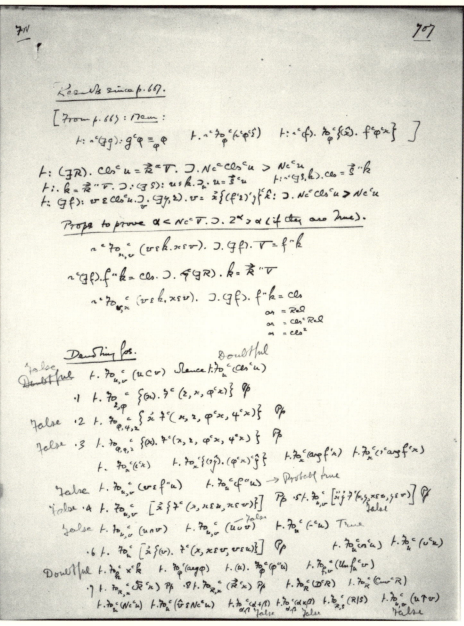

PLATE 4. A striking folio from Russell's working text 'FN' of 1904 (RA). He reviewed the efforts of the previous 40 folios to solve his paradox, with especial reference to Cantor's (correct) power-set theorem. The sheet is shown also as Russell *Papers 4* (1994), Plate 6.

have been inspired by the alternation between functions and their negatives. However, although he entertained the theory for some time (§7.4.4),
he never found a criterion of complication sufficiently simple to be
practical.

Another important notion arising from these considerations was the
relation of identity. Using Schröder's symbol '1'', Russell first used it for
classes with this assumption (*not* definition) from equivalent propositions:

$$'\vdash \;:.\,(x): \phi x .\equiv .\, \psi x :\, \supset .\, \{x \ni (\phi x)\} 1'\{x \ni (\psi x)\} \; \text{Pp.}' \quad (732.1)$$

(*m1903c*, 5: the equivalence came out as a theorem on p. 15). The feeling
here is intensional: equality of classes *a* and *b* came out extensionally on
p. 17 as

$$'a = b .\equiv .\, a \supset b .\, b \supset a \; \text{Df}', \text{ with } 'a \supset b .\equiv\, : x \in a .\supset_x .\, x \in b \; \text{Df}'.$$

$$(732.2)$$

By the time of writing to Couturat the identity relation was defined

$$'\ast 11{\cdot}1 \quad x = y .\equiv .\, (\phi)\phi'x \supset \phi'y \quad \text{Df}', \quad\quad (732.3)$$

where apparently *x* and *y* were terms (*m1904a*, 81). This (Leibnizian)
form of definition was to be adopted hereafter, although changes of theory
altered the range of entities over which ϕ was quantified. One relevant
factor here was the (lack of) restriction to propositional functions; for he
was well aware that mathematical functions were different in kind, which
needed to be treated with great care.

7.3.3 *The mathematical importance of definite denoting functions.* (Russell *Papers 4*, Part 3)

> When we are speaking of an Individual, it is usually an abstraction that we
> form; *e.g.* suppose that we are speaking of the present King of France; he
> must actually *be* either at Paris or elsewhere [...]
> Whately *Logic*₉ (1848), 84 (from the third edition (1829) onwards)

We saw in §6.7.4 that Russell was much engaged with six little words, such
as 'a', 'any' and 'some'. Among them 'the' rose to prominence, partly for its
intrinsic interest but also for its central place in mathematical analysis and
set theory. In particular, Cauchy's insistence that mathematical functions
had to be single-valued (§2.7.2) was upheld by all successors, including the
Weierstrassians; thus Russell had to be able to express them within his
logical system. They were an especially important case of 'denoting functions', which he defined as any function which, unlike a propositional
function, did not take a proposition as its value; others included, for

example, the class of all arguments satisfying a propositional function. A good theory of these functions should also help to explain the use of such phrases in natural languages as well as in mathematics.

Russell occupied himself not only with the referentiability of denoting functions but also with allied issues such as their relationships with propositional functions and in turn to propositions, sentences and statements; the status and occurrence of universals and particulars; the ontological consequences of quantification; the effect of the paradoxes upon the generality of logic by imposing restrictions in the ranges of variables; and the place of truth-values. The intricacy of the resulting systems, and the intensity of Russell's endeavours both published and unpublished, allow for much interpretation:[7] this section picks out some principal features which bear most closely upon logicism.

'If I say "I met a man," the proposition is not about a *man*', wrote Russell in *The principles*, 'this is a concept which does not walk the streets, but lives in the shadowy limbo of the logic-books' such as his own (*1903a*, 53). In a suite of manuscripts he tried to disentangle the relationships between concepts, proper names, and phrases and especially between propositional functions and denoting. In an early one, written in the summer of 1903, he reflected 'On the meaning and denotation of phrases' by noting of the current Prime Minister that 'A proper name, such as *Arthur Balfour*, is destitute of *meaning*, but *denotes* an individual. On the other hand, verbs and adjectives have meaning but no denotation'; and soon he considered the case of 'the present King of France' (*m1903d*, 284–285). Whether or not Russell took the case of Monsieur le Roi from Whately, he treated Him differently; instead of pondering upon the current Royal Residence, he found a phrase with meaning but no denotation.

The relationship between names and descriptions was a very important issue, since truth values could change. A very nice arithmetical example came from the proposition, assumed true, that 'the number of people at the meeting was greater than any one expected'; 5,432 came, but '5,432 was greater than any one expected' is false (p. 317). Whitehead and Russell discussed this case in more detail the letters of the following April (RA).

With such examples Russell entered the labyrinth of reference, and wandered around it looking for a credible exit. As guide he worked with 'complexes', well-formed formulae in the predicate calculus including functions of functions, relations and quantification (*m1904c*, a summary written in October to enlighten Whitehead). One source was the philosophy of reference of the Austrian psychologist Alexius Meinong (1853–1920),

---

[7] Among the profitable portions of recent secondary literature, note Rodriguez-Consuegra *1988a* and *1991a*, and Landini *1988a* and *1998b*, although they take differing positions on the importance of the substitutional theory (§7.4.6–7). Other studies include Dejnožka *1996a* on identity theories, and de Rouilhan *1996a* on concepts and objects. I do not attempt to appraise the total literature.

especially a theory of complexes and 'assumptions' which Russell popularised in Britain with a long essay *1904e* in *Mind* (J. F. Smith *1985a*). Meinong granted reference to 'the present King of France' and even oxymorons such as 'round square', and permitted them to refute the law of contradiction by existing and not existing at the same time—both too luxuriant and illogical for Russell. However, in other respects, such as the notion of complexes, Russell received Meinong's ideas positively, though he did not use them in his logicism (see the mixed reception of Meinong on zero in ch. 42 of *The principles*). The later, converse, influence is noted in §8.7.2.

In June 1905 an escape route began to appear when Russell laid down a list of principles in an experimental text 'On fundamentals'. 'A complex *C* has both *being* and *meaning*', he decided (*m1905a*, 369), importing the former property from *The principles* as 'that which belongs to every conceivable term, to every possible object of thought' (*1903a*, 449). Being occurred with the assertion of *C*, while meaning was linked to its truth; similarly, each component of *C* had a corresponding '*entity*-position' or '*meaning*-position' (*m1905a*, 361). However, 'The manner of occurrence of "the author of *Waverley*"' in 'People were surprised that Scott was the author of *Waverley*' was 'peculiar', for the kinds of reason pertaining to the arithmetical substitution above (p. 370).

Russell analysed at length types of occurrences of components in complexes and propositions in compound propositions, and especially the difference between denoting and propositional functions. For notation 'We may use $(C \wr x)$ for a general complex' (p. 366); the unusual symbol was possibly taken from Arthur Cayley's sign for polynomial forms in his theory of invariants. On identity, the form (732.1) was preserved, but ϕ 'may be a denoting function, or a propositional function, or a propositional function of a denoting function, or a denoting function of a propositional function, but must not be a propositional function of a propositional function' (p. 371).

7.3.4 *'On denoting' and the complex, 1905.*   Russell later wrote at the head of this manuscript that the list of its principles contained 'the reasons for the new theory of denoting' (*Papers 4*, 358). This was written in the next month, July 1905, as the paper 'On denoting', which appeared in *Mind* the following October (*1905d*). Perhaps his most famous essay, it was written in difficult personal circumstances, especially the death of a close friend, which must have contributed to the poor exposition (Urquhart *1995a*). Understandably, editor G. F. Stout had been very reluctant to publish it (Russell *1959a*, 83); but Russell never placed a major philosophical paper there again.

Russell began by stressing 'the distinction between *acquaintance* and *knowledge about*', of which the latter came through denoting. In a later paper he expressed this distinction in terms of 'knowledge by acquaintance and knowledge by description', and introduced the name 'definite descriptions' for his theory (*1911c*, 151–154). Working here with his six little words, he covered some of the corresponding complexes, such as '*C* (a man)' as 'It is false that "*C*(*x*) and *x* is human" is always false'. But the bulk of his analysis fell upon phrases using 'the', which led to 'by far the most interesting and difficult of denoting phrases' (*1905d*, 417). After a succinct and organised resumé of the June list of principles, with Scott and Monsieur le Roi again as main examples, he came up with these criteria under which a definite denoting phrase actually denoted, within the context of a proposition (p. 423):

> [... T]he proposition 'Scott was the author of *Waverley*' (*i.e.*, 'Scott was identical with the author of *Waverley*') becomes the proposition 'One and only one entity wrote *Waverley*, and Scott was identical with that one'; or, reverting to the wholly explicit form: 'It is not always false of *x* that *x* wrote *Waverley*, that it is always true of *y* that if *y* wrote *Waverley y* is identical with *x*, and that Scott was identical with *x*'.

Russell did not point out that his trio of criteria for the existence of a referent for a denoting phrase were exactly those which Peano *1897b* had proposed (§5.4.3) in the special context of a single-valued mathematical function. Presumably Russell had forgotten Peano's text, which he must have read in 1900; the first paragraph of the section is line-marked in the margin of his copy (RA). But he never referred to it, or to Peano at all, in the manuscripts preceding this paper.

Another difference is that Peano's criterion was given nominally whereas Russell's came *contextually within a proposition*. This basic feature was not emphasised in the paper, and is often overlooked; but Russell made it quite clear to Jourdain on 13 January 1906 (my *1977b*, 70):

$$\psi\{(\imath x)(\phi x)\} . = : (\exists b) : \phi x . \equiv_x . x = b : \psi b \text{ Df.} \qquad [(734.1)]$$

We put [for existence]

$$E!(\imath x)(\phi x) . = : (\exists b) : \phi x . \equiv_x . x = b \text{ Df.} \qquad [(734.2) \dots]$$

If *u* is a class, we write $\imath' u$ for $(\imath x)(x \in u)$, that is,

$$\psi(\imath' u) . = : (\exists b) : x \in u . \equiv_x . x = b : \psi b \text{ Df.} \qquad [(734.3)]$$

Contextual definitions were to be centrally important in *PM*, as we shall see in §7.8.4–5. They show one reason for distinguishing descriptions from names: the former, but not the latter, involve the *scope* within the defining formula.

Connected to contextual definition was 'denoting complex', a phrase composed of more than one word and working via the denotations of its constituent words, which Russell also emphasised in his paper. But later he played down its importance; when Ronald Jager *1960a* argued plausibly that this complex denoted a meaning, as opposed to a 'denoting phrase' which expressed it, Russell wrote to him on 28 April 1960 (RA) that he had seen his paper as a '*reductio ad absurdum* of the view that a denoting complex has a meaning as well as (sometimes) a denotation', was 'surprised by your view that the concept of a denoting complex seems to you an essential part of my theory' and judged that 'I came later to think all that stuff about denoting complexes is unnecessary and in no degree essential to my argument'.

This self-criticism of the paper reflects the small reaction to it at the time. But G. E. Moore welcomed it in a letter of 23 October 1905 (RA); the ontological parsimony of Russell's criteria would have warmed him. However, he acutely queried Russell's claim that '*all* the constituents of propositions we apprehended are entities with which we have immediate acquaintance' (italicising in a clause near the end of Russell's paper), and wondered if the variable permitted such acquaintanceship. In his rapid reply Russell claimed that this was the case, but he admitted himself uncertain as to whether the variable was an entity or not (*Papers 4*, xxxv). The philosophy was still wanting; indeed, the feature highlighted by Moore was to concern Russell deeply after *PM*.

7.3.5 *Denoting, quantification and the mysteries of existence.* Russell's account above to Jourdain involves a rather muddling aspect of his logic, which becomes noticeable from 1905 onwards: his multiple uses of the word 'existence', and the symbols that went with them (my *1977b*, 71–74). There were two senses for individuals ('I' in Table 735.1), and three for classes ('C'):

On individuals, the two senses do not necessarily interact; for example, the present King of England would not have been ignominiously placed at the bottom level in the type theory of *PM*. However, proved there is

$$\text{'} *14\cdot201 \ \vdash \ : \text{E}!(\imath x)(\phi x) . \supset . (\exists x) . \phi x \text{'}, \qquad (735.1)$$

with a comparable proposition involving relations at $*53\cdot3$.

TABLE 735.1.   Russell's congeries of existences

Case	Notation(s)	Sense of existence
I1	$\exists x \ (\exists x)$	As in existential quantification
I2	$\text{E}!(\imath x)(\phi x)$	Of a referent of a denoting phrase
C1	$\exists u \ (\exists u)$	As in existential quantification
C2	$\text{E}u$	Abstractable from a propositional function
C3	$\exists\text{'}u \ \exists u \ \exists!u$	Non-emptiness (Peano's (541.3))

The relationship between the senses for classes is harder to determine, especially as Russell used C2 and C3 rather informally; but, for example, the empty class exists (or may do so) in senses C1 and C2 but not C3 while conversely the class which generates Russell's paradox exists only in C3. In some theories existence was allowed under conditions; in particular, C2 obtained in *PM* only within contextual definition (734.2). The zigzag theory tried to deny C2 to classes when the propositional function was too complicated.

In addition, some relationships obtained between the I and C senses. I1 might permit any of the Cs, while I2 implied C3 thus:

$$\vdash \, : E!(\imath x)(\phi x) . \supset . \, \exists! \hat{x} \phi(x). \tag{735.2}$$

Russell did not prove this proposition in *PM* because by then he had largely dropped C3.

Russell did not always have these different senses of 'existence' under control. We saw his exaggerated claim in 1900 that only his definition of irrational numbers delivered the existence theorem (§6.4.7); thus he granted Peano's sense C3 a primacy which it did not deserve. It appeared in another round of exchanges at this time.

7.3.6 *Russell versus MacColl on the possible, 1904–1908.* (Bibliography, Rahman *1997a*) We saw in §2.6.4 that Hugh MacColl had proposed in *1877a* that Boole's algebra could be used to formulate the propositional calculus. Over the years he had continued with his researches, and with a paper *1897a* of a series on 'Symbolic reasoning', published in *Mind* as he entered his sixties, he made an innovation for which he is now best remembered: modal logic (not his name), in which 'possible' and 'impossible' were basic notions.

Russell was one of MacColl's principal targets for discussion. Already in a long letter of 6 October 1901 MacColl had queried a passage in Russell's recent paper on order in *Mind* (§6.4.2) on the relationships of implication between various theorems in Book 1 of Euclid's *Elements* (Russell *1901a*, 296). The issue for MacColl was not geometry but implication itself; for propositions A and B 'A implies B' meant for him 'It is impossible that A can be true and at the same time B false'; while 'A does not imply B' meant 'It is possible that A can be true and at the same time B false'. He gave further examples from arithmetic, and from modal relations such as the certainty 'a whale can swallow a herring';[8] he gave variants on this example in *1902a*, 357–358. For him a proposition was 'certain' if it

---

[8] This letter by MacColl is one of a large number which he sent up to his death in 1909 and which Russell kept (RA); nevertheless, Russell's covering note to the batch reads merely: 'A writer on mathematical logic with whom I disagreed'. There seems to be no MacColl *Nachlass*. A group including myself has surveyed his life and work in the *Nordic journal of philosophical logic* (1999), incorporating his letters to Russell.

followed from the pertaining assumptions and rules, and 'impossible' if not. While Russell's reply to MacColl's letter is not extant, he probably did not warm to these suggestions; indeed, it may have helped prompt his note *1902b* on 'the teaching of Euclid' (§6.2.2), when he stressed the lack of rigour in Euclid's *Elements*.

Their published exchanges date from 1904. MacColl had another string of (short) papers on 'Symbolic logic' running in *The Athenaeum*, and in a pair *1904a* he raised similar doubts about the logic of non-Euclidean geometry, as presented in Russell's book *1897c* on geometry. In his reply in the October issue, Russell *1904g* detached questions of our ignorance, such as whether or not 'our actual space may be non-Euclidean', from logic, where 'all propositions are merely true or false. I should not now divide true propositions into necessary and contingent, or false propositions into impossible and possible'. He also stressed 'the difference between geometry as the science of actual space, and geometry as a branch of pure mathematics'. His logicism asserted that 'In pure mathematics, as such, we do not consider actual objects existing in the actual world, but hypothetical objects endowed by definition with certain properties'; however, this was a not a proper reply in that MacColl had been discussing Russell's pre-logicist book on geometry.

MacColl pursued his line in his *Mind* series, especially in *1905a* in the January issue with a short April addendum *1905b* on 'Existential import'. Russell's reply *1905b* was again short but instructive, for it shows that he saw existence in logic *only* in the Peano sense C3 above: 'To say that A exists means that A is a class which has at least one member'. He made this point in order to confront MacColl's unsatisfactory definition of the empty class as 'our universe of non-*existences*'. Russell opposed this admission of possible objects with his own theory of denoting; but in May 1907 Jourdain was to reprove him for not having explained also existence in the I2 sense, and Russell accepted the criticism (my *1977b*, 102).

Russell sent MacColl a proof of his reply, and MacColl added a paragraph *1905c* pleading for a broader philosophy in which 'Symbolic Logic has a right to occupy itself with any question whatever on which it can throw any light', especially existence. In a book *1906a* on *Symbolic logic and its applications*, based upon several of his articles, MacColl presented his modal system in detail (including a range of rather unhelpful notations, not reproduced here). In a review in the April 1906 issue of *Mind* Russell noted that he dealt 'always with whole statements or propositions, not, like most writers, with classes', so that 'he is primarily concerned with *implication*, not with *inclusion*' either in part-whole theory or in Russell's preferred Cantorian way (Russell *1906c*, 255). He also repeated his disagreements about possibilities and existence, and in another review *1908a* of the book for *The Athenaeum* he rejected MaColl's empty class of unreal members in a witty remark on 'the present king of France': 'Thus republics have kings, who only differ from the kings of monarchies by being unreal'.

Perhaps motivated by MacColl's work, Russell surveyed the various senses of possibility and necessity used in logic and epistemology, in a lecture *m1905f* delivered to the Oxford Philosophical Club in October. However, finding none of them to be definitive, he recommended 'that the subject of modality ought to be banished from logic, since propositions are simply true or false'—a hostile *non sequitur* followed by a declaration which for MacColl begged the question. He also misinterpreted his adversary as linking possibility and necessity respectively with existential and universal quantification of a propositional function; later he promoted this sense of modality himself.[9]

In this lecture (p. 518), the *Mind* review and elsewhere Russell also deplored MacColl's failure to distinguish a proposition from a propositional function. As MacColl had pointed out to Russell in a letter of 24 February 1906 (RA), back in his early days he had interpreted the universally affirmed proposition 'All *X* is *Y*' as 'if an individual belongs to the class *X* it belongs to the class *Y*' (*1877a*, 181). However, he treated propositions not as designators but as forms of words, sometimes true and sometimes false; for example, 'Mrs. Brown is not at home' (*1906a*; 18–19). In a longer reply in *Mind* to Russell, he insisted that a propositional function, which 'I should prefer calling a *functional proposition* [, . . .] must, from my point of view, be classed as a proposition' (MacColl *1907a*, 470).

The last exchange ran through the 1908 volume of *Mind*. Starting out from A. T. Shearman's review *1907a* there of *The principles* (§6.8.1), MacColl concentrated this time on implication (*1908a* and *1908b*, with Russell *1908b* in between). Again the principal divide was the status of propositional functions; for example (due to MacColl) 'he is a man' was a proposition that could possibly be true, while for Russell it was a propositional function if 'he' was indeterminate.

MacColl deserves great credit for his innovations. Both in his work from the late 1870s and now, he emphasised *propositions* in logic over and above classes or terms. He also made a good distinction between the senses of truth-values attributable to, say, '3 > 2' or '3 < 2' and to 'Mrs. Brown is not at home'. His stress on knowledge rather than truthhood is good, as is his claim that we need logics, as (though not just as) we need algebras and need geometries. But the modal brands received an unclear start here, with his views on propositional functions and variables, absurd

---

[9] See especially Russell *1918–1919a*, 231 and *1919b*, 163. In an attack of Russell's criticisms of MacColl, Rescher *1947a* slides over MacColl's unclarities and seems to be unaware of Russell's use of modality. On this and other mistakes see Dejnožka *1990a*, 406–412, whose own advocacy of modality in Russell, however, is heavily tempered in Magnell *1991a*. Dejnožka presents his evidence in detail in *1999a*. A greater obstacle than Russell's objections to the recognition of MacColl's proposals was the minimal acknowledgement in C. I. Lewis (§8.3.3).

conceptions of empty classes and of infinitude, and unintuitive notations. So their later development had to wait some years for C. I. Lewis (§8.3.3), and in many respects far longer than that (Rescher *1969a*, ch. 1).

7.4 FROM MATHEMATICAL INDUCTION TO LOGICAL SUBSTITUTION, 1905–1907

7.4.1 Couturat's Russellian principles. While Russell was searching for solutions and axioms, and puzzling over denoting, his publications were gaining attention in Paris. As in the late 1890s (§6.2.3), the main venue was the *Revue de métaphysique et de morale*, with Couturat and Poincaré among the authors.

The news started well for Russell. Up to the appearance of *The principles*, Couturat's interest in logic had been dominated by the algebraic tradition; for example, in an article on 'Symbolic logic or algebra of logic', written with C. S. Peirce's former student Christine Ladd-Franklin at her request[10] for an American dictionary of philosophy and psychology, they gave merely two passing mentions to Peano and none to Russell (Couturat and Ladd-Franklin *1902a*). But we saw in §6.8.1 that Couturat published a lengthy review *1904a* of *The principles*, and the book then inspired him to write in the *Revue* a series of five long articles *1904–1905a* with the same title in French. The publisher of the journal quickly put out a lightly revised version, together with some other material, as a book *1905b*. Citations are to this version; a German translation *1908a* soon appeared.

Although Couturat's book was less than half the length of Russell's, he presented logicism pretty comprehensively, and for the first time in French. As usual, he did not criticise Russell much, rendering not only the title but much of the content of *The principles*: the basic logic and set theory; the finite and transfinite numbers; order and continuity; real numbers and magnitudes; and finally dimensions and the three branches of geometry. But he ignored mechanics without explanation, and did not dwell on the underlying philosophical issues. He cited a wide range of literature, with a notable enthusiasm for the recent work on axioms and models (§4.7.3) by Oswald Veblen and E. V. Huntington (pp. 168–174); he even added a postscript on p. 308 on Huntington *1905a–b* on the continuum (§7.5.6), and wrote warmly about it to Russell on 21 October (RA). Among recent developments, he sided with Keyser (§7.2.7) on the need for an axiom of infinity (p. 60).

The other material included two new notes, on set theory and group theory; perhaps aware of Russell's (644.2) or Whitehead (§6.4.9), Couturat had subsumed the latter under the logic of relations as 'a branch of the

---

[10] Couturat told Russell on 7 May 1905 that Ladd-Franklin had been 'mécontante' with the article he had written and so she wrote another one, 'ce qui fait le melange le plus bizarre' when the two were combined (copy in RA).

science of order' (p. 208). Finally, an appendix reprinted a long piece *1904b* from the *Revue* on Kant's philosophy of mathematics: he doubted the legitimacy of *synthetic a priori* judgements and the formulation of hypothetical ones, and concluded that 'the progress of Logic and of Mathematics in the 19th century has invalidated the Kantian theory and given right to Leibniz' (p. 303).

Later that year Couturat published with Gauthier-Villars a short complementary guidebook *1905a* to *L'algèbre de la logique*—a much older subject but still little known in France. In 1914 Open Court published an English translation, with a preface by Jourdain: a Polish translation is noted in §8.8.2. He ran through the theories in their classial and propositional forms roughly at the level of Schröder's first volume, though unfortunately with only a mention at the end of the logic of relations. There and in art. 11 he cited, as a work in preparation, a 'Manual of logistic'; it had arisen from a course given at the *Collège de France* where he was substituting for Henri Bergson. For some reason this piece of *vulgarisation* never left his busy study (§8.6.2);[11] but in his inaugural lecture *1906b* for the course, on 'logic and contemporary philosophy', he aired his oppositions to psychological and to sociological foundations for logic. With Russell, and maybe Poincaré also, he was at one here. However, his advocacy of logic earned derision in a reply Borel *1907a* in the *Revue* stressing the need for intuition in mathematics—which doubtless Couturat did not deny.

7.4.2 A second pas de deux with Paris: Boutroux and Poincaré on logicism. (Sanzo *1976a*) Borel is typical of the reception of Couturat's writings in France; indeed, on 18 December 1904 Couturat had written to Russell about his failure to interest Borel and Lebesgue in Peanese (text in §11.1). The hostility soon became public: from Poincaré once again, and his nephew Pierre Boutroux (1880–1922), a mathematician and philosopher. 'Poincaré is an oracle for the readers of the *Revue de métaphysique et de morale*, and Boutroux also', Couturat told Russell on 11 February 1904;[12] the high reputation of the nephew in his lifetime is difficult to explain at this historical distance. The sequence in the *Revue* this time is as follows:

Boutroux *1904a* (November) and *1905a* (July), Russell *1905g* (November);

Poincaré *1905c* (November) and *1906a* (January), Couturat *1906a* (March);

Poincaré *1906b* (May), Russell *1906h* (September), Poincaré *1906c* (November).

---

[11] Lalande hoped to publish Couturat's manual (*1914a*, 676), but he never did. See Couturat's letter of 9 November 1905 to Huntington about his lecture course (Ladd-Franklin Papers, Box 3).

[12] Copy in RA: 'Poincaré est un oracle pour les lecteurs de la *Revue de métaphysique et de morale*, et Boutroux aussi'.

We can deal with Boutroux quickly. His two papers were devoted to showing that the notion of correspondence between elements under a function could not be expressed by a theory of relations, as the logicists claimed, since it was not a 'logical notion' but 'an intuitive fact analogous to physical law' (*1905a*, 620–621). The two views are in fact not contradictory: in his reply Russell *1905g* contented himself with a tutorial on relations as propositional functions of two independent variables, and on distinguishing a class as such from a listing of its members. Boutroux's second paper elaborated upon a lecture *1905b* given at the International Congress of Philosophy at Geneva in August 1904 (§7.5.1). He wrote the text at Cambridge in the following December, when he visited Whitehead; the chats may not have been too fruitful, for Whitehead had to add two paragraphs *1905a* to Russell's reply denying that he thought that functions did not belong to logicism but that the symbolism 'is not *practically* useful in the development of Analysis'. He was also sorry that a remark on functions 'made in the course of a conversation would be presented as a carefully considered definition'!

7.4.3 *Poincaré on the status of mathematical induction.*   (Schmid *1978a*, ch. 6) The collision between Russell and Poincaré started from a collection of partly reworked articles on the philosophy of mathematics and science which Poincaré had published as the book *La science et la hypothèse* (*1902a*). An English translation *1905a* came out three years later, which Russell reviewed in the July issue of *Mind*. He found fault with Poincaré's position that MI was 'a means of passing from the particular to the general: it is merely a means of passing from one general proposition to another' (Russell *1905c*, 590). We saw in §6.5.3 that his logic of relations, partly enhanced by Frege's ancestral relation, demoted MI from the prime place which Poincaré assigned to it. In addition, as during the last dance, he rejected again the view that a geometry is not entirely conventional. Editor Stout invited Poincaré to reply, which he duly did in a short letter *1905b*, repeating his own stances on both issues and announcing that an essay on MI was soon to appear in the *Revue*.

Poincaré had been motivated by Couturat's Russellian essays, perhaps by two of the three references to himself; for one found against him on MI and another only partially accepted conventionalism in geometry (Couturat *1905a*, 62–63, 205). Called 'Mathematics and logic', Poincaré's paper appeared in the *Revue* in two parts; then he added a third part in response to new work from Russell. Later he reworked the trio in his next popular book, *Science et méthode* (Poincaré *1908b*). They formed less of an advance over nephew Boutroux than one might expect.

The first part, Poincaré *1905c*, comprised a survey of mathematical logic as he understood it, which was not very deeply: for example, a propositional function '$\phi(x)$, x being the variable. The proposition $\phi(x)$ can be true or false' (p. 827), like MacColl but probably inspired by deliberate obtuseness. Again, he was amused that Couturat's presentation 'contains numeral adjectives, cardinal as well as ordinal', such as ' "A relation holds between two terms" ' (p. 830); he might as well have criticised the use of page numbers in the *Revue*. Some reading of, say, Dedekind on *Zahl* and *Anzahl* (§3.4.2) would have helped him.

Poincaré treated MI better; indeed, *'Thus it is only here that the true debate commences*' (p. 832). For him specifying the integers by the Peano axioms or some equivalent procedure has to be primitive because any effort to justify or prove it assumed the integers in the first place (pp. 832–835). In his second part *1906b* he treated David Hilbert's work on the foundations of arithmetic (§4.7.5) with much greater warmth, but still found the definition of MI to be wanting since the talk of finite and infinite numbers embodied it (p. 23).

Couturat's long reply to these two papers also reads somewhat like a tutorial; for example, on the different senses of number (*1906a*, 216). He also ridiculed the implication that Russell was the first to (claim to) advance logic beyond Aristotle. He ended by hoping that Poincaré 'will come to a more just and more favourable appreciation of Logistic when he will have studied it' (p. 250). However, in one important respect he was on the same limited ground as Russell (§7.3.5); he spoke of the existence of a class only in Peano's sense of its non-emptiness (pp. 232–233).

7.4.4 *Russell's position paper, 1905.* (Moss *1972a*) The next stage of Russell's progress was motivated not by Paris follies but by a paper recently published in the *Proceedings* of the London Mathematical Society (hereafter, 'LMS') by the Cambridge mathematician E. W. Hobson (1856–1933). In his attempt to solve especially Burali-Forti's paradox, and also to appraise Zermelo's introduction of the axiom of choice, he had proposed 'a law, or set of laws, forming the *norm* by which the aggregate [class] is defined' (Hobson *1905a*, 173). He seems to have intended a constructive approach in building up these norms, although without using propositional functions. After mutual discussions and correspondence, critical replies came in from Hardy, Jourdain and Russell; we are concerned with the last, which was submitted to the LMS in November 1905 and published the next March as his *1906a*.

Reformulating Hobson's norm as a propositional function, Russell separated issues surrounding the paradoxes from those related to Zermelo's axiom. He compared the latter with the multiplicative axiom (which he still thought less general), and gave theorems from set theory when needed. He also published for the first time his charming illustration about the need for infinite selections to show that $\aleph_0$ boots divide into pairs as any reasonable owner would desire (pp. 47–48).[13]

---

[13] Russell's first use of this illustration seems to be in his letter of 31 July 1905 to Jourdain, repeated in December (my *1977b*, 55, 64). On explaining it later in terms of a millionaire's possessions 'to a German mathematician' he received the response 'Why a millionaire?' (Russell *1959a*, 93), which may not have been the stupid *Frage* that he obviously thought; for $10^6 \neq \aleph_0$.

Most of Russell's paper was devoted to the paradoxes of classes. He gave his first extended listing of them, including the general relational version described in §7.2.1, and showed that Hobson's norm was not sufficiently restrictive to avoid them. Curiously, he did not reinforce his point by mentioning also the paradoxes of naming (§7.2.2), although he thanked Berry for another detail (p. 36).

Russell also surveyed three candidate Solutions, all based on abandoning class comprehension in some way. He treated first 'The zigzag theory', which came closest in form to Hobson's own approach; 'we define a *predicative* propositional function as one which determines a class (or relation if it contains two [or more] variables)' (p. 38). But he warned that in this theory the 'axioms as to what functions are predicative have to be exceedingly complicated', and that he had not found them (p. 39). Next, 'The theory of limitation of size' was inspired by the greatest number paradoxes, but seemed less convincing under his functional generalisation (pp. 43–44). He might have added that the class producing his own paradox was not that large; as he had put it excellently to Jourdain in June 1904, it is 'only half way up' (my *1977b*, 35), for if a class did not belong to itself then its complement did.

In the final candidate, 'The no-classes theory', 'classes and relations are banished altogether', and propositional functions avoided, leaving only propositions p and their constituents a. When x was substituted for a in p, the resulting proposition was written '$p\frac{x}{a}$', and the theory rested largely on these objects, which replaced classes (Russell *1906a*, 45–47). (Relative to the type theory to come in *PM*, this one was not ramified.) He noted their single-valuedness, and confessed to difficulties in defining some transfinite numbers. Meanwhile Poincaré got wound up again.

7.4.5 *Poincaré and Russell on the vicious circle principle, 1906.*   Poincaré's addition *1906b* to his paper on 'mathematics and logic' was partly provoked by Couturat's reply *1906a*, and also by a treatment by Peano's follower Mario Pieri in the *Revue* of the consistency of the axioms of arithmetic (Pieri *1906a*). But Russell's new paper also fell within his sights. Poincaré fixed upon the naming paradoxes (which Russell had not treated), especially the version due to his countryman Richard (§7.2.2); he diagnosed the illness as lying in the impredicative manner of defining a number by a condition which itself involved naming. Therefore his 'TRUE SOLUTION' lay in avoiding such a 'vicious circle', and he wondered if Russell's 'zigzaginess' had the same aim (*1906b*, 307–308).

Poincaré also returned to MI, but in the new spirit. In a discussion of the axioms of choice Russell had mentioned in his paper their need for proving the equivalence of the inductive and reflexive definitions of cardinal finitude (*1906a*, 49); Poincaré now regarded as Russell's first defining clause ('a cardinal number which obeys mathematical induction starting from 0') as impredicative, and so the former treatment of finite and infinite numbers in Russell *1902* (§6.5.3) as 'vicious' (*1906b*, 310, attributed to Whitehead). After attacks on the axiom of choice for its lack of obviousness (pp. 311–315) he came to radical conclusions: '*logistic is no longer sterile, it engenders antinomy*', and even more generally, '*there is no actual infinity*; the Cantorians forgot that, and they have fallen into contradiction' (p. 316).

Russell could not let pass such dismissals *sous silence*; so a muse by him on 'The paradoxes of logic' appeared in the *Revue* four months later, in September. As usual, there were errors to correct; for example, the liar paradox had nothing to do with the infinite (Russell *1906h*, 633). But he responded to the naming paradoxes positively by taking seriously the vicious circle principle (hereafter, 'VCP') as a key to a solution, while pointing out that it too had no intrinsic link to the infinite (p. 634). He expressed it in Peanese as 'All that contains an apparent variable must not be one of possible values of that variable' (p. 634), and he began to think out how the '*universe of discourse*' (p. 641) would have to be divided up into different sorts of things (pp. 640–646).

On MI, Russell recalled that his remark on equivalent definitions had been directed towards the need of the axiom of choice, not a confession that finitude was indemonstrable (pp. 646–648). Poincaré's brief rejoinder *1906c* dwelt mainly on this last point, including a new and succinct statement of the basic issues involved:

Definition A. A *finite* number is a cardinal number n such that $n < n + 1$.

Definition B. An *inductive* number is a number which is part of all the recurrent classes [of numbers, that is, which contained 0 and also $(n + 1)$ if they contained n].

Proposition C. Any finite number is inductive.

For him MI was C; for Russell it was *B*, with *C* as an alleged theorem.

The dispute lay largely upon philosophical differences. Poincaré compared it as the Kantians (himself and his nephew) against the Leibnizians; Couturat dismissed the analogy in his reply *1906a* to Poincaré, but there were evident cross-purposes of a similar kind. The Kantian in Poincaré distinguished mathematical from logical knowledge and saw the role of the mind as unavoidable; the logicist in Russell/Couturat located mathematical within logical knowledge and sought to leave out the mind (Detlefsen *1993a*). Further, Poincaré seems to have been loyal to syllogistic logic, which Russell had recently described in his lecture on modalities (§7.3.6) as 'a subject scarcely more useful or less amusing than heraldry' (*m1905f*, 516).

The place of intuition in mathematical thought was another issue: Poincaré stressed it, while logicists wanted to avoid it in their foundations although presumably they would admit intuition in the creation of mathematics, and surely it played a role in choosing indefinables. In a review of the exchange for the American *Journal of philosophy* W. H. Sheldon *1906a* made such points; and indeed Russell had some creation of his own to do, for he thought that at last he had constructed a general logical system to Solve the paradoxes.

7.4.6 The rise of the substitutional theory, 1905–1906. In his reply to Poincaré Russell also announced that the no-classes theory was 'the most satisfying' Solution (*1906h*, 636), and he gave some details of its apparatus. The presence of the 'constituent' *a* in proposition *p* was the basic '*matrix*' of the substitution, written 'p/a'; the result of substituting *b* for *a* in *p* to produce proposition *q* was symbolised '$p_a^b!q$'. Simultaneous substitutions were used, to replace relations, as in '$p_{(c,d)}^{(a,b)}$' (pp. 636–638). The existence of an infinite class was proved on p. 639 by inductively generating from two distinct 'constituents' *a* and *u* the sequence of (true or false) different propositions

$$'p_0 . = . a = u' \quad \text{Df}, \quad 'p_{n+1} . = . p_n = u' \quad \text{Df.} \qquad (746.1)$$

Russell worked intensively on this theory from the autumn of 1905 and through 1906. Early in February 1906 he added a note to the proofs of his LMS paper *1906a* that 'the no-classes theory affords the complete solution of all the difficulties' about the paradoxes. He prepared a successor *m1906d*, which he submitted to the LMS on 24 April; it was accepted, but he withdrew it during the autumn, for he was losing faith in the theory. Hundreds of folios on it survive, though none saw print;[14] *m1906d* was published in the collection Russell *Analysis* (1973). The funeral is recorded in the next sub-section; here I summarise the theory, using that manuscript and a shorter one *m1905e* which he had written out in December for Hardy's benefit.

'We must distinguish between *substitution* and *determination*', Russell began to Hardy. Variables in the usual sense having disappeared, the proposition 'Plato is a man' was obtained by substituting 'Plato' for (say) 'Socrates' in 'Socrates is a man' instead of determining constituent *x* with the value 'Socrates'. Thus '$\phi!x$' was now an 'expression' (proposition?) containing *x* rather than a propositional function, and one of the axioms was

$$'\vdash . \phi!x . \supset \vdash . \phi!\text{Socrates Pp.'} \qquad (746.2)$$

---

[14] These manuscripts include the miscellaneous files catalogued as Mss.220.010950 and 230.031000, which I identified in 1984 as one connected text Russell *m1905e* of 256 folios 'On substitution'. Another collection of 100 folios contains a substitutional 'list of propositions' corresponding to Part I of *PM* (230.031260). The manuscript *m1906k* on the multiplicative axiom (§7.2.6) was also so developed. On the mathematical aspects of this theory, see my *1974d*, 389–401; on its logistic potential, try Landini *1998b*, pt. 3.

Universal quantification was understood thus: '$(x) . \phi!x$ is the proposition 'any value of $\phi!x$ is true"'.

The result of substitution had always to be unique; so Russell stressed that definite denoting phrases played a central role in the theory. After quoting their contextual definition (734.1) from 'On denoting' for the expression $\psi!x$, he gave the corresponding sense I2 of existence (fol. 4):

$$\text{'E}!(\imath x)(\psi!x) . = : (\exists b) : \psi!x . \equiv_x . x = b \text{ Df'}. \tag{746.3}$$

Russell then gave a string of definitions concerning the presence or absence of a from p (fol. 6), with substitution now also written '$p\frac{x}{a}!q$', where they were based upon

$$\text{'}a \text{ out } p . = . (x) . p\frac{x}{a}!p \text{ Df' and '}a \text{ in } p . = . \sim (a \text{ out } p) \quad \text{Df'}. \tag{746.4}$$

At this stage negation was an indefinable in Russell's theory; by the time of the manuscript *m1906d* for the LMS it was defined on p. 169 for proposition q as

$$\text{'} \sim q . = . q \text{ is false'}. \tag{746.5}$$

Although this proposition was not presented as a formal definition, it rendered a profound difference, for now the truth *and falsehood* of propositions were admitted as indefinables in the theory. Further, and more explicitly than before, propositions themselves were objects as much as were entities (the new word for 'constituents'); 'p should be the name of a genuine entity, and not a mere phrase like "the King of France" or "the King of England"' (p. 168).

Russell did not stress differences between propositions and entities; for example (an important one), he defined identity between x and y simply as

$$\text{'}x = y . = . x(y/x)!x \text{ Df'}. \tag{746.6}$$

without discussing any relationship between them (p. 169). It is hard to understand what the former property might describe, since this was decidedly *not* a theory about names (or about universals); however, he allowed propositional and non-propositional substitutions together (p. 175).

This theory solved the paradoxes by splitting matrices into 'types', starting with p/a and then moving to $p/(a, b)$ for the substitution of two individuals, $q/(p/a)$ which 'gives rise to classes of classes' (or rather to their analogue), '$q/\{p/(a, b)\}$' 'which is a matrix of the third type', and so on (pp. 176–177). Here was another profound change of policy; for the first time in a major theory Russell had stratified his logical universe. After writing his reply to Poincaré a few months later, he was to strengthen this approach by adopting the VCP (§7.7.1).

The construction of mathematics began naturally enough with the cardinal integers, to be defined as the analogue to classes of similar classes. For the lowest type

$$0_{p,a} \,.\!=\, . \left\{(x)\,.\sim (p/a)^{'}x)\right\}/(p,a) \quad \text{Df,} \qquad (746.7)$$

$$1_{p,a} \,.\!=\, . \left[(\exists\, c)(x)\{(p/a)^{'}x\,.\!\equiv\, . \,x = c\}\right]/(p,a) \quad \text{Df,} \qquad (746.8)$$

and so on (my symbolic version of a rather unclear informal discussion on pp. 175–177). The procedure went up to the transfinite numbers, which were defined by imitating Cantor's principles of generation (§3.2.6). Then he set in well-order a sequence of entities by inductively defining a sequence of two-place relations from propositions; the proof (746.1) that infinitely many objects existed, in the later reply to Poincaré, was to be a simplified version. The resulting numbers ω and $\aleph_0$ were respectively 'entity-ordinals' and '-cardinals', ω_1 and $\aleph_1$ 'ordinal-ordinals' and '-cardinals', and so on (pp. 180–183). But the continuation created only a finite number of these types (p. 177), which would not allow ω_ω or $\aleph_\omega$ to be reached: thus Cantor's and Burali-Forti's paradoxes were brutally solved in that *many* numbers prior to the offending ones could not be reached anyway. Similarly, for the naming paradoxes, '*definable* is relative to some given set of fundamental notions' I, and ' "definable in terms of I" ' is never itself definable in terms of I' (p. 185); and his own paradox was solved by banning a matrix from being substituted within itself (pp. 171–172). The liar paradox seemed to be banished by distinguishing a proposition from its truth-value.

7.4.7 The fall of the substitutional theory, 1906–1907. Russell's manuscript was refereed for the LMS by A. B. Kempe (§4.2.9). In his report he was generally praising, but he did not understand the reductionist flavour, either concerning the senses in which classes no longer existed or the meaninglessness of 'the present King of England' (Kempe Papers, Packet 39; copy in RA). Sending this report to Russell, the editor of the *Proceedings*, A. E. H. Love, wrote to him on 12 October 1906 saying that another referee was quite critical but that nevertheless the paper was accepted, and suggesting that some revision might be effected (RA). Russell's response was immediate; he withdrew the paper. The reasons were probably not due to Kempe's criticisms, which largely concerned presentation and unfamiliarity with his philosophical procedures. A few days later he reported the withdrawal to Jourdain as effected because 'there was much in it that wanted correction, and I preferred to wait till I had got things into a more final shape' (my *1977b*, 93).

It seems that during the year Russell saw the various difficulties, mathematical and philosophical, grow to unacceptable levels; at all events, late in 1906 he abandoned the substitutional theory, although not the notion of substitution. The status of entities as 'constituents' of propositions was in any case vague; the best move is to make them individuals, but their place in logic still needed clarification (§7.9.3). In addition, the technical difficulties of the theory were formidable. Could it really provide sufficient apparatus to deliver the mathematics envisaged, especially something to imitate the functional hierarchy? The outer regions of Cantor's empire, such as ω_ω, were out of reach anyway.

A philosophical difficulty concerned the status of propositions and their truth-values, especially concerning falsehoods; for what sort of object could correspond to a false proposition? Russell prepared a short manuscript *m1906e* on 'Logic in which propositions are not entities', where he converted propositional quantification '$(p).p$' into functional and individual quantification '$(\phi x).\phi x$'; but it is hard to see how meaning could be preserved.

Russell tackled this question in two philosophical papers: *1906g* in the October issue of *Mind*, and especially in a lecture read on 3 December to the Aristotelian Society and published as *1906j*. His target was the monistic theory of truth as advocated by his relative Harold Joachim in *The nature of truth* (1906); one of his main criticisms was based upon rejecting the internalist construal of relations (§6.3.1). In developing his own reductionist position he held that beliefs held of some thing such as a fact; but then 'objective falsehoods' had no referent. For some reason he did not then entertain the alternative option that there was not a fact to that which a false proposition corresponded, rather than a non-fact to which it did. But, as he told Jourdain the next June, 'Consideration of the paradox of the liar and its analogies has led me to be chary of treating propositions as entities' (my *1977b*, 105); so 'p is false' in (746.5), to take one example from the substitutional theory, had to go.

A further fear was that 'some contradiction should be found to result from the assumption that *propositions* are entities', as Russell put it in his LMS paper, though he had not found one (*m1906d*, 188). But it soon arrived (*m1906a*, fol. 7): I express it in terms of the propositions

$$F(a, p) = p\frac{b}{a}!q \text{ Df and } p_0 . = \; : (\exists p, a) : a_0 . = \; . F(a, p) : \sim \left(p\frac{a_0}{a} \right) \text{ Df,}$$

$$(747.1)$$

where apparently all letters symbolise propositions. The negation in the last clause exhibits kinship with the liar paradox, and indeed he showed that essentially

$$F(a_0, p_0) = F(a, p) . \supset \; : p(F(a_0, p_0)/a) . \equiv \; . p_0(F(a_0, p_0)/a_0), \quad (747.2)$$

an unacceptable equivalence in the consequent. He communicated it on 22 January 1907 to Ralph Hawtrey (text in §11.3), a former student who was reading the manuscript of *PM* (§7.8.1). He did not consider his paradox stated at the end of *The principles*, based upon associating a proposition with the proposition 'every member of a class *m* of propositions is true' (§6.7.9); but it may also be constructed within the theory.[15] It is a pity that he seems to have forgotten about this paradox, both at this time and later (§7.9.1–2). He also did not react to the observation in Grelling and Nelson (§7.2.3), that the theory assumed the self-contradictory proposition '*A class may not be subject of a sentence*' (*1908a*, art. 9).

Despite being abandoned, the theory left a noteworthy mark upon Russell's logical career, and not only for its prime place in his thought for well over a year. A close link between his theory of definite descriptions and his logicism, in the succeeding theories substitution was used—for example, in a published paper on a basic feature of logic, which we now consider.

7.4.8 *Russell's substitutional propositional calculus.*

Soon after publishing 'On denoting' Russell wrote a paper on 'the theory of implication'. On 23 July 1905 he sent it to Frank Morley (RA), who accepted it for the *American journal of mathematics*, where it appeared in the following April as a 44-page paper. He concentrated upon 'material implication' between propositions, construed as 'the theory of how one proposition can be inferred from another' (*1906b*, 159). Thus he jumbled together implication and inference Peano-style; in particular, the theorem

$$'\vdash \; : p \,.\, p \supset q \,.\, \supset q' \qquad (748.1)$$

was 'an important principle of inference, which I shall call the "principle of assertion"' (p. 180) and often used as the *modus ponens* rule of inference.

In other respects, however, 'the ideas are more those of Frege' (p. 160), with a reference to the *Grundgesetze*. Russell noted four features. Firstly, he continued to use the notion of assertion of a proposition *p* and the sign '$\vdash \,.\, p$', here with the "explanation" that 'it may be read "it is true that" (although philosophically this is not what it means)' (p. 161). Secondly, 'The essential property that we require of implication is this: "What is implied by a true proposition is true"' (p. 161), which is *not* a tidy reading of Frege. Thirdly, he distinguished implication between (variable) propositions and over quantification, as in 'the formula

$$\vdash \; : .\, \sim p \,.\, \supset \; : p \,.\, \supset \,.\, (q) \,.\, q' \qquad (748.2)$$

---

[15] Contrast de Rouilhan *1996a*, 178–194 and Church *1984a*, 516–522 with Landini *1998b*, 227–230; the differences of reading arise from additional assumptions and axioms made to Russell's incomplete formulation.

where '$(q).q$' was 'an absolute constant, meaning "everything is true"'' (p. 193). Finally, he defined his logical connectives from 'implication and negation as our primitive ideas' (p. 160); for example (p. 176),

$$\text{'}p \equiv q . = . p \supset q . q \supset p \text{ Df.'} \tag{748.3}$$

Russell also used his notation '$(C \, \wr \, x)$' from denoting (§7.3.3); but, in a ghastly choice of name, he called it a 'propositional function', where, for example, '"$p \supset q$" is a propositional function of p and q' (p. 163), not the usual denotation!

Substitutional influence was evident not only in the prominent place of truth-values and propositional quantification but also in the 'principle of substitution' and the fraction symbol to separate substituent from substituand (p. 165). However, perhaps because that theory was at an early stage, Russell allowed variables to vary, and quantification to occur over them, in normal ways. These procedures formed the second of his ten 'primitive propositions' for the calculus (pp. 164–168, with no *2·4 for some reason), fairly similar to the ten presented in *The principles*, 16–18. There he had warned that 'the method of supposing an axiom false, and deducing the consequences of this assumption' as in normal independence proofs 'is here not universally available' (p. 15); now he stressed that that caveat 'concerning primitive propositions applies with even greater force to primitive ideas' (*1906b*, 160). He compared his treatment of formal rules with that of the 'algebra of logic' presented in Huntington *1904a* (§4.7.3); but he showed no American sensibility to model theory (pp. 183–192).

For propositional quantification Russell added the 'primitive idea', written '$(x).(C \, \wr \, x)$', meaning 'the truth of $(C \, \wr \, x)$ for all values of x' (p. 194). This introduced 'formal implication' with quantification over x; but now the reference of 'propositional function' seemed to revert to normal, or maybe covered functions of propositions also. The three new assumptions did not enlighten: two of them stated rather Biblically that 'What is true of all is true of any' and its converse, while the third announced that 'If it is true, for all values of x, that p implies $(C \, \wr \, x)$, then p implies that $(C \, \wr \, x)$ is true for all values of x' (pp. 194–195). He ended by noting that the paradoxes required limitations on the ranges of the quantifiers, but sensibly suspended consideration of the matter.

Russell introduced his paper as 'the first chapter of the deduction of pure mathematics from its logical foundations' (p. 159), with doubtless a large substitutional text in mind. This was not to be; before examining his next phase, however, in the next section we review others' reactions to his and Whitehead's mathematical logic in the mid 1900s, and in §7.6 Whitehead's concurrent researches.

7.5   REACTIONS TO MATHEMATICAL LOGIC AND
       LOGICISM, 1904–1907

By 1905 Whitehead and Russell each had several papers in print on logicism, and Russell also *The principles*. This work began to take its place among the corpus of foundational studies of the time, especially among mathematicians. We note the reactions in this section, starting internationally, and then taking in turn Germany, Italy, the U.S.A. and Britain. They form only a part of a considerable concern with foundations at that time: in particular, the intense discussion of the axioms of choice is not described here.

7.5.1 *The International Congress of Philosophy, 1904.* This assembly succeeded the Paris Congress of 1900 and launched a four-year series, like the mathematicians (and some other disciplinary communities). However, unlike last time there was no closer link; for the mathematicians met at Heidelberg (§7.2.2) while philosophy was prosecuted at Geneva, in September 1904.

Another difference was that neither Whitehead nor Russell attended; however, their concerns were well represented, as Couturat showed in his report in the *Revue* of the Section on 'Logic and philosophy of science'. For example, he reported Peano's reservations over Russell's logic of relations thus: 'One can (as Mr. Russell has done) define mathematical functions by means of the Logic of relations; one can also, inversedly (as Mr. Peano has done in the *Formulaire*), define relations by means of functions, on the condition of taking for variables, no more numbers, but classes of numbers' (Couturat *1904c*, 1046). Again, to Boutroux's contribution (§7.4.2) Peano commented that 'the relation xRy of Russell is not identical to the function of Mathematics [...] the relation of Russell is always invertible: the function of Mathematics does not have this property' (Boutroux *1905b*, 719). Thus he did not understand Russell's papers on the logic of relations that he had recently accepted for the *Rivista*! (§6.5.2, 4).

Among other talks, the President of the Section, the Genevan mathematician Henri Fehr (1870–1954) envisaged 'the progressive fusion of logic and of mathematics'. Unlike many mathematicians, he enthused over 'the magisterial studies of Mr. RUSSELL' and the 'remarkable papers of Mr. COUTURAT' in the short extract *1905a* published in the proceedings. Further, Couturat recorded him as seeing 'the union, if not the unity, of these two disciplines,' of which 'Logic is a mathematical science by its form' while 'Mathematics is a purely logical science by its methods and by its principles' (*1904c*, 1037).

Couturat's own talk *1905c* reviewed some basic principles of logic. He started by discussing its title, 'On the utility of algorithmic logic': 'I prefer this name to that of *symbolic logic*' for stressing symbols too much, or to '*Algebra of Logic*, which presents it as an Algebra, and not as a Logic; [or] to that of *Mathematical logic*, which is equivocal'. He then recalled that the word 'Logistique' had been used to characterise arithmetical calculations by Augustin Cournot, in the volume *1847a* on algebra and geometry which had influenced his own book on the infinite (§4.2.3); indeed, he had mentioned Cournot's use of the word there (*1896a*, xxi).[16] Now he suggested 'logistique' as a name for mathematical logic with mathematical intent.

Couturat also noted with pleasure that two Congress colleagues had thought of this word independently: the French philosopher André Lalande, and a participant, Gregorius Itelson (1852–1926). He is a fascinating outsider in our story; he seems to have published only one paper, on psychophysics (Schröder *1890b*, 704), but he could make good one-line comments at conferences (§10.1.1). Russell was to meet him at the International Congress of Mathematicians at Rome in 1908, and tell his wife of a Russian living in Berlin who was 'very poor, and lives in an apartment without any servant', possessing ' "eine noble Passion" for old books, of which he has a fine library'; apparently by then Itelson saw Couturat as a plagiariser of the word 'logistique' (Russell *Letters 1*, 318).

Couturat reported Itelson's two talks at Geneva at generous length (*1904c*, 1037–1042). One on 'The reform of logic' covered Renaissance and 17th-century history, doubtless gleaned from his old books. In the other talk, perhaps under the influence of reading Bolzano, Itelson made the prophetic suggestion that logic be construed as the '*study of objects in general*', whether constant or not, rather than the tradition of analysing modes of reasoning. He also examined the relationship between 'Logic and mathematics' in a spirit distant from Couturat's. Logic was 'the Logic of extension' while 'pure mathematics' was defined as 'the *science of ordered objects*'; both disciplines deployed 'the same method' in handling '*ensembles of objects*', but the differences were significant. He also claimed that, in its 'pure' rather than transcendental form, 'Logic is distinguished from ontology in that it does not inquire of the existence of objects' (p. 1041):

> Moreover, no science, no theory can be prior to or higher than Logic, which is the foundation of any science and of any theory; one can say, in parodying the word of Pascal: that which surpasses Logic surpasses us; thus there cannot be *metalogic*.

What a pity that Russell was not there, to hear this early use of 'metalogic'; he would have concurred with the opinion expressed about it.

---

[16] The word 'logistical' was also used after Cournot by Cayley *1864a* to refer to the execution of algebraic operations; but he may well not have known of Cournot. Similarly, Couturat could have missed Cayley's short (and poorly phrased) paper. For some reason he did not mention that Leibniz had called his proposed theory of reasoning 'logistica'.

Itelson's remarks were quoted by authors in the *Revue*: in Boutroux *1905* (§7.4.2), and in a piece on 'Metaphysics and mathematical logic' by the mathematician Maximilien Winter. Obviously well familiar with the Peanist literature—he cited their lectures at the 1900 Congress—Winter took an idealist stance, and felt himself unable to tell in the writings of Couturat and Russell whether he was in philosophy or in 'the most abstract branch of mathematics' (*1905a*, 602–603); they could have replied that they were in both at the same time. Later he reworked his paper and two others in the *Revue* as a book *1911a* on 'method in the philosophy of mathematics', which contained a fairly wide survey of proofs and derivations in arithmetic and algebra.

By then the *Revue* had published two more notices. Henri Dufumier *1909a* gave a warm and well-referenced survey of Russell's and Moore's new philosophy. The Swiss philosopher Arnold Reymond (1874–1958) wondered in *1909a* whether the definition of cardinals as the class of all similar classes could embrace indeterminate classes such as that of living men, and proposed that zero could be defined as the class of such classes (an unhappy idea, similar to MacColl's empty class in §7.3.6). This paper followed Reymond's doctoral thesis *1908a* at the University of Geneva on the supposed history of the infinite; although he had corresponded with Russell and Frege among others, he still made some rather basic mistakes, as reviewers Russell *1909a* and Sheffer *1910a* pointed out.

7.5.2 German philosophers and mathematicians, especially Schönflies. Now we catch up on German reactions. The general question of 'Kant and modern mathematics' was treated by Ernst Cassirer (1874–1945) in a long paper in *Kantstudien* inspired by *The principles* and Couturat's *Principes*. His neo-Kantian background led him to emphasise the distinction between analytic and synthetic judgements (Smart *1949a*). After summarising Russell's programme, he regarded as a 'basic lack of Couturat's critique' that from the 'purely l o g i c a l nature' of his system 'its a n a l y t i c a l character is in no way proven' (Cassirer *1907a*, 35). He also noted that Couturat stressed that Kant recognised 'the simply analytic significance of mathematical propositions', while Russell appraised as a weakness the denial of the same status to logical judgements (p. 37). He also shared the dislike of Winter *1905a* for sharply separating mathematical and empirical judgements (p. 46), and he criticised Russell's definition of mathematics as the subject where we do not know what we are talking about (§6.5.1), on the grounds that a mathematician would have some idea of (say) a point or a surface when analysing it (pp. 47–48). Surprisingly, he did not discuss the paradoxes, or Part 7 of *The principles* on mechanics.

The German translation of Couturat's *Principes* led Joseph Geyser *1909a* to consider, with a similar Kantian reserve, 'Logistic and Relation-logic' in the *Philosophisches Jahrbuch*. He gave more attention to the status of judgements than to the logicist thesis, which he did not specifically mention; and he objected to inclusion differing from identity, which he seemed to regard as the dominant relation. This paper was cited in an essay on 'the significance of Couturat's researches in logic' by Joseph Schnippenkötter *1910a* in the next volume of the journal. Among the topics which Couturat had treated, he covered especially set theory (not, however, the paradoxes) and Leibniz's view of logic. While unoriginal, the tone was noticeably warm.

Among mathematicians, Cantor's fervent follower Artur Schönflies, in his early fifties, addressed the *Deutsche Mathematiker-Vereinigung* (hereafter, '*DMV* ') on 'the logical paradoxes of set theory'. Labouring the point that they were caused by 'inconsistent concepts' which did not satisfy the law of excluded middle (hereafter, 'LEM'), Schönflies *1906a* invalidly concluded that Russell's class 'is thereby nothing other than the "class of all classes" '. The same level of logical acuity was evident in a reply by Alwin Korselt; rather than appraising senses of existence, he held that non-existent objects did not satisfy the LEM, and could be valuable in developing mathematics (undoubtedly the case, of course). Thus 'all' and 'nothing' were valuable concepts 'in the works of exact logic by P e a n o and S c h r ö d e r' (*1906a3*, 217; a remarkable pairing!). He could not 'find in the Russell paradox any reason for doubt in the basic truths of *Mengenlehre*' "because" 'logicians and mathematicians operate only with a *finite* number of inferences' (p. 218); and he mistakenly thought that since Russell's definition of 'number of the class *a* names the totality of classes, then ' "to have the equal number with *a*", gave 'an interpretation of the word "number" ' (p. 219). In a succeeding piece *1906b* 'On logic and set theory' he rather confused the name '*A*' of a set with a pertaining 'Attribute *α*' and so allowed the latter only to name numerals in the context of arithmetic.

The Austrian philosopher Benno Urbach criticised Schönflies in a paper seeking 'the essence of logical paradoxes'; for him self-reference was admissible, and one should deny the LEM to predicates which generate paradoxes. Of Russell's solutions, he liked the 'zick-zack theory' (Urbach *1910a*, 102). In a long paper on 'the place of definitions on mathematics' with the *DMV* Schönflies disagreed with Urbach that the paradoxes were difficult to solve (*1910a*, 254), and stressed that the naming paradoxes cast doubt only on certain concepts and not on finite sets (p. 236). Throughout he affirmed his enthusiasm for *Mengenlehre*, where definitions could be used as elsewhere in mathematics. Noting the logical character of mathematical deduction, he even suggested that mathematics might be defined as the only branch of knowledge admitting indirect proofs, which involved concepts with no corresponding mathematical objects (p. 254). However, he regarded Russell's logic as mistaken, for self-(non-)membership infringed the distinction between subject and predicate (p. 253). So he ended with a polemical declaration: '*For Cantorism, but against Russellism!*'. Korselt *1911b* replied, reviewing forms of definition and noting definitional equivalence, disliking Dedekind's definition of an infinite set (§3.4.3), and praising Frege and above all his hero Bolzano (§4.5.9). This profile placed him closer to Russell than to Cantor.

In the second part of his report on *Mengenlehre* (§4.2.7), which concentrated upon point-set topology and some applications, Schönflies gave the paradoxes just five pages. Starting with the footnote 'It must suffice, to go into this in all shortness', quickly but competently he went through the cardinal, ordinal, Russell's and Richard's (Schönflies *1908a*, 26–31). His solution was to avoid 'contradictory concepts', a view which all would share; for him '*Comparability* and *well-ordering* constitute the basic problems of *Mengenlehre*' (p. 31). Fearing that the paradoxes would help 'a nearly increasing *scepticism*' against the subject, he hoped that it 'will not let itself go astray into scholastic dead-ends' (p. 39).

However, Schönflies also considered the paradoxes of naming in a paper *1909a* in *Acta mathematica*. Objecting to Richard's claim that the class of objects definable in a finite number of words was denumerable, he offered among his counter-examples a constant-valued function, which had a non-denumerable range of values available; he took Richard's argument as proof by *reductio ad absurdum* of the contrary theorem. In a reply Poincaré *1909a* defended his countryman on the grounds that in this and indeed all of his examples Schönflies 'defines an object *A* as having a relation *B* with another object *C*. This relation *B* does not suffice to define *A*; one must equally define the object *C*', such as the constant values themselves.[17]

7.5.3 *Activities among the Peanists.*

The Italian campaign was still in progress, with the fourth edition of the *Formulaire mathématique* (current title) coming out in 1902–1903 at 420 pages; but the next one, *Formulario mathematico* (1905–1908, around 500 pages[18]), was to be the last. The aim and style of this compilation (§5.4) was not much affected by English logicism, although its works were duly cited. No logic of relations was shown; functions were given all the work. This increased to encompass the calculus and differential geometry, some elementary theory of functions, and more point-set topology.

Elsewhere the Peanists continued to popularise their movement. Alessandro Padoa, in his late thirties, addressed a Venetian cultural

---

[17] Poincaré had recently repeated his view on naming at the end of a widely published survey *1908a* of 'The future of mathematics' delivered at the International Congress of Mathematicians in Rome. Both Russell and Zermelo had attended the meeting, where Zermelo *1909b* spoke about proving MI (§4.7.6); in his *Acta* paper Poincaré *1909a* also objected to Zermelo's recent treatment *1909a* of his topic there, for its use of the impredicative Dedekindian definition of the chain of a member of a class.

[18] A nearly complete edition (440 pages) of *Formulario mathematico* was published in 1906 in 100 copies; E. H. Moore used it in the context to be described in §7.5.5. I have consulted the copy in the Library of the Department of Mathematics in the University of Milan, which also holds a copy of the definitive version annotated by Peano.

society on 'logical ideography'. Giving simple examples of arithmetical and set-theoretic propositions in partly symbolic form, Russell was mentioned only for having adopted his own notation 'Elm' (§5.4.7) of the class of unit classes (Padoa *1906a*, 337).

In the same year Pieri *1906b* performed a similar task with a lecture on the 'new logico-mathematical direction of deductive sciences' at the University of Catania, where he currently taught. The relationship between the new symbolic logics and mathematics was a main theme, and he claimed to quote *The principles*, 9 that 'Logic constitutes the most general part of Mathematics, and Mathematics consists in the application of logical principles to certain special relations' (p. 403); but this is *far* too free, and waters down logicism. However, he emphasised one aspect of logicism, for he had affirmed it himself earlier (§5.5.5): the hypothetical character of mathematics (pp. 425–431). Mentioning Fehr's lecture at Geneva (§7.5.1), he saw a 'progressive fusion [...] in the works of Boole, Schröder and C. Peirce on one side, and of Weierstrass, Cantor and Peano on the other' (p. 435). On the paradoxes, he referred to a 'a most ingenious and happy solution of the paradox of Richard' (p. 424), but for some reason he did not give the details: it was Peano's remark quoted in §7.2.4 that that paradox belonged to language rather than to mathematics, recently put forward in the *Rivista* as a 'NEW SOLUTION', in reaction to ' "THE TRUE SOLUTION" of M. POINCARÉ', namely the VCP (*1906b*, arts. 5, 4). Unfortunately, as was mentioned, nobody took it further at the time, including its creator and followers.

Outside the Peano school Federigo Enriques (1871–1946) caused quite a stir with his book *Problemi della scienze* (*1906a*); for a German translation (by Grelling) appeared in 1910, and an English one four years later. His long ch. 3 treated 'problems of logic', followed by one on 'Geometry' (his own mathematical speciality). The contrast with the Peanists continues, for he treated symbolic logic only briefly, and saw it '*as a part of psychology*' (pp. 106–109), which must have appalled his Turin compatriots and their Cambridge allies. Earlier in the book he presented Russell's paradox; but he concluded merely that the process which produced the generating class was 'transcendental', and so should be banished from mathematics, like all metaphysics (pp. 16–17). His compatriot Beppo Levi (1875–1961) was more definite: in a long paper *1908a* on 'Logical antinomies?' he favoured limitation of size and imposed restrictions on permissible forms of definition and on correspondence between elements and sub-classes of a class.

Italian interest in mathematical logic was continuing, but on a much smaller scale. We pick up the next fragments in §8.6.1.

7.5.4 *American philosophers: Royce and Dewey.* The English translation of Enriques's book would have appeared soon after Grelling's in German, had the husband of the translator not been slow in checking the technical parts; when it came out in 1914, he apologised in his preface. He was the philosopher Josiah Royce (1855–1916), and the edition was put out by the Open Court Publishing Company. By then Americans had been writing for over a decade on foundational issues in mathematics, such as model theory (§4.7.3); here we note views in and around, or avoiding, logicism.

Royce's first book had been a short primer *1881a* on applying (post-) Boolean algebra to the teaching of English. In the late 1890s he deepened his concern with logic, and wrote papers of various levels of technicality during the rest of his life. The most significant item here is a long and difficult paper *1905a* sent to the American Mathematical Society (hereafter, 'AMS'), in which he related 'the principles of logic to the foundations of geometry'. He produced the first major treatment of Kempe's theory of multisets (§4.2.8), which he reworked by replacing Kempe's 'between' relation by analysis of the relationships between the 'elements' of an '*O*-collection' of mutually disjoint multisets which filled a universe, and between collections which were *O*- and those which were not. Laying down six principles for the existence and basic properties of an *O*-collection (p. 367), he produced a very general theory of collections, emphasising the various kinds of 'series' into which the elements could be ordered; he regarded as 'probably quite superficial' Russell's emphasis on the difference between the logics of relations and of classes and propositions (p. 355). The word 'logic' of his title should have been 'algebra', perhaps; he treated Boolean algebra as a special case, with single membership of elements to a collection (pp. 401–404). One of his main concerns, inspired by Kempe *1890a*, was to reduce all relations to combinations of symmetrical ones, and over the next decade he devoted enormous energy to the study of such expansions, and representation by Venn diagrams.[19] He cited Veblen and Huntington, and used the same postulate techniques to show that his six principles were independent (pp. 412–415).

In an extensive review of Royce's paper for the *Journal of philosophy* Theodore de Laguna *1906a* claimed that the title promised more than the text delivered; he attacked Kempe's enterprise rather than Royce's version of it, which he liked (Kuklick *1972a*, esp. ch. 10). The issue rests more on whether the theory is properly logical. Later Lewis *1914a* reworked it in terms of ordered elements.

We recall from §4.7.3 that Veblen accepted the term 'categorical' around 1904 from John Dewey while he was studying under E. H. Moore at the University of Chicago. Dewey edited a volume of *Studies in formal logic* in 1903 (his 45th year), written by himself and colleagues and

---

[19] Royce Papers, Boxes 1–6 of notebooks on symbolic logic. There are some rather rambling manuscripts of this time on set theory and logic, including a review-like piece on Russell's *The Principles* (volumes 72–75 *passim*). Kempe read Royce's paper 'with the greatest interest' in October 1905 (Incoming Correspondence, Box 2); Peirce discussed relationships in terms of relations (Miscellaneous Correspondence File; compare Royce *Letters*, 488–492).

associates of his Department of Philosophy. Despite the adjective of its title, no symbolic logic was used; only Venn and Jevons of that ilk received mentions, and then in passing. After his four-part examination of 'thought and its subject-matter' (Dewey *1903a*, 1–85), his colleagues treated logic as an empirical search for norms somewhat like J. S. Mill (§2.5.8) but more concerned with the social and ethical aspects (Sleeper *1986a*, ch. 3).

In 1904 Dewey moved to Columbia University in New York, where he became a colleague of Keyser. In a popular lecture on mathematics delivered there in October 1907 Keyser praised the 'creators of modern logic'; in particular, logicism showed that 'mathematics is included in, and, in a profound sense, may be said to be identical with, Symbolic Logic' (*1907a*, 13). So in his good wishes Keyser captured one of the confusions which pestered Russell's logicism.

Keyser was both philosopher and mathematician (and also historian of sorts). Let us turn to his mathematical colleagues, who became interested in Russell's work without adopting logicism.

7.5.5 *American mathematicians on classes.* In his Presidential address to the AMS, delivered in his 45th year, E. H. Moore *1903a* prefaced a review of mathematics education with a survey of 'abstract mathematics' of all kinds, including Hilbert's and Veblen's. The Peanists were duly noted, but he judged that 'this symbolism is not an essential part of their work', and he wondered in general 'whether the abstract mathematicians [...]' are not losing sight of the evolutionary character of all life processes, whether in the individual or in the race' (p. 405). His own research on 'general analysis' (§4.2.7) used many of the Peanist 'logical signs', listed as taken from the recent edition of the *Formulario* (Moore *1910a*, 150); but they were there for utility rather than philosophy. However, letters during 1908 to his former student Veblen show him developing the symbolism extensively (Veblen Papers, Box 8).

Then in his late twenties and just appointed to Princeton University, Veblen had become more engaged in logicism. Flattered by Whitehead's and Russell's interest in his own work, he carefully read Russell's reply *1906a* to Hobson (§7.4.4) and wrote at length on 13 May 1906 (text in §11.2). As usual among mathematicians, set theory itself was his main concern, and he made a proposal like Padoa's of 1899 (§5.4.7) but doubtless conceived independently; that membership of objects to a class should be distinguished from that of classes to the class of all classes, so that 'The paradoxes are all gone, and all real mathematics remains'. Russell might have pointed out that neither claim could be affirmed so categorically (as it were).[20]

---

[20] Russell's letters to Veblen of 30 May and 7 October 1905 (Veblen Papers, Box 11) do not address this point; they show that Veblen visited Oxford that summer but that the two men did not meet.

Royce's colleague at Harvard, the mathematician Maxime Bôcher (1847–1918), also corresponded with Russell. The contact came from an address on 'The fundamental concepts and methods of mathematics' which Bôcher gave to an International Congress of Arts and Sciences at St. Louis in September 1904 and published as *1904a* in December with the AMS. At the end he stressed the presence of 'fundamental' in his title; he did not try to tackle the use of intuition or the creative aspects of mathematics. He contrasted three positions within his chosen remit, with a representative figure for each. First was the standpoint enigmatically expressed in 1870 by Benjamin Peirce (§4.3.2) about mathematics as *'the science which draws necessary conclusions'* (p. 117). Bôcher extrapolated from it the position that mathematics 'does not deal directly with reality', to him a nicely unifying view (p. 124, where he also cited C. S. Peirce *1896a*); but he regretted the unclarity of Peirce's phrase and the ignoring of the creative side of the subject. He contrasted this emphasis on methods with the position of Kempe, based on a very general conception of mathematical objects and relationships between them. Bôcher liked the way that here 'mathematics is not necessarily a deductive science', but he wondered, vaguely, if these two positions were 'co-extensive' for axiomatised theories (pp. 129–131).

The third position considered by Bôcher was Russell's logicism as expressed in *The principles*, which he saw as a mixture of the other two in treating both objects and methods. On the former, he noted the unclarity with which Russell handled existence theorems. However, his own under-standing of Russell's position was defective, for he thought that Russell did *not* proffer a Peircean stress on hypotheses for pure mathematics (pp. 131–132). Russell understandably objected in a letter (now lost), and in reply on 21 April 1905 (RA) Bôcher apologised, and hoped feebly that clarity had flowed from his next sentence in his paper, about establishing the existence of a mathematical system from logical principles.

In his paper Bôcher also deplored Russell's 'unequivocal repudiation of nominalism in mathematics' (p. 132), and in this letter he made his grounds more precise:

> The central point at issue is your 'class as one'. Your attitude towards this term is that of the realist, if I understand you correctly; mine is that of the nominalist. I cannot admit that a class is in itself an entity; it is for me *always* many entities (your 'Class as many') [...] If you were to accept my position here, which of course you will not do, your remarkable paradox would crumble to pieces.

Presumably Russell defended his realism over classes, for in a letter of 16 June (RA) Bôcher took the somewhat oddly named institution 'The Rugby Cricket Eleven' as merely 'a name (sign, mark, what you will)' when he wished to say of them (rather than of any members) that they had won a game. '[I]t may happen that at some school exactly the same boys formed the foot-ball eleven as the cricket eleven. In this case we should need two marks each associated with these same boys'. Presumably Russell replied that this was realism under some other guise; at all events, Bôcher did not understand his reply.[21]

Another difference between the two men concerns Kempe's work. Whereas Bôcher found it valuable (like Royce), Russell told Couturat on 4 July 1905 that he knew of it only through those two commentators, whereas he had received offprints from Kempe the previous October (Kempe Papers, Packet 37).

7.5.6 *Huntington on logic and orders.* Bôcher's Harvard colleague Huntington addressed such questions for logic in general when he teamed up with Ladd-Franklin to write an article on 'Logic, Symbolic' for *The Americana* encyclopaedia, published by the Scientific American Compiling Department (Huntington and Ladd-Franklin *1905a*). He seems to have been largely responsible for it, incorporating her 'suggestions, but I fear with rather tame results in places' he confessed to her on 30 July 1905.[22] They gave a reasonable survey of the topic; the algebraic tradition gained rather more space than the mathematical one, fairly reflecting the balance of published material at that time. But their opening definition of the subject shows how unclear even to experts was the ensemble of logics which used symbols:

> Symbolic Logic, or Mathematical Logic, or the Calculus of Logic,—called also the Algebra of Logic (Peirce), Exact Logic (Schröder), and Algorithmic Logic or Logistic (Couturat),—covers exactly the same field as Formal Logic in general, but differs from Formal Logic (in the ordinary acceptation of that term) in the fact that greater use is made of a compact symbolism—the device to which mathematics owes so largely its immense development.

At this time Huntington also published in two parts a study *1905a–b* of 'the continuum as a type of order', taking up around 60 pages of the *Annals of mathematics*. Harvard University quickly put out the offprints as a separate publication; and in 1917 its Press issued a slightly revised version as a book, reset and with the conventional pagination. It became a standard source in English for Cantorian order-types developed from postulates.

In the first part Huntington began with the basic notions of set theory, and then he treated simple order (ch. 2), 'Discrete series: especially the type of the natural numbers' (ch. 3) and 'Dense series: especially the type

---

[21] Bôcher to Ladd-Franklin, 24 December 1913 (Ladd-Franklin Papers, Box 3). I have found no *Nachlass* for Bôcher.

[22] Ladd-Franklin Papers, Box 4. Couturat told Russell on 28 June 1905 that Keyser was in charge of the section for mathematics in this encyclopaedia (RA). I have also found no *Nachlass* for Huntington.

of the rational numbers', the latter including a '*Theorem of mathematical induction*' proved from the postulates of simple order and the Dedekind cut (art. 23). The second part, which he read to the AMS in September 1905, treated 'Continuous series: especially the type of the real numbers' in one and several dimensions (chs. 5–6), and in an appendix he summarised transfinite arithmetic. Russell's *The principles* had covered this material in prosodic form with far fewer explicit theorems and an elaborate philosophical framework; Huntington largely avoided the latter (he did not mention the paradoxes), but he cited Russell's book regularly on various details and renderings in prose of Cantor's technical terms. In art. 63 he presented Russell's definition of irrational numbers but also noted the priority in Pasch *1882a* (§6.4.7).

Huntington may have sent an offprint to Russell; at all events he received a letter from Russell (now lost) to which he replied on 6 January 1906 (RA). He mentioned some intended cuts (which in the end were not carried out): one was the example, due to Royce, of a 'self-representative system' by a map of London laid out on a pavement in London (art. 28). 'I am also quite disconsolate' over Russell's scepticism about various proofs in set theory: the use of an axiom of choice seems to have excited Russell's doubts over, for example, the equivalence of the reflexive and inductive definitions of finitude (art. 27), and the theorem that a dense series was dense-in-itself (art. 62); Cantor's diagonal argument and the basic covering inequality (663.1) were also causing qualms which Huntington did not share. This postulate theorist made a good general point with which Russell should have concurred:

> In fact the central problem in all the recent discussion about the transfinite ordinals seems to be this: How shall we prove the consistency of a set of hypotheses in the cases where no finite or denumerable example can be exhibited?

7.5.7 Judgements from Shearman.   Finally we consider reactions to logicism from an Englishman. In a lecture on 3 April 1905 to the Aristotelian Society, A. T. Shearman *1905a* reviewed 'Some controverted points in symbolic logic'. These included the primacy of classes (Venn, say) or of propositions (MacColl), the place of 'Modals' in view of 'MacColl's very ingenious system', inclusive or exclusive disjunction, and the worthwhile role of inversion procedures and of the logic of relatives. The published version does not contain the discussion: this is a pity, for on the last topic he mentioned only Peirce, and Russell was present in the audience.

Shearman used this lecture in a 'critico-historical study' of *The development of symbolic logic*, which was published the following year, his 41st. Related to his teaching at University College London, where he was a lecturer, his coverage was quite wide: all the algebraic logicians, MacColl, Frege and Peano (but unfortunately not his disciples, or Couturat) as well as Whitehead and Russell. They came mainly in a passage on 'The new treatment of mathematical conceptions' (Shearman *1906a*, 196–220). After quoting Russell's definition of logicism from the head of *The principles*, he stressed its quantitative as well as qualitative character (p. 200). As well as discussing Russell's and Peano's choices of indefinables, he gave Russell's definition of cardinals and even mentioned the multiplicative class (though not its current disputes), and also emphasised the place of relations in logicism. He included some guarded remarks on Whitehead's *Universal algebra*, and five appreciative pages on Frege, the latter seemingly drawn from Russell's appendix in *The principles* (pp. 213–218, with 'Kerry' (§4.5.4) misspelt as 'Kelly').

By and large Shearman gave a good impression to the general reader of the range of theories within symbolic logic without, however, bringing out sufficiently the great differences between them. On the contrary, he began his preface by indicating that he followed Johnson, who had written to him in 1903 of the 'error' of thinking of the various symbolic systems as being radically distinct (p. v)—which indeed they are! In addition, he did not present much set theory (Cantor was never mentioned), and so he could not contrast it with part-whole theory. Thus his study was still less critical than historical.

Shearman followed with a sequel volume *1911a* in which he described Frege, Peano and Russell in roughly equal measure, including their notations (Frege's praised but its consumption of paper disliked). But he did not analyse logicism or even the paradoxes, or explain sufficiently the mechanics of set theory, and gave a fair-minded but not penetrating 'survey of symbolic logic'.

At least Shearman *had* an attitude: at Cambridge itself logic was still dominated by the syllogistic tradition with a few splashes of the algebraic tradition. Such at least is the impression conveyed by the fourth edition of J. N. Keynes's *Studies and exercises in formal logic* (1906), where no mathematical logician was mentioned. Whitehead was extremely isolated in the city.

7.6 WHITEHEAD'S ROLE AND ACTIVITIES, 1905–1907

> [*Principia mathematica*], which was of great importance in its day, doubtless owed much of its superiority to Dr. (after Professor) Whitehead, a man who, as his subsequent writings showed, was possessed of that insight and spiritual depth so notably absent in Russell; for Russell's argumentation, ingenious and clever as it is, ignores always those higher considerations that transcend mere logic.   Russell *1936b*

7.6.1 *Whitehead's construal of the 'material world'*.   (Lowe *1985a*, ch. 14)
On 18 August 1904 he spoke on 'Peano's symbolic method' to the British Association for the Advancement of Science (notice in their *Report* (1904),

440); no text survives. In September 1905 he sent a long paper to the Royal Society, to which he had been elected two years earlier (§6.8.2). In 1899 he had had to withdraw his last paper, for lack of originality (§6.2.4); but this one was accepted, although he had some tussles with his referees, largely on presentation and use of logical symbols.[23] It was published in their *Philosophical transactions* as *1906b*. Russell, who had not been much involved in it, made several pages of notes after its appearance (RA).

In 61 pages Whitehead examined 'mathematical concepts of the material world' by laying out in axiomatic form elaborate constructions from basic 'entities' which 'constitute the "stuff" of space' (p. 465). For general background he cited Kempe *1890a*, Veblen *1904a* and Royce *1905a* (p. 469). It reads like logicism, and not just logical notations (especially relations), but applied to physical science; it lay within his planned treatment of geometry in *PM* (§11.5), and also presaged some later philosophical concerns (§9.4.3).

Five 'Concepts' were presented (pp. 478–484). The first was 'the classical concept of the material world', composed of points of space, instants of time and particles of matter. The next Concept treated physical particles statically in space; Whitehead defined particles in terms of two-place relations between points, acknowledging *The principles*, 468 for the idea. 'Concept III' handled them in motion, so that three-place relations were need in order to include time. The last two Concepts dealt correspondingly with the aether in static and dynamic states; we recall from §6.2.4 that his first mathematical interest had lain in electromagnetism.

'In what sense can a point at one instant be said to have the *same position* as a point at another instant?' (p. 481). To provide a framework Whitehead used the four-place relation '$S'(uvwt)$', where u, v and w were rectangular 'kinetic axes' at instant t, from which velocity and acceleration were definable. One of his reductionist aims was to avoid the predicament of treating a physical point both as a basic notion and as a complex (Russell in *The principles*, ch. 51, for example) by defining it from lines, a primitive notion which he called 'objective reals'. He offered two strategies, one using a theory of dimensions and the other working via 'interpoints'. This latter notion, an abbreviation of 'intersection points', was an extension to space-time of relationships of order in projective geometry with the notion of intersection taken as primitive. He assumed that any given entity a determined the order of three more entities b, c and d at time t in a relation '$R'(abcdt)$', defined two entities x and y to be 'in a similar position' if they could be exchanged at any of the three central places in R (and so have the same position in the order), and specified the 'class of interpoints' of a at t to be composed of a and all points which

---

[23] The referees of Whitehead's paper were G. B. Matthews, N. D. Niven and W. Burnside; their reports are kept in the Royal Society Archives, RR vol. 16, nos 399–403 (my *1986a*, 64–67).

were similar to any x substitutable in R (pp. 484–488). This gave him a general means of treating points in space-time.

Although deploying quite different methods, Whitehead's foray is not unlike Royce *1905a* in its generality and concern with geometry. But some of his notations were not friendly: the segment of between points a and b was symbolised '$R^;(a;b)$', its prolongations beyond a and beyond b were respectively '$R^;(ab;)$' and '$R^;(;ab)$', and the class of entities on a with positions similar to x was '$R^;(\frac{a???t}{x})$'—a horrible notation. This forbidding appearance must have deterred even symbolically literate readers; and naming the Concepts merely by Roman numerals hardly excites the intuition, either. Like Royce's, Whitehead's paper made little impact—apart from upon his own philosophy when it flowered in the 1920s after he had finished with logicism (Lowe *1962a*, chs. 7–8).

7.6.2 *The axioms of geometries.* Whitehead soon presented some of this material on geometry at a more elementary level when in 1905 Cambridge University Press launched an important series of short 'tracts in mathematics and mathematical physics'. He was one of the first authors, producing the fourth and fifth volumes as his *1906a* and *1907b* (the prefaces are dated October and March of their respective years). Conceived as a pair, he presented the axioms of projective and of descriptive geometries, distinguishing them by the property that in projective geometry 'two coplanar lines necessarily intersect. Thus Euclidean Geometry is not projective, but becomes so' when the points of infinity were adjoined; 'A non-projective Geometry will be called a Descriptive Geometry' (*1906a*, 5–6), for which he concentrated upon transformations (including Sophus Lie's theory), congruence and distance.[24]

Whitehead's treatments were quite converse to Russell's; he gave elementary but technical presentations, and not much philosophy (for which he referred the reader to *The principles*). Unlike the 1905 paper, he used no logical symbols, although he came close to talking propositional functions when formulating the 'general theory of correspondence' (*1907b*, 34–36). However, the philosopher appeared in a few places; in particular, the first tract began with an extensive discussion of the nature of axioms and of definitions (a part of it was translated in the *Revue* as Whitehead *1907a*). For projective geometry he cited Pieri *1898a* (§5.5.5) as his main source; for descriptive geometry he relied similarly on Peano *1888a* (§5.2.4). However, although he had been in contact with Veblen and cited his *1905a* (*1907b*, 7) he did not allude to categoricity.

7.6.3 *Whitehead's lecture course, 1906–1907.* While Whitehead worked on the second tract he also gave a substantial course on 'The principles of

---

[24] In *1907b*, 14, Whitehead thanked Berry for correcting a detail in a proof in *1906a*, 58. On 9 July 1897 Berry had submitted to Felix Klein a short article on synthetic geometry (Klein Papers, 8:82); it seems to have been rejected.

mathematics' at Trinity College in Lent (Spring) term 1907. No manuscripts survive, as usual, but luckily we have the notes taken by H. W. Turnbull (1885–1961), who later made his career as a mathematician and historian of mathematics. Below I have silently expanded some abbreviations; and most of the folios are not numbered.

Whitehead began by sketching various views on numbers, including that of 'Frege (following Herbart)', which was 'practically that of B. Russell' (Turnbull *m1907a*, fol. iv). This latter naturally dominated the subsequent discourse; for example, 'the present king of France is $\{{}^{\text{bald}}_{\text{not bald}}$ are both false and one contradicts the other' (fol. 3). But he treated mainly the mathematical side, from the integers produced by MI (the 'nerve of the whole thing') through rational and real numbers to transfinite arithmetic, including cardinal exponentiation and uses of the multiplicative axiom. His current understanding of model theory was that the basic notions (number, zero, successor) in the Peano axioms 'might be fitted on to anything. B. Russell doesn't seem to object to this'. In the last part of the course he summarised the material of his new tracts, and then also parts of his recent Royal Society paper in his symbols (all those quoted in §7.6.1, for example).

In the middle part Whitehead explained some features of the logical system that he and Russell were developing. No mention was made of substitution; instead, he presented propositional functions, their 'ranges of significance' divided into types, functions of functions, truth-functions and functional quantification; curiously, he seemed not to mentioned the paradoxes. Referring to Russell *1906h*, he reported that 'Russell says a proposition is an entity, in contrast to the view that p ⊃ q had meaning when p and q are any entities whatever. This must be dropped and propositions must be looked on as entities distinctly by themselves'. What was going on at Bagley Wood?

7.7 THE SAD COMPROMISE: LOGIC IN TIERS

7.7.1 *Rehabilitating propositional functions, 1906–1907.* Meanwhile Whitehead's collaboration with Russell continued, now in a definitive form. They had referred to 'volume II' up to at least November 1905;[25] but Russell told Couturat on 21 August 1906 that 'We are thinking of doing an independent book, which we call "Principia mathematica"'. The new title was chosen perhaps not in imitation of Isaac Newton but in line with G. E. Moore's book *Principia ethica* (1903), which treated ethics within the new

---

[25] See Russell's letters in November 1905 to Donnelly (his *1967a*, 180) and to Jourdain (my *1977b*, 44). The quotation soon to come reads: 'Nous pensons en faire un livre indépendant, que nous appellons "Principia mathematica"' (copy in RA).

philosophical framework within which they wished to site logic and logicism.

After abandoning the substitutional theory, Russell reverted to propositional and denoting functions with free and apparent variables. From the late summer of 1906 through 1907 he wrote many manuscripts, focusing especially on functions but also considering their relationships to classes. The main common factor was the VCP and some stratification into types, although more than one version was put forward. Much could be said on the changes which Russell rang on the various possibilities; a small selection of points is made here, with the texts awaited in *Papers 5*.

One of the most striking transitional manuscripts is a diary-type text *m1906i* of 108 folios called 'the paradox of the liar', which Russell started in September of that year. He started out with that conundrum, stressing the place in it of negation. Granting existence to propositions led to great difficulties with MI, and he wondered about taking 'and so on' as a primitive idea (fols. 15–17). But this was abandoned, and a propositional hierarchy was proposed; 'propositions' had no quantifiers, and the location of a 'statement' was determined by its apparent variables (fols. 30–34). Later he used substitutional theory before rejecting it for its complications (fols. 66–87); although he added a note in June 1907 doubting this reservation, this was its last appearance in full regalia. Otherwise he deployed propositional functions; a type theory involving the VCP was imposed, with (lack of) quantification again determining the bound and free variables. The base was occupied by individuals, defined as 'any single existent', so that Dedekind's proof of an infinitude of objects by a sequence of thoughts (§3.4.2) had to be abandoned (fol. 65). Near the end he wondered whether the axiom of infinity was an empirical question about the universe, and saw the need of both it and the multiplicative axiom for proving existence-theorems in mathematics (fols. 101–104).

In another long journey through 'Fundamentals' Russell *m1907b* started with the distinction between a propositional function and the class that could be formed from it. A rumination on functions included worry that the difference between predicative and non-predicative functions was not clear, whether truth-functions could be values of apparent variables, and the place of belief predicates in the overall structure (fols. 8–12). As regards his paradox, he noted that non-self-membership could sometimes lead to true or to false propositions and that self-membership was 'not always nonsense'; so he mooted the idea of restricting the use of the law of excluded middle (fols. 24–25). For classes (fol. 38), 'we are to have

$$x \in \alpha . \equiv_x . x \in \beta : \supset : \alpha \in \kappa . \supset_{[\kappa]} . \beta \in \kappa'. \tag{771.1}$$

However, despite (or because of?) such flights, no substantial theory resulted.

A group of manuscripts on functions, which seem to belong to this period and are cited as *m1907c*, included in the second text a different approach drawing on the notion of the form of an expression X containing a variable x, called 'function' and written '$\dot{x}(X)$', specified as that 'which may be preserved constant [in X] while x varies'. Conversely, when x was introduced into the form F, then the expression 'F $|x$' was produced. These converse indefinable notions were based on the following assumptions

$$\text{'}\vdash : \dot{x}(X) = \Phi . \supset . \Phi | x = X \text{ Pp'. and '} \vdash \dot{x}(X) = \Phi \text{ Pp.'.} \quad (771.2)$$

The approach was ambitious, even encompassing implication:

$$\text{'} \supset . = . \dot{p}\dot{q}(p \supset q) \text{ [Df] \& } (\supset | q) | p . = . p \supset q \text{ Df'.} \quad (771.3)$$

While these ideas were not durable, the notion of logical form was to appear later in his philosophy of logic (§8.2.4).

7.7.2 *Two reflective pieces in 1907.*   On 9 March Russell addressed the Cambridge Mathematical Club on 'The regressive method of discovering the premises of mathematics'. Although he discussed an important aspect of logicism which otherwise was rather ignored, he never published his manuscript (it appeared in the collection Russell *Analysis*). He was concerned with the philosophical consequences of analysis (in his narrower sense of §6.6.1); that 'in the logical theory of arithmetic' a theorem such as '2 + 2 = 4 [. . .] is more certain than the premises, and the supposed proof therefore seems futile' (*m1907a*, 272). His justification was surprising, and hard to articulate: the 'logical simplicity' of a proposition, 'measured, roughly speaking, by the number of its constituents', the fewer the better (p. 273). He argued that aiming at simpler premises 'gives a greater chance of isolating possible pervading elements of falsehood', and that they 'have many more consequences than the empirical premises'. Then he contrasted Peano's axioms for arithmetic with the regresses to Frege's and his own definitions of integers as classes of similar classes (pp. 276–278, where he noted also the paradoxes). Admitting that some of Frege's logical premises were 'more intrinsically obvious' than others, he assessed similarly his own axioms for the propositional and predicate calculi. He also mentioned the need for an (empirical) axiom of infinity (p. 282), but passed over the issue of its obviousness!

In October Russell published an introductory essay *1907d* of similar intent on 'the study of mathematics'. He had written it five years earlier, as part of a general consideration of educational questions (the little piece *1902b* on teaching Euclid (§6.2.2) may belong to the same concern); but it had been rejected by the original journal, so he offered it to his friend Desmond McCarthy to help launch *The new quarterly*. Demanding that textbooks be written in a more systematic manner than was customary, he proffered 'Symbolic Logic' as 'the fundamental science which unifies and systematizes the whole of deductive reasoning', and argued against the psychological interpretation of logic (*1907d*, 90, 91).

7.7.3 *Russell's outline of 'mathematical logic', 1908.*

Between the views on logic presented in these two essays lay a difficult area of heuristics which Russell addressed in July 1907, when he wrote an important paper outlining the form which *PM* was to take. Published ten months later in the *American journal of mathematics* as his *1908c*, it is one of his most famous; for in 41 pages he sketched out the apparatus of 'Mathematical logic as based on the theory of types'—an early use by him of Peano's phrase 'mathematical logic' (§5.3.2). He started in art. 1 with seven 'contradictions' (the liar, his own, the relations version, Berry's, the least indefinable ordinal, Richard's and Burali-Forti's, in that peculiar order), diagnosed 'self-reference or reflexiveness' as the illness, and prescribed as medicine the VCP, which he cast in the form 'Whatever involves *all* of a collection must not be one of the collection'. In each of the seven cases some totality no longer belonged or referred to itself; but various harmless features of logic and mathematics, such as the law of excluded middle and MI, also required modification (art. 3). In the rest of the paper he sketched out some details.

The exposition started in art. 2 with 'all and any' in quantification; the VCP showed the importance of 'The distinction between asserting ϕx and asserting $(x) . \phi x$ [, which] was, I believe, first emphasized by Frege'. The predicate calculus split into a 'hierarchy of types' and also orders; the presentation (art. 4) is too brief, and often misunderstood, but the theory is *different* from that to be used in *PM* (compare §7.9.1–2). 'A *type* is defined as the range of significance of a propositional function'; that is, a class (of humans for 'x is a man', say); at the bottom are individuals, "defined" as 'something destitute of complexity'. Quantification over x, to make it an 'apparent variable' in a hitherto unquantified 'elementary proposition', produced '*first-order propositions*. These form the second logical type', that is, the class of such propositions. Quantification over these propositions produced new ones belonging to the third type, and so on. Some examples would have been helpful; for example, that the second type contained both

$$\vdash . (x) . \phi x . \supset . p \text{ and } \vdash . (x, y) . \psi(x, y) . \supset . p. \qquad (773.1)$$

Since each quantification was effected independently of others, a function might have only some of its lower-order functions quantified; for example, $(\exists \phi) . \phi a$. Substitution remained in that Russell briefly related his theory of functions to matrices within propositions (for an elaboration, see Landini *1998b*, ch. 9).

Functions which were 'of the order next above' that of at least one of their variables were called 'predicative', and marked with an exclamation mark as in §7.3.1: for example (Russell's own), in '$f!(\psi!\hat{z})$' f contained only $\psi!\hat{z}$ as apparent, within which only z was apparent. The importance of this kind of function resided in the repair needed to render to mathematics; the VCP forbad, say, subtractions such as $26 - \sqrt[3]{7}$ because the latter number was irrational, defined in terms of a class of classes of rationals and so determined by a function of an order different from that for 26. To save arithmetic and mathematics in general, he assumed 'the *axiom of classes*, or the *axiom of reducibility*' (art. 5), also 'the axiom of *relations*' for functions of more than one variable. It claimed that every propositional function had a logically equivalent mate among the predicative sub-class: for example for 'double functions',

$$\text{`(14)} \quad \vdash \; : . \; (\exists f) : . \; (x, y) : \phi(x, y) . \equiv . \, f!(x, y)\text{'}. \tag{773.2}$$

It seems strange that he (and Whitehead) did not realise that this assumption minimised, maybe even nullified, the hierarchy of orders; evidence in the same article is given by the Leibnizian definition (732.3) of the identity relation, now quantified over predicative functions. However, he stressed that asserting a function was not equivalent to asserting its predicative mate.

Formula (773.2) occurred in art. 6, where Russell launched a systematic and symbolic account of the logical system: the indefinables (propositional function, the truth of any and of all its values, negation and disjunction, predicative function in any type, and assertion); axioms for the propositional and predicate calculi; the determination of classes and relations from predicative functions by means of contextual definition (art. 7); his form (734.1) of definite description, and intensional and extensional functions (art. 8); and the definitions of cardinal and ordinal numbers up to ω_n and $\aleph_n$ for any finite ordinal n, and the need in the arithmetics for the multiplicative axiom (arts. 9–10). The axiom of an infinitude of individuals was adopted reluctantly in art. 9, again made as an empirical assumption. Curiously, he did not explicitly state logicism, although the nature of the enterprise should have been clear to his mathematical audience.

7.8 THE FORMING OF *PRINCIPIA MATHEMATICA*

7.8.1 Completing and funding Principia mathematica. (My *1975b*) By the time that that article appeared in May 1908, Russell and Whitehead were well into writing the book which it presaged; they more or less completed it during the winter of 1909–1910. Not much record survives of the details of the labour; Whitehead discussed his current progress and reported various matters in letters passing between Grantchester and Bagley Wood during 1908 and 1909. An outside reader was Ralph George Hawtrey (1879–1975), then early into his distinguished career in Treasury; an undergraduate in mathematics at Trinity of the same year as Jourdain, he had retained sufficient Tripos mathematics to make valuable comments in long letters (RA and copies).

While Whitehead and Russell revised each other's work (Russell *1948a*), Russell seems to have written out the final version for publication. He reported progress to Ivy Pretious (1881–1958), whom he had come to know in connection with his work on Free Trade (§6.1.2). On 25 May 1908 he told her that 'On average days I do 9 or 10 hours' work at my book [. . .] I have written about 2000 pages of the manuscript of my book since last September; there will be about 6000 or 8000 altogether' (RA). On 18 October 1909 he reported to Lucy Donnelly (RA) that

> tomorrow I go to Cambridge taking with me the manuscript of the book for the printers. There is a certain amount at the end that is not yet finished, but over 4000 pages are ready, and the rest can be finished easily. [. . .] the manuscript is packed into two large crates [. . .] It is amusing to think how much time and trouble has been spent on small points in obscure corners of the book, which possibly no human being will ever discover.

The reaction of the Press to the crates was financial shock. Whitehead received from them a letter of 29 October announcing that 'the manuscript sent would make 1648 pages, and the cost of 750 copies, without binding or advertising, would be £920'. An application for financial support from the Royal Society was suggested and acted upon. Russell had been elected Fellow in May 1908, after nomination in October 1906 by Whitehead (my *1975a*): so they were able to apply jointly for £300, in a fascinating document written together (text in §11.5). On 22 January 1910 £200 was agreed by Council of the Society, so that, as Russell put it in his autobiography, 'we thus earned minus £50 each by ten years' work' (*1967a*, 152).

Russell's memory of the financial loss was once again selective, though this last time probably intentionally. Composition was 'slow work, as there is only one compositor who can read our queer symbols', he told R. B. Perry in April 1910 (Sheffer Papers, Correspondence Box). But the first Volume appeared without apparent hitch in December 1910, in 750 copies at £1.5s. each for around 670 pages; they sent copies to the Society, Trinity College, Berry, Couturat, Forsyth, Frege, Hardy, Hawtrey, Hobson, Johnson (the Press's reader), Jourdain, Peano and Royce. However, during 1911 Whitehead found that he had made a serious mistake in the second Volume by using the axiom of infinity without restriction (§7.9.3). Several passages had to be rewritten, and a special preface prepared for the Volume, which did not appear until the spring of 1912. The extra cost was considerable: Russell seems to have borne it alone. The third Volume came out a year later, like its predecessor in 500 copies; they sold respectively at £1.10s and £1.1s each for around 800 and 490 pages.

The period of publication straddled Whitehead's 50th birthday and Russell's 40th; and the two were together for a time, since in October 1910 Russell started a five-year lectureship in 'logic and the principles of mathematics' at Trinity College. By 1920 all three Volumes were nearly out of print, and a second edition was being mooted between the Press and Russell (§8.4.4).[26]

In fact, the first edition was still not complete. In January 1913 the Press had agreed that the third Volume be split into two; Whitehead was to write the fourth Volume, on geometry. However, while he prepared quite a lot of it, he was to abandon it during (and partly because of) the Great War, and the material was destroyed with all his effects after his death in 1947. It joined its predecessors, for they had burnt the contents of the crates soon after publication; only one folio survives, for $*208 \cdot 4$, in a letter of 3 November 1911 to Ottoline Morrell in which Russell said that he wrote out the book on a little oak table which had belonged to his mother (copy in RA). However, Russell kept over 680 folios of discarded or replaced folios and also a 250-folio concordance of cross-references in the work. They show that various passages had been omitted from the final printed version, whose design will now be summarised.[27]

7.8.2 *The organisation of Principia mathematica.*   After a short preface, the book opened with a general introduction of 88 pages comprising three chapters on basic 'ideas and notations', type theory and 'incomplete symbols'. The rest was divided into six Parts numbered by Roman numerals, each one going into 'Sections' ordered by letters; I shall cite a Section as, say, 'IVB'. Each Section contained 'numbers' (of all names to choose!) with definitions or propositions numbered in Peano's way as cited already (say, '$*35 \cdot 812$' of number 35); as mentioned in §1.3.1, citation by page number is to the second edition. Table 782.1 summarises the contents of each Section, indicating also the asterisked numbers and the numbers of pages in the first edition; '$+$' indicates that a quantity of the replaced or

---

[26] This paragraph draws also on information from the correspondence between Russell and the Press (RA) and that kept in the Press's own archives at Cambridge. No proof sheets of *PM* survive for either edition (on Russell's side, see footnote 43 in §8.9.2); during the Second World War, the Press disposed of mounds of stored sheets in response to a drive to save paper.

[27] The discarded folios are kept at RA, mss. 220.031210–1250. The concordance (230.031270) shows that the following numbers were finally omitted from *PM* (I use Peano's method of concatenating reference numbers): $*2 \cdot 22 \cdot 23 \cdot 34 \cdot 35 \cdot 434 \cdot 51 \cdot 66 \cdot 7 \cdot 71 \cdot 72 \cdot 84$ $*10 \cdot 211$ $*11 \cdot 121–123$ $*13 \cdot 197 \cdot 2$ $*37 \cdot 403$ $*40 \cdot 28 \cdot 29$ $*51 \cdot 39$ $*52 \cdot 34$ $*71 \cdot 3 \cdot 301 \cdot 302$ $*75–*77$ $*83 \cdot 91 \cdot 911 \cdot 912$ $*84 \cdot 15–17 \cdot 44$ $*85 \cdot 82 \cdot 821 \cdot 822 \cdot 83$ $*90 \cdot 361 \cdot 92 \cdot 122–125 \cdot 4 \cdot 6 \cdot 7$ $*93 \cdot 43 \cdot 432 \cdot 44$ $*94 \cdot 15 \cdot 151 \cdot 311 \cdot 32 \cdot 321$ $*96 \cdot 6 \cdot 61–63$ $*97 \cdot 25 \cdot 251$ $*111 \cdot 17 \cdot 171 \cdot 312$ $*113 \cdot 153 \cdot 17 \cdot 171 \cdot 312 \cdot 7 \cdot 71$ $*114 \cdot 49–497 \cdot 561$ $*116 \cdot 144 \cdot 422$ $*117 \cdot 2$ $*274 \cdot 24$.

TABLE 782.1.   First edition of *Principia mathematica* (1910–1913).

The numbers of pages are for the first edition. Volume 2 started at Section IIIA, Volume 3 at Section VD. The titles of the Parts, and numbers of pages (omitting ithe introductions) were I. 'Mathematical logic' (251); II. 'Prolegomena to cardinal arithmetic' (322); III. 'Cardinal arithmetic' (296); IV. 'Relation-arithmetic' (210); V. 'Series' (490); VI. 'Quantity' (257).

Section; pages	(Short) 'Title' or Description: other included topics
IA: *1–*5; 41	'Theory of deduction': Propositional calculus, axioms
IB: *9–*14; 65	'Theory of apparent variables': Predicate calculus, types, identity, definite descriptions
IC: *20–*25, +; 48	'Classes and relations': Basic calculi: empty, non-empty and universal
ID: *30–*38, +; 73	'Logic of relations': Referents and relata, Converse(s)
IE: *40–*43; 26	'Products and sums of classes': Relative product
IIA: *50–*56; 57	'Unit classes and couples': Diversity; cardinal 1 and ordinal 2
IIB: *60–*65; 33	'Sub-classes' and 'sub-relations': Membership, marking types
IIC: *70–*73; 63	'One-many, many-one, many-many relations': Similarity of classes
IID: *80–*88, +; 69	'Selections': Multiplicative axiom, existence of its class
IIE: *90–*97; 98	'Inductive relations': Ancestral, fields, 'posterity of a term'
IIIA: *100–*106; 63	'Definitions of cardinal numbers': Finite arithmetic, assignment to types
IIIB: *110–*117; 121	'Addition, multiplication and exponentiation' of finite cardinals: inequalities
IIIC: *118–*126; 112	'Finite and infinite': Inductive and reflexive cardinals, $\aleph_0$, axiom of infinity
IVA: *150–*155, +; 46	'Ordinal similarity': Small 'relation-numbers' assigned to types
IVB: *160–*166; 56	'Addition' and 'product' of relations: Adding a term to a relation, likeness
IVC: *170–*177; 71	'Multiplication and exponentiation of relations': Relations between sub-classes, laws of relation-arithmetic
IVD: *180–*186; 38	'Arithmetic of relation-numbers': Addition, products and powers
VA: *200–*208, +; 97	'General theory of series': Generating relations, 'correlation of series'
VB: *210–*217; 103	'Sections, segments, stretches': Derived series, Dedekind continuity
VC: *230–*234; 58	'Convergence' and 'limits of functions': Continuity, oscillation
VD: *250–*259, +; 107	'Well-ordered series': Ordinals', their inequalities, well-ordering theorem
VE: *260–*265, +; 71	'Finite and infinite series and ordinals': 'Progressions', 'series of alephs'
VF: *270–*276; 52	Compact, rational and continuous series: Properties of sub-series
VIA: *300–*314; 105	'Generalisation of number': Negative integers, ratios and real numbers
VIB: *330–*337; 58	'Vector families': 'Open families', vectors as directed magnitudes
VIC: *350–*359; 50	'Measurement': Coordinates, real numbers as measures
VID: *370–*375; 35	'Cyclic families': Non-open families, such as angles

discarded ones belongs to the Section, not necessarily with the same numbers. I shall refer to 'the authors' when 'they' is ambiguous.

The book reveals itself as a *Principia Peaniana atque Cantoriana* rather than a veritable *Principia mathematica*: meticulous detail on these mathematical topics, but complete silence on the rest. The coverage was far less than in *The principles*: the absence of mechanics clarified logicism (§6.7.6), but the silence over the calculus and all its consequences, mathematical topics then in very vigorous development, is hard to understand, especially as *PM* advanced as far as continuous functions (§7.9.8). The rather scrappy design of the third Volume was much guided by the scope of the fourth one; but that does not explain the silence over, say, statistics. The four-page preface made no reference to any omissions; but then the logicist thesis was not stated there either! Maybe Whitehead and Russell had become too engrossed with the endless details; or perhaps the three dubious axioms (reducibility, infinity, choice) held them back.

As a technical exercise *PM* is a brilliant virtuoso performance, maybe unequalled in the histories of both mathematics and logic; the chain-links of theorems are intricate, the details recorded Peano-style down to the last cross-reference, seemingly always correctly. However, one has to pass beyond the unclear introductory material—an eccentricity these days—before these virtues emerge, especially in the second Volume, which is easily the best of the trio. The rest of this section deals with the general principles of the system; the next one treats the type theory and the handling of mathematics.

In addition to the book, some French articles are available. Firstly, in the *Revue* Poincaré *1909b* expressed doubts about logicism as presented in Russell *1908c*; so the author brought him up to date in a reply *1910b* based upon a translation of much of the last two chapters of the general introduction. Secondly, there are the products of a hectic time in Paris when Russell gave three lectures: *1911d* on 22 March 1911 on the axioms of infinity and of choice, *1911b* the same evening on the philosophical aspects of mathematical logic (published later in the *Revue*), and *1911a* the next day on 'analytic realism'. The occasion of his visit had an unexpected but profound consequence for his life: on the way to Paris he stayed overnight in London at the home of Morrell, and fell in love with her to a depth which he had never previously experienced.

7.8.3 The propositional calculus, and logicism. The first chapter of the introduction of *PM*, and the opening Sections IA and IB of the main text, are unfortunately the most unclear in the entire work. In them the authors laid out the propositional and predicate calculi (not their names), but in a bizarre manner. For example, as in Russell's paper on implication (§7.4.8) they used the phrase 'propositional function' and notation 'ϕx' at first to refer to a logical combination of propositions before turning to the usual sense; the explanation was buried on p. 15, to back up the explanation of 'elementary proposition' as a proposition free of quantifiers and variables and containing only a finite number of connectives (p. 91). The 'primitive proposition'

'∗1·11. When ϕx can be asserted, where x is a real variable, and $\phi x \supset \psi x$ can be asserted, where x is a real variable, then ψx can be asserted, where x is a real variable. Pp.'

has to be understood in the *first* sense; the example following, which concerns the proof of a theorem (∗2·04) in the propositional calculus, makes this tardily clear.

Another ambiguity is also present here; ∗1·11 concerns inference, following

'∗1·1. Anything [*sic*] is implied by a true elementary proposition is true. Pp'.

on implication. (Their definition of truth is quoted in §7.9.2.) The difference of the latter from inference was given as between

$$\text{'}\vdash p \supset q\text{' and '}\vdash p \supset \vdash q\text{',} \qquad (783.1)$$

with the former omitted for convenience as the intermediate proposition of the latter. The explanation 'An inference is the dropping of a true premiss; it is the dissolution of an implication' (p. 9) makes nice English but muddy logic. In ∗1 the difference was expressed more simply as between 'it is true that p implies q' and 'p is true; therefore q is true' (p. 92). However, in the predicate calculus,

'∗9·12. What is implied by a true premiss is true. Pp.'

was held to deal 'with *inference* to or from propositions containing apparent variables, as opposed to implication', which muddied the logic again.

All this talk of implication belies the fact that it was not a logical primitive. The authors assigned this status to negation and disjunction, specifying them informally via truth-values of propositions p and q as 'p is false' and 'p is true or q is true' respectively (pp. 93–94). Since the latter clause uses 'or' (presumably inclusively), then more muddiness is evident; the earlier account on p. 6 is better. They then defined implication, conjunction and equivalence for propositions (∗1·01, ∗3·01, ∗4·01); these definitions locate their connectives contextually within the propositions involved, although they may also be construed as rules of substitution.

Apart from these general axioms, they offered straightforward symbolic ones, giving them short names for convenient referencing: for example,

$$\text{'}∗1·4. \vdash : p \vee q . \supset . q \vee p \quad \text{Pp'.} \qquad \text{('Perm.')} \quad (783.2)$$

and $\text{'}∗1·5. \vdash : p \vee (q \vee r) . \supset . q \vee (p \vee r) \quad \text{Pp.'} \qquad \text{('Assoc.').} \quad (783.3)$

Since they chose to have no means of testing the independence of the axioms of the calculi (p. 91), it is not surprising that ∗1·5 turned out to be redundant (Łukasiewicz *1925a* and Bernays *1926a* (§8.7.4)). On 22 June 1908 Russell told Hawtrey that 'I consider the proof of the [Pps] to be inductive' (copy in RA), obviously in the scientific sense of induction rather than that of MI; and, as in science, the research was found not to be foolproof. They required their system to 'embrace among its deductions all those propositions which we believe to be true and capable of deduction from logical premises alone', and also to 'lead to no contradiction' (pp. 12–13); but they had no means of *establishing* completeness or consistency.

In these Sections the authors ran through many of the basic theorems. The treatment of equivalence was rather slight, and lacked theorems such as

$$\vdash\ :p \equiv q\,.\,r \equiv s\,.\,\supset\,:p \equiv r\,.\,\equiv\,.\,q \equiv s \text{ and } \vdash\ :p \equiv q\,.\,q \equiv r\,.\,\supset\,:p \equiv r;$$

$$(783.4)$$

the first is needed at least once (in ∗20·51), while the lack of the second was pointed out to Russell by G. H. Crisp on 7 October 1919 (RA). Some other proofs in Part 1 have slips of this kind.

Since the substitutional theory had been abandoned, substitution was handled rather casually, unlike the explicit principle in Russell's paper *1906b* on implication (§7.4.8). The fractional notation of (746.4)$_1$ was used; for example, the proof of

$$\text{`}\ast 2\cdot 11.\ \vdash p\ \vee\ \sim p\text{' involved the move `}\left[\text{Perm}\,\frac{\sim p,p}{p,q}\right]\text{'}. \quad (783.5)$$

This theorem was 'the law of excluded middle', a metalaw to us; others of this status include the laws of contradiction and of double negation (∗3·24 and ∗4·13 respectively).

Related to these unclarities is the logicist thesis itself. The short introduction to Section IA announced that the ensuing 'theory of deduction' would 'set forth the first stage of the deduction of pure mathematics from its logical foundations, where deduction was 'the principles by which conclusions are inferred from premises' and 'depends upon the relation of *implication*' (*PM 1*, 90). This sounds like the implicational logicism of *The principles*; however, pure mathematics was mentioned *for the first time* in the book here: elsewhere 'mathematics' was mentioned, and an inferential standpoint about its relationship to logic seems to have been intended. Only admittedly doubtful principles such as the axiom of infinity or the multiplicative axiom were stressed as antecedents 'so that our propositions, as enunciated, are true even if the axiom of infinity is false' (to quote a typical example from *3*, 234). But this view is surely too easy an option:

hope that the premisses are true, but if not, the truth of the theorems will not be affected. The ambiguities of implication and inference strike at the heart of their logicism.

7.8.4 *The predicate calculus, and descriptions.* After stating a propositional function $\phi\hat{x}$ as being of a variable, the authors rather unfortunately asserted that 'the essential characteristic of a function is *ambiguity*' (p. 39), whereas of course the function *itself* is quite determined, or should be. Both here and elsewhere the status of variables was, well, ambiguous: early on 'variables will be denoted by single letters' (p. 5), but pretty often and surely correctly they *were* letters. Moreover, they ranged over not only 'entities' but also non-entities such as 'propositions, functions, classes or relations' (p. 4). Here and elsewhere x usually ranged over individuals—an important mystery of their own, explored in §7.9.3.

Quantification was defined veridically in terms of truth values. For $\phi\hat{x}$ 'there is a range, or collection, of values of x yielding true or false propositions': then 'the symbol "$(x) . \phi x$" may be read [...] ϕx is always true', while ' "$(\exists x) . \phi x$" may be read "there exists an x for which ϕx is true"' (p. 15): better, surely, there exists a *value a* of x for which ϕa is true.

The other axioms of the calculus were

$$\text{'}*9\cdot1. \ \vdash \ : \ \phi x . \supset . (\exists z) . \phi z \quad \text{Pp.' and}$$

$$\text{'}*9\cdot11. \ \vdash \ : \ \phi x \lor \phi y . \supset . (\exists z) . \phi z \quad \text{Pp.'.} \tag{784.1}$$

(McKinsey *1935a* showed that the first one was redundant.) Rather oddly, they then *defined* '$\exists x$', thus;

$$\text{'}*10\cdot01. (\exists x) . \phi x . = . \sim (x) . \sim \phi x \ \text{Df'.} \tag{784.2}$$

'Descriptions' were based in $*14$ upon definition (734.1) and its brother for two independent variables, with the companion sense 'E!' of existence ($*14\cdot02$). They emphasised the need to watch the 'scope' (their word) of descriptive terms within formulae, so as to avoid ambiguities; for example, the basic contextual definition was now written

$$\text{'}*14\cdot01. [(\imath x)(\phi x)] . \psi(\imath x)(\phi x) . = : (\exists b) : \phi x . \equiv_x . x = b : \psi b \ \text{Df'},$$

$$\tag{784.3}$$

where the square-bracket expression launched the scope.

As we have seen, a main purpose of this theory was to specify single-valued mathematical (that is, denoting) functions $f(x)$ in terms of propositional functions ϕx; they might have stressed the relationships more. For example, in the equation $f(x) = 0$ 'x' is an unknown constant taking the zeroes of the mathematical (denoting) function as its values; but in the corresponding propositional function

$$\phi x . = : f(x) = 0 \quad \text{Df}, \tag{784.4}$$

'x' is still a free variable, determining a true proposition when x takes a zero of f for its value and a false one otherwise.

Most of the basic notions for one variable were repeated for two; for example, for (784.3) at $*14 \cdot 111$. The role within logicism of functions of several variables emerged in the companion theories of classes and relations, which we now consider, postponing type theory until §7.9.1.

7.8.5 *Classes and relations, relative to propositional functions.* In the rest of Part I the authors laid out these calculi as pure *façon de parler* in that they avoided 'the assumption that there are such things as classes' but subsumed them under predicative propositional functions, again written '$\psi ! x$' (p. 187). This was the sense of the title 'incomplete symbols' of the last chapter of the general introduction, a name perhaps chosen in imitation of Frege's 'unsaturated' signs (§4.5.5).

The status of functions themselves had to be made clear. While keeping to an intensional view of functions as properties, they stressed the extensional relationship that two were 'formally equivalent' if they always took the same truth-value, and defined a function of a function as '*extensional* when its truth-value is the same as with any formally equivalent [functional] argument' (p. 72): symbolically (p. 187),

$$`\phi ! x . \equiv_x . \psi ! x : \supset_{\phi, \psi} : f(\phi ! \hat{z}) . \equiv . f(\psi ! \hat{z})'. \tag{785.1}$$

Thus the 'extension' of a function, the class of arguments satisfying it, did not need to be taken as an object; the only requirement was the equivalence between same ones. In this way, and in line with 'the mere principle of economy of primitive ideas', 'an extension (which is the same as a class) is an incomplete symbol, whose use always acquires its meaning through a reference to intension' (p. 72); that is, contextually again:

$$`*20 \cdot 01. \ f\{\hat{z}(\psi z)\} . = : (\exists \phi) : \phi ! x . \equiv_x . \psi x : f(\phi ! \hat{z}) \ \text{Df}'. \tag{785.2}$$

Further, 'intensional functions of functions only occur where some non-mathematical ideas are introduced, such as what somebody believes or affirms' (p. 74). But are the needs of mathematics suitably met? For example, if belief functions are admitted under the quantifier, is the equivalence of classes guaranteed? Is it not sly to 'have our classes and delete them too' (Hill *1997b*, 101)?

This definition of a class drew upon universal quantification over its (potential) members x which satisfied the corresponding propositional function, which in turn bears upon the relationship between identity and equality. Identity of two terms was introduced in ∗13, based upon the Leibnizian definition (732.3); and equality, especially as used in mathematics for classes, numbers and so on, was to be construed as various special cases of it. Identical objects belonged to the same classes while equal classes had the same objects as members, as stated in two strikingly adjacent propositions

$$ \text{`} ∗20·34. \;\vdash\; :.\, x = y .\equiv\; : x \in \alpha .\supset_\alpha .\, y \in \alpha \text{'}, \tag{785.3}$$

$$ \text{`} ∗20·43. \;\vdash\; :.\, \alpha = \beta .\equiv\; : x \in \alpha .\supset_x .\, x \in \beta \text{'}, \tag{785.4}$$

which also compare interestingly with (732.2–3).

The identity relation was repeated for relations in ∗21·43 as part of 'The general theory of relations' (∗21), which closely followed the same philosophy using functions of two variables (rarely were functions of more variables discussed). The universal class V was defined by self-identity as $\hat{x}(x = x)$, with the empty class defined as its complement (∗24·01·02). But far more machinery had to be developed concerning their domains, converse domains and fields. Whitehead devised the notations, with rather excessive enthusiasm; several combinations are only rewrites of each other. Thus we have, unnecessarily,

$$ ∗33·11. \quad \vdash\; . R\text{“}V = D\text{‘}R = \hat{x}\{(\exists y).\, xRy\}, \tag{785.5}$$

$$ ∗33·111. \;\vdash\; . \check{R}\text{“}V = \mathbb{Q}\text{‘}R = \hat{y}\{(\exists y).\, xRy\}, \tag{785.6}$$

$$ ∗33·112. \;\vdash\; . C\text{‘}R = \hat{x}\{(\exists y): xRy .\lor. yRx\} \tag{785.7}$$

('*C*' for 'campus'). Further, for the classes of referents and of relata of a given term under *R*,

$$ ∗32·11. \quad \vdash\; . \vec{R}\text{‘}y = \mathrm{sg}\text{‘}R = R\text{“}(\iota\text{'}y) = \hat{x}(xRy), \tag{785.8}$$

$$ ∗32·111. \;\vdash\; . \overleftarrow{R}\text{‘}x = \mathrm{gs}\text{‘}R = \check{R}\text{“}(\iota\text{'}y) = \hat{y}(xRy). \tag{785.9}$$

In Section IIC, which dealt in great detail with 'ONE-MANY, MANY-ONE, AND ONE-ONE RELATIONS', they used respectively '$1\varepsilon \to \mathrm{Cls}$', '$\mathrm{Cls} \to 1$' and '$\mathrm{Cls} \to \mathrm{Cls}$' (p. 419)—unhappy extra roles for '1'. In general, if the domain and converse domain were sub-classes respectively of α and β, then *R* was classified as '$\alpha \to \beta$' (∗70).

But many notations were unique, and symbolised notions used frequently in the book. A typical example is this one for a relation *P*:

$$ \text{`} ∗36·13. \;\vdash\; : x(P[\alpha)y .\equiv. x, y \in \alpha .\, xPy \text{'}. \tag{785.10}$$

Classes of classes κ were treated carefully, in view of the forthcoming definitions of numbers: for example, the intersection and union of classes became respectively

$$`*40\cdot01.\ p`\kappa = \hat{x}(\alpha \in \kappa.\supset_\alpha.x \in \alpha)\quad \text{Df}', \qquad (785.11)$$

$$`*40\cdot02.\ s`\kappa = \hat{x}\{(\exists\alpha).\alpha \in \kappa.x \in \alpha\}\quad \text{Df}'. \qquad (785.12)$$

A companion pair of definitions dealt in the same way with classes of two-place relations ($*41\cdot01\cdot02$).

One major definition for the theory of relations was that of the 'ordinal couple' of x and y, based upon the relation between members of two classes

$$`*35\cdot04.\ \alpha\uparrow\beta = \hat{x}\hat{y}(x \in \alpha.y \in \beta)\quad \text{Df}',$$

$$\text{as } `*55\cdot01.\ x\downarrow y = \iota`x\uparrow\iota`y\quad \text{Df}'. \qquad (785.13)$$

But the defining expression is a conjunction, and so commutative; thus the order itself is determined by that of the class abstraction operator, which is no real advance. Couturat had raised this matter with Russell in letters of May and June 1904 (RA) without clear resolution; perhaps the best solution is to take propositional functions in more than one variable as primitives. Soon after its publication, *PM* was to inspire Norbert Wiener to define the ordered pair in (827.1).

To link relations to mathematical needs, the 'descriptive function' for a relation R was defined as the denoting function

$$`*30\cdot01.\ R`y = (\imath x)(xRy)\quad \text{Df}'. \qquad (785.14)$$

In the theory of 'double' functions structural similarity obtained between several 'double descriptive functions' such as the union and intersection of two relations (or of two classes), the relative product of two relations, various relations between relations and the (restricted) classes of its relatives and referents, and the addition and subtraction of cardinals. They proved the required results in $*38$ simultaneously by using a two-place connectival schematic letter '$\dagger$', which took "values" such as '$\cup$', '$|$' (the compounding of two relations) and '$+$' from these definition schemae:

$$*38\cdot01.\ x\dagger = \hat{u}\hat{y}(u = x\dagger y)\quad \text{Df' and}$$

$$`*38\cdot02.\dagger y = \hat{u}\hat{x}(u = x\dagger y)\quad \text{Df}'. \qquad (785.15)$$

This strategy was Whitehead's idea, and the ensuing simplifications pleased him (Quine *1985a*, 84).

It is surprising that these definitions were nominal, for important in logicism is the central role of *contextual* definitions of classes and relations, and of definite descriptions. In Moore's reductionist spirit, over and above the reductions to propositional functions or the criteria of referentiability, the class, relation or description may not exist anyway. As we saw at (734.1), Russell was well aware of the central role of these definitions, and he was to stress it again to Wiener in 1913 (§8.2.7).

7.8.6 *The multiplicative axiom: some uses and avoidance.* As we saw in §7.2.6, Russell had realised in the summer of 1904 that the theory of infinite products required a new 'multiplicative axiom', which was named here 'Mult ax'. They presented it in the account of set theory in Section IID, in the form

$$' *88 \cdot 03. \text{ Mult ax}. = :. \kappa \in \text{Cls ex}^2 \text{ excl.}$$

$$\supset_\kappa : (\exists \mu) : \alpha \in \kappa . \supset_\alpha . \mu \cap \alpha \in 1 \quad \text{Df'} \quad (786.1)$$

for each type; they stated various equivalent axioms. Sharing the doubts of many contemporaries of their legitimacy, in the rest of the book they gave an excellent and careful display of their need and use as then understood. However, they did not mention the special difficulty raised by Jourdain concerning the definability of such a class by a (finite) propositional function; the specifying expression,

$$\text{Cls}^2 \text{mult} = \hat{\kappa}\left\{\exists! \left[(1 \rightarrow \text{Cls}) \cap \hat{R}(R \in \in) \cap \sqcap \text{`}\kappa\right]\right\} \quad (786.2)$$

($*88 \cdot 02$ with $*80 \cdot 01$ and $*61 \cdot 12$), is finite—the members are selected by the many-one membership relation applied to each class in a class κ of mutually exclusive classes—but it was not shown how the defining clause would be reducible to a propositional function.

Among major properties of classes, the Schröder-Bernstein Theorem 425.1 was proved, twice. Firstly, they obtained it as the last result ($*73 \cdot 88$) in the number on similar classes, following Zermelo precisely because Mult ax was not needed; Russell had praised his proof in a letter to him of 23 May 1908 (Zermelo Papers, Box 1). Then they imitated Felix Bernstein's use of transfinite induction in the context of the relative product of relations ($*94 \cdot 53$, $*95 \cdot 71$), giving an excellent diagrammatic representation on pp. 589–590. They presented the theorem *as* a result concerning similar classes as such rather than about cardinal numbers; to the latter, and the type theory within which they were located, we now turn.

7.9   TYPES AND THE TREATMENT OF MATHEMATICS IN *PRINCIPIA MATHEMATICA*

7.9.1 Types in orders. The theory was again based upon the VCP, phrased in various ways such as 'Whatever involves *all* of a collection must not be one of the collection' (*PM 1*, 37); as before, the authors also spoke of (in)definability. Somewhat precariously, they seemed to regard the formulations as equivalent; but common to all was the requirement that $\phi(\phi\hat{x})$ be forbidden for all variables, and for any finite number of them (p. 40). It amounted to a distinction of grammatical categories (a class cannot be named by a noun and a verbal clause at the same time) although unfortunately Russell did not exploit or develop such a reading.

The hierarchy was presented both in the introduction (ch. 2) and in $*12$, with some differences; the latter version will be followed here. Propositional functions were stratified by the systematic consideration of quantifiers. 'A 'matrix' (the old word from the substitutional theory, with a similar meaning) had no quantifiers, and its arguments were individual variables; it was given the special notation '$\phi\hat{x}$' and also the additional name 'predicative'. If some *or* all of the arguments were quantified, then a 'first-order' function was produced (pp. 161–162). Then the VCP required that one move up to 'a *second-order* matrix[, which] is one which has at least one first-order matrix among its arguments, but has no arguments other than first-order matrices and individuals'; quantification led to functions and propositions again in this order (p. 163). So one could continue, finitely many times; the order of a function was one above that where all variables were quantified. Quantification was restricted to predicative cases; in class terms, the VCP permitted its members to belong only to the order one above. At any order, 'A function is said to be *predicative* when it is a matrix. [...] "Matrix" or "predicative function" is a primitive idea' (p. 164, tardily on an important point!).

In addition to orders, there were types, where 'a "*type*" is defined as the range of significance of some [propositional] function' (p. 161). After quantification of some or all the variables of a matrix, the remaining free ones determined the type of the function, and it has become customary to regard a type as divided into orders, each one specified by its quantifiers. However, the prior place given by the VCP to quantification shows that the relationship is reversed: an order is divided up into types, each one specified by the free variables left *after* quantification, and a propositional function is specified by *both* its order and type indices.

But the underlying strategy remained unclear, in that $*9$ developed the theory constructively from 'elementary propositions' and successive quantification of its variables, a 'purely philosophical' exercise which was complemented in $*10$ by an 'alternative method' based upon taking negation and implication as primitive and proceeding more formally.

Russell clearly had different 'reasons for ramification' (Goldfarb *1989a*), but he did not fully explain the variety, either here or in use in the later numbers.

The stratification of functions led to a companion hierarchy of propositions when the free variables were given values. On 26 May 1909 Whitehead had queried with Russell this feature of the type theory in Russell's 1908 paper, puzzled by the relationship between the (propositional) values of functions of different types and the hierarchy of propositions itself (RA). Presumably Russell explained that the VCP stratified propositions by propositional quantification, in order to Solve the liar paradox: 'p. (p is false)' is a proposition whose 'second falsehood' does not belong to the 'first falsehood' of the original propositions, and so on up finitely (pp. 41–43). After this explanation in the introduction one reads in disbelief in *9 that 'we never have occasion, in practice, to consider propositions as apparent variables' (p. 129).

Further on they gave the axioms (784.1) of the predicate calculus, and interpreted the first as 'if ϕx is true, then there is a value of $\phi \hat{z}$ which is true' (p. 131)—which is surely not a correct reading of a propositional function. In both hierarchies 'typical ambiguity' was deployed, where theorems applied to *any* appropriate collection of orders and/or types (for example, not only individuals and classes but also classes and classes of classes).

In the context of type theory the authors introduced their definition of truth: 'When we judge "a has the relation R to b," our judgement is said to be *true* when there is a complex "a-in-the-relation-R-to-b," and is said to be *false* when this is not the case' (p. 43). While it was framed again in terms of correspondence, the old uncertainties of Russell *1906g* over objective falsehoods (§7.4.7) were over; indeed, when he reprinted that article in a book anthology of *Philosophical essays* (§8.2.4), he wrote a new piece *1910b* in this vein to replace its uneasy final pages.

The motivation for the theory of types was, of course, the need to construct mathematics logicistically while Solving all the paradoxes. However, the latter were not presented until after the summary of type theory in the introduction. This time the paradoxes reported were, in order, the liar, Russell's, the relations version of it, Burali-Forti's, Berry's, the least indefinable ordinal, and Richard's; the list was followed by a summary of each Solution as furnished by the VCP (pp. 60–65). Notable omissions include the cardinal paradox(es), and especially the one stated at the end of *The principles* arising from associating a class m of propositions with the proposition 'every member of m is true', which once again (§7.4.8) is constructible within the simple theory (de Rouilhan *1996a*, 223–230). Grelling's recent paradox on heterology (§7.2.3) was also ignored.

7.9.2 *Reducing the edifice.* (Hill *1997a*, ch. 9) As in Russell's 1908 preview, this Byzantine partition of propositional functions into orders and

types sabotaged the reconstruction of mathematical theories and operations; hence was introduced 'the axiom of reducibility' (sometimes named
'Reduc ax'), also 'axiom of classes' and 'axiom of relations' as at (773.2). It
claimed that any propositional function had a logically equivalent quantifier-free predicative function with the same free variables (*PM*, *12); thus
it restored Leibnizian identity (732.3) in full form. However, it seems to
destroy the structure of orders, in that any propositional function up there
had a mate on the ground floor. (They stated on p. 58 that the axiom could
equally well assert the existence of the function at *any* same order.)
Somehow, however, the semantic paradoxes to not seem to be restored;
Myhill *1979a* presents a modern type theory in which avoidance is assured
even with such an axiom in place.

One of the more useful queries about logicism made in the *Revue* by
Poincaré *1909b* (in reaction to Russell *1908c*, we recall from §7.8.2)
concerned the place of this queer axiom; in particular, whether it was more
or less general than MI, and whether it could be justified psychologically.
Russell's reply *1910b*, based on the introduction to *PM*, argued for greater
generality, in that MI was derivative; as for the second point, he included a
somewhat isolated passage on judgement which we shall consider in §8.2.6
when it took a more developed form after *PM* appeared.

When reading the manuscript of the book Hawtrey had expressed his
puzzlement over the axiom on 3 September 1908 (RA):

> You have adopted a hierarchy which is not the most general possible and which
> is not consistent with the Science of Pure Mathematics. You have then made
> your hierarchy consistent with the Science of Pure Mathematics, by adopting as
> an axiom a statement about the functions existing within your hierarchy, which
> there is no ground for believing to be true.

While not quite the criticism that the axiom amputated the hierarchy of
orders, his point was well put; Russell's reply of 5 September (RA) only
recalled the raising of types in the definition of identity, and the text in
PM does not really answer it. He also gave the definition of inductive
cardinals as another example, whereupon two days later Hawtrey worked
through the theorem that the sum of two of them was also inductive
(compare *110·02). Russell accepted the proof on 23 September, and in a
discussion of identity he mooted, not for the first time (§6.3.1), that
'Speaking as a philosopher rather than a mathematician, I should say that
"diversity" is an ultimate notion, & that identity may perhaps be correctly
regarded as the negation of diversity' (copy in RA).

A final question is the status of the theory in another sense; whether
objects of some sort are being stratified, or their symbols. In later writings
Russell preferred the latter view, regretting that he had not adopted it
when developing the theory (*1918–1919a*, 267; *1944a*, 691); we see his
tendency to conflate symbols and their referents. Possibly the type theory
was a central feature of his *calculus* of logic rather than of the logic itself,

which remained universally general although its formulae had to be split up into orders and types. Variables and constants surely have to be taken as symbols, and the former distinguished from schematic letters—another conflation in their calculus, especially grave when typical ambiguity was deployed.

Hawtrey raised but attacked this position in an undated letter probably written, like most of his others, in the autumn of 1908. He saw Russell as having 'felt bound to make all variables unrestricted' especially because 'A restriction on the variable can only take the form of a hypothesis about the variable'; however, 'the use of the unrestricted variable led to the contradiction, and you now use the [VCP] to show that, in order that a proposition containing an apparent variable may have any meaning at all, there must be a certain restriction on the variable implicit in it'. He opined that asserting propositionhood as such 'was in reality fundamentally different' a proposition from the proposition concerned, so that non-restriction did not apply anyway. Russell's reply, if written, seems not to have survived; once again Hawtrey's query was penetrating.

7.9.3 *Individuals, their nature and number.*   One consequence of abandoning the substitutional theory was that the generation (746.1) of an infinite class was also lost: so Russell was forced, reluctantly, to adopt an axiom of infinity (named 'Infin ax'). Due to the return to propositional functions and orthodox variables, it took a quite different form, which we now summarise. The issues arising, very significant for logicism, are rarely recognised.

Firstly comes the question of the nature of individuals. *PM* was not a happy text; the "definition" as 'neither a proposition nor a function' (p. 132; see also p. 51) does not exclude, for example, logical connectives or assertion. A further explanation as 'something which exists on its own account' (p. 162) is perhaps worse given Russell's many senses of existence. The phrase 'destitute of complexity' in the 1908 paper (*1908c*, art. 4) is more specific; but the status of structureless entities in logicism is still unclear (Prior *1965a*).

Type theory bears upon this issue. In *PM* it was held to be relative in that 'In practice, we never need to know the absolute types of our variables, but only their *relative* types' (p. 165); ∗65 was devoted to typical ambiguity, where examples were given of propositions and definitions (in cardinal arithmetic, in that case) which applied across any collection of neighbouring types. However, in one of his lectures in Paris in March 1911 (§7.8.2) he explicitly took the absolutist position: 'Here the word *individual* contrasts with class, function, proposition, etc. In other words, *an individual* is *a being in the actual world, as opposed to the beings in the logical world*' (*1911d*, 23). But this empirical status for individuals entailed that logic was *a posteriori*—surely a mistake, requesting the physicists to decide about a basic feature of logicism! *Was* Russell influenced by current speculations about molecules? It is worth noting his talk soon afterwards of 'atomic' and 'molecular' propositions in his epistemological writings (§8.3–4).

Secondly is the question of the number of individuals assumed. Infin ax took the form of asserting that there existed a non-empty class of any finite number of members ($*120\cdot03$). The structure of type theory required such a class, 'Indiv' (*2*, vii, 18), to exist at the lowest level of types in order to allow logicism to ascend to(wards) the Cantorian transfinite clouds. However, the reductionist spirit decreed that it be used only when absolutely essential, especially when, 'as monists aver, there is only one individual' (*2*, 325: compare $*22\cdot351$, $*24\cdot52$, $*50\cdot33$ and $*254\cdot431$); they had even named its unit class as '1 (Indiv)' (*1*, 345). However, the identity relation is thereby trivialised, for $x = y$ becomes $x = x$.

Perhaps for this reason, Whitehead's mistake (§7.8.1) was to have forgotten this restriction when preparing Part III of Volume 2 on cardinal arithmetic. During 1911, while publication was suspended, he altered proofs and text in various places. A letter to Russell of 19 January (text in §11.6) shows him rethinking $*100\cdot35$ on the equal cardinality of non-empty classes being equivalent to their similarity, and worrying about $*126$ on 'typically indefinite inductive cardinals' (§7.9.4) and other passages; Russell noted theorems in $*117$ and $*120$ on the letter. On 20 May Whitehead reported that $*118$ and $*119$, respectively on substitution and the subtraction of cardinals, were redone.

To explain his new procedures Whitehead prefaced Volume 2 with a long and very difficult 'Prefatory statement of symbolic conventions'. Following Russell's definitions as classes of all similar classes and starting from 0 defined as the unit class of the empty class, he tried to lay down ways of using only 'adequate' types T for a cardinal c; that is, those Ts within which c was definable. For 0 and 1 occurred in all types, but not 2,... in Indiv, or 3, 4,... in 'Indiv, Cl'Indiv, Rl'Indiv; and so on' (*2*, x). He asserted that there were three kinds of hierarchy: a 'functional' one of the various types of propositional function; a 'propositional' one built up from elementary propositions (without quantifiers) by means of substitution of non-elementary functions; and an 'extensional' one, which started from any point in the functional hierarchy and relative to it generated classes, classes of classes, and so on. There were also two kinds of 'formal number': 'constant' ones identical with a cardinal number in the given type; and 'functional' ones, defined 'by enumeration' and using the standard arithmetical operations (p. xiv). The type of such a number was its 'actual type', whose adequacy was established by an existence theorem for that type (pp. xv–xvi). Compound formal numbers could contain components from other types, which had to be 'normally adjusted' to ensure non-empty classes, and 'arithmetically adjusted' for adequacy. 'Symbolic forms' could be *equations* or *inequations*, and formal numbers might make *argumental*, '*arithmetical*', '*equational*', '*attributive*' or '*logical*' occurrences in them (pp. xvii–xix).

Further, two conventions limited the choice of types for formal numbers: one concerned the relationships between different kinds of occurrence, while the other insisted that a type be adequate, leading to restrictions on the substitutivity of numbers in arithmetical equations (pp. xxi–xxiv). Ambiguous types in a symbolic form were identifiable with types, ambiguous or specified, in any other one. Failures of adequacy were removed by conventions that all equations involving arithmetical formal numbers were arithmetical, and that inductive cardinals should be taken in a type adequate to ensure the existence of all cardinals and an avoidance of non-empty classes (pp. xxviii–xxxi). Whitehead concluded, somehow, that 'all discrimination of the types of indefinite inductive numbers may be dropped; and the types are entirely indefinite and irrelevant' (p. xxxi).

Their failure to spot this elementary slip must have been traumatic for both men. Russell referred later to the assumption of any individual at all as a 'defect in logical purity' (*1919b*, 203), and in 1912 he had become agitated about it. In a recollection made to me in 1972 by a mathematics undergraduate of that time who wished to remain anonymous,

> McTaggart and Bertrand Russell held periodic coffee parties on Sunday evenings in the latter's rooms attended by a dozen students or so from various colleges; the 'audience' sat in silence in a half ellipse with McTaggart to the left and B.R. to the right [...]
>
> At the last conference [...] Bertrand Russell called me aside as a mathematician I suppose and likely to appreciate the gravity of his statement—'I have just realised that I have failed—it is easy to establish the unit one but I have omitted to establish a second like unit'—(I won't guarantee the precise wording but it's not far off). He went on to say 'I have finished'.
>
> With this ringing in my mind and the dreadful distress of a great scholar I was glad I was alone.

7.9.4 *Cardinals and their finite arithmetic.* This was developed, within each type, in two somewhat distinct forms. Throughout the treatments the authors indicated carefully the roles of both Mult ax and Infin ax, and Whitehead added on proof some remarks on his new procedures to minimise the role of the latter.

Firstly, the definite description 'the cardinal number of a class α', written 'Nc'α', was defined as the class of all classes β similar to α; its type lay one above that of α itself (*100·1), but the β were taken from all types. When β was restricted to the type 't'α' (*63·01) of α, then the 'homogeneous cardinal $N_0 c$'α' (*103·01) was defined. As an exercise, they showed in *105 how to define the cardinals relative to any class of lower type.

So far so Peanist, in that the cardinals were tied to one-one correspondences although defined within logicism; their "direct" definitions of numbers started from the initial definition of 0 as the unit class of the empty class Λ ($*54\cdot01$), 1 and 2 as the classes of unit classes and of unordered couples respectively ($*52\cdot01$, $*54\cdot02$), and successively on upwards ($*100$–102). None of these integers was an object, since as classes each of them relied on the contextual definition (785.2) of classes—a major difference from Frege's *nominal* definitions of cardinals (§4.5.3).

Cardinal arithmetic was defined within each type, and moreover needed classes of disjoint classes to work, which in general does not obtain. We recall Whitehead's emphasis (653.2) on such classes; now he devised in $*110$ a way of defining the 'arithmetical sum' $\alpha + \beta$ of *any* two classes α and β by forming two disjoint classes α' and β' each of the type (β, α) and respectively similar to α and β: essentially,

$$\alpha' = (\Lambda \cap \beta, \iota\text{‘}a) \text{ Df with } a \in \alpha, \text{ and } \beta' = (\iota\text{‘}b, \Lambda \cap \alpha) \text{ Df with } b \in \beta,$$
(794.1)

where each empty class Λ came from the type of its partner class.[28] Such restrictions did not obtain in defining the 'arithmetical product' $\beta \times \alpha$ of α and β ($*113$), but they showed why and how 'The arithmetical product of a class of classes' ($*114$) required Mult ax. They also gave a treatment of 'double similarity' between two classes of classes κ and λ, which obtained not only between the member classes but also between their own members:

$$\text{‘}*111\cdot44. \ \kappa \text{ sm sm } \lambda . \supset . \kappa \text{ sm } \lambda . s\text{‘}\kappa \text{ sm s‘}\lambda\text{’}.$$
(794.2)

It was needed in arithmetical properties such as equality of sums or products of cardinals.

The connective variable '$\overset{+}{}$' (785.15) was used to define, via arithmetical addition, the relation '$(+_c \nu)$' between two homogeneous finite cardinals μ and ν:

'$*110\cdot02. \ \mu +_c \nu$

$$= \hat{\xi}\{(\exists\alpha, \beta) . \mu = N_0 c\text{‘}\alpha . \nu = N_0 c\text{‘}\beta . \xi \text{ sm } (\alpha + \beta)\} \text{ Df'}.$$
(794.3)

---

[28] Whitehead's explanation in *PM 2*, 63 would have benefited from stating the definitions of intermediate notions: check successively through $*55\cdot23\cdot231$, $*37\cdot01$ and $*51\cdot11$.

where '$+_c 1$ is the relation of a cardinal to its immediate predecessor' (2, 203). Similarly, subtraction of a cardinal from a greater one was defined thus:

$$'*119·01. \quad \gamma -_c \nu = \hat{\xi}\{Nc'\xi +_c \nu = \gamma . \exists!Nc'\xi +_c \nu\} \, Df'. \qquad (794.4)$$

The extension to negative cardinals is described at (799.1).

7.9.5 *The generalised ordinals.* There now followed, as Part IV, Russell's most substantial contribution to the mathematics of logicism: his 'RELA-TION-ARITHMETIC', built upon his insight of 1901 (§6.5.4) of generalising from well-order to order-types. Russell told the Press on 27 May 1961 (RA) that when Johnson had read the manuscript of *PM* for them, he had stressed the importance of this Section.

Two relations P and Q were ordinally similar ('smor') under a 'correlator' relation S (*151) when

$$\text{if } tPw \text{ and } uQv, \text{ then } tSu \text{ and } v\check{S}w, \text{ so that } P = S|Q|\check{S}. \qquad (795.1)$$

Then the relation-number of P was given by

$$'*152·1. \quad Nr'P = \hat{Q}(Q \text{ smor } P) = \hat{Q}(P \text{ smor } Q)' \qquad (795.2)$$

with the corresponding class NR of such numbers as its domain (*152·02). In *164 they described the 'Double likeness' of two classes of relations by a 'double correlator'; the needs corresponded to those in (794.2) for the double similarity for cardinals.

Much of the Section was taken up with the "arithmetic" of these numbers, imitating that of cardinals already presented and ordinals still to come: homogeneous numbers, addition (but not subtraction), multiplication, exponentiation, laws such as associativity, and inequalities. The greater generality made the exegeses much more extensive, and Mult ax still more ubiquitous (for example, to choose correlators in certain circumstances). From the mathematical point of view the generality was especially evident in *170−*171, where they presented in detail Felix Hausdorff's exposition of various kinds of order-types *1906a* and *1907a* (§4.2.7) in what they described as 'brilliant articles' (2, 391; see also 3, 171). The procedure worked in terms of ordering the sub-classes of a relation by taking away members from classes until they became different from each other; adapting his name of it to 'the principle of first differences', they handled sub-types similarly.

Notations were again rather over-prolific; and, since much structure similarity was evident, the schematic connective letter '$\dotplus$' of (785.15) might have been deployed earlier than *182. The symbols for operations with cardinals were adopted, with some graced by overdots. Links to cardinal arithmetic were exhibited; for example,

$$`*180 \cdot 71. \; \vdash \; : \mu, \nu \in \text{NR} . \supset . C``(\mu \dot{+} \nu) = C``\mu +_{\text{c}} C``\nu'. \quad (795.3)$$

7.9.6 *The ordinals and the alephs.* The ordinal numbers were rather unclearly presented in *PM*: some small ones in Part II, and the rest in Part IV on Relation-Arithmetic and Part V on Series, where they appeared as special cases of two theories that were more general in different ways. Here the portions are gathered together.

The theory started out in *250 from the notion of a well-ordered relation, 'when every existent sub-class of its field has one or more minima' (*3, 4*); from the class Ω of these relations the class NO of ordinals formed in *251·01 the domain of the relation-number relation when Ω was placed in its converse domain ('NO = Nr``ω Df'). They showed how to produce Cantor's series (326.2) of ordinals, and how Burali-Forti's paradox was avoided by the raising of types (p. 74); Zermelo's proof of the well-ordering theorem was treated in *258.

In order to generate the required domains they drew on Frege's ancestral relation (452.1) in *90 to launch Section IIE on 'INDUCTIVE RELATIONS'. They worked with Frege's own 'proper' non-reflexive version, written 'R_{po}', but started with the reflexive relation 'R_*':

$$`*90 \cdot 01. \; R_* = \hat{x}\hat{y}\left\{ x \in C`R : \breve{R}``\mu \subset \mu . x \in \mu . \supset_{\mu} . y \in \mu \right\} \text{Df'}, \quad (796.1)$$

so that ' $*91 \cdot 52. \; \vdash . R_{\text{po}} = R_* \,|\, R = R \,|\, R_*$ '. $\qquad (796.2)$

This theory was also used to distinguish finitude from infinitude, where they drew upon both the reflexive and inductive definitions of infinity. The former was provided via the ancestral relation; for inductive cardinals, 'those that obey mathematical induction starting from 0' (*2, 200*), were defined by successive additions of 1 to 0, so that their class, 'NC induct', was the posterity of 0 relative to the relation '$(+_{\text{c}} 1)$' (*120·01·02: these cardinals were used in (794.3–4)). But the 'Cls infin' of infinite classes was defined not as non-inductive but reflexively, 'when there is a one-one relation which correlates the class with a proper part of itself' (*2, 270*). Then an infinite series was defined as one with a reflexive field, so that its class was given by

$$`*261 \cdot 02. \; \Omega \text{ infin} = \Omega \cap \breve{C}``\text{Cls refl} \quad \text{Df'}. \quad (796.3)$$

The treatment of reflexive and inductive definitions of infinitude was handled with a care that is still rare (Boolos *1994a*). Mult ax haunted the account; at *124·01·02·03 they defined together Cls refl, the class 'NC refl' of such cardinals, and the class 'NC mult' of classes satisfying the axiom.

The transfinite ordinals and the alephs were defined by their versions of Cantor's definitions. His well-order-type ω was changed into the class 'Prog' of 'progressions' generated by ancestral relations:

$$\text{'} *122\cdot01.\ \text{Prog} = (1 \rightarrow 1) \cap \hat{R}\big(\text{D'}R' = \overleftarrow{R}_* \text{'}B\text{'}R\big)\ \text{Df'}, \qquad (796.4)$$

$$\text{with '} *123\cdot01.\ \aleph_0 = \text{D''Prog Df'}, \qquad (796.5)$$

where $\vec{B}\text{'}R$ was the complement of the domain of R with respect to its converse domain ($*93\cdot101$), so that its member(s) could launch the progression ('B' for 'begin'). Similarly,

$$\text{'} *263\cdot01.\ \omega = \hat{P}\{(\exists R).\ R \in \text{Prog}.\ P = R_{\text{po}}\}\ \text{Df'}. \qquad (796.6)$$

Higher alephs were defined from the corresponding initial ordinals of the number-classes, starting with

$$\text{'} *265\cdot01.\ \omega_1 = \hat{P}\big\{\overrightarrow{\text{less}}\text{'}P = (\aleph_0)_r \cup \Omega \text{ fin}\big\}\ \text{Df'}, \qquad (796.7)$$

where 'less' was the relation of ordinal dissimilarity between two well-ordered series ($*254\cdot01$), and '$(\aleph_0)_r$ the class of well-ordered series whose fields have $\aleph_0$ terms' (3, 169, using $*262\cdot03$). Then

$$\text{'} *265\cdot02.\ \aleph_1 = C''\omega_1\ \text{Df'}; \qquad (796.8)$$

the field of ω_1 was taken rather than its domain because of Cantor's way of defining the number-classes into terms of rearrangements of order of the members (§3.2.7).

The further pairs of definition followed this pattern, with the additional requirements for cardinality added into the defining clauses of the ωs. However, as in the 1908 preview (§7.7.3), since only a finite number of types was permitted, $\aleph_\omega$ and its successors could not be defined; ω_ω and beyond were also inaccessible (as it were) (*PM 2*, 183–184; *3*, 170, 173). Perhaps for this reason, they did not include Cantor's alephs paradox (663.2) in the construction of 'the series of alephs' ($*265$); for the Solution of both it and of Burali-Forti's paradox were rendered rather pointless since they could advance so far up the series of ordinals anyway—significant lacunae in this highly Cantorian logicism.

The authors handled beautifully Cantor's theory of cardinal exponentiation, forming the class '$\alpha \exp \beta$' of $(\text{Nc'}\alpha)^{\text{Nc'}\beta}$ ordered couples $x \downarrow y$ chosen from α and β respectively such that each y took only one x ($*116$). They also connected the operation with multiplication in $*116\cdot361$

(thanks to MI), again with a fine diagrammatic representation to clarify the proof.

7.9.7 *The odd small ordinals.* The ordinal numbers started out surprisingly (∗153, with parts of an overture curiously placed as ∗56 in the Section on cardinal arithmetic). '0_r' ('r' for 'relation') was defined as the unit class of the diversity relation '$\dot\Lambda$' and so was the correlator of relations ordinally similar to it:

$$` *56\cdot03.\ 0_r = \iota`\dot\Lambda\ \text{Df'}, \text{ so that } ` *153\cdot11.\ \vdash\ .0_r = \text{Nc}`\dot\Lambda', \quad (797.1)$$

$$\text{where, from } *25\cdot01\cdot02,\ \dot\Lambda = \hat{x}\hat{y}(x \neq x\ .\ y \neq y). \quad (797.2)$$

Similarly, 2_r was the class of ordered pairs:

$$` *56\cdot02.\ 2_r = \hat{R}\{(\exists x, y)\ .\ x \neq y\ .\ R = x \downarrow y\}\ \ \text{Df'}. \quad (797.3)$$

But, since 'series must have more than one member if they have any members' (*1*, 375), the identity relation, being symmetrical, was not serial. Hence the relation-number 1 was not an ordinal; so they offered

$$` *153\cdot01.\ 1_s = \hat{R}\{(\exists x)\ .\ R = x \downarrow x\}\ \ \text{Df'} \quad (797.4)$$

as 'the nearest possible approach' ('s' for 'series'?). They also defined another cousin number:

$$` *56\cdot01.\ \dot2 = \hat{R}\{(\exists x, y)\ .\ R = x \downarrow y\}\ \ \text{Df'}. \quad (797.5)$$

Another relation-number '$\dot1$' arose, for the purpose of adding one term to a relation-number μ to produce '$\mu \dot+ \dot1$' or '$\dot1 \dot+ \mu$' (∗181·02·021). It was granted the property

$$` *181\cdot04.\ \dot1 \dot+ \dot1 = 2_r\ \ \text{Df'} \quad (797.6)$$

in order 'to minimize exceptions to the associative law of addition' (*2*, 467).

7.9.8 *Series and continuity.* Part V dealt with 'SERIES' in general, based upon a non-identical relation P which was transitive and connected:

$$\vdash :.\ x, y, z \in C`P\ .\ \supset_{x,y,z}\ xPy\ .\ yPz\ .\ \supset\ .\ xPz, \quad (798.1)$$

$$*202\cdot103.\ \vdash\ :\ x, y \in C`P\ .\ \supset_{x,y}\ :\ xPy\ .\lor.\ y = x\ .\lor.\ yPx. \quad (798.2)$$

The former property was not stated; instead '$P^2 \subset\ \cdot P$' was used (∗201·1). The class of series was notated 'Ser' (∗204·01).

Among the basic notions of series given in Section VA, the most significant were the 'minimum points of a class α with respect to a relation P', namely 'those members of α which belong to $C'P$ but have no predecessors in α'; and similarly the 'maximum' points of α with no successors in α (2, 541). *205 gave a variety of properties of these points for various kinds of relations (for example, connected ones). Then the authors treated the 'sequent points' of α relative to P in *206; both the class '$\overrightarrow{\text{seq}}_p$ 'α' of 'its immediate successors' whether or not α had a maximum, and "$\overrightarrow{\text{prec}}_p$ 'α' of its immediate predecessors (p. 559). The key notion in much of the entire Section was the 'limit or maximum': 'A term x is said to be the "upper limit" of α in P if α has no maximum and x is the sequent of α', symbolised 'lt$_P$ 'α' (p. 575). Similarly, 'the *lower* limit of α will be the immediate predecessor of α when α has no minimum; this we denote by tl$_P$ 'α'. Thus the key umbrella notions, 'limit or maximum' and 'minimum', were given thus:

$$\text{'} *207 \cdot 4. \;\vdash\; :.\, x \,\text{limax}_P\, \alpha \,.\equiv\, :\, x \,\text{max}_P\, \alpha \,.\vee.\, x \,\text{lt}_P\, \alpha\text{'}, \qquad (798.3)$$

with 'limin$_p$' handled via $\check{P}$ (*207·401).

With these and earlier notions, especially ancestral relations, the authors were able to treat in Sections VB–C several features of real line theory. It was guided by the notion of dividing a series into two parts after the manner of a Dedekind cut of the continuum (*210–*215), and Cantorian derived sets (323.3) as 'Derivatives' of series with limit points (*216). From there they proceeded into mathematical analysis as far as defining the oscillation and the continuity of a mathematical function (itself expressed as a relation), its upper and lower limiting values, and other delicate tools from the Weierstrassians' surgery (§2.7.5). As in *The principles* (§6.4.7), Ulisse Dini's textbook in its German edition *1892a* was the principal source (mistitled in 2, 724), and in connection with his definition of continuity in both the (ε, δ) and sequential forms they noted 'that practically nothing in the theory of continuous functions requires the use of numbers' (p. 725)—a feature of many of the later Sections of *PM*.

The theory of ordinal numbers described in §7.9.5 was contained mostly in Sections VD–E. It included transfinite induction, then by no means routine, in contexts such as 'The transfinite ancestral relation' (*257), where the 'transfinite posterity' of a relation was generated by a method imitating Cantor's principles (§3.2.6). Some other material was described in §7.9.6.

The final Section VF dealt with three of Cantor's other principal order-types. His dense sets became 'compact series' in their hands, 'in which there is a term between any two, *i.e.* in which '$P \subset \cdot P^2$, where P is the generating relation' (3, 179), a formulation of compactness (*270·01) delightfully converse to their definition of connectivity stated after (798.2). 'Rational series' was 'ordinally similar to the series of all rational proper fractions (0 excluded)' (p. 199) following Cantor of *1895b* (§3.4.7): compact and containing a progression within its field ($*273 \cdot 01$).

The Part ended with continuous series, where the authors again followed Cantor, using his second definition, also of 1895. It was preferred over Dedekind's cut procedure because two Cantor-continuous series were ordinally similar ($*275 \cdot 3 \cdot 31$); but in $*214$ they had worked out in detail 'Dedekindian relations', where 'every class has either a maximum or a sequent with respect to it'. They used Hausdorff's theory to determine the sub-classes of such a series, and used it to show that its cardinality was $2^{\aleph_0}$($*276 \cdot 43$). They did not mention Cantor's continuum hypothesis—unanswered mathematical questions were not their concern—but the German heritage was well in evidence in this Part.

7.9.9 Quantity with ratios. Part VI, on 'QUANTITY', was largely Whitehead's work; his surviving letters to Russell suggest that much of it was done during 1909, especially in the autumn when they were also applying to the Royal Society for a grant and he was scheduled to give a two-term course on 'The principle of mathematics'.[29] The importance of the theory 'grows upon further consideration', he opined to Russell on 14 September, 'the modern "<u>arithmeticisation of mathematics</u>" is an entire <u>mistake</u>'—of course a useful mistake as turning attention upon the right points'. However, it 'leaves the whole theory of applied mathematics (measurement etc) unproved'. He then advised that

> You will have to devote some attention to my ms, since their results will come as a shock to the current orthodoxy. In fact mathematicians will feel much like Scotch Presbyterians who might find that a theological professor in one of their colleges had dedicated his work to the Pope.

In fact, as he might have expected, the mathematical community has always treated this Part as if it was, say, 'The Assumption of Moses'—not among anyone's normal reading.

By 12 October Whitehead found that 'The whole part on quantity is naturally rather long—for it embraces the whole theory of ordinary [real-variable] mathematical analysis—The comforting thing is that our previous ideas and notations are exactly adapted for the exposition', although much attention had to be paid to typing of notions and to proving existence theorems (in the Peanist sense of classes being non-empty). He reported that the Part had two 'subdivisions': ratios, and then 'the Quantitative Relations (or, quantity proper)'. But in print it began with a theory of 'positive' and 'negative integers', extending the treatment (794.3–4) of the addition and subtraction of cardinals. In $*121$ on 'intervals' the authors had treated various relations of 'the class of terms between x and

---

[29] *Cambridge University reporter* (1909–1910), 66.

y with respect to some relation P', specifically the case 'P_μ' where $(\mu + 1)$ terms were counted from x to y inclusive ($*121\cdot013\cdot02$, with 'μ' for 'ν'). Then they introduced in $*300\cdot232$ these integers in terms of addition and subtraction of two positive inductive cardinals by μ:

$$\vdash . U_\mu = (+_c \mu) \restriction \alpha . \breve{U}_\mu = (-_c \mu) \restriction \alpha, \text{ where } \alpha = \text{NC induct} - \iota'\Lambda \quad \text{Df},$$

$$(799.1)$$

and from $(785.15)_2$ for the connectival schematic letter '$\pm$' '$(\pm_c \mu)$' were the relations of $(\mu \pm_c \nu)$ to μ (compare *PM 2*, 181). In order to include 0, they appealed to the identity relation (785.3) ($*300\cdot03$). The change over the previous theory was that these integers were relations (effectively counting forwards and counting forwards) rather than classes of classes, and so more suitable for the theories of ratios and quantity to come. As before, subtraction was still of the lesser from the greater; but 'negative ratios' were handled in $*307$ by using the converses of the relations furnishing the positive ones.

They needed a definition of ratios '(1) narrow enough to preserve all the algebraic properties (2) wide enough to include all the applications', as Whitehead nicely put it to Russell on 12 October. In an undated letter of this time he found Russell's proposed definition of the power R^σ of a relation R to any inductive cardinal σ to be 'excellent, and must have a chapter for its own sake'; this was duly prepared as $*301$. Then ratios were defined as follows: distances lay in the ratio μ/ν of co-prime cardinals if their corresponding relations R and S possessed at least one pair of terms x and y such that $xR^\nu y . xS^\mu y$ ($*303\cdot01$). 'The series of real numbers, positive and negative' was defined Dedekind's way by specifying an 'irrational number' by the condition that a class of ratios had neither a maximum nor a minimum point with respect to the 'less than' relation between ratios ($*310$). Again, negative numbers came via relations converse to those for positive ones ($*312\cdot01\cdot02\cdot1$); maybe they were influenced by Frege's similar strategy in his theory (458.1) of real numbers. The treatment was rather disappointing, however, for their reductionist philosophy led them to be 'mainly concerned with just those few simple properties which are independent of the axiom of infinity' (3, 316); but they did present the arithmetic operations ($*311$-$*314$).

Quantitative relations drove the other three Sections, which concentrated the notion of 'vector'. This was the somewhat restricting choice of name for 'conceiv[ing] a magnitude as a vector, *i.e.* as an operation, *i.e.* as a descriptive function in the sense of $*30$'. It was specified as a one-one relation that 'shall be capable of indefinite repetition' (p. 339: Russell's definition once again), and a 'vector family' as a class of those relations which commuted ('Abel', $*330\cdot02$). Much of the rest of Section VIB dealt with various special kinds (for example, connected relations); but throughout they assumed that the family took no upper bound over the class of permitted magnitude. Norbert Wiener was to find here a fruitful source for modifications (§8.2.7).

Section VIC, 'MEASUREMENT', wedded the theories of ratios and real numbers with those of vector families. Ground was prepared for Volume 4 by a means called 'rational nets' for the 'introduction of coordinates in geometry', in which a class of rational multiples of a given vector family was set up in a way that it also formed a group ($*354\cdot14$–17). Thus Whitehead's hopes to use algebra in logic (§6.4.4, 9) were revived, albeit in a different context.

The final Section VID treated 'CYCLIC FAMILIES', where a vector could have more than one multiple: for example, angles between two straight lines, for which the ratio α/β could also be $(\alpha + 2N\pi)/\beta$ for any cardinal N. The definition included the condition that 'it might contain a non-zero member which is identical with its converse' relation, such as π in this example (p. 458); the other conditions ensured that this member was unique ($*370\cdot23$). Thus in Benjamin Peirce's terms (§4.3.2) cyclic families formed an algebra with one idempotent element, and the theory was similar to defining complex numbers as ordered pairs; they mentioned neither feature.

The modifications to the theory were not great, so that 'we have given proofs rather shortly in this Section', since many were 'perfectly straightforward, but tedious if written out at length' (p. 461). It is surprising to read such a statement in the Bible of logicism; doubtless it was motivated by the role of the theory as a tool for use in the Volume 4 to come. In the end Whitehead was to abandon it; we note this disappointment in the next chapter, which begins with their work immediately following *PM*.

CHAPTER 8

The Influence and Place of Logicism, 1910–1930

8.1 PLANS FOR TWO CHAPTERS

The reception of *PM* from its publication to around 1940 is covered in this and the next chapters, with the break coming around 1930. There was a wide range of reactions both to the logical calculus of *PM* and to logicism; some striking similarities arose from different backgrounds. I associate each main philosophy with a 'school', in contrast to the 'traditions' of algebraic and mathematical logics.

This chapter falls into two roughly equal parts, with Anglo-Saxon attitudes followed by reactions elsewhere. §8.2 surveys Whitehead's and Russell's very different transitions from logic to philosophy till around 1916. Whitehead did not adopt empiricism, but Russell's empiricist philosophy built closely upon logicism. §8.3 describes an influential visit made by Russell to the U.S.A. in 1914, and his further work in epistemology during the decade. §8.4 is dominated by Cambridge, with the reactions to logicism of Wittgenstein and Ramsey, and Russell's own revision of *PM* in the mid 1920s. §8.5 notes the new responses until around 1930 of Britons and Americans to logicism, especially the second edition of *PM*.

Then the focus falls upon Continental Europe. After the limited interest in Italy and France (§8.6), German-speaking countries take over. §8.7 continues the story of parallel processes from §4 by noting Frege's late rejection of logicism; Hilbert's second phase of proof theory; developments in set theory, and the rise of intuitionism with Brouwer; and the reactions of various mathematicians and philosophers.

Next, §8.8 records the remarkable rise of interest during the 1920s in the new country of Poland; a superb group of logicians emerged under the leadership of Łukasiewicz and Leśniewski. Finally, §8.9 describes the emergence in Vienna of a circle of mathematicians and philosophers, among whom logic was a major concern. The two dominant figures are Carnap and Gödel; the chapter stops with the latter's completeness theorem of 1930. The next chapter starts with its successor on incompleteness published in the following year, and then concentrates upon the place of logicism in the 1930s, with only summary note taken of parallel developments.

Among the bibliographical sources, the reviewing *Jahrbuch* is still the best single source, with the literature under headings such as 'Philosophie', 'Grundlagen', and 'Mengenlehre' and 'Logic'—itself a sign of emergence of the field. Herbertz *1912a* is a useful 'study guide' to philosophy which not only informs on logic and foundations of mathematics but also shows all the main tendencies in philosophy of its time; understandably it is skewed towards German work.

For the U.S.A. the Open Court Publishing Company comes into its own, with many articles and reviews published in its philosophical journal *The monist* and also new editions, translations and original books by Russell and others (McCoy *1987a*). Furthermore, from 1905 the *Philosophical review* published almost every year until the Second World War a review of developments of philosophy in France; the author was usually Couturat's friend André Lalande (1867–1963). A companion series from Germany started in 1907 with Oscar Ewald; halted in 1914 by the Great War, it was resumed in 1927 by Arthur Liebert. While valuable for general context, it never treated work in symbolic logics or the philosophy of mathematics in much detail; so it is not cited here. But use is made of the articles and book reviews there, and also in the older *Journal of philosophy*.

8.2 WHITEHEAD'S AND RUSSELL'S TRANSITIONS FROM LOGIC TO PHILOSOPHY, 1910–1916

8.2.1 *The educational concerns of Whitehead, 1910–1916.* For Russell *PM* completed his philosophical programme for mathematics (geometry to come and the three dubious axioms excepted, of course); but for Whitehead it was an (important) stepping-stone for a broader vision of mathematics, which included its creative aspects. Their application for a grant to the Royal Society (§7.8.1) shows the difference in a small but striking way: in three places Whitehead had described their intention of deriving 'mathematics' from logic, and each time Russell added the adjective 'pure' (§11.5).

At that time Whitehead resigned from his Trinity College Fellowship, in protest over the treatment of A. R. Forsyth concerning a *scandale d'amour*. He moved to London, and a readership in applied mathematics at University College London became available in 1912 when Karl Pearson moved over to the new Galton Chair in Eugenics. In his application Whitehead presented a 'large scheme of work, involving the logical scrutiny of mathematical symbolism and mathematical ideas', and having 'its origin in the study of the mathematical theory of electromagnetism' (Lowe *1975a*). He secured the post, in his 52nd year; but two years later he moved to a chair at Imperial College, which he held until 1924.

The scope of Whitehead's ambitions is suggested in a little *Introduction to mathematics* (*1911c*) which he wrote for the Home University Library. After opening chapters on 'The abstract nature of mathematics' and 'Variables' (an interesting choice of starters) he treated some mechanics, real and complex numbers, trigonometry, and aspects of the calculus, series and functions. His 180 pages show a mixture of nice heuristics and pop history; but he never mentioned logicism, logic or even set theory. However in a substantial article *1911b* on 'Mathematics' for the new 11th edition of *Encyclopaedia Britannica* he gave logicism disproportionate space, in order to stress that mathematics was more than the study of number and magnitude. He compensated for the long article 'Logic' and its history, written by two Oxford philosophers, which ignored all symbolic logics, even Boole's!

In his article, and with acknowledgement to Benjamin Peirce (§4.3.2), Whitehead defined mathematics as 'the science concerned with the logical deduction of consequences from the general premisses of all reasoning' (*1911b*, 880). It is surely too broad in scope: he may have been influenced by questions of mathematical education, which loomed large in his new career in London. His lecture *1913b* to the Fifth International Congress of Mathematicians at Cambridge in 1912 (§8.2.4) dealt with the education of boys. He focused upon the formation of 'abstract ideas' and on the 'logical precision', but he did not advocate teaching logic itself to promote the latter; interestingly, he supported instruction in the history of mathematics. That year, as President of the London branch of The Mathematical Association, he followed the same line in a more discursive piece *1913a*; logic was now linked mainly with mathematical functions. In his later Address *1916a* as President of the full Association he pleaded for 'reform' in mathematics education, though in the direction of teaching fewer topics thoroughly rather than emphasising logic or reasoning. But in his Presidential Address *1917a* to Section A of the British Association for the Advancement of Science, on 'the organisation of thought', he gave much space to logicism, outlining its four 'departments': 'arithmetic', dealing with connectives between propositions; 'algebraic', using propositional functions and relations (A. B. Kempe praised); 'general-function theory', where mathematical functions were introduced via definite descriptions; and finally 'analytic', where he summarised the mathematical contents of *PM*.

Although his later career was dominated by philosophy and relativity theory, Whitehead maintained an interest in education. He included these four essays in an anthology *1917b* entitled *The organisation of thought*, and the last three again in a later collection *1929b* on *The aims of education*.

8.2.2 *Whitehead on the principles of geometry in the 1910s.*   Whitehead also contributed to the *Encyclopaedia Britannica* a substantial survey *1911a* of the 'axioms of geometry'. Concentrating on the projective and descriptive parts, he broadly followed his two Cambridge tracts of the mid 1900s (§7.6.2). He also wrote with Russell the article *1911a* on non-Euclidean geometry to replace Russell *1902c* in the previous edition (§6.2.3); he was largely responsible for the changes, which consisted in leaving the historical part almost intact, eliminating much of the philosophy, and adding in technical matters such as metrics (Russell *Papers 3*, 472).

In 1914 Whitehead participated in an event stimulated by the *Encyklopädie der mathematischen Wissenschaften* and its French counterpart (§4.2.4). Both projects were to be completed with a final seventh Part on the history and philosophy of mathematics, and for the French version Federigo Enriques (1871–1946) had been appointed editor. To help fulfil his aims he initiated 'The First Congress on Mathematical Philosophy', held in Paris early in April 1914 immediately after one on mathematics education (Reymond *1914a*). He planned to create an 'International Society' to further these efforts, but they foundered with the start of the Great War in August, and the Part was never produced for either version of the *Encyklopädie*. The Congress proceedings were to appear in the September issue of the *Revue de métaphysique et de morale*; but only the opening address appeared there (a scribble *1914a* by Pierre Boutroux on mathematicians and philosophers working together), though a few other talks were published in later issues, including Whitehead *1916b* on the relational theory of space.

This study was related to the fourth volume of *PM*, which Whitehead continued to prepare after the earlier ones had appeared. However, he stopped work on it in 1918, seemingly after the death in War action of his younger son Eric.[1] After his own death in 1947 his widow followed his instruction to burn all his manuscripts; so we can never know the extent to which it was written. But the application document to the Royal Society (§11.5) shows that he had four Sections in mind. Those on the projective and descriptive branches presumably drew upon his two tracts of the mid 1900s (§7.6.2); then came the metrical branch, where much of the material in the third volume, especially concerning quantity and measurement, was already waiting to be used; finally was promised the 'Constructions of space', probably following the scheme of his paper *1906b* (§7.6.1) and maybe the Paris paper cited above. Surviving letters to Russell up to 1914 suggest that he was making considerable progress, so that one can form a good impression of the intended product (Harrell *1988a*). Clearly several branches would be missing; for example, the absence of the calculus excluded differential geometry. So our sceptical query about the amount of mathematics in logicism (§7.8.2) remains.

For Whitehead *PM* ended in sad circumstances. Further, in a letter of January 1917 he objected to Russell's use of his ideas on the construction of space in *Our knowledge of the external world* (Russell *1968a*, 78), despite Russell's full acknowledgement in the preface; so their collaboration ceased. Let us turn now to the initial reception of their great effort.

---

[1] Shortly before this tragedy Whitehead contributed his War effort: a short paper *1918a* sent to the Royal Society the previous November, in which he gave graphical estimates for calculating the paths of shells projected at high angles.

8.2.3 *British reviews of* Principia mathematica. Philip Jourdain was the most assiduous reviewer, with seven pieces. Three appeared in the *Jahrbuch*: *1913c* on the first volume, mostly citing other literature with little discussion of its contents; then *1915a* summarising the treatment of arithmetic (more detail was given in a review *1913b* elsewhere); and by far the longest, an extended summary *1918a* of the last volume. Throughout the decade he both wrote much on Russell's logic and philosophy and also secured him publication in *The monist*, of which he became the English Editor in 1912 (my *1977b*, ch. 23). The most significant work was a pair of articles on 'The philosophy of Mr. B∗rtr∗nd R∗ss∗ll' (*1911a*, *1916a*), to which Russell himself contributed a few chapters; with great wit they satirised many of Russell's main concerns, such as paradoxes and definite descriptions. The asterisks followed a practise of the Cambridge University humorous magazine *The granta*, where two chapters had first appeared in 1907. Further, in imitation of Russell's appendix of original texts to his book on Leibniz (§6.3.1), Jourdain showed in appendices that Lewis Carroll had anticipated many of the topics, principally in his *Alice* books (though not in the ones on logic). A book version Jourdain *1918b* was put out with Russell's blessing by Allen and Unwin, his publisher from 1916.

Jourdain also published in a mathematical journal a long and valuable series of studies on logicians from both traditions: *1910a* on Leibniz and Boole; *1912b* on MacColl, Frege (§8.7.3) and Peano; and *1913d* on Jevons. Apart from the first one his accounts were good, with many references and perceptive comments; and those on Frege and Peano were graced by important remarks on the drafts made by their subjects and added as footnotes. Russell read all the drafts, though he had few remarks to add; one wonders about the reaction had Jourdain fulfilled his intention of writing on C. S. Peirce. Among related activities, Jourdain also published with the Company an English translation of Cantor's last two-part paper with his own long introduction (Cantor *1915a*).

Apart from Jourdain, *PM* did not receive many other British reviews of substance (some American ones are recorded in §8.3.2); and unlike his, the majority dwelt upon the first volume. An interesting one appeared in *The spectator* in July 1911. The anonymous reviewer was James Strachey (1886–1967), a younger brother of editor John, and of Oliver of the liar paradox in §7.2.2.[2] He gave a good survey *1911a*, ending with an analogy worthy of Russell himself: the book 'seeks to establish the immensely complex structure of mathematics upon a basis more solid that the universe itself, because it is independent of the universe, namely, upon a set of axioms that would remain true if the universe were swept away, and that are true now, even if the universe is merely a delusion'.

---

[2] I am grateful to Kenneth Blackwell for identifying the reviewer as James Strachey, from the publisher's file copy of *The spectator*. Also a younger brother of the famous writer Lytton, James became well-known later as an editor of the writings of Sigmund Freud.

Another anonymous review of the first volume appeared, in the *Times literary supplement*. After lamenting mathematicians' lack of interest in such studies, G. H. Hardy *1911a* urged it to their attention, and of its components he stressed that 'mathematics, one may say, is the science of propositional functions', and 'The theory of "incomplete symbols" is one of the authors' triumphs'. On type theory he was more reserved, although in wondering if there could be an infinitude of types he had not read the book carefully enough.

8.2.4 *Russell and Peano on logic, 1911–1913.* In 1910 Russell prepared a collection of his general essays, including *1907d* on the study of mathematics (§7.7.2), as a volume *1910a* for Longmans, Green entitled *Philosophical essays.*[3] In October he joined Hardy again when he was appointed to a lectureship at Trinity College, in effect taking over the teaching of foundations of mathematics from Whitehead (§7.9.9). He normally gave a two-term course on 'The fundamental concepts of mathematics' for a 'Fee 10*s*. 6*d*.', together with free three-term courses on 'The principles of mathematics' and some years one term on 'mathematical logic'.[4] Rather little is known about Russell's teaching (§7.9.3 contains a hint), but G. E. Moore took the paying course in the winter of 1911–1912; his notes (Papers, 10/4/3) show that Russell dealt with all the main notions and intentions of *The principles*, including geometry and mechanics, without the technical contents of *PM*. Mathematics, 'more certain than philosophy', was analytic in the sense of being derivable from logic alone, not in the traditional sense that the predicate be part of the subject. 'Logic in some sense = pure form', that which remained constant when all constituents had been changed. Russell became quite keen on this approach, and wrote a short manuscript *m1912b* answering 'What is logic?' with 'Logic is the study of the forms of complexes', with 'form' specified as above. 'A complex is *logical* if it remains a complex whatever substitutions may be effected in it. Df.', so that 'Logic = the class of logical complexes'. In line with his current avoidance of objective falsehoods (§7.4.7), he associated complexes only with true propositions. The text peters out inconclusively.

At the request in January 1912 of organiser E. W. Hobson, Russell invited speakers to a section on 'philosophy and history' of the Fifth International Congress of Mathematicians; it took place at Cambridge in September, with the proceedings coming out the following year. Couturat and Frege declined, but of the Peanists Peano *1913b* contrasted the existence of definite description with that of the non-emptiness of classes,

---

[3] For the production schedule and costs of the book, see Reading University Archives, Longmans, Green Papers, Impression Book R34B.

[4] The details for Russell's courses may be retrieved from the lists of courses in mathematics published in the *Cambridge University reporter*. It stated the intended courses; insufficient support would cause their cancellation.

Cesaro Burali-Forti *1913a* treated functional and operational symbols, and Alessandro Padoa *1913a* discussed the status of mathematical induction (and also argued at length with Russell: *Papers 6*, 444–448). Other contributors to this or other sessions included Ernst Zermelo, Emile Borel and Jourdain. Among visitors from the U.S.A. were Maxime Bôcher and E. H. Moore—and also Paul Carus (1852–1919), editor of *The monist*, who met Jourdain and recruited him as the English Editor for Open Court. Russell himself did not give a paper, but in his Chairman's Address *1913b* for his section he regretted the absences of Cantor and Frege, and the recent death of Henri Poincaré.

In his paper Peano had noted logicism without comment; but soon afterwards he reviewed the first volume of *PM* at length in an Italian journal (not his *Rivista*, which had stopped in 1908). Peano *1913a* contrasted his own use of 'logic-mathematics' as an 'instrument' with its role in *PM* 'for science in itself'. Using classial rather than logical formulations, he quoted many of the symbolic definitions and properties of classes and relations (of which the logic was 'all new'), and the construction of cardinal arithmetic as far as the need for the multiplicative axiom. In three lines at the end he noted the contents of the second volume, promising to write at greater length later; but he never did so, for his passion had switched to his international language of Latin without inflection, in which he had written the review.

8.2.5 *Russell's initial problems with epistemology, 1911–1912*

I found Matter a large and fruitful theme, and I think very likely I shall work at it for some years to come. I have done the philosophy of *pure* mathematics, and this would be the philosophy of *applied* mathematics.
                    Russell to Lucy Donnelly, 19 December 1912 (*Letters 1*, 444)

Like his master Peano, Russell was leaving logic behind at this time. He devoted most of 1911 to philosophical writing, of which the best-known product was a short book on *The problems of philosophy*, written for the Home University Library series at the request of editor Gilbert Murray. It appeared early the next year as Russell *1912a*, soon after Whitehead's *Introduction to mathematics*. As he made clear in his opening note, he presented his own preferred selection of problems, so that epistemology was prime, with Moorean reductions and techniques from logic prominent. Indeed, his title was misleading, since he was always a systems philosopher rather than a problems philosopher. While he did not describe logicism, he asserted that 'All pure mathematics is *a priori*, like logic' (p. 43, surely recalling *The principles*!), and alluded to the definability of arithmetic from logic (p. 65). He also announced that the laws of identity, contradiction and excluded middle were 'self-evident' (p. 40), without reference to the inductive processes of finding basic principles which he had mentioned to Ralph Hawtrey (§7.8.3).

Russell's epistemological principles were guided by the search for certainty, which he mentioned three times in the opening two paragraphs. In ch. 5 he distinguished 'knowledge by acquaintance', drawn from sense-data presented directly before the mind, from 'knowledge by description', given by physical objects and presented indirectly for expression via definite descriptions; the account was partly based upon, and even copied from, a lecture of this title delivered to the Aristotelian Society in March 1911 and published by them as his *1911c*. Again, in chs. 9 and 10, built upon his first Presidential Address *1911e* to the Society, he stressed the importance of universals, and assumed that relations played a major role. But soon after the book appeared Oliver Strachey suggested on 4 January 1912 that he rethink the distinction between universals and particulars and demote relations in some ways (RA); Russell noted on the letter that it 'influenced me considerably', so it is transcribed in §11.7. (Strachey was to make similar points in *Mind*, not so sharply, in *1915a*, 20–23.) In ch. 12 on 'Truth and falsehood', Russell appealed again to the correspondence theory, and discussed the status of judgements, which were to come to the fore in his next epistemological phase.

8.2.6 *Russell's first interactions with Wittgenstein, 1911–1913.*   1911 saw two important changes in Russell's life: the start in March of his affair with Ottoline Morrell; and the beginning in October of his relationship with Ludwig Wittgenstein (1889–1951), then a young Austrian graduate engineer from Manchester University. Wittgenstein had been aware of Russell's work already in 1909, when he corresponded with Jourdain about Russell's paradox and a solution proposed in terms of regarding paradoxes as meaningless limiting cases of meaningful propositions (see Jourdain's text in §11.4). Now in personal contact, they experienced an intense exchange until the Great War started in 1914 (McGuinness *1988a*, ch. 5).

One major consequence concerned the fate of Russell's first effort at a book on epistemology, which he wrote at speed during May 1913. He intended to work out in detail his theory of knowledge by acquaintance, firstly with a Part 'On the nature of acquaintance' (with predicates, sensation, time and so on), and then Parts on 'atomic' and 'molecular propositional thought', the latter formed by linking up examples of the former with logical connectives such as 'or' and 'unless' (Russell *m1913a*). His rejection of objective falsehoods led him to replace a two-place theory relating a judgement to a proposition *P* to assess its truth-value by a 'multiple relation' theory (his name), where the judgement was made of the various constituents of *P*; typically, 'person *S* understands that *a* is in the relation *R* to *b*'.

However, Wittgenstein criticised this theory on some fundamental grounds. The exact details are hard to make precise, for the main surviving evidence comes only in June letters from Wittgenstein to Russell (fragments drawing on doubtless long chats) and from him to Morrell.[5] But clearly Russell's old and new philosophical concerns were involved, since the epistemology *of* logic was a major issue. One criticism was that Russell's theory could not handle asymmetrical relations and discriminate between '*S* believes that *a* precedes *b*' and '*S* believes that *b* precedes *a*'. This 'direction problem', as it has become known, struck at the heart of Russell's epistemological aims; for example, he gave prominence to logical forms and complexes, and his third part would have contained a taxonomy of complexes.

Another dart was aimed at Russell's logic, where already in June 1912 Wittgenstein was convinced that 'The prop[osition]s of Logic contain ONLY APPARENT variables', so that 'there are NO *logical* constants' (presumably meaning that they were not objects: *Letters*, 10). He must have realised that Russell's logic was muddled up with logicism and so needed its own characterisation; he may have been led to his view of variables by noting that a logical order was specified by its quantified variables. He seems to have accepted type theory; but he did not subscribe to logicism, since he concluded this letter that 'Logic must turn out to be of a TOTALLY different kind than any other science', presumably including mathematics.

For logicist Russell, however, such issues were still more serious, and confidence in his new book gradually disintegrated; as with his substitutional theory (§7.4.6), several different factors may have been involved. At all events, he abandoned it in June after writing only the first Part and much of the second one. He did publish the first six chapters of the first Part during the first half of 1914, sending them to Jourdain as a quartet of articles for *The monist*; but their appearance may reflect his financial difficulties more than any intellectual conviction. (The text appeared in full only in Russell *Papers 7* (1984).) However, the *general* thrust of his philosophy—empiricist and reductionist epistemology drawing upon techniques from logic—was unimpaired, and a more successful successor was soon written (§8.3.1).

8.2.7 *Russell's confrontation with Wiener, 1913.* (My *1975b*) Three months later, in October 1913, Russell had another acquaintance with a foreigner, less consequential than that with the Austrian but tricky enough. This time it was an American, still younger than Wittgenstein but already the possessor of a doctorate from Harvard University: Norbert Wiener (1894–1964). He came over to Cambridge with his father, who 'looks like a Hindoo, but I think it would come off in water', Russell told Morrell on 26

---

[5] For this correspondence see Wittgenstein *Letters*, 23–27; and Russell's letters 760–820 *passim* to Morrell (copies in RA). Historical interpretation has become an endless exercise, in which Nicholas Griffin's pioneering efforts are impressive (*1980a*, 152–169, and *1985a*); among other studies, see Iglesias *1984a*.

September 1913 (letter 877, copy in RA). 'The son is fat, bland and smug', and

> after a period of dead silence—suddenly woke up and began an equal torrent, on the subject of his doctor's thesis—pulling out books from my shelves and pointing out crucial passages, pointing out, kindly but firmly, where my work is one-sided and needs his broad view and deep erudition to correct it [...] I believe the young man is quite nice and simple really, but his father and teachers have made him conceited.

Partly through the influence of Josiah Royce's lectures at Harvard (§7.5.4), Wiener had found a nice topic for his teenage investigations: 'A comparison between the treatment of the algebra of relatives by Schröder and that by Whitehead and Russell' (Wiener *m1913a*). It was the first, and still the only substantial, contrast between the two traditions of algebraic and mathematical logic, drawing upon an especially interesting context. For Wiener showed that from a structural point of view there were very many similarities between the two theories, even down to specific kinds of relation defined in each. However, on their respective foundations, he was much less sure. In particular, while he noted that Schröder treated collections as part-whole theory whereas Whitehead and Russell deployed Cantorian set theory, he did not grasp the consequences; he even praised Schröder's conflation of membership and inclusion as subsumption. Curiously, he used Schröder as his only algebraic representative; fellow Harvard graduate Peirce, still alive though in isolation, was not mentioned.

The 'infant phenomenon is staying here until I go to America', Russell opined to Morrell on 28 September (letter 879, copy in RA). 'I have read his Dr. thesis, and think him more infant than phenomenon. Americans have no standards'. Some of his written comments to Wiener are of great historical value; in particular, he recorded the cause of his initial attachment to Peano in 1900 (quoted in §6.4.1), and stressed the central importance of contextual definitions in logicism (§7.8.4–5). Throughout he was hard-hitting, especially on the inferior power of part-whole theory. Presumably as a reaction, Wiener kept these comments but never published a line of his thesis, not even a summarising paper (my *1975b* serves as a partial substitute). This is a great pity, as the vast differences between the two traditions have never been properly recognised historically or even philosophically, and he had plenty of good material and examples to present.

However, Wiener made other useful contributions to logic then. One was a note *1914a* on 'A simplification of the logic of relations', where he showed how the calculus of relations could be reduced to that of classes by a proper definition of the ordered pair to replace $(785.13)_2$:

$$'(\exists\, x, y)\,.\,\varphi(x, y)\,.\,\alpha = \iota'(\iota'\iota'x \cup \iota'\Lambda) \cup \iota'\iota'\iota'y', \qquad (827.1)$$

for the class α contained only the ordered pair (x, y), as provided by the respective clauses in the union. His motive was to reduce the axiom of reducibility (§7.9.2) for relations to that for classes. The definition is somewhat clumsier than needed; the job can be done by $(\iota'x \cup \iota'y) \cup \iota'x$, as Kuratowski *1921a* was to show later.

Wiener's note was published by the Cambridge Philosophical Society after its presentation in February 1914 by Hardy, with whom Wiener had also become acquainted. (827.1) was only a technical device, albeit interesting, without consequences for the *philosophy* of relations, such as possible relationships between x or y; naturally he had made no such claim for it, although he showed how several definitions in *PM* could be simplified. Perhaps for that reason, or for residual annoyance, Russell did not react at all; but he had noticed the infant's prodigious gifts. He suggested that Wiener apply the logic of transitive and unconnected relations to non-overlapping intervals and the theory of instants of time; it soon led to a succeeding paper Wiener *1914b* which was to influence Russell himself much later (§9.5.4), and to an extension *1914c* on relations of many variables with application to the theory of intensities of sensation. At the end of the decade he returned to this topic, modifying the theory of magnitudes by vector-families given in Part VIB of *PM* (§7.9.9) to allow for those which have a finite upper bound. He submitted this paper to Hardy for the London Mathematical Society, and Russell wrote a very warm report (*Papers 9*, 469–470); so it duly appeared as Wiener *1919a*.

Maybe inspired by this extra exposure to foundational questions, from 1914 to 1919 Hardy himself offered a free course in the Easter term (summer to everyone else) on 'Elements of mathematics (for non-mathematical students)'. G. E. Moore took it, seemingly in 1915; his notes reveal a course of set theory influenced by *PM*, and including variables, finite and transfinite cardinal and ordinal arithmetic, mathematical induction, the multiplicative class, continuity, and some elementary geometry.[6] But he published none of it, and returned in print to foundational questions only a decade later (§8.5.2).

8.3 LOGICISM AND EPISTEMOLOGY IN AMERICA AND WITH RUSSELL, 1914–1921

8.3.1 *Russell on logic and epistemology at Harvard, 1914.* Another foreign product came into Russell's hands late in 1913: the English edition of a recent German compendium of logic, intended to launch an 'Encyclopaedia of the philosophical sciences' but prevented by the Great War

---

[6] For details of Hardy's course, see the lists of courses in mathematics in the *Cambridge University reporter* from 1913–1914 to 1918–1919. G. E. Moore's notes are in his Papers, 10/6/1–4.

from proceeding further (Windelband and Ruge *1913a*). Royce, Couturat and Enriques were among the sextet of authors, surveying various kinds of logic or its applications to other disciplines, especially philosophy. The whole is a mish-mash: even Couturat's long survey *1913a* of symbolic methods, mainly Peano's and *PM*'s, lacks life. Arguably the most interesting piece is Royce *1913a* on the notion of order as he was developing it from Kempe's theory (§7.5.4). In a short review Russell *1914a* was wittily sarcastic: 'The book, in fact, resembles a compendium on the British Constitution composed during the Civil War, with an introduction by King Charles and an epilogue by Oliver Cromwell'.

During that autumn and winter Russell had the regular company of Wittgenstein, Wiener and Jourdain, and occasionally he used Jourdain's secretary to take down dictation of various papers and also Wittgenstein's notes on logic. But his main concern was the visit to America mentioned to Morrell above: he had negotiated to visit Harvard between mid March and mid May 1914 to give two lecture courses in the Philosophy Department. Early in April Peirce died; a student of the courses, Victor Lenzen (*1971a*), went to collect Peirce's *Nachlass* at the end of the year (§4.3.1).

One of Russell's courses comprised a highly technical account of *PM*, whose three volumes were now all published. The audience included T. S. Eliot (1888–1965), a post-graduate student specialising in the philosophy of F. H. Bradley, who was much taken with Russell's very different philosophical world: further, his notes on the course on logicism are much the more extensive. They record one innovation since *PM* appeared: the presentation of logical connectives as tables of the truth-values, which Russell and Wittgenstein seem to have thought up between them in 1912 or 1913.[7]

Russell's other course, on epistemology, had been partially delivered at Cambridge as a course on 'The fundamental concepts of physics' in the Lent term of 1914.[8] Russell had largely written a book version during the previous autumn, and much later he claimed that he dictated it to a secretary at the beginning of 1914 in one session (*1956a*, 195–196; *1967a*, 210); but the faulty memory was on form again, for the torrent of letters to Morrell tell a far more prolonged story. If any work of that time was produced this way, it may have been a related paper on 'sense-data and physics' prepared with the help of Jourdain's secretary and published as Russell *1914b* (Blackwell *1973a*).

---

[7] Eliot's notes on both courses are held in his Papers at Houghton Library, Harvard University, File Am. 1691.14.(3); the truth-tables are written on fol. 43, for inclusive disjunction, implication, and the Sheffer stroke. For some reason the notes of Russell and Wittgenstein were left out of Russell *Papers 8*; but their substance is summarised in McGuinness *1988a*, 160–162, and they appear, with good discussion, in Shosky *1997a*.

[8] *Cambridge University reporter* (1913–1914), 483; for an auditor's notes, see G. E. Moore Papers, 10/4/4. Russell gave, or at least offered, the course a year later ((1914–1915), 91).

The book appeared from Open Court in August 1914, just before the Great War (Russell *1914c*). A somewhat revised edition was published by Allen and Unwin in 1926 in Britain, and a different one with rather fewer changes three years later in the U.S.A.; both versions were reprinted, with the British one of 1993 being mis-described as of the original edition!

The long title strikingly encapsulates Russell's philosophical hope: 'Our knowledge of the external world as a field for scientific method in philosophy'. Unfortunately, some of the reprints only give the first six words of its title, under which it is generally known. In an opening chapter he summarised the prevailing philosophies, idealism and evolutionism, which he wished to replace: Bradley and Henri Bergson were the respective prime targets. The positive doctrine was displayed across six chapters, prefaced by his creed of 'Logic as the essence of philosophy' (ch. 2); he made a quick historical survey of modern versions, especially his own, though without the symbols. Then in ch. 3 using those first six words, he summarised a version of knowledge by acquaintance, although he neither used that name nor analysed judgement or truth-values of propositions: the failed book of the previous year seems to have reduced his ambitions. The main novelty was a passage on 'perspective space', effectively the class of all individual perspectives of a physical object, which served like an invariant for them. This passage closely resembled part of his paper *1914b* on sense-data; maybe it was added later, causing him to mis-remember the manner of composition of the book. In a succeeding ch. 4 on 'the world of physics and the world of sense' he displayed his reductionism in 'the maxim that inspires all scientific philosophising, namely "Occam's razor"'.

Much of the next three chapters was concerned with continuity and infinity, where Russell discussed Zeno's paradoxes at length and the views of Cantor. At their end he considered the relationship between classes and mathematical logic, and mentioned Wittgenstein's recent unpublished idea that logical constants were not objects. The final ch. 8 dealt with cause and free-will, giving him chances to extol certainty and describe induction. Overall the book has a patchy scenario, more mathematics than necessary and not really enough science. But the important role of logic was clear; and while he avoided a symbolic treatment, the fusion of logic with epistemology was to give the book a warm reception, with long complimentary reviews such as Jourdain *1914a* in the *Mathematical gazette* and C. D. Broad *1915a* in *Mind*.

Back home from Harvard, Russell rehearsed his position in a lecture on 'the scientific method in philosophy', delivered on 18 November 1914 at Oxford University. He also drew upon his recent thoughts on the logical forms of complexes (§8.2.4) to distinguish between two 'portions' of logic: one handling 'general statements' and the other 'concerned with the analysis and enumeration of logical *forms*' (Russell *1914d*, 65).

By contrast, reactions in America, both to the book and to *PM*, were also substantial but more mixed, as we now see.

8.3.2 *Two long American reviews.* The *Journal of philosophy* carried a fine critical piece on 'the logical-analytic method in philosophy' by Theodore de Laguna, who taught at Bryn Mawr College. While praising the merits of the method in its mathematical contexts (*1915a*, 451), he greatly doubted its utility in Russell's new book, stressing the epistemic dangers of using Ockham's razor (p. 453) and sensing a vicious circle in constructing space via perspective spaces: 'Mr. Russell has *deduced* his conclusion from his knowledge of physical space; nobody ever *induced* it' (p. 460). He concluded that 'Mr. Russell's philosophy is as complete and radical a failure' as a theory of ethics which Russell had recently abandoned (p. 462).

Logicism was discussed also in mathematical journals, especially those published by the American Mathematical Society (hereafter, 'AMS'). One of the best reviews of the first volume of *PM* was 25 pages written by J. B. Shaw in its *Bulletin*. We last met him in §4.3.2, summarising knowledge about Benjamin Peirce's linear associative algebra five years earlier; here, after reviewing some of the main techniques used, such as types, relations and descriptions, he stressed how little mathematics seemed to fall with the logicist purview, and among notable absentees he noted 'structure and form' (Peirce in mind?), 'invariance', 'functions as functions' and 'inversions' (Shaw *1912a*, 410). 'A Principia Mathematica should cover the field, or it ceases to justify its title': regretting the limited amount of mathematics covered, he noted that *PM* 'examines the rules of the great mathematical game. But it does not play the game or undertake to teach its strategy' (pp. 389, 411). Among other authors, he mentioned Kempe for the 'whole consideration of mathematical form' (p. 406).

In a sequel paper in *The monist* Shaw *1916a* continued in the same vein, concluding that 'Logistic has a right therefore to exist as an independent branch of mathematics, but it is not the Overlord of the mathematical world' (p. 414). He reprinted this review in a book of *Lectures on the philosophy of mathematics* given at the University of Illinois, published by Open Court as Shaw *1918a* after some delay (Open Court Papers, Boxes 32/16–18 *passim*). While not penetrating philosophically, he gave a nice survey of the *variety* of such concerns: he also considered possible reductions of mathematics to algorithms, algebra, and 'transmutations' such as Royce on order (§7.5.4). Other topics included form in the spirit of Kempe, and the theories of number, functions and equations.

8.3.3 *Reactions from Royce students: Sheffer and Lewis.* When Royce died in September 1916 in his early sixties, four years after William James, the Department of Philosophy at Harvard University was at a low ebb (Kuklick *1977a*, ch. 21). In June and July the Chairman J. H. Woods had corresponded with Russell about lecturing there again, but his pacifist activities prevented him from obtaining a visa (Russell *1968a*, 65–66). On

23 September, a few days after Royce's death, Woods invited Russell to edit some of Peirce's logical manuscripts which they had acquired (§8.3.1); Russell declined this obviously inappropriate proposal (RA).

Despite the languor, the best remembered American contribution to logic from this time came from a junior member of the Department, Henry Maurice Sheffer (1882–1964). His philosophical training under Royce had drawn him strongly to logic (they may have been writing a textbook together, though it was not finished) and especially to logicism. Sheffer made a tour in Europe during the winter of 1910–1911, when he met Peano and Russell, among others; and back home he published with the AMS a short paper Sheffer *1913a* on 'Boolean algebras' (the origin of this phrase), in which he showed that they could be defined solely from an operation written '|' and four laws, assuming two elements in the algebra. Then he showed that the propositional calculus could be produced with this operation applied to two propositions as their joint denial ('neither-nor', or 'nand'); he called it 'rejection'. In a footnote he appealed to duality to show that alternative denial, '*either not-p or not-q*', could play the same role; this reading has became known as the 'Sheffer stroke'.

This paper eclipsed a recent one by the Polish mathematician Edward Stamm (d. 1940), who had accomplished in *1911a* a similar reduction using 'nand' and 'or'. Sheffer seems not to have known of Stamm; and neither author could have been aware of the anticipation of 'nand' in a manuscript *m1880a* by his Harvard predecessor Peirce (§4.3.5).

This paper has left Sheffer's name on the philosophical scene; thereafter he became notorious for publishing virtually nothing. He seems to have become paranoid about the printed page, or even the hand-written one; most of his *Nachlass* consists of collections of lectures and research notes cut into small rectangles, so that the texts are very hard to reconstruct.[9] In 1919 he sent to Russell a manuscript which he claimed would 'make a great portion of Principia mathematica superfluous, meaning, I think, that his was much a much simpler method of getting your results', as his colleague R. F. A. Hoernlé reported to Russell on 28 November (RA).

Maybe this work was a version of a substantial manuscript Sheffer *m1921a* describing 'The general theory of notational relativity', in which he combined the American penchant for postulational studies (§4.7.3) with tableau-like 'grafs' showing which individuals did (not) satisfy a given two-place relation (more like the displays of Benjamin Peirce than truth-tables, which presumably he had come to know during Russell's visit). He sent copies of his manuscript to Russell and Hardy among others, but

---

[9] There are 50 boxes in confetti-ish state in the Sheffer Papers; I confess to have rummaged in only a few of them. In compensation, less mangled parts include some good correspondence with various figures, mainly after 1921 but also including a letter to Peano of 1911 while on his European trip. On his Harvard doctorate, see the documents in the University Archives, file UA V 687.235, no. 39. Michael Scanlan is studying Sheffer, confetti included.

published only a few details in a congress paper *1927a*. On July 1928 he told Heinrich Behmann (§8.7.8) that he had written 'about one-third of a book' on notational relativity before being held up by a nervous breakdown (Behmann Papers, File I 68).

An exact contemporary at Harvard and fellow student under Royce was Clarence Irving Lewis (1882–1964); during the 1910s he taught at the University of California at Berkeley. His review *1914b* of the second volume of *PM* in the *Journal of philosophy* was comprehensive; he noted details which were often missed, such as the assumption of only one individual and homogenous cardinals, and he found relation-arithmetic to be 'a miracle of patience and ingenuity' (p. 501). He admired the enterprise: 'The "Principia" is to intellect what the pyramids are to manual labor. And the "Principia" has the added wonder that the whole structure is balanced on the apex of logical constants' (p. 502).

But the main influence of *PM* on Lewis was negative: he was repelled by Russell's all-purpose use of implication, especially that 'a false proposition implies any proposition' (*PM*, $*2\cdot21$), which for him betrayed the construal of implication. In a sequence of papers from 1912 he developed an alternative logic of 'strict implication', based on impossibility as a primitive notion. 'I am quite convinced now', he told Royce on 15 October 1911, 'of the possibility of modifying the calculus of propositions so as to bring its meaning of implication into accord with that of ordinary inference and proof'.[10] He also mentioned a 'preliminary paper' *1912a* sent to *Mind*, where he introduced his version; in a succeeding paper there he described impossibility as 'intensional disjunction' (symbolised ' $\vee$ ') such as no Tuesday being a Wednesday, in contrast to the traditional extensional case (written '+': *1914a*, 241–242). He presented calculi for both strict and material implication, laying out each one American style as a system of postulates, and comparing the status of the new calculi with that of non-Euclidean geometries relative to their Euclidean parent. In a companion presentation in the *Journal of philosophy* he introduced substitution as one of his 'primitive ideas' (*1913a*, 434), thus making explicit a common assumption of symbolic logicians.

Lewis *1914c* developed his symbolism of 'The matrix algebra for implications' in *The philosophical review* by writing ' $\sim p$ ' for '[proposition] p is impossible', in contrast to ' $-p$ ' for ' p is false'. He stressed the role of semantics, indicating that five truth values were now available: these two, truth itself, and also those due to ' $-\sim p$ ' and ' $\sim -p$ '; further ones were imaginable. He worked through his systems in more detail, including rules of substitution, and closed by claiming that 'the consequences are important not only for logic, but also for epistemology and metaphysics'.

The reaction to Lewis's innovation was mixed. In *Mind* Oliver Strachey felt that strict implication belonged to 'applied' logic while Russell had

---

[10] Royce Papers, Incoming Correspondence, Box 1.

properly concerned himself with the more general 'pure' logic (Strachey *1915a*, 26–28). In the *Journal of philosophy* Wiener *1916a* took Lewis's criticism of Russell as guilty of the fallacy of denying the antecedent (that if the postulates delivered an incorrect logic, then they must be incorrect themselves); but he saw strict implication as a theory worth developing in its own right.

Lewis produced a book *1918a* of over 400 pages, published by his University as part of its semicentennial celebrations. Entitled *A survey of symbolic logic*, it included a chapter on strict implication containing his latest version. Some of the previous versions had been faulty; and this one was also to be found wanting, so that the systems with which his name is now associated came only in the 1930s (§9.4.1). When the book was reprinted in 1960 he had this chapter omitted.

Lewis then also left out the final chapter, which had reviewed the relationships between 'Symbolic logic, logistic, and mathematical method'. He contrasted in detail three 'types of logistic procedure': the 'simple' one of the Peanists, of translating mathematics into logical language; 'the hierarchic method, or the method of complete analysis, exemplified by *Principia mathematica*'; and 'the method of Kempe and Royce', dominated by order (these titles on pp. 367, 368). Assessing their strengths, he imposed his own formalist definition of a *'mathematical system'* as *'any set of strings of recognizable marks in which some of the strings are taken initially and the remainder derived from these by operations performed according to rules which are independent of any meaning assigned to the marks'* (p. 355). Perhaps he realised later the unsatisfactory nature of this characterisation, leading him to omit the chapter from the reprint.

The rest of Lewis's book may surprise: very little on mathematical logic, but a good review, partly historical, of *algebraic* logic from his hero initiator Leibniz through Boole, De Morgan and Peirce to Schröder, followed by a detailed account of the methods such as expansion theorems and elimination, and of ancillary techniques such as Euler and Venn diagrams and the Gergonne relations. All this material was reprinted, although by 1960 much of it was out of fashion; indeed, the account had been an unintentional tombstone in 1918.

Within this surprise was another one: the small space assigned to Hugh MacColl, seemingly Lewis's predecessor in modal logic (§7.3.6). However, to Lewis MacColl's procedures only 'suggest somewhat' (p. 108) his own strict implication; the differences are considerable, especially because between the two logicians lay the recognition of Peano, Frege and *PM* (Parry *1968a*).

A curious feature of the book is Lewis's use of 'class'. When he finally reached mathematical logic, he clearly explained Cantorian properties such as membership differing from inclusion (pp. 260–265); but he did not stress the differences from part-whole 'class' theory used in the earlier account of algebraic logic (especially pp. 184–189). Similarly, when he introduced quantification there, he used the Peirceian symbols 'Σ' and 'Π' and defined the quantifiers as infinite disjunctions and conjunctions respectively (p. 236); thus he admitted horizontally infinitary language without qualms, maybe without noticing. Again, his account of Kempe and Royce did not include a definition of multisets. In these respects Lewis's survey was rather undiscriminating; but his bibliography is still superb.

8.3.4 *Reactions to logicism in New York.* In January 1913 Lewis had written to Christine Ladd-Franklin, fearing that 'similar difficulties' over implication to those in Russell's logic obtained in her system, which was based on her inconsistent triads of propositions (§4.3.7).[11] He was thanking her for an offprint of her recent paper *1912a* in *The philosophical review* on 'Implication and existence in logic'. Although rarely publishing, she had continued her interest in logic, largely in the algebraic style of her master Peirce. In this paper she used her method to interpret (rather inconclusively) the case of consistency as involving possibilities and thereby existence: Russell's notions of material and formal implications were held to be over-rated in importance (pp. 642–643, 656–657).

Late in 1917, her 71st year, Ladd-Franklin delivered a series of lectures on 'symbol logic for the logician' at Columbia University in New York, where she held a lectureship. Presumably in this context, she drafted papers on Russell's logic; among a mass of notes in files sometimes called 'Bertie' (Ladd-Franklin Papers, Boxes 10 and 38) the most developed account is *m1918a?*, of 18 folios. Like Lewis, Ladd objected to Russell's all-purpose use of implication; to her the *lack* of paradox lay in the negation of the antecedent, *not* in distinguishing implication from inference. But she also disliked Lewis's alternative, on the grounds that it could not properly handle qualities.

Ladd's main disagreement with Russell concerned the arrangement between propositional functions and classes chosen for *PM*; she preferred the no-classes theory of Russell *1906a*, which she had recently read (fol. 5). As we saw in §7.4.6, Russell had developed it as the substitutional theory but had published virtually none of it; Ladd understandably misread the (small) use made of it in *PM* as the foundation of that calculus. She liked it for its avoidance of propositional functions and (as she saw it) the reinstatement of classes, despite the presence of paradoxes. She also did not grasp the significance of membership to Cantorian classes, regarding its non-transitivity as merely a fallacious use of the copula (fols. 11–12). While furnishing a nice sideways look at logicism, her paper might not have been well received if published, although perhaps the lecture course was successful.

---

[11] Ladd-Franklin Papers, Box 1; there seems to be no reply in the meagre Lewis Papers at Stanford University Archives (letters to me from Polly Armstrong).

New Yorkers also reviewed the first volume of *PM*. The first was Morris Cohen (1880–1947), a Russian-born immigrant and Harvard student of Royce (like Sheffer on both counts) and newly promoted to professor at his *alma mater*, College of the City of New York. In *The philosophical review* Cohen *1912a* admired the logicist thesis while being critical of repetitions in the presentation and sceptical of type theory. In an article *1918a* in the *Journal of philosophy* he advocated, rather lamely, the new approaches to logic, though he took logicism as an identity thesis (pp. 679–680). An interesting detail was his recalling Russell saying that 'all inference is deductive' (p. 686).

Next came Cassius Keyser, Ladd's colleague at Columbia and the victor over Russell a decade ago on the need for the axiom of infinity (§7.2.7). Then in his fifties, he wrote a praising though waffly review *1912a* in *Science* of the first volume of *PM*, showing general sympathy with logicism. More importantly, in 1917 he organised a seminar on *PM*, and excited the interest of Emil Post (1897–1954), who then wrote a most remarkable doctoral thesis which appeared in the *American journal of mathematics* as Post *1921a*. In art. 2 he presented 'truth-tables' (his name), with ' + ' and ' − ' symbolising the two values. We saw in §8.3.1 that Russell had publicised them at Harvard six years earlier; Post seems to have been independent, for he unconvincingly cited as precursors *PM* itself, and texts before that in Jevons and Venn. He also made far greater use of the tables, to prove that the propositional calculus was consistent and complete (art. 3). Further novelties came when he generalised the tables to n values, and thereby introduced many-valued logics, at least from a postulational point of view (arts. 5–7). Finally, in a 'generalisation by postulation' he presented his logics as systems of inference (arts. 8–16).

8.3.5 *Other American estimations.* In The *philosophical review* de Laguna *1916a* attacked the theory of types. After querying the status of the variable as a symbol or with some kind of referent, he considered three paradoxes. The liar held 'very little significance', since apparently it could be solved by not asserting propositions to be true (p. 23). Much more important was Russell's, and his own variant in terms of a property of a property not holding of itself: types failed because inclusion of sets was *not* 'a constant property' across types, and self-membership was quite acceptable, especially if construed in terms of properties (p. 27).

The Pennsylvanian scholars Robert P. Richardson and Edward H. Landis published an article *1915a* in *The monist* on 'Numbers, variables and Mr. Russell's philosophy' as handled in *The principles*; they seem not to have seen *PM*. Some points were silly, such as preferring the word 'group' to 'class' (p. 324). But others were perceptive; for example, Russell's confusion over 'term' and 'object' as the most general philosophical notion (p. 331), and over symbols and their (possible) referents (pp. 349–351). Taking variables to be 'things represented by symbols' (p. 350), they found

Russell's use of 'variable' to be 'in a sense peculiar to himself' in denoting 'either a *general class name* [...] or else the object represented by the distributed class name'; thus his definitions of numbers and of variables were a 'failure [...] complete and utter' (pp. 362–363). While detecting several cases of conflation in Russell, they did not appreciate that he gave variables a much wider remit than that of the ordinarily mathematical; they also failed to discuss propositional functions or quantification.

Their article was reprinted as a booklet by Open Court, who soon put out also their book Richardson and Landis *1916a* on *Fundamental conceptions of modern mathematics*. Again variables were to the fore; and this time mathematical logic gained some attention, though not happily. Frege's definition in the *Grundlagen* of 0 as 'the number which belongs to "unequal to itself"' was sloppy enough a translation (compare §4.5.3) to lead them to conclude that his 'zero is simply nothing' (p. 59); thus 'As a logician he cannot be ranked above the level of Schröder and Peano', whose 'treatises', however, were 'a hindrance rather than a help to precision of thought and speech' (p. 152). Russell was more summarily treated this time, but his definition of cardinals as classes of similar classes was 'such as though one were to define whiteness as the class of all white objects', an absurdity due to his failure, shared with Peano, 'to recognize the important distinction between equality and identity' (pp. 152–153). While again ignoring propositional functions and quantification, they gave a wide selection of examples of conflation of symbols with their referents from writings in algebras and mathematical. So their choice of hero is very surprising: 'It is a crying shame that the University of Cambridge, which has recently stood sponser [*sic*] for so many treatises of dubious value, has not yet set her press to the work of issuing an edition of the collected works of Augustus De Morgan, one of the greatest of her sons' (p. 121).

This volume was the first of 13 planned parts 'dealing with 'Algebraic Mathematics', summarised at its end; to come were most parts of mathematical analysis, Cantorian infinitude and the theory of functions—but not algebraic or mathematical logics. A vaster collaboration even than that for *PM* was envisaged, but no more parts appeared.

While no author in this sub-section made a durable contribution, they all touched nicely on issues relating to the distinction between logic and metalogic. We shall hear more and better on it from the Americans in §8.5.5.

8.3.6 *Russell's 'logical atomism' and psychology, 1917–1921.* The start of the Great War in August 1914 changed Russell's priorities completely, and he devoted most of his energies to a personal 'make love not war' campaign, with pacifist lectures and writings mixed in with a not-well-ordered series of sexual relationships. The former activity led to his dismissal from his lectureship at Trinity College Cambridge in July 1916 (Hardy *1942a*) and imprisonment two years later (§8.3.7). But during 1917 he prepared a collection of his general essays on mathematics and science, partly overlapping with that in *Philosophical essays* (§8.2.4); including *1901d* and *1907d*, it appeared as *1918a* from Longmans, Green, under the title *Mysticism and logic*.[12]

In the autumn Russell's interest in philosophy began to revive, leading to two public courses delivered between October 1917 and March 1918 at Dr. Williams' Library in London. As at Harvard three years earlier (§8.3.1), one course treated logic and the other epistemology. The latter course was recorded by a stenographer and typed up for publication, including some of the questions from the audience and Russell's answers. Thanks to Jourdain, the eight lectures appeared in four consecutive issues of *The monist* between October 1918 and the following July, around 130 pages in all (Russell *1918–1919a*).

The title of the course, 'The philosophy of logical atomism', reflected Russell's use of 'analysis' in both senses (§6.1.1), breaking a complex into its basic components and then synthetically reconstructing. As in his pre-War writing, he mulled over the (non-)existence of objective false-hoods, propositions and facts as complexes, and the relationship to all of beliefs; but he acknowledged at the head a greater influence of Wittgenstein. In particular, '*propositions are not names for facts*', since, for example, 'Socrates is dead' and 'Socrates is not dead' correspond to the same fact, one truly and the other falsely (p. 187). This finding by 'a former pupil of mine' led him to rethink his own reductionist enterprise, in which analysis (in the narrow sense) led him at one point to revive the approach of his abandoned book of 1913 (§8.2.6) by associating 'atomic propositions' with simple facts and 'molecular' ones to combinations formed from these atoms by means of logical connectives (pp. 203–208).

This procedure brought logic most explicitly to the fore in Russell's course. He included the truth-table (§8.3.1) for 'or', in a horizontal layout, probably for convenience of printing:

$$\begin{matrix} `TT & TF & FT & FF \\ T & T & T & F\text{'}; \end{matrix}$$

(836.1)

he then explained the Sheffer stroke, calling it 'incompatibility' but not mentioning its creator (pp. 209–210). He presented his own definition of (im)possibility in terms of the (non-)existence of a values satisfying a propositional function (§7.3.6); he may have had Lewis as a target, since a few pages earlier he had used a similar Tuesday/Wednesday example (pp. 231, 223). The old substitutional theory (§7.4.6) made a brief appearance, with the Socrates/Plato example again (pp. 237–239); and he exhibited the

---

[12] For the production schedule and costs of the book, see Reading University Archives, Longmans, Green Papers, Impression Book R34B. The publisher issued a short review in its house journal *Notes on books* (1918), 38.

limitations of logicism by stating the axioms of infinity and choice as cases of non-logical propositions (pp. 239–241). Definite descriptions had the sixth lecture to themselves, followed by one on type theory and classes. The last lecture dealt with 'what there is' as he saw it: as usual, he brandished Ockham's razor, and he stressed his methodology of preferring 'logical' constructions to inferred entities, much like his use of definitions in logicism to reduce assumptions.

Just as the last part appeared in *The monist*, in July 1919, Russell followed up with a lecture *1919c* to the Aristotelian Society on propositions, which were defined at the head as '*What we believe when we believe truly or falsely*'. Truth and falsehood were prominent: he adhered to the correspondence theory but dithered (as in the earlier lectures) over the status of 'negative facts', in face of a recent essay in *Mind* on 'negative propositions' by Raphael Demos (1892–1968), who had studied with Russell at Harvard in 1914. Demos *1917a* argued (rather naively) that 'not' introduced an opposition to positive propositions similar to that between truth-values. Russell, who helped in the publication of this essay, supported William James's view (which he had previously rejected) that the distinction of mind from matter lay only in the causal laws involved. Admitting that 'Logicians, so far as I know, have done very little towards explaining the nature of the relation called "meaning"' (*1919c*, 290), he transferred the problem to psychology, where he broadly followed the American psychologist J. B. Watson on behaviourism, the reductionist philosophy of psychology.

Between May 1919 and the following March Russell gave three public courses on 'the analysis of mind' at Dr. Williams' Library (Russell *Papers 9*, 477–484), and completed a book under this title which appeared from Allen and Unwin as Russell *1921a*. Thanking Watson for reading the draft, his approach largely adopted behaviourism; not wishing to reify 'consciousness' or 'introspection' (lectures 1 and 6), sense-data became non-entities, and sensations the chief human epistemic source. Truth and falsehood occupied lecture 13; as usual they were predicated of a belief, and negative falsehoods were eschewed. Notably absent from the book was logic, although most of his July lecture *1919c* ended up in it; in particular, its traditional links with psychology were abandoned.

G. E. Moore, who had chaired Russell's lecture, addressed the Aristotelian Society himself in December 1919 on 'External and internal relations', with Whitehead chairing. Arguing as previously (§6.2.4) for the former interpretation, Moore *1920a* gave the issue a welcome fresh airing; he tried to isolate entailment out of Russell's all-purpose notion of implication, and also to link up with Lewis's strict version.

8.3.7 *Russell's 'introduction' to logicism, 1918–1919.*   Russell's first London course in 1917 had been called 'Introduction to modern logic', and he planned to make a book of it also. His publisher for a year, Allen and

Unwin, had seen an announcement of the course, and on 17 October 1917 asked about a published version (RA). The writing was achieved thanks to Russell's unpatriotic pacifist behaviour, for which he was convicted to six months in prison in 1918. In the end he spent $4\frac{1}{2}$ months in Brixton, with six weeks' remission for good conduct; the privileged status of Category A allowed him time for much reading and writing (*Papers 8*, 312–328). Some of the book on mind was prepared there, but the most substantial outcome was the book on logic. Upon release Russell's manuscript was typed up by one Mrs. Kyle; on the verso of one of her bills he later recalled her as 'an admirable typist but very fat. We all agreed that she was worth her weight in gold, though that was saying a great deal' (RA).

Allen and Unwin proposed to publish it in their series 'Library of philosophy' edited by J. H. Muirhead. The book did so appear, in March 1919, as *Introduction to mathematical philosophy* (Russell *1919b*); but Muirhead inserted a note at the front, apologising for the book to those sophisticated enough to observe 'the distinction between Mathematical Philosophy and the Philosophy of Mathematics'. Despite this aggressive sales talk and relatively few reviews, it sold well enough for a new printing (misnamed 'second edition') to appear in the next year, and several more later. It was also translated into German as *1923a* (one of the translators, Emil Gumbel, praised the original version in a review *1924a* in the *Jahrbuch*), and later into French (Russell *1928a*).

In 18 short chapters and just over 200 pages Russell reviewed, in prosodic manner, all the main features of the three volumes of *PM* (geometry was avoided): definitions of integers, mathematical induction, order relations and ordinals, Cantorian transfinite arithmetic, limits and continuity, and the axioms of choice and of infinity. This latter chapter covered also type theory and the paradoxes; near the end of the book he confessed that his assumption of the existence of at least one individual (§7.9.3) was 'a defect in logical purity' (p. 203).

By contrast, logic was curiously fugitive. The propositional calculus appeared only in chapter 14; in a review of connectives Russell gave prime place to the 'incompatibility' newcomer, with Sheffer credited (p. 148). In a footnote he introduced a 'non-formal principle of inference' to sanction substitution, and lamented its omission from *PM* (p. 151). The next three chapters treated propositional functions (including his treatment of modality), definite descriptions, and classes (oddly including a more extended account of type theory, with the axiom of reducibility).

In the final chapter, on 'mathematics and logic', Russell started unhappily by stating logicism as an identity thesis instead of inclusion (p. 194). But he explained clearly the inferential character of both subjects, and reviewed the connectives. He also sketched out the notion of forms of propositions in terms of logical constants; but he spared his reader the anguish of negative falsehoods and the conundrums of belief.

Russell referred several time to Frege, introducing him as one 'who first succeeded in "logicising" mathematics' (p. 7). Apart from the mis-representation (which Russell partly rectified by explaining his own view of the role of arithmetic in mathematics), the passage is notable for the word which he put in quotation marks; but their presence suggests nervousness, and he never used the word again, so that 'logicism' did not emerge until the later 1920s (§8.7.6, §8.9.2).

8.4 REVISING LOGIC AND LOGICISM AT CAMBRIDGE, 1917–1925

8.4.1 *New Cambridge authors, 1917–1921.* In his *Introduction* Russell cited a paper by another new pupil, the Frenchman Jean Nicod (1893–1924). In October 1916 Nicod had submitted to the Cambridge Philosophical Society, through Hardy, a paper applying Sheffer's Boolean algebra (§8.3.3) to *PM*; it appeared as Nicod *1917a*. Noting that Sheffer had interpreted his operator both as alternative and as joint denial, he adopted the former for its simpler definition of implication. He also noted that it corresponded to the 'disjunctive relation' of W. E. Johnson (*1892a*, 19), although not given the same role there.

With this connective Nicod reconstructed the propositional calculus of *PM* with two 'Non-formal' axioms: that the stroke operator created a new elementary proposition '$p \mid q$' for propositions p and q; and *PM*, $*1\cdot1$ about anything implied by a true proposition being true. He added

$$\text{'Formal III. } p \mid q/r \mid t \mid t/t . \mid . s/q \mid \overline{p/s}\text{'}, \qquad (841.1)$$

where the stroke took the forms '$/$', '$\mid$' and '$\mid$' in rising order of bracketing, and the overbar indicated negation (itself defined in terms of the stroke). He worked out some of the main features of the calculus; the derivation of the axioms of *PM* was to be executed later in Quine *1932a*.

In an appendix to his paper Nicod commented on a strikingly similar one. By curious coincidence, two months before his paper arrived Hardy had communicated to the Society by one C. E. Van Horn (b. 1884) of the Baptist College in Rangoon, India, and the two were printed together. Van Horn *1920a* also used the stroke (symbolised 'Δ'), proposed the same two axioms as Nicod's non-formal duo, and added a third which however merely gave the truth-value of '$p \, \Delta \, q$' in terms of those of p and q. He derived the axioms of *PM*; however, as Nicod commented, he had not clarified the status of the law of excluded middle, and his contribution has been eclipsed.

After the War Nicod returned to France. Perhaps at Russell's suggestion, he wrote a new article *1922b* for the latest edition of the *Encyclopaedia Britannica* on 'Mathematical logic and the foundation of mathematics', a rather parochial review of Peano, Frege and *PM* without mention of, for example, Hilbert. His main concern had switched to the philosophy of science, and he prepared doctoral *thèses* at the University of Paris on geometry and the perception of space and on induction in science; they appeared in 1924, after his premature death in February. Rare cases of serious French interest in Russellian epistemology, a joint English translation soon appeared (Nicod *1930a*), to which Russell wrote a very warm prefatory tribute *1930a*.[13]

Quite different were Russell's relations with Jourdain in the late 1910s. The creeping paralysis from which Jourdain suffered seems to have crept to his brain, for his desire to prove the axioms of choice became an obsession. Russell's failure to recognise his achievement led to sad letters and bitter remarks in *The monist* and elsewhere which Russell had formally to repudiate. The tension continued after Jourdain's death in 1919 over financial issues—a very unfortunate end to a fine relationship (my *1977b*, chs. 24–25).

Involved in this sorry affair was another new follower of Russell, Dorothy Wrinch (1894–1976). In 1917 this student at Girton College travelled down to London weekly to join Nicod and also Victor Lenzen in a study group on *PM* run by Russell (Lenzen *1971a*). Later she helped Russell as secretary, setting up some of his lecture courses at Dr. Williams' Library, going regularly to Brixton prison with material that he wished to read, and negotiating about his books with Allen and Unwin. Her mathematical interest centred on transfinite arithmetic and order-types, usually deploying *PM* versions in terms of relations. For example, in *1923a* she picked up the question posed in $*124 \cdot 61$ about what properties 'mediate' cardinals, those which are defined neither inductively nor reflexively, might possess if they existed at all. She also continued Russell's application of logic (especially relations) to epistemology, with papers *1919a* on judgement and *1920a* on memory in *Mind*. Then she switched to relativity theory in the 1920s and to mathematical biology in the 1930s, although she wrote on logic and mathematics from time to time later.[14]

Finally, we note *A treatise on probability* (1921) by John Maynard Keynes (1883–1946). Much influenced by Russell (Dejnožka *1999a*, ch. 10), he took probability to measure the logical relevance between a premise and a conclusion: a notable consequence was a 'definition of *inference* distinct from *implication*, as defined by Mr. Russell', with the latter placed within probability theory (Keynes *1921a*, 117–119).

---

[13] A fresh translation appeared much later as Nicod *1969a*; Russell gave money to help publish it, supporting an anonymous donor (the poet W. H. Auden, according to letters of the time in RA from R. F. Harrod).

[14] For an excellent survey of Wrinch's careers, see Abir-Am *1987a*. Wrinch spent her later career at Smith's College in Massachusetts; her papers there contain nothing of note on logic or mathematics, although there are good letters in RA and also in the Ogden Papers (about to be used in §8.4.2).

8.4.2 *Wittgenstein's 'Abhandlung' and Tractatus, 1921–1922.* When acknowledging Wittgenstein's influence at the head of his lectures on logical atomism in 1919, Russell confessed that 'I do not know if he is alive or dead'. In fact, this combatant in the Austrian army had prepared a manuscript called 'Der Satz' in moments of tranquillity, and he showed it to Russell when they met again at The Hague in December 1919. He tried to publish it with various German houses but without success; in addition, Frege was mystified by it (*1989a*, 19–26). He sought Russell's help, and Wrinch became a key figure; after it was rejected by the Cambridge University Press (was Johnson a reader?) and by a publisher and some journals in Germany, she placed it with the *Annalen der Naturphilosophie*, edited by the chemist and zoologist Wilhelm Ostwald—an improbable venue, secured by Ostwald's high opinion of Russell and the promise of an introduction from him. The essay appeared as Wittgenstein *1921a*, entitled 'Logisch-philosophisch Abhandlung', including a German version *1921b* of Russell's introduction.

For the next stage a main role was played by C. K. Ogden (1889–1957). Of considerable wealth, by profession he was a (minor) philosopher (Gordon *1990a*); his main importance lay as editor from 1912 of the *Cambridge review*, and from 1919 the founding editor and subsidiser of the journal *Psyche*. He published it with Kegan Paul, for whom he also launched a book series in philosophy and psychology in 1921; Nicod *1930a* was to appear in it. To organise an English translation of the 'Abhandlung', he recruited another new Russell student, Frank Ramsey (to be introduced in §8.4.5), to dictate one; it took $10\frac{1}{2}$ hours. After typing out, it was revised by Ogden and Russell before despatch to Wittgenstein, who went over it line by line and suggested many revisions to Ogden, especially over the rendering of technical terms. It appeared as Wittgenstein *1922a* in Ogden's series under the title 'Tractatus logico-philosophicus' (which may have been suggested by G. E. Moore), set page by page opposite the original; Russell's introduction *1922a* had been somewhat revised (Iglesias *1977a*).[15]

Following the Peanist numbering system of *PM*, and maybe also a similar one used in the structured numbering of laws in the Austro-Hungarian Empire, the main text was organised in a sequence of short numbered clauses, where the fewer digits in the number indicated the greater importance of the text (thus 5 was important, 4.04 less so, and

---

[15] Wittgenstein's correspondence with Ogden, and a discussion of the origins of the *Tractatus*, are contained in Wittgenstein *Ogden* (1973). An important source was the papers of Ogden's solicitor Mark Haymon, whose *Nachlass* has just become available in University College London Archives: Box 4 contains Wittgensteiniana, including the rebuffs by Russell and Wrinch in the early 1960s of claims by the disciples that The Master played no role in the translation. Wittgenstein's correspondence with Russell is included in Wittgenstein *Letters* (1974), with some already in Russell *1968a*, 116–121; later finds are published in McGuinness and von Wright *1991a*. The Ogden Papers is a vast collection, including correspondence with almost everybody described in this section in alphabetically ordered Boxes.

6.1233 still less). Sometimes the disclosed order of importance is interestingly unexpected. We shall consider first Wittgenstein's conception of logic, and then the bearing of his views on mathematics; the translations are mine, as usual. Russell's introduction (in its English original) is also noted where appropriate.

8.4.3 *The limitations of Wittgenstein's logic.* From their earlier discussions Wittgenstein must have realised that Russell had mixed logic and logicism together (§8.2.6), and so he sought to characterise logic separately. In his preface he stated as an aim 'to set a limit to [...] the expression of thought: for in order to be able to set a limit to thought, we should to be able to think on both sides of this limit (thus we should be able to think, what cannot be thought)'. This passage may have stimulated Russell to counter 'that every language has, as Mr. Wittgenstein says, a structure concerning which, *in the language*, nothing can be said, but that there may be another language dealing with the structure of the first language, and having itself a new structure, and that to this hierarchy of languages there may be no limit' (*1922a*, xxii).

To Russell this proposal extended his propositional hierarchy of types (*1940a*, 62). But to us a feature of *capital* importance to philosophy was proposed here: a *general* and *fundamental* distinction of language from metalanguage, and by implication (as it were) of logic from metalogic, and of theory from metatheory. But Wittgenstein rejected it totally. 'The limits of my language marks the limits of my world' (*1922a*, clause 5.6); 'The world and life are one' (5.621), so that 'Whereof one cannot speak, thereof must one be silent' (the famous closing clause 7, which followed a Viennese philosophical tradition). Thus, as a main consequence, 'What can be shown, cannot be said' (the astonishingly minor 4.1212). Similarly, 'Logic fills out the world; the limits of the world are also its limits' (5.61), so that 'It is clear: the logical laws may not themselves fall again under logical laws' (6.123). Likewise, 'Philosophy should set limits to the thinkable and thereby to the unthinkable' (4.113–114).

Wittgenstein was a metaphysical monist, believing all physical and mental entities to be unified; so that 'There is no thinking, representing subject' (5.631) with his own private mental products (Cornish *1998a*, ch. 5). (Such passages may have pleased Ostwald, who was a monist for science.) Thus, as a special case, he affirmed Russell's logical monism, affirming the all-embracing generality of logic, in its bivalent form. Now Russell had just refuted this stance with his hierarchy, but did not recognise the gold in his hands: although he mentioned it in some later writings, he never gave it a major place (§10.2.3). In particular, it played no role in his revision of *PM*, prepared only two years later (§8.4.4). Those equipped with metalanguage and -logic know better, that speaking then comes into its own. For example, Wittgenstein's distinction between showing and saying is itself said; so where is *it*?

Wittgenstein's monistic way ahead was to base his epistemology on 'facts' ('*Tatsachen*', 1.1) which showed 'the existence of states of affairs' ('*Sachverhalten*', 2). He then proposed a rather naive metaphysics founded upon the notion of a picture ('*Bild*') theory of the meaning of a proposition relative to a state of affairs, true *or* false when corresponding to a correct *or* an incorrect one (2.21–2.225), and involving 'positive' or 'negative facts' respectively (2.06). Thus he adopted Russell's position before 1906 of admitting objective falsehoods (§7.4.7), in the form of admitting possible as well as actual states of affairs (Kreisel *1968a*).

The link between a picture and reality was the 'logical form' (2.18), which can depict the world' (2.19), allowing logic to be 'a mirror-image of the world' (6.13). 'Each statement about complexes can be resolved into a statement about its components and into the propositions that the complexes completely describe' (2.0201, 'complex' unexplained). Thus the link was like an isomorphism, down to 'An elementary proposition[, which] consists of names' (the unbelievable 4.22) but which nevertheless 'asserts the existence of a state of affairs' (4.21).

The corresponding atomicity of propositions was handled in terms of 'proposition-signs', the truth-tables thought out by Wittgenstein with Russell around 1912 (§8.3.1), which treated a compound proposition as a truth-functions of its components (4.31, 5.54). His example connective was implication (4.442), in both a tabular form and a horizontal one like Russell's (836.1):

p	q	
T	T	T
F	T	T
T	F'	
F	F	T

'(TT-T) (p, q)',

[...] or, more clearly,   (843.1)

'(TTFT) (p, q)'.

In addition, assertion was abandoned (4.124, 6.2322).

From such tables Wittgenstein took the 'two extreme cases': where it is 'true for the entire possibilities of truth value of the [component] elementary propositions', when it was 't a u t o l o g i c a l' and the converse cases of non-stop falsehood, when it was 'c o n t r a d i c t o r y' (4.46). Moreover, 'There is one and only one complete analysis of the proposition' (3.25). His view resembles Russell's logical atomism (§8.3.6), but he must have thought it out independently in the trenches, following the bipolarity of propositions which he had suggested to Jourdain back in 1909 (§8.2.6): 'Tautologies and contradictions lack sense. (Like a point from which two arrows go out in opposite directions to one another.)' (4.461); they are limiting cases' of the logical 'combination of signs' denoting the constituents (4.466). Now he could characterise logic thus: 'the propositions of logic are tautologies' and 'say nothing. (They are the analytic propositions.)' (6.1–6.11). Both cases 'are not pictures of reality. They do not represent any possible situations' (4.462), so that logic was detached from reference, and tautology and contradiction lost their more fruitful connotations of being respectively true or false by virtue of the meanings of their constituents. Moreover, his doctrine was applied only to the propositional calculus.

Under this regime Wittgenstein devoted a whole section to the consequences of the principle that 'The proposition is a truth-function of elementary propositions' (5), whether tautological, contradictory, or in between. He used the horizontal layout in (836.1) to present all 16 possible connectives between two propositions (5.101), and went through the usual ones. He stressed that the proposition was prior to any logical combination that it may contain: 'The occurrence of an operation [or connective] does not characterise the sense of the proposition' (5.25), so that any two logically equivalent propositions were identical. In particular, 'The operation can vanish', as in the case ' $\sim \sim p = p$' (5.254)—unhappily written, given his abandonment of identity.

Wittgenstein treated the predicate calculus in terms of 'formal concepts' (4.126), but rather briefly. 'The existence of an internal relation between possible situations expresses itself linguistically by means of an internal relation between the propositions representing them' (4.15): a claimed answer to the internalist/externalist question (4.1251) which however only transfers it. On quantification, 'I disassociate the concept all from truth-functions', but he falsely asserted that 'Frege and Russell have introduced generality in association with the logical product or the logical sum' (5.521).[16]

So logic was trivialised to a mere algorithm, though now specified without reference to logicism; but what about mathematics? While allegedly 'Mathematics is a logical method' (6.2) or 'a method of logic' (6.234), 'A proposition of mathematics does not express a thought' (6.21), the latter being 'A logical picture of facts' (3). In such ways did Wittgenstein reject logicism. But he also seems to have abandoned mathematics (or forgotten what he knew when an engineer) when characterising its propositions as 'equations, and therefore pseudo-propositions' (6.2) and decreeing that 'The theory of classes is quite superfluous in mathematics' (6.031).

On *PM*, Wittgenstein rejected the theory of types, since Russell 'had to mention the referent of signs when [making] the line-up of the sign-rules' (3.331); similarly, though in a weak analogy, Russell's paradox was solved by preventing a propositional function from being an argument of itself (3.333). The axiom of reducibility was also banished because 'It is possible to imagine a world in which [it] is not valid'; however, the underlying

---

[16] According to possibly faulty notes taken by G. E. Moore (*1955a*, 297), Wittgenstein had read quantifiers this infinitary way in the *Tractatus* but later rejected it because 'grammar' was needed to grasp their ranges of values.

reason was that 'logic has nothing to do with the question of whether our world really is so or not' (6.1233).

Another casualty was Russell's definition (732.3) of identity: 'not satisfactory; because according to it one cannot say, that two objects have all properties in common. (Even if this proposition were never correct, it still has s e n s e.)' (5.5302). Presumably universal classes, defined by Russell via self-identity (§7.8.5), also had to go. However, the concept of 'numerical equality' survived, as 'the general form of all special cases' (6.022). This "definition" is hardly lucid, but presumably as a non-logicist Wittgenstein did not have treat equality as a derivative of identity; after all, 'there are no numbers in logic' (5.453).

A further victim of Wittgenstein's abandonment of identity was the axiom of infinity, which 'is intended to express in language, that there be infinitely many names with different referents' (5.535). But he did not explore this possibility as a philosophical issue; thus, while it is too much to say that his 'whole theory overlooks the distinction between finite ranges and infinite ranges and is, therefore, quite irrelevant to the foundations of mathematics' (Wang *1968a*, 21), its utility is very limited. While accepting the attack on identity, Russell had queried this feature in August 1919 when reading the manuscript, in connection with Wittgenstein's definition of only finite ordinals (6.03) (McGuinness and von Wright *1991a*, 108), and he raised it again in his introduction (Russell *1922a*, xx).

After the publication of the *Tractatus*, Wittgenstein rejected the possibility of an academic career and became a school-teacher in Lower Austria. Russell soon become very disappointed with his star. 'He was very good', he told Sheffer perhaps in 1923,

> but the War turned him into a mystic, and he is now quite stupid. I suspect that good food would revive his brain, but he gave away all his money, and won't accept charity. So he is an elementary schoolmaster and starves. I do not believe his main thesis; I escape from it by a hierarchy of languages. He wrote his book during the War, while he was at the front; hence perhaps his dogmatism, which had to compete with the dogmatism of bullets.[17]

Nevertheless, Wittgenstein's thought was much on his mind when he came to revise *PM*.

8.4.4 *Towards extensional logicism: Russell's revision of* Principia mathematica, *1923–1924*. After the popular *Introduction* of 1919 Russell worked little on logic. In October 1920 he recommended his lectureship in logic and the principles of mathematics at Trinity College Cambridge. In

---

[17] The original of this letter is lost; Sheffer included this passage in a letter of 27 October 1923 to Hoernlé (Sheffer Papers, Correspondence Box; copy in RA). In a letter of 16 May 1960 to A. Shalom, Russell defended his reading of the *Tractatus*, stressing its syntactical side and playing down the metaphysics (RA).

1921 he visited China and lectured on logic and set theory among other topics; his *Introduction* was translated into Chinese the following year. He was accompanied by Dora Black, to whom he had been introduced by Wrinch. Later in the year they married, and he resigned his lectureship. This marriage yielded children; in January 1925 he asked Ogden to be the guardian of his children and executor of his will should he and Dora die (RA). His main interest switched to education, although he maintained a stream of philosophical writing and reviewing.

One occasion to consider logic arose when Muirhead asked Russell to wax autobiographical in a survey of *Contemporary British philosophy* under his editorship. Russell responded with a summary *1924a* of 'Logical atomism'. After characterising logicism in terms of 'pure mathematics' in the manner of *The principles* rather than *PM* (p. 325), he turned to Ockham's razor, a 'very important heuristic maxim', expressed as 'Wherever possible, substitute constructions out of known entities for inferences to unknown entities' (p. 326), and gave examples of its use in logicism as well as in epistemology (for example, defining integers as classes of classes, definite descriptions, and Whitehead's construction of points). Unusually for him and showing the impact of Wittgenstein (as in other writings of the time), he stressed the importance of language in philosophy, reviewing in some detail syntax, relations and paradoxes (pp. 330–336). He was much more positive than usual about type theory: it had led him to 'a more compete and radical atomism than any that I conceived to be possible twenty years ago' (p. 333), and he characterised 'an ideal logical language' as not only avoiding paradoxes but preventing Ockham-like illegitimate inferences 'from the nature of language to the nature of the world' (p. 338). The essay is a beautiful cameo of Russell's life's ambitions in logic and philosophy to date. An early sentence set his scene perfectly: 'I hold that logic is what is fundamental to philosophy, and that schools should be characterized rather by their logic than by their metaphysic' (p. 323).

With *PM* now out of print, Russell agreed with Cambridge University Press to prepare a new edition. Most of the work was done during the summers of 1923 and 1924 at his summer home near Land's End in Cornwall. The outcome was a new introduction *1925a* of 34 pages in the first volume, and three new numbers reworking certain theories at its end. (Two short manuscript notes are printed in *Papers 9*, 155–160.) Another addition was an index of symbolic definitions; maybe Wrinch prepared it, for she had so offered in an undated note to Russell (RA). Whitehead was initially interested in the venture; on 29 June 1923 he promised to send some notes, including on the 'various meaning of "function"' in the second volume, 'entirely due to the niggly criticisms of Johnson' (§7.8.1) as the Press reader (RA, Dora Russell Papers). However, in the end he played no role; when he found no mention of this fact in the new first volume, he sent a testy note *1926a* to *Mind*. Nevertheless, Russell still referred to 'we' in later reminiscence! (*1959a*, 122).

The first two volumes appeared in 1925 and 1927; for some reason their texts were reset by Cambridge University Press, at about 4% more pages. Some typographical errors were corrected, but who knows if changes in text are not buried in odd places?[18] Sadly, not I, though some random checking did not reveal any changes. Mercifully the third volume was photo-reprinted for publication in 1927.

A major alteration, stimulated by Wittgenstein but which Russell had himself been following, was to adopt an extensional view of logical notions. The old idea of elementary propositions was now replaced by 'atomic' ones, which might contain individuals and propositional functions and relations but not quantifiers; 'molecular' ones were constructed as usual by linking up atoms with logical connectives (*PM*$_2$ *1*, xv–xix). However, he did not refer to tautologies, doubtless feeling that mathematics could not be grounded in them; nor did he present truth-tables.

The new Appendix C, on 'Truth-functions and others', was based upon such functions f of propositions p and q preserving equivalence:

$$\text{`} p \equiv q \,.\, \supset \,.\, fp \equiv fq \text{'}. \text{ In consequence, `} p \equiv q \,.\, \supset \,.\, p = q \text{'}, \quad (844.1)$$

so that identity was still present, although some of Russell's uses of ' = ' seem to be equality by definition (see especially pp. xxxiii–xxxviii). Thus he had dropped Wittgenstein's abandonment of identity, accepted in his introduction to the *Tractatus* (Russell *1922a*, xvii); doubtless he realised meantime that Wittgenstein had not handled numerical equality well.

Russell also doubted that belief predicates could be satisfactorily handled. Concerning the status of judgement, he also noted 'the difference between propositions considered factually and propositions as vehicles of truth and falsehood', according to which apparently 'The paradoxes rest on the confusion between factual and assertive propositions' (pp. 664, 666). This was his closest approach to his new idea of a hierarchy of languages each talking about the one(s) below. However, he explicitly stressed again '*implication* (as opposed to *inference*)' (p. xxiv; note also p. xxviii).

Among the connectives, the Sheffer stroke was given prime place, along with Nicod's single axiom (841.1); the *modus ponens* role of inference was suitably restated, and substitution also allowed for (pp. xv–xix). Russell hailed these two innovations as 'the most definite improvement [. . .] in mathematical logic' since the first edition (p. xiii), which shows how out of general touch he had become; even Wiener's definition (827.1) of the ordered pair was omitted. However, in a bibliography at the end of the

---

[18] In July 1923, on completing four years of study at Göttingen University, the Hungarian logician Alfred Boskowitz (1897–1945) sent to Russell a detailed list of errors and possible changes for the first edition (RA; see also Behmann Papers, File I 08); it seems that only the former category was acted upon. Boskowitz turned to a career in banking in Hungary (information from C. Thiel and P. Boscowitz).

introduction he did include some weightier literature, for example by Brouwer and Hilbert (§8.7.4, 7).

The predicate calculus was reconstructed similarly; a propositional function without quantifiers taking elementary propositions as values was still called a 'matrix' (p. xxii). Concerning quantification (which, presumably following Wittgenstein, Russell unusually named 'generalisation'), the quantifiers were even defined as infinite conjunctions and disjunctions (p. xxxiii), with no concern over the horizontally infinitary logic invoked. Again after Wittgenstein, 'there is no need of the distinction between real and apparent variables' or of assertion (p. xiii); yet one page later, when subsuming propositional quantification under functional and individual quantification, 'in place of "⊢ . (p) . fp" we have "⊢ . (ϕ, x) . $f(\phi x)$"'!

Propositional functions of functions were also presented extensionally:

$$\text{`}\phi x \equiv_x \psi x . \supset . f(\phi \hat{z}) \equiv f(\psi \hat{z})\text{'} \tag{844.2}$$

(p. xxxix). Russell wrote a new ∗8, as Appendix A, to replace the old ∗9 on the basic features of the predicate calculus. In addition, from this premise '$\phi \hat{x} = \psi \hat{x}$' (identity again), so that 'there is no longer any reason to distinguish between functions and classes' (p. xxxix). Further, the deployment of truth-functions dispensed with the troublesome axiom of reducibility. However, it was still necessary to distinguish the orders of classes which might have members from the same order, a point illustrated by a nice discussion of examples such as Zermelo's proof of the Schröder-Bernstein Theorem (§7.8.7) and mathematical induction (pp. xxxix–xliii). He reworked the latter, together with the ancestral relation, in Appendix B as the new ∗89; however, the theory faltered in the proof of a theorem about ancestral relations (∗89·29), seemingly irreparably (Myhill *1974a*). Further, several parts of mathematics, such as the theory of real numbers, were not captured anyway, so that a replacement axiom of some kind was needed (p. xlv).

As this summary suggests, the second edition is hardly a philosophical advance upon the first, at least for clarity; for example, as before, it is still uncertain whether or not the calculus is consistent (Fitch *1974a*). The influence of Wittgenstein was in some ways unfortunate: he was not a logicist, but in adopting some of his positions Russell had now to assume that mathematical propositions said nothing about the world—pure mathematics, indeed, but in the traditional sense rather than the implicational form of *The principles*. Again, the emphasis on extensionality hardly fits well with a logicism in which, for example, non-denumerability is central. But a trend was now in place, and would soon be taken still further.

8.4.5 *Ramsey's entry into logic and philosophy, 1920–1923.* Russell was helped in the proof-reading of the first two volumes of *PM* by a new star in his circle, Frank Ramsey (1903–1930). Son of the applied mathematician

A. S. Ramsey, then President of Magdalene College Cambridge, Frank centred his short life on the University, entering Trinity as an undergraduate, moving to King's as a Fellow in 1924, and holding a university lectureship two years later until a perforated ulcer in his fat frame took him away. A selected edition of his writings soon appeared in 1931 (Ramsey *Essays*), in Ogden's series; for some reason two more and different selections appeared in 1978 and 1980. He left an interesting *Nachlass* (Ramsey Papers), from which a nice selection, largely centred on probability theory, was published in 1991 (Ramsey *Notes*).

Ramsey's other interests lay in logic(ism), Russell-style philosophy and mathematical economics. He seem to have come to this unusual quartet early on: letters of 1920 sent from school to Ogden (a Fellow of his father's college) deal with them all. Concerning Russell's work, he reported on 17 June reading the *Introduction*, and a month later the German translation *1908a* of Couturat's book on Russellian mathematical logic (§7.4.1), which 'is always referring to Russell's Principles of Mathematics vol 1 which I hadn't heard of before' (Ogden Papers, Box 111, File 1).

Ramsey published first on logic and philosophy in book reviews, mostly in *Mind*. In a long piece *1923a* on the *Tractatus* he was largely descriptive, emphasising the picture theory of meaning; he also tried to understand the mystical passages. But he doubted that Wittgenstein's rejection of identity was adequate for mathematics, and also the decree that mathematics consisted of equations (pp. 279–280). In addition, he disagreed with Russell's construal of the book as concerned primarily with a logically perfect language (p. 270); curiously, he passed over Russell's proposal of a hierarchy of languages. On the first volume of the new *PM* he wrote two similar short reviews, *1925a* in *Nature* and *1925b* in *Mind*, picking out the Sheffer stroke and Nicod's axiom, and the attempt to eliminate the axiom of reducibility; he judged Appendix C on truth-functions to be the most interesting innovation.

Privately Ramsey was much more critical of Russell. '[I] am reading the manuscript of the new stuff he is putting into the Principia', he wrote to Wittgenstein on 20 February 1924 (Wittgenstein *Ogden*, 84),

> You are quite right that it is of no importance; all it really amounts to is a clever proof of mathematical induction without using the axiom of reducibility. There are no fundamental changes, identity is just as it used to be. I felt he was too old: he seemed to understand and say 'yes' to each separate thing, but it made no impression so that 3 minutes afterwards he talked to me on his old lines.

8.4.6 *Ramsey's recasting of the theory of types, 1926.*   The major revision of theory was published in two papers: a lengthy account *1926b* of 'The foundations of mathematics' with the London Mathematical Society, and a shorter version *1926a* for the *Mathematical gazette*. In the latter piece he stated very clearly the need to unwrap Russell's logic from its logicism (p. 75):

> When Mr Russell first said that mathematics could be reduced to logic, his view of logic was that it consisted of all absolutely true general propositions, propositions, that is, which contained no material (as opposed to logical) constants. Later he abandoned this view, because it was clear that some further characteristic besides generality was required. [. . .]
>
> If, then, we are to understand what logic, and so on Mr Russell's view mathematics is, we must try to define this further characteristic which may be vaguely called necessity, or from another point of view tautology.

The details were presented in Ramsey's main paper. His commitment to extensionality carried through to the extent that he used Wittgenstein's theory of truth-tables for an infinity of argument places, adding as weak explanation that 'an infinite algebraic sum is not really a sum at all, but a *limit,* and so cannot be treated as a sum except subject to certain restrictions' (*1926b,* 7). Similarly, he followed Russell's new interpretation (844.2) of quantification of propositional functions in terms of infinite conjunction and disjunction (p. 8).

Sceptical of always subsuming classes under propositional functions, Ramsey mooted the option of restricting the action to specific kinds: 'although an infinite indefinable class cannot be mentioned by itself, it is nevertheless involved in any statement beginning "All classes" or "There is a class such that", and if indefinable classes are excluded the meaning of all such statements will be fundamentally altered' (p. 22). He emphasised here the difference between restricting classes or functions to particular kinds or allowing *all* possible kinds to be quantifiable; it has become known as the distinction between 'non-standard' and 'standard' interpretations (Hintikka *1995b*). Distinguished predecessors include Kronecker disliking talk of "any" function (§3.6.5), and discussants of the legitimacy of axioms of choice (§7.2.5–6).

Next Ramsey reconstructed the theory of types. Adopting Russell's definitions of 'elementary proposition' and 'elementary function', he gave Russell's old term 'predicative function' a new reference. Starting out from atomic functions, a 'predicative function of individuals' was a truth function of a (finite of infinite) number of atomic functions of individuals or of propositions. Similarly, a 'predicative function of predicative functions of individuals and of individuals' was such a function of propositions or of 'atomic functions of functions of individuals and of individuals' (which had only one functional but many individual arguments), and so on. The value of a function was a proposition of the corresponding type. Any function occurred in a predicative function through its values, and quantification was interpreted in terms of truth-functions of an infinity of arguments; thus *every* function was predicative in Ramsey's sense, so that the axiom of reducibility was not needed (*1926b,* 38–42, 46–47; compare Quine *1936b*).

Ramsey also criticised the vicious circle principle for its denial of harmless descriptions presupposing a totality, such as 'a man as the tallest in a group' (p. 41). In the old *PM* this principle had prohibited any value taken by an intensionally specified class from presupposing that intension; but with functions now defined extensionally, their values could be defined separately and thus involve objects from higher types. Thus he also abandoned this principle. Since he preserved the theory of types, he might have concluded that the principle was independent of it (compare §9.3.5, §10.2.5).

Ramsey's new theory of types yielded a calculus which hopefully was equivalent to that of the new *PM* and therefore avoided the paradoxes. Yet his best remembered suggestion entailed that part of his reconstruction was unnecessary anyway. For he pointed out that the paradoxes divided by subject matter into two groups which we now call 'mathematical' and 'semantic', to be solved respectively by the 'simple' and the 'ramified' theories of types. (Ramsey used no names: on their origins, see §8.8.4 and §9.4.5.) In the first group of paradoxes he listed Russell's, its relation version, and Burali-Forti's, formulated in terms of classes; they were solved by the simple theory of types. But within each type the paradoxes of the second group were constructible—he listed the liar, Berry, the least indefinable ordinal (König), Richard and Grelling—and they needed the ramified theory.[19] While not distinguishing paradoxes such as the liar involving truth-values of propositions from those of definability based upon properties of sentences in a language, he noted that their mathematical character was contaminated by linguistic or semantic elements; but he regarded the mathematical element as sufficiently significant to proceed with his complete restructuring of the theory of types. But later opinion has sided with the remark in Peano *1906b* that they had no place in logic (§7.2.4); and it is surprising that Ramsey did not draw the same conclusion, for he cited Peano here, and also recognised the distinction for the two roles for logic involved: the paradoxes 'would not be relevant to mathematics or to logic, if by "logic" we mean a symbolic system, though of course they would be relevant to logic in the sense of the analysis of thought' (*1926b*, 21). We return to this matter in §10.2.2, among the general conclusions.

8.4.7 *Ramsey on identity and comprehensive extensionality.* Perhaps Ramsey pursued his full restructuring because he saw for it a role greater than merely avoiding (or Solving) some paradoxes; for another of his proposals concerned the definition of identity. While not adopting Wittgenstein's rejection of it, he did accept the view that two objects being identical is nonsense and one object being identical is vacuous (p. 50). So

---

[19] Ramsey *1926b*, 20, 42–46; compare Ramsey Papers, 2–20 to 2–22, 5–12. He attributed Grelling's paradox to Weyl *1918b* (§8.7.7).

he defined the identity of terms x and y as a *non-predicative* combination of a predicative contradiction and a predicative tautology (pp. 51–52):

$$\text{For } x \neq y, \; x = y := . \,\text{'}(\exists \phi).\phi x.\sim \phi x : (\exists \phi).\phi y.\sim \phi y\text{' Df.;} \quad (847.1)$$

$$\text{For } x = y, \; x = y := . \,\text{'}(\phi):.\phi x. \vee .\sim \phi x : (\phi).\phi y. \vee .\sim \phi y\text{' Df.}$$
$$(847.2)$$

This need for non-predicative functions also led Ramsey 'to drop altogether the notion that ϕa says about a what ϕb says about b; to treat propositional functions like mathematical functions, that is extensionalize them completely' (p. 52). He called his new primitive notion a 'function in extension', symbolised 'ϕ_e'; under it and the interpretation of quantification the Leibnizian form of identity

$$x = y := . \,\text{'}(\phi_e).\phi_e x \equiv \phi_e y\text{'} \quad (847.3)$$

was acceptable, for it covered all possible associations of proposition and individual and so would be a tautology if x were identical with y and a contradiction otherwise (pp. 52–53).

Classes were subordinated to functions in extension. Among their roles, their identity sufficed for a condition of identity in higher types, so that only predicative functions of classes (and therefore predicative functions of functions) were required there (p. 54). In addition, the multiplicative axiom was legitimised, since no difficulty arose in positing a function in extension to deliver the choice class (pp. 57–59).

Ramsey ended by revising the axiom of infinity.[20] Following Russell's empirical interpretation of it in *PM* (§7.9.3), and also Wittgenstein's retort that it must be a tautology or a contradiction, he proposed that any proposition of the form 'there are at least n individuals' be taken as one about diversity, starting out from '$(\exists x, y).x \neq y$' for $n = 2$. As n increased, either such propositions would become contradictory when some

---

[20] In the Ramsey Papers there are some manuscripts relevant to this axiom and related issues, seemingly concurrent: they are published in *Notes*, especially pp. 131–148, 178–194. Wittgenstein rejected Ramsey's definition (847.1) of identity, largely on the grounds that the negation of nonsense was also nonsense, rendering unacceptable the switch between tautology and contradiction (see, for example, Carnap's notes of a discussion in 1927 between Wittgenstein and Ramsey in Carnap Papers, 102-77-01). A late paper Ramsey *1930a* on the partition of infinite and finite classes was to stimulate much mathematical research in combinatorics and some aspects of proof theory.

Another mathematician with Trinity connections was H. T. J. Norton (1886–1937), who knew Russell (and Hardy) around the time of his doctorate in 1910. Perhaps in the late 1920s, he wrote several manuscripts on aspects of PM_2 such as Nicod's axiom, quantification and the axiom of reducibility; unpublished at the time, they do not seem worthy of revival now (University College London Archives, London Mathematical Society Papers, Norton Manuscripts).

finite value N of n was reached, in which case there were only N individuals; or tautologies would continue to arise up to $\aleph_0$ or even $\aleph_1$ individuals. While it may never be known which of the two cases obtained, the axiom had a logical character (pp. 59–61).

What Ramsey meant by 'necessity' in the opening quotation above was a total commitment to extensionality, especially as represented by predicative functions in extension, enlarging Wittgenstein's characterisation of logic in terms of tautologies. This extensionality resembled Frege's system in some respects (Chihara *1980a*). Ramsey did specify logic independently of the logicist thesis (which he upheld); and moreover, as he noted, it used functions in a way commonly attributed to mathematics anyway. However, conceptual misgivings arise, especially concerning infinitary language, and the handling of non-denumerable classes and their defining functions. At the end of his shorter paper Ramsey confessed to some disquiets, although they concerned the status of certain axioms rather than his general strategy (*1926a*, 79–81). But it launched an enterprise of great and original promise, whose completion was prevented by his death a few years later.

After the mid 1920s Ramsey concentrated upon probability theory and philosophical questions, the latter largely within Russell's approach. For example, in 1927 he and G. E. Moore conducted a symposium at the Aristotelian Society on 'Facts and propositions'. Preferring Russell's view of judgement as a multiple relation, he suggested that believing proposition p and disbelieving not-p were equivalent judgements (Ramsey *1927a*, 147–148). But his paper was not very original or even well thought out, as Moore *1927a* gently indicated in an intricate analysis of the distinction between judging a fact and the fact that one is making (or not) such a judgement at some given place and time.

When the posthumous edition of Ramsey *Essays* appeared in 1931, Russell *1931a* reviewed it at length in *Mind*. Lamenting the early death of his student, he accepted two of Ramsey's criticisms of logicism: always subsuming classes under propositional functions, and not distinguishing between two kinds of paradox. But he did not discuss Ramsey's hint of removing the semantic ones from logic; his own conception of logic as completely general would have prohibited such a move. Further, he was doubtful about the revisions of identity, pointing out that 'take *two* things, a and b' already takes diversity as primitive; he preferred to read it as saying that a and b 'have different properties', which does not solve this difficulty. Much later and against Wittgenstein, he went further, treating it as a primitive notion (Russell *1959a*, 115).

8.5 LOGICISM AND EPISTEMOLOGY IN BRITAIN AND AMERICA, 1921–1930

8.5.1 *Johnson on logic, 1921–1924.* In §2.6.4 we noted Johnson *1892a* in *Mind* as a significant paper for diffusing algebraic logic. Over the years

he lectured on logic regularly at Cambridge, and often read manuscripts of other logicians (for example from §7.8.1, of *PM* for the University Press), but he published rarely. His *Logic* finally appeared from the Press at about 700 pages in three Parts in 1921, 1922 and 1924, when he was in his mid sixties. While he still linked logic with laws of reasoning and stressed the importance of syllogistic logic, he also noted the symbolic traditions; a few features are noted here.

In the first Part Johnson *1921a* saw the difference between mathematics and logic as lying in that 'between form and matter' (pp. xxii–xxiv)—another traditional distinction (§2.4.4). But pure mathematics was an application of 'pure logic', with applied mathematics marked out from both by its reference to reality (p. xxv). Finite classes were handled in the part-whole way in ch. 8, with a ponderous explanation of parts of parts of.... The only mention of *PM* in this Part was a disagreement over existence: for him the definition there of non-emptiness in terms of existential quantification should be reversed (p. 172), although he did not really handle quantifiers anywhere. He noted, with approval, Russell's theory of definite descriptions, but in the context of identity and substitution rather than referentiability (p. 198).

Johnson had lectured on mathematical logic in the winter of 1916–1917, with G. E. Moore present (Moore Papers, 10/5/1), and he engaged with it most closely in the second Part. The start was unhappy, for he connected 'the mental act of inference and the relation of implication' as 'analogous to that between assertion and the proposition' (Johnson *1922a*, 1), which is no better than Russell himself. He proposed another reversal from Russell's procedure, that propositional functions be derived from descriptive ones (pp. 66–68)—or rather from mathematical functions in general. However, he thereby masked one consideration underlying Russell's logicism, which he did not discuss. But later on he quite nicely distinguished Boole's algebraic logic from the work of Peano, Cantor and *PM* (pp. 135–138); not a logicist of any kind, he felt that 'The current phrase *mathematical logic* is ambiguous inasmuch as it may be understood to mean either the logic of mathematics or the mathematics of logic' (p. 137). He even described some aspects of transfinite arithmetic (pp. 173–180), although he misunderstood the continuum hypothesis as 'the number [of points in a line] may be assumed to be $2\exp\aleph$' (p. 174). He also reworked Russell's principle of abstraction for symmetrical and transitive relations (§6.4.4), but by an odd method stressing adjectives (pp. 145–150). In a survey of methods in syllogistic logic he described the antilogism (pp. 78–79) without mentioning Ladd-Franklin (§4.3.7), to which she objected bitterly in a note *1928a* in *Mind*.

Johnson's third Part was largely taken up with inductive logic and the philosophy of science, especially causality. However, in the introduction he pointed to identity as a mark of 'transition from pre-mathematical logic to mathematical logic' (*1924a*, xv).

Johnson's book has many nice touches; but its view of logic is a heterogeneous mixture of old and new, with no clear direction emerging either technically or philosophically. Ramsey *1922a* reviewed the second Part very warmly, but was sad that Johnson had not really engaged in some of his points of disagreement with Russell (for example, the existence of classes in terms of that of individuals). Although the book was used for years in the basic logic course at Cambridge, most logicians and mathematicians, such as Russell and Ramsey, largely ignored it in their own work.

8.5.2 *Other Cambridge authors, 1923–1929.*   One newcomer was the mathematician Max Newman (1897–1984). For his fellowship thesis *m1923a* at Saint John's College he chose 'The foundations of mathematics from the standpoint of physics', and presented an axiom system broadly similar to that of *PM* but with a more subjective cast in treating logic as an investigation of means whereby 'acts of belief may be rendered possible' (fol. 15). Treating such as acts as 'performances', he denoted the corresponding beliefs by 'performance functions', and regarded as a sequence of them as a 'process', starting out from 'primitive performances' which 'are believed possible without any proofs' (fol. 27). Many of his notations were based upon those of *PM*, and among notions he used ordinal similarity (795.1) between two processes (fol. 49). Aware of other current work on foundations, he adopted Hilbert's version $(874.1)_1$ of the axioms of choice as a 'selection function' (fol. 40). Among 'mathematical processes' he defined the integers in a manner akin to Peano's axioms, referring the reader to the *Formulaire* for full details about the real-number system; and he presented long division as an example of a process which he could express (fols. 85, 68). As for the physics promised in his title, he treated the measurement of quantities on scales (fols. 102–120), but little else.

It is a pity that Newman published none of this essay, which bears some similarity to Carnap's later formalised epistemology (§8.9.4); but some years later he assisted in Russell's latest epistemological enterprise, a study of *The analysis of matter* which appeared in Ogden's series as Russell *1927a*.[21] As usual, Russell replaced inferred entities by logical constructions, but he also paid attention to recent developments such as relativity theory and quantum mechanics. Logic and set theory were rather more prominent than in the book on mind for Muirhead (§8.3.6); in particular and thanks to Newman's advice, in ch. 28 he presented recent developments in topology, including some due to Felix Hausdorff (§8.7.6), and new definitions of dimension (on which see D. Johnson *1981a*). In ch. 24 he even drew on his idea of relation-number (§7.9.5) to characterise similar spatio-temporal 'structures' for which '*all their logical properties are identi-*

---

[21] In 1927 Russell also published a popular *Outline of philosophy* (entitled 'Philosophy' in the U.S.A.) in which logicism was not mentioned, and no mathematics seriously addressed.

cal'. Newman himself sent to *Mind* an adumbration *1928a* of Russell's ch. 20 on the 'causal theory of perception'. Russell thanked him for the essay on 24 April, accepting the criticism on aspects of his theory of perception (*1968a*, 176–177). Thereafter, like Russell, his interest in foundations decreased; it also changed to metamathematics and set theory (§9.5.3).

In the front material of Russell's book the publishers had listed a forthcoming volume from Hardy on 'mathematics for philosophers' for the series; but it never appeared, and none of it seems to survive. Then in his early fifties and professor at Oxford University, in December 1928 he delivered at Cambridge a lecture on 'mathematical proof' which appeared in *Mind* as *1929a*. Complaining about the dubious three axioms in logicism, he welcomed Wittgenstein's revision of Russell's theory of judgement (§8.2.6), which for him had lacked the feel for correspondence which a decent theory of truth should possess.

Ogden himself and his younger Magdalene colleague I. A. Richards (1893–1979) had published as *1923a* in his series a volume on *The meaning of meaning*. The title suggests a descent into essentialism, but in fact they were primarily concerned with linguistics and semiotics. In an agreeable if rather superficial way they considered words as signs and their relationships to thoughts and things (ch. 1). Concerning 'The meaning of philosophers' (ch. 8), itself a double meaning, they quoted Russell's admission about the small contribution of logicians quoted in §8.3.6. 'The meaning of meaning' itself was treated in a wry ch. 9 where, for example, Russell's theory of definite descriptions was held to 'have only led to further intricacies which logicians are once more endeavouring to unravel' (p. 190). Again, 'all definitions are essentially *ad hoc*' (p. 111), which revealed their limited familiarity with mathematics—contextual definition, for example, which cannot be so dismissed. The only general contentual remark on mathematics concerned clause 6.2 of a recent book in the same series: 'Some, like Wittgenstein, have been able to persuade themselves that "The propositions of mathematics are equations, and therefore pseudo-propositions"', a view which 'can be presented without the background and curtain of mysticism which this author introduces' (p. 89).

The comments most pertinent to logic came at the end of the book. In Appendix E 'On negative facts', the authors attacked the notion of opposition stressed in Demos *1917a* (§8.3.6). Appendix D treated 'Some moderns': Frege and Russell received routine summary, but Husserl and Peirce had several pages each, the latter covering both aspects of his algebraic logic and his existential graphs (§4.4.7) and including some letters, then unpublished, to his main British follower Lady Welby (1837–1912).

In a warm and quite long review Russell *1926b* recalled that he had once 'become acquainted with Lady Welby's work' on meaning and symbolism 'but failed to take it seriously'. However, a few lines later he expressed the astonishing opinion 'that a logic independent of the accidental nature of space-time becomes an idle dream'. In *Mind* Ramsey *1924a* dismissed the book, rating as 'valueless' their conception of sensory perception, although he was glad that they presented 'a lot of amusing writings on the use of words by plain men, savages and philosophers'. However, the book became well known, mainly in a reduced second edition which appeared three years later; the several later editions show only minor revisions.

An unusual application of mathematical logic was made by W. W. Greg (1875–1959) in a book on *The calculus of variants* (Greg *1927a*). This was not a new kind of mathematics, but 'An essay on textual criticism' by a scholar and also the former head librarian at Trinity College when Russell and Whitehead had been there.[22] Greg used the logic of relations, especially the ancestral relation, to confect a symbolic representation of the relationships between an original text and its various later versions. While the display of graph-like notations is at first startling, the calculus is rather impressive, especially in its clear generality. By such means he became known as a founder of the notion of the copy-text, though presumably he did *not* know of the habits of the younger Russell of transferring folios from one version of a manuscript to the next one! But in a later discussion of symbolic systems, he gave *PM* as an example of one where the symbols 'form a silent language of their own' (Greg *1932a*, 248).

8.5.3 *American reactions to logicism in mid decade.*   The second edition of *PM* received a good deal of notice in the U.S.A. A surprising place is the history of science journal *Isis*, run by the History of Science Society, which Russell joined for some years in 1927. On the first volume Sheffer *1926a* modestly noted the role of his own connective (which he called 'non-conjunction'). Writing admiringly of the project which produced the first edition, he contrasted it with postulate theory; but he puzzled that '*In order to give an account of logic, we must presuppose and employ logic*', and saw no way out of this '«logocentric predicament»' (p. 228).

The logician C. H. Langford (1895–1965) was assigned the last two volumes; but he found them unchanged and so devoted his review to the new material. Sceptical of the extensionality that he found, such as infinite conjunctions, he worried if a 'notational pantomime' had taken place (*1928b*, 516). His review framed two papers in *Mind*: *1927a* against Wittgenstein's emphasis on tautologies and in favour of existential import and individuals, and *1929a* attempting a theory of 'general propositions' based upon functions which took 'non-general facts' as values. In Langford

---

[22] RA possesses the copy of an undated letter to Greg where Russell announced that the third volume of PM_1 was about to appear (in 1913): 'it will be thin, & there will be a final thin Vol. IV'. The letter was found by a bookseller in Greg's copy of this volume, which was later owned by Karl Popper.

1926–1927a he studied the semantic completeness of some axiomatised theories.

Langford was later a collaborator with Lewis, who himself produced a review *1928a* of the second edition of *PM* for the *American mathematical monthly*. Also concentrating on the changes, he too noted the 'Sheffer-Nicod "stroke-function" ' but disliked Russell's name 'incompatibility' for it, on grounds similar to his reading of 'impossibility' (§8.3.3). He gave a truth-table for material implication, using Post's '$\pm$' notation (§8.3.4). But he did not read the later volumes carefully, for he claimed that the preface on symbolic conventions to the second one (§7.9.3) 'was considerably expanded', whereas it had been reprinted without change.

For the AMS the first volume was reviewed by B. A. Bernstein (1881–1964) of the University of California. While lamenting the small impact of the first edition upon mathematicians, Bernstein *1926a* pointed to the prominence of notions such as assertion and intensionality which they would find 'not sufficiently convincing'. Summarising the postulatary approach, he applied it later to the propositional calculus of *PM* (§.9.4.1).

The other two volumes of the second edition were reviewed by Alonzo Church (1903–1995). He proposed that the definition of cardinal numbers as classes of similar classes could be replaced, via Russell's principle of abstraction, by defining 'the class determined by a propositional function ϕ as the class of propositional functions equivalent to ϕ', and also assuming the existence of the abstraction: an approach like Peano's (§5.3.3), as he noted (Church *1928b*, 238–239). Russell would not have been keen; and he would also have been surprised by Church's acceptance of the Bolzano-Dedekind "proof" of the axiom of infinity (p. 240), with his old *The principles* (§7.2.7) cited as the source.

Church wrote the review soon after graduating at Princeton University under Oswald Veblen with a thesis *1927a* on alternatives to the axiom of choice (Aspray *1991a*). Then he spent part of 1929 at Göttingen, where Haskell Curry (1900–1982) was writing his *Dissertation* under Hilbert on combinatory logic; it duly appeared in German in the *American journal of mathematics* (Curry *1930a*), thus continuing the German influence.[23] Later Church *1932a* reviewed the edition of Ramsey's writings for the *Monthly*, in general warmly but with objections: 1) to abandoning identity, on the usual grounds that if x and y had all properties in common, then one of them was being identical with x, so that $y = x$; 2) to reading universal quantification in terms of infinite conjunction, since every component

---

[23] On the early career of Church, see E. H. Moore Papers, Box 4, Folder 5, where Veblen described him in 1927 as 'one of our very strongest doctors'; and his own letters to Veblen in Veblen Papers, Box 3, including mention of Curry in Göttingen (a file on Curry is held in Box 4). Curry had thought of combinators in 1922 from trying to simplify substitition procedures in the propositional calculus of PM_1. In the mid 1930s he annotated in detail his copy of PM_2 without however publishing anything. Thanks for this information are due to Jonathan Seldin (University of Lethbridge, Canada).

proposition had to be understood to start with; and 3) to dropping the subsumption of a classes under propositional functions, since surely class α was determined by the function '$\hat{x} \, \varepsilon \, \alpha$'.

An amusing review was produced for the *Journal of philosophy* by Harry Costello, who had started off Russell's Harvard logic course in 1914 before he arrived (§8.3.1). Harking back to that time, Costello *1928a* recalled the publication of the first edition of *PM*, with 'its pages that look something like hen-tracks on the barnyard snow of a winter morning', and also Royce's emphasis on the importance of foundational studies. Following a tradition which Russell was trying to replace, Costello wondered if the system should be called 'logic' since it was not concerned with reasoning. Like his compatriots, he was not happy with the changes; for example, that type theory was now held to be about symbols whereas paradoxes such as the liar dealt with things (of some kind), or that atomic propositions were as general as claimed (pp. 442–445).

8.5.4 *Groping towards metalogic.* Bernstein's review *1926a* of *PM* ended thus:

> As a mathematical system, the logic of propositions is amenable to the postulational treatment applicable to any other branch of mathematics. As a language, this logic has *all* its symbols *outside* the system which it expresses. This distinction between the propositional logic as a mathematical system and as a language must be made, if serious errors are to be avoided; this distinction the *Principia* does not make.

This distinction is close to that between logic and metalogic; we also saw Sheffer wanting for it with his 'predicament'. Several other American authors approached it at this time, as we shall soon see.

At Cornell University Harold Smart presented much of his Ph.D. as a short book on *The philosophical presuppositions of mathematical logic*. Reviewing the recent development of the subject, with Russell and Royce as principal figures, he judged it to be 'a complete failure' (Smart *1925a*, 97) for misunderstanding the vision of hero Leibniz, for not using part-whole theory, and for ignoring the creative aspects of mathematics; curiously he hardly mentioned algebraic logic, which was innocent of the first two criticisms. Lewis *1926a* was cool in his review for the *Journal of philosophy*, criticising the author for muddling 'what is essential' with 'what is accidental', especially an excessive emphasis on extensionality, which MacColl and Lewis himself had shown could be avoided. In reply Smart decreed that 'formal or abstract truths' were incompatible with extensionality (*1926a*, 297). Later he broadened his treatment into a book on the 'logic' (philosophy) of the sciences: two chapters were devoted to mathematics. A rather disappointing survey, perhaps the most interesting feature was his appeal to Russell *1914d* on 'the scientific method in philosophy'

(§8.3.1) to stress the need of showing the validity of logical principles (Smart *1931a*, 98).

Two American commentators on logicism discussed the theory of types in *Mind*. Firstly, Suzanne Langer *1926a* was understandably concerned by the 'Confusion of symbols and confusion of logical types', and wondered if the ban on $\phi(\phi\hat{x})$ by the vicious circle principle might also exclude cases of $\phi(\phi x)$, which could be harmless. Perplexed by the linguistic rendering of $\phi\hat{x}$ ('is a cat'? 'The being-a cat-*ness* of'? ...: p. 225), she carried over her doubts to the status of truth-values (p. 229); for the ϕx 'this proposition is a lie',

> our proposition, now $\phi(\phi x)$, becomes: ' "This proposition is a lie" is a lie', which is no longer—this proposition. 'This proposition is a lie' and ' "This proposition is a lie" is a lie' are, as a matter of fact, two discrete propositions, and cannot be denoted by the same symbol in the same complex.

Secondly, Paul Weiss *1928a* queried the epistemological status of type theory in view of its apparent inability to appraise itself; in the same spirit he analysed the liar and 'Weyl's' heterological paradoxes. In a companion piece in *The monist* he proposed 'Relativity in logic' based upon various interpretations of implication, including Lewis's, with some emphasis given to entailment; he used a large truth-table to represent the various cases (*1928b*, 545). He deployed these features also in a lengthy sketch of logical and mathematical 'systems', and means of comparing them; following Wittgenstein, he took logical systems to be tautological but mathematical ones not (*1929a*, 440).

Publishing with the AMS, Church *1928a* considered discarding the law of excluded middle without necessarily making an alternative assumption; he doubted that a contradiction could be obtained, and felt it to be 'meaningless to ask about the truth' of the law. In reply Langford held that 'logical facts' were analytic and that 'logical principles' were based upon them; hence no alternative logics were possible; further, 'π is transcendental' was allegedly 'just as much a logical principle' as the law (*1928a*, 581).

The next year the philosopher Kurt Rosinger published in *The monist* a study of 'The formalization of implication'. Taking the contrast between the central concern of 'traditional logic' with inference and the modern interest in implication, he saw Boole as a herald of the change, and noted also the importance of quantification of the predicate (§2.4.6), where *'logic and mathematics are joined'* (Rosinger *1929a*, 277). The symbolic traditions had led to the normalisation; he cited Johnson *1922a* on 'functional deduction' as a source, although he saw neither *PM* nor Lewis as giving 'the real meaning' of implication, if indeed one existed. In a succeeding note Rosinger *1930a* followed Langer in puzzling over the difference between '$\phi\hat{x}$' and 'ϕx', deciding that in terms of Frege *1891a* (§4.5.5) they were respectively 'belongingness' ('*Zusammengehörigkeit*') and 'necessity of completion' ('*Ergänzungsbedürftigkeit*'). He mentioned, correctly, that Russell had sometimes used 'propositional function' to refer to both; but usually it was clear that the former was the function as such and the latter the template for its values, as with mathematical functions.

8.5.5 *Reactions in and around Columbia.* These American sensitivities were partly stimulated by their tradition of postulate theory. We recall from §4.7.3 that John Dewey had suggested the adjective 'categorical' to student Veblen in the early 1900s; while he did not study postulate theory, he worked on logic from time to time. His philosophy was a type of pragmatism more comprehensive than that of his mentor Peirce in that he saw *all* aspects of thought, including (a) logic, as influenced by human activity: this member of the older generation had arrived at alternative logics long ago. In particular, in 1916, when in his mid fifties and at Columbia University, he published a volume of *Essays in experimental logic*, including his own pieces in the *Studies* of 1903 (§7.5.4). His title would have puzzled many logicians, but he made his position quite clear: claiming that logic was based upon action (Dewey *1916a*, ch. 6) and, for example, warning against Locke's mistaking 'logical determinations' for 'facts of psychology' (pp. 404–405). He made a delightful analogy of the difference between idealists and realists as that between 'eaterists' and 'foodists' (pp. 270–271). Applying his view to epistemology, especially in ch. 11 to Russell's recently published *Our knowledge of the external world*, he found the priorities there the wrong way round: world first, patches of colour (or whatever) afterwards, not constructions from the latter to the former (Sleeper *1986a*, ch. 4).

In a long review for the *Journal of philosophy*, written in his temporary Brixton residence (§8.3.7), Russell *1919a* agreed with many of Dewey's points but clashed over the role and place of logic, which Dewey had spread far too widely. Anxious to keep time out of its remit, Russell distinguished temporally from logically primitive notions. He also defended his recent book by explaining the different referents which he and Dewey assigned to the word 'data', and protesting at the end that indeed 'it was not I who made the world'.

Dewey's Columbia colleague Keyser produced a large book *1922a* on *Mathematical philosophy* as he entered his sixties. He rambled agreeably though without much penetration around a good range of topics, including chapters on invariance, group and postulate theories, the 'psychology of mathematics' and the 'mathematics of psychology', and even engineering. Logic did not gain a chapter, though *PM* was praised and some of its concerns appeared in the chapters on limits, variables and infinitude. The next year he wrote a rave review *1923a* of Wittgenstein's *Tractatus* for the *New York evening post*, estimating the author as 'a really great genius of the philosophic order', a 'logical mystic' like Spinoza and Pascal with a 'deep and beautiful' theory of inference, who however 'has not taken sufficient pains to be clear' on all matters. To give the book more publicity, he abridged his review for the AMS in 1924.

Dewey sustained a large correspondence in addition to prolific publication.[24] One regular contact was with Scudder Klyce (1869–1952), a naval officer whose passion lay in rather eccentric metaphysical writings based on a concern with 'the One and the Many'. He applied it to mathematics in an article *1924a* in *The monist* on the 'Foundations of mathematics'; there he praised Keyser *1922a*, like Keyser judged Wittgenstein to be 'a mathematical star of the first magnitude' (p. 622), and felt that there was no real foundation for mathematics. In 1932 he expanded in similar style in a short and self-published *Outline of basic mathematics*, in which he preferred a postulary approach and found logicism useless: the notion of class of classes confused the One with the Many, for example, and the defining number in terms of them was confused since a number was 'a word pointing towards an element' (Klyce *1932a*, 19, 68). The book is deservedly obscure.

One of Dewey's most intense correspondents was another amateur, Arthur Bentley (1870–1957), who made enough money to retire in early middle age and devote himself to philosophy. He was an enthusiastic letter-writer himself, continuing many extensive exchanges. His publications were less impressive; in particular, an essay *1931a* on 'mathematical consistency', published in Ogden's journal *Psyche*, contained an obviously mistaken proof of the denumerability of the real numbers, as Black *1931a* pointed out. Next year Bentley put out a book-length *Linguistic analysis of mathematics*, in which he concentrated on current foundational movements, especially their use of words. An interesting chapter dealt with 'Word-clusters lacking consistency', including German examples such as 'Menge', 'Auswahl' and 'Zahl' (*1932a*, ch. 5). He did not evaluate logicism as such, but *PM* was praised (the frontispiece marked out the sentence in $*54\cdot43$ where '$1 + 1 = 2$' was proved); he also noted Russell's stress on bivalency and on the role of operation. As with Keyser's book, the range was refreshingly wide, but a price was paid in shallowness.[25]

We have seen a considerable body of American commentary, with the informed students far outweighing the nonentities and eccentrics. Now we

---

[24] An electronic edition of Dewey's correspondence is currently in preparation from the originals in the Dewey Papers, University of Southern Illinois at Carbondale. Klyce's side of the correspondence with Dewey is held in his Papers, Box 4. His main book is *Universe* (1921: drafts in Box 16), again self-published after Open Court rejected a version of it in 1915–1916 (Open Court Papers, Boxes 32/17, 19).

[25] The Bentley Papers, at Indiana University at Bloomington, deserve to be *far* better known; the letters are filed in alphabetical order of writer. On his paper and book, see also his correspondence with Ogden in Ogden Papers, Box 3. Like Klyce (another correspondent, incidentally), his book was rejected by Open Court (Bentley Papers, Box 3). His *Philosophical correspondence* with Dewey was published in 1964; during their lifetimes they had jointly issued a book on *Knowing and the known* (1949).

switch to Continental Europe for the rest of the chapter, starting with two countries of minor importance.

8.6 Peripherals: Italy and France

8.6.1 *The occasional Italian survey.* Italian interest faded during the 1910s as Peano switched interest steadily more to international languages, though we saw his work of 1913 in §8.2.4. But one lingering worry concerned the clash between Peano's method of definition of cardinals by abstraction and Russell's nominal definition by classes of classes (§6.5.2). Burali-Forti *1909a* had proposed, as a third option, definitions of function(al)s of ordered pairs '$(a;b)$' of 'simple entities' a and b, such as arithmetical operations on numbers, or (Russellian) classes of ordered pairs of integers. One of Peano's minor followers, Eugenio Maccaferri (1870–1953), took up this idea to suggest that if members x and y of a class u satisfied an equivalence relation α, then each member of the corresponding equivalence class ν was a function ϕ of some member of u, so that

$$\text{`}\phi x = \phi y . = . x \alpha y\text{'} \tag{861.1}$$

(Maccaferri *1913a*, 165; on p. 167 he specified the second ' = ' as logical equivalence). Then this kind of definition could be based on appropriate α and ϕ; for example, for integers m and n under equality of rational numbers m/n or n/m as ordered pairs, or of numbers of the form $\sqrt{m/n}$. Further, the classes defined this way were clases of classes, so that Russellian definitions could be expressed; in particular, Russell's definition of irrational numbers (p. 169).

The most substantial Italian item of the 1910s was the second edition *1919a* of Burali-Forti's textbook on 'mathematical logic', published in his late fifties. We saw in §5.3.7 that in 1894 the first edition had been a slim but pioneering work; now at over 480 (small) pages it had more than trebled in length. The five chapters covered, in turn, 'Ideographical symbols' for mathematical logic and set theory, still basically Peanist; 'Operators and connectives', including some attention to the axioms of choice; 'Ideographical algorithms in general', with syllogistic logic and the calculus of classes; 'Definitions', nominal (maybe under hypothesis), or by abstraction or induction; and finally and briefly 'Some applications', to real and complex numbers, and point-set topology. Theories not Peanist were not well received; already in the introduction, 'the method of *relation*, that, like foreign goods, is received at once in Italy' such as 'the chaotic and imprecise geometrical system of Hilbert' (p. xxxii) despite the existence of the 'superior' version of Mario Pieri (§5.5.5). He disparaged 'such empty discussions' of non-self-membership of classes (p. 84), incorrectly gave

Russell's definition of cardinal 0 (pp. 355–357) and dismissed as 'impossible' that of 1 (p. 167), and found it 'more *simple* and more *common*' to take relations as operators rather than a Russell-style primitive (pp. 226–227).

The book did not circulate well, perhaps due to the poor economic state of Italy after the Great War, which may have also inspired the bitter tone. However one reader was Enriques, who published a suite of comments *1921a* in *Periodico di matematico*; Burali-Forti *1921a* responded, exciting a reply Enriques *1922a*. Among the issues, Enriques preferred to distinguish a class from an abstract concept, rather than membership from inclusion in order to cope with 'apostle ⊃ twelve' (in Peanese): Burali-Forti rejected the latter on the ground of simple entities not being classes, and, attending to second-order predicates, he related apostlehood to dozen-ness. He also retained the non-identity of an object with its unit class, which did not assuage Enriques's puzzlement over the role of identity in Burali-Forti's book. Enriques also questioned the form of definition by abstraction, citing Maccaferri while regretting the use of disjunction to specify class abstraction; Burali-Forti cited his paper *1912a* on abstracting a class *without* characterising the members. On the status of mathematical induction and the legitimacy of the axioms of choice, Enriques's nominalism split him from his colleague's contented realism.

At this time Enriques was completing a history of logic, deductive and inductive, from antiquity to modern times. Much broader a book but shallower than Burali-Forti's, it became far better known, with translations into French, German and (American) English within the 1920s. He rather mixed algebraic and mathematical logics together, with Peirce and Schröder under-represented. Russell's contribution was found to be 'refined and profound (perhaps in places too subtle)', a judgement which Enriques verified by mistaking a proposition for a propositional function (*1922a*, art. 18).

8.6.2 *New French attitudes in the* Revue. While interest in logicism remained sporadic, the situation among French authors changed: the polarity between sneers from Poincaré (died 1912) and applause from Couturat (died 1914) was replaced by more neutral positions, even in the *Revue de métaphysique et de morale*. Its co-founder, the idealist philosopher Léon Brunschvicg (1869–1944), preferred in *1911a* intuition among current philosophies of mathematics, especially as handled by Poincaré (pp. 165–169); he distrusted dogmas such as Ockham's razor, and did not welcome logicism (p. 146). But a softer line came through the following year in his large book surveying various stages in the development of philosophies of mathematics. Of its seven Books the sixth was devoted to 'the logistic movement'. Its 58 pages began with a short treatment of Boole's methods before passing to Frege's programme and especially Russell's, where he clearly indicated the prime influences of Cantor and

Peano (Brunschvicg *1912a*, 381–383). But at the end he was sceptical of Russell's empiricist 'realism' in both epistemology and logic: 'it is on the terrain of positive science that the positive mathematical philosophy must be placed in future' (p. 426). To stress his own position he devoted his long last Book to the need for intuition and psychology in mathematics when, for example, understanding complex numbers (pp. 542–550). As in many such writings, there is an unresolved tension between the philosophies of grounding mathematical theories and of creating them.[26]

In his survey Brunschvicg drew upon the first volume of *PM*, which was reviewed in a mathematical journal by Henri Dufumier *1911a*. Far from the Poincaresque sneer, he praised the enterprise, even pointing out the need for the axioms of choice, unusual among the reviews. He continued more amply in this manner in a 'critical essay' for the *Revue* of both the volume and Russell's *Philosophical essays*; applauding both the 'courageous precision and solid originality' of the logic and 'the originality and profundity' of the epistemological enterprise (*1912a*, 539, 564), he underlined the respect which he had already shown in that journal in 1909 (§7.5.1).

Padoa renewed his crusading efforts for Peano's programme with a septet of lectures at the University of Geneva, which the *Revue* published at over 100 pages and also in a separate book version (Padoa *1911–1912a*). To the latter Peano *1912a* contributed a preface reporting Padoa's lecture courses since 1898, not only Brussels (§5.4.6) and Rome (§5.5.4) but also in Pavia, Padua and Cagliari. Padoa's title promised 'Deductive logic in the latest phase of development', thereby updating the (scrappy) survey Vailati *1899a* there (§5.4.6); but in fact he largely told the story as known in Turin for a long time. He explained most of the symbolism (though rather little on quantification), and made applications to syllogistic logic rather than to logicism or even Peano's programme. But he gave an able and clear account; and its length and manner of publication shows the maintained French interest.

Some years later another rehearsal appeared there: a posthumous piece Couturat *1917a*, taken from his uncompleted 'manual of logistic' (§7.4.1). Written seemingly around 1906, it covered so much of the same ground as in Padoa's lectures that the publication seems unnecessary. In the same volume another Swiss connection appeared, when Arnold Reymond (1874–1958) (§7.5.1) considered 'the transfinite ordinals of Cantor and their logical definition'. He did not cite Padoa's article, which would have helped his chatter; his logic depended only upon the notions of individuality and plurality, so that *PM* was absent. He mentioned the 1914 Paris

---

[26] Brunschvicg's consideration of logistic soon received some unintended further publicity when it was "heavily used" by the Argentinean philosopher Camilo Meyer in a paper *1918a* on this topic. Meyer had already helped himself to chunks of the third and fourth Books in a survey *1916a* of 19th-century views on continuity and arithmetic.

Congress, upon which he had reported (§8.2.2), and he also cited Richardson and Landis (Reymond *1917a*, 695, 700).

Reymond also mentioned a recent essay on 'geometric proof and deductive reasoning' by Louis Rougier (b. 1889). Rather unclearly, Rougier relied upon syllogistic logic if possible; but he also presented many features of mathematical logic, including relations, and stressed the role of definitions and '*formative principles*' such as axioms to furnish existence theorems (*1916a*, 613–619). He thus opposed the position of Poincaré on geometry, a topic to which he later devoted a book; but there he was in line with his master, with the dispute with Russell of the late 1890s (§6.2.3) receiving only a few lines at the end (Rougier *1920a*, 197–198).

8.6.3 *Commentaries in French, 1918–1930*.   In his *Revue* essay Rougier also differed from the Lyon philosopher Edmond Goblot (1858–1935), who published a large *Traité de logique* in 1918, after delay caused by the War. In his preface Goblot explained that 'logistic' would not be addressed, since he wished to treat 'reasoning in general and not just mathematical proof' (*1918a*, xix), which for some logisticians begged questions. The chapters on 'deductive reasoning' were driven by syllogistic logic and Kantian concerns, together with doubts over Poincaré's claims for the generality of mathematical induction (pp. 257–272).

Goblot's silence on logicism reflected the decrease in Francophone interest, over several years. However, the volume for 1922 of the *Revue* contained both Nicod highly praising both Russell's logic and philosophy as 'This glory of logical and mathematical reasoning' (*1922a*, 84) and, in an issue devoted to 'American thought', Lewis *1922a* surveying both traditions of symbolic logic up to and (briefly) including his own modal version.

Belgium came into the story in the mid 1920s, when Robert Feys (1889–1961) published two 85-page three-part articles in the Catholic *Revue néo-scholastique de philosophie*. In the first one, *1924–1925a*, he treated 'the logistic transcription of reasoning', but rather under-stated the differences between the algebraic and mathematical traditions; he touched upon modalities (art. 9). Not surprisingly, he concluded that logistic 'does not seem to have revolutionised logic' because it was 'founded upon essentially different PRINCIPLES' (art. 14). The second article, *1926–1927a*, dealt specifically with the bearing of 'Russellian logistic' upon reasoning; here he was more detailed and positive, and also more up to date in noting Sheffer, Wittgenstein, Nicod (arts. 9–12), Lewis, and the first volume of the second edition of *PM* (especially in arts. 17–20 the new theory of predicative functions). But he said little on the paradoxes, logicism or Frege; and nothing on Russell's epistemological writings, though they bore closely upon his interests. Thus the impression was inconclusive.

The Polish mathematician Stanisłav Zaremba (1863–1942) contributed a short book *1926a* on 'the logic of mathematics' to a Gauthier-Villars series of short monographs similar to the Cambridge tracts for which Whitehead had written (§7.6.2). He largely followed the Peanist/*PM* lines, except for muddling in algebraic logic in places; and in a section on the theory of proof he outlined the algebra of classes but judged the logic of relations to be irrelevant (pp. 45–47). The most original contribution, inspired by Poincaré's warning over impredicativity (§7.4.5), was to interpret Russell's paradox as proving the theorem that to any class *E* of classes there corresponded another class which did not belong to *E*; thus there was a notion broader than class, which he called 'category' and to which this theorem itself applied (pp. 11–15). Further, type theory was 'unsustainable' for stratifying only classes (p. 44). We shall find in §8.8.4 something more substantial from a former student of Zaremba, and in §9.6.6 that he gained at least one brilliant reader.

Intuitionism rather than logicism inspired various authors to short exchanges, especially in the 1926 volume of the *Revue*; Borel reproduced them two years later in the new edition of his textbook on the theory of functions (§4.2.2), as an appendix 'for and against empirical logic' (*1928a*, 254–278). In Switzerland Ferdinand Gonseth (1890–1974) produced a book on 'the foundations of mathematics', concentrating on geometries, general relativity and intuitionism. Logicism was rejected because numbers needed experience for their ground whereas logic did not (Gonseth *1926a*, 186); so it was almost entirely ignored in the final chapter on 'mathematics and logic'.

Among new authors Albert Spaier (1883–1934), a phenomenologist by inclination, produced two long books on 'thought' in 1927. One, his doctoral thesis *1927a*, was a survey of psychological notions, such as consciousness and intuition. A short section near the end dealt with 'construction of mathematical concepts', but none too valuably, since his claim that numbers formed 'the basis for all mathematics' was followed by examples from plane geometry (pp. 383–400). The second book was a lengthy discussion of number and quantity as handled not only by Russell (more *The principles* than *PM*) but also by Frege and Hilbert among others. Anxious to give experience an important role in a study of 'existential questions', he regarded Russell's logical definition of cardinals as 'some illusion' since no concept was provided for the intuition to grasp (*1927a*, 12); however, he misunderstood it in stating that all numbers were 'equally infinite' in using classes of classes, and also for assuming order when successively deploying operations (p. 184).

Finally, Jules Tricot produced a 'treatise' handling 'formal logic' quite nicely, based largely on traditional logics (including quantification of the predicate), with Goblot a main cited author. But 'Logistic and the algebra of logic' appeared in only the last 10 of his 317 pages, and was judged as 'a very special discipline' of which 'their very foundation is vicious' (*1930a*, 306, 312). He did little more than list its principal authors (Whitehead not at all), and none of their works appeared in his bibliography—not even the recent French translation *1928a* of Russell's popular *Introduction*.

Both French and Italian authors took some note of recent German writings on logic and the foundations of mathematics. We shall now do the same.

8.7 GERMAN-SPEAKING REACTIONS TO LOGICISM, 1910–1928

Different countries, different scale of reaction, and quite a different philosophical climate. This section briefly notes some of the reactions to (or ignoring of) mathematical logic and logicism, in an environment where Kantian and/or Hegelian traditions still reigned strongly, so that logic was usually regarded as analytic and mathematics synthetic, and intuition played a significant epistemological role. For some philosophers symbolic logics were irrelevant; for example, Wilhelm Koppelmann in the 700 pages of his two-part *Untersuchungen zur Logik der Gegenwart* (1913–1918). In a book on 'Symbols' Richard Gätschenberger (b. 1865) regarded the 'giant work' *PM* as a 'formula-cemetery' involving an 'elimination of the calculus of classes' (*1920a*, 121). The young Martin Heidegger (1889–1976) noted the first volume of *PM* at the end of a review *1912a* of recent literature in logic: after mis-stating logicism as an identity thesis, he reasonably judged that 'the deeper sense of its principles remain in the dark' for the lack of a theory of judgement, and he did not bother with it later, including not in his various courses on logic.

In the account to come the words '(neo-)Kantian' and 'phenomenology' are broad characterisations of the positions of the authors described, although often they are pretty exact. Logicism features explicitly only in §8.7.3 with Frege, and then primarily with his rejection of it; however, the logic of *PM* appears quite regularly, and some mathematical features such as the definitions of integers. Several of the papers cited, especially by Hilbert and Brouwer, are translated into English in Ewald *1996a* and/or Mancosu *1998a*.

8.7.1 *(Neo-)Kantians in the 1910s.* Paul Natorp (1854–1924) produced a book-length survey *1910* of 'the logical foundations of exact knowledge'. He followed a brand of neo-Kantianism in which 'pure thought' and logic were distinguished from mathematics; but he was not hostile to Russell or predecessors such as Cantor and Dedekind, who were treated in some detail (chs. 3–4). Patriotically he found Frege's work 'essentially reproduced in Russell and Couturat' (p. 114). The range of mathematics treated was wide, including foundational, pure and applied branches; but the insights into logic and set theory were shallow, as Jourdain *1911b* showed in a review for *Mind*.

Much more impressive a neo-Kantian was Ernst Cassirer (1874–1945) (Ferrari *1996a*). Also in 1910, he published an extensive study of *Substance and function*, to quote the title of the English translation which Open

Court published in 1923 (and cited here). He saw science as structured by concepts drawn actively from reality rather than imposed by it. His title took up the balance between substance, 'to which the purely logical theories of Aristotle have reference', and 'logical relations', to be expressed by functions and avoiding defective 'psychology of abstraction' (*1910a*, 7–21); in the Marburg school, to which he belonged, thought was preferable to sensibility wherever possible. Mathematics was an essential component of the processes required, especially for creating general laws and concepts (ch. 1); his first example was 'the concept of number' (ch. 2), where he stressed the logic of relations as embodied in Dedekind's definitions and brought out by Frege and Russell. He then proceeded through geometry and space to physics and chemistry, and in a supplement of 1921 he added 'Einstein's theory of relativity'. The rest of the book was philosophical, treating (scientific) induction and 'the concept of reality', and relations in more detail.

We noted Cassirer in §7.5.2 as a commentator on Russell's logic; here Russell appeared mainly though rather incidentally in the chapter on numbers, as treated in *The principles* (*PM* had not yet appeared), alongside Frege's *Grundlagen* and Dedekind's booklets. Russell was cited mainly for the logic of relations (on which, curiously, Cassirer did not draw) and the definitions of cardinals (especially on pp. 50–54 over the relationship between cardinals and concepts, with particular reference to 0 and 1). On real numbers Dedekind gained prime place (pp. 58–62). More attention was paid, however, to Cantor, and to Frege for attacks on Mill and on (symbol-writing) formalists. Surprisingly, he did not cite any Peanist or algebraic logician.

Less inspiring is Hugo Dingler (1881–1954). After studying mathematics at Göttingen with Hilbert and Klein, he wrote at length on *Mengenlehre*. In a pamphlet *1911a* on 'Burali-Forti's antinomy' he made a rather fine distinction between 'contradictory' concepts to which no objects corresponded, and non-contradictory ones for which however allied assumptions led to trouble. Regarding *reductio ad absurdum* proof methods as typical of the second kind, he argued, for example, that that antinomy was such a proof against the assumption of the existence of the series of all ordinals (pp. 5–9); he also analysed Zermelo's recent two proofs of Cantor's well-ordering principle (§4.7.6) from this point of view (pp. 10–17).

Dingler rehearsed some of this material in his *Habilitation 1912a* written at Munich University (where he was to make his career for twenty years), in which he discussed well-ordered and 'scattered' sets in general. His analyses were largely based upon Gerhard Hessenberg and Hausdorff respectively; 'scattered' ('*zerstreut*') came from Hausdorff *1908a*, defining a point set which contained no dense subsets (see §7.9.5 for context). Dingler made papers out of both parts of the thesis; for the *Deutsche Mathematiker-Vereinigung* (hereafter, '*DMV*') he took many of the paradoxes, including Russell's and Richard's, to be *reductio* proofs of impossible situations within a 'logical building' of axioms, theorems and rules of inference (*1913a*, 308).

So far so fairly good, though not always competent; Dingler inserted a sticker into his pamphlet disclaiming its last section, on consequences of his position for Cantorian limit ordinals. A short book *1915a* on 'logical independence in mathematics' looked no better; he explored the foundations of arithmetic and 'infinite processes' and a few aspects of mathematical analysis, as 'an introduction to axiomatics'. However, while he dedicated it to Hilbert, he followed the paper-formalism attributed to Johannes Thomae (§4.5.9), as Löwenheim *1922a* noted in a very cool review for the *Jahrbuch*. Some *Mengenlehre* featured in the account, and consistency was deployed; but no mathematical logic appeared, even though he outlined 'relation-arithmetic' (pp. 73–75). A later book on the foundations of physics included a 70-page chapter on logic (*1923a*, 43–113); but the main notion was that of concept, with much space given to simplicity (such a complicated idea). Algebraic and mathematical logics were taken together; set theory, as used by Frege and Russell, formed an 'astonishing complication', with classes difficult to characterise as logical—including his own error over the definition of 1 (p. 83).

Dingler *1915a* also warmly reviewed a recent volume proposing 'New foundations for logic, arithmetic and set theory'. He liked many of the details involved, especially (without stressing the point) its Kantian leanings. The author was not German, but the Hungarian Julius König, with an incomplete posthumous book *1914a* published soon after his death by his mathematician son Dénes with help from Hausdorff. We saw König last in §7.2.2, failing to prove Cantor's continuum hypothesis but instead finding paradoxes of naming; they were treated here also, but in a much wider canvas. The book contains an interesting fusion of elements of logicism, metamathematics and axiomatic set theory.

König showed his philosophical inclination by starting with 'experiences of my consciousness' as his principal notion (pp. 1–3); but the ensuing narrative was not purely idealist or solipsist. Of major importance was 'the experience' of naming object B as A, written 'A nom. B', within a 'thought domain'; collections of such namings led to the class-concepts' (pp. 24–26), each one associated with 'property-', 'relation-' and 'order-concepts' (pp. 50–57). They were all the products of synthetic judgements, another main notion; his first title for the book had been 'synthetic logic' (p. iii).

Upon this framework König rebuilt set theory upon 'logical basic concepts' based upon the relation between experiences A and B of 'not different from'; called 'Isology', it and its name were symbolised respectively 'A id. B' and $x \stackrel{\smile}{=} y$' (pp. 70–74). He built up a calculus like the propositional, using '$\smile$' over the corresponding algebraic symbols for connectives, such as '$\stackrel{\smile}{\times}$' for conjunction (pp. 70–81). Truth values 'ʋ' and its contrary '$\mathfrak{v}''$' were associated with experiences by the notations such as '$A \frown v$'; tables of them were given for simple compound propositions (pp. 81–89), but not as an equivalent for truth-tables, which had just been born (§8.3.2). The law of excluded middle came out as

$$[x \frown \mathfrak{v}] \mathbin{\bar{+}} [x \frown \mathfrak{v}']',\tag{872.1}$$

where x was the name for the 'logical form' X (pp. 122–125), of which several were exhibited. He explicitly gave rules for substitution, using the column notation from permutation theory (pp. 92–98).

In ch. 6 König attempted an axiomatisation of set theory, after Zermelo's but less formal. He was especially concerned to show the axiom of choice to be 'proved evident, like any other fundamental intuition of logical-mathematical knowledge' (p. 163); the reason was that the choice function created a new domain of thought and so was a synthetic judgement (pp. 166–172). Earlier in an unconvincing footnote he stated that Dedekind's construction of an infinite set (§3.4.2) by iterated thought processes was meant only to show the legitimacy of such generation, not the existence of the set as such (pp. 62–63). The book ended with an incomplete survey of key results of set theory such as the Schröder-Bernstein Theorem 425.1 (chs. 8–9).

On arithmetic, finite and transfinite, Cantor and Dedekind were the stars; König's theory of real numbers used dyadic expansions not unlike Frege's (458.1), though seemingly independently (p. 206). Formalism arose in the assumption, following Hilbert, that consistency implied existence (p. iv); in the text he sought to show that it was possessed by his calculus, and by set theory in order to avoid the 'antinomies'. They were oddly treated, in various places: Russell's first (pp. 28–32), then the naming ones (pp. 211–214), and finally 'the antinomy of the set of all things' (pp. 223–229, after a warm-up on pp. 149–151). His solutions broadly followed Poincaré on avoiding impredicative definitions (§7.4.5) rather than limiting size.

In addition to Dingler's review, König was also noticed by a newcomer. In a *Dissertation 1918a* defended at Kiel University, Georg Behrens (b. 1892) compared König, Schröder and Russell, the latter apparently a 'Professor in London, and Professor of Logic in Trinity College Cambridge' (p. 5); Peirce and Frege were omitted. Grouping the trio under 'mathematical logic', he did not fully bring out the differences between Schröder and the other two in a largely descriptive and oddly ordered comparison; thus no meticulous analyses *à la* Wiener (§8.2.7) appeared, although he made some nice contrasts. König featured the least (pp. 44–46, 53–54) with the logical calculus and the paradoxes, this latter topic occupying the third and last chapter. Behrens never published again in this area, not even making a paper out of the thesis; so his work was virtually without influence.

Finally, Aurel Voss (1845–1931) contributed in 1914 a long essay to a volume of mathematics edited by Felix Klein for a series on 'Contemporary culture'. Long interested in such questions in addition to his specialism in differential geometry and mechanics (Reich *1985a*), Voss considered both psychological and epistemological aspects of mathematics from a broadly Kantian perspective, with a strong historical component and fine bibliography. But his treatment of 'the newer l o g i s t i c' was brief and disappointing, muddling the algebraic and mathematical traditions together. He quoted and rejected both Russell's implicational definition of mathematics from *The principles* and the short version about not knowing what is being talked about; but he construed logicism as an 'abstract formalism', which smudged a distinction which Russell wished to address (*1914a*, 29–31). Concerning the paradoxes, he favoured limiting the size of classes, but with no discussion (pp. 88–89).

8.7.2 Phenomenologists in the 1910s. The Austrian Alexius Meinong (1853–1920), a former student of Franz Brentano, had developed a theory of reifying non-existents such as 'round square', to the cautious interest of Russell *1904* (§7.3.3). A main source for Russell was a book of 1902 'On assumptions', and in later articles and the second edition Meinong extended it to objects of thought of 'higher order' (*1910*, 253–266). This work inspired his own former student Ernst Mally (1879–1944), who was interested in symbolic logics, especially the algebraic tradition; a short book Mally *1912a* on 'logic and logistic' was quite close to Schröder. In a paper 'on the independence of objects from thought', Mally considered 'a thought, which does not happen to itself' and came to a paradox akin to Russell's (*1914a*, 39–43). Soon afterwards, in a long article on 'emotional presentations', Meinong avoided the 'Russell-Mally paradox' by the principle that no higher-order object such as a class could contain 'its own inferior' ('*Inferius*'), so that the paradox-forming clause referred to a 'defective object' (Meinong *1916a*, 12, 10, 27). Langer was to draw a similar conclusion, from a different background (§8.5.4).

Meinong's solution was hailed as new by a fellow Brentanian, the Viennese Alois Höfler (1853–1922), in the second edition *1922a* of a treatise on logic, which was dedicated to the recently deceased Meinong. He showed much more sympathy for mathematics and even physics, with several long footnotes on their 'logic' and teaching. His logic was still based upon judgements and their contents within this tradition; taking the laws of contradiction and of excluded middle as basic (pp. 542–549: not exactly early on!), he relied on syllogistic procedures, including Euler diagrams and Gergonne relations (pp. 202–207, 454–460, 626–638). In a discussion of Russell's paradox and the vicious circle principle, he noted with regret the British departure from Kant's sense of analyticity (pp. 564–577). Mally added four passages, including detailed summaries of the axioms and principles in Schröder and in *PM*, and a contrast of their treatments of implication between propositions (pp. 577–592, 886–892). The range and limits of interests of this over-long but fine book are striking: Bolzano's *Wissenschaftslehre* (§2.8.2) among Höfler's influences, but nothing on Frege; concern with relations but not with their logic; attention to mathematical logic, but not to logicism.

In contrast to most contemporaries, the Halle philosopher and psychologist Theodore Ziehen (1862–1950) took Frege as a central figure in a short book *1917a* published by the *Kantgesellschaft* on 'the relationship of logic to set theory'. The title suggested that logic came first, but he gave much more space to set theory, through various forms from Cantor through Artur Schönflies to a recent treatment by Hausdorff to be noted in §8.7.6; his coverage included cardinality, infinitude and transfinite induction, well-ordering and the need for axioms of choice, order-types, and continuity. No logicism was offered: Russell featured only in details such as defining cardinals from similar classes (p. 23) and solving his own paradox (p. 38, paradox mis-attributed to Burali-Forti). Instead, 'Set theory then starts, as it were, where logic stops' (p. 24), and, as his final remark, 'Set theory is no part of logic but its preferred daughter science, from whose inspiration many more results are to be awaited' (p. 78). He took this view because set theory contradicted basic logical laws such as the whole being greater than the part (p. 61–63); but in a review for the *Jahrbuch* Paul Bernays *1922b* rightly retorted that this law was not logical, at least not in the sense of being tautological, but only one obeyed by 'extensive magnitudes'.

Ziehen returned to this topic in much greater length but rather less merit in a huge textbook *1920a* on logic and its history. The 240 pages of the latter included only a short review of both traditions, though with a good bibliography (pp. 227–236). His own approach used a little formal logic in connection with *Mengenlehre* (pp. 410–416), but basically it was psychological and cognitive, using some strange notions which A. E. Taylor *1920a* assessed as 'almost unreadable' in a review for *Mind*.

8.7.3 *Frege's positive and then negative thoughts.*   Ziehen *1917a* exemplifies the higher level of attention which Frege was gaining in the 1910s, especially because of Russell's publicity. In particular, for his series of articles on the history of logic (§8.2.3) Jourdain sent his account of Frege to its subject in April 1910 and received detailed comments which he incorporated into the published version (Jourdain *1912a*, 237–269).[27] No

---

[27] The history of the manuscript of Frege's comments to Jourdain is curious. At Russell's request Jourdain had forwarded those folios that he could find in August 1911 (now RA); but he did not find the rest in his own notebook! (Russell *1967a*, 217). I was able to reconstruct the whole from these two sources (my *1977b*, 141), and they are published in full, with Jourdain's letters, in Frege *Letters*, 114–124; Jourdain's article is reprinted on pp. 275–301. The history of Frege's *Nachlass* was recorded in §4.5.1.

basic change of view was presented, but Frege stressed the central place in his system of the notion of thought, independent of thinkers and their mental acts or beliefs in its truth-value (§4.5.2).

However, Frege published nothing. He drafted a reply to Schönflies's proposed solution *1906a* of the paradoxes (§7.5.2) but did not complete it (*Manuscripts*, 191–199); and the failure may have led him to despair of finding one. When the young Wittgenstein visited Frege in 1911 for advice about studying logic, he was recommended to go to Russell. At that time and also two years later Rudolf Carnap took Frege's course on logic (Carnap *m1910–1913a*), and in later reminiscence he described more a wake than a lecture course, with the paradoxes never mentioned (*1963a*, 4–5). The material of the lectures included irrational numbers, and seemed to come from the third volume of the *Grundgesetze*; Jourdain asked after it when corresponding about his article (Frege *Letters*, 124–125), but it never appeared.

In 1918 Frege retired from Jena University in his late sixties, and moved the next year to the spa town of Bad Kleinen. Then he published three articles on 'Logical investigations' in a philosophical journal. In the first one he dealt with 'The thought', stressing its objective existence of a 'third realm' ('*drittes Reich*') separate from those of 'things' and of 'ideas' (*1918a*, 353) and relating it to the judgement of its truth-value (only two of those available, of course). The second article dwelt upon 'Negation', where he stressed that philosophical concern sometimes failed to recognise its place as an artefact of language: of the propositions 'Christ is mortal' and 'Christ is immortal', for example, 'Where do we now have here an affirming, where a denying thought?' (*1919b*, 369). He played down the action of negation in various contexts, especially in his final statement that 'a thought clothed in double negation does not alter the truth-value of the thought' (p. 378)—a position which Brouwer was challenging (§8.7.7). The third article appeared some years later, as Frege *1923a*: failing powers are suggested by the rather pedantic run through 'Compound thoughts', showing how the various logical connectives produce different compound propositions to which thoughts related but not making clear the relationship of the whole to its component parts. He did not mention the Sheffer stroke, or tackle the predicate calculus.

As with his notes for Jourdain, apart from the new prominence of the concept of thought, Frege exhibited no major changes of position. However, the draft *m1924b*? of a fourth article on 'Logical generality' was much more promising, and it is sad that he was not able to complete it before his death. Wondering how language translated compound thoughts into laws of physics, he focused upon the appearance of sentences and the letters of which words were composed. This led him to distinguish the 'auxiliary language' ('*Hilfssprache*') in which the physical discourse is conducted and the 'explanation language' ('*Darlegungssprache*') in which the propositions of the auxiliary language were studied. This is the distinction between object and metalanguage, like Russell's hierarchy (§8.4.3) but individuated mainly by Gödel and Tarski later (§9.2.3, §9.6.7).

The journal where these articles were published or intended was of a right-wing persuasion; Frege's contact was with his younger Jena colleague (and anti-Semite) Bruno Bauch (1877–1942), a neo-Kantian philosopher. In the mid 1920s, with Germany in economic collapse, Frege enthused in his diary *m1924a* over the rise of young Adolf Hitler. He also then wrote a sequence of short essays and diary entries which death also prevented him from developing into a book for a series co-edited by Bauch (Frege *Letters*, 9, 83–87). Maybe in a mood of negativism and professional disappointment, he *rejected* his life's aim: 'My strivings to bring light to the question as to what has been attached to the word "Number" by its particular number-words and number-signs, seem to have ended in a complete lack of success' (*m1924–1925a*, 285). Surprisingly, the main reason lay not in paradoxes but in continuity and infinitude. Proposing three distinct 'sources of knowledge' ('*Erkenntnisquellen*')—sense perceptions, logic and geometry with time—he claimed that only in the last source 'the infinite flows' (*sic*) where 'infinite' was used 'in the proper and strongest sense of the word', not the common sense which may mean merely 'very many'. Apparently 'The both-ways ['*beiderseits*'] infinite time is equal to a both-ways infinite line' (pp. 293, 294). Thus 'Counting, coming psychologically from a requirement of trading life, has misled scholars' (p. 297).

Frege's appeal to geometry, space and time recalls Russell's empiricism, such as his interpretation of the axiom of infinity (§7.9.3); but such analogies cannot be pressed hard between that committed reductionist and this enthusiastic Platonist. In any case, the consequences of Frege's new stance for his *general* position are not clear; for example, the difficulty over the paradoxes remain unassuaged. The reasons for it are also murky, and surely he exaggerated a fruitful philosophy to say that '*all* mathematics is properly geometry' (p. 297, my italics). If he had read the third volume of *PM*, he should have found a treatment there of continuity (§7.9.8) which could have been adapted to his system without drastic new assumptions. Was an old man, bitter over the lack of reception of his fine work, *trying* to throw it away now that his life would surely soon end?

Apart from Bauch and his colleague, nobody seems to have heard about the change of mind. Gottlob Frege died in July 1925, still largely a footnote to logic and philosophy.

8.7.4 *Hilbert's definitive 'metamathematics', 1917–1930*

The procedure of the axiomatic method, as it is expressed here, thus comes equally to a *deepening of the foundations* of the individual sciences [...]

Hilbert *1918b*, 148

From obscurity to fame; for we take now Hilbert's revival of interest in the foundations of mathematics. I avoid the common name 'formalism' for his position, because he never used it; we note its origins in §8.7.7. The sources for his work are of two kinds: unpublished lecture courses delivered at Göttingen University and usually edited and typed up by a follower; and papers, normally based upon public lectures. The story is very rich: only general features and aspects related to logic and logicism are treated here, though also some contributions by followers. More details may be found especially in Sieg *1990a* and *1999a*, Hallett *1995a* and G. H. Moore *1997a*.

After a dozen years largely elsewhere in mathematics (§4.7.5–6) Hilbert came back to foundations in the summer of 1917 with a course on set theory (§8.7.6); the last chapter dealt with the 'application of set theory to mathematical logic', of which 'the Russell-Zermelo paradox' was an example (*m1917a*, fol. 132). In September he spoke in (politically neutral) Zurich to the Swiss Mathematical Society on 'axiomatic thought', a lecture *1917b* which became better known (including the motto above) as a paper *1918b* in *Mathematische Annalen*. Emphasising syntax thoughout, he covered the four main features of axiomatisation: independence, completeness and consistency of systems, and to 'the *decidability* of a mathematical question through a finite number of operations' (p. 153). He mentioned a wide range of applications to both pure and applied mathematics.[28] One was to logic, where Frege's 'profound investigations' were followed by Russell's 'magnificent enterprise of the *axiomatisation of logic* as the crowning achievement of the work of axiomatisation as a whole' (p. 153), Bernays *1922b* reviewed the paper enthusiastically and at length in the *Jahrbuch*, without reference to logicism.

The scale of the enterprise emerged in Hilbert's course on the 'Principles of mathematics' delivered in the winter of 1917–1918; it was to be the most original of the sequence. After exemplifying the axiomatic method in detail with Euclidean geometry, Hilbert *m1918a* turned to 'mathematical logic'. In a novel division, he handled the four calculi separately: propositional, 'predicate and class' (a short and incomplete survey of syllogistic modes, but not quantification of the predicate), 'narrower functional' (first-order) and 'extended functional' (higher-order). However, for some reason he omitted the logic of relations. Perhaps inspired by Schröder, he put forward that kind of axiomatisation of the propositional calculus, including symmetries and duality. He also drew upon 'normal forms',

---

[28] Some of Hilbert's examples from applied mathematics were in fact not fully axiomatised theories; his choice was led by his current concern with the foundations of physics (Corry *1997a*). He showed his lack of patriotism by inviting Russell to lecture at Göttingen in 1917; Russell, without a passport because of subversive pacifism, had not been able to accept (Hilbert Papers, 339).

conjunctive or disjunctive, to which any formula could be reduced (especially fols. 176–182); this move simplified analysis of the main features. Consistency of the axioms was proved by assigning 0 or 1 to each proposition and its negation and showing that each axiom evaluated at 0 (fols. 150–151). 'Completeness', no longer carrying his model-theoretic connotation of the early 1900s (§4.7.3), conveyed two other senses: that every well-formed formula or its negation was provable, and that the addition of any non-derivable formula to the calculus rendered it inconsistent (fols. 151–153).

The paradoxes gained due attention, with a theory of 'levels' ('*Stufen*') of predicates proposed as the solution; but instead of the stratification of types of *PM*, each level contained all of its lower predecessors (fols. 213–226). In an original passage Hilbert used Cantor's diagonal argument to show that the number of one-place predicates '$F_n(x)$' in each nth level was not denumerable by considering the set S of those that were satisfied by only one value a of x, and treating that property itself as a predicate of the next level up; by definition, it could not belong to S (fols. 227–230). The procedure resembles that which Gödel was to use in his incompletability theorem (§9.2.3); Hilbert interpreted it as a difficulty in grounding set theory and mathematical analysis in logic (fol. 229). He also proved that the definition of identity at any given level could be proved only from the corresponding one at any lower level (fols. 230–235).

This lecture course was typed up by Paul Bernays (1888–1976), who was to became Hilbert's most important follower on foundational studies. He used it in his *Habilitation m1918a* on 'the axiomatic treatment of the logical calculus'. In fact he treated only the propositional calculus, but in a manner quite different from that of the course; in particular, he replaced Hilbert's symmetries with the axiom system from *PM*, an orientation of logic towards *PM* which was to become standard in the Hilbert school. However, after its manner he showed that the fourth of its five axioms for propositions was provable from the others; he also distinguished axioms from rules of inference (*modus ponens* and substitution), and tried some other systems by replacing certain axioms by additional rules. Unusually and perhaps due to wartime conditions, this important thesis was not printed; but several years later he published a paper containing some of its main features and results, including the new definition of completeness and the non-independence result (Bernays *1926a*, cited in §7.8.3).

Hilbert gave various courses on logic, foundations and/or set theory in the early 1920s, not all noted here. Bernays edited most of them, including one on the 'logical calculus'. Based much on the course *m1917a* on set theory, the main novelty was a 'start of a new founding of the theory of numbers' (Hilbert *m1921a*, ch. 3), where he considered the signs used to present arithmetic. More details were given in various public lectures in 1922, from which two papers resulted.

In the first paper, Hilbert *1922a* stressed the importance of axiomatics and secure foundations for mathematics, and of distinguishing signs from their referents. Using arithmetic as his case study, 'number-signs [. . .] are themselves object[s] of our consideration, but otherwise they have no *meaning* ['*Bedeutung*'] at all'. The theory of them was developed by rules of formation, such as setting up the compound sign '1 + 1 + 1' and (maybe) abbreviating it by 'the sign 3' (pp. 163–164). This was the beer-mugs stance of the 1890s (§4.7.2), now formulated in terms of mug-signs and clearly meaning by 'sign' its ideographic sense, not any inscriptive instance of it; '3', say, without concern for its font, size or printed colour (compare §4.5.9). In contrast to signs lay the properties of arithmetic proper, such as the 'principle of complete induction', a 'higher level of pertaining principle' (pp. 164–165); and to make the distinction still more important, he introduced into print the word 'a *metamathematics*' ('*Metamathematik*'), which 'serves for the securing of that' mathematics (p. 174).[29] The rest of the paper was taken up with rules for the signs, and axioms and rules of inference for arithmetic (including the propositional calculus); Gothic and italic letters were used respectively, although the same sign ' = ' acted in each theory. While not referring to *PM*, Hilbert must have known that many of its unclarities were removed by invoking metamathematics. But his own treatment was not always clear; he sometimes let the signs stand for their referents, such as '1' for 1 (Hallett *1994a*).

The second lecture-paper, Hilbert *1923a* in *Mathematische Annalen* on 'the logical grounding of arithmetic', was related to a recent lecture course *m1922–1923a* with the same title. Here he went further, in recognising that *both* the sign theory and its metatheory could have axioms, and he stressed more explicitly that the latter was finitary. In addition, he grounded both quantifiers and the axioms of choice in a new 'transfinite axiom', which announced that if a one-place predicate '$A(a)$' was satisfied by some object 'τa', then it was also satisfied by all the other applicable ones: Zermelo's form of the axiom of choice was proved (Hilbert *1923a*, 183, 191). He concluded by claiming that mathematical analysis could now be grounded 'and that of set theory opened up'.

Hilbert's later writings were largely individual lectures and papers, although he also gave two more lecture courses. As a spectacular case study, he sketched (but failed to complete) in *Mathematische Annalen* a proof of Cantor's continuum hypothesis from the axioms of arithmetic. Among these he included the transfinite axiom in a converse form and

---

[29] In an astonishing coincidence of timing, at this time the British applied mathematician Sir George Greenhill (1847–1927) introduced the word in his Presidential address *1923a* to the Mathematical Association on 'Mathematics of reality and metamathematics'. He referred to 'the Tyranny of Mathematical Rigour, converting our subject into one which Aristotle might have called *Metamathematics*'.

with a new Greek letter (*1926a*, ax. 3): for a propositional function $A(a)$,

$$\text{instead of } `A(\tau a) \rightarrow (a)A(a)\text{'}, \text{ now } `A(a) \rightarrow A(\varepsilon A(a))\text{'}. \quad (874.1)$$

He also outlined a means of constructing an infinitude of 'levels' (lemma 2).

Hilbert's crusade was popularised in 1928 by a 120-page textbook written with his former doctoral student Wilhelm Ackermann (1896–1962). The title, 'Founding of theoretical logic', was new, but the text was based upon three of Hilbert's earlier lecture courses, especially *m1918a*; so the book carried both men's names (Hilbert and Ackermann *1928a*). A few features will be noted: their parentages are not always recorded, though *m1918a* was especially influential.

The separate treatment of the four calculi was maintained: the propositional calculus was based upon the *PM* version as modified by Bernays, with the Sheffer stroke merely a 'curiosity' (p. 9). Two rules of inference were given: *modus ponens*, and the substitution of propositional variables (p. 23). The four main tasks of metamathematics dominated the later account, including both senses of completeness. The status of metalogic was stressed, but sometimes oddly; in particular, two propositions were 'equal-valued' ('*gleichwertig*', symbol ' $\sim$ ') if both were 'correct' or both 'false', while they were 'equal-referring' ('*gleichbedeutend*', symbol 'äq') if logically equivalent (pp. 4–5), so that the first notion rather than the second seems to be metalogical. Duality and normal forms were prominent, and the bibliography of principal writings included Schröder's *Vorlesungen* as well as (the first edition of) *PM* (p. 116). Curiously, for the narrower calculus separate axioms for existential and universal quantification were given (p. 53), although negation was freely used thereafter; the names 'free' and 'bound' variables were introduced (p. 46).

The main appearance of logicism came in the chapter on the extended calculus. Some of the paradoxes were stated; surprisingly in view of Zermelo's role, and moreover at Göttingen (§4.7.6), Russell was said to have 'first discovered' the one named after him (p. 93). As a means of solving them, the authors rehearsed their cumulative version of type theory, though only for functions of one variable and with the propositional hierarchy omitted (pp. 98–106). The axiom of reducibility now distanced them from logicism: a good survey of its unwelcome consequences included the theory of identity, and the manner of constructing non-denumerable classes and of defining real numbers (pp. 106–115). However, they did not seem to notice its effective dismemberment of orders.

In the same year, 1928, the latest International Congress of Mathematicians was held at Bologna. Hilbert led the German delegation at their first such attendance since the Great War, and spoke on foundational issues in mathematics in general; a revised version appeared in *Mathematische*

Annalen as *1929a*. Once again he exalted axiomatics and proposed various analyses for metamathematics, including the consistency of axiom $(874.1)_2$, the axioms of choice, impredicative definitions with the 'very problematic axiom of reducibility' of Russell and Whitehead (p. 3), and the completeness 'of the logical system of rules' including identity (p. 8). Mis-quoting Cantor's motto about freedom in mathematics (§3.6.2), he repeated from the 1900s (§4.7.5) that 'in mathematics there is no Ignoribamus' (p. 9), a hope for completeness which Gödel was soon to rebuff (§9.2.3).

Hilbert's metamathematics was then taking prime place among the competing philosophies of mathematics, both inviting conceptual questions and stimulating mathematical techniques. His followers prosecuted the doctrine; for example, Bernays gave an elementary course on mathematical logic at Göttingen in the winter of 1929–1930 (Bernays Papers, 973:212). But he retired in 1930, and his own health went into steep decline during the 1930s; from the mid 1920s he had been suffering from pernicious anaemia. Thus, although he was still to appear in print as author (§9.6.2), his career as a researcher was largely over.

8.7.5 Orders of logic and models of set theory: Löwenheim and Skolem, 1915–1923. (G. H. Moore *1980a*) Hilbert may have been influenced to stress normal forms by the use made of them in a remarkable paper in *Mathematische Annalen* written by Leopold Löwenheim (1878–1957). Like Eugen Müller (§4.4.9), Löwenheim passed most of his career as a schoolteacher (in Berlin), and was inspired to research by Schröder's algebra of logic (Thiel *1975a*, *1977a*). After some papers on the solvability of class equations in that calculus, Löwenheim *1915a* continued Schröder's way (446.8) of treating quantification as infinite con- or disjunctions; dual normal forms were even used with an infinitude of quantifiers of various orders.[30] Aware that this calculus was infinitary, he used infinite matrices (art. 3); he might have been drawn to them by their recent development, especially by Hilbert in connection with integral equations (Bernkopf *1968a*).

Löwenheim's main result stated that if a well-formed formula was not valid in any domain, it was also not valid in a denumerably large one (thm. 2). His proof amounted to a model-theoretic demonstration of the compactness theorem; but he did not seem to envision it as such (he cited only Schröder *1895a* on the logic of relatives), although he mentioned E. V. Huntington. The proof was simplified and the consequences considered in the early 1920s by the Norwegian Thoralf Skolem (1887–1962). None of

---

[30] In 1909 Löwenheim had been stimulated to study the foundations of arithmetic as a result of the non-discussion between Frege and Thomae (§4.5.9). His correspondence with Frege was destroyed before Scholz had transcribed it (§4.5.1), but he mentioned his motivation to Bernays in a letter of 25 May 1937 (Bernays Papers, 975:2938–2939; noted in Frege *Letters*, 161).

TABLE 875.1.   Views on the character of the predicate calculi

Figure	First order	Higher order	Infinitary
Löwenheim	Yes	No	Yes, horizontally
Skolem	Yes	No	No
Hilbert	Yes	Yes	No
PM	Yes	Yes	No
Zermelo	Yes	Yes	Yes, vertically

this work drew specifically on logicism, although the formulae involved lay within the calculus of *PM*; the closest contact came in Skolem *1923a*, where, after a weak criticism of Russell's theory of definite descriptions (art. 1) he founded arithmetic upon recursive procedures.

These researches heightened the role of model theory, because of the seeming paradox of the satisfiability of axiom systems due to Löwenheim's theorem. They also emphasised the differences between kinds of logic (first-order, higher-order, finitary and infinitary), and also quantification (Goldfarb *1979a*); no agreement was found, though the desire of Skolem to confine logic to first-order and finitariness gradually became common. The various positions on the use of a logic in set theory are shown in Table 875.1. Zermelo is there for an initiative of around 1930 (§9.2.5); and *PM* is included, although neither Whitehead nor Russell took part in the discussion.

8.7.6 *Set theory and Mengenlehre in various forms.*   We saw in §4.2 how Cantor's *Mengenlehre* became widely adopted in the 1900s, although not normally accompanied by his philosophy. It gradually spread beyond mathematical analysis to other branches of pure mathematics and into physical applications.[31] We pick up a few major texts from 1913 concerned with the foundational sides, with the main focus on logicism as usual. Most of the writing was in German.

In §4.2.4 and §5.7.2 we noted the two parts of Schönflies's report on *Mengenlehre* for the *DMV*, and their influence. In 1913, his 61st year, he published a heavily revised edition of the first part *1900a*; the second, on the theory of functions and related topics, was reworked later by Hans Hahn.

As before, Schönflies *1913a* covered both general and set-topological aspects, but he took 400 pages for the purpose. The new edition was better organised than before and the references were pretty complete; but, like

---

[31] Fraenkel's massive bibliography of set theory in *1953a* deliberately does not cover physical applications. Among a rather spotty literature, see F. Bernstein *1912a* on perturbation theory; Zermelo *1913a* on chess; van Vleck *1915a* on dynamics; and Bouligand *1928a* on potential theory, *1931a* on hydrodynamics, and *1932a* on differential geometry. Later literature includes Brush *1967a* on statistical mechanics.

last time, he deliberately left out the paradoxes and the relationship to logic (p. v). No reason was given; and in any case it does not excuse his virtual silence over Russell in the description of well-ordering and the axioms of choice (pp. 170–184). In a similar prejudice, the Peanists were dismissed in a footnote (p. 257).

Much more significant was a treatise on 'the basic characteristics' ('*Grundzüge*') of *Mengenlehre* by Hausdorff, published the following year, his 47th (Hausdorff *1914a*). We saw him praised and used in *PM* for his treatment of order-types (§7.9.5); here, like Schönflies, he covered both general and topological aspects of the subjects but left out paradoxes almost entirely—curious in a mathematician whose work often involved paradoxes in the colloquial sense of surprises (with some caution, Czyż *1994a*). Even Zermelo's axiom system received only a passing mention (Hausdorff *1914a*, 2–3, 450, this latter one of two citations of Russell's *The principles* in a rather spare appendix on literature). Dedicating the book to Cantor, Hausdorff followed him in working with an intuitive conception of sets, including the empty set (p. 3). He felt no apprehensions over the axioms of choice, mainly because he grounded Cantor's well-ordering principle in transfinite induction (pp. 133–139); so the painstaking analysis in *PM* was ignored. In his preface he described his book as a textbook—a fine but daunting one, where 'no difficulties remain and none but mild climaxes are reached', as Henry Blumberg *1920a* concluded in a long and admiring review for the AMS. Later Hausdorff produced a second edition *1927a*, with title reduced just to 'Set theory' and text also cut down by a third, and the authority lessened (as noted in H. M. Gehman *1927a*, also for the AMS); paradoxes and Russell were now absent.

However, in 1917 a new paradox had been introduced in the Swiss journal *L'enseignement mathématique* by Dmitry Mirimanoff (1861–1945), a Russian-born mathematician long on the staff of the University of Geneva. Taking any well-ordered class E containing an 'indecomposable element' e, he formed the class E' of all its 'segments'; then E' was similar to E, and had e as a member. Iteration of this procedure led him to consider descent of membership of classes in nesting segments, and especially to notice that the class α of all classes descending to e in a finite number of steps was paradoxical; for it admitted an infinite descent of self-membership, in inverse well-order: $\ldots \varepsilon \, \alpha \, \varepsilon \, \alpha$ (Mirimanoff *1917a*, 45–48).[32] He saw his paradox as in 'a form a little different' from Burali-Forti's. In a following paper, Mirimanoff *1917a* reworked his approach partly under the influence of König *1914a*, which he had reviewed

---

[32] The adjective 'grounded' has become attached to this kind of class (Hallett *1984a*, 185–194); Mirimanoff used no special name. In the course of expounding his idea of nested segments, he used the forms for his 'types 1, 2 and 3' which von Neumann *1923a* was to give in defining the ordinals 1, 2 and 3 in his own foundations of arithmetic (Mirimanoff *1917a*, 46). In a letter of perhaps 1908 to Russell G. G. Berry had named as 'epsilonic classes' those showing infinite ascent of membership (Garciadiego *1992a*, 177).

admiringly in *1914a*. Perhaps due to its appearance in wartime, neither Russell nor Ramsey seems to have noticed this new entry to their list of paradoxes.

A major new entrant into set theory at this time, after a first career in ring theory, was Adolf Fraenkel (1891–1965), as he signed himself at that pre-Nazi time. As well as papers, he wrote an introductory textbook while serving in the German Army during the Great War. Rejected by Teubner as part of their declining commitment to publishing mathematics (Fraenkel *1968a*, 135–136), it came out in 1918 from the house of Julius Springer, as part of their *rising* commitment; eventually there appeared three greatly expanded editions (the inverse order of revision from Hausdorff's practise!).

As indicated in the sub-title of the first edition, 'a comprehensible introduction in the realm of the infinitely large', Fraenkel *1919a* concentrated on the general aspects. Its 156 pages started out rather unhappily with an extensional conception of classes, so that the empty one was only 'so-called' (pp. 9, 13). The bulk of the 23 pages on paradoxes was devoted to Zermelo's axiom system, but among the paradoxes, he left out (or did not know) Mirimanoff's. However, he included Russell's in both its class and impredicativity forms, the latter as one 'with which mathematics has nothing to do', apparently (pp. 132–133). Logicism was not discussed.

In an important paper Fraenkel *1922a* clarified the notion of 'definiteness' in Zermelo's separation axiom by restricting the function involved to a finite number of logical operations; untypically of this ardent bibliographer, he had not (yet) noticed the anticipation by Weyl *1910a* (§4.7.6). This was one of his modifications to Zermelo's system which has led his name to be attached to it, and the acronym 'ZF'.

Fraenkel's second edition *1923a* was now called an 'elementary introduction'. At 251 pages, and larger ones, it followed the same structure as its predecessor, but with amplifications throughout, a much better subject index and now a name index, and more references (including Mirimanoff *1917a* on p. 152). Among the solutions of the paradoxes, the axiom system was preferred, and influence from Hilbert was also evident in a new passage on the completeness of Zermelo's system and more material on its consistency (pp. 226–241). But other solutions were noted, including eight pages on logicism and related attempts. Type theory was (too) briefly described and not welcomed, partly for its 'arbitrariness' and especially the 'weak point' of the axiom of reducibility; König's book was more warmly received (pp. 182–184).

Next year Kurt Grelling published an introduction to set theory in a Teubner series of short books on mathematics and physics. Confining himself to basic properties of classes, order and transfinite arithmetic, he finished with some of the paradoxes (including Russell's and his own); Zermelo's axiom system was the preferred solution, with just a few lines on type theory (Grelling *1924a*, 48).

Three years later Fraenkel published, also with Teubner, a short book *1927b* containing 'ten lectures on the founding of set theory' which, at the invitation of the logician Heinrich Scholz (1884–1956) (§9.6.3), he had delivered at Kiel University over five presumably hectic days in June 1925. Logicism appeared in various places: in connection with impredicativity (pp. 26–43); and with the relationships between mathematics and logic, where all main standpoints were noted (pp. 50–56). In a careless footnote he asserted that numbers were presupposed in type theory (p. 144). Axiomatisation naturally took the main place; Russell was excluded from the lengthy discussion of the axioms of choice (pp. 88–97). Presumably the lectures were well received; in 1928 he was to move to Kiel from Marburg when Scholz took a chair at Münster (Fraenkel *1968a*, 180–181).

In that book Fraenkel collected his references together in an excellent bibliography. Soon he followed this format in the third edition of his introduction, Fraenkel *1928a*, authoritative if rather long at 424 pages. On the axioms of choice Russell was now noted, for his example of infinite pairs of boots (p. 345).

Fraenkel divided the paradoxes into 'logical' and 'epistemological' ones, but with Grelling's placed in the first group; the latter was confined to those concerned with naming (pp. 210–218). Without comment he used the name 'logicism' to characterise the Whitehead/Russell position (in the title of the section on p. 244, explanation on p. 263): Carnap also introduced it around this time (§8.9.3), apparently independently and with a greater influence on others later. Fraenkel's account of logicism was much fuller than before (pp. 244–268), especially the place of both types and orders in type theory; presented as an exercise in non-predicativity, he mentioned commentators on the role of the vicious circle principle, especially Poincaré. Fraenkel's attitude remained cool, including to the role now given to tautologies, which to him seemed satisfactory for logic but not for mathematics (pp. 263); but when reviewing the three main schools, he inclined towards logicism (p. 384). However, he ended the book 'on the significance of set theory' (pp. 388–393).

Some years earlier Fraenkel had encountered an amazing newcomer to mathematics, the Hungarian schoolboy Janos Neumann von Margitta (1903–1957). Beginning in *1923a* with a new definition of ordinals using iterated nestings from the empty set, von Neumann next gave a new axiomatisation *1925a* of set theory; Fraenkel refereed the first paper and proposed the second one after receiving an astounding 14-page letter dated 26 October 1923 from the lad.[33] Logicism did not feature in von Neumann's concerns, which focused on axiomatisation and models, and so

---

[33] For recollections, see Fraenkel *1968a*, 166–169. I saw this letter in the presence of his widow in 1982, at her flat in Jerusalem; it is now in his Papers. It includes also letters from Carnap, Gödel and Quine. No copy of this letter exists in the von Neumann Papers, held at the Library of Congress, Washington.

were oriented towards Hilbert's proof theory, on which he also soon wrote a long paper *1927a* attempting to prove that 'mathematics' was consistent. *PM* was mentioned right at the end, where he claimed that consistency had been proved for that system 'without' the axiom of reducibility.

Another and older newcomer was Paul Finsler (1894–1970), supplementing a noted career in differential geometry. In his inaugural professorial lecture of 1923 at Cologne University, published by the *DMV* as Finsler *1925a*, he wondered 'are there contradictions in mathematics?'; after a review of all foundational traditions of that time, he concluded that circular definition was the true cause of malady, not only Russell's paradox but even Cantor's 1895 general definition of a set. However, he followed Cantor in the consistency-implies-existence kind of formalism, and Platonism over sets (§3.6.1). He wished to confine them to 'pure' ones containing only their own kind as members with the empty set as basis (hence a combinatorial aspect came in), and began his construction in a paper *1926a*.

Like von Neumann, Finsler took no special notice of mathematical logic or of logicism. Unlike von Neumann, his approach was not well received, and it has been neglected until recently, with editions and discussions of these writings (Finsler *Sets*, *Essays*). No set theorist, not even Fraenkel, bothered about Kempe's multisets.

8.7.7 Intuitionistic set theory and logic: Brouwer and Weyl, 1910–1928. (van Dalen *1999a*, ch. 8) However, Fraenkel took notice of another kind of set theory, introduced by the Dutch mathematician L. E. J. Brouwer (1881–1967) as a central part of his rethinking of the foundations of mathematics. His mathematical career shows two main and largely distinct parts: a) topology and dimension theory, to which he contributed brilliantly, partly by largely ignoring b) his philosophy of mathematics. Great mathematician but ghastly philosopher (and difficult person); we note here the latter two aspects.

The origins of Brouwer's philosophy lie partly in poor understanding of certain mystical texts, and partly on a naive reading of Kant's views on the place of intuition. He outlined his position in his doctoral thesis *1907a* 'on the foundations of mathematics' defended at the University of Amsterdam; being printed but not really published (and moreover in Dutch), it made little impact, although it received an unusually long review in the *Jahrbuch* from his friend and doctoral opponent Johan Barrau *(1910a)*. His position developed in various stages. A basic stand was to reject the law of excluded middle (hereafter, 'LEM'), once he had disentangled it from the propositional implication $A \supset A$ (van Dalen *1978a*, 300).

Much concerned with point-set topology, partly in connection with topology in general, Brouwer saw the Cantor-Bendixson Theorem 329.1 as 'fundamental' for perfect sets (van Dalen *1999a*, ch. 8; Hesseling *1999a*). However, he wished to reprove it: Cantor's proof had used the Ωth derived set of a set, where Ω was the initial ordinal of the third number-class, which young Brouwer had decided did not exist because the transfinite ordinals from ω onwards could not form a set. So with him the theorem became: '*If we destroy in a closed set an isolated point, in the rest set again an isolated point, and so on transfinitely, this process leads after a* denumerable *number of steps to an end*' (*1910a*, art. 2, his English, non-italics inserted). In *1918–1919a* he reworked *Mengenlehre* as a set theory with the LEM banned from the comprehension of sets but the transfinite ordinals still allowed. Then, in a three-part paper *1925–1927a*, published in *Mathematische Annalen*, he offered a general re-founding of mathematics, based upon set theory, order and ordinals.

Brouwer sometimes allied his position to the 'old intuitionism' of Poincaré and Borel, though not to Kronecker (for example, *1930a*, art. 1). His basic principles, some going back to 1907, were 1) mathematics was languageless, to be distinguished from any literal or symbolic presentation of it, of which there was no preferred version; 2) it needed for source only the 'primordial intuition' (hence the name) of time;[34] and 3) logic was part of mathematics, allied especially to arithmetic.

More valuably, in a footnote of his thesis Brouwer 4) distinguished between mathematics and 'mathematics of the second order, which consists of the *mathematical consideration of mathematics or of the language of mathematics*' (*1907a*, 61; see also p. 101). Sadly, he rarely mentioned this distinction later; but he publicised his position(s) regularly, even in book reviews. For example, despite having helped Schönflies much with the new edition of his report, he wrote a long nominal review *1914a* of it for the *DMV* in which he concentrated on his own position. (Later, in *1930b* he treated there Fraenkel's book of ten lectures in similar manner.) He polemicised against Hilbert's position, which he came to call 'formalism'; presumably it was intended as a criticism, for he caricatured it as the marks-on-paper-only brand of which Frege's opponents may have been guilty but which cannot be maintained against Hilbert. In response Hilbert was vigorous (see especially his *1922a*), seeing intuitionism as a dangerous successor to Kronecker's pessimistic constructivism which he had attacked in the 1900s (§4.7.5). He tried to turn the tables by *also* using only finitary methods, though at the metamathematical level rather than in the mathematics itself.

One factor which greatly helped Brouwer's cause, and probably also increased the intensity of friction with Hilbert, was that Hermann Weyl (1885–1955) became an ally after he met Brouwer in Switzerland in 1919.

---

[34] From this naive conception of time Brouwer infused his intuitionism with 'two-ities' of various kinds, such as before and after. Whitehead was then developing his 'process philosophy', especially in his *1929a*, using a far more elaborate conception of time as a spread-like interval where experience was 'anticipation tinged with reminiscence' (Lowe *1990a*, chs. 10–11 *passim*).

Weyl had been a doctoral student with Hilbert, working on integral equations; but he had shown early interest in foundational questions. An important influence was a lecture on 'Transfinite numbers' delivered in Göttingen by Poincaré on 27 April 1909 and published as *1910a*. Richard's paradox suggested that classes were denumerable, whereas Cantor had shown that the continuum was not so; following his usual line (§7.4.5), Poincaré had (dis)solved the apparent contradiction by pointing out that Richard's argument used an impredicative definition while Cantor's did not.

This lecture seems to have encouraged Weyl's interest in foundational issues. (As we shall see in §8.8.4, he was not the only such listener to be influenced.) Soon he published the paper *1910a*, cited in §4.7.6, in which he clarified Zermelo's separation axiom; it was based upon the lecture delivered at the defence of his *Habilitation*. He continued this line with a short book *1918b* on *Das Kontinuum*. Motivated by the importance of the paradoxes, early on he stated the paradox of heterologicality (p. 2); however, he did not mention Grelling although he must have known of its Göttingen genesis (§7.2.3) when he was a student there, and it was and is often mis-attributed to him. For solution/avoidance he followed Poincaré in adopting the vicious circle principle as fundamental and using mathematical induction to define the cardinals, and he attempted to rework in predicative form all the basic components of mathematical analysis, including Zermelo's axiom system and the continuum, up to (but not including in detail) measure theory (Feferman *1988a*). Hilbert's system was the main guide for his logical calculus, including some notations; as far as logicism was concerned, he accepted the principle and the simple theory of types but not the axiom of reducibility, so that predication was restricted to the first order (pp. 35–37). Rademacher *1923a* described the book at unusual length in the *Jahrbuch*.

Weyl's opinion that 'To formalise is indeed the mathematician-illness', expressed in a paper on vicious circles published by the *DMV* (*1918a*, 44), and his silence over *PM*, show that he sought salvation in neither formalism nor logicism. Seemingly impressed by Brouwer's construction of the continuum, more sophisticated than his own, he switched to Brouwer's position while avoiding most of the philosophy; for example, like Poincaré he took mathematical induction as primitive (and in its normal form), and considered some consequences for applications (Beisswanger *1966a*).

In a succeeding paper Weyl *1921a* coined the phrase 'foundational crisis' ('*Grundlagenkrise*'), with principal reference to the conflict between Brouwer and Hilbert; however, he did not describe a crisis as such but presented his own and Brouwer's versions of the continuum. (Perhaps sensing weakness in the camp, Fraenkel *1925a* noted differences between the two versions in a long piece for the *Jahrbuch*.) Nevertheless, the paper was read widely, and his phrase became popular to the extent that Hasse and Scholz *1928a* even planted a 'foundational crisis' upon ancient Greek mathematicians, for their alleged treatments of irrational numbers; this claim has long been a source of historical misunderstanding.

Among Weyl's own occasional later contributions, a 40-page essay on 'the current state of knowledge in mathematics' did include logicism; while he praised the effort put into *PM*, he found intuitionism to be 'an achievement of the greatest epistemological weight' (*1927a*, arts. 3–4). In a contemporary book on 'the philosophy of mathematics and natural science' Weyl *1927a* began with two chapters on logic and foundations, favouring 'intuitive' over 'symbolic mathematics' (arts. 9–10); but he ignored logicism, although type theory had been mentioned in the preceding passage on set theory. This book also received an unusually long description in the *Jahrbuch*, from Helmut Grunsky *1935a* (by the time of its tardy publication, the editor of the journal); he did not mention the silence over logicism.

Thus the impact of intuitionism upon logicism was much less than that upon metamathematics. Brouwer mentioned *PM* very occasionally, and only as a system sadly using the LEM; Weyl usually referred only to Russell's paradox, and to type theory and the regrettable axiom of reducibility. Conversely, Russell *1925a* merely listed Brouwer and Weyl in the bibliography of his new introduction to *PM*. Ramsey referred to them occasionally and dismissively in his writings (for example, *1926a*, 65–67), and made notes on Brouwer's notion of 'absurdity of absurdity' in Brouwer *1925a* and elsewhere, used to found intuitionistic logic (Ramsey Papers, 6-06–07). In his paper on mathematical proof (§8.5.2) Hardy *1929a* dismissed intuitionism, and also mis-represented 'formalism' as mere marks-on-paper philosophy (he twice lectured on 'Hilbert's logic' while on tour in the U.S.A. early in 1929).

The conflict between metamathematics and intuitionism was sharper than would now be recognised because then intuitionism was often identified with finitism in general. There were also personal factors: in particular, Brouwer had been a member of the editorial board of *Mathematische Annalen* since 1914, but Hilbert removed him in 1929 by the device of sacking everybody and inviting back everybody else. (Not everybody accepted: see van Dalen *1990a*.) The "betrayal" of Hilbert's former graduate student Weyl must have raised the temperature; nevertheless, when Hilbert retired in 1930, Weyl succeeded him.

These personal aspects, ideal for coffee-time chats, increased the notoriety of the conflict, helping the eclipse of logicism. However, the number of *professing* intuitionists remained small. On a trip in Europe, Huntington told E. H. Moore on 28 October 1928 that Brouwer 'has no large following in Europe, even in Holland'. Again, during the spring of 1929 Bentley spent six weeks 'in the neighbourhood of Brouwer' but, probably in a *double entendre*, reported that intuitionism 'can easily be reduced to a patchwork of absurdities';[35] in the chapter on intuitionism in his book on mathematical philosophy (§8.5.5) he was to be untypically sarcastic (*1932a*, esp. ch. 9). However, when outlining his philosophy in Vienna in 1928, Brouwer made some impact (§8.9.5).

8.7.8 *(Neo-)Kantians in the 1920s.* By far the most substantial contribution by a mathematician came from Otto Hölder (1859–1937), who had been trained in Weierstrassian mathematical analysis but had since worked notably also in abstract algebra and in geometry. In addition, he treated philosophical questions, including an extended essay *1914a* on the real number system concentrating upon compatriots' theories but with Frege's as well as Russell's ignored: Weyl had sub-titled the paper *1918a* on vicious circles as an extract from a letter to him.

Hölder presented his standpoint in 1924 in a large book on *Die mathematische Methode*. The title was unfortunate; as his impressive coverage shows, mathematics *has* method*s*, several of them. He included mathematical analysis, and some aspects of mechanics and physics. Set theory was quite prominent in the treatment of analysis, and in the second of the three Parts he considered the 'Logical analysis of methods'. Largely guided by Kantian and some phenomenological concerns, he saw method as concerned with treating given concepts and proof methods as objects for study by higher-stage 'synthetic concepts'. Not to be confused with synthetic judgements in Kant's sense (*1924a*, 361), these concepts were built up from given ones by some mathematical construction; for example, the centre of mass of a given collection of mass-points, or the creation of a square upon a given line (pp. 292–297). This structuring sounds like metamathematics, and he noted similarities with Hilbert on some contexts (pp. 319–326), although his distinctions were not so clearly drawn or technically developed.

On logic itself, Hölder dealt with some older questions, such as Lewis Carroll's worry (§6.7.4) over the paradox of implication (pp. 269–272). He followed mathematical logic without deploying much of the machinery; the predicate calculus was construed in terms of 'species-concepts' ('*Gattungsbegriffe*'), relations and judgements (pp. 247–253, 272–292). He discussed 'Russel' a certain amount, mainly citing *The principles* and Couturat *1908a*, but only on specific issues such as abstraction and the continuum (pp. 260, 138). Agreeing with Natorp *1910a* (§8.7.1) that 'logistic' was 'a comparably subordinate province' of mathematics (p. 5), he saw it mainly as a 'calculus', mixed with the algebraic tradition (pp. 272–274, including one of only two citations of Schröder).

Hölder closed with a nice appendix distinguishing 'Paradoxes and antinomies' (pp. 533–556). The former were just mysteries, such as dividing by

[35] Huntington letter in E. H. Moore Papers, Box 2, Folder 1. Letters from Bentley of 29 April 1931 to the Open Court Publishing Company, and of 21 March 1930 to E. D. Grant, in Bentley Papers, Box 1.

zero or the odd integers being isomorphic with all integers, and Kant's antinomies were also only 'so-called'. The real ones were those of Burali-Forti, Richard and Russell; for this non-logicist the last was 'at its base not mathematical' but concerned difficulties over handling species (the rather disappointing pp. 551–552).

Hölder's book is now undeservedly neglected; then it was well recognised, with several reviews. Wrinch *1925a* in *Mind* found it to be too daunting to say anything useful, A. R. Schweitzer *1926a* in the *Monthly* liked it in general but was surprised that *PM* was ignored, and Fraenkel *1927a* in the *Jahrbuch* admired it at some length while querying the treatment of set theory.

Other readers included two fellow senior German mathematicians with philosophical interests. We saw in §6.4.7 that Max Pasch (1843–1930) had anticipated Russell's theory of irrational numbers: in later life he wrote also on philosophical questions in mathematics, but his concerns lay with formation of concepts, axiomatic set theory and proofs as practised by mathematicians (hence overlapping with Hölder). However, he did not consider logic explicitly, not even in a short book *1924a* on 'Mathematics and logic', reprinting four earlier articles. He was answered by Eduard Study (1862–1930), a student of Klein who had made his mathematical mark in line geometry and invariant theory but was also concerned with cultural (including historical) aspects of mathematics. In a short book musing upon 'thought and presentation in mathematics and science' he took issue with Pasch on ignoring the creative sides of mathematics (Study *1928a*, 5, 27). Surprised that Pasch had not mentioned 'l o g i s t i c i a n s', he referred to Russell in a footnote when acknowledging the great extensions made to syllogistic logic (p. 7); however, he regarded the paradox of implication (§6.7.4) as a 'thought-error' (pp. 32–33). He praised Hölder's book (p. 36), and also Whitehead's little introduction *1911c* (§8.2.1) on mathematics (pp. 45–46).

Hilbert's former doctoral student Heinrich Behmann (1891–1970) published a short introduction *1927a* to 'Mathematics and logic', in the same Teubner series as Grelling *1924a* (§8.7.6). He placed a sketch of the recently deceased Frege on the cover; but he ran closer to *PM* in his presentation of logic and set theory, including some paradoxes and cardinal arithmetic. His innovation was notations: nice combinatorial designs to display kinds of relation, but a rather clumsy system of overmarks, sub- and superscripts for the predicate calculus (which he construed in terms of 'concepts'). Some exercises were given at the end. In a review for the AMS A. A. Bennett *1930a* preferred E. H. Moore's notations in general analysis (§4.2.7), which led Behmann in 1931 to seek and receive enlightenment from their creator.[36]

---

[36] E. H. Moore Papers, Box 16; Behmann Papers, File I 49. Behmann's correspondence with Russell is in File I 60, and in RA.

Behmann's friend Walter Dubislav (1895–1937), a Berlin philosopher with mathematical training including with Hilbert, first expressed views in an unusual setting; his entries for a 'systematic dictionary of philosophy' prepared with a medical colleague (Clauberg and Dubislav *1922a*). Under 'Logic, mathematics', mainly quotations from Kant, he did not distinguish 'logistic' from the 'algebra of logic' (pp. 275–279); and the latter suffered, for he followed Cantor in some detail over both 'Menge' in mathematics and 'Klasse' in logic (pp. 292–299, 249–251), giving Cantor's definition of integers in 'Number' (pp. 545–557) and summarising Zermelo's axiom system in 'Sets, measurable' (299–302). Even 'Part' was handled Cantor's way (p. 462), with part-whole theory treated briefly elsewhere (pp. 177–178, 191–192). However, he contrasted Kantian 'Antinomy' (pp. 26–28) with 'Paradox' (pp. 325–326) such as Russell's and Cantor's. Propositional functions appeared under both that name and 'Function' (pp. 175–176), contrasted in the latter entry from mathematical functions.

Logicism fared quite well, with an accurate summary of 'Hierarchy of types' including a warning that 'Russel' did not always distinguish a sign from its referent (pp. 218–220). But in a paper 'on the relationship of logic to mathematics' Dubislav *1926a* rejected logicism on the grounds that logic and mathematics needed different axioms. However, he considered *PM* in a short book 'On definition', building upon a careful entry in the dictionary; I take the second edition, *1927a*. After a list on p. 6 of much of the relevant literature (where he acknowledged advice from Scholz), he ran through four prevailing theories of definition: specification of an essence, of a concept, 'establishment (not stipulation)' ('*Feststellung* (*nicht Festsetzung*)') of the referent of a sign, and *vice versa* of the latter. Then he drew upon Gergonne *1818a* (§2.4.6) to discuss implicit definitions, cited Burali-Forti *1919a* (§8.6.1) on definition by abstraction, and accepted Frege's strictures (§4.7.4) against creative definitions (pp. 51–57). However, he left out both contextual definitions and definite descriptions even though, as a case study in predicativity, he listed the primitives and axioms of the propositional calculus of the first edition of *PM* (pp. 37, 42–46). The much revised third edition (1931) of his book is described in §9.3.5.

Lastly, in one of his finest books, Cassirer provided a massive study of 'The philosophy of symbolic forms' as a historico-philosophical study of signs.[37] In the introduction of the first volume, published in 1923 when he was in his early fifties, he explained that he wished to extend his earlier philosophy of mathematics and science (§8.7.1) to cultural studies in general. One principal device was to merge Kant's distinction of regulative forms for knowing the world into the constitutive ones used in achieving that knowledge, calling them all 'symbolic' in order to emphasise the central place of signs. To meet this end he took a wide-ranging look at

---

[37] An edition is in preparation of Cassirer's manuscripts, which are held mainly at Yale University (Krois *1997a*); many of them relate to this book.

languages, including a survey of 'the linguistic development of the concept of number' which contrasted the absence of intuition in the definitions of Dedekind, Frege and Russell (but hardly *linguistic* definitions in the first place) with number systems in African tribes (*1923a*, ch. 3, sect. 3). His principal treatment of mathematics came in the third volume, devoted to 'The phenomenology of knowledge'. In its third part, on 'the building up of scientific knowledge', he gave an intelligent survey of all the main foundational positions for mathematics (*1929a*, pt. 3). His Kantian inclinations (§7.5.2) led him to find fault with all of them; for Russell, the issue of intension versus extension, and the lack of grounding of the theory of types. He also considered phenomenology, a tradition to which we now turn.

8.7.9 *Phenomenologists in the 1920s.* Husserl retired from Freiburg University in 1928, to be succeeded by Heidegger; but his phenomenological school continued vigorously, although with diminished connections to arithmetic and set theory. This is well shown by two book-length articles of that time in the *Jahrbuch für Philosophie und phänomenologische Forschung*, which he had founded in 1913. His own account of 'formal and transcendental logic' included 'apophantic analytic', his hardly limpid name for that part of formal logic which sought conditions under which judgements could be made 'clearly' (that is, consistently). In this programme set theory and 'ordinal numbers of various levels' were regarded as 'not apophantic mathematics' (*1929a*, art. 24); so Cantor, Frege and *PM* were out (though Boole appeared in art. 25), and the treatment was almost entirely prosodic.

Husserl's article finished with a short note on tautologies by his assistant and former doctoral student at Freiburg, Oskar Becker (1889–1964), who had written on 'Mathematical existence' two years earlier. However, Becker's chief concern was the conflict between formalism and intuitionism viewed in phenomenological terms.[38] Thus both Cantorian *Mengenlehre* and later set theory featured quite well; but logicism was rarely mentioned, and then mainly on the shattered hopes for a logical foundation for mathematics because of the paradoxes, and on König's reading of type theory as the objectification of thought processes (Becker *1927a*, 446, 556–559). He gave time a high status, which Cassirer queried in a note appended to the end of his own consideration of scientific knowledge (*1929a*, pt. 3).

The Rostock *Privat-Dozent* Wilhelm Burkamp (1879–1939) published in the same year as Becker a study of 'concept and relation' for the 'founding

---

[38] In an interesting exchange of letters in the autumn of 1930 following the receipt of Becker's article in book form, Hilbert recalled to him the impact of Dedekind's booklet on integers and saw Becker's reservations over metamathematics as a source of strength. In reply Becker wondered about the obviousness of the ε-axiom $(874.1)_2$ (Hilbert Papers, 457).

of logic'. His innovation was to take individuality as a basic notion with concept as its opposite. Thus he construed logic in terms of form rather than existence or meaning; for example, he lamented Lewis's failure to argue for strict implication (§8.3.3) rather than its material counterpart on grounds of the 'conceptually logical sense of the calculus' (*1927a*, 173–175). Nevertheless, when explaining quantification and relating the universal to the existential he used necessity (pp. 169–172).

Burkamp considered in some detail the constructions of arithmetic by Frege (to whom he dedicated his book) and in *PM*. But he took Russell's new extensionality too far in mistakenly attributing to him instead of Schröder the view that 'Brown and Jones together are a class, and Brown alone is a class' (p. 185). Yet when discussing the definitions of cardinals he clearly distinguished an object from its unit class, and preferred the empty class to Schröder's empty manifold (pp. 224–229). His study of part-whole relationships included cases where parts existed only because of a pertaining whole (husband in a married pair, for example), and he also deployed Euler diagrams and Gergonne relations (§2.4.6) when discussing Schröder (pp. 190–196). A thoughtful book which gained positive reviews for its philosophical acuity (A. C. Benjamin *1927a* in the *Journal of philosophy*, J. N. Wright *1928a* in *Mind*), nevertheless its treatment of collections is difficult to grasp, and it did not make a large impact.

In a sequel analysis of 'the structure of wholes', conceived very generally with flames and ocean-flows among the many examples mentioned, Burkamp examined parts and moments. While he touched upon counting and well-ordering, he hardly used either Cantor or Russell, and took counting to be the epistemological basis for both cardinal and ordinal arithmetic (Burkamp *1929a*, 37–41). Again, he was not widely read.

One welcoming reader of Burkamp's first book was Cassirer *1929a*, with several statements of praise. Another was Gerhard Stammler (b. 1898) at Halle, who soon published his own book on 'Concept judgement inference'. Distinguishing in his foreword 'logistic' from 't h e   a l g e b r a i s i n g   t e n-d e n c y   o f   t h e *logical calculus*' (Stammler *1928a*, ix–xi), he recalled as his original aim 'to introduce the logic calculus into philosophy', but then saw it as only 'an exact technique of logic'. So, partly under Burkamp's influence, he switched to a study of logicism, partly Frege's but mainly the new edition of *PM*; regrettably, the Peanists were almost entirely ignored. In the second part he contrasted the two traditions as between 'the calculation of ranges' ('*Umfangsrechnung*') in the part-whole sense and 'the calculating with truth-values'; and to the axioms of the latter's propositional calculus he added one stating '*Each axiom is true*' (p. 209). To us this newcomer is a meta-axiom, and indeed Stammler moved in the same direction, such as noting that Russell had not explicitly stated rules of substitution (p. 210). The most startling manifestation came at the end, where he noted that determining the limits of logic was an 'extralogical problem' (p. 294) and so raised the possibility of 'Metalogic'. However, he rejected it, on the grounds that logic could be appraised by the trio of 'thoughts' of his title, which were handled in broadly Kantian styles: thus 'There is no metalogic as extralogical grounding of logic. Logic stands for itself' (p. 317). Stammler reached a major issue in logic at his time, but offered a conservative response to it. Perhaps for this reason, like Burkamp he has never gained the attention that he deserved; the excerpts above do not do justice to his book.

Among other groups, the Fries school (§4.2.5) maintained some interest. Gerhard Hessenberg came back to philosophical questions in his late forties when he took a chair at Tübingen University. He devoted his opening lecture *1922a* to the 'sense of number', including foundational questions; but he ignored logicism and metamathematics completely, appealing to Cantor and Dedekind (p. 28) and to group theory rather than relations for handling transformations (pp. 38–45). He even spurned the notions of function and variable in handling numbers, citing Thomae's sequence (459.1) of '3's of different sizes as a counter-example (p. 37), though that had hardly been Thomae's own point.

Some years later Grelling surveyed German 'philosophy of the exact sciences' for *The monist*; in the section on mathematics, he treated intuitionism, set theory, axiomatics such as with Pasch, and above all metamathematics (Grelling *1928a*, 99–106). Typically of German authors on the foundations of mathematics at this time, the logic of *PM* was noted but logicism gained least attention, here none at all.

8.8 THE RISE OF POLAND IN THE 1920s: THE LVÓV-WARSAW SCHOOL

8.8.1 *From Lvóv to Warsaw: students of Twardowski.* New country, new initiatives: one consequence of the defeat of Germany in 1918 was the re-creation of Poland to its east, and very soon major schools of mathematicians and logicians emerged. Moreover, many of the mathematicians worked in set theory, so that links with logic could be close. This impression was confirmed in 1920 when the Poles launched in Warsaw a journal, *Fundamenta mathematicae*, to cover both disciplines (Lebesgue *1922a*). But in fact the union was not successful; little logic appeared there, and the two co-editors for logic resigned in 1928, with little regret from the mathematicians (Kuratowski *1980a*, 33–34).

Those two had become the leaders of the new school; and their background goes back to a common teacher, the Austrian-born philosopher Kazimierz Twardowski (1866–1938). Trained at Vienna largely in phenomenology under Brentano's influence (Twardowski *m1927a*), he moved in 1895 to a chair at the university at Lvóv, which was then the town named 'Lemburg' in the Austro-Hungarian Empire. While not a major philosopher (B. Smith *1994a*, ch. 6), he did much to raise the professional status of philosophy in Poland to a level comparable with his home country and Germany. The university came to have a fine seminar room for philosophy, incorporating his own personal library.

Although not a specialist in logic, Twardowski taught it from 1898, including the algebraic tradition; thus it came to the attention of his many students. Two especial beneficiaries were the future leaders, Jan Łukasiewicz (1878–1956) and Stanisłav Leśniewski (1886–1939). After taking doctorates at Lvóv, they were given chairs at Warsaw University in 1915 and 1919 respectively. They had begun to publish before the re-creation of Poland, especially in philosophical journals which Twardowski had helped to create; but the main flowering dates from 1920.

During the 1920s, Twardowski's students and grand-students formed the largest community in the world working on logic and related topics. Partly under his influence, they tended to start out from specific problems rather than build up large systems Russell-style, although Leśniewski became such an architect later. Many of them published in their own language (which, I am told, contains many properties relevant to logic and set theory, which may have helped stimulate the interest in the first place). But, knowing that the consumption would be almost entirely local, they also published in standard languages, and abroad. Thus their achievements gradually became known, especially during the early 1930s.

However, their country came to an end with Hitler in 1939; Leśniewski had died in May, and of the others several lived or died at home, or went away (Tarski to the U.S.A. just before the War, Łukasiewicz to Ireland just before its end). Many manuscripts were lost, including Leśniewski's during the Warsaw insurrection and Łukasiewicz's later by bombing; some publications also stopped, especially the journal *Studia philosophica* which had started in 1935 to replace the *Fundamenta* for logic, and Łukasiewicz's new *Collectanea logica*.

But historical interest has been considerable. The survey Z. Jordan *1945a*, in English, quickly brought information, and much of it was reprinted in the English edition McCall *1967a* of main papers by various authors. Since then, a new generation of students has examined the history of modern Polish philosophy in general; among products not in Polish, Wolenski *1989a* is especially important (ch. 2 here), and Szaniwaksi *1989a* more generally but also on the Vienna Circle. Editions of many of the works of the leading Warsaw trio (and others) have also been published, and further material is being prepared under the general direction of Jan Srzednicki. The trio occupies the next two sub-sections; in §8.8.4 we consider Leon Chwistek, not a member of the school, but upon whom Russell's influence was the closest.

8.8.2 *Logics with Łukasiewicz and Tarski.* Łukasiewicz's first book, published in his 33rd year, was an analysis *1910a* of the status of the law of contradiction in Aristotle's logic; efforts to prove it brought him to consider paradoxes, including Russell's. As a result he tried to construct a non-classical propositional logic; success came in 1918, which he advertised twice, in a Polish journal. Łukasiewicz *1920a* was a note stating a 'three-valued logic' in which propositions could take the values 0, 1, and 2, the latter named 'possible' (and later symbolised '1/2'). His innovation was contemporary with that of Post (a fellow Pole, incidentally: §8.3.4), but independent. In a longer successor Łukasiewicz *1921a* presented 'two-valued logic' (a main origin for this name, by the way), so that the three-valued case could be seen as a natural extension. His use of propositional quantification followed the algebraic way as infinite con- and disjunctions. He drew upon both Peirce and Schröder, as presented in the survey of algebraic logic in Couturat *1905a* (§7.4.1), which had recently been translated into Polish as Couturat *1918a*; but he also adopted truth-values and assertion from Frege.

At that time Łukasiewicz met Alfred Tarski (1902–1983), Leśniewski's doctoral student (in the end his only one). Taking his doctorate in 1923 (his 22nd year), Tarski quickly published a version of it as two papers in French in *Fundamenta mathematicae*. Like Bernays (§8.7.4), he examined the propositional calculus, and started out from *PM*; like Łukasiewicz, he worked with the equivalential form (which is now known as the 'equivalential calculus'). In the first paper, Tarski *1923a* defined all the other connectives from the bi-conditional, and rebuilt the calculus including universal propositional quantification. Interestingly, he cited Peirce *1885a* for the origin of the word 'quantifier' (§4.3.8), and also interpreted the resulting formula as 'involving logical multiplication' (pp. 15–16). He could also replace equality by definition, a primitive notion in *PM*, by stating the equivalence of propositions appropriate for the defined term and the defining clause: for example, for negation, he offered

$$'[p] :. \sim (p) \equiv : p \equiv . [q] . q ' \text{ instead of } \sim p . = : p . \supset . (q)q \text{ Df.}$$

$$(882.1)$$

in the style of *PM*, where I quote his use of square brackets to mark quantification (p. 16).

In the second paper Tarski *1924a* was concerned with truth-functions f of propositions, which he defined by a new '*law of substitution*'

$$'[p, q, f] : p \equiv q . f(p) . \supset f(q)';$$

$$(882.2)$$

in the usual formulation f was not quantified. He found various necessary and sufficient conditions for f to be truth, noting the similarity of some to Boolean expansion theorems (255.5) (p. 29). He extended the conditions to functions of several propositions (pp. 35–37).

Tarski became Łukasiewicz's assistant, but for the rest of the decade he published little on logic, concentrating on transfinite arithmetic and point-set topology. Łukasiewicz himself published very little, though in the note *1925a*, cited in §7.8.3 with Bernays *1926a*, he showed that the propositional calculus in *PM* contained a redundant axiom. But in 1930 professor and student came together with three papers placed with the Polish Scientific Society. Firstly, Tarski *1930a* treated 'some fundamental concepts of metamathematics', formalising logical consequence in a Hilbertian frame-work; one outcome was the deduction theorem, as it was to become known. The calculus itself was deductive, not necessarily a logic, an enlargement which he was soon to stress (§9.6.7). Then Łukasiewicz and Tarski *1930a* presented the propositional calculus both without and with quantification, and admitting any number of truth-values even up to $\aleph_0$. They also used Łukasiewicz's bracket-free notation for connectives based upon '*Np*' for '$\sim p$' and '*Cpq*' for '$p \supset q$', with compound propositions expressed by concatenation and rules of binding; Łukasiewicz had developed it in 1924, following a suggestion by Chwistek, and this paper was an early appearance. Finally, Łukasiewicz *1930a* defended these new calculi philosophically, using quotations from Aristotle to help. He also displayed their truth-values in 'matrix' form, an extension of bivalent truth-tables after the style of Benjamin Peirce's tables (432.4).

Other students of Łukasiewicz, such as Jerzy Słupecki (1904–1987) and Mordechai Wajsberg (1902–1939), took up these new logics (Wolenski *1989a*, ch. 6); but the advance beyond two-valuedness was achieved at some pain for Łukasiewicz. News of the innovation had been circulating, but usually both he and Tarski were credited as its originators. Upon being told the truth by Łukasiewicz, Lewis *1933a* corrected his own impression in a note in the *Journal of philosophy*; but the error generally remained in place, and clouded the rest of Łukasiewicz's life.[39]

8.8.3 *Russell's paradox and Leśniewski's three systems.* (Wolenski *1989a*, ch. 7) Leśniewski's main stimulus to mathematical logic came when he learnt of Russell's paradox in Łukasiewicz's book on Aristotle. He published a few papers in Polish journals up to 1916: in particular, in

---

[39] Lewis *1933a* corrected the assumption made in Lewis and Langford *1932a*, a textbook to be described in §9.4.1. When G. E. Moore mentioned 'Tarski's "3-valued" Logic' in an article on Wittgenstein's lectures of the early 1930s (*1954a*, 300), Łukasiewicz wrote to him on 21 January 1955, bitter that 'Tarski does nothing to rectify' the situation (Moore Papers, 8L/26). Moore published a correction when reprinting his article, while making clear that his records seemed to record Wittgenstein's understanding accurately (*1959a*, 324). In 1955 a selected English edition of Tarski's papers was being completed, and Łukasiewicz's letter to Moore suggests that he himself wrote the correcting note at the head of the translation of their joint paper (Tarski *Semantics*, (1956), 38). In 1977 his widow wrote an equally bitter letter to the editor of an Italian journal in logic, which was published there as Łukasiewicz *1990a*. On the other hand, the Łukasiewiczes had been regarded during the War as Nazi sympathisers.

Leśniewski *1914a* he tackled Russell's paradox by considering the relation-ship of the name of a class to its members (Sinisi *1976a*). If K were the class of bs, then it was also both the class of classes of bs, and (the class of bs or the class of bs), and so on; he deduced that *every* class was subordinated to itself, Russell's paradoxical one included. 'Russell's "paradox" is "slain"', he ended, 'Let it rest in glory!'. However, he must have realised, as Russell had long before, that this association of K with its defining property b was too simple; for example, the class of cardinal integers may be taken as the union of even numbers and odd ones, but the sub-class of, say, multiples of 7 cannot be so specified.

Another durable early reaction to *PM* was negative. Leśniewski found the notion of assertion unsatisfactory, because as a truth-functor it mud-dled up a proposition p with its name; further, '$\vdash p$' was not a proposi-tion, so where did it belong? The consequences of avoiding assertion were profound: 1) abandon assertion and instead distinguish a language from its metalanguage; 2) shun psychological predicates such as asserting and believing, and also the abstract objects that pertained to them, and instead adopt nominalism (this in tune with a predilection for J. S. Mill's philoso-phy of logic (§2.5.7) from his youth); and 3) for the same reason, treat logic extensionally as much as possible.

Leśniewski deepened his analyses with a subtlety hitherto unsurpassed, and gradually evolved three doctrines. His disenchantment over assertion did not prevent him from planning a Polish translation of *PM*; he wrote to Russell about his intention on 12 October 1923 (RA), though nothing seems to have been done in the end. He regularly taught logic and foundations of mathematics after receiving his chair at Warsaw in 1919; but he did not go to print again until 1927, when in a long paper in several chapters, written in Polish and dedicated to Twardowksi, he repudiated several early papers (though not *1914a*) in the course of rejecting asser-tion, appreciating the superiority of Frege over *PM* in the handling of quotation marks, and reminiscing about his researches generally (Leśniew-ski *1927–1931a*, 182, 189–193, 198). The reception of his mature trio of doctrines was more gradual than that for Łukasiewicz or Tarski, but interest has increased in recent decades, partly inspired by the general study Luschei *1962a*; Polish non-readers have been much helped by recent English translations of his writings, especially Leśniewski *Works*.

'Protothetic' was described in a long essay Leśniewski *1929a* on the 'Fundamentals of a new system of the foundations of mathematics', written in German and published in *Fundamenta mathematicae*. Słupecki later published an extensive account *1953a* based upon notes made by a colleague student of a course given in the semester of 1932–1933. Wishing to replace the propositional calculus of *PM*, Leśniewski adopted the equivalential calculus, and extended it to quantification over (his version of) predicates; quantifiers seem again to be interpreted as con- and disjunctions. Inspired by a suggestion made by Chwistek in 1920, he developed a semiotic notational system for connectives based upon little wheels with four spokes. He could have symbolised each of the 16 possibilities, but he concentrated upon the two most important: '⌀' for the conditional, and '⌀' for the bi-conditional (pp. 456, 441). He gave various axiom systems for the calculus; it became an obsession in Warsaw to find the shortest one(s) for all calculi and to compare different versions, and in a review of this paper for the *Jahrbuch* Skolem *1935a* noted the 'huge set of terminological explanations' required.

Leśniewski also recorded the importance of his theory of 'semantic categories', created in 1922 (pp. 421–422). An extension of the theory of types in *PM* (but without orders) to classify parts of speech and partly developed by Husserl, it is notable at a time when logic was still usually dominated by syntax.

On 'Ontology', the 'calculus of names' based upon Protothetic, Leśniewski published, again in German, only a paper *1930a* of 22 often obscure pages; in compensation, Słupecki later published an extensive account *1955a*, in English, based upon various student notes of Leśniewski's lectures of 1929–1930. Reacting to a question of Łukasiewicz, Leśniewski developed a calculus based upon singular predication, where A 'is' the sole b; akin to the 'is' ('*jest*') in Polish, he saw it also as a refinement of part of syllogistic logic. It also resembled a special form of set theory, which he over-emphasised by using Peanese to write '$A \, \varepsilon \, b$'. Finding it possible to assume only one axiom, in the Polish way he sought the shortest form, and found it in 1929. Placed at the end of his paper, it stated the transitivity of 'is': for all names A and b, A 'is' b if and only if there exists at least one C such that A 'is' C and C 'is' b.

'Mereology', originally called 'general set theory', was an extensional treatment of non-empty collections of objects, and so more in tune with Leśniewski's philosophical inclinations; he published on it in most detail, especially much of his long paper (*1927–1931a*, chs. 4–10). This paper was called 'On the foundations of mathematics', so that he imitated logicism though offering a very different underlying theory of collections. His 'collective' treatment of collections of objects $P, Q, \ldots$ followed the part-whole tradition, but handled with a new level of sophistication. The one relation, 'part', was given axioms such as transitivity and the assumption that if P were a part of Q, then Q were not a part of P. Collections were not objects; a class of objects was the same as the class of classes of those objects, and the empty class was no thing (pp. 215, 211–214).

Leśniewski seems to have first thought of linking mereology to ontology, but then he split them; one advantage was that he could now Solve Russell's paradox (Sobociński *1949–1950a*, arts. 6–7). It arose because the collective formation of classes was mixed with the Ontological; when separated, at least one axiom of each theory was not satisfied, so that its construction was blocked.

In the course of developing his systems Leśniewski advanced beyond Dubislav in understanding the roles of definitions in formal calculi (Rickey *1975a*). In a paper written in German, Leśniewski *1931a* formalised the procedures used in developing the propositional calculus; for example, rules for making legitimate substitutions within well-formed formulae. He also conceived of a kind of definition which became known as 'creative'; it aided in the proof in a calculus of at least one otherwise unprovable theorem which contained neither the defined term nor any term dependent upon it. But he published nothing on his theory (not even in this paper), Łukasiewicz mentioned only occasionally, and their students rarely discussed it (for example, in Słupecki *1955a*, 61–66); so it remains an important but shadowy part of both logic and axiomatics.

8.8.4 *Pole apart: Chwistek's 'semantic' logicism at Cracov.*   While his Warsaw colleagues used *PM* for its logic and/or set theory, Leon Chwistek (1884–1944) attempted to reconstruct its logicism. Trained in philosophy at Cracov, mainly under Zaremba (§8.6.3) and the philosopher Jan Sleszyński (1854–1931), he was appointed a lecturer there. He was also a talented painter, with many canvasses displayed in Polish galleries.

Chwistek spent several months in 1909 and 1910 at Göttingen. Hilbert's work presumably made an impact, although foundations were not then his main concern; more significant was the lecture of April 1909 by Poincaré on (im)predicativity and Richard's paradox, described in §8.7.7 in connection with Weyl. As Chwistek recalled later (*1948a*, 78–79) both he and Weyl were in the audience, and in November he sent Russell a manuscript on specifying mathematical induction impredicatively and asking for Russell's opinion (Jadacki *1986a*, 245–251). There was then a pause in contact of a decade, during which time Chwistek rethought logicism. In a Polish paper on paradoxes he joined the chorus against the axiom of reducibility, but for a new reason: it admitted a version of Richard's paradox within simple type theory (*1921a*, art. 1). However, he was mistaken; when he repeated the claim in a paper in German (*1922a*, 239–241), Ramsey was to point out that even if a propositional function were finitely definable in terms of some given symbols, the equivalent elementary function allowed by the axiom may not be (*1926a*, 28–29). Neither man noticed the paradox at the end of Russell's *The principles*, which does seem to be constructible in the simple theory (§6.7.8).

There are various similarities between the theories of Chwistek and of Ramsey, with priority resting with the former. They became clearer in Chwistek's main production, which he sent in manuscript to Russell: a long essay in English on 'the theory of constructive types', published in two parts by the recently formed Polish Mathematical Society after rejection by *Fundamenta mathematicae* (*1924a*, *1925a*).

One similarity has been foreshadowed: in his earlier papers Chwistek spoke of the 'simple' theory, meaning the types only, distinguishing it from the full version. We saw in §8.4.6 that Ramsey was to make the same distinction; maybe he was inspired by Chwistek's allusion. We also noted that he did not name the kinds; by contrast Chwistek was bountiful, for the simple theory was also 'simplified' or 'primitive', while the full version was variously 'pure', 'branched' or 'constructive' (*1924a*, 12–13, 20; *1925a*, 92, 95–98, 110). The last name occurred in his title, indicating his intention to confine propositional functions Richard-style to finitely constructible ones and so avoid the axiom of reducibility. In addition, his simple theory differed from that in *PM* in that types were defined from a given class by its sub-classes, sub-classes of sub-classes, and so on (*1924a*, 26–28, 36–40); further, classes were not incomplete symbols relying upon contextual definitions in terms of propositional functions but treated on an epistemological par with them, so that 'there is no difference at all between a function with I variable and a c l a s s, or between a function with II variables and a r e l a t i o n' (p. 40). The most interesting and influential change was to alleviate the problem of identity by defining it only for classes and relations, since they sufficed for all mathematical purposes (p. 19). Thus for classes α and β (p. 42, here in *PM* symbols):

$$\alpha = \beta . = . (x).(x \, \varepsilon \, \alpha . \equiv . x \, \varepsilon \, \beta) \, \text{Df}. \qquad (884.1)$$

Chwistek also discarded the theory of definite descriptions in favour of a surely inadequate collection of nominal definitions (pp. 45–47). He laid out his 'Directions concerning the meaning and use of symbols' in a very formal way, claiming that it 'may be called the real "Metamathematic"', because it was 'more precise' than Hilbert's (p. 22). However, his general rule of definition (p. 28) was to be shown by Leśniewski to lead to a contradiction (*1929a*, 488–491).

The second part of Chwistek's paper was largely devoted to a reconstruction of 'cardinal arithmetic'. He used his simple type theory to imitate the theory of homogeneous cardinal arithmetic (§7.9.4) in *PM* (Chwistek *1925a*, 110–137). But the theory was limited, both mathematically and philosophically. For example, the new definition of identity gave his calculus an extensional flavour, although he was guarded about its measure (pp. 93–95). Again, the restriction to constructive types and their functions prevented him from defining any transfinite cardinals, even though he adopted an axiom of infinity (pp. 133, 137). In the introduction to the second edition of *PM* Russell noted that 'a great deal of ordinary mathematics' was lost (*1925a*, xiv).

Chwistek promised to publish a 'complete system of Logic and Mathematics' (*1925a*, 99); the closest approximation during that decade was a paper *1929a*, written in German. The 'and' in his phrase is significant: logic no longer grounded mathematics, but both had some common basis. This lay in 'semantics', which for him referred to the study of signs and their means of manipulation. It was a development of his 'Metamathematic', which differed from Hilbert's in that its language incorporated both symbols and their referents. In this paper he presented his calculus in which all formulae were symbolised by a string comprising five letters and 'special signs' when necessary, and concatenations thereof; thus it was not unlike Łukasiewicz's notation. For example, the Sheffer stroke '$p \mid q$' was expressed as '$* cpccq$' (where 'c' was a place-marker), and 'H is the result of the substitution of G for F in E' became '$* EFGHc$' (pp. 705, 707). He laid out a long table of logical and mathematical notions, including a number-coding system strikingly similar to a list of characters in a modern computer manual (pp. 714–717).

At the time Chwistek's enterprise must have seemed mysterious; but it was sufficiently well regarded for him to be preferred over the young Tarski for the chair in logic at Lvóv in 1930. Russell had supported the candidature in a letter of 23 December 1929 to the University, while adding that he did not know Tarski's work.[40] In the 1930s Chwistek's 'semantics' was to gain some currency (§9.3.2), and he was to extend it (§9.6.7). But the main centre of activity for Polish logic remained in Warsaw, with Tarski and his colleagues and seniors.

8.9 THE RISE OF AUSTRIA IN THE 1920S: THE SCHLICK CIRCLE

8.9.1 *Formation and influence.* In this final section we move south of Poland to consider another group circle of philosophers and mathematicians, which formed in the mid 1920s and came to have great influence world-wide. I avoid their name 'Vienna Circle' here, since it was introduced in 1929, in a phase of work heralded in §9.2.1; here the earlier period is described.

The origins goes back to 1900 when Hans Hahn (1879–1934), Otto Neurath (1882–1945) and Philipp Frank (1884–1966), students together at Vienna University, began to discuss philosophical questions about mathematics and science. Hahn was to make his career as a mathematician, especially in the calculus of variations and also in the theory of functions; his achievements bought him back to a chair in Vienna in 1921, after five years at Bonn. By then his interest in philosophy had begun to develop, and he started off a discussion group with his old chums and some others. He also sought for the chair of 'inductive philosophy', which had been established in 1889 for Ernst Mach, to be given to the German philosopher Moritz Schlick (1882–1936).

Schlick had started as a physicist, writing a *Dissertation* in 1904 under Max Planck at Berlin (his home town); but then he transferred to philosophy, specialising in epistemology and ethics. He devoted his *Habilitationsschrift 1910a* at Rostock University to a search for 'the essence of truth

---

[40] Russell's letter, kept in the Archives of Lvóv University with a copy at RA, is printed in Jadacki *1986a*, 243; his correspondence with Chwistek is on pp. 251–259.

according to modern logic', and plumped for one in which judgement '*univocally designates a specific state of affairs*' (pp. 94–95)—strikingly similar to Russell's theory then in embryo (§8.2.6), though without the apparatus of relations. Professor at Rostock from 1917, he developed this approach in a book *1918a* on epistemology, where he followed Hilbert in relying upon implicit definitions (§4.7.2). Widely admired, but with reservations from Weyl *1923a* in the *Jahrbuch*, its appearance helped Hahn's hope to be fulfilled in 1922.

Thereafter the Vienna discussions fell under Schlick's leadership; a philosophical circle was created, with invitation from him as the criterion of membership. New figures joined from time to time, including our two major figures: Rudolf Carnap (1891–1970) and Kurt Gödel (1903–1978). The circle met quite regularly, usually in certain Vienna coffee-houses; a member gave a talk, and discussion followed. Often notes were taken by a stenographer, and sometimes also by those members who had learnt a shorthand. A collaborator from the Mathematics Department was Karl Menger (1902–1985), who also organised a mathematical colloquium of his own, which ran a yearly *Ergebnisse* of its results from 1928 to 1937 (Menger *1998a*); co-editors included Gödel.

The circle wished to bring to philosophy the standards of rigour and exactness that were (presumed to be) found in mathematics, logic and science. Positivism and reductionism were held to govern both epistemology and ontology, and metaphysics was disliked; father figures included Mach and especially Russell, and Wittgenstein's *Tractatus* exercised both positive and negative influence. The members used phrases such as 'scientific philosophy', surely inspired by the full title of Russell's *Our knowledge* (§8.3.2), which had recently appeared in a German translation as *1926d*. However, on many specific issues and special interests they differed quite widely.

The influence of the circle gradually grew, especially from late in the decade. A book series on 'the scientific world-view', co-edited by Schlick and Frank, started in 1928, when also Schlick and Hahn led a *Verein Ernst Mach* to hold meetings and conferences; and a journal was taken over two years later (§9.2.1). But after 1933 many members left the influence of the ever-widening Hitler Circle, going mainly to the U.S.A.; and thereby the influence increased further, leaving a heavy mark on philosophy for decades, especially in Anglo-Saxon countries. The scale of this influence has led in the last thirty years to great historical interest, including editions and translations of the writings of several members.

In addition, and in contrast to the poor Poles, several large *Nachlässe* and other collections of manuscripts have been conserved; they include not only protocols but also notice-books of lecture courses, versions of lectures, and vast collections of correspondence (often typed, with carbon copies). I cannot do justice to this range of published or manuscript material, not even to my own partial knowledge of it; the survey here and in §9.2–3 is strictly confined to the concerns with logic and the philosophy of mathematics, though it takes note of some peripheral members.[41]

8.9.2 *The impact of Russell, especially upon Carnap.* (My *1997c*) During 1924–1925 Hahn conducted a year-long seminar on *PM*; it must have been detailed, but no records seem to survive. It may have suggested emulation to Schlick, who, although publishing little on logic, took a deep interest in the subject. At all events, for the winter semester of 1925–1926 he gave a general course on the philosophy of mathematics (Schlick Papers, 23-B1), and for the summer he chose Russell's *Introduction*, presumably in the German translation *1923a*, as the text for a seminar on 'philosophy of number'. Following normal practise, a student was assigned to prepare a chapter or two and lead its presentation, while another student took notes for the record book (Papers, 52-B32-2). No new insights emerged, but the students will have gained a good impression of the scope and limitations of logicism. Curiously, while Gödel was one of the students, he neither made presentations nor took notes; maybe at *this* stage he was not too interested.[42] But the effect on Schlick must have been great; for when the German translation of Russell's *The problems of philosophy* appeared (Russell *1926c*), he wrote an ecstatic review *1927a*:

> The method of his philosophising can hardly be estimated highly enough. In my strong conviction it is the method of the future, the only method through which LEIBNIZ's ideal can be realised, and will introduce the rigour of mathematics (Russell is a recognised mathematician) into the treatment of philosophical questions.

Maybe Schlick had been counselled by Carnap, who had been appointed *Privatdozent* to Vienna University in 1926. After attending Frege's lectures at Jena in the early 1910s (§8.7.3) and Army service throughout the

---

[41] The most valuable single source on the c(C)ircle is Stadler *1997a*, a Baedecker guide to the whole story, together with transcriptions of the surviving meeting protocols, bio-bibliographies and teaching records of all members and peripheral figures, and a bibliography of historical literature. Among the latter, Dahms *1985a* and Haller and Stadler *1993a* comprise relevant collection of essays; Menger's posthumous reminiscences *1992a* are also well worth noting. The most extensive source in English is the 'Vienna Circle' collection of translations and editions published by Reidel (now Kluwer); among the commentaries, Coffa *1991a* takes some nice sideways looks, and Giere and Richardson *1996a* stands out among the collections of essays.

The various *Nachlässe* are held in several centres, with microfilm copies available from the *Sozialwissenschaftliches Archiv* at the University of Konstanz. Since correspondents quite often made carbon copies of texts or letter, there is duplication across collections; I may well not have registered all cases in my citations. In addition, the Behmann Papers is a rich but little-known source, catalogued in Haas *1981a*.

[42] A fellow student was Olga Taussky, who later recalled Gödel leading one seminar and being generally active (Taussky-Todd *1987a*, 35–36). But in general her recollections are inaccurate.

Great War, he returned to Jena for a few years; then, from 1921 until his call to Vienna he lived under his own means near Freiburg. His initial philosophical forays were neo-Kantian studies of space and time, following especially Natorp and also Dingler to some extent; but then, as he recalled later, he moved towards Russellian positions and remained there for many years (Carnap *1963a*, 16–17). After reading *PM* in 1919–1920 (Carnap Papers, 81–39) and then *Our knowledge*, he wrote to Russell on 17 November 1921, explaining that his own conception of space was closer to that in Russell's neo-Hegelian *Essay on the foundations of geometry* (§6.2.1) than in the positivistic *The problems of philosophy*. He also listed in seven pages 'the most important signs, definitions, axioms and theorems' in *PM*, perhaps to check his catalogue with the master. The next June he enquired about buying the first volume, and Russell responded in July by writing out *all* the main definitions and results of the work in a manuscript of 35 folios.[43] This extraordinarily generous effort by a famous and mightily busy figure for a promising but little-known philosopher in Germany at the time of its massive inflation of currency must have impressed young Carnap, over and above the philosophical kinship.

Carnap also sent to Russell news of a proposed meeting for 1923 on the 'Construction of reality (structure theory of knowledge)' and on the 'Doctrine of relations' ('*Beziehungslehre*'). He invited the participants: they included fellow philosophers such as Hans Reichenbach (1891–1953), Schlick, Dubislav, Paul Hertz (1888–1961), Behmann (1891–1970, an exact contemporary), and Fraenkel (Thiel *1993a*). Two meetings took place in March at Erlangen, under the auspices of the *Kantgesellschaft* (Reichenbach Papers, 15-50-03). No publications were produced, but the stage was set for the kind of philosophy which Carnap wished to pursue: one of his talks at the second meeting dealt with Russell's logic of relations.

Around then, when in his mid thirties, Carnap began to write books on logicism largely following *PM*, and on epistemology much based upon *Our knowledge*. (The influence of Russell thus extended from subject matters to concurrent writing!) They appeared in different countries in reverse chronological order, which we follow here.

8.9.3 *'Logicism' in Carnap's Abriss, 1929*. Carnap's account, published in Vienna in the Schlick-Frank book series as *Abriss der Logistik*, was much more substantial than Behmann's recent little book (§8.7.8). It begins with a durable innovation, not reflected in its title. As we saw in

---

[43] Carnap's November letter is in RA, Dora Russell Papers; copy in his Papers, 102-68-32. Russell's manuscript is at 111-01-01, his thanks of 29 July 1922 at 102-68-31. The later letters in the latter file include news of obtaining a copy of the first edition of *PM*, seemingly from Ogden, in 1926 (68-28), soon after the first two volumes of *PM*₂ were available. He lent Russell's notes to Behmann, asking for their return on 19 February 1924 (Behmann Papers, File I 10). On October 1923 Russell had given the proofs of the first volume of *PM*₂ to Reichenbach (Papers, 16-40-05; and RA).

§7.5.1, the French word 'Logistique' was introduced by Couturat and others at the 1904 International Congress of Philosophy, and was used by Russell and others from then on, in versions appropriate for various languages. However, it covered both the views of the Peanists and the different position adopted by Russell and Whitehead. It seems that in the manuscript he had used 'Logistik'; for, upon receiving a copy, Behmann told Carnap on 29 December 1927 that he found it to be an 'unlucky bastard word', like 'logical product' and 'propositional function'.[44] Carnap retained the word in the published *Abriss*; but, perhaps for this reason, he also proposed there 'Logizismus' (*1929a*, 2–3) to name positions like that proposed in *PM*. As we saw in §8.7.6, at this time and seemingly independently Fraenkel had adopted the word for the same purpose; but the spread was mainly due to Carnap, from 1930 onwards (§9.2.2). 'Logizismus' had been used before (sometimes as 'Logismus'), mostly in connection with phenomenological logic; in line with the German practise of '. . . ismus' indicating a negative stress, it connoted a *rejection* of a reduction of the non-logical to the purely logical, granting the latter an independent existence.[45] There was also a positive sense, such as grounding metaphysics in logic alone; maybe Carnap had seen that use in Dubislav's recent book on definition (§8.7.8: *1927a*, 67).

Carnap packed a lot of material into his 114 pages. The first part, 'System of logistic', ran through the calculi, with the Sheffer stroke and other connectives defined from truth-tables. For the latter he missed Post and cited only Wittgenstein; however, he redefined tautology as 'c o n - t e n t l e s s (but not meaningless)' (pp. 8–9). Then he treated paradoxes, with only Russell's discussed; Russell's theory of definite descriptions; and type theory, including its 'verzweigte' part involving orders and the pertaining paradoxes of naming (pp. 15–23, 30–33). 'Verzweigte' reads like a translation of Chwistek's 'branched' (§8.8.4); but he did not cite Chwistek and may not have known his work. The word may have come back later into English as 'ramified' (§9.4.5), now the standard name.

Carnap also presented Euler diagrams, together with the 'controversial' issue of classes conceived extensionally (pp. 23–25). However, he missed Wiener's definition (827.1) of the ordered pair. The logicist construction of arithmetic went through the finite cardinals and 'relation-numbers', and touched upon the infinite, limit points and continuity (pp. 45–50). The second part, on 'applied logistic', included the Peano axioms, bits of topology and projective geometry, and the analysis of language. The tight links forged between notions gave an overall structure to the parts of

---

[44] Carnap Papers, 28-07-05; Behmann Papers, File I 10.

[45] See Wundt *1910a* for this use of 'Logizismus'. Ziehen deployed it similarly in his textbook on logic (§8.7.2) to characterise logics opposed to psychologistic, sensualistic or inductive trends, such as those of Bolzano, Brentano, Husserl and Meinong (Ziehen *1920a*, 172–173, with some historical notes on the word).

mathematics handled; indeed, his version of Russell's ordinal similarity (795.1) between two relations was called a 'structural property' of a relation (p. 54). Curiously, he did not discuss model theory, although it had featured in a recent logicist article *1927a* in which he preferred 'proper' ('*eigentlich*') concepts over 'improper' ones, which were specified only by implicit definitions and so lay open to non-categoricity (Howard *1996a*). The end matter of the *Abriss* included some of the most popular current symbols and terms then in use. In the index Russell and Whitehead were the longest entries; Frege was cited in the text three times, and only for references.

Although it sold respectably into the 1930s,[46] Carnap's *Abriss* is now largely forgotten, even by historians of the period. Its sub-title, 'with especial consideration for relations theory and its applications', indicated that the logic of relations linked it together. This feature was prominent also in Carnap's companion volume.

8.9.4 *Epistemology in Carnap's Aufbau, 1928.* The publisher of the *Abriss*, Julius Springer, had rejected the other book, which, by contrast has become a classic.[47] Apparently based upon part of his *Habilitationsschrift* at Vienna in 1926,[48] *Der logische Aufbau der Welt* was published as Carnap *1928a* in Berlin by the Welt-Kreis Verlag, with a print-run seemingly of 600 copies. The house was closely linked to the circle; it had recently published the translation *1926c* of Russell's *Problems* mentioned in §8.9.2, which had been prepared by Hertz.

Again Carnap's sub-title, 'Attempt at a constitution-theory of concepts', is instructive, for it reflects the fact that while preparing the book he often used 'Konstitutionstheorie' as the intended title. The change is instructive: from a neo-Kantian noun to the catch-word 'construction', in Germany and Austria, in the 1920s, where post-war aspirations were to be fulfilled in fields social, architectural (such as the *Bauhaus*) and philosophical (Galison *1996a*). Maybe also the publisher had asked for a catchier title.

Carnap provided a quite formalised construction, drawing upon logicist techniques and methods; prominent was the logic of relations, which he stressed briefly in the opening sects. 7 and 11 and in detail in the formal

---

[46] Springer told Carnap that 951 copies were sold by 1937 (Carnap Papers, 27-40-109). He had told them in the summer of 1927 that Bruno Cassirer had rejected the *Abriss* as financially too risky (29-37-02). Russell received a copy from Carnap, but he did not slit many pages (RA).

[47] Springer told Schlick on 25 May 1926 that they rejected Carnap's *Aufbau*, because of sales prospects (Schlick Papers, File 119); the following March he gave the manuscript a very warm testimonial (for Welt-Kreis Verlag?), and discussed the title with Carnap later in the year (21-A89, A85).

[48] On this somewhat obscure point, see Vienna University Archives, Philosophy Faculty Board Minutes, 13 March and 9 June 1926; and Schlick Papers, 85-C29-4 and -9.

outline in Part 3, where various different kinds were used (sects. 69–79). Algebraic logic was entirely absent.

Carnap imitated type theory by assuming a 'basis', so that 'objects of each level were "constructed" from the objects of the lower levels' by creating '*basis, ascension form, object form*, and *system form*' (sect. 26, launching Part 3). His processes of construction, worked out in Part 4, drew heavily upon '*structure descriptions*', which comprised 'all formal properties' of a relation (sect. 12). He associated his concept of structure with 'relation-number', Russell's generalisation of ordinal arithmetic in *PM*; Russell had made this link himself in the *Introduction*, and Carnap cited both passages here.

Russell's theory of definite descriptions was also used to handle uniquely specifiable objects (sects. 13, 50). It led Carnap to 'the general possibility of structural definite descriptions' (title of sect. 15), an epistemological notion since 'all scientific statements are structure statements' (title of sect. 16). In addition, it helped him avoid considering the metaphysical essence of objects with his requirement that 'the indication of the nominatum of the sign of an object, consists in an indication of the truth criteria of those sentences in which the sign of this object can occur' (sect. 161). He noted that Russell had used the contextual definition of definite descriptions (and also of classes and relations) in the same spirit of avoiding assuming existence (sects. 38–39).

The technical details of Carnap's 'constructional system' were based upon such notions from *PM*, together with many of the symbols (sects. 106–122). A notable technique was 'quasi-analysis', where primitive elements were examined by the pertaining relations between them (sects. 71–83). As with Russell and like the restricted compass of mathematics in their logicisms, Carnap treated almost entirely only the physical sciences; a few nods were made towards biology (sect. 137), but none to, say, medicine or geology. Also following Russell in the second edition of *PM*, Carnap treated classes and relations extensionally (especially sects. 40–45 in constructing ascension forms); but as a result universals could not be handled, and generality was compromised (Popper *1935a*, art. 14). Like his *Abriss*, the tight links between notions made his epistemology highly structured.

While Carnap's overall system resembled Russell's type theory, his broader aims made the choice of the basis more tricky than Russell's assumption of structureless (and physical) individuals. The neo-Kantian in Carnap came through in his discussion of 'possible psychological bases', where he decided upon the 'autopsychological basis' of the single person; objectivity was secured through the 'forms of experience' started by members of a community, and thus was markedly structural (sects. 61–68).

Wittgenstein's *Tractatus* was only cited three times, though warmly (sects. 43, 180 and 183). It had become a major talking point in Schlick's circle, read 'page by page', Schlick wrote to Carnap on 29 November 1925 (Carnap Papers, 29-32-34); but Carnap chose from it warily. We noted his different definition of tautology, and he also defined analytic theorems as 'deduced from the definitions alone' (sects. 106–107). A major difference concerned the hierarchy of languages, which Russell had proposed in his introduction to the *Tractatus* without recognising its significance (§8.4.3). The circle split over its merits. Orthodox Wittgensteinians such as Felix Kaufmann and Felix Waismann followed Their Master in rejecting it; but Carnap, and also Hahn and Gödel, recognised its major importance—with momentous consequences for Gödel's contributions.

8.9.5 *Intuitionism and proof theory: Brouwer and Gödel, 1928–1930.* During the 1920s Menger had been working on the topological definition of dimension, and stayed for some time in Amsterdam with Brouwer (Menger *1979a*, ch. 21). Their relationship was to flounder in 1929, due to Brouwer's conduct; but before that Brouwer gave two lectures on intuitionism in Vienna in March 1928.

The first lecture was published in the Viennese *Monatshefte für Mathematik und Physik*, then edited by Hahn, as Brouwer *1929a*. He discussed 'mathematics, knowledge and language', with his mixture of weird psychology and sociology. But he argued his case for languagelessness clearly, and near the end he distinguished 'correct theories', in which the law of excluded middle was forbidden, from the broader collection of 'non-contradictory' ones, in which it was allowed but his other restrictions still applied. In the second lecture, which for some reason appeared only as a pamphlet, Brouwer *1930a* contrasted the classical and the intuitionistic continua with more clarity than usual. Logicism was mentioned only once, when Russell was classified as a formalist, along with Peano and Dedekind... .

The impacts on the audiences of these lectures, especially the first one, were unexpectedly great. Wittgenstein was drawn back to philosophy after several years' absence, perhaps by Brouwer's stress, congenial to a monist, upon personal will and the state of reverie (*1929a*, art. 1); and both Carnap and Gödel were impressed by the pessimistic idea that there might be unsolvable problems in mathematics (Köhler *1991a*, drawing upon Carnap's diary). The fruits of this seed, if sown then, were to blossom surprisingly at the end of 1930; but meanwhile Gödel continued with the optimism inherent in the circle's position, and also in Hilbert, by establishing another important case of completeness.

In February 1930, during his 25th year, Gödel defended his *Dissertation*, written under Hahn's supervision; a re-written version appeared in the *Monatshefte* as Gödel *1930b*. Noticing in Hilbert/Ackermann that the syntactic completeness of the 'narrower calculus of functions' was not proved or disproved (*1928a*, 68), he established the former; that is, assuming it to be consistent, each of its well-formed formulae was either 'refutable or satisfiable'. He used metamathematical and model-theoretic notions and notations to work out the details, with Hilbert/Ackermann as his main source, and also some kinship to Skolem (§8.7.5). *PM*, as modified by Bernays *1926a* (§8.7.4), provided the axioms and rules of inference, and the theorem partially vindicated its authors' assumptions about completeness; but their broader aims were soon to suffer from Gödel a fundamental rebuff, as we shall see in the next chapter.

CHAPTER 9

Postludes: Mathematical Logic and Logicism in the 1930s

9.1 PLAN OF THE CHAPTER

As noted in §8.1, this chapter contains rather less detail than before; for example, most details after 1935 round off old stories rather than launch new ones. Other philosophical schools, especially formalism (where I follow authors who used the word) and intuitionism, are treated lightly.

To start, §9.2 focuses upon Gödel's incompletability theorem and corollary (as his second result is often called), and the first reactions to them. §9.3 takes the Vienna Circle: Carnap dominates, but we find two new figures and two others already met. §9.4 covers the U.S.A., whither several Circle members had to emigrate after 1933, and also W. V. Quine emerged; in addition, the *Journal of symbolic logic* was founded. §9.5 returns to Britain, including the late views of both Whitehead and Russell, and continuing skirmishes over the merits of mathematical logic. Developments in Europe outside Vienna concern §9.6, with Germany still prominent and Poland flourishing.

In addition to the annual records of German and French philosophy in the *Philosophical review* (§8.1), the *British Journal of philosophical studies* (from 1931, *Philosophy*) treated in shorter order these two countries, and also Italy and 'Russia' (where some attention was paid to logicism (Anellis *1987a*), though seemingly with little effect abroad). In general, the treatments of symbolic logic and the philosophy of mathematics were rather cursory.

A new German publication needs to be mentioned. The reviewing *Jahrbuch* for mathematics continued even more tardily than before (§8.1), with volumes sometimes completed years late and not even in chronological order; it finally closed with the 1942 volume (Siegmund-Schulze *1994a*). Meanwhile a rival had arisen: the *Zentralblatt für Mathematik und ihre Grenzgebiete*, under the general editorship of the historian of mathematics Otto Neugebauer. The first volume appeared in 1931, and was far more prompt than the *Jahrbuch* in its reportage.

On general literature, the bibliographies in Church *1936a* and *1938a* (§9.4.5) fill most gaps. My *1981b* briefly surveys logic between the world wars, with an extensive basic bibliography.

9.2 GÖDEL'S INCOMPLETABILITY THEOREM AND ITS IMMEDIATE RECEPTION

9.2.1 The consolidation of Schlick's 'Vienna' Circle. In 1929 Schlick rejected a call to Bonn University, and in celebration his discussion group met more formally to prosecute their philosophy. Otto Neurath gave them the name 'Vienna Circle', which most of the other members disliked; but as it became standard, I shall use it, normally in the acronym 'VC'. The principal members concerned with logic continued to be Rudolf Carnap and Kurt Gödel; other members active in neighbouring areas (including the philosophy of mathematics) included Felix Kaufmann (1895–1949) and Freidrich Waismann (1896–1959), and we find again Heinrich Behmann and Walter Dubislav. From the late 1920s until his death in 1934 Hans Hahn popularised logic(ism) and empiricism in mathematics and science in several papers and pamphlets (now included in his *Papers*).

Carnap became *ausserordentlich* professor in Vienna in 1930, but moved the next year to a similar chair in the philosophy of science at the German University in Prague; however, he maintained regular contacts with Vienna and indeed probably exercised his main influence there. The protocols of the Circle meetings were recorded and typed out more systematically than before—or least, most survive from 1930. In that year the Circle joined with the *Gesellschaft für empirische Philosophie* in Berlin to take over the *Annalen der Philosophie* as a journal in which to publish its views; renamed *Erkenntnis*, Carnap and Hans Reichenbach in Berlin were appointed editors. Logic was quite prominent in its pages, especially because of Carnap; other philosophies of mathematics were also aired (and Neugebauer once wrote on history). A book series was started in 1931, and one of short books entitled 'Unity of knowledge' ('*Einheitswissenschaft*') in 1933, to complement the Circle series edited by Schlick and Philipp Frank. From 1934 until 1939 the groups held a series of annual congresses in cities in various countries, with a final gathering at Chicago in 1941. Contacts also continued with the *Verein Ernst Mach* for lectures and Karl Menger's mathematical colloquium. The French became quite interested in this movement, especially Louis Rougier (§8.6.2); he organised the congress for 1935 held in Paris. Its proceedings were published by the house of Hermann in their fine series of short books 'Actualités Scientifiques et Industrielles', which also included some editions or translations of works by VC members and associates.

This and the next section chart work effected until around the mid 1930s. Then the arrival of Hitler in 1933 began to scatter the members to several countries (Dahms *1985b*); Schlick was murdered in June 1936 by an apparently deranged former student, and the "relaxed" attitude of the authorities to the loss finished off an already broken Circle. Stadler *1997a* continues to be our admirable documentary guide, including sections on the contents of *Erkenntnis*, the Menger colloquium, the congresses, and Schlick's death.

9.2.2 *News from Gödel: the Königsberg lectures, September 1930.* The *Verein Ernst Mach* was involved in a symposium on the 'exact knowledge' held at Königsberg early in September 1930 (Wang *1987a*, 85–87). Three speakers spoke on the main philosophies of mathematics: written versions appeared the next year in *Erkenntnis*, followed by the transcription of a remarkable discussion session (Dawson *1997a*, 68–72).

Carnap *1931c* spoke on 'The logicistic grounding of mathematics': it is from this publication rather than his *Abriss* that the word 'logicism' slowly but gradually gained currency. (His draft material for this lecture (Papers, 110-03) was entitled 'The basic thoughts of logicism'.) For 'the derivation of mathematical concepts' (art. 1), he ran though some connectives and quantification, although not contextual definition despite giving the definitions of cardinal integers. He then sketched the treatment of real numbers, and also mentioned the differential (derivative?) and the integral even though neither *PM* nor *Abriss* had provided any details. Concerning 'the derivation of theorems' (art. 2), he noted the three doubtful axioms (infinity, choice, reducibility) in *PM*, and spent most of the rest of the paper on the 'simple' and 'branched' theories of types, and Ramsey's attempt to reconstruct them (arts. 3–4). He did not mention Gödel's new theorem, although seemingly its author had told him of it.

Brouwer's former student Arend Heyting (1898–1980) argued the intuitionistic cause; but his paper *1931a* is notable for the limited treatment of Brouwer's philosophy, with no mention made of two-ities or of primordial intuitions of time. The details were restricted to the theory of real numbers, and praise rather for Becker *1927a* (§8.7.9). Von Neumann *1931a* was really too brief on formalism, but he stressed consistency and the decision problem.

Like the lectures, the published version of the discussion (cited as 'Vienna Circle *1931a*') was much abbreviated, although it ended with a valuable bibliography of recent publications on the foundations of mathematics. Hahn began with a long advocacy of his empiricism as the route to certainty in mathematical knowledge; but he also accepted Wittgenstein's characterisation of logic as tautologous and so rejected Russell's empirical interpretation of individuals while supporting logicism (pp. 135–138). He also mis-attributed to Russell the view that the axiom of choice said 'something about the world' (p. 138: compare §7.8.6). In a later comment he announced that 'mathematics has a purely extensional character' and rejected the axiom of reducibility on this ground (p. 145). Carnap saw (his version of) physicalism as a way of bridging the gap between the three philosophies, and hoped for progress in the interpretation of axiom systems (pp. 141–146 *passim*).

Gödel made two short contributions, of which the second recorded his finding of the incompletability theorem (p. 148). He may not have had the definitive version then, and spoke briefly; it is not known if the audience noted its significance, but the published account of the discussion shows no trace. He did not mention the corollary, and probably had not yet found it. By the time of publication of this discussion, his paper had appeared; so, at the request of the editors, he added a short 'Postscript' *1931a* stating both the theorem and its corollary.

Curiously, Reichenbach did not mention Gödel in his report *1930a* on the Königsberg meeting for the general science journal *Die Naturwissenschaften*. Gödel *1932a* himself reviewed the three papers for the new *Zentralblatt*, giving each one about equal space. He described accurately the contents of each one, but at the end of his notice of von Neumann he mentioned his corollary as a new development. Carl Hempel (1905–1997) reviewed the whole ensemble for the tardy *Jahrbuch* in *1938a*, dwelling most on Carnap's paper, especially type theory and Ramsey's revision. Although he did not mention Gödel's bombshell in his report of the discussion, he drew attention to the postscript. It is time to record the arrival of the main paper.

9.2.3 *Gödel's incompletability theorem, 1931.* (Dawson *1997a*, ch. 4) Around August 1930, a few months before Gödel's doctoral thesis on completeness was to appear in Hahn's *Monatshefte für Mathematik und Physik*, he conceived a metatheorem of a quite opposite character; the resulting paper was also published there, as Gödel *1931a*. Hahn had presented an abstract Gödel *1930a* to the Vienna Academy in October, and in the following January Gödel had addressed the VC himself (Stadler *1997a*, 278–280) and Menger's colloquium (*1932b*); both theorem and corollary were given in these versions.[1]

Apparently Gödel found his theorem when he represented each real number by an arithmetical propositional function $\phi(x)$ and found that, while '$\phi(x)$ is provable' could also be so treated, '$\phi(x)$ is true' landed him in liar and naming paradoxes (Wang *1996a*, 81–85). Maybe because of Viennese empiricist doubts over truth, he recast the paradox in terms of unprovability and 'correct' ('*richtig*') propositions; his main paper began with this intuitive version. He took *PM* as the axiomatic layout of first-order

---

[1] The various reprints and English translations of Gödel *1931a* are given in the bibliography. I cite by page number the original as reprinted in Gödel *Works 1*, drawing upon the translation on the opposite pages; it follows a fine account of the theorem and corollary by S. C. Kleene on pp. 126–141. Numerous papers and books have been written, especially in the last 20 years; I cite only a few items from the historical portion. Among general introductions in English, Braithwaite *1962a* can be recommended. On the early reception of the theorems with the VC and friends, see Mancosu *1999a*.

arithmetic (that is, with quantification only over cardinals) though for brevity he used the Peano axioms; in similar mood I shall still call his system '*PM*'. He began by well-ordering all well-formed formulae in one free variable by some relation '*R*' (lexicography, say); the *n*th one was symbolised '*R(n)*'. 'Let α be any class sign; by $[\alpha ; n]$ we denote the formula that results from the class sign α when the free variable is replaced by the sign denoting the natural number *n*' (p. 148), attending to the difference between sign and referent with a degree of care *which itself* was then novel. With this device he could defined a class *K* of integers by the condition

$$n \in K \equiv \sim Bew[R(n) ; n], \qquad (923.1)$$

where he used ' $\equiv$ ' for equality by definition and '*Bew*' (from '*beweisbar*') for the property 'is a provable formula'; I have replaced his overbar for negation by the ' $\sim$ ' of *PM*. Noting that (923.1) itself was a formula with a free variable, it was *R(q)* for some cardinal *q*; and 'we now show that the proposition $[R(q) ; q]$ is undecidable in *PM*'. For if it were provable then it was true, but contradicting the definition of *q* via (923.1); however, if its negation were provable, then so would be $[R(q); q]$, as well as its own negation. So neither proposition was provable (p. 175 and thm. 6). He associated his argument with the Richard and liar paradoxes, though it is really closer to (benign) Cantorian diagonalisation.

Hence *PM* was syntactically incomplete—and moreover incomplet*able*, since adjoining either that proposition or its negation as a new axiom only excites a repeat of the argument and the exhibition of another renegade proposition. Thus I shall refer to the theorem as proving 'incompletability', for it characterises its force better than the 'incompleteness' that is usually used.

Gödel assumed that *PM* was consistent; but examining that issue led him to the corollary. It stated that any set *S* of consistent formulae of *PM* could not include the formula *F* asserting its consistency (p. 196 and thm. 11). As is stands, the result is not surprising, for if *S* were inconsistent, it would contain *F* as one proposition among all of them: the point is that *S* belonged to a metatheory *embracing* that of *PM*, not to a more primitive one as Hilbert had presumed (§8.7.4). Thus the corollary affected meta-mathematics as seriously as the theorem touched logicism; however, Gödel politely denied any such consequences since 'it is conceivable that there exists finitary proofs [of consistency] that *cannot* be expressed in the formalism' of *PM* (p. 198). In fact, in this one and only respect his paper was not as general as he had hoped, for he had to assume that *PM* was '*ω*-consistent' (his name, alluding to Cantor's ordinal), a special kind of consistency which forbade both $\sim (\nu)R(\nu)$ over cardinals ν, and $R(m)$ for each cardinal *m*, from being provable (p. 172). He had not mentioned this detail in his Königsberg postscript.

Both proofs required that the metatheory could be expressed in the language of *PM* in the first place. Gödel met this requirement without paradox by the principal technical feature of the paper (art. 2), which gained great importance in its own right under the name 'Gödel numbering'. To each primitive logical and arithmetical symbol of his system, and the brackets, he assigned a (different) cardinal, choosing $1, 3, 5, \ldots, 13$; each variable of the nth type was given a cardinal of the form p^n, where $p > 13$ was a prime (p. 179). Then any proposition was converted to the form F using only these symbols, and assigned the cardinal $\Pi_r \, p_r^{n_r}$, where p_r denoted the rth prime in order of magnitude and n_r was the cardinal assigned to the rth symbol in F. Similarly, a proof, understood as a sequence of propositions, received a cardinal of the same form, where this time 'n_r' was the cardinal for the rth proposition. In this way, each theorem and proof of *PM* took a unique cardinal relative to the chosen numbering. Conversely, assuming prime factorisation (taken to be proved in *PM*), any formula or proof could be reconstructed from its factors.

With this apparatus Gödel formally defined every desired meta-notion for *PM*, such as being negation, being a variable of order n in the type theory, or substitution. He ended with the 46th definition, on being 'a provable formula' (p. 186), and the crux of his theorem lay in its defining condition, that a proof existed; for no upper bound was set upon its (finite) length, and thus not on the corresponding cardinal. Thus this definition stood apart from the others, which were expressed in terms of arithmetical functions which he called 'recursive' (p. 180); to both theorem and corollary he added various results on classes of such functions. Far more clearly than with Paul Finsler a few years earlier (§8.7.6), he characterised the notion of arithmetical and algebraic operations, and indicated their scope as well as their limitations (Webb *1980a*). The functions associated with the first 45 definitions became known as 'primitive recursive', and the wider world opened up by the 46th led in the mid 1930s to theses and meta-theorems on undecidability studied especially by Church, his student S. C. Kleene and others (Davis *1965a*).

9.2.4 *Effects and reviews of Gödel's theorem.* The main features of Gödel's paper are best appraised in four categories:

1) the theorem, affecting especially logicism, and formalism to some extent;
2) the corollary, affecting especially formalism, and logicism to some extent;[2]
3) the use and potential of recursive functions, and of finitary proof methods on general; and

---

[2] I refer to Gödel's corollary as it was then understood; Detlefsen *1986a* has queried it, on the ground that it assumes that *only* Gödel's consistency proposition can express the consistency of the theory, and no alternative propositions are available.

4) the necessity to distinguish *rigidly* meta- from object-, be it logic, language or theory—and the difficulty often of so doing.[3]

Both logicism and formalism now had to be set aside in their current forms, although *PM* still provided a main source for many basic notions in mathematical logic. However, in assuming bivalency, the theorem did not affect intuitionism although, as we saw in §8.9.4, both Gödel and Carnap had been impressed by Brouwer's pessimistic opinions in his lecture of 1928. Further, it had no major effect on mathematicians; apart from their general uninterest in foundations, it used a far more formal notion of proof than even their most "rigorous" practitioners entertained, so that it would not have seemed to bear upon their concerns.

Both reviewing journals reviewed Gödel's paper. The second volume of the *Zentralblatt* was launched by Dubislav *1932a*, only a dozen lines but highly praising and clearly stating the two results. He saw the theorem affecting not only *PM* but also the axiomatic set theory of Zermelo and Fraenkel; and it was the latter who reviewed for the *Jahrbuch*. Fraenkel *1938a* begun 'this momentous work', and over three paragraphs he described theorem and corollary (seeing the latter as refuting Hilbert's hopes), and summarised Gödel numbering.

The senior set-theorist of the day also reacted, but from a different point of view. We now record a remarkable case of cognitive dissonance.

9.2.5 *Zermelo against Gödel: the Bad Elster lectures, September 1931.*
(My *1979a*) From its launch by Cantor in 1890 the *Deutsche Mathematiker-Vereinigung* (hereafter, '*DMV*') held its main annual meeting somewhere each September. In 1931 the chosen venue was Bad Elster, a spa town in Southern Saxony near to the border with Bohemia; and for the occasion Ernst Zermelo, then entering his sixties, organised some lectures on foundational issues. Imitating a marvellous recent foray *1930a* in *Fundamenta mathematicae* deep into Cantor's transfinite ordinals, he offered in Zermelo *1932a* a theory of *infinitely* long proofs and thereby hoped to show that all true mathematical propositions were provable in this extended sense.

In his proof Zermelo followed the algebraic line in reading quantification in terms of infinite con- and disjunctions, attacked model theory, and left truth-values unexplained. The details are not our concern; the point is that he rejected the normal preference for finitude in proof, and especially the message of the lecture 'On the existence of undecidable arithmetical theorems in formal systems of mathematics' which was given by Gödel. The audience must have had any scepticism of foundational studies rein-

---

[3] On this aspect of Gödel's paper I rely in part on the reminiscences made to me in 1973 by J. Barkley Rosser (1907–1989), who felt himself to be speaking not only personally. Entering the field in the mid 1930s as a student of Church, in Rosser *1936a* he replaced Gödel's ω-consistency by normal consistency.

forced by this brute clash of theorems! Of the two speakers, the young man was little known (perhaps Hahn had suggested his participation). Zermelo wrote to him on 21 September, doubting that the assertion (923.1) that $R(q)$ was unprovable belonged to his system, and wondering about a whiff of paradox 'analogous to Russel[l]'s antinomy' in the argument. In a long reply on 12 October, Gödel emphasised that the opening pages of his paper were only a sketch, and then explained carefully the role of class signs and their referents. He pointed out that Zermelo had incorrectly defined the class of not 'correct' formulae, and that the proper version did not determine a concept, unlike his own class of unprovable formulae.[4] Unlike his paper, his reply involved truth-values, and it may have led him to break his self-imposed silence on truth and increase his interest in semantics, which was usually "under cover" in positivist the VC.

In a reply of 29 October Zermelo thanked Gödel for the elaboration but contrasted their approaches, finitary versus infinitary. Similarly, when he wrote up his lecture for the *DMV*, he included a rather scathing paragraph on his young competitor, claiming 'the inadequacy of confirming *any* "finitisitic" proof theory, without yet having at one's disposal a means for the elimination of this [finitary] inconvenience' (*1932a*, 87). However, while he published a little more on his theory until the mid 1930s, it did not gain much attention, and was presumably too sketchy to influence the re-emergence of infinitary logics after the Second World War.

Three figures had quickly reacted to Gödel's work, with two positive reviews and one negative exchange. The responses of many others were less rapid or even non-existent; for example, most authors in *Erkenntnis* in the following years did not mention it at all. The patchiness of the response will run as a theme in this chapter.

9.3 LOGIC(ISM) AND EPISTEMOLOGY IN AND AROUND VIENNA

9.3.1 *Carnap for 'metalogic' and against metaphysics.* (My *1997d*) The place of logic in *Erkenntnis* was registered at once with a position paper *1930a* on 'the old and the new logic' by Carnap, who was then in his 40th year. The old logic, based upon both algebraic and mathematical logics, was 'The method of philosophising' (p. 12), and it helped to

---

[4] This correspondence was published in reverse order of writing: Zermelo's letter from the Gödel Papers in Dawson *1985a*, and Gödel's reply (and an answer from Zermelo soon to be noted) from the Zermelo Papers in my *1979a*; I was denied access to the Gödel Papers when preparing that article. Some amateurish type-setting of the letter needs the following corrections ('L' denotes 'line', 'd'own or 'u'p'):

p. 298, L11d: *read* eines	p. 300, '7', L1d: *read* Implikationsregel
p. 299, L12u: *read* viel-	p. 301, at '8': *read* $B \subset W$
p. 300, '6', L5d: *read* Formeln,	p. 301, '10', L7u: *read* vergleiche

answer Hilbert's call (head of §8.7.4) for the 'deepening of foundations' (p. 14). Carnap presented logicism in its analytic/tautological form, with the paradoxes solved by type theory. He also published a tardy but warm review *1931a* of the second edition of *PM*, generally welcoming the changes but doubting the generality of Wittgenstein's extensional reading (844.1) of functions as appearing through their truth-values.

One consequence of Carnap's position was the 'elimination of metaphysics' (*1930a*, 25). He became strident: in the next volume of *Erkenntnis* he wrote on the 'Vanquishing of metaphysics through logical analysis of language', emphasising that the meaninglessness of metaphysics was meant 'in the strongest sense' of the impossibility of verifying its propositions (*1931e*, 220). Since he mentioned Hegel and Heidegger as examples of metaphysicians, his repulsion is understandable; but one may see his own stance as an opposite stupidity, which came to weaken VC philosophy. It is strange that a philosopher so attuned to metalogical notions such as consistency should be close to oxymoronity in asserting that a proposition was meaningless *after* one has had to understand it in the first place.

Carnap also returned to model theory (§8.9.3) in a short 'report' *1931a* on 'general axiomatics'. He offered a suite of model-theoretic and metalogical definitions for an axiom system, and indicated relationships between them. The system was 'fulfilled' (*'erfüllt'*) if it had a model; 'monomorphic' (categorical) if there was only one of them; also 'decision-definite' (the Hilbertian *'entscheidungsdefinit'*) if either any other compatible system *g* or not-*g* was a consequence; and 'bifurcatable' (*'gabelbar'*) if compatible with both *g* and with not-*g* under certain restrictions. His report is too short to be really clear; it was based upon a much larger manuscript which he seems to have largely abandoned[5]—traces of Russellian influence again, perhaps.

Another large project which remained largely in manuscript was a study of 'Metalogic', upon which Carnap spoke to the Circle on three successive Thursdays in June 1931; the text *m1931a*, including some discussion, has been published twice recently.[6] In the first talk he defined metalogic as 'the theory of forms, which appear in a language' (p. 314). Thereafter he much followed Gödel, setting up his symbols for a treatment of first-order arithmetic with integers defined as the successors of zero and quantifiers taken as infinite con- and disjunctions, presenting rules for forming well-formed formulae and checking them to be such (pp. 315–317). Taking identity as a relation between names, he mis-stated Russell's definition (785.3) by omitting quantification over predicates (p. 317). He criticised

---

[5] This manuscript is in Carnap Papers, 80-34 and 81-01; with comments by Kaufmann at 28-23-19.

[6] This manuscript is in Carnap Papers, 81-07-17 to -19; and VC Papers, 186/9, fols. 58–89 and 187/10, WK 12–14. The first recent publication is graced by a useful explanation by the editor (Padilla Galvez *1995a*); I cite the second one, from Stadler *1997a*, as it is the more accessible.

C. I. Lewis for claiming that *PM* had only one sign for implication, and asserted that inference was 'a metalogical concept' (p. 317).

The rest of Carnap's first talk and the second one were mostly devoted to a Gödel-like list of metalogical notions. Hahn interrupted to seek clarification of rules of inference, whereupon Carnap stated four: not only substitution and *modus ponens* but also, and surprisingly, identity and mathematical induction (p. 323). Later in the talk he stated that 'the method of <u>Arithmeticising</u> <u>metalogic</u>' handled 'not the empirically presented, but the possible structures' (p. 325). In the third talk he applied Gödel numbering to his chosen collection of symbols, and included a table of (un)decidable theories (p. 329). The discussion focused more on the distinction between intensionality and extensionality, in view of his replacement of Russellian intensional proposition 'A believes p' by an extensional syntactical sentence about believing 'p' (pp. 327–328). This distinction was to become more important in his exegesis of metalogic.

9.3.2 *Carnap's transformed metalogic: the 'logical syntax of language', 1934.* Carnap's culmination of this line of work was a book *1934b* of around 275 pages on *Logische Syntax der Sprache*, published in the Schlick-Frank series; a revised English translation appeared three years later, in C. K. Ogden's series (Carnap *1937a*, not quoted here).[7] His finest contribution to formal systems, justice cannot be done to it here; but a few points and details will be noted.

Carnap's title reflected a great extension of his metalogic to 'the f o r m a l   t h e o r y of language forms' of a given language (p. 1), especially when it was itself formally specified. He presented three of these: a 'definite language I' (chs. 1–2, hereafter 'SI'), sufficient just to construct the arithmetic of cardinals; an 'indefinite language II' (ch. 3, 'SII'), properly containing SI and (hopefully) broad enough to cover the needs of mathematics and physics; and 'General syntax' (ch. 4, 'AS'), really a schema for handling a language and its 'syntax language' (p. 46). In a final chapter he related 'Philosophy and syntax'. In all three exegeses he used sequences of various kinds, especially for logical consequence (for example, sect. 48 for AS).

For each language Carnap distinguished axioms from rules of inference. Quantifiers were now taken as 'operators' rather than infinite con- and disjunctions in each language; however, he did not discuss them. Using Hilbert's terms (§8.7.9), quantification converted 'free' variables into 'bound' ones and produced 'open' or 'closed' formulae according as

---

[7] For correspondence with Ogden and others on the translation and also the books noted in §9.3.3, see Carnap Papers, especially 81-13 and 89-43; and Ogden Papers, Box 7, File 10 and Box 81, File 2. The translator of *Syntax* was the Countess Zeppelin; previously Max Black (whose own correspondence is in Box 4) and J. H. Woodger (§9.5.2–3) had been mooted.

whether free variables were left or not (sects. 6 for SI, 26 for SII, and 54–55 for AS).

Carnap gave credit to Frege for distinguishing a sign from its referent, and for emphasising the importance of discriminating a language from its syntax (pp. 110–111), and followed him by laying out corresponding propositions of various kinds in parallel columns. For example, the 'object sentence' '5 is a prime number' was set alongside its 'pseudo-object' 'Five is no thing, but a number' and 'syntactic sentence' ' "Five" is no thing-word, but a number-word' (p. 212). Again, stressing sentences rather than propositions, in AS he contrasted the intensional cases 'Karl says/thinks A' with their extensional syntactic counterparts 'Karl says/thinks "A" ' (p. 191). The most elaborate exercise of this kind was a trio of rather weak analogies for geometry and SI: 'arithmetical', between coordinate geometry and the prime factorisation of integers; 'axiomatic descriptive', between Hilbert on geometry and himself on the syntax of SI; and 'physical', between applying geometry to space or physics and arithmetical place-markers in SI being replaced by the (correct) signs (pp. 69–71).

Russell (with Whitehead) was still prominent; indeed, as in his books of the late 1920s (§8.9.2–3), Carnap cited him more than anyone else. But he praised Frege for the historically important transfer from the logic of 'concept-ranges' to that of 'concept-contents' (p. 202), and he devoted sect. 38 of SII to a summary history of the paradoxes and the 'elimination of classes' by Russell and Frege via their type theories. Using the word 'logicism' when needed, he saw the purpose of 'a l o g i c a l g r o u n d i n g o f m a t h e m a t i c s' as a reduction, in contrast to the formalist association of mathematics and logic as 'constructed in a common calculus'; it 'w i l l n o t b e s o l v e d t h r o u g h a m e t a m a t h e m a t i c s, i . e. a s y n t a x o f m a t h e m a t i c s, a l o n e, b u t f i r s t t h r o u g h a s y n t a x o f t h e o v e r-a l l l a n g u a g e, which unifies logico-mathematical and synthetic sentences' (sect. 84). Another pair of columns was used to contrast Russell's 'contentual' ('*inhaltliche*') definition of cardinals as objects with Hilbert's 'formal' view as expressions, as Carnap took the latter (p. 227). SII was endowed with a Ramseyan (simple) theory of types (sect. 27). However, not all was pleasure; Carnap found Russell's all-purpose treatment of implication to be 'very u n h a p p i l y e f f e c t e d' (p. 198).

The place of Gödel was recognised in various places, such as the 'arithmeticisation of syntax' (the title of sect. 19) for SI, and undecidable propositions in arithmetic for SII and AS (sects. 36, 60). Early in the final chapter Carnap explained his rejection of Wittgenstein's decrees on the impossibility of exhibiting logical form and of formulating metatheory (sect. 73), and on the latter point he propounded 'knowledge-logic instead of philosophy' (the title of sect. 72) in both its general and Wittgensteinian senses. The new word denoted the 'question-complex of the logical analysis of knowledge' for a given theory, seen as part of the VC's philosophy of science replete with protocol sentences and (hopefully) effecting the 'unity of knowledge' by reducing all scientific sentences to that form (pp. 248–250, with some intriguing remarks on Karl Popper's book *1935a*, about to appear). Such pursuits were not only precise but also avoided the detested metaphysics (pp. 204–205); thus 'l o g i c   i s   s y n t a x' (p. 202), so that 'knowledge-logic is syntax of the knowledge-language', the title of the anti-Wittgensteinian sect. 60, where the former notion was 'logical methodology' (p.7).

Such fervour sits ill with another, well-remembered, feature of the book: Carnap's 'tolerance principle of syntax', according to which 'I n   l o g i c there   is   no   norm['*Moral*']. Each may construct his logic, i.e. his language-form, as he wants' (sect. 17). He adapted it from a similar approach which Menger had proposed for mathematics in order to calm the formalists and the intuitionists (Menger *1979a*, 11–16); denying preferences for any logical system, it introduced an element of conventionalism. Thus it let him consider intuitionism in connection with SI in the preceding sect. 16, and handle some of Lewis's modal logics as examples of intensional languages (Carnap *1934b*, sect. 69 and p. 230).

Like most logicians and philosophers of the time, Carnap assumed that the semantics of a (bivalent) language followed from its syntax (Coffa *1991a*, ch. 17). Thus semantics gained only two references in his book, and moreover was taken from Leon Chwistek (§8.8.4) to be the study of symbols and the means of their manipulation (Carnap *1934b*, 191–192). However, in sect. 60 on 'the antinomies' in AS he noted Ramsey's division of the paradoxes and admitted that ' "t r u e"   a n d   "f a l s e"   a r e   n o g e n u i n e   s y n t a c t i c   c o n c e p t s'. Much later in reminiscence, he recognised that his stance had been too narrow, and now preferred 'metatheoretical' to 'syntactical' (*1963a*, 56). For later in the 1930s he was to realise that semantics (in the general sense) enjoyed independence from syntax; this finding is briefly explained in §9.4.6, when we note his first years in the U.S.A.

The book was well and warmly reviewed. In the *Jahrbuch* Dubislav *1935a* judged it as an 'epoch-making work', effected 'with extraordinary success'. In the U.S.A. Ernest Nagel *1935b* emphasised in the *Journal of philosophy* the distinction between languages, noted the divergences from Wittgenstein, and doubted the generality of the tolerance principle. In the *Philosophical review* Quine *1935a* also stressed languages, and noted the use of Chwistek's sense of semantics and of Gödel numbering; he had been through the manuscript when visiting Carnap in Prague two years before (§9.4.4). There was also a review in *Mind*: we note it in §9.5.1, in connection with a visit which Carnap paid to London in 1934.

9.3.3 *Carnap on incompleteness and truth in mathematical theories, 1934–1935.*   Apparently to save production costs, Carnap omitted some material from his book, and converted it into two papers. Much concerned with Gödel's theorem and corollary, they appeared in the same journal, the

Monatshefte, as *1934c* and *1935b*; he sent offprints to Russell (RA), and put back most of the original material into the English translation of the book (*1937a*, 211–222 and 260–270, 98–129).

Starting the first paper with the distinction between 'O b j e c t   l a n-g u a g e' and 'S y n t a x   l a n g u a g e', Carnap used Ramsey's distinction of kinds of paradox, with Russell's and Grelling's as one example of each kind (*1934c*, 264–266). Still under the dominance of syntax and following Chwistek's sense of semantics, he noted again that truth-values were not syntactic notions (p. 268). Then he formulated 'syntactic antinomies' by imitating the liar paradox for two pairs of notions: a proposition was 'analytic' in a language if it were derivable by logical means alone, and otherwise 'contradictory'; and 'provable' if capable of proof and otherwise 'refutable' ('*widerlegbar*') (pp. 269–272). These moves brought Carnap to his main concern; that, because of Gödel's theorem, which imitated these paradoxes, 'm a t h e m a t i c s   i s   n o t   e x h a u s t i b l e   b y   O n e   s y s t e m, but demands an infinite sequence of ever richer languages' (p. 274).

In the rest of his paper Carnap reworked his theory in terms of two-place relations, syntactically defining properties such as symmetry and distinguishing the 'L- . . .' (logical) version in the object language from its mate 'G- . . .' ('*gültig*', valid) in the syntax language (pp. 277–278). He stressed the 'Isomorphism' between two relations in both languages (pp. 278–281), a variant of Russell's ordinal similarity (795.2) between two relations resembling his own 'structure property' of the *Abriss* (§8.9.2) but now defined for each language. As a mathematical application he considered transfinite cardinals and non-denumerable classes in set theory, taking Fraenkel's version (§8.7.6) as reference; he showed that the paradox (663.1) of cardinal exponentiation was avoided by distinguishing equinumerousness in the two languages (pp. 281–284).

At the end of this paper Carnap mentioned the Löwenheim-Skolem paradox (§8.7.5); and in its successor he sought, as a 'main task of the logical grounding of mathematics', a 'c o m p l e t e   c r i t e r i o n   o f   t h e   v a l i d i t y   o f   m a t h e m a t i c s' (*1935b*, 163, 165), necessary and sufficient to show that a theorem was 'valid (correct, true)' (p. 166) while of course allowing for Gödel's theorem. Thereafter, however, truth disappeared from the syntactic exercise, though the two languages were separated: the two pairs of notions from the previous paper were still in place, and joined by other pairs using sequences (as in his book), such as 'derivable' mated with 'sequence' (the double column of p. 166). Broadly following Gödel, Carnap built up a formal system for first-order arithmetic, using different fonts to distinguish its symbols from their mates in the syntax language; for the (meta)logic he deployed normal forms of propositions, and formed a type theory (pp. 167–180). Oddly, 'We shall not explicitly define the concept "analytic" ' (p. 173), for he followed the version given in the previous paper; further, it furnished the sought criterion of validity, covering not only logical staples such as the law of excluded middle but even mathematical induction (pp. 182–188). However, the status of the axiom of choice was unclear, because of the uncertainty of reducing it to logical form (pp. 188–190)—his version of Russell's dilemma (§7.8.7).

A footnote in Carnap's paper is very striking (p. 165): Alfred Tarski has recently published a short note *1932a* in German on the notion of truth 'in the language of deductive disciplines', summarising work in Polish which 'is not available to me' but seems akin to Gödel. But there was an important difference: at that time Carnap gave logic, language and mathematics a syntactic and thereby structural character with semantics defined in Chwistek's limited sense, whereas, as we shall see in §9.6.7, in Tarski's theory semantics took prime place.

9.3.4 *Dubislav on definitions and the competing philosophies of mathematics.* The rest of this section treats contemporary work by companions or other members of the VC. Wondering in *Erkenntnis* about 'the so-called objective ['*Gegenstand*'] of mathematics', Dubislav rejected Kant's position for its incorrect emphasis on intuition and plumped for Hilbertian formalism as the preferred alternative; logicism could be maintained only as an implicational thesis, not for deriving formulae from 'the initial formula' by 'operational rules' (Dubislav *1930a*, 47–48).

The next year, his 37th, Dubislav issued the third edition of his book on definitions, now just called 'Die Definition' (Gabriel *1972a*, 52–55); it launched, and seemingly also ended, a book series linked to *Erkenntnis*. More than twice the length of the last edition of 1927 (§8.7.8), again he started with four prevailing theories: essence, concept, establishment, stipulation (*1931a*, pt. 1). Then in pt. 2 he described other or related kinds, at much greater length than before: definitions as rules for substitution, both as associated with Frege or those influenced by him, and of a 'formalistic' sub-kind for calculi such as *PM*; 'as coordinations of signs with objects'; 'the specification of concepts'; 'the explanation of signs'; and 'the explanation of situations'. Again he treated special forms such as inductive and implicit definitions (with Russellian contextual definition rather buried at p. 39 of the latter), especially in formal systems where (preservation of) consistency was an important factor. His account of *PM* was relatively detailed, including the 'so-called branched theory' of types (p. 91, citing the second edition and Ramsey). In a substantial review *1932a* for the *Zentralblatt*, Hilbert's graduate student Arnold Schmidt (1902–1967) noted that Dubislav assumed that the consistency of the system containing these definitions was known, a hope which Gödel's corollary had dashed for many cases.

This oversight is surprising, for Dubislav was quick to appreciate Gödel's paper. He considered 'contemporary' philosophy of mathematics in another short book, published in 1932 in a series of 'philosophical research reports' for which Wilhelm Burkamp and Kaufmann wrote on other topics. Into 88 pages he managed to include empiricism and phenomenology as well as the main three schools. He also not only stated Gödel's theorem but also sketched out the proof and mentioned the corollary (Dubislav *1932c*, 20–22)—the first account outside the VC, it seems, apart from his own review (§9.2.4). But even he did not recognise all its consequences: for example, he did not mention the assumption of completeness that underlay *PM*. He found all schools unsatisfactory or silent in some aspect or other of mathematical existence (pp. 48–66). On logicism (so named on p. 38, without explanation of the word but with Frege's version mis-identified as reducing mathematics to logic), he noted the three doubtful axioms and Ramsey's reconstruction (pp. 41–42), and failed the second edition of *PM* for relying upon tautologies to image reality (pp. 62–63).

On the whole Dubislav gave an able review of the varied current situation. The bibliography was also excellent; nevertheless, when that notable historian Heinrich Scholz *1933b* reviewed it for the *DMV*, the most substantial notice that it received, he pointed out some omissions of literature, especially the latest edition (1930) of Hilbert's book on geometry where several foundational papers had been reprinted (§9.6.2). But in general he welcomed Dubislav's book, noting the account of Gödel's theorem which, like Fraenkel (§9.2.4), he called an 'epoch-making work'. Schmidt *1933a* was also generally warm in the *Zentralblatt*, stressing the treatment of consistency but not mentioning Gödel.

Also in 1932, Dubislav spread the word in a French annual launched by Albert Spaier (§8.6.3) with Alexandre Koyré as a collaborator. To a special section devoted to philosophy abroad he summarised current German work in the philosophy of mathematics in 13 dense but highly informative pages graced with fine references. He devoted the first part to 'metamathematical problems', including a statement of Gödel's theorem and corollary (Dubislav *1932b*, 303)—perhaps the first non-German presentation. In the second part, on 'epistemological problems', he covered the three main schools and also work on infinitude. For logicism he mentioned Ramsey's efforts to clear up difficulties; but neither there nor on formalism (pp. 306–310) did he make any connection to Gödel!

9.3.5 *Behmann's new diagnosis of the paradoxes.* In a paper *1931a* published by the *DMV*, Heinrich Behmann, then entering his forties, proposed a new solution to the paradoxes based upon banning impredicativity from nominal symbolic definitions. The correction took two forms. The first stated that all signs in the defined term had to be eliminated when it was replaced by the defining expression (p. 41). Russell's paradox arose, in its properties version, because the propositional function 'ϕ does not come within itself' (p. 40); for from the nominal definition

$$\text{`}F(\phi) = \sim \phi(\phi) \text{ Df' it followed that `}F(F) \equiv \sim F(F)\text{'.} \qquad (935.1)$$

The second correction required that a quantifiable variable could be used in its range of significance in a given context only if the results could be expressed without deploying the defined terms. For example, for three propositional functions ϕx, ψx and χx, if the value a of x were 'insertable' ('*einsetzbar*') in the first two but not the third, then any formula using them would have to be restricted by a 'metalogical' property 'χx!' stating that 'x is insertable in χx' (p. 44); thus the usual

$$'(x).\ \phi x \supset \psi x' \text{ was replaced by } '(x): \chi x! . \supset . \phi x \supset \psi x'. \quad (935.2)$$

To us this procedure is akin to free logic; Behmann contrasted it with both type theory and axiomatic set theory (pp. 43–44).

The response to these corrections was somewhat confused. Dubislav had criticised the first one in his book on definition, reporting that Gödel had given a counter-example to it (*1931a*, 94, 96). In reply in *Erkenntnis* Behmann *1932a* noted that his second correction had not been discussed, and in correspondence with Dubislav Gödel clarified that he had not refuted Behmann but had shown that unrestricted type theory would generate paradoxes independently of the forms of nominal definitions.[8] This finding vindicated a review of the paper in the first issue of the *Zentralblatt*, where Schmidt *1931a* had doubted that rules about signs in definitions were sufficient.

Later in *Erkenntnis* Behmann *1934a* posed the question 'Are mathematical judgements analytic or synthetic?', but answered enigmatically. Taking 'pure mathematics' as undefined, he compared it with logic (whose basic notions, taken from his recent short book *1927a* (§8.7.8), included necessity and possibility), and decided that both were analytic and grounded in the same axioms, but with no sharp distinction between them. As usual, he was unhappy with logicism because of the three doubtful axioms.

Behmann brought his views on paradoxes to English attention in a short paper *1937a* in *Mind*, in response to a discussion there involving Kurt Grelling. He gave Russell's paradox in 'the brief formulation in language: "Does the quality 'not applying to itself' apply to itself or not?"'—an example which is usually attributed to its presumably independent reappearance in Gödel *1944a* (§10.2.5).

9.3.6 Kaufmann and Waismann on the philosophy of mathematics. In a survey in *Erkenntnis* of 'current contrasts in the founding of mathematics' Fraenkel dwelt on set theory and logicism, noting difficulties in the latter caused by the doubtful axioms (*1931a*, 298–302). But Kaufmann went further when considering 'the disagreement on foundations in logic and mathematics', for he regarded these axioms as causing 'unsurmountable difficulties' to logicism (*1931a*, 290). He was continuing a line presented in

---

[8] Gödel's letter to Dubislav, and related letters, are held in Behmann Papers, File I 23; the documents are to appear in Gödel *Papers 4*.

a recent book *1930a* on 'the infinite and its elimination'. Much less keen on empiricism than his colleagues, he drew upon Husserl's phenomenological emphasis on acts of thought and judgement (pp. 41–44), and allowed only a denumerable number of them (pp. 18–19). Brouwer and Hermann Weyl were seen as fellow thinkers, though with a different philosophy (pp. 57–68); and Wittgenstein was followed on tautology and identity (pp. 38–40).

Kaufmann banned infinitude from the foundations of mathematical analysis and set theory. For example, theories of irrational numbers (including Russell's) were re-interpreted as treating sequences of (sequences of) rational numbers converging towards 'condensation-intervals' of arbitrarily small rational lengths (pp. 121–128). Ordinals came before cardinals because, when laying out n objects in *any* order, the last one would always be the nth (p. 78); thus cardinals were invariants under orderings, and not classes of classes. In set theory, following Brouwer, he found the comprehension principle to be unreliable beyond the denumerable (pp. 61–64), and invoked the 'Löwenheim-Skolem antinomy' (§8.7.5) to effect suitable reductions in size (pp. 163–166). Thus, defining integers as classes of (well-ordered) classes was not permitted, because classes were not susceptible to second-order comprehension (p. 99); axioms such as infinity, choice and power-set were not welcome in unrestricted form (pp. 173–179); and Cantor's diagonal argument showed only that any *finite* list of decimal expansion of real numbers could be extended (pp. 139, 145).

Logic was a secondary factor for Kaufmann, but he sketched the first-order predicate calculus, following Hilbert (pp. 46–49). He noted the differences between the various current senses of completeness and categoricity (not his terms: pp. 71–73). On (un)decidability he mooted the 'unprovability of the [arithmetical] assertion, whereby the unprovability itself is provable' (p. 186), a possibility which Gödel was to fulfil the next year. He did not address logicism directly, but quoted Russell's own reservations over Chwistek and Wittgenstein from the second edition of *PM* (p. 180). The paradoxes of logic and set theory arose from illegitimate comprehension (pp. 193–194), while the semantic ones were solved by claiming that naming, say, the least indefinable ordinal treated 'the thinking of the number' than the number itself (p. 196); similarly, the vicious circle principle was 'arbitrary and perhaps too far-reaching a ban for thinking' (pp. 192–193).

Kaufmann's book was widely and well received: ten notices are listed at the end of the welcome given by Hempel *1933a* in the *Jahrbuch*.[9] Another

---

[9] Interest in Kaufmann's book has continued recently in translations into English (1978) and Italian (1990), the latter with a nice long introduction on 'the infinite by etcetaration' by the translator, Liliana Albertazzi. I have not use Kaufmann's *Nachlass*, which is housed at Waterloo University, Canada, and is catalogued in Reeder *1991a*; much of his work lay outside our topic and period.

warm review, and the longest, was written for the *Journal of philosophy* by Nagel; finding the book 'very interesting though by no means convincing', he was glad to see Brouwer's 'extreme and mistaken views avoided' (*1932a*, 402–403).

The book was not intended for the Frank-Schlick series, which was supposed to have been launched with a volume on 'logic, language, philosophy' by Waismann. However, although he fiddled with his manuscript until 1939, it appeared in full only posthumously (Waismann *1976a*). Similarly, he did not publish a talk on the Wittgensteinian 'language-criticism' philosophy of mathematics which he gave at the Königsberg meeting in 1930 (§9.2.2). Indeed, he published little before an introductory book *1936a* on 'mathematical thinking', also not in the series though put out by the same house, Julius Springer. His coverage was similar to Kaufmann's, but he concentrated more on the formation of concepts rather than their justification; for example, he did not define integers at all in his 'rigorous construction' (ch. 4) but instead laid down laws which pairs of them should satisfy (such as equality) and properties that they should obey (such as addition). The most advanced treatment came in ch. 8, on mathematical induction. He did not discuss philosophical issues in detail, even in ch. 9 on two main schools, though the outline of formalism did include a clear statement of Gödel's theorem; for logicism he covered only isomorphism between classes. However, both there and elsewhere he credited Frege, although not in ch. 6 on definitions of real numbers, where also Russell was left out. While he praised Brouwer occasionally, he did not appraise intuitionism. Unlike Kaufmann, he ignored phenomenology: like Kaufmann, he was much influenced by Wittgenstein, and in a brief epilogue he noted a manuscript by Wittgenstein on the philosophy of mathematics which was published only after its author's death.

Waismann developed his relationship with Wittgenstein at Cambridge in 1937 after immigrating to Britain; two years later he moved to Oxford. The VC had been forced to split up, members going elsewhere in Europe or to the U.S.A. To the latter country the next section is devoted, starting with work of some compatriots.

9.4 LOGIC(ISM) IN THE U.S.A.

9.4.1 *Mainly Eaton and Lewis.*
One continuing interest in *PM* for American mathematicians was as a postulate system: forays include B. A. Bernstein *1931a* and Huntington *1933a*. But, as in other countries, logicism was becoming eclipsed by the competing philosophies. It did not feature at all in Tobias Dantzig's introductory account of *Number, the*

language of science, although Cantor's *Mengenlehre* was described in some length, with striking historical carelessness (Dantzig *1930a*, ch. 9). Again, in a long paper in the general journal *Science* on 'Tendencies in the logic of mathematics', based upon a lecture given at the end of 1932 to The American Association for the Advancement of Science, E. R. Hedrick (1876–1943) devoted only about an eighth of his space to 'logistics', and then as much for its postulational interest as for its logic (Hedrick *1933a*, 341); logicism was ignored. Exactly a year later Saunders Mac Lane, finishing a *Dissertation 1934a* at Göttingen under Weyl, summarised it at the end of 1933 in a lecture to the American Mathematical Society on the 'logical analysis of mathematical structure'. A printed version *1935a* appeared in *The monist*; drawing upon the mathematical logic of *PM*, he examined the partitioning of long proofs into parts and their reorganisation to bring out the overall strategy. In the thesis he also presented 25 rules for abbreviating sub-proofs or steps in derivations; several amounted to deduced rules of inference. He did not consider Gödel's theorem because, as he told me recently, nobody at Göttingen ever mentioned it to him. His work made very little impact, but it set some of the lines of his own later structuralist philosophy of mathematics.

The reception of *PM* among philosophers was much warmer. Shortly before his death, R. P. Eaton (1892–1932) at Harvard University, another former student of Josiah Royce, published a long introduction to *General logic*. While he 'represents no school and has no philosophical axe to grind' (*1931* 2, vii), his preferences came through in the text, with nearly 300 pages on 'The Aristotelian tradition' and 120 on 'Mathematical logic'; and for the latter the first edition of *PM* dominated so much that Hilbert gained only two footnotes and Brouwer nothing. Following the tradition of such books, a final part covered inductive logic, for 140 pages.

Noting from his friend Henry Sheffer the 'logocentric predicament' (§8.5.3) that logic was needed to discuss logic (pp. 38, 385), Eaton gave epistemological priority to propositions over sentences, facts and judgements; he defined logic as the theory of valid inference among them, and relative only to their form (p. 22). The propositional function was defined as a general '*scheme of structure for a proposition*', and quantification in terms of generality (pp. 391–393). He saw algebraic logic only as 'an important stage in the approach, from the side of logic, to the unification of mathematics and logic' (p. 361). The latter clause was influenced by his unhappy construal of logicism that 'mathematics is the later portion of logic' and *vice versa* (p. 359); he sketched it up to the definitions of finite cardinals. His account of *PM* was straightforward and quite detailed, though he (intentionally) left out the logic of relations (p. 408). Failure to consider the second edition of *PM* and reactions in the 1920s made his book rather *passé*.

In the following year a very different picture was conveyed by Lewis, who published with C. H. Langford a 500-page account of *Symbolic logic*

(Lewis and Langford *1932a*). Set at a similar level to Lewis's own *Survey* of 1918 (§8.3.3), it contained far less on algebraic logic and history and more on the modal logics (interestingly, not a name used, although they wrote of 'modal functions'). Most of the material on those topics came in the first eight chapters of the book, for which Lewis was chiefly responsible. They also included the propositional calculus as a Boole-Schröder 'two-valued algebra' (ch. 4), a name nicely mixing Polish logic (they cited Łukasiewicz and Tarski in the preface) and American postulate theory. The method of truth-tables was extended to 'the matrix method' to accommodate the values taken by many-valued logics (ch. 7 and app. 2). In a paper of the same year in *The monist*, Lewis *1932a* gave an introductory account of his modal logics and contrasted them with *PM*.

The last five chapters of the book, mainly due to Langford, started with the bivalent calculi but included the modal operators (chs. 9–10); then 'postulational technique' was developed in detail (chs. 11–12). *PM*, featured at times, arrived in the final ch. 13. Entitled 'The logical paradoxes', it began with a survey of several (with the Grelling attributed to Weyl on p. 449); however, most of its 48 pages were devoted to a detailed account of type theory, mainly that of the first edition of *PM* but with the later modifications noted (pp. 458–461). While warning of ambiguities, their account was not critical; for example, the vicious circle principle was not queried, the three doubtful axioms were passed over, and logicism was not much considered. Modality crept in when they read 'Men exists' as 'Men *might* exist' (p. 472), which was not the way that Russell construed possibility (§7.3.6).

The book was generally well received, most notably by Scholz *1935b* with the *DMV*. For him 'the most interesting piece of the work', indeed 'a landmark in the history of the logistic propositional calculus', was the 'System of strict implication'. Oddly claiming that it '*cannot* be constructed by the matrix method', he defended logical pluralism: each consistent logical system was 'a representative of a possible logic', with truth values to be taken conventionally rather than absolutely (p. 89). The new logics were now coming out from under the skirts of bivalency.

9.4.2 *Mainly Weiss and Langer.* By contrast, a modified orthodoxy was proposed by Paul Weiss, then in his late twenties. He suggested in *The monist* that 'The metaphysics and logic of classes' be handled not by the normal abstraction via predicates such as ' … is red' but using the 'indexical rhema' such as 'this book is …' (Weiss *1932a*, 120). He made much play of building up classes from unit classes, such as the extensional indexical class 'Columbus is …' specified by the intension 'discoverers of America' (p. 127, and allowing for some dubious history of navigation). While his variant is appealing, its efficacy is uncertain; for example, his treatment of classes of classes and of definitions of cardinals using double abstraction (pp. 142–149) is not convincing.

Weiss credited the notion of indexical rhema to C. S. Peirce (p. 116), whose works he was then editing for publication in collaboration with Charles Hartshorne. Their six volumes came out from the Harvard University Press between 1931 and 1935 (Peirce *Papers*). Logic featured most prominently in volumes 3 and 4 (1933), for which Weiss was largely responsible. Despite the many faults in the preparation, the edition launched the general recognition of Peirce. It was widely reviewed, and not only in the U.S.A.; for example, R. B. Braithwaite wrote a long piece *1934a* on volumes 1–4 for *Mind*, concentrating on the semiotics and probability theory but also making clear that a remarkable logician was waiting to be discovered.

In the *Philosophical review* Weiss *1933a* considered 'alternative logics', especially Lewis's, and argued against them by granting priority to entailment, which he took to be 'a relation between groups or propositions or single propositions such that the antecedent has the consequence as a necessary result' (p. 523). Missing metalogic, he concluded that logical pluralism 'will remain a position stated and not a doctrine proved' (p. 525). Not surprisingly, Lewis wrote a terse reply *1934a* reaffirming the plurality of notions involved in implication.

Also in 1934 there appeared another textbook, from Morris Cohen (§8.3.4) and his former student Ernest Nagel. Following the tradition of largely philosophical works, they presented parts on deductive (bivalent) logic and on 'scientific method'. Selling well from the start, an abridged edition appeared in 1939, and the first part of the original version was reprinted as a separate volume in 1962. Concentrating there on proof, they carefully distinguished several related notions, not only implication and inference, and they astutely treated traditional logic (Cohen and Nagel *1934a*, chs. 2–5, all omitted in the abridged edition). But the selection of topics from mathematical logic was rather odd, though showing postulatary traits: the notion of relation and an algebraic account of 'the calculus of classes' (ch. 6); then 'The nature of a mathematical or logical system', with the ambiguous 'or' heralding mostly a survey of basic aspects of axiomatics (ch. 7). Logicism was not discussed, and throughout *PM* received only a handful of passing references; Eaton and Lewis/Langford were just cited, for 'the algebra of logic'; Hilbert's metamathematics and Gödel were absent.

In 1937 an elementary *Introduction to symbolic logic* was published by Suzanne Langer, then in her early forties and teaching at Harvard's female counterpart, Radcliffe College; we noted her in §8.5.4 a decade earlier, spotting Russell's conflations of symbol and referent. In the preface she warned against teaching only technicalities, and her text was laced with nice everyday examples; in addition, each chapter was furnished with simple exercises, and the index was unusually rich.

Langer emphasised logical form, citing a passage from ch. 2 of Russell's *Our knowledge* as supporting evidence (*1937a*, 32); she even preferred a propositional function to be called 'propositional form' (p. 91). Postulate theory turned up in the handling of Boolean algebra and the propositional calculus (chs. 5 and 7), but *PM* dominated her treatment; for example, much of the exegesis was handled in terms of classes, though with ' < ' inaptly symbolising improper inclusion (p. 135). She discussed *PM* in detail at the end, presenting its 'assumptions' (ch. 12) and then its aims as 'Logistics' (ch. 13); but the impression is strangely incomplete. For example, she did not make clear the differences between the two editions. Again, she did not advance beyond the definitions of (only finite?) cardinal numbers (pp. 314–321); but she then announced that thereby 'Whitehead and Russell rear the whole edifice of mathematics [sic], from Cardinal arithmetic to the several types of Geometry', mentioned the paradoxes for the only time, and stated that type theory was too complicated to be described (p. 331; compare p. 324). In her account of the propositional calculus she arrived at the border between logic and metalogic, quoting Sheffer (pp. 278–281) but passing over Hilbert and Gödel. As Nagel *1938a* judged in his generally positive review for the *Journal of philosophy*, the book was rather *passé* in being so dominated by *PM*. Let us turn now to its senior author.

9.4.3 *Whitehead's new attempt to ground logicism, 1934.* Whitehead, since 1924 at Harvard University, seems to have set logicism aside after abandoning the fourth volume of *PM* (§8.2.2). However, in his philosophical account of *Process and reality* he outlined a theory of indicating the subjects of a proposition, including relations (*1929a*, 274–279), and five years later he furnished the details in a difficult, and badly proof-read, paper *1934a* in *Mind*.

Whitehead started out from the new notion 'Ec!x', which referred to the 'indication' of a unique object denoted by x (p. 282):

> [...] 'Ec' stands for the latin word 'ecce' meaning 'Behold'. [...] Ec!x is a proposition about x—the object, not the symbol—which ascribes to x no intension other than the intension derived from purely logical notions. [...] it is a proposition involving x in pure extension. It expresses that recognition of individuality which is involved in counting.

Upon this epistemologically obscure foundation he built a theory of classes, using also the notion of 'togetherness' of indicated objects; 'Ec!x' itself delivered the unit class of x, 'Ec!$x \cup$ Ec!y' the class with members x and y, and so on (pp. 284–285). Classes were 'special instances of true propositions' (except for the empty class Λ), and their algebra was developed as a branch of the propositional and predicate calculi. For example, 'Ec!x' was defined in terms of the existence of a propositional function ϕ, and Λ from a contradiction using propositional quantification:

$$\text{'Ec!}x := \text{Df} . (\exists \phi) . \phi x \text{' and '} \Lambda := \text{Df} : (p) . p . \sim p \text{'} \qquad (943.1)$$

(p. 286, with an important misprint corrected). However, to avoid paradox there was no universal class, because he did not assume that every propositional function was 'associated with a unit entity which in some way is derived from the totality of objects satisfying it' (p. 286). He also abandoned the propositional hierarchy of types (p. 284).

Whitehead criticised the construction of arithmetic in *PM* on two grounds: its formulation within each type; and the dependence upon 'shifting accidents of factual existence', where 'a new litter of pigs alters the meaning of every number, and of every extension of number, employed in mathematics' (p. 288). Thus he still adhered to the empirical status assigned to individuals (§7.9.3). His new theory was based upon 'Ec' and Λ, defining from 0 in each case:

$$\text{ordinals: } \Lambda, \text{Ec}!\Lambda, \text{Ec}!\text{Ec}!\Lambda, \ldots;$$
$$\text{cardinals: } \Lambda, \text{Ec}!\Lambda, Ec!\Lambda \cup \text{Ec}!\text{Ec}!\Lambda, \ldots. \tag{943.2}$$

The latter definitions are not unlike von Neumann's a decade earlier using iterated nesting (§8.7.6). He also defined $\aleph_0$ from $(943.2)_1$ as the union of the preceding finite ordinals (pp. 288–289).

Noting the inadequacy of the definition $(785.13)_2$ of the ordered pair in *PM*, Whitehead replaced it by

$$\text{`}(x \downarrow y) . = \text{Df.} \quad \text{Ec}!x_1 \cup \text{Ec}!x_2\text{'}, \tag{943.3}$$

with the suffices defined from $(943.2)_1$. Then he rebuilt the foundation of the logic of relations extensionally by treating them as classes of ordered pairs, noting also Wiener's 'analogous definition' (827.1) of them (pp. 290–291). With these modified gadgets he could set up finite cardinal arithmetic. He ended with some cryptic remarks about forms of propositions, rejecting the obligatory quantification of all propositions executed by Russell in the second edition of *PM* (pp. 294–297).

It is tricky to assess the potential of Whitehead's proposal, for he was often obscure and he only sketched out consequences. In particular, the range of deployable propositional functions is not clear, so that his abandonment of type theory is enigmatic. Further, the status of 'Ec!x' as logical is questionable (although Ackermann *1938c* accepted it in his review for the *Jahrbuch*); maybe a theory of logically proper names or of haecceity could be confected. In addition, his system would be vulnerable to Gödel's theorem, which he ignored both here and in all of his writings after 1931.

Whitehead had consulted his junior Harvard colleague Quine, who gave a lecture *1934a* upon the proposal to the American Mathematical Society. He never returned to mathematical logic; quite the opposite happened with Quine.

9.4.4 *The début of Quine.* (Dreben *1990a*) Studying at Harvard University with Whitehead nominally as advisor, Quine presented a doctorate there in March 1932, when 23 years old, on 'The logic of sequences' (*m1932a*). Spending the following autumn and winter in Europe, he met the Viennese and Polish logicians, especially Carnap in both Vienna and Prague (Quine *1985a*, 92–104). Upon his return he made a radical revision (but not purpose) of the thesis into his first book, which was published by his University Press in 1934 as *A system of logistic*. The account below is largely confined to the latter.

Taking the second edition of *PM* as his inspiration, Quine emulated it but used quite different primitives. The first one was 'ordination', a non-associative operator which made a 'sequence' 'x, y' out of terms x and y (which might be already existent sequences), then both '$x(y, z)$' and '$(x, y), z$' out of three terms, and so on (Quine *1934b*, ch. 1). The extension was very much the point; iterations could produce 'n-ads' (this name independent of its use by A. B. Kempe in §4.2.8) and thereby allow the calculi of classes and of n-ary relations to be developed simultaneously instead of the separate treatments in *PM* of classes and of (only) binary relations. Quine claimed that ordination could also ground the propositional calculus, since he defined a proposition as the form 'x is a member of [class] y' (p. 26), presumably only for the purposes of mathematical logic.

Quine's other two primitives were presented in ch. 4. One was the operation of 'congeneration' on a class x to produce its 'superclass' '$[x]$' of classes (his square brackets) within which it was properly included. This notion allowed set theory to be developed, based largely on the relationship of inclusion of class α within β (the sequence '$[\alpha], \beta$') and their 'implexion' as the class $((U - \alpha) \cup \beta)$, where U was the local universal class (pp. 91, 85). The other primitive was the operation of 'abstraction' on the bound variable x to produce the class '$\hat{x}y$' of true propositions y (from which x may be absent: p. 36). The class abstraction (785.2) of *PM* was the special case $\hat{x}(\alpha, x)$ of terms x belonging to the class α. He credited the name 'abstraction' to Weiss *1932a*, although his system followed different lines (p. 7).

With these primitives Quine was able to eliminate truth and falsehood, as reducible respectively to U and to its complementary empty class (ch. 12). Classes also went; for the class of true propositions was $\hat{p}p$, so that membership could be defined as the converse sequence (p. 168). The logical connectives were also definable: pride of place was given to material implication '$\supset$' between propositions p and q as $[\hat{x}p], \hat{y}q$ (p. 45); thus, for example, negation '$\sim p$' came out as '$p \supset . U, \hat{q}q$' for any proposition q (p. 96). The *modus ponens* rule of inference was also formulated using material implication (but incurring a Russellian conflation): it joined the

TABLE 944.1.   Primitives in Quine's First Two Logical Systems

Notion in the thesis	Status or situation in Logistic
'Catenation' (pairing of sequences)	Extended to 'ordination'
'Superplexion' (related to inclusion)	Changed to 'congeneration' (inclusion)
'Predication' (membership)	Defined
'Assertion'	Dropped
'Abstraction' (spuriously defined on p. 20)	Taken as primitive

rules of substitution, and of universal quantification on a free variable in 'any theorem or postulate' containing it (ch. 5).

Among other features of *PM*, Quine modified Russell's theory of definite descriptions, giving prime place to the descriptive function '$R'x$' of *PM* (786.1) rather than the general contextual definition (785.2) (ch. 14). He also adopted the simple theory of types (as another primitive?), without appeal to the vicious circle principle (p. 19); and he argued that the ramified structure and the axiom of reducibility were irrelevant to the logicist programme (pp. 186–187: he used neither of these adjectives). Much of the book was taken up with developing a system logically equivalent to that of *PM*, with a summary given in the final ch. 18.

The relationships between the primitive notions in the thesis and in the book are summarised in Table 944.1. One important difference concerned the notion of assertion: primitive in the thesis, he abandoned it in the book (p. 5) following Carnap's argument in *Abriss* (§8.9.2). But several features of the book had been in its predecessor: for example, the status of the axiom of reducibility, and the careful distinction of a symbol from its referent (where on p. 4 of the book he thanked Sheffer for suggesting the systematic use of 'X variable' as opposed to 'X' for various notions X). An important common factor was extensionality: in the book it was based upon ordination, which produced 'A class of sequences[, which] is called a *relation in extension*' (p. 18), and membership of classes was also thus construed (p. 93). But then the usual reservations about generality (§8.4.7) arise again.

There are some surprising omissions. For example, even though he had been to Vienna on his trip, Quine did not mention Gödel's incompletability theorem or its corollary; indeed, completeness and consistency were nowhere discussed. In a recent letter he told me that he may have felt that the theorem had no *direct* bearing on the reconstruction effected; but he had launched a similar reduction of arithmetic, especially in ch. 18. Further, while he attended well to the axiom of reducibility, its companions for infinity and choice were passed over. His remodelling of type theory recalls Ramsey, whose work he seemed also not to know. D. J. Bronstein *1936a* noted some of these lacunae in a review for the *Philosophical review*.

In his thesis Quine had not even cited the 1914 definition (827.1) of the ordered pair due to Norbert Wiener, long resident at MIT down the road (although not then working on foundational questions). He recalled in *1989a* that Harvard's logicians and philosophers were not so *au fait* with the literature; but we saw after (943.3) that Whitehead knew this definition, and Wiener's paper was in the bibliography of Lewis's *Survey* of 1918 (§8.3.3) though not mentioned in the text. Who knew or remembered what, and when? There can be surprising answers in history!

In a preface Whitehead *1934b* welcomed the book as a 'landmark', an opinion endorsed by Ackermann *1938b* in his review for the *Jahrbuch*. Black *1935a* was also praising in his review for *Mind*, although he wondered about the potential 'wider objects' of the system. In similar vein, upon receiving a copy of the book in the summer of 1935 Russell told Quine on 6 June that 'I think you have done a beautiful piece of work' (1968a, 213–214). He raised various points, such as the status of the axiom of reducibility, and Quine replied at length in July; the exchange, beautiful, is transcribed in §11.8.

By contrast, Quine's book was to be mentioned only in the bibliography of Langer's *Introduction*, as 'A difficult but significant little book' (*1937a*, 357). It has been eclipsed by his own later achievements, which came forth in a steady stream of papers, and marked out more clearly than any other work the place of logicism after *PM* and Gödel (Ferreirós *1997a*). His next volume was called *Mathematical logic* (Quine *1940a*); and the title is instructive, for while it began like *PM* with the logical calculi and went though arithmetic, no logicist thesis was put forward. On the contrary, classes were not subsumed under propositional functions but stood as 'a second level of mathematics' (sect. 23), and their paradoxes were avoided by a 'stratification' of classes under a criterion of membership different from the vicious circle principle (sect. 28); Wiener's definition of the ordered pair was now used to simplify the calculus of relations (sect. 36). Furthermore, metalogic was prominent, from the early explanation of 'statements about statements' (sect. 5) and introduction of 'quasi quotation' as a means of symbolising the logical contexts of well-formed formulae (sect. 6) through to the version of Gödel's incompletability theorem pertaining to his system (sect. 59). At the time of the publication of this book he also had the opportunity to assess logicism, as we shall see in §10.2.5.

9.4.5 *Two journals and an encyclopaedia, 1934–1938.* The increase in interest in philosophy and logic led to the formation of new organisations in the U.S.A., with evident influence of the VC. When the Philosophy of Science Association was set up in 1934, Carnap was on the editorial board, and his paper *1934a* 'On the character of philosophical problems' launched its journal *Philosophy of science*. Its early volumes contained several papers on postulate theory.

Two years later the Association for Symbolic Logic was established. At its opening session Whitehead recalled the work on *PM*, and criticised the current tendencies towards positivism in logic (Weiss *1936a*). The Association launched the *Journal of symbolic logic*, edited by Church and Langford, seemingly in that order; Church, then in his early thirties, exercised transfinitely meticulous control over all aspects of its production. For example, one of the consulting editors was Paul Bernays, who during the early years received from his younger colleague dozens of letters not only discussing logic and *Journal* submissions but also instructing on the use of genitives in English and the need to apportion the postage costs of any letter sent to a correspondent which included both personal and Association matters.[10] Church also completed the first volume with a magnificent bibliography *1936a* of symbolic logic of over 100 pages, followed by a supplement *1938a*. He was largely responsible for editing the section of book and article reviews, a service exercised on such a scale that volume 26 (1951) consists solely of indexes. The VC was not involved in the formation, but several of its members and followers were soon writing or reviewing there. Topics related to *PM* featured in the *Journal*, not only various pieces by Quine but also, for example, F. B. Fitch *1938a* on a system logically implying that of *PM* which was provably consistent, but which Gödelian arguments revealed as incapable of prolonging transfinite induction up to ω^ω. Fitch also introduced the word 'ramified' into type theory here, without reference to earlier adjectives (§8.9.3).

The *Journal of philosophy and Philosophical review* continued to publish papers in and around logic and epistemology on occasion, including reports on the VC Congresses. In addition, then exiled in Turkey, Reichenbach *1936a* described in the *Journal* current 'Logistic empiricism' in both the VC and his Berlin group; as in 1930 (§9.2.3), he did not mention Gödel. In the same issue Nagel *1936a* described his impressions of a recent visit to Europe, not only to the VC (where Carnap took pride of place) but also to the Polish group, which he compared with 'Russellized Cambridge' (p. 49), another port of call. In addition, he noted the uninterest in history evident at all centres (p. 6).

Also in 1936, his 46th year, Carnap emigrated from Prague and moved to the University of Chicago. This arrival pushed forward a project that Neurath had launched two years earlier; a grandiose 'International Encyclopedia of Unified Science' under his editorship, in which many authors would produce books of around 100 pages on specific topics across the sciences and humanities, grouped into eight volumes. Another key figure was the philosopher Charles Morris (1909–1979), also at Chicago; through his advocacy and the arrival of Carnap, the University of Chicago Press took on the publication. But the production was fraught with difficulties: in

[10] See Bernays Papers, 975:774–818 for letters from Church to early 1940; the full sequence, to no. 1004 (1972), contains much important material.

particular, Neurath had moved to Amsterdam, far away from both the publisher and two main collaborators.[11]

The books for the first volume began to appear in 1938. The very first one was a collection of short essays on 'Encyclopedia and Unified Science' by eminent writers (Neurath *1938a*). Divergence is very evident: the pieces by Neurath himself, Russell (a note *1938a* 'On the importance of logical form') and Carnap satisfied VC rules on philosophising, but those by Morris and Dewey followed the American pragmatist tradition. Although Morris had been influenced by the VC since 1933, he was primarily concerned with the philosophy of signs and indeed had coined the word 'semiotician'. Despite his efforts to join the two philosophies (Rossi-Landi *1953a*), such as in a short book *1937a* of reprinted articles in Hermann's series (§9.2.1), Carnap's principle of tolerance (§9.3.2) was being stretched to extremes, and the title of the encyclopaedia became a name without a referent.

9.4.6 *Carnap's acceptance of the autonomy of semantics.* One of the early individual Encyclopedia booklets to appear was Carnap *1939a* on the 'Foundations of logic and mathematics'.[12] An abbreviated mixture of his *Abriss* and *Syntax*, he ran through the construction of a language using basic German as example (arts. 1–9), and then the propositional and predicate calculi and the Peano axioms for arithmetic (arts. 13–17). But then knowledge of mathematics seems to desert him: 'On the basis of a calculus of the arithmetic of natural numbers the whole edifice of classical mathematics can be erected without the use of new primitive signs' (art. 18). Similarly, he did not mention Gödel's theorem in a very brief description of logicism, although the corollary was noted in a (better) notice of formalism (art. 20).

More happily, in art. 12 Carnap noted the arrival of other logics and appealed to logical pluralism; and this position may relate to a major modification from his earlier writings, the replacement of 'syntax-language' by 'semantical system'. The change was not merely in name: syntax had *ceased* to dominate his philosophy. A clue occurs in his phrase '*normal interpretation*' used for the calculi from art. 13 onwards; for he had found

---

[11] There are masses of letters and other materials on the Encyclopedia in (at least) the Papers of Carnap, Neurath, Morris (this a rich but little-known source in general) and Reichenbach; full references cannot be attempted here. On its publication, see the University of Chicago Press Records (University Archives), especially Box 345, Folders 6–9. After the War Carnap and Morris assumed joint editorship for the project, each representing a philosophy which it embodied. A moderate history will be found in the doctoral thesis Reisch *1995a*; an excerpt dealing with Neurath is given in his *1994a*.

[12] This booklet had originally been assigned to Gödel, but he failed to produce. He finally settled in the U.S.A. in 1940, after some years of apparent personal undecidability (Dawson *1997a*, chs. 5–6). Most of his work in the 1930s was related to formalism, intuitionism and axiomatic set theory; type theory was sometimes used (*Works 2-3 passim*).

other ones. The occasion may have been a graduate seminar on logic which he ran in 1937–1938 at the University of Chicago with financial backing from the Rockefeller Foundation, in which he treated 'syntax and semantics' first before passing on to the theory of types, metamathematics, and other topics.[13] Presenting each topic in a pair of volumes which launched his own series of 'Studies in semantics', the manner of their publication has masked the origins.

First to appear was *Introduction to semantics* (Carnap *1942a*), a systematic treatment of syntactic-semantic 'systems', with semantics understood much more widely than in Chwistek's sense adopted earlier (§9.3.1); in a final section he showed how various notions in *Syntax* had to be rethought as semantics, at least partially. But this book, which became well-known, was *preceded* in preparation by *Formalization of logic*, published the following year (*1943a*). In this forgotten volume Carnap presented counter-examples as 'non-normal' interpretations. For example, if the truth of a proposition were associated with its derivability from axioms in a logical calculus by the assigned rules of inference, then neither a basic proposition p nor its negation was a theorem, so that each proposition was valued 'false'—thereby disobeying the law of excluded middle (sect. 16). Further, he stated at the end of its preface that this book was drafted in the autumn of 1938 (not long after the University course), that non-normal interpretations of the propositional calculus had been among his findings, and that *Semantics* was conceived out of this draft.[14] Proper historical understanding of the development of Carnap's semantics requires the order of these two books to be reversed. Further, the counter-examples not only raised the status of semantics but also compromised the dominating status of classical bivalent logic; after the War non-classical logics of all kinds began to develop rapidly.

The changes to logic initiated by Carnap were not always accepted: Quine continued to adhere to logical monism, while Sheffer became so opposed to the activities of 'Carnap and Co.' that he felt 'If any work of mine has done anything to stimulate this development, I had rather not have been born' (Berlin *1978a*, vii–viii). *PM* had become just one source,

---

[13] Carnap sent to Ogden a summary of this course (Ogden Papers, Box 7, File 10). Carnap recalled his seminar and growing interest in semantics from about 1935 in *1963a*, 35, 60–64.

[14] On this chronology, and on the significance of non-normal interpretations, see Belnap and Massey *1990a*. The two books were due to appear with the University of Chicago Press; but they deferred *Formalization* in 1940 (Records, Box 21, Folders 9-10), and both books were transferred to Harvard University Press after Carnap taught there in 1941 (see his letters to Neurath of 1941–1942 in Neurath Papers, File 222). Like the *Abriss*, *Formalization* is usually ignored by Carnap specialists, who in my view underestimate the importance of this change in position on semantics from Chwistek's theory (§9.6.7). Several otherwise excellent articles in Giere and Richardson *1996a* seem to exhibit these features.

albeit important, among a growing number; let us return to the land of its birth.

9.5 THE BATTLE OF BRITAIN

9.5.1 *The campaign of Stebbing for Russell and Carnap.*   The most active advocate of mathematical logic in Britain was not Russell but L. (for 'Lizzie'—honestly) Susan Stebbing (1885–1943). After studying at Cambridge University and then teaching there, from the early 1920s she passed her career at Bedford College, a women's college in the University of London. She came across Russell at Cambridge, and she attended his London course of 1918 on logical atomism (§8.3.6);[15] but her main involvement with logic, at least as far as her publications are concerned, dates from the late 1920s.

For the new (14th) edition of the *Encyclopaedia Britannica* Stebbing wrote a substantial piece *1929a* on 'Logistic'. After a short survey of the algebraic tradition, she concentrated upon mathematical logic, explaining the main notions and associated notations, and coming up to date with Ramsey's contributions; the only notable defect was a failure to distinguish part-whole theory from set theory. The last part treated logicism itself, with the background in Frege and Peano; having begun the piece by defining logistic as 'the realization of the ideal of logic, the exhibition of form', she ended with a sketch of how 'all the propositions of arithmetic are shown [in *PM*] to follow from the analysis, in purely logical terms', so that 'mathematics is reduced to pure logic, and the achievement of the ideal of form is complete'—a nice literary touch, though effected at the heavy cost of identifying mathematics with arithmetic.

By then Stebbing had been teaching courses in logic for some years, and in the following year she went public with a 500-page 'modern introduction to logic' which she had been invited to write by the house of Methuen. As she explained in the preface, the book was written for two levels of undergraduate study, resulting in some repetitions of material.

In the first of the three parts Stebbing treated deductive logic, both 'in ordinary life' and in formal versions. The first 100 pages or so were dominated by syllogistic logic, though she hoped that quantification of the predicate had had its day (*1930a*, 80); the account was modernised with a few Russellian infusions such as distinguishing knowledge by acquaintance from knowledge by description (pp. 22–27) and particulars from universals (pp. 52–54, including the phrase '*propositional forms*'). But then mathemat-

---

[15] See Stebbing's letter of 1 January 1919 to G. E. Moore (Moore Papers, File 8S/39); other letters here are sources for my account to follow of her book. There seems to be no *Nachlass* for her, but valuable career documents are held in the Archives of the Royal Holloway College, Englefield Green, Surrey, especially Files AR150/D381 and PP33/1.

ical logic largely took over, for 100 pages: relations (pp. 111–114, 166–174), *then* propositional functions, with a nice comparison with mathematical functions (pp. 121–133); *PM* notations in all the calculi (pp. 133–149, 188–196); postulates (though not model theory) prefacing some set theory (pp. 147–149, 180–188, with Peirce's '–<' (§4.3.4) symbolising improper inclusion); and ordinal similarity (pp. 201–207).

However, Stebbing did not welcome every innovation. For example, on Russell's theory of definite descriptions she disliked the name 'incomplete symbol' because of its use elsewhere such as for 'd/dx' in the differential calculus (pp. 152–155, including some correspondence with her mentor G. E. Moore), and she noted the muddles over implication and inference (pp. 212–221): she even quoted a passage from Russell's *The problems of philosophy* (*1912a*, ch. 2) as an example of an argument with suppressed premises (p. 110, with an incorrect page number for Russell). Among other authors, she cited Frege and Wittgenstein for some details; but she ignored features such as the Sheffer stroke, Nicod's axiom system, and Wiener on the ordered pair. More seriously, there was no mention of Hilbert or of Brouwer, though Lewis gained a footnote (p. 222).

In tradition, the second part of Stebbing's book treated inductive logic, for nearly 200 pages. But her coverage was unusual in incorporating historiography and elements of mathematical statistics (chs. 19, 18). The short third part comprised a miscellany of general topics, including definitions (but with the contextual kind omitted, though it had arisen on pp. 144–146 in connection with definite descriptions), and a sketch of the history of deductive and inductive logic (with Peirce omitted, though father noted).

Overall the book is impressive but demanding, similar to Eaton's of the following year in the U.S.A. but rather more astute, and anticipating Langer's for some terms and symbols (§9.4.1–2). Reviews were generally welcoming, the philosopher-psychologist C. A. Mace writing a long descriptive piece *1931a* for *Mind*.

A slightly revised second edition appeared in 1933, the main changes to the section of deductive logic being a suite of short new appendices on various Russellian matters. The most significant one is the third, in which Stebbing saw logicism as a 'directional system' where the postulates are really primitive and the consequences sought, in contrast to the 'postulational system' of Huntington and Oswald Veblen where simplicity of derivation of theorems was the prime criterion for their choice (p. 506). Less happily, she also associated Hilbert with this approach, and took formalism to be only the marks-on-paper philosophy (Stebbing *1933a*, 506–509). In the historical chapter, she repaired her omission of C. S. Peirce with some admiring lines (p. 458).

In the same year Mace published his own introduction to logic, with Longmans, Green. Like Stebbing (who had read his manuscript) he covered both deductive and inductive logic, but on the former he was largely concerned with the syllogistic and related traditions. Mathematical logic arrived in a chapter of 40 pages on 'the general theory of deduction', where he went though all the basic components of the first edition of *PM* except for relations, which were intentionally omitted. While generally quite welcoming, he felt that the reduction of classes to propositional functions 'reverses the historical order, and in a sense the psychological order, of exposition' (Mace *1933a*, 177).

Such enthusiasm for the new could not occur without demur from the British establishment. It came from H. W. B. Joseph (1867–1943), one of the traditional Oxford philosophers with whom Russell had long argued (my *1986c*). Already in a muse *1928a* on 'logic and mathematics' he had attacked the thesis that they were identical (which of course only Russell in careless moments had advocated); as an example he opined that Dedekind's "proof" of the existence of an infinite class, based upon thinking about thinking about... (§3.4.2), seemed to be good logic in being involved with thought but was mathematically illegitimate. Now, inspired by Stebbing's first edition, he launched a non-discussion with her in *Mind* between 1932 and 1934 which became three rambling papers (50 pages in all) with two replies by her (28 pages).

The differences rested upon conflicts over symbols and variables, nouns as opposed to demonstratives, and the various kinds of implication. Joseph did not understand propositional functions at all, denying any analogy with mathematical ones and not grasping quantification (*1932a*, 425–427); on the other hand, he correctly objected to Stebbing's construal of an un-known constant as a variable (*1933a*, 430), and opined that logistic as a calculus was not logic any more than 'the construction of reckoning machines is the science of number' (p. 443). Stebbing would not have disputed this view, but she concentrated on trying to clarify various features. She also admitted that Russell often conflated symbol or word with its referent; in particular, she thanked Moore for taking Russell's theory of definite descriptions to deal with '*how an expression is used*' rather than '*a form of words*' as Russell had stated (Stebbing *1933b*, 342). Perhaps inspired by this exchange, Joseph's pragmatist contemporary F. C. S. Schiller (1864–1937), then based in California, wrote an ironic piece *1935a* on the various current schools in logic; to him symbolic logic was an odd mixture of traditional logic, the new many-valued logics of Łukasiewicz, mathematics and pragmatism.

By contrast, Stebbing formed strong links with the VC, forming her own group to study its work,[16] and also becoming the British representative for the Congresses and of the Encyclopaedia. Like Carnap, she was a founder member of the editorial board of *Philosophy of science* (§9.4.5); and she made another opportunity to further the VC cause by arranging for

---

[16] See Stebbing's letters in Neurath Papers, File 303; and the Moore file in the previous footnote.

Carnap to come over and deliver a trio of lectures at the University of London in October 1934. They soon appeared as a short book on *Philosophy and logical syntax* in Ogden's series.[17] The titles of the lectures capture well the purpose: 'The rejection of metaphysics', 'Logical syntax of language' and 'Syntax as the method of philosophy'. His recent *Syntax* was the main text, and its main features were rehearsed, especially the two levels of language and parallel notions defined for each one; however, apart from noting the different definitions of cardinals (Carnap *1935a*, 76–77), he treated little mathematics. As he stated as his close, the aim was to show that 'The method of logical syntax, that is, the analysis of the formal structure of language as a system of rules, is the only method of philosophy' (p. 99).

Stebbing *1935a* reviewed this and three other books by Carnap in 13 enthusiastic pages in *Mind*. Giving almost all space to *Syntax*, she provided an excellent short survey of all its main features, including Gödel numbering and modality; but she queried the utility for science of the reduction of ordinary sentences to formal versions, and wondered if 'perhaps too great tolerance was permitted by the principle' (p. 501).

9.5.2 *Commentary from Black and Ayer.* Another book by Carnap included in Stebbing's review was a short survey *1934d* of *The unity of science*, handled VC style. It was a translation of a recent *Erkenntnis* article, prepared by the Russian-born philosopher Max Black (1909–1988). A graduate in mathematics from Cambridge, in the early 1930s he taught at a school in Newcastle-upon-Tyne. We saw him review Quine in §9.4.4.

In 1933 Black published with Ogden his own treatment of *The nature of mathematics*. He restricted himself to the three main schools; like Stebbing (who read his proofs), he took 'logistic' as the main concern, giving in 130 pages quite an extensive account of the logic. The treatment of incomplete symbols was rather long-winded, with no proper explanation of contextual definitions (Black *1933a*, 76–84). He also noted the complications of type theory, but 'no attempt will be made to distinguish between orders and types' (p. 104) is not the best remedy. For him a main difficulty with the axiom of reducibility was 'understand[ing] what is meant by asserting the existence of propositional functions', while the axiom of infinity asserted 'the existence of infinitely many' of them (p. 112). He also quoted Russell stating logicism as an identity thesis, but offered an unclear alternative based upon 'the fact that mathematics must be used in the systematic development of logic' and claiming that logic was 'the syntax of all possible state of affairs' (p. 144). The latter remark suggests influence from Carnap, though Black listed only the *Aufbau* in his bibliography. Outside *PM*, he treated Ramsey, Weyl (on vicious circles), Chwistek (said on p. 118 to be

---

[17] On Carnap's lectures and the publication of Carnap *1935a* see Carnap Papers, 28-28 and 81-03; and Ogden Papers, Box 7, Folder 10.

deceased) and Wittgenstein, for the last giving an '(unauthorized) report' of his 'thorough repudiation of the logistic thesis' (p. 129). Like Stebbing, he ignored Wiener, Sheffer and Nicod (though Nicod *1917a* was in his bibliography). 'We conclude that the logistic thesis is not proven', and that the required 'elaborate reconstruction' would destroy 'that method's ambitions' (p. 144); but, as usual, he did not wonder at the limited amount of mathematics covered by *PM*.

Black granted formalism only 20 pages, although he listed the main axioms. This section ended with a 'note' on Gödel's theorem and corollary, the latter unfortunately phrased as that 'a contradiction could be deduced from any proof that the *entire* calculus of propositional functions could be formalized' like the first-order predicate calculus (pp. 167–168)! Intuitionism received nearly 40 pages, though they included a 'digression' on set theory to note the controversy over the axioms of choice (pp. 178–185); given Russell's role there, the passage would have been better placed within logicism.

The book ended with no final comparisons or conclusions. Although more advanced in level than Stebbing's volume, it did not have the same calibre: reviews were cool, John Wisdom *1934a* in *Mind* being especially frosty. Nevertheless, in 1939 Black successfully submitted it to the University of London as part of his doctorate.

The philosophy of the VC gained much publicity in mid decade when A. J. Ayer (1910–1989) published an introduction called 'Language, truth and logic'. He stressed epistemology, starting out from 'the elimination of metaphysics' (*1935a*, ch. 1) by the 'verification criterion of significance' of a proposition. Logic indeed came last of the three categories in the title, mainly in ch. 4 on 'The *A priori*'. There he characterised both logic and mathematics (how much of it?) in terms of tautology, which he took as synonymous with 'analytic proposition', itself defined 'when its validity [truth?] depends solely on the definitions of the symbols it contains' (p. 78). While noting *PM* as a 'deductive system', he cited Lewis's and Langford's recent book (§9.4.2) for the possibility that it 'is probably only one among many possible logics, each of which is composed of tautologies as interesting to the logician as the arbitrarily selected Aristotelian "laws of thought"' of identity, excluded middle and non-contradiction (p. 81). Like Quine (§9.4.4), he was in Vienna in 1932–1933, and later he recalled the Circle's silence over Gödel's theorem (*1977a*, 113: he also described this to me in 1983). Perhaps for this reason, as in Quine's *Logistic*, he did not mention the theorem.

9.5.3 *Mathematicians—and biologists.* *PM* still did not attract or retain British mathematicians. In particular, we saw in §8.5.2 Max Newman writing on formal systems and helping Russell late in the 1920s; but then he followed practise in changing foundational interests towards formalism and set theory, which he taught at Cambridge. A student on his 1935 course was Alan Turing (1912–1954), who addressed the Cambridge Moral Science Club on the theme that 'a purely logicistic view of mathematics was inadequate', according to Braithwaite's minutes (Hodges *1983a*, 85–86). Turing came to specialise in recursion theory and computability, with momentous consequences during the War when he led the design at Bletchley Park of the decryption machine of the German 'Enigma' encoder (with Oliver Strachey as a colleague)—and Newman directed the preparation of the 'Colossus' computers.

More interest was excited among life scientists. In 1931 John Butler Burke (b. 1873) published a long book on 'The emergence of life, being a treatise on mathematical philosophy and symbolic logic by which a new theory of space and time is evolved'. This unusual combination of disciplines was whisked together with Hegelian philosophy in order to state philosophical theories about The Real, Time, and Kant's conception of space. The main ingredients came from Boolean algebra, especially propositional calculus, the expansion theorems (255.5) and even Taylor's series; however, *PM* contributed especially the basic predicate calculus (J. Burke *1931a*, chs. 2–3). On paradoxes, he described Russell's and Grelling's (pp. 73–75), exactly Carnap's choice in *Syntax* three years later (§9.3.2)! The manipulations were competently done, but the underlying grasp was shaky; for example, he seemed not to notice the difference between part-whole and set theories. Not surprisingly, the book did not make much impact; but, like W. W. Greg on textual criticism (§8.5.2), it is an interesting example of the spread of mathematical logic into other areas.

A much more significant contribution came from J. H. Woodger (1894–1981), Reader in Biology at the University of London. Deeply concerned with the philosophy of his discipline, he reviewed options in a long survey *1929a* of *Biological principles* in Ogden's series. A central issue was 'Judgement in biological explanation' (the title of ch. 6). The two most cited philosophers were Russell on logical empiricism, and Whitehead on organisation and constructions in space-time.

In the 1930s Woodger decided in favour of adapting and applying *PM* for his purposes, and fulfilled his aim in a volume *1937a* on *The axiomatic method in biology*, published by Cambridge University Press. He began with accounts of the calculi of propositions, classes and relations, but with a nominal and classial definition of the existence of a definite description (p. 31). A novelty was his attractive layout (Figure 953.1) of the 16 basic connectives (p. 24). The proposition at any node was the conjunction of those propositions *to* which arrows ran, and the disjunction of those propositions *from* which arrows issued; they ran from contradiction Z to tautology U. The negation of any proposition was located at the opposite corner of the other cuboid. The Sheffer stroke '$p|q$' was 'not defined in the text because it is not used in subsequent pages'.

Then Woodger introduced his own modifications, such as identity 'with respect to a certain class of [classes α determined by a class of chosen]

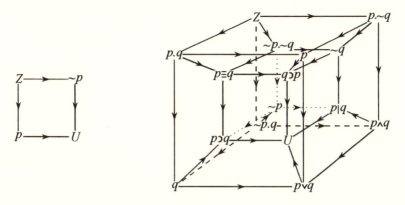

FIGURE 953.1. Woodger's presentation of logical connectives.

properties λ' (p. 99):

$$\text{'7.1.1} \quad I_\lambda =_{\text{Df}} \hat{x}\hat{y}\{(\alpha) : \alpha\varepsilon\lambda . \supset . x\varepsilon\alpha \equiv y\varepsilon\alpha\}'. \tag{953.1}$$

The most significant innovation was a relation R called 'hierarchy', defined as asymmetrical, one-many, containing only one initial member a in its field, and with a converse domain identical to the class of terms in relation R^n to a for some integer n (p. 42): it helped him to characterise generations and other properties of biological organisation.

With this and other devices Woodger handled theories such as gender and embryology; his system deployed notions in type from Cls'Indiv to Cls^5'Indiv, and various kinds of relation (pp. 147–148). Much of his exegesis was as wallpaper-like as *PM* itself, with some proofs postponed to appendices. Although he wrote for biologists, using their terms without explanation, his presentation must have deterred them; conversely, he was also too biological for the logicians. Hence his enterprise 'may fail to develop because of a paucity of investigators', as Kurt Rosinger *1938a* put it in an admiring review for the *Journal of philosophy*. Later Woodger *1958a* saw his book as '*constructing the metatheory* of a scientific theory'; and he maintained his philosophical interests, even writing on *The technique of theory construction* in 1939 for Neurath's encyclopaedia.

Woodger had read well in the literature; he began *The axiomatic method* with a long (unreferenced) passage from Whitehead's recent *Adventure of ideas* (*1933a*, 176–177) on the treatment in *PM*, *93 of the field of the ancestral relation (796.1) as an example of the qualitative aspects of logicism (Woodger *1937a*, iv). Other authors included Kempe (§4.2.8) on form and multisets (pp. 1, 9). He had also secured the help of Carnap and Tarski, and also Dorothy Wrinch (p. x); in particular, Tarski contributed a 12-page appendix *1937a* to the book, in which he used Leśniewski's mereology (§8.8.3) to present properties of part-whole theory suitable for

Woodger's purposes. Curiously, Woodger seems not to have contacted Russell, although his concern with temporal succession (which featured also in Tarski's appendix) overlapped with Russell's current interests, as we now see.

9.5.4 *Retiring into philosophy: Russell's return, 1936–1937.* During the 1930s Russell continued his mid life career of popular writing and lecturing, regular carousing,[18] and also transferring the property of wifehood from Dora Russell to Peter Russell *née* Spence. This action occurred in 1936, when he also marked a return to mathematical logic with a paper placed with the Cambridge Philosophical Society 'On order in time' (Russell *1936a*). Picking up a thread from his *Our knowledge* on defining instants in time Whitehead-style as certain kind of events (*1914c*, ch. 4), he noted the search in Wiener *1914b* for conditions for instants to be well-orderable (§8.2.7) to wonder how instants could occur at all. He deployed his logic of relations in a sophisticated way, showing that the techniques had not left him (or else had returned). Sadly, the paper has been largely ignored.

In the following year Russell reprinted with Allen and Unwin two books which Cambridge University Press had published around the start of the century: the study of Leibniz, and *The principles*. To each reprint he added a new introduction, and the one for the latter volume revealed that his grasp on the philosophy of logicism was weaker than that on its technicalities. He began by stating that the book showed that 'mathematics and logic were identical' (*1937a*, v), whereas the inclusion thesis had clearly been argued (§6.5.1). Discussing in detail his implicational definition of logicism, he recorded its origins in geometry (p. vii). Among changes made since, he noted the theory of definite descriptions and the consequent 'abolition of classes', queried the existence of logical constants, and noted the need for the axioms of infinity and of choice (p. vii–xi). On paradoxes, he adopted Ramsey's distinction, but overlooked the propositional variant of his own paradox in the second appendix (§6.7.9). For solutions he recommended on p. xiii Carnap's *Syntax*, just out in English (§9.3.2); but he ignored Gödel's theorem and its corollary. On the other schools, he considered formalism only in the marks-on-paper version; intuitionism was taken more seriously, including its destructive consequences for mathematics (pp. vi–vii). Scholz *1938a* welcomed the reprint in the *Deutsche Literaturzeitung* (where he was a regular contributor), but as much for its initial publicity for Frege as anything else (compare §4.5.1).

Thus revived into mathematical logic, Russell hoped to renew his philosophical career in his 66th year, a common age for retirement. After

---

[18] Some of the most amusing files of letters in RA belong to this context. Try especially those with Miss Joan Follwell around 1927 (Russell the pursuer) and Mrs. Amber Blanco White in 1931 (Russell the pursued).

checking with Moore that chances at Cambridge were hopeless (Russell *1968a*, 214–215), he asked Veblen about the new Institute of Advanced Study at Princeton in the U.S.A.; but he learned that the Director, Abraham Flexner, did not wish to encourage philosophy (at least involving Russell[19]). Succeeding Whitehead at Harvard University was not possible either. But he secured the academic year 1937–1938 at Oxford University, and then took his family to the U.S.A. until 1944 (ch. 6). His next main philosophical work was *An enquiry into meaning and truth* (*1940a*), where various logical techniques were used.

Russell also reacted negatively to John Dewey's new account *1938a* of logic as 'the theory of enquiry'. When invited by the philosopher P. A. Schilpp (1897–1993) to contribute to a volume on Dewey which launched a book series entitled 'Library of living philosophers', Russell wrote a critical piece *1939a*, to which Dewey replied in the volume. As in the 1910s (§8.5.5), incompatibility reigned: Russell did not grasp the range of Dewey's notion of enquiry, or his view that one judged the choice of action for an enquiry to be pursued, not truth-values of propositions; conversely, Dewey never appreciated symbolic logics at all (T. Burke *1994a*). So we leave Russell in controversy again, and conclude with a review of European work lying outside that of the VC.

9.6 EUROPEAN, MOSTLY NORTHERN

9.6.1 *Dingler and Burkamp again.* German interest in logicism was evident in a translation of the introductory material of both editions of *PM* (Whitehead and Russell *1932a*). But in mainstream German philosophy (neo-)Kantian or phenomenological positions continued to be prominent.

The 50-year-old Hugo Dingler came up with another book *1931a*, 200 pages of *Philosophie der Logik und Arithmetik*. Naturally he took Hilbert's approach as the most significant, and claimed priority for himself over various details. In the opening chapter, on the 'philosophy of logic', he adopted the term 'logistic' but sadly took it as synonymous with both mathematical and algebraic logics (p. 17); in several later places he linked it with mere calculation. As before (§8.7.1), he avoided the Kantian *a priori* and saw the choice of axioms as a (somewhat) conventional matter of decision rather than based upon self-evidence. Among recent literature, he also noted and contrasted both Wittgenstein and Carnap, relating them to

---

[19] See the exchange between Russell and Veblen of February and March 1937 in RA and Veblen Papers, Box 11. Flexner's reluctance may have been motivated by his friendship 20 years earlier with a gynaecologist in Chicago, whose daughter had then been seduced by Russell but had not gained the wifehood which she expected. Russell described some of this affair in his autobiography (*1967a*, 213–214) without releasing her name, 'Helen Dudley'; for more details see Monk *1996a*, chs. 12–13 *passim*. He dragged his estranged wife Alys into it, as she recalled in her old age (my *1996a*).

'arithmeticisation' (pp. 65–66). The presentation of arithmetic was based upon axioms and mathematical induction, where he allowed relations a major place with acknowledgement to Russell (p. 92). In a remark intriguing to appear just before Gödel's theorem, he claimed that on ordering series by relations 'complete axiomatisation is impossible', although the reasons appear to lie in practice rather than in principle (p. 101). Reviews of the book were reserved; in particular, Nagel *1932a*, 406 found it of 'much less merit' than Kaufmann's (§9.3.6).

A rather better grasp of the relationship between symbolic logic and Kant was shown by Burkamp, first met in §8.7.9. In his mid fifties, he published as *1932a* a general survey of logic(s). After a short history of logic from antiquity to recent times (chs. 1–3), he compared the Kantian theory of concepts and judgements with the more modern versions on topics such as relations, individuals and negation, giving due credit to symbolic logics (chs. 4–10); citations included not only Russell and Hilbert but also Peirce, Schröder, Bolzano, Cassirer and Becker. But his preference came out in the last chapters, such as the extensive treatment of syllogistic logic in the account of deduction (pp. 122–132). Only a few paradoxes were listed, with not much discussion of solutions (pp. 142–144); and the final ch. 13 was divided into sections on 'Logic and consciousness' and 'Logic and [the] person', topics abhorred in 'The formalist-logicist conception (Russell, Carnap)' (p. 153). While somewhat incoherent in overall impression, the book exemplified well the perception of mathematical logic in the broad scope of logic at the time; Ackermann *1938a* nicely balanced the strengths and weaknesses in the *Jahrbuch*.

9.6.2 German proof theory after Gödel. Hilbert did not attend the meeting in September 1930 in Königsberg (his home town) where Gödel announced his theorem, although he spoke at one there immediately following it. He would have been shocked by Gödel's news, as his own lecture *1930a* ended with the call that 'we must know, we shall know'; what did audience member Gödel think? However, soon afterwards in a paper on 'the founding of elementary number theory', after recalling the impact of Dedekind's booklet on integers in 1888 and sarcastically rejecting Kronecker's constructivism, Hilbert *1931a* introduced a new rule of inference in which if a formula $A(z)$ were correct for each numerical value of z, then $(x)A(x)$ was available as a premise. The connection with Gödel's procedure is clear, although Hilbert did not mention him. Similarly, Gödel *1931b* ignored the link in his review of the paper for the *Zentralblatt*.

Hilbert's health was declining, and it was decided to publish an edition of his papers with Springer. Those on foundations were rather unfortunately handled, for some were placed in the seventh edition of his book on geometry in 1930 and the others included in the third volume of the edition five years later. The latter was edited by Otto Blumenthal, who told

Bernays in 1938 that Hilbert's mathematical activity had ceased and he lay 'half dreaming on the sofa'.[20]

Hilbert's decline was sad, for metamathematics continued to lead foundational studies. Among new figures Gerhard Gentzen (1909–1945) was especially significant: he completed his *Dissertation* on 'natural deduction' at Göttingen under Weyl in 1933, publishing it two years later as *1935a*, when he became Hilbert's assistant. The influence on him of *PM* was negative, for he began by contrasting his new approach with the formalised proofs to be found there, and also in Frege and Hilbert (p. 68). He also introduced the symbol '∀' for universal quantification (p. 70). In *1936a* he studied a Russellian simple type theory without the axiom of infinity, and elsewhere found a proof of the consistency of number theory (using transfinitary methods), examined aspects of intuitionism, and also reviewed much for the *Zentralblatt*. The war interrupted his career, for he was called to military service as a radio operator. In an extraordinary irony, one of the very few photographs taken by the Germans of the Enigma encoder shows it being operated by him in uniform ... ;[21] luckily nobody seems to have asked him to think about its underlying logic, for he might have become the German Turing. Instead, he was invalided out in June 1942, and taught at the German University in Prague (where Carnap had held his chair) until the end of the War, when he was interned and died of malnutrition.

Gentzen joined the Göttingen group soon after Bernays had been dismissed by the Nazis and had moved to Zürich. Then in his mid forties, Bernays completed and published the first volume of the canonical statement of proof theory. Although 'Hilbert and Bernays' *1934a* was entitled 'foundations of mathematics', it covered only the logical (meta-)calculi and first-order arithmetic with recursion, though treated in great detail. Gödel featured more for his completeness than the incompletability theorem, which featured was treated in the sequel volume *1939a*, along with topics such as the ε-axiom (874.1), transfinite recursion, and decision procedures. Naturally Whitehead and Russell were little cited (nor Frege or Peano), but *PM* was present implicitly in the axiom systems deployed.

Also in 1934, Heyting contributed a comparison between intuitionism and 'proof theory' to a series of short books associated with the *Zentralblatt* and published by Springer. As in his 1930 lecture (§9.2.2), he concentrated upon Brouwer's mathematics and largely ignored the philosophy; he also stepped delicately around the relationship between Brouwer and Hilbert (Heyting *1934a*, 52–57). In the introduction he promised a sequel

[20] Blumenthal to Bernays (then in Zurich), 17 February 1938: Hilbert's 'geistigen Interessen haben ganz nachgelassen' and he largely spent the intervals between meals and walks 'halb träumend auf dem Sofa' (Bernays Papers, 975:414). Hilbert was then in his mid seventies; maybe a degenerative disease had also begun to develop.

[21] I owe this information to Eckart Menzler-Trott, who found the photograph in the course of writing a biography of Gentzen, as yet unpublished.

volume on other topics, including logicism; but it never appeared, again a sign of priorities of the time. His text exhibited a further change, somewhat evident already in his and the other Königsberg lectures (§9.2.2): perhaps in reaction to Gödel's theorem, he was ready to consider *both* of the other schools, and thereby entertain more modest philosophical tasks for mathematics than Handling All Mathematics Somehow as in the previous generation (Franchella *1994a*).

9.6.3 *Scholz's little circle at Münster.* Heyting's book included a fine bibliography, for which he thanked access to Scholz's library. After producing various lecture courses and related material at Münster University, the 50-year-old Scholz launched in 1934 a series of short typescript books undertaking 'Investigations on logistic', published at his (or the university's) expense by Meiner at Leipzig. They seemed to gain little circulation; I confine myself to the two least unknown ones. The first, written by Friedrich Bachmann (1909–1982), won for the academic year 1932–1933 a prize proposed by the university through Scholz on the relationship between logic and mathematics; it served also as his *Dissertation*. Bachmann *1934a* worked meticulously through the foundations of arithmetic, principally as handled by Dedekind, Frege and Russell, though Peano was also noticed. While he picked up some nice details, such as the different ways in which Frege and Russell defined successorship of cardinals, little original was said—and nothing about Gödel's theorem. Overall the analysis is somewhat underwhelming.

The other book was a joint survey Scholz and Schweitzer *1935a* of the related topic of definition by abstraction, incorporating another Münster *Dissertation*, by Hermann Schweitzer. Once again the treatment was careful but unexciting; for example, Frege's notion of value-range, which had allowed Russell's paradox to slip into his system, did not easily fit there (pp. 94–95). The texts examined included Carnap's *Abriss* and Huntington's writings for axiom systems, and Maccaferri *1913a* (§8.6.1) on defining cardinals. In a review for *Mind* Ayer *1937a* praised the treatment in general; however, he thought that, as in Carnap's *Aufbau*, Schweitzer's extensional interpretation of classes blocked the handling of universals, thus reducing the generality of this form of definition.

A fervent Christian, Scholz also wrote on religious matters: in a paper *1935a* he linked the two together, on the Leibnizian position that the axioms and propositions of logic were valid in all possible worlds; relative to them logic served as an invariant. In his historical work he acquired the *Nachlässe* of Frege and of Schröder (§4.4.9–5.1) and began to transcribe the former collection, briefly reporting on its contents in 1935 at the VC Paris Congress (Scholz and Bachmann *1936a*). On 2 December 1937 he told 'Prof. Dr.' Russell of his progress over Frege (copy in RA); a book series of 'Frege-Studien' began in 1940, but soon had to stop. He also told Russell of his intention to write an article on mathematical logic for the new edition of the *Encyklopädie der mathematischen Wissenschaften* (RA), a topic not covered in the first edition; but it appeared only after the War, as Scholz and Hermes *1952a*, written with his former student Hans Hermes. It was one of the very few articles published before the new edition was abandoned.

9.6.4 *Historical studies, especially by Jørgensen.* Scholz's historical writings included a general book *1931a*; while too brief, he ended the main text by noting Gödel's new theorem and relating it (only) to formalism. Much longer that year was an historical 'treatise of formal logic', written by the Danish philosopher Jørgen Jørgensen (1894–1969). In 1925 the manuscript won him a prize on the history and philosophy of logic offered by the Royal Danish Academy of Sciences and Letters. The published version, Jørgensen *1931a*, contained regrettably few additions in its three volumes and 830 pages: in particular, virtually nothing was said about the new edition of *PM*. But the coverage up to the time of completion was generally good, and well indexed at the end; the most notable absentees were Poles (including Post), where only Chwistek gained attention. A bibliography would have emphasised his wide reading of both primary and historical literature evidenced in the many footnotes; not only principal logicians but also, for example among many, Bôcher and Keyser, Hölder and Pasch.

Jørgensen's first volume comprised a general history in four chapters. In the opening one he contrasted 'classical', algebraic and mathematical logics, with the latter two involving 'the difference between *algebra of logic, which is a special deductive theory, and logistics, which is the theory of deduction itself*' (p. 25)—another good example of the time when metalogic was still fugitive. After a chapter on antiquity and the Middle Ages, the third one treated 'the algebra of logic' in detail, including quantification of the predicate and all the main figures. The last chapter featured mathematical logicians, including Frege and Lewis, although too little was said about Cantor. Jørgensen noted Russell's tendency to conflate a proposition as object and as word-string, and noted the propositional hierarchy in the theory of types (pp. 196, 222).

The second volume comprised a trio of 'outlines'. 'Classical logic' was largely devoted to syllogistic modes, described in much detail. Then Jørgensen introduced the algebraic tradition, following Schröder in scope but with the Sheffer stroke and truth-tables presented early, duality almost absent, and part-whole theory mostly taken for granted. The rest of the volume, nearly half of it, was devoted to 'logistics', following Russell's implicational logicistic thesis; the main axioms and definitions, and many theorems from Part I of *PM* were transcribed, but the subsequent theory of (Cantorian) classes and definitions of numbers of various kinds was treated more summarily.

The 'discussion and criticism' came in the last and longest volume. Clearly mathematical logic won the contest for foundations, especially

Russell's version, although Hilbert's 'axiomatic method' was quite nicely reviewed (pp. 141–160); intuitionism was very coolly received, largely because of Brouwer's unintelligibility (pp. 50–55). Much of Jørgensen's appraisal rested on the competing needs of the creation and the justification of mathematics, where he did not seem fully to grasp that logicism aimed at the latter need while of course requiring originality to achieve fulfilment. Most aspects of *PM* were considered, including some changes in the new edition (pp. 288–289), although he overlooked the problem of the status of individuals (§7.9.3). While noting the three doubtful axioms, he optimistically thought that 'logic has become more and more mathematical' and 'mathematics more and more logical' (p. 296).

On the whole Jørgensen's book gave a fair impression of the development of mathematical logic and logicism up to the mid 1920s, and a better one of the range of pertinent literature. His own philosophy of 'formal logic', outlined in the last two chapters, was traditionally grounded in a link with thoughts, and the various 'forms' in which it might manifest. At the end he too arrived at the edge of metalogic: 'Logistics [...] differs from all other deductive theories in utilising the principles of deduction themselves as its primitive propositions' (p. 293), followed by a list of issues concerning forms, connectives and principles which needed to be addressed (pp. 297–298).

Stebbing was pretty cool in her review for *Mind*, especially on the 'serious demerit' of associating logic with thinking, and moreover unclearly (*1932a*, 241). Interestingly, she also wished for more on Kempe and Royce, whose work was 'far too little known to English students' (p. 238); but she welcomed the account of Frege. By contrast, for the *DMV* Scholz *1933a* was disappointed about the treatment of Frege but liked those of the Peanists, Russell and Lewis; and he called for deeper studies of Russell on existence and on forms of definition, and on the logicistic construction of real numbers. Different countries, different perceptions again!

Jørgensen became closely associated with the VC and the Encyclopedia. In a superficial survey in *Erkenntnis* 'on the aims and problems of logistic' he followed Carnap in connecting logic with the syntax of language, and split it from mathematics on the grounds that only logic consisted of tautologies (*1932a*, 75, 99); he mentioned Gödel's theorem only when noting that not all formulae could be converted into normal forms (p. 98).

9.6.5 *History-philosophy, especially Cavaillès.* In 1932 two important editions of works appeared. One was of Cantor's writings, rather sloppily prepared by Zermelo (Cantor *Papers*). It included a shortened version Fraenkel *1932a* of his biographical article *1930a* cited in §3.1.2. Fraenkel also pointed out that Cantor's correspondence with Dedekind was conserved in the latter's *Nachlass*; the letters of 1899 were transcribed by Jean Cavaillès (1903–1944), but messed up by Zermelo when publishing them in his edition (§3.1.2).

The other edition was the last volume of Dedekind's, including his two booklets and some other writings on foundational questions (Dedekind *Works 3*). One of the editors was Emmy Noether, who joined Cavaillès in editing all the earlier letters, written between 1872 and 1882; their edition was ready in 1933, but her departure to the U.S.A. that year and death two years later delayed publication. It appeared in 1937 in Paris with Hermann (Cantor-Dedekind *Letters*).

Cavaillès also completed his doctoral thesis that year, which Hermann reprinted as a sequence of three short books, nearly 200 pages in all (Cavaillès *1938b*). This history of the growth of axiomatisation in mathematics in the 19th century and its crystallisation in metamathematics is still the best study of its topic: *PM* played a minor role, arising principally in connection with the logical calculi (pp. 105–113). He also produced a fine historical commentary on set theory from its background in mathematical analysis to the axiomatic versions to the 1920s, although too little was included on Whitehead and Russell (*1938a*, two more short books). One philosophical theme in his writings was the place of structure in mathematics (Benis-Sinaceur *1987a*); quite keen on the VC, in Cavaillès *1935a* he had reported in the *Revue de métaphysique et de morale* on the Congress held the previous year in Prague.

When the War started Cavaillès largely ceased his research and became a leader of the French resistance (Ferrières *1950a*). He sacrificed himself to the cause in 1944, when he falsely confessed to be a leader sought by the Gestapo and was shot. After the War his body was re-interred in the chapel of the Sorbonne in Paris, near to that of another philosopher—René Descartes.

9.6.6 *Other Francophone figures, especially Herbrand.*   Cavaillès was exceptional among French historian-philosophers in the intensity of his interest in logic and related topics; their concern continued at its modest level (§8.6.2–3), with nobody embracing logicism. The veteran Emile Meyerson (1859–1933) published in 1931 over 1,000 consecutively numbered pages *Du cheminement de la pensée* in three volumes, the last one comprising extended notes to the text of the other two. Much of his philosophy focused upon the tension between identity as exhibited by rationality and diversity as presented by experience; he was concerned with both creation as well as justification, though not sufficiently distinguishing the two for logic. In the opening chapter he considered symbolic logic rather superficially (*1931a*, 18–30); the syllogistic tradition was clearly preferred. The most relevant part was the third Book on 'Mathematical reasoning', 160 pages of text and clearly the result of wide reading even though confined to arithmetic, algebra and geometry; he included, for example, Hermann Grassmann (pp. 227–228, 893–896). However, no clear conclusion emerged, not even in the 50-page chapter on 'Mathematics and logic', where on pp. 442–443 he also showed his ignorance of (458.1) in stating that Frege had not considered negative or real numbers. Russell and Whitehead were cited several times, but always on particular points rather than in a general appraisal of logicism; and algebraic logic was very lightly treated.

The following year the Swiss philosopher Arnold Reymond, then in his late fifties, published a book on 'The principles of logic and current criticism', based upon guest lecture courses at the Sorbonne. We came across him in §7.5.1 and §8.6.2, not impressively; here he was somewhat more substantial. Influenced by the recent surveys of mathematical logic by Robert Feys (§8.6.3), he concentrated upon *PM*, describing with approval many of the techniques. He doubted the logicistic thesis, but for various corrigible reasons; for example, the definitions of ordinal numbers depended upon order, which 'envelopes' the notion of collection (Reymond *1932a*, 178). Perhaps for similar reasons, he listed logicistic writings in his bibliography in the section on 'Algorithmic logic' rather than that on 'Mathematics and logic'. Ignoring phenomenology entirely, he noted other schools in less detail, and cited only the Hilbert-Ackermann textbook (§8.7.4) for formalism (pp. 235–245).

Reymond also described (pp. 245–254) some recent work on logic produced by a Frenchman: Jacques Herbrand (1908–1931). Like Gentzen a few years later, Herbrand developed proof theory and sought to show the consistency of arithmetic; but he took *PM* as his principal source, with its axioms, notions and notations, all seasoned with a dressing of use-mention muddles. He explained his approach in a general paper *1930b* in the *Revue*.

The main product of Herbrand's short life (he died in a skiing accident) was his thesis, in which he found two ways of proving that tautologies are provable. One was based upon a means of matching any quantified formula with a quantifier-free mate and proving that each was derivable (*1930a*, ch. 2); it reversed the handling of quantification in *PM*, ∗9, and also its systematic application in the second edition. The other method drew on model theory and normal forms, as developed by Löwenheim and Skolem (§8.7.5). A highlight was a result which became known as 'the deduction theorem'; it took the form that if the premises of a theory were stated as a single conjunction H, then a proposition P was true within it if and only if '$H \supset P$ *be a propositional identity*' (ch. 3, sect. 2.4). In effect though not in intention, he clarified some of Russell's conflations and implication and inference, and also removed a standard sloppiness among mathematicians when (not) relating a proof to its theorem. While several proofs were unclear and even defective, the thesis inspired important new lines of research.

The indifference of French mathematicians to mathematical logic is particularly marked in Herbrand's case: with some difficulty a jury was found to read and approve the thesis, which he then published in Poland. But this completed a circle of contact; for he had partly been motivated to study logic by the 1926 survey of 'logic in mathematics' published in Paris by the Pole Stanisłav Zaremba (§8.6.3). Let us now turn to that important country.

9.6.7 *Polish logicians, especially Tarski.* The work of Polish logicians (§8.8) continued to flourish, with some new figures emerging. Several of their papers are translated into English in McCall *1967a*, which also contains part of the historical survey Z. Jordan *1945a*.

Of the senior figures, Łukasiewicz had sugurested in 1926 that deductions be effected "locally" from given premises, and his student Stanisłav Jaśkowski gave in *1934a* a version which resembles Gentzen's (independent) theory of natural deduction (§9.6.2). Łukasiewicz also studied the earlier history of logic, and defended the mathematical tradition: in particular, in *1936a*, art. 2 he rejected Carnap's association of it with the overthrow of metaphysics (§9.3.1). Stanisłav Leśniewski continued to develop his systems; in addition and perhaps motivated by his nominalism, he is said to have doubted Gödel's incompletability theorem and sought a mistake in its proof.

Their younger colleague Tarski, in his early thirties, made the greatest mark, not only by his work but also through external contacts, especially with the VC (Wolenski *1989a*, ch. 8; Wolenski and Köhler *1999a*). Like Herbrand, he concentrated on proof and model theory, but he used *PM* as an important source or test case. For example, in *1933a*, published in the *Monatshefte*, he formed a system like Gödel's and thus like that of *PM* for first-order arithmetic, and studied its ω-consistency and -completeness (§9.2.3); thm. 2 is a version of the deduction theorem, formulated in terms of classes of propositions. Again, in a lecture given to the VC Prague Congress in 1934 and published in *Erkenntnis* as *1935a*, he noted the analogy between two groups of concepts pertaining to deductive theories, such as 'axiom' or 'theorem', as against 'definable concept' or 'definition'; he studied them in detail for a logical system resembling *PM* without ramified types. He noted that his method underpinned the ideas launched in Padoa *1901a* at Paris in 1900 (§5.5.4).

Tarski's main contribution of the decade was his definition of truth. He claimed to have found the essential components by 1929, and they were stated without proof in the short paper *1932a* communicated to the Vienna Academy in January 1932 which Carnap had seen (§9.3.3). The first long version appeared in Polish as a book in 1933: I take the extended German account *1936a* published in the new Polish logic journal *Studia philosophica*, for it made most impact; but I quote from the English translation in Tarski *Semantics* (1956) made under the inspiration of Woodger. Later Tarski *1944a* nicely rehearsed his theory; and a useful guide is provided in Sofroniou *1979a*.

Acknowledging the work of Leśniewski on semantic categories, Tarski began by pondering the definability of truth for natural languages, and decided against it, especially because of unavoidable paradoxes; he stated a version of the liar paradox due to Łukasiewicz based upon giving the sentence '*c* is not true' the name '*c*' (*1936a*, 158). But he saw a chance for a definition in a formal language by distinguishing it from a 'second language, called the *metalanguage* (which may contain the first as a part)' and belonging to a 'second theory which we shall call the *metatheory*' (p. 167). This is seemingly the origin of those names: Carnap, to whom 'object language' is due (§9.3.2), mistakenly credited himself with 'metalanguage' much later (*1963a*, 54). The distinction was essential to Tarski's theory, since truth was a property in the metalanguage of a sentence correctly expressing some state of affairs in the object language: ' *"it is snowing" is a true sentence if and only if it is snowing'* (Tarski *1936a*, 156).

Making use of recursive definitions, Tarski constructed a predicate calculus for the metalanguage, imitating the structure of the one in the object language (pp. 175–187). In order to ease the use of recursion, he worked with sentential functions rather than sentences: '*for all* [objects] *a, we have a satisfies the sentential function "x is white" if and only if a is white'* (p. 190). The crucial property was '*satisfaction of a sentential function by a sequence of objects'* in some domain (p. 214), for from it he defined truth for any formal language with a finite number of orders of semantic category in terms of satisfaction by any sub-sequence in that domain (pp. 200–202). The background influence of *PM* was explicit in his analogy between categories and simple types (p. 215), and maybe in his decision to work with sentential functions.

For languages of infinite order, Tarski went back to impossibility, on the grounds that the formal structure admitted liar-type paradoxes via diagonalisation (pp. 247–251), and followed *PM* in confining himself to a finite number of orders (p. 244). However, in a postscript he found a way of handing transfinite orders in both language and metalanguage, extending his definition of satisfaction and truth under the condition that the metalanguage contained still more orders than the language itself (pp. 268–274).

Tarski cited Carnap's *Syntax* for details, many of which in fact go back to *PM* (§7.9.8); and a comparison with Carnap is worth making. Both men highlighted the central importance of metalanguage and -theory; but Tarski's emphasis on semantics (p. 252) contrasted with Carnap's adherence to syntax. As Scholz *1937a* put it in a review of Tarski's paper in the *Deutsche Literaturzeitung*, Carnap had thought in *Syntax* that the problem of defining truth was unsolvable—'and now it is yet solved!', so that 'the classical criterion of truth is saved'. Moreover, Tarski's use of correspondence via satisfaction did *not* appeal to belief, judgement or any of the paraphernalia that had perplexed Russell, Carnap and so many others; thus his theory was *neutral* with respect to those epistemologies which drew upon a realist ontology, and hence was significant to their various adherents. Finally, he was more concerned than Carnap with model theory (for example, categoricity on p. 174); indeed, later it became one of his major concerns.

Comparing Tarski with Gödel, some of his techniques, and the impossibility result, correlate with incompletability and numbering; hence he was anxious to emphasise the independence of his own work (Murawski *1998a*), pointedly so in his Vienna note *1932a*. However, his proof allowed for denumerably infinite sequences, while Gödel's was finitary. Another contrast lies in Russell's understanding: Gödel's theorem always escaped him (§10.2.3), but Tarski's definition was described in his *Inquiry* (*1940a*, 62–65), although it largely disappeared thereafter.

Tarski's Leśniewskian version of semantics supplanted that proposed by Chwistek in terms of formal analysis of symbols and their manipulation (§8.8.4), which Carnap had adopted for some time (§9.3.2). Chwistek had advocated his definition again in a presentation *1933a* of 'the nominalistic grounding of mathematics' in *Erkenntnis*, where he noted that uncontrolled use of impredicative definitions led to paradoxes. He also reworked some logicist principles in his own logico-mathematical system; the full details came in a book *1935a* on 'the limits of science', to quote the title of the heavily revised English version *1948a* published in Ogden's series. While symbolically quite different from *PM*, he constructed arithmetic and a type theory; but he also formally examined various meta-properties. One was a Gödelian incompletability theorem (given already in Chwistek *1939a* in the *Journal of symbolic logic*); philosophically he always advocated pluralities, and in a paper *1926a* he had already hinted at the possibility of various forms of set theory.

Overall the ambitions of Chwistek's enterprise surpassed the clarity of its expression; it has been little noted, and then guardedly.[22] So he has become a somewhat marginal figure among the constellation of Polish logicians. Tarski moved centre stage, especially with an introductory book in logic and applications to arithmetic, which appeared in Polish in 1936, in German the year later, and in English in 1941 after delivery of a course at Harvard University (Tarski *1941a*).

9.6.8 *Southern Europe and its former colonies.* In Italy, once so important, interest had fallen very considerably. Alessandro Padoa was most active: in particular, he opened an Italian multi-volume 'Encyclopaedia of elementary mathematics' with an article *1930a* of 79 pages. It may have been written much earlier, as the project had been seriously delayed. He covered the usual Peanist ground, with a few references to others, including Russell; the technical level was quite high for the volume, and the material may have been new for many readers.

Peano's reputation remained high. For example, his last student was

---

[22] On Chwistek's system, see Myhill *1949a*; and J. Russell *1984a*, 300–312. His *1948a* contains a helpful introduction by Helen Brodie, one of the translators.

Ludovico Geymonat (1908–1991), who travelled around Europe sampling logical positivism and mathematical logic in the early 1930s; his connection with Peano 'opened many doors', as he put it to me in 1982 (compare his *1986a*). Among his early publications was a 20-page review in a philosophical journal of 'the philosophical significance of some recent directions of logic-mathematics' especially the three schools; on logicism he noted the doubtful axioms in *PM* (*1932a*, 265–266), and he also cited later authors such as Behmann, Dubislav, Carnap and Hahn. As with Padoa, nothing was new apart from the news, but useful as such for his compatriots.

Peano died suddenly on 20 April 1932 in his 75th year, after teaching in the university in the morning. Russell's prompt tribute *1932a* was quoted at the head of §6.4; among other homages, Ugo Cassina (1897–1964), later to become his editor (§5.1) and main historian, recalled in *1933a* the importance of the *Formulaire*, and reviewed at length in *1933b* his contributions to mathematics and logic. At the invitation of the *Revue* he also recalled for the French 'the philosophical work of G. Peano', both in mathematics such as the space-filling curve (§5.2.4) and in logic (Cassina *1933c*); however, he mentioned neither Cantor nor Russell. At the VC Paris Congress in 1935 Padoa *1936a* paid tribute to Peano, stressing his contributions to the study of exact reasoning and to distinguishing particulars from universals. Then in his late sixties, Padoa himself died in 1937, six years after Cesare Burali-Forti, who seems not to have published on logic in his last decade.

By contrast, in Spain attention began slowly to grow from its previous slight level (§4.4.4, §5.3.2). In particular, the philosopher and scholar Juan García Bacca (1901–1992) was very productive in the 1930s. In a remarkably extensive encyclopaedia article *1933a* on 'Symbolics (logical)' his coverage of the propositional calculus included truth-tables and the reduction of connectives to the Sheffer stroke, while for the predicate calculus and 'set-theoretic logic' ('*Lógica conjuntal*') he ran through many basic features of *PM*, notations included, even up to Dedekind cuts and a touch of the alephs. In a concluding short historical survey he gave more space to the algebraic tradition than to the mathematical, and finished by noting Lewis on modal logics. Then, two volumes *1934a* of 'Introduction to logistic and applications to philosophy and to mathematics' in Catalan were followed by a shorter Spanish version *1936a*.[23] The coverage was wide, with *PM* guiding much of the treatment of the calculi, followed by short surveys of intuitionistic, modal and many-valued logics, and ending with some aspects of phenomenology.

As a socialist, García Bacca then had to exile himself to the former

---

[23] In astonishingly close timing, the Chinese philosopher Jin Yuenlin taught symbolic logic at Beijing University from 1932, and published a book similar in range to García Bacca's also in 1936; then a book 'On different logics' in 1941 (information from J. W. Dauben).

Iberian colonies in South America, where interest had begun in foundational studies. In Brazil the mathematician Manuel Amoroso Costa (1885–1928) became concerned, especially after studying in Paris in the early and mid 1920s (Silva da Silva *1997a*). A year after his premature death in an air crash, there appeared his book on 'the fundamental ideas of mathematics', which included logic and definition, set theory, and elements of the differential and integral calculus and of geometries. For him symbolic logic 'is not a reform of traditional logic [...] but an enlargement' (*1929a*, 204); nevertheless, he showed the difference in a survey of the second edition of *PM* which explained the calculi and stated the logicistic 'thesis of the English school' (p. 216). In Argentina Claro Dassen (1873–1941) wrote in French on paradoxes and especially on intuitionism in *1933a*, and then a biography *1939a* in Spanish of Louis Couturat.

Gradually the foundations of mathematics began to be taken up around the world, with *PM* acknowledged both as a source of logical techniques and as an exercise in axiomatisation; however, logicism was becoming part of history. This is the cue for the final chapter.

CHAPTER 10

The Fate of the Search

A STORY as rich and interconnected as this one could generate masses of meta-consideration; but I avoid the temptation, especially as various "local" conclusions have been drawn and summaries made *en route*. After a general comparison of algebraic and mathematical logics, the focus falls mainly upon Russell, and is mainly organised before, during and after *PM*; it ends with several appraisals of logic(ism) in the U.S.A. in the early 1940s. The chapters ends with a flow-chart for the whole story and some notes on formalism and intuitionism,[1] before locating symbolic logic in mathematics and philosophy in general, and emphasising the continuing lack of a definitive philosophy of mathematics.

10.1 INFLUENCES ON RUSSELL, NEGATIVE AND POSITIVE

10.1.1 *Symbolic logics: living together and living apart.* (My *1988b*)

Boole, Schröder, etc., have made the Algebra of Logic, while Mr. Peano and his school have made the Logic of Algebra.

Itelson (Couturat *1904c*, 1042)

While algebraic logic has not been a main theme of this book, its main thrusts have been conveyed: the separate initiatives from de Morgan and Boole (§2.4–5) and their fusion by Peirce and Schröder (§4.3–4). The many differences from mathematical logic also emerged: mathematics applied to logic rather than the other way round, as Gregorius Itelson nicely put it at the Congress of Philosophy at Geneva (though his remit to algebra was too narrow); laws rather than axioms, and algebraists' exploitation of duality; rapid derivation of consequences, not the detailed exposure of proofs; part-whole rather than set theory; the language of nouns and adjectives rather than of particles; and some algebraists' concern with model theory, especially with Boole on (un)interpretability. The roots of the differences go back to the 1820s, when Cauchy grounded mathematical analysis on the theory of limits and rejected Lagrange's foundation of the differential and

---

[1] It is strange that the names for the three main philosophical schools were already in use in ethics (Clauberg and Dubislav *1922a*, 161). Ethical grounds for exercising the will were 'logicistic' if their consistency was held to be morally sufficient; ethical norms were 'intuitionistic' if they were held to be inborn rather than acquired, and 'formal' if they came through general principles rather than individual objects.

integral calculus in power series (§2.7.2). The later mutual lack of sympathy is especially clear when Russell virtually ignored the algebraists when developing his mathematical logic of relations (§6.4.4), and Peirce dismissed *The principles* in a few lines (§6.8.1).[2] Reactions to Kempe are another good contrast: Peirce's enthusiasm (§4.2.9) came through to Royce's reworking *1905a*; (§7.5.4); Russell received an offprint of that paper, but his marginal notes consist of rewriting various results in Peanese (RA).

Despite the (sporadic) advocacy of Peirce's work by Ladd-Franklin (§8.3.4) and the approval of Schröder by Löwenheim and Skolem (§8.7.5), the algebraic tradition was heavily eclipsed by mathematical logic, especially after *PM* appeared (Anellis and Houser *1991a*). One reason was its seeming lack of purpose: we may admire the insights of Peirce and check the catalogues of Schröder, but where are they going? (Similarly, the initial influence of Grassmann upon Schröder, Peano and Whitehead diminished.) The aims of Peano and Russell, and of Frege when noted, were much more specific: directional rather than postulational logic, as Stebbing put it (§9.5.1). This difference seems to have underlain the change of allegiance for Whitehead: publishing *Universal algebra* in 1898 (§6.2.4) but then collaborating with Russell on logicism. When algebraic logic began to revive in the 1940s (due to Tarski among others), Boolean algebra remained prominent; but otherwise it was a much altered subject, concerned with metalogic and model theory, and deploying abstract algebras and set theory.

The incompatibility between the two traditions came back to tax Russell in the 1960s, when G. Spencer Brown sought his support for an unorthodox manuscript with the Boolean title *The laws of form*. It contained in effect the propositional calculus based upon the Sheffer stroke but without brackets, a great simplification of the procedures of *PM* (compare Hoernlé on Sheffer in §8.3.3). The theory of types was discarded, although the predicate calculus was not developed sufficiently to refound logicism. However, the anti-logicistic cast of the theory was clear, as Russell realised when praising 'a new calculus of great power and simplicity' for Allen and Unwin, with whom it first appeared as Spencer Brown *1969a*, in the first of many printings.[3]

10.1.2 *The timing and origins of Russell's logicism.* We have just noted negative influences; now for positive ones. Although Russell could be

---

[2] On 11 February 1899 Russell told Couturat of Peirce that 'je ne l'ai vu qu'une seul fois, de sorte que je ne le connais guère' (copy in RA). Presumably this meeting occurred during Russell's time in the U.S.A. in 1896 (*1967a*, 130–133).

A fine bibliography of and on algebraic logic is provided in *Modern logic 5* (1995), 1–120.

[3] Russell's praise appears on the dust-jacket or front matter of the editions; letters to him from Spencer Brown in 1961 are kept at RA. I am grateful to Brown for discussions in 1976 about the book; apparently the patent cited in app. 2 covered a circuit to operate lifts.

amazingly in error in his recollections (§6.1.2), there is no doubt over his repeated assertions of the three most important forces: G. E. Moore, for a turn from neo-Hegelianism to empiricism; Cantor, for envisioning a foundation for mathematics in *Mengenlehre*; and above all Peano, for providing logical and mathematical techniques for such a vision to be effected. Thus armed, Russell came to his conception of an empiricist philosophy of mathematics grounded in Peanist logic enhanced with relations and expressed in set theory (§6.5.1).

The timing of Russell's creation of logicism in 1901 is striking. Not merely did an ambitious and intelligent young man in his late twenties set out a programme, but also an ambitious and intelligent member of the British aristocracy, whose Inheritance Of The World Was A Large Responsibility, sought an empire of his own, colonising (pure) mathematics within mathematical logic. The consonance of situations is worth noting; for example, recall from §6.1.2 his concurrent work in the mid 1900s on logic and on the Free Trade question. Hilbert's proof theory of that time also has similar global aspirations, and there are other cases for the period (Stump *1997a*); for example, International Congresses in various disciplines, especially those in connection with the 'Universal Exhibition' held in Paris in 1900 (§6.4.1). The word 'universal' then had a *cachet* often granted today to 'international'; Peano is a nice example, with his 'universal' class V (§5.2.3) and his advocacy of 'universal' languages, especially his own Latin without inflexion. This interest was shown by some other logicians: Couturat for Esperanto (which Carnap was also to learn) and then Ido (which also attracted Ladd-Franklin).

On the paradoxes, when recognised they played a large role in all foundational theories, as "mistakes" to be avoided or even Solved; *flaws* of thought, as it were. However, the well-known story that they *promoted* the 'foundational crisis' (following Weyl's dubious phrase in §8.7.7) has been heavily undermined by historical research of recent decades, and has gained no support here. Indeed, it is doubtful that, with the possible exception of Zermelo (§4.7.6), set theory motivated anybody to take up foundational studies; in particular, Cantor created *Mengenlehre* to tackle technical problems in mathematical analysis (§3.2), the foundational side only coming later (§3.4). 'Away with the myth of the crisis!', to paraphrase the title of a recent historical survey of foundational studies (Borga and Palladino *1997a*).

10.1.3 *(Why) was Frege (so) little read in his lifetime?*   It is also clear, and correct, from Russell's recollections that Frege's influence on him began to assert itself (as it were) only when his main lines had been formed from Cantor and Peano. Sadly, it is "common knowledge" among many philosophers that Frege was Russell's main source; their writings point to the close similarities between the systems of the two men, and often marginalise or even ignore completely all other figures. However, historians who have mastered enough logic to distinguish conjunction from implication know that the similarity of some *B* to an earlier *A* does not necessarily entail influence from *A* to *B*, and Frege-to-Russell is a clear case in many respects. The first known influences of Frege were recorded in §6.7.7, and further examples were given later; but in scale they do not at all match those from Cantor and Peano. It is a great pity that Russell did not digest Frege earlier; at the very least, his logical calculus and logicism could have been more clearly stated. But even then, Frege's logic was notably Platonic while Russell's tried to be empirical; and Frege's logicism was explicitly confined to arithmetic and (some) mathematical analysis, whereas Russell's ambitions extended to much more (pure) mathematics.

How well was Frege's work known before Russell's publicity? His claim to be the 'first reader' of the *Begriffsschrift* has been shown to be absurd (§4.5.2). Further early readers of Frege included Kerry (an excellent start in §4.5.4), Cantor, Schröder, Peano, Husserl and Hilbert, and reviews in the *Jahrbuch* by Michaelis; among others not recorded here are Brentano and some of his followers (Linke *1946a*). Russell's appendix in *The principles* spread the news considerably; but the book also reported the paradox to which Frege's system was susceptible, and he reduced his publishing afterwards, an embittered victim (§8.7.3). Thus, unlike *PM*, the reception of his logic was *not* separated from that of his logicism. After applause in the 1910s from figures such as Jourdain (§8.2.3) and Ziehen (§8.7.2), the revival of interest in the 1920s led by Wittgenstein and then Carnap tended to focus on the linguistic and semantic aspects of Frege's system and lead to the invention of Frege′ (§4.5.1), a philosopher of language and of all mathematics, who has not been a concern here.

Thus, while the reception of Frege was far from the silence which Russell imagined, it never attained a deserved level. The usual reason given for repelling readers is the strange notations; but in §4.5.9 additional sources were sought in Frege's unfortunate use of standard terms such as 'function', and especially in the level of foundation proposed, extreme even in a country where in general such studies were favoured. Furthermore, his claim in the 1900s that the Euclidean was the only geometry was not good advertising (§4.7.4). This last factor will return soon, as Russell's own influence is considered.

10.2 THE CONTENT AND IMPACT OF LOGICISM

Logical and mathematical judgements are true only in the world of ideal entities. We shall probably never know whether these entities have any counterparts in any real objects.   Łukasiewicz *1912a*, 12

10.2.1 *Russell's obsession with reductionist logic and epistemology.* A major motivation for Russell was the improvement of rigour, especially in mathematical analysis already before Cantor. One of its main features was *reduction*: not only of real numbers to rationals and thence to integers, but also of other branches of mathematics to arithmetic. Whitehead and Russell effected these reductions in their logicistic construction of (some) mathematical analysis, but went deeper down to classes and then to propositional functions and relations.

These mathematical reductions were twinned with the philosophical empiricism which Russell adopted from G. E. Moore. As he recalled later, in the 1900s 'my universe became less luxuriant. [...] Gradually, Occam's razor gave me a more clean-shaven picture of reality' (*1959a*, 62). Logicism was his first detailed foray into reduction; not just the mathematical and logical cases just mentioned but also of mathematical logic to its 'primitive propositions', and of many philosophical categories to relations. A related term was 'meaningless', which however came close to self-reference with his many uses: denoting phrases without referents, classes or names or propositions which generate paradoxes, non-membership of an empty class, and ill-formed phrases or formulae (my *1977b*, chs. 13–17 *passim*).

Russell's desire to eliminate abstract objects also led him to construe individuals as basic elements of the physical world (§7.9.3). But then his logic became a *posteriori*, and he was forced to assertions such as 'Logic, I should maintain, must no more admit a unicorn than zoology can; for logic is concerned with the real world just as truly as zoology, though with its more abstract and general features' (*1919b*, 169). But one does not have to follow Leibniz, or Scholz for that matter (§9.6.3), on logic holding in all possible worlds to wonder why it should worry about unicorns, or bullocks either: Russell's conflations of various senses of existence (§7.3.5) has reared its ugly head (or horn) again, and his word 'concerned' above is hopelessly ambiguous. Logic is not so much an exact science as a *subtle* science.

Further, reification slips back into this threadbare empire, especially when quantification is applied: then propositions have surely to be taken as abstract objects, propositional functions seem to be 'attributes' (to use the term of Quine *1969a*, ch. 11), and the relationship between the predicate calculi and set theory is unclear (G. H. Moore *1980a*). Russell's gradual change from intensionality to extensionality, especially in the second edition of *PM*, was also reductionst; it raises similar doubts (§8.4.4), and Ramsey's extreme version (§8.4.7) still more. Later Russell queried the measure of extensionality deployed by Carnap (Russell *1940a*, ch. 19), and Carnap himself became doubtful (*1967a*, ix).

These last cases concern reductionist epistemology. During Russell's logicist phase he was doubtful about the utility of his logical techniques in philosophy in general. For example, while making a point on 6 February 1905 to his philosopher relative Joachim about how a complex is not wholly determined by its constituents, he made clear both his indebtedness to Moore and the limited philosophical scope of logicism:

> Mathematical pre-occupations can hardly be the cause of my views, since they are derived from Moore, who only crammed up enough Mathematics to get through the Little-Go [entrance examination]. As a matter of fact, it was mainly through Ethics that Moore reached his views. As for me, I admit that the philosophy of Arithmetic seemed to me to require something like the theory of entities. But I consider Aesthetics or Ethics or Theory of Knowledge just as good a field for proving the necessity of my views. [...] My Symbolic Logic is only intended to apply within a certain sphere. I hold it to be absolute truth, but not all truth. But I never apply it to properly philosophical problems. As to 'man is mortal', I am aware that there are dozens of other valid interpretations besides 'if x is a man, x is a mortal'; but this is for me the <u>most convenient</u> .[4]

But by the early 1910s Russell was applying his logical techniques and principles to reductionist epistemology; prominent were the logic of relations, the avoidance of abstract objects, and the tying of true propositions to facts by correspondence. In the late 1910s he came to call his position 'logical positivism' (§8.3.6); but the earlier *Our knowledge of the external world as a field for scientific method in philosophy* (1914), to quote its title in its important full, made perhaps the greatest impact, especially on the VC. The enthusiastic conversion of Schlick was noted in §8.9.1. The major effect on Carnap has been recorded in detail, and personal evidence is also available. When in London in July 1965 Carnap gave Russell a copy of Schilpp *1963a* on his philosophy, and wrote in it: 'To Bertrand Russell, from whom I learnt the scientific method in philosophy, I give this token of my admiration and gratitude'; he also marked the passage in his autobiography mentioning Russell's book (RA).

The word 'scientific' hides another reduction, made by Russell and his followers: of science to mathematics and physics. This narrowing has blighted much philosophy of science ever since, by retarding the philosophies of the life sciences and medicine, and also of technology. The effort of Woodger to logify biology (§9.5.3) was most ingenious but mistakenly conceived or at least very limited in scope.

The policy of reduction continued when Russell's (and Frege's) philosophies were mutated by others, especially philosophers of language, to 'analytic philosophy', an umbrella term which became popular in the late 1940s; it embraced especially Moore's and also parts of Wittgenstein's later philosophy, and helped to invent Frege'.[5] But it often degraded into language games, to Russell's disgust: 'Bad philosophy is an Oxford

---

[4] Russell's letter to Joachim is held at the Bodleian Library (Oxford), Eng.c.2026. I am grateful to Richard Rempel and Albert Lewis for communicating it to me.

[5] There is a mass of literature on the history of analytic philosophy; among the better parts related to Russell are the collections of articles in Russell n.s. *8* (1988), Irvine and Wedeking *1993a*, and Monk and Palmer *1996a*.

speciality', he exclaimed in 1956, 'and bad philosophy is still philosophy!' (*Logic*, 322).

Finally, to resume a theme from §1.2.4, despite the stress laid on the links between mathematical logic and language during our period, its impact upon linguistics was very limited. The adherence to bivalency was one reason, but take even the case, say, of classical negation: what is the positive counterpart to 'I have not eaten even an apple today'?

10.2.2 *The logic and its metalogic*

It has been a great misfortune to logic that universities have considered the necessary training of a logician to consist of a knowledge of Greek rather than of science.   Russell to J. O. Wisdom, 18 August 1959 (RA)

We have found the foundations of all systems studied here to be somewhat obscure. One reason was the lack of recognition of axioms as schemata; in particular, in *PM* schematic letters were conflated with real variables, and quantifier words such as 'all' and 'some' were used multiply. This unclarity is especially marked because both the propositional and the predicate calculi incorporated quantification in the normal sense. Frege had made some distinctions by using different kinds of letter (§4.5.2); but even his ordinary words can be ambiguous.

Another feature of logic was its universality, which was upheld by most adherents to all versions into the 20th century: one, and only one kind. Itelson used the word 'metalogic' at Geneva in 1904, but he rejected the notion involved (Couturat *1904c*, 1041); Stammler acted similarly much later (§8.7.9).[6]

Russell always held to logical monism; he took it first came from Peano, along with the term 'primitive proposition' to cover both axioms and rules of inference (§6.4.4). It prevented both men from clarifying the foundations of their logic: Russell's seems to have been inductive, in the scientific sense (§7.8.3). But the simple claim that he muddled logic with metalogic needs some refinement. The theory of definite descriptions shows that he knew the difference between a phrase and the object which it may be denoting (if any); and, like any one else, he wrote *about* a symbol if necessary. But some important demarcations *are* lacking; for example, whether type theory classified symbols or their referents (§7.9.2).

A very serious case is implication. While Russell identified it with $\vdash p \supset q$ between propositions p and q and inference with $\vdash p \supset \vdash q$, often he did not separate them, or individuate logical consequence or

---

[6] This was new sense for 'metalogic'; previously it denoted 'a science regulating the processes and symptoms of thought which are not universal', as Thomson put it (*1842a*, 23), like the somewhat later sense of 'metageometry' (§3.6.2).

entailment; laws such as excluded middle and contradiction were placed in the logic rather than stated of it (§7.8.3). Ponder his reaction in a letter of 15 May 1919 to Lewis upon receiving the *Survey* (§8.3.3): 'I have never felt that there was any very vital difference between you and me on this subject, since I fully recognize that there is such a thing as "strict implication", and have only doubted its practical importance in logistic' (RA).

It is not even clear whether the logicism of *PM* was implicational or inferential. But its logic became viciously encircled by its logicism, in that (some) mathematics came out of mathematical logic, which itself was taken to be whatever was required to deliver this mathematics—an ironic feature of a calculus supposedly based upon the vicious circle principle.

While the logical calculi of *PM* became well established (including the logic of relations for more advanced readers), the universalist conception behind it became a growingly negative influence from the late 1910s. Rules of inference were gradually recognised as distinct in status from axioms. Brouwer admitted 'mathematics of the second order' by 1907, although giving it little publicity (§8.7.7). Later, Wittgenstein and Ramsey defined logic without involving logicism, and for the second edition of *PM* Russell made some changes under Wittgenstein's influence; however, he did not use the hierarchy of languages which he had proposed in 1922 in reaction to the *Tractatus* (§8.4.3). Frege also spotted it at that time (§8.7.3), but kept it to himself. Several Americans groped towards metalogic and metalanguage (§8.5.3–5), while Hilbert's coterie developed metamathematics (§8.7.4). But the full distinctions were effected principally by Gödel (§9.2.3) and Tarski (§9.6.7), insights which mark the logic of the 1930s most clearly from that of the 1920s. This point leads us to the next topic.

10.2.3 *The fate of logicism.* While not exhaustive, the survey in the preceding two chapters shows clearly that *a far greater variety of positions was held and uses made of mathematical logic* than is conveyed by the traditional history about three competing philosophies of logicism, formalism and intuitionism. This feature belongs to a full study of the development of logics and epistemologies between the wars—a fine topic not yet explored comprehensively, which deserves a book the size of this one.

Logicism competed not only with formalism and intuitionism but also with (neo-)Kantian philosophy, phenomenology, conventionalism, and axiomatic set theory. It maintained a fairly good status in Britain, the U.S.A. and Italy, and even picked up a little from the mid 1920s among the French. But then it became overshadowed, especially by formalism and the attendant techniques of metamathematics and model theory; Russell, and also Whitehead and Frege, were blind to the latter, and so misinterpreted formalism as marks-on-paper philosophy (a mistake which has been made endlessly ever since). *PM* is replete with *proof* theory in all its details; but it lacked proof *theory* in the sense which Hilbert and his colleagues were individuating, thus helping the eclipse to occur.

There was also a difficulty of discipline: logicism, whether in Russell's or Frege's version, was too mathematical for philosophers to understand and too philosophical for mathematicians to appreciate. Even E. H. Moore opined to the historian Florian Cajori on 17 November 1926 that 'Certainly the extreme tendencies of Peano, Schröder, and Whitehead and Russell are beyond the approval, or at least the adoption, of working mathematicians' (Moore Papers, Box 1, Folder 16). In consequence, many nice mathematical features of *PM* were largely unnoticed, especially the axiomatic set theory embodied within the logical calculi, and the point-set topology and especially transfinite arithmetic developed in its latter Sections. Other features were taken for granted; for example, the tri-distinction (already with Frege) between zero, the empty class and nothing (§6.7.2), which cleared away much ambiguity.[7] The point can be generalised: mathematicians are usually dismissive of logic and philosophy, and thereby correspondingly careless from logical and philosophical points of view (Corcoran *1973a*).

For the same reason, while Stebbing and especially Ramsey and Carnap were distinguished followers, neither Whitehead nor Russell had doctoral students in Hilbertian quantity, nor corps of successors. Neither man seemed to have such aspirations; and in any case logicism was only part of Whitehead's philosophy of mathematics, and after abandoning the fourth volume of *PM* (§8.2.2) he went in other philosophical directions, apart from his strange recasting of *PM* in 1934 (§9.4.3). Thus he is often demoted in the history of logicism—an understandable but unfortunate situation which I have tried to repair with the information available.

As part of the modest reception of logicism, there was little discussion of the limited range of mathematics covered in *PM* (Table 782.1): only set theory, finite and transfinite arithmetic, and some mathematical analysis and (had the last volume been completed) geometry. Russell had put forward a broader vision in *The principles*, even if the exposition was incomplete: onwards to the differential and integral calculus, then through geometries to handle space, and thus to some mechanics; the indefinables 'alone form the subject-matter of the whole of mathematics; no others [. . .] occur anywhere in Arithmetic, Geometry or rational Dynamics' (*1903a*, 11). However, the range of mathematics to be unified was not mentioned in *PM*; seemingly the authors had become so bogged down in the details of the early segments that they lost sight of the rest. In addition, potential followers would have been discouraged from attempting

---

[7] An eloquent appreciation of this distinction, itself not recognised as such, is John Cage's composition 'Four minutes, 33 seconds', where any *number* of musicians play *nothing* for this duration of time. It cannot be a coincidence that the title indicates 273 seconds—that is, $-273°$ Centigrade, the absolute *zero* of temperature.

the trek by the recognised presence of the three doubtful axioms (infinity, choice, reducibility).

This point can be related to a general feature of implication which is often misunderstood. In $A \supset B$, proposition A is sufficient for proposition A, and A is necessary for B; so *an implication moves from the less general to the more so*, with the conclusion B embodying less knowledge than the premise A. From 'Socrates is a man' it follows that 'Socrates is mortal'; but the conclusion follows also from, for example, 'Socrates is a cat'. Similarly, logicism takes a relatively narrow base of mathematical logic and seeks (pure) mathematics as conclusion; but it fails, not only because of Gödel's incompletability theorem and corollary but also in the many branches of mathematics that slipped through the net anyway.

Gödel's results caused the logicist aim to be reformulated, especially by Quine (§9.4.4), in terms of possible relationships between mathematical logic, set theory and (some) mathematics. We noticed that, except for a few figures like Carnap and Dubislav, the reception of the results after publication in 1931 was rather slow; and Russell never absorbed them. At some stage he acquired the offprint of Gödel's paper previously owned by Countess Zeppelin, the translator of Carnap's *Syntax* (§9.3.7); but he marked only footnote 19, on the class of elementary formulae (RA). It is highly ironic that he had proposed the notion of a hierarchy of languages, essential for Gödel, for he never properly understood it (§9.5.4); for example, in a survey of logical positivism he mentioned his hierarchy and within a few lines said that Gödel's theorem applied to 'any formal system' (*1950a*, 371), while in his philosophical recollections he judged that the hierarchy merely 'disposes of Wittgenstein's mysticism and, I think, also of the newer *puzzles* presented by Gödel' (*1959a*, 114, italics inserted). In his last decade he was still wondering about the significance of the theorem (text in §11.9).

The balance between *PM* and metamathematics in the early 1940s is well exemplified by Church. Then in his early forties, he published as *1944a* some of his undergraduate course on mathematical logic given at Princeton University: due to war conditions, the book is not well known, at least not in Europe. While he covered the same logic as in *PM*, the treatment was much closer to Hilbert-Bernays (§9.6.2), from the separation of the first- and higher-order 'functional' calculi to the stress on consistency, completeness, deduction theorems, normal forms and decidability. *PM* was not much mentioned, and of its ingredients only type theory was presented in detail, and then with the 'ramified' theory omitted (pp. 109–112). Later Church greatly revised the parts of the book up to the second-order predicate calculus into a new edition *1956a*. Whitehead and Russell were mentioned dozens of times in the many new historical footnotes, but always on details. Although a type theory was constructed, logicism was not discussed: apparently it would have appeared in a second volume (p. 332), which was never published.

Similarly, when in the mid 1960s the Polish logician Andrzej Mostowski (1913–1975) delivered 16 lectures on 'thirty years of foundational studies' from 1930, he covered axiomatic set theory, intuitionism, formalism and model theory, but not logicism. In the introduction he mentioned its conversion into 'a reduction of mathematics to set theory', which was 'unsatisfactory' because of the difficulties facing the latter subject (*1966a*, 7); Quine appeared only for the stratification used in his book *Mathematical logic* (p. 143). However, the silence rebounds in part on Mostowski; for Quine and others were keeping alive the relationship between logic and set theories (Quine *1969a*), and in recent years various aspects of *PM* have been freshly studied.[8]

10.2.4 *Educational aspects, especially Piaget.*   As with all philosophical schools, logicism paid no attention within arithmetic to 'goes-into' integers. They arise in contexts such as the Euclidean algorithm: for example, 7 goes into 23 thrice, with 2 over. Words like 'thrice' show that a special vocabulary applies to these integers, which are neither cardinals nor ordinals; and the mention of Euclid shows that their history is long. Yet they usually escape the attention of mathematicians and philosophers— and, despite their heuristic utility, educators also.

The Swiss educational psychologist Jean Piaget (1896–1982) came eventually to the borders of logicism. An early book *1923a* in his studies of the child dealt with reasoning, with syllogistic logic providing the main basis to the extent that in ch. 3 he examined understanding of the 'logic of relations' without resource to any recent authors on the topic. However, by the late 1930s, while deploring Russell's separation of logic from psychology, he was imitating *PM* in giving prominence to (order-)isomorphism between classes, and even their 'additive composition', when studying the child's supposed 'conception of number' (*1941a*, chs. 3–6, 7). But not only did he ignore negative and (ir)rational numbers (and goes-into numbers), but more broadly he seems not to have grasped the distinction between

---

[8] For example, Bostock *1974a* and *1979a* has attempted a Fregean construction of finite arithmetic in which the insight that non-zero integers are the counterpart to non-existence is used to define them as quantifiers of propositional functions; however, the exegesis faces various philosophical and technical difficulties, well summed up by Resnik *1982a*. Rescuing *PM* from use-mention muddles and the excessive extensionality of its second edition, Church *1974a* and elsewhere has offered revised theories of types and of definite descriptions. Cocchiarella *1987a* has reformulated the logistic theses of Russell and Frege using more modern techniques, especially Church's λ-calculus. Hintikka *1996a* has 'revisited' *The principles* with the (absence of) quantifier order as a main concern, constructing a non-bivalent logic and deploying model theory to emphasise the distinction between the standard and non-standard interpretation of higher-order quantifiers (that is, their unrestricted or restricted ranging over the types of object falling within their range of values). Landini *1998b* has constructed a logico-mathematical system similar in scope to *PM* based upon a version of the discarded substitutional theory (§7.4.6–7).

creating or appreciating mathematics and justifying it epistemologically. These two books, and others, appeared in English in Ogden's series.

Piaget next wrote a volume *1942a* on 'numbers, classes and relations', a plod through the algebra of classes and relations with a touch of type theory on pp. 76–79. Publication of the book during the War greatly limited its impact, though the author sent Russell a copy in 1946 (RA). The influence of *PM* came through Piaget's belief that rationality resembles mathematical reasoning, which in turn was captured by (mathematical) logic. But such a position is hardly credible, especially for the creative sides of mathematics itself. Later his work played a role in the 'new mathematics' educational idiocy of the 1960s onwards.[9]

10.2.5 *The role of the U.S.A.: judgements in the Schilpp series.* The importance of Americans in this story, only partly recognised, deserves separate consideration. The initial stimulus did not come from C. S. Peirce, although both he and his father helped to foster an interest in algebras. The key mathematician was E. H. Moore, with his own interest in set theory and the development of postulate theory by his student Veblen, and also by Huntington; and from philosophy came Royce, who publicised Kempe and especially stimulated mathematical logic through his students, especially Lewis, Sheffer and Wiener, and also Cohen and Eaton. Thereafter interest continued steadily to the emergence of Veblen's student Church and Whitehead's sort-of student Quine, and the immigration of Carnap and Gödel in the 1930s.

Then Royce's influence arose again. The founding of *Journal of symbolic logic* (§9.4.5) owes much to C. J. Ducasse (1881–1969), who was the founder President of the sponsoring Association. Although mainly an analytic philosopher, he was deeply interested in logic—because, like Sheffer, Lewis and Eaton, he took Royce's courses around 1910 (Ducasse and Curry *1962a*).

Soon after the founding, opportunities to appraise Whitehead and Russell were provided by a major new American enterprise. We recall from §9.5.4, in connection with Dewey, that Schilpp launched in 1939 a series of volumes entitled 'Library of living philosophers', where extended articles on a philosopher were published together with his replies. In 1941 Whitehead was the third philosopher to be treated (though he was not well enough to reply). His principal follower, Victor Lowe (1907–1988), provided a long survey *1941a* of his philosophy to the mid 1920s, including the *Universal algebra* and the construction of space in 1905 (§7.6.1). Lowe deliberately left the mathematical logic to Quine who, then in his early thirties, contributed a piece *1941b* on 'Whitehead and the rise of modern logic'. Reviewing briefly but comprehensively the entire contents of *PM*,

---

[9] On this theme, and historical and philosophical aspects relating to it, see my *1973a*. Around that time Quine told me that when he had heard that set theory was being used in mathematical education, he had thought that he was being told a joke.

he drew upon his recent *Mathematical logic* and other work (§9.4.4) to point out imprecisions caused by conflating theory with metatheory; for example, the conditional and implication (pp. 140–141), and propositional functions as abstract attributes and as linguistic expressions (pp. 144–146). He also noted the excess multiplicity of notations in the logic of relations, and Whitehead's return to logicism in 1934, both with ennui (pp. 152–153; compare *1985a*, 113).

Quine did not try to distinguish Whitehead's contributions to *PM* from those of Russell, who was Schilpped three years later (after G. E. Moore, incidentally). Three authors considered mathematical logic. Reichenbach reviewed it pleasantly, but inaccurately when attributing to Russell the notion of propositional function (*1944a*, 25). His most thrusting query concerned the status of intuitionistic and three-valued logics, which he cautiously supported (pp. 40–44, including also a mention of Gödel). Russell replied that 'I agree, of course, that a three-valued logic is possible', as to Lewis in 10.2.2; but he argued for bivalency on the ground that it 'embraces unverifiable truths', with non-true encompassing both false and unknown (*1944b*, 682).

In response to Russell's introduction *1937a* to the reprint of *The principles* (§9.5.4), James Feibleman (1904–1987) felt that 'the old Russell is to be defended against the new Russell' by advocating 'modified realism' for logic; for example, disjunction 'is *logical* because it can neither be successfully contradicted nor shown to involve self-contradiction', while definite descriptions and classes were justified on the grounds that 'real existence means possibility of actualization, expressed in propositional functions' (*1944a*, 158–159, 161). Defending his current position, Russell merely stressed preferring 'minimum vocabularies' in formal systems (*1944b*, 687).

Logicism was reviewed by Gödel, then in his late thirties, in an article which has become well known in its own right. Like Quine on Whitehead, he noted the imprecisions of *PM*, 'in this respect a considerable step backwards as compared with Frege' (*1944a*, 126), and he deployed metatheory thereafter. Focusing upon the vicious circle principle, he pondered various senses of 'applied to itself', and found that when applied to itself a contradiction arises; so he concluded that it was false (pp. 132–133: we saw Behmann's priority in §9.3.4). The Platonist in Gödel emerged in doubting the location of logicism in the extensional logic of the second edition of *PM* (pp. 143–146); he also pointed out that the reconstruction of mathematical induction effected there (§8.4.4) was unsuccessful. Curiously, he did not drive home the consequences for *PM* of his theorem (p. 139). As usual, he struggled to produce the final version of his contribution (Dawson *1997a*, 162–166), so it came to Schilpp too late for Russell to write a reply—a great pity.

The appearance of the volume in 1944 motivated Weyl *1946a* to meditate in the *American mathematical monthly* on the relationship between mathematics and logic. In 1935 he had opposed the founding of the Association for Symbolic Logic, on the grounds that the subject was part of mathematics anyway;[10] now he addressed Russell's opposite view, that mathematics was part of logic. He reviewed the three main schools (including a detailed outline of type theory), and also his own position in *Das Kontinuum* (1918) (§8.7.7). While noting that the distinction between classes and predicates, so important to logicians, 'leaves the mathematician cool' (p. 268), he concluded (p. 279) that

> we are less certain than ever about the ultimate foundations of (logic and) mathematics [...] we have our 'crisis'. We have had it for nearly fifty years. Outwardly it does not seem to hamper our daily work, and yet I for one confess that it has had a considerable practical influence on my mathematical life [...].

10.3 THE PANOPLY OF FOUNDATIONS

Figure 103.1 is a summary flow-chart for the principal and largely positive influences of branches of mathematics upon algebraic logic, mathematical logic, formalism with model theory, and axiomatic set theory. The emphasis on influence means that achievements which did not make much impact for some time are omitted (principally Bolzano, the Grassmanns and Frege's creative phase). The main traditions or schools left out are intuitionism, phenomenology, Polish logic (too varied in range to be captured here), and (neo-)Kantian philosophies. The thick black line separating algebraic and mathematical logics largely concerns the predicate calculi and theories of collections; in other respects, such as the propositional calculi, there was more common ground.

Also not explicitly indicated are national differences. Having created most of the elements of algebraic logic, Britain showed some interest in the mathematical version in the new century. But much suspicion of symbolic logics remained (my *1986c*); for example, J. S. Mill (§2.5.8) and the neo-Hegelian tradition were taught at Oxford University until after the Second World War. The U.S.A. was far more sympathetic, as we have just seen. German mathematics and philosophy were both concerned with foundations, especially via axioms or assumptions; but mathematicians preferred formalism and/or axiomatic set theory, while many philosophers maintained the (neo-)Kantian or phenomenological traditions, which involved factors such as intuition and judgement which logicism hoped to avoid or re-interpret. The VC became an important force in Central Europe and to some extent in Britain, and then in the U.S.A. after emigration of key members; so logic and Russellian epistemology gained

---

[10] Church to Veblen, 19 June 1935 (Veblen Papers, Box 3). For context, see Aspray *1991a*, 61-63.

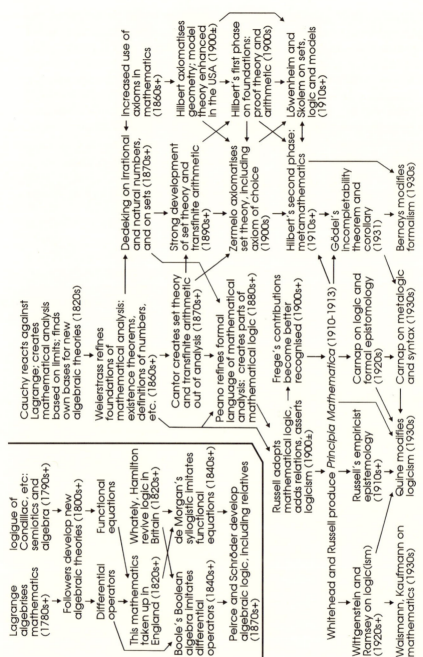

FIGURE 103.1. Connections between mathematical and algebraic logics, set theory and formalism.

considerable currency there. Poland rapidly became an important country after its creation in 1920, and *PM* was a major source for Leśniewski and Chwistek, and quite significant for Łukasiewicz and Tarski. France was hostile and then mildly interested, but Italian vigour was located almost entirely with the Peanists. None of these interests seems to manifest national*ism*, although the growth of Polish logic with its new country is striking.

Among the philosophies, intuitionism created much publicity in various countries but gained few converts; however, one was Weyl. As he judged above, the search for mathematical roots had not produced any internationally accepted position. One reason was the continuing failure to find a definitive solution (or Solution) of the paradoxes of logic and set theory; for at least some of them should fall within any philosophy of mathematics.[11] Similarly, the concern with form persisted from De Morgan's distinction of form from matter (§2.4.4) through Russell's attempt to define mathematical logic as 'the study of forms of complexes' (§8.2.4) to Carnap's structuralisms (§8.9.3, §9.3.2) and Stebbing on 'the ideal of form' (§9.5.1); but no version was definitive even for bivalent logic. Again, De Morgan also pioneered the logic of relations, which both traditions developed considerably; but it cannot carry a philosophy of mathematics on its own.

Weyl's judgement is still valid: the following half century has still not produced a victor. On the contrary, many more logics have been developed (Mangione and Bozzi *1993a*); and variant or new philosophies of mathematics have appeared (Tymoczko *1986a*), including anti-foundationalism, which incidentally Neurath pioneered from the 1920s for the philosophy of science. The variety of these developments underlined an observation made long before by Condillac in his *Logique* (§2.2.7) that 'the study of a well-treated science is reduced to the study of a well-made language' (*1780a*, pt. 2, ch. 9); for the science of logic needed self-study—Sheffer's predicament, from which hierarchies rescue the enlightened.

In the same spirit of pluralities, the writing of this book was inspired by the *historical importance* of the matters described, especially for their lovely mixture of logic, mathematics, set theory and philosophy. No *philosophical defence* of either logicism or logical positivism is offered or intended; for, if done properly, history is metatheoretic, looking at the situations of its historical figures and of the past *without* having to affirm or deny any stance found there.

The place of symbolic logic in mathematics and philosophy may be broadly described as follows. Algebraic logic came from a readiness in mathematics to entertain new algebras; it gave back especially Boole's

---

[11] In an ingenious feat Priest *1994a* has claimed that all the paradoxes share a common form, and so should receive the same solution. In my reply my *1998b* I find renegade paradoxes, and especially argue for a plurality of solutions even for those paradoxes embraced by Priest by giving priority to the *contexts* in which they arise.

algebra of logic, a sophisticated logic of relations with quantification based upon part-whole theory, and Schröder's logicising catalogue of mathematical theories. Mathematical logic grew out of the desire in mathematical analysis for rigour, and came back with a sophisticated logic of predicates and relations with quantification based upon set theory, and a detailed method of framing definitions and working out proofs. It also took Cantor's vision of *Mengenlehre* as a foundation for mathematics, and handed back a detailed logicistic exposition of the links between set theory, finite and transfinite arithmetic, and some mathematical analysis and geometry. But it also led to surprising news about the limitations of axiomatisation with Gödel's theorem and corollary.

Both traditions took from logic and philosophy a long concern with valid and invalid arguments, and came back with clear indications of the limitation of syllogistic logic. They also provided new insights about the connection between relations (including predicates) and collections, part-whole theory in one case, set theory in the other. Eventually mathematical logic helped to give to philosophy the centrality of the distinction between theory and metatheory—its most precious gift to knowledge of this century. But it also warned about the central place of paradoxes of various kinds, and failed to provide a definitive solution of them.

In this general context, Russell's aspirations show one main negative and one main feature. Both are independent of Gödel's incompletability theorem while reinforced by it:

−) A leading motivation for Russell's logic and philosophy, but also a source of his difficulties, was his quest for *certainty* in (mathematical) knowledge. But maybe he was mistaken to ground fairly clear objects like 1, 2 and even $(\aleph_0 + 371)$ in the uncertainty of classes, especially when he found paradoxes at their centre. His search for the logicist roots of mathematics was unsuccessful; moreover, the logical roots which he proffered were enigmatic. Logic(s) and mathematics are overlapping disciplines, with (axiomatic) set theory in the intersection, where the Peanists had (unclearly) located it. Hence logicism founders: logic may clothe the body mathematic, but it cannot provide the body itself, at least not all of its many limbs. Not only is there far more to logic than mathematics (as a logicist would affirm), but also there is far more to mathematics than logic, even the mathematical brand. Mathematics exhibits an amazing range of interconnections within and between its branches, and some of them inspire quite large-scale reductions, among which *PM* was one of the most extensive. But even there most mathematics 'gets away'; *many links, yes, but no unifier of the whole lot.*

+) While *PM* failed in its main objective, theories were developed and insights made in the prosecution of this great enterprise which endured long after the original vision had become clouded. In Russell's hands it also inspired an influential tradition in empiricist epistemology. All the traditions and schools described in this book greatly deepened understanding of the (meta-)theoretic questions posed; by 1940 the foundations of mathematics enjoyed a far richer range of treatments than ever before, and the searches of Whitehead and of Russell had played major roles in the progress.

10.4 SALLIES

It was long before I got the maxim, that in reading an old mathematician you will not read his riddle unless you plough with his heifer; you must see with his light, if you want to know how much he saw.
De Morgan to W. R. Hamilton, 1853 (Graves 1889a, 438)

Symbolic logic [...] has been dismissed by many logicians on the plea that its interest is mathematical, and by many mathematicians on the plea that its interest is logical.   Whitehead *1898a*, vi

[...] philosophers subsequent to Kant, in writing on mathematics, have thought it unnecessary to become acquainted with the subject they were discussing, and have therefore left to the painful and often crude efforts of mathematicians every genuine advance in mathematical philosophy.
Russell, close of 1900a

Accordingly, we can say that the algebra of logic is a *mathematical* logic by its form and by its method, but it must not be mistaken for the logic of *mathematics*.   Couturat, close of *1905a*

Historians whose purpose is simply to understand the meaning and drift of mathematics will do better to leave [*Principia mathematica*] alone.
Sarton 1936a, 54

CHAPTER 11

Transcription of Manuscripts

MOST OF the manuscripts transcribed here are (parts of) letters to or from Russell, both ways in §11.8: they are ordered chronologically. Unless otherwise stated, the originals are kept in the Russell Archives; permissions to publish have been recorded in §1.6. Orthography has been followed, including the rendering of underlinings and capital letters, and the presence or absence of punctuation and inverted commas (single or double). The layout of formulae has been followed as closely as practicable. But deletions and slips in writing have been ignored when they are insignificant; and opening flourishes and signatures are omitted. Editorial insertions or comments are placed between curled brackets because Russell used square ones in §11.3–4; all footnotes are mine.

11.1 COUTURAT TO RUSSELL, 18 DECEMBER 1904

This letter belongs to the time of Couturat's first major advertisement of Russell's logicism (§7.4.1–2), especially in his articles in the *Revue de métaphysique et de morale* and the book version *Les principes des mathématiques* (*1905b*). The letter was written just before the two friends faced hostility from Poincaré and Boutroux, the latter a recent visitor to Britain (§7.4.2). Couturat tells Russell of his current writings, his failure to popularise logicism among mathematician compatriots such as Borel and Lebesgue, and the latest news on proving the well-ordering theorem (§7.2.2, 5). The letter also contains some details on two lost works: Couturat's own 'manual of logistic', and the first version of *PM* as the second volume of Russell's *The principles*. In a postscript he adds some remarks comparing some symbols in mathematical and algebraic logics. Apart from the last paragraph, the letter is typescript, from Paris.

Voilà bientôt un mois que je vous ai laissé tranquille; j'ai terminé mon petit volume sur l'Algèbre de la Logique, et je vais me remettre à ma "Logique mathématique", que je vais publier chez Alcan sous le titre: "Manuel de Logistique". J'ai reçu bien des fois, surtout depuis mes articles de la R.M.M. et le Congrès, des demandes d'explications sur la "Logique nouvelle", et cela me décide à publier ce manuel, qui contiendra le résumé du système de Peano et du vôtre,...en attendant que vous ayez réformé celui-ci. Je ne me dissimule pas le caractère contingent et provisoire de ce travail, que votre 2e volume rendra peut-être caduc; mais il ne faut pas se flatter de travailler pour l'éternité, et il vaut mieux faire oeuvre imparfaite, mais utile pour le moment présent.

Qu'il soit utile de vulgariser la Logistique, ou même de la faire connaître, c'est ce dont j'ai chaque jour la preuve. J'ai fait la petite expérience que voici: mon ami M. Borel, mathématicien très distingué, m'ayant donné des ouvrages qu'il vient de publier pour en rendre compte dans la R.M.M., j'ai rédigé quelques ''Remarques de logicien'' que j'ai envoyées à lui, à M. Baire et à M. Lebesgue (ses deux principaux collaborateurs) pour leur apprendre l'existence de la Logistique. Je vous envoie la copie de ces Remarques, et de la lettre que j'ai répondue à M. Baire; vois devinerez aisément les objections de celui-ci. Borel et M. Lebesgue m'ont fait des réponses analogues, c'est à dire tout à fait sceptiques, sous une forme plus ou moins aimable. Il est clair qu'ils dédaignent et ignorent les travaux de Peano et de son école, et les croient absolument inutiles et stériles. ''Ils n'ont pas besoin de cela pour raisonner juste, etc.'' Je crains bien que ce ne soit là l'attitude de tous les mathématiciens à l'égard de la Logistique; ces gens qui vivent de symboles (au point de réduire toute leur science à un pur jeu de symboles) ont une aversion étrange et irréfléchie pour tout symbole qu'ils ne comprennent pas. Ils ne savent pourtant pas l'Algèbre de naissance!—C'est pourquoi j'ai trouvé et trouve très fâcheuse l'attaque de P. Boutroux, si inoffensive qu'elle soit au fond. Le seul fait qu'un mathématicien de profession paraisse déclarer la Logistique inutile ou incompétente[1] en mathématiques confirmera dans leur préjugé paresseux bien des gens qui ne savent ni les mathématiques ni la Logistique; et les professeurs de philosophie, que j'ai assayé de secouer dans leurs préjugés, continueront, rassurés, à rabâcher les lieux communs de l'épistemémologie kantienne.—Je viens d'apprendre que M. P. Boutroux voyage en Angleterre, et qu'il a dû aller vous voir; on me dit en même temps qu'il prépare un article sur l'idée de correspondance en mathématique {§7.4.2}. Je souhaite que votre conversation l'ait instruit et converti; mais à vrai dire, je crains plutôt qu'il ne s'imagine que le fait de vous avoir parlé une heure ou deux lui a permis de pénétrer à fond dans votre système, et le dispense de lire et d'étudier votre ouvrage; il se permettra peut-être de porter un jugement sommaire, à priori, sur des doctrines qui ont coûté des années d'élaboration à leurs auteurs, et que je ne me suis assimilées que par des années d'étude. Comme je le disais à Borel, on a beau avoir du génie, on risque fort de ne dire que des bêtises quand on parle de ce qu'on ne connaît pas.

Je vois remercie vivement de l'offre de me communiquer votre ms.; mais, d'abord, j'ai toujours peur qu'il arrive quelque malheur à vos

---

[1] Over this word Couturat has written '(irrelevant)'.

papiers; ensuite, je crois qu'il vaut mieux que je me contente de ce que je connais de votre 2e volume. Vous savez pour quelles raisons d'ordre pratique et pédagogique je ne crois pas pouvoir adopter vos principes dans mon exposé élémentaire, destiné à des "commençants". Je tâche toutefois de le faire profiter de vos théories; par exemple, j'y introduis dès le début les principes de déduction et de substitution. J'appelle pr. de déduction le pr. 4 du §18 de votre vol. I, ou la Pp. 2.1 de votre vol. II (dont vous m'avez communiqué le 1er chapitre). Je réserve le nom de principe d'assertion à l'axiome spécial de Schröder, à savoir: a = (a = 1): "Affirmer une proposition, c'est affirmer qu'elle est vraie". Il me semble que ces deux noms sont bien appropriés au sens des deux principes dénommés.

Mon ami Brunschvicg (que vous devez connaître par la Revue de Métaphysique) est plongé en ce moment dans la lecture de votre ouvrage, qui lui plaît extrêmement par la profondeur des analyses logico-grammaticales de la 1re Partie. Il est plus métaphysicien que mathématicien (il a fait une thèse sur la modalité du jugement); il met, à mon gré, trop de métaphysique dans la logique; mais c'est un excellent esprit, et j'espère que vous en ferez la conquête. Il occupe la chaire la plus importante des lycées de Paris (au lycée Henri IV), et il forme la moitié des jeunes gens qui entrent à l'Ecole normale, et qui deviendront ensuite professeurs de philosophie. Il vient de publier une belle édition des Pensées de Pascal chez Hachette.

Je suis bien aise de pouvoir vous apprendre une importante nouvelle (que je dois à mon ami Borel): M. König s'est trompé dans sa refutation du théorème de G. Cantor, et il l'a reconnu lui-meme. D'autre part, M. Zermelo publie dans les Math. Annalen une démonstration de ce même théorème {Zermelo *1904a*}. Mais comme elle repose sur le calcul des "alephs", certains mathématiciens ne lui attribuent pas plus de valeur qu'à la réfutation de M. König. Quand vous aurez le temps de l'étudier, je serai curieux de savoir ce que vous en pensez, et si vous pouvez résoudre la question dans un sens ou dans l'autre au moyen de votre calcul.

P.S.—A propos de ce que vous m'avez répondu au sujet des inférences de la forme: $ab \nleqq a$, il me vient à l'esprit une question: puisque le fait que la thèse est contenue en facteur dans l'hypothèse est insignifiant, comment définissez-vous la distinction des jugements ou des inférences analytiques et synthétiques, que vous paraissez admettre? J'avais cru voir une différence essentielle entre l'Algèbre de la Logique, où toute conséquence (formelle) est contenue dans les prémisses, en vertu de la formule:

$$(A + B = 0) \nleqq (A = 0)$$

et votre Logistique, qui n'est pas un calcul formel, une "algèbre". Dans l'Algèbre de la Logique (surtout comme elle est présenté par M. Poretsky[2]), on peut obtenir pour ainsi dire automatiquement toutes les conséquences d'un système de prémisses données, et de même toutes ses "causes". Il n'en est pas de même dans la Logistique de Peano: ce n'est plus une machine à raisonner. Je vous prie de réflécher à cette question, et de me dire, d'abord, si la différence que je vois vous paraît juste, ensuite, d'où elle vient selon vous. C'est cette différence que j'avais essayé de caractériser par les expressions classiques "analytique" et "synthétique"; mais les mots n'ont pas d'importance; l'essentiel est de savoir si la distinction est juste; on verra après comment on doit l'appeler.

Dans mon Algèbre de la Logique, j'emplois les signes de Schröder (à cause de l'analogie si commode avec l'Algèbre ordinaire) sauf le signe $\in$, que je remplace par $<$, et le signe de négation, que je remplace par l'accent (x').—En ravanche, dans mon Manual de Log. et dans les Principes des math. j'adopterai le signe de négation de Peano, parce qu'il est le seul uniforme (le même pour les Cls. les P{ropositions} et les Rel.)

11.2 VEBLEN TO RUSSELL, 13 MAY 1906

This letter is an early example of American interest in Russell's developing logicism, from a young mathematician at Princeton University early in a major career (§7.5.5). Reacting to Russell's paper *1906a* on the paradoxes recently published by the London Mathematical Society (§7.4.4), Veblen outlines a nominalistic approach to classes in an attempt to "save" as much set theory and mathematics as possible while avoiding them; he also comments cautiously on the axioms of choice, newly revealed. The notion of 'norm', due to Hobson *1905a*, corresponds to a propositional function (§7.4.4). While his ideas do not constitue a major contribution, they show well the kind of response which Russell's approach to foundational questions stimulated in a sympathetic and informed mathematician of the time. The letter is entirely typescript, from Princeton.

I have been reading your valuable paper on "The Theory of Transfinite Numbers and Order Types" and should like to indicate a way somewhat similar to your "no-classes theory", in which one may look at the so-called paradoxes about classes. My point of view is in a sense a nominalistic one and I think does not require the introduction of any new hypotheses into logic.

---

[2] The Pole P. S. Poretsky (1846–1907) brought the algebra of logic to Russia, and developed some of its methods (Styazhkin *1969a*, 216–253).

A class is defined in terms of its elements, either by enumerating the elements separately, or by stating a norm or characteristic property of the elements. The existence or non-existence of a class depends on the existence or non-existence of all its elements. Any norm, therefore, which requires a class a to be an element of a is subject to the dual fault that a is defined in terms of itself and that any attempted proof of its existence must involve a vicious circle.

Now, all the "paradoxes" you consider involve in their statement the use of this kind of a norm. They, therefore, involve processes which are as invalid as any other kind of defining in a circle or inferring in a circle.

I think, therefore, that if the existence of a class is determined by the existence of its elements the terms "all classes", "all well-ordered classes", etc. cannot be used to define classes. On the other hand, "all rational numbers", "all continuous functions" and "all well-ordered classes of cardinal number $\aleph_\alpha$", are designations of classes to whose existence theorems there is no valid objection. The paradoxes are gone, and all real mathematics remains. This does not mean that one may not use the phrase "all classes" with great freedom, or that there is anything wrong with the proposition, "a is a class." It does mean that Peano's ε in "$N_0 \varepsilon Kls$" is different from his ε in "$a \varepsilon N_0$".

I presume that there may be possible extensions of logical theory that will admit another than a nominalistic view of classes, and that will determine in a more searching way the significance of terms like "all classes". One thinks at once of the wellknown definitions by postulates, by abstraction, etc., but it must be remembered that such definitions are always to be supplemented by existence theorems which are effected by giving possible nominal definitions of the same objects. A logic in which the norms "all well-ordered classes" and "all classes" determine aggregates which are regarded as individuals, I think will also have to regard the norm "a class whose only element is itself-regarded-as-individual" as determining a class. My own feeling is that "all classes" can be treated as many, but not as one. Moreover, even if the extension is made, it cannot affect the validity of what has already been achieved in mathematics and logic.

I take a like view of "Zermelo's axiom". It may or may not be true in general that "given a class of classes a there exists a function f such that f(a) is always an element of a". In some special cases this can be proved true. In the important case of the continuum it must either be false, or be a consequence of known theorems, since these theorems characterize the continuum completely. Therefore, if the Zermelo proposition is to be taken as an independent axiom, it must not be stated so as to apply to the continuum or any known set of objects. Hence, I do not see any particular use of stating it at all, as a postulate.

I trust that you and Mr. Whitehead are making satisfactory progress on the second volume of your Principles of Mathematics. The volume is being awaited with interest.

11.3   RUSSELL TO HAWTREY, 22 JANUARY 1907 (OR 1909?)

In this letter Russell presented the paradox to which he found his substitutional theory of classes and relations susceptible (§7.4.7). He often did not distinguish propositions from individuals in this theory; presumably a and a_0 are propositions, available for substitution in compound propositions p and p_0. Negation is used to define falsity: ' ~ p' means 'p is false'. This paradox, a substitutional variant on the liar paradox using negation to transmit falsity, led to the decisive abandonment of the theory, seemingly early in 1907. The letter is hand-written, from Bagley Wood, but the year date is unclear; '07' is my preferred reading, as it belongs better to the time of their correspondence. The original is kept at Churchill College, Cambridge in the Hawtrey Papers, File 10/81, with a copy at RA; it has recently appeared in facsimile as the frontispiece of Landini *1998b*.

I forgot to send you the paradox which pilled this substitution-theory. Here it is. Put

$$p_0 . = : (\exists p, a) : a_0 . = . \, p \frac{b}{a} !q : \sim \left(p \frac{a_0}{a} \right) \quad \{Df\}$$

$$\left[\text{where } "p \frac{b}{a} !q" \text{ means } "\underline{p} \text{ becomes } \underline{q} \text{ by substituting } \underline{b} \text{ for } \underline{a} \right].$$

Then

$$p_0 \frac{p_0 \frac{b}{a} !q}{a_0} . = : (\exists p, a) : p_0 \frac{b}{a_0} !q . = : p \frac{b}{a} !q : \sim \left(p \frac{p_0 \frac{b}{a_0} !q}{a} \right)$$

Hence

$$p_0 \frac{p_0 \frac{b}{a} !q}{a_0} . \supset : (\exists p, a) : p_0 \frac{b}{a_0} !q . = . \, p \frac{b}{a} !q : p_0 \frac{p_0 \frac{b}{a_0} !q}{a_0} . \sim \left(p_0 \frac{p_0 \frac{b}{a} !q}{a} \right)$$

(1)

$$\left(p_0 \dfrac{p_0 \dfrac{b}{a}!q}{a_0} \right) . \supset : . \, p_0 \dfrac{b}{a_0}!q . = . \, p\dfrac{b}{a}!q : \supset_{p,a} . \, p \dfrac{p_0 \dfrac{b}{a_0}!q}{a} . :$$

$$\supset : . \, p_0 \dfrac{p_0 \dfrac{b}{a_0}!q}{a_0} \tag{2}$$

$$\vdash (1) . (2) \dots : . \, p_0 \dfrac{p_0 \dfrac{b}{a_0}!q}{a_0} : . \, (\exists p, a) : p_0 \dfrac{b}{a_0}!q . = . \, p\dfrac{b}{a}!q : \sim \left(p \dfrac{p_0 \dfrac{b}{a}!q}{a} \right) \tag{3}$$

But if $p_0 \dfrac{b}{a_0}!q$ is the same as $p\dfrac{b}{a}!q$, it seems plain we must have $p = p_0 . a = a_0$, whence

$$p \dfrac{p_0 \dfrac{b}{a_0}!q}{a} . \equiv . \, p_0 \dfrac{p_0 \dfrac{b}{a_0}!q}{a_0} .$$

Thus it is impossible that

$$p \dfrac{p_0 \dfrac{b}{a_0}!q}{a} \text{ should be false while } p_0 \dfrac{p_0 \dfrac{b}{a_0}!q}{a_0} \text{ is true,}$$

which, by (3), is shown to be involved.

In trying to avoid this paradox, I modified the substitution-theory in various ways, but the paradox always reappeared in more and more complicated forms.

11.4   JOURDAIN'S NOTES ON WITTGENSTEIN'S FIRST VIEWS ON RUSSELL'S PARADOX, APRIL 1909

This text is taken from notes made by Jourdain of a conversation held when Russell visited him on 20 April 1909 at his home at Broadwindsor, Dorset. He records Wittgenstein's initial manner of solving Russell's paradox, seemingly in correspondence which is now lost; later Wittgenstein writings, including the *Tractatus*, followed similar lines in various ways. The date is surprising, since Wittgenstein came into personal contact with Russell only 30 months later (§8.2.6). The text up to 'which are not meaningless.' was printed in my *1977b*, 114; I give here the whole paragraph. Jourdain went on to discuss some particular solutions, especially Schönflies's (§7.5.2); Wittgenstein was not mentioned.

As very many words are abbreviated, I have made the expansions silently. The original text is in Jourdain Papers, notebook 2, fols. 205–206, with a copy at RA.

Russell said that the views I gave in a reply to Wittgenstein (who had 'solved' R's contradiction) agree with his own. These views are: The difficulty seems to me to be as follows. In certain cases (e.g. Burali-Forti's case, Russell's 'class' $\hat{x}(x \sim \varepsilon x)$, Epimenides' remark) we get what seem to be meaningless <u>limiting cases</u> of statements which are not meaningless. Thus there may be certain <u>x</u>'s such that the statement (propositional function) ϕx,—which is a (true or false) proposition for other <u>x</u>'s, —is meaningless: thus if $\phi x = '\underline{x}$ is a liar', ϕx becomes meaningless if I put x = myself. Analogously, we see no contradiction in thinking of $\hat{x}(x \sim \varepsilon x)$, where <u>x</u> is <u>restricted</u> to (say) finite integers; but when we <u>drop</u> this restriction this class w is such that $w\varepsilon w \equiv w \sim \varepsilon w$. Thus the 1st problem is to find a principle to exclude those & only those limiting cases.

11.5 THE APPLICATION OF WHITEHEAD AND RUSSELL TO THE ROYAL SOCIETY, LATE 1909

This is the text which Whitehead and Russell prepared to request for financial support from the Royal Society to publish *PM* (§7.8.1). It seems to have been written around November 1909. Naturally they stressed to the Society the mathematical parts of *PM* rather than the philosophical ones—even the paradoxes receive only a short paragraph in the middle— but the text also has philosophical interest.

I follow the original hand-written version; presumably a typed version went in to the Society, but no such copy has survived. It appeared, with historical elaboration, in my *1975c*; it should have been included in *Russell Papers 6*, but in view of its omission I repeat it here. The authorship alternates: I mark the new writer by inserting '{W}' and '{R}' respectively. Note Russell's three-fold restriction of Whitehead's 'mathematical' to 'pure'. I have also silently inserted stray missing punctuation and expanded the occasional abbreviation.

{W} The proposed title of the book is 'Principia Mathematica' and its object is to comprise a complete investigation of the foundations of every branch of mathematical thought.

The book commences by stating the logical principles and ideas which govern the course of all demonstrative reasoning. Then all the fundamental ideas which occur in {R} pure {W} mathematics are considered in detail: Each idea is stated in its most general form and is shown to be definable in terms of the fundamental logical notions which are considered at the commencement of the book. It is also proved by exact reasoning that the fundamental propositions from which the various branches of {R} pure {W} mathematics start are deducible from the logical principles stated in the book, without the aid of any other undemonstrated axioms.

Thus the definitions form an analysis of all mathematical ideas, and the demonstrations are (1) proofs of the 'axioms' from which the various branches of mathematics start, (2) guarantees of the adequacy of the analysis of ideas, effected in the definitions, by exhibiting the ways in which the ideas as thus analysed occur in reasoning.

The proofs are so arranged that it is possible to trace through the whole work the occurrences of any idea or any proposition.

{R} We have in each case sought the utmost generalization of the various mathematical ideas involved which is compatible with the truth of the properties usually assumed in their mathematical treatment. Thus for example in Cardinal Arithmetic we seek the most general definitions of cardinal number, and of the addition and multiplication of cardinal numbers, which secure the truth of the associative laws of addition and multiplication and of the distributive law, and we then show that these laws are applicable without distinction to the infinite and to the finite, and are themselves capable of generalized forms which include their usual forms as special cases. Similarly, in dealing with limits of functions and the continuity of functions, we so generalize the usual definitions as to make them independent of number, thereby greatly extending their scope. In short, given the usual "axioms" of any subject, our problem is to find the most general hypotheses from which these axioms follow, in other words, the largest set of objects to which they are applicable.

The proofs are so arranged that it is possible to trace through the whole work the occurrences of any idea or any proposition. {W} For example, the relevance of 'mathematical induction' to various parts of mathematical reasoning can be traced: also mathematical induction itself is analysed in its most general form, and in its various particular shapes. By this procedure the falsehood of the opinion that mathematical induction is essential to all mathematical reasoning concerning the infinite is at once evident. But the reason for the growth of such an opinion is also clear, for mathematical Induction is shown to be necessary for most of the most interesting deductions. This is only one example of the type of question which is considered in the work. Other examples of ideas which are analysed, generalised and traced in their mathematical uses, are Cardinal Number, Ordinal Number, Rational Number, Real Number, Complex Number, Quantity.

It is shown that by the aid of the logical principles considered at the commencement all the mathematical contradictions, which have formed the subject of so much recent discussion, vanish: so also do the logical paradoxes and puzzles {R} some of {W} which have come down to us from the Greeks. At the same time {R} our principles do not make {W} it necessary to abandon any important propositions in which contradictions have not been found to occur.

{R} Apart from the philosophical interest possessed by the analysis of the fundamental ideas of mathematics (including such logical notions as 'class' and 'relation'), the main {W} value of the book lies in its exactness, its particularity of reasoning, and its completeness. It is in fact an encyclopaedia which professes to exhibit all the ideas of {R} pure {W} mathematics and the various ways in which they are related to each other.

The book contains some thousands of propositions and many thousands of references. Apart from their preparatory studies both authors have worked at it directly since the autumn of the year 1900: one author has devoted to it {R} nearly {W} his whole time, the other such time as he could spare after performing his duties as a College lecturer.

It has been found to be absolutely necessary to conduct the proofs by the aid of a logical symbolism. Only by this means could the necessary brevity and exactness be obtained, combined with an almost pictorial aid to the imagination, very necessary for the abstract ideas which occur. The basis of the symbolism is the logical notation of Boole, Schröder, and Peano. But the authors have continuously modified it, in the light of their ten years' experience of its use, so as to make it a simple and direct instrument. Accordingly the symbolism is in fact created to suit the ideas and to express them in the simplest way possible, {R} thereby securing the advantages of mathematical symbolism in regions of thought which have hitherto been hampered by the ambiguity, obscurity, and prolixity of verbal reasoning. The Purport {W} of every important proposition and section is explained in ordinary language. These explanations are sufficiently full to form an abstract of the whole work.

The Pitt Press reckons that the book will occupy nearly 2,000 pages, if printed uniformly with its mathematical series, containing (for example) Forsyth's Theory of Functions {Forsyth *1893a*}. This would require the book to appear in three such volumes. The Syndics consider that the book would have to be sold at 2 guineas and that after allowing for sales a probable loss of £500 would result. They are willing to publish the book and to incur a considerable loss upon it; but £500 is more than they can undertake. Accordingly we ask the Council of the Royal Society whether they will cooperate with the Cambridge University Press by making a grant to enable the book to be published.

TABLE 115.1.   Plans and Fulfilment of the Last Sections of *PM*

Section	Application title	PM title
VIA	Generalizations of Number. (Rational Numbers, Real Numbers, etc.)	Generalization of Number
VIB	Non-numerical Theory of Quantity	Vector-Families
VIC	Measurement	Measurement
VID		Cyclic Families
VIIA	Projective Geometry	
VIIB	Descriptive Geometry	
VIIC	Metrical Geometry	
VIID	Constructions of Space	

Russell concluded the application with an abridged table of contents (my *1975c*, 98–100); most of it is closely similar to that of *PM* summarised in Table 782.1. However, Whitehead seems to have somewhat reworked Part VI on 'Quantity', and he abandoned Part VII on 'Geometry' in the 1910s (§8.8.2); so the titles of the Sections involved are reproduced here, in Table 115.1.

11.6   WHITEHEAD TO RUSSELL, 19 JANUARY 1911

This letter belongs to the period when Whitehead discovered his mistake in his construction of cardinal arithmetic in Part III of *PM*, where he had assumed the axiom of infinity whenever required instead of only one individual whenever possible (§7.9.3). He describes the existence theorems which fail (that is, where the corresponding classes are empty, symbolised '$\sim \exists!$') when this reductionist principle was applied, and the sections of Part 3 which need reworking. The letter is hand-written, in London, the first of two sent to Russell that day.

Please consider this with the view of determining accurately what we mean as distinct from how to put it best.

On p 278 it is shown[3] (rightly, I think) that if values are determined before types,

$$\vdash \; : \mathrm{Nc}'\alpha = \mathrm{Nc}'\beta . \equiv . \; \alpha \; \mathrm{sm} \; \beta \qquad (1)$$

Now also

$$*2 \cdot 02. \; \supset \vdash \; :: \mathrm{Nc}'\alpha = \mathrm{Nc}'\beta . \equiv . \; \alpha \; \mathrm{sm} \; \beta : \supset :. \sim \exists!\mathrm{Nc}'\alpha . \sim \exists!\mathrm{Nc}'\beta . \supset :$$

$$\mathrm{Nc}'\alpha = \mathrm{Nc}'\beta . \equiv . \; \alpha \; \mathrm{sm} \; \beta \qquad (2)$$

---

[3] Presumably this page number is that of the manuscript folio. The closest corresponding theorem in *PM*, $*100 \cdot 35$.

$$\vdash . (1) . (2) . *1\cdot1 . \supset \vdash ~\sim \exists!Nc'\alpha . \sim \exists!Nc'\beta .$$

$$\supset : Nc'\alpha = Nc'\beta . \equiv . \alpha \text{ sm } \beta \qquad (3)$$

$$\vdash . (3) . *24\cdot51 . \supset \vdash ~\sim \exists!Nc'\alpha . \sim \exists!Nc'\beta . \supset . \alpha \text{ sm } \beta : \vdash . \text{Untrue Prop}$$

$$\{(4)\}$$

Notice that in (2) $\sim \exists!Nc'\alpha . \sim \exists!Nc'\beta$ may be true in some types and false in others but (2) is always true. Yet (3) and (4) are both false.

Thus with 'values before types' no proposition whose truth value can vary with type must ever be <u>employed</u> in any deduction.

But no such limitation has been adhered to before *126. Hence all the proofs of *126 (where 'values before types' is the rule) are fallacious because they appeal to results proved in previous numbers.

Am hoping to have some positive suggestions soon—but the subject is very difficult.

I am inclined to think that 'values before types' must go, as too difficult to work—but rather think that all we want can be secured, if we can get a good symbolic theory of the dependent variable on its legs

Can you tell from your list of references where *103·16 is subsequently used?[4] We may have a fine crop of fallacious proofs on our hands from it. But I hope not.

Whitehead followed with a flurry of letters for the rest of the month, outlining attempts to get over his problem of existence; *126, which dealt with the sequence of cardinal numbers, continued to be an especial target. While the aim of the exercise is clear, these letters cast little light on the obscure 'Prefatory statement' which he added as front matter for the second volume of *PM*. On 20 May 1911 he announced that 'Hurrah! At last all the stuff is off' to Cambridge University Press.

11.7 OLIVER STRACHEY TO RUSSELL, 4 JANUARY 1912

This letter came in the context of a discussion of the relationship between universals, particulars and relations at the time when Russell was launching his first epistemological programme after the completion of his role in preparing *PM* (§8.2.5). On an attached sheet Russell recorded that the letter 'influenced me considerably'. It is hand-written, from Agra, Egypt.

You see my notion is that the difference between Particulars & Universals may after all be simply that the former can only be subjects,

[4] At this point Russell added: '*117·107·108·24·31·54, *120·311 [In some, only the implication wh. holds anyhow is required.]', presumably drawing upon his concordance of cross-references for *PM*. The theorem states that the homogeneous cardinal of class α equals the cardinal of class β if and only if their cardinals are equal.

the latter either subjects or predicates. Admitting external relations we have to agree that in 'A loves B' nothing is predicated of A or B; but if my analysis is right the proposition may be resolved into a predication about the relation.

It seems to me that if you admit particular & universal relations, it would dispose of the only difficulty you appear to raise against all propositions being reducible to subject-predicate affairs—viz that it involves internal relations.

Though this is perhaps no argument, it would be a considerable simplification surely. A complex such as 'this before that' would then consist only of particulars; and could we not then say that a thing 'is' the sum of its predicates, as long as it is clearly understood that 'is' is the mark of predication; so that a thing does not consist of its predicates at all, but only is them; it consists of the sum of the particulars in it.

There may be difficulties about universals used as subjects; I spoke last week of Relations relating universals not being particulars, but this seems doubtful. When we say Honesty is better than Success, I suppose it means 'If A is honest and not successful and B is successful and not honest, then A is better than B'— a function in which the relation is certainly (in any case where the value of A & B is given) a particular relation in my sense. Thus there is really no relation of 'better' between the universals Honesty & Success. In cases where there really is a relation between two universals—such as 'Universals are different from Particulars' I think the relating relation is also a particular. So that I see no reason why all relating relations should not be taken to be particulars.

I have so far been able to see no objection to particular relations, though you seem always to assume their universality. It occurs to me that possibly your previous views about the difference between 'being' & 'existence' may have led you into putting down all relations as universals. Yet by inspection it seems to me there is every bit as much in favour of particular relations as of particular qualities. Moreover the 'in-ness' that relates you to your room no doubt is the same as that relating me to my room; but then 'in-ness' is clearly a universal. Surely the actual relation 'in' relating you to your room is different from the in between me and my room?

11.8 QUINE AND RUSSELL, JUNE–JULY 1935

These letters launched the correspondence between two philosophers, one famous at the time and the other to become so pretty soon. Quine had sent to Russell a copy of his first book, *A system of logistic* (*1934b*: §9.4.4). Obviously pleased with it, Russell discusses Quine's way of handling definite descriptions; the status and specification of classes, especially as related to the proposition 'α, x' (Quine's way of saying that x was a member of class α); and the axiom of reducibility in *PM*. In a postscript he raises two specific issues about notations, including '$\hat{p}p$', the relation of all true propositions p, or α, x (p. 168). Quine's long reply deals with these issues, especially his way of using Russell's theory of descriptions but without the notation, and his argument for dropping the axiom from a logico-mathematical system akin to *PM*.

Russell's letter appeared in his autobiography (*1968a*, 213–214), but with minor errors of transcription and no explanation of notations; I have included it here in order to set the scene for Quine's reply. Quine alluded to the exchange in his own autobiography (*1985a*, 84); it beautifully captures the transition between classical and modern mathematical logic at the time.

11.8.1 *Russell to Quine, 6 June 1935.* The letter is hand-written, from Petersfield. It belongs to Professor Quine, with a copy in RA.

Your book arrived at a moment when I was overworked & obliged to take a long holiday. The result is that I have only just finished reading it.

I think you have done a beautiful piece of work; it is a long time since I have had as much intellectual pleasure as in reading you.

Two questions occurred to me, as to which I should be glad to have answers when you have time. I have put them on a separate sheet.

In reading you I was struck by the fact that, in my work, I was always being influenced by extraneous philosophical considerations. Take e.g. descriptions. I was interested in "Scott is the author of Waverley", & not only in the descriptive functions of PM. If you look up Meinong's work, you will see the sort of fallacies I wanted to avoid; the same applies to the ontological argument.

Take again notation (mainly Whitehead's): we had to provide for the correlators in Parts III & IV. Your α_β for our R | S would not do for three or more relations, or for various forms (such as R‖S) we needed.[5]

I am worried—though as yet I cannot put my worry into words—as to whether you really have avoided the troubles for which the axiom of reducibility was introduced as completely as you think. I should like to see Induction & Dedekindian continuity explicitly treated by your methods.

---

[5] In *PM* '$R \mid S$' denotes the compound of two relations R and S ($*34 \cdot 01$) and '$R \parallel S$' that holding between any relation Q and $R \mid Q \mid S$ ($*43 \cdot 112$, using the value '$\mid$' for the operator "variable" '$\dagger$' (785.15) in $*38 \cdot 11$). Quine introduced his symbols 'α_β' for relations α and β in *1934b*, 157.

I am a little puzzled as to the status of classes in your system. They appear as a primitive idea, but the connection of "α" with "$\hat{x}(\phi x)$" seems somewhat vague. Do you maintain that, if $\alpha = \hat{x}(\phi x)$, the prop. "α, x" is identical with "ϕx"? You must, if you are to say that all props are sequences. Yet it seems obvious that "I gave sixpence to my son" is not the <u>same</u> as "my son is one of the people to whom I gave sixpence".

And do you maintain that an infinite class can be defined otherwise than by a defining function? The need of including infinite classes was one of my reasons for emphasising functions as opposed to classes in PM.

I expect you have good answers to these questions.

In any case, I have the highest admiration for what you have done, which has reformed many matters as to which I had always been uncomfortable.

p. 16.   If an ordered couple is taken as a primitive idea, we have the difficulty (I think urged by Sheffer) that in distinguishing x, y from y, x we are surreptitiously appealing to spatial intuition. This is urged as an objection to PM's distinction of "xRy" & "yRx", &, if valid, applies equally to you.

p. 168.   Props are sequences: is (x, y) of the same type as (x, y, z)? You seem usually to imply that it is not. But, if so, your p in $\hat{p}p$ is limited to one sort of prop. Moreover your DF D1 implies that p & q are of the same type. If, therefore, there are props of different types (apart from bound prop-variables) you will have a great complexity of which nothing appears in your Chap. VII.

       This may be a mere misunderstanding on my part but on the face of it it looks awkward.

11.8.2 *Quine to Russell, 4 July 1935.*   The letter (RA) is hand-written, from Cambridge, Massachusetts.

I am happy that my book interested you to the extent that your comments and questions indicate, and am extremely pleased to learn your reactions.

Logical analysis of the singular definite article is a matter of first importance, and I recognize that it must be so carried out as to cover not only purely logical instances such as "s'Λ",[6] but instances in general, e.g. "the author of Waverley". This has been successfully accomplished, I am convinced, by your theory of descriptions. In abandoning the description notation in my book I do not abandon your theory of the

---

[6] This symbol denotes the union of the empty class (for some given type), which is 'the' empty class itself (*PM*, ∗40·21).

singular "the"; what I do abandon is the use, within the symbolic language, of a special notation imitative of the "the"-locution of common speech, and this I abandon in favor of one or another paraphrase equivalent to your definiens of that imitative notation. My reasons for this abandonment are, I believe, clear; besides simulating a form which it does not logically possess, the "the"-locution and its descriptive parallel "$(\imath x)(\phi x).\psi(\imath x)(\phi x)$" entail a complex technique, and are avoidable, in their logico-mathematical applications at least, without prolixity, as I have shown. But my departure is confined to the notational level; theory is unaffected. Asked how "Scott is the author of Waverley" could be translated into my system given "Scott", "author of" and "Waverley" as primitive, I should reply in effect with your theory of descriptions, merely withholding your definienda in favor of their definientia, and imposing such further uniform modifications as the divergences between my system and PM demand.

Despite my reason (pp. 158–159) for adopting the subscript notation for relative multiplication, I have come to regard its adoption as unfortunate. As your remarks illustrate, it is useful in binary operations to keep the two operand signs on the line, with an operation sign uniformly interposed (or, as with the Poles, uniformly prefixed[7]). Still Sheffer's use of the vertical stroke, though antedated by use in PM for relative multiplication, has become so familiar as to make some new notation desirable for the latter purpose.

As to the troubles for which the axiom of reducibility was introduced in PM, I am convinced that I avoid them completely, and that they are indeed gratuitous complications for PM itself. This conviction is explained in the three typewritten pages which I am enclosing. These pages are from a MS of the summer of 1932, an earlier stage of my book; the same argument appears in my Ph.D. thesis The Logic of Sequences (1932; Harvard Library).[8] In these pages I say "function" instead of "class", but the reference is the same. The argument may be framed as a refutation of the statement on p. xxxix of PM that if we assume that "all functions of functions are extensional," so that "there is no longer any reason to distinguish between functions and classes," still we "have to distinguish classes of different orders composed of members of the same order." So directed, my argument takes the following form.

Case I. The axiom of reducibility is in fact false. This case may be passed over, as it is equally unfortunate for everybody.

---

[7] This is the notational system introduced by Łukasiewicz after a suggestion by Chwistek (§8.8.2), and now known as the 'Polish notation'.

[8] Attached to Quine's letter are fols. 36–38 comprising the opening of 'Chapter 2. Signification and type', where such arguments are rehearsed; they were repeated, partly verbatim, in the thesis (m1932a, 5–8). He will repeat his position in a paper 1936b in Mind.

<u>Case II.</u> The axiom of reducibility is in fact true. Then, for every non-predicative function there is a predicative function of the same extension; and only one, by the principle of extensionality assumed above. Let us then always reinterpret the symbolism of a non-predicative function as denoting, instead, the correlated predicative function. Non-predicative functions thereby vanish from the subject-matter of the system, become in effect non-existent, and the complications underlying the axiom of reducibility drop out of the system, carrying the axiom with them. The system is the same as if we had ignored the distinctions of order at the start.

Note that the above argument is independent of Ramsey's argument to the same end (<u>Foundations of Mathematics</u> {*Essays*}, pp. 24–49). Perhaps the above argument leaves a doubt; I should like to have your criticism of it.

You ask whether I identify the propositions α, x and if ϕx if $\alpha = \hat{x}(\phi x)$. I cannot answer this directly, since "ϕx" does not occur in the language of my system, and the translation of it into that language is in danger of begging the question. It is true that where $\alpha = \hat{x}(\beta, x)$ I identify the propositions α, x and β, x. On the other hand where "$\cdots$" is any propositional expression containing "x", and $\alpha = \hat{x}(\cdots)$, I do not necessarily identify the propositions α, x and $\cdots$; e.g., though $\jmath y = \hat{x}(y = x)$, i.e., by D3, $\jmath y = \hat{x}([\ni y], \ni x)$, still $\jmath y, x$ and $[\ni y], \ni x$ are for me distinct sequences, therefore distinct propositions. At the same time, independently of the accidents of my system, I see no difficulty in identifying the respective propositions expressed by "I gave sixpence to my son" and "My son is one of the entities to which I gave sixpence". Obviously the two <u>sentences</u> are distinct; likewise the two sentences "α, x" and "β, x" are distinct symbols, as also are "2" and "1 + 1". What manner of symbolic diversity on the part of two sentences is to be regarded as precluding their synonymity, however, i.e. their denotation of one and the same entity ("proposition"), strikes me as within certain bounds an arbitrary matter. It is a question of what a <u>proposition</u>, as the <u>denotation</u> of a sentence, <u>is</u>; and for my part I should not be averse even to identifying propositions in this sense with truth-values and thus eventuating with but two, although that is far from the course adopted in my book. I enclose an offprint of a paper in <u>Mind</u> {*1934c*} in which I have expressed views on this matter.

I agree that an infinite class can be defined only by a defining function. Indeed, I should say the same of a finite class, for even simple enumeration embodies a defining function which happens to have the form of an alternation of identities. But I do not see how my system conflicts with these considerations. My device of "abstraction"—the $\hat{x}(\cdots)$" borrowed from PM—makes for the definition of a class in terms of a "defining function", i.e. in terms of a propositional expression

" ··· " containing the variable "x". Functions, in this sense of propositional expressions containing variables of which they are said to be functions, occur in the notation of my system just as in that of PM. What I do suppress are "functions" as elements of the system (rather than grammatical components of the language of the system), such as are denoted by the expressions "ϕ", "ψ", "$\phi\hat{x}$", "$\psi(\hat{x}, \hat{y})$", "$\psi(\hat{x}, y)$", "$\hat{x} = y$", etc. in PM; or it may be said that I do not even suppress functions in this sense, but identify them rather with the classes which they "define". Seen from this latter point of view, the only way in which the procedure differs from that of PM is in imposing upon "propositional functions" the principle of extensionality.

As to your comment regarding p. 16, I disagree with Sheffer's claim that the spatial order of symbols smuggles a primitive idea of "order" into the "xRy" and "yRx" of PM, and likewise I hold that in distinguishing x, y from y, x I have recourse to no such concealed primitive. In xRy we have the result of a ternary operation; in x, y we have the result of a primitive binary operation; and I regard the order of application of any plurary operation to its operands as integral to the operation itself. Consistently with Sheffer's point of view, on the other hand, it would be incumbent upon us always to disregard typographical order, and hence to abandon all commutative laws as trivial identities; and it is difficult to see how Sheffer's primitive idea of "order" could thereupon be added explicitly to restore the wanted ordinal determination, for any special symbolism for that purpose would have to be exempt from the above rule that typographical order be disregarded. Sheffer's stand is an example, to my mind, of what might be termed the "introspective fallacy"—the same fallacy, e.g., whereby students erroneously object to Sheffer's own stroke-function that it involves two primitives, "either-or" and "not", in view of the verbal explanation of the stroke as "either not —or not". In either case the fallacy depends upon venturing too far from the pole of the formalist, who asks, regarding primitives, only: "What notational devices are not introduced by abbreviative conventions in terms of previous notational devices?"

Regarding your comments on p. 168, I agree, first, that x, y is of different type from x, y, z; of course x, y may be of the same type as x, u, v, for y may itself be u, v. But I do not see that "p" is limited, in the context "$\hat{p}p$", to one sort of proposition; I should say rather that "p" here has complete typical ambiguity except for the fact that it has propositional position, i.e. must represent a proposition, and that correspondingly "$\hat{p}p$" has complete typical ambiguity except for the fact that it must represent a class of propositions. Again, I disagree when you say that the definition D1 implies that p and q are of the same type. It is true that D1 implies that $\hat{x}p$ and $\hat{y}q$ are classes of the same type; however, $\hat{x}p$ can be a class of any desired type independently of the type of p, and the analogous is true of $\hat{y}q$, as becomes clear upon translating "x̂p" and "ŷq" into the language of PM as $\hat{x}(x = x \, . \supset . \, p)$" and "$\hat{y}(y = y \, . \supset . \, q)$".

When you find time I hope you will let me know in what state this letter leaves the various problems. I should like to add that I am far from contented with the present state of mathematical logic, either as of PM or as of my book; and that the sore points for me are types and apparent variables. I am at work now on a theory dealing with these matters, and the first step, viz. a unified algebraic systematization of just so much of mathematical logic as does <u>not</u> depend upon types and apparent variables, is already at press. You shall have a copy when it appears;[9] meanwhile, I enclose a somewhat obsolete abstract of the paper.

11.9   RUSSELL TO HENKIN, 1 APRIL 1963

Henkin published in the American journal *Science* an article *1962a* on logicism and its history from *PM* onwards. He sent an offprint to Russell, who replied with some nice reflections, in his 91st year, of his time on *PM* with Whitehead. The letter, which is typed at Penrhyndeudrath, Wales, belongs to Professor Henkin, with a copy at RA.

Thank you very much for your letter of March 26 and for the very interesting paper which you enclosed. I have read the latter carefully and it has given me much new information. It is fifty years since I worked seriously at mathematical logic and almost the only work that I have read since that date is Gödel's. I realized, of course, that Gödel's work is of fundamental importance, but I was puzzled by it. It made me glad that I was no longer working at mathematical logic. If a given set of axioms leads to a contradiction, it is clear that at least one of the axioms must be false. Does this apply to school-boys' arithmetic, and, if so, can we believe anything that we are taught in youth? Are we to think that 2 + 2 is not 4, but 4.001? Obviously, this is not what is intended.

I should like to make a few general remarks about my state of mind while Whitehead and I were doing the <u>Principia</u>. What I was attempting to prove was, not the truth of the propositions demonstrated, but their deducibility from the axioms. And, apart from proofs, what struck us as important was the definitions.

You note that we were indifferent to attempts to prove that our axioms could not lead to contradictions. In this, Gödel showed that we had been mistaken. But I thought that it must be impossible to prove that any given set of axioms does <u>not</u> lead to a contradiction, and, for

---

[9] The paper is Quine *1936a*; no abstract seems to have been published, so presumably Quine sent a manuscript version.

that reason, I had payed little attention to Hilbert's work. Moreover, with the exception of the axiom of reducibility which I always regarded as a makeshift, our other axioms all seemed to me luminously self-evident. I did not see how anybody could deny, for instance, that q implies p or q, or that p or q implies q or p.

Both Whitehead and I were disappointed that the Principia was almost wholly considered in connection with the question whether mathematics is logic. In the later portions of the book, especially, there are large parts consisting of what would be called ordinary mathematics if the subject matter had been more familiar. This applies especially to relation-arithmetic. If there is any mistake in this, apart from trivial errors, it must also be a mistake in conventional ordinal arithmetic, which seems hardly credible.

More generally, Aristotelian logic is almost exclusively concerned with propositional functions having only one variable. The philosophies of Spinoza, Leibniz and Hegel are entirely dependent on this limitation. We wanted to construct a logic of functions containing two variables— i.e. logic of dyadic relations. The fourth volume, which was never completed, was to deal with triadic and tetradic relations. Whitehead defined a space as the field of a triadic or tetradic relation, and was going to work our geometry on that basis. All such work still seems to me to stand on the same level as the more familiar parts of mathematics.

I had fondly imagined that we were making the kind of advance that Descartes made in geometry by the use of co-ordinates.[10] Unfortunately, while algebra was familiar, what was needed for the logic of relations was unfamiliar, and therefore the advantages of our symbolism were not evident to readers.

If you can spare the time, I should like to know, roughly, how, in your opinion, ordinary mathematics—or, indeed, any deductive system —is affected by Gödel's work.

Russell's appraisal that the logicism of *PM* made far more impact than the logic is refuted by this book, which shows quite the opposite. Further, once again he misunderstood Gödel's theorem; Henkin gave the correct version in his reply of 17 July 1963 (RA), which was mainly concerned with current developments in logic. But Russell was still struggling with the theorem at the end of his life when he wrote an addendum to his replies for a new edition of the Schilpp volume (*1971a*, xviii–xix). His continuing difficulties encapsulate a principal theme of this book in which he has been the central figure.

[10] Russell's history was shaky here. Descartes invented analytic geometry, with no privileged directions such as furnished by coordinate axes; this extension soon came from successors such as Leibniz.

BIBLIOGRAPHY

List of archives

We begin with the locations of the archives where major collections are held, usually the surviving *Nachlass* of a figure; unless otherwise indicated in brackets, a collection is held in the University library or archive of the institution named. Collections are not noted for figures whose roles in the story have been marginal (for example, Bolzano and Lagrange).

Europe

Cambridge (England): Hawtrey (Churchill College); G. E. Moore, Turnbull (University); Newman (St. John's College); Venn (Gonville and Caius College); University Press records.

Chichester (England), West Sussex Record Office: Kempe.

Djursholm (Sweden), Institut Mittag-Leffler: Jourdain, Mittag-Leffler.

Englefield Green (England), Royal Holloway College: Stebbing.

Erlangen (Germany), University: Behmann.

Freiburg (Germany), University: Zermelo.

Göttingen (Germany), University: Cantor, Dedekind, Hilbert (also at the Mathematics Institute), Hurwitz, Klein.

Haarlem (The Netherlands), State Archives of North Holland: Schlick, Neurath, Vienna Circle files.

Jerusalem (Israel), The Jewish National and University Library: Fraenkel.

Konstanz (Germany), University, Philosophical Archive: microfilms of many collections, especially for members of the Vienna Circle.

Lausanne (Switzerland), University: Gonseth.

London (England): Boole, Society archives (Royal Society); College archives (University College).

Manchester (England), Manchester University: Jevons.

Münster (Germany), University: Scholz.

Vienna (Austria), Vienna University: University files.

Zurich (Switzerland), Technical High School: Bernays, Weyl.

North America

Bloomington (Indiana), University of Indiana: Bentley, Pearsall Smith (family of Russell's first wife).

Cambridge (Massachusetts): B. and C. S. Peirce, Royce, Sheffer (Harvard University); Wiener (MIT).

Carbondale (Illinois), Southern Illinois University: Dewey, Open Court Publishing Company.

Chicago (Illinois), University of Chicago: E. H. Moore, University Press records.

Hamilton (Ontario), McMaster University: Ogden, Russell.

Indianapolis (Indiana), University of Indiana, Peirce Project: C. S. Peirce (copies), Morris.

New York (New York), Columbia University: Keyser, Ladd-Franklin.

Northampton (Massachusetts), Smith College: Wrinch.

Pittsburgh (Pennsylvania), University of Pittsburgh: Carnap, Ramsey, Reichenbach.
Princeton (New Jersey), University: Gödel.
Stanford (California), University: Lewis.
Washington (D.C.), Library of Congress: Klyce, Veblen, von Neumann.
Waterloo (Ontario), Wilfred Laurier University: Kaufmann.

ORGANISATION

The entry for each author named above begins with 'Papers:', followed by the location of the materials and sometimes a comment on their use here. For every author his entry may start with items given a catchword; if so, then they begin with an edition of his '*Works*' or '*Writings*', say, and may continue with editions of correspondence or translations or special selections of writings. The entry is completed by the items given dating codes: those for manuscripts left unpublished are prefaced by '*m*', and for the authors named above the original is in the collection specified under 'Papers' unless otherwise indicated.

The ordering within a year is determined as far as possible by the chronology of the first appearances of publications and the completion or abandonment of manuscripts; but it is coarse, since often no accurate information on dating is available. When the year assigned to an article is that, or one of those, for the volume of the journal concerned, it is not repeated in the reference details. However, otherwise the nominal year *is* recorded; always so with the *Jahrbuch*, which usually covered a year in a volume which was necessarily published some years later. A few items relating to an author but lacking an author or editor are listed within his entry, with the code suffixed by a prime (for example, 'Couturat *1983a*').

The entry for an item includes where appropriate the main (photo-)reprints and translations (usually into English); if just a date is given, then the original publisher so acted. Reprints cited in the text have their own codes. No attempt has been made to record *all* reprints or translations, or even all British and American publishers of a book; the full list for Russell alone would be enormous. In the past more than today publishers would call a reprint some *n*th 'edition', presumably as an attack on customers' wallets.

Any information or comments not properly part of the reference details are enclosed within square brackets. They include cross-references between an item and its review, and the location in a library of a copy of a rare item.

ABBREVIATIONS

For brevity, the following symbols are used:

T	translation *or* translated by (maybe with editorial apparatus added)
E	English T
G	German T
F	French T
I	Italian T
‡	this printing cited *by page number* in the text
n.s.	*new series*

R   review (of book or paper following)
≈   original reprinted
=   original photoreprinted
n   nth edition of item

To save more space, titles are often shortened, the usual abbreviations of words in names of journals are deployed ('*J.*' for 'Journal', '*math.*' for 'mathematics' or mathématiques', and so on), and the following acronyms are used for certain organisations, publications, publishers and journals:

AMS   American Mathematical Society
DMV   *Deutsche Mathematiker-Vereinigung*
LMS   London Mathematical Society

A & U   Allen and Unwin
GV   Gauthier-Villars
PU(F)   Presses Universitaires (de France)
RA   Russell Archives
(R)KP   (Routledge and) Kegan Paul [or related names]
UP   University Press

C & G   Cassinet and Guillemot *1983a 2*
EMW   *Encyklopädie der mathematischen Wissenschaften*, Leipzig (Teubner)
PM   *Principia mathematica*

AHES   *Archive for history of exact sciences*
AM   *Acta mathematica*
HM   *Historia mathematica*
HPL   *History and philosophy of logic*
JDMV   *Jahresbericht der Deutschen Mathematiker-Vereinigung*
JFM   *Jahrbuch über die Fortschritte der Mathematik*
JSL   *Journal of symbolic logic*
JP   *Journal of philosophy*, psychology and scientific methods
MA   *Mathematische Annalen*
PR   *Philosophical review*
RdM   *Ri(e)vista di matematica* or *Revue de mathématiques*, edited by Peano
RMM   *Revue de métaphysique et de morale*
ZfM   *Zentralblatt für Mathematik*

A few further acronyms are introduced under individual entries.

ITEMS

Aarsleff, H.
  *1982a. From Locke to Saussure.
  Essays on the study of language
  and intellectual history*, London
  (Athlone).
Abel, N. H. (1802–1829)
  1826a. "Untersuchungen über die
  Reihe [...]", *J. rei. ang. Math. 1*,
  311–329. [F: *Oeuvres complètes 1*,

  1881, Christiania (Göndahl) =
  1965, New York (Johnson), 219–
  250.]
Abir-Am, P.
  1987a. "Synergy or clash [...] the
  career of mathematical biologist
  Dorothy Wrinch", in Abir-Am and
  D. Outram (eds.), *Uneasy careers
  and intimate lives*, New Brunswick

and London (Rutgers UP), 239–280, 342–354.

Ackermann, W. (1896–1962) *See also* Hilbert and Ackermann

1938a. R of Burkamp *1932a*, *JFM 58* (1932), 54–55.

1938b. R of Quine *1934b*, *JFM 60* (1934), 845–846.

1938c. R of Whitehead *1934a*, *JFM 60* (1934), 848.

Adamson, R.

1878a. R of Schröder *1877a*, *Mind 23*, 252–255.

Albury, W. R.

1986a. "The order of ideas: Condillac's method of analysis as a political instrument in the French Revolution", in J. A. Schuster and R. R. Yeo (eds.), *The politics and rhetoric of scientific method*, Dordrecht (Reidel), 203–225.

Alexandrov, P.

1971a. (Ed.) *Die Hilbertschen Probleme*, Leipzig (Geest und Portig).

Amoroso Costa, M. (1885–1928)

1929a. *As idéas fundamentaes da mathematica*, Rio de Janeiro (Pimenta de Mello) ≈ (Ed. M. Reale), *As idéias fundamentais da matemática a outros ensaios*, 1971, Sao Paolo (Grijalbo), 171–326‡.

Anderson, D. E. and Cleaver, F. L.

1965a. "Venn-type diagrams for arguments of n terms", *JSL 30*, 113–118.

Andrews, G.

1979a. (Ed.), *The Bertrand Russell memorial volume*, London (A & U).

Anellis, I.

1987a. "The heritage of S. A. Janovskaja", *HPL 8*, 45–56.

Anellis, I. and Houser, N.

1991a. "Nineteenth century-roots of algebraic logic and universal algebra", in H. Andréka and others (eds.), *Algebraic logic*, Amsterdam (North-Holland), 1–36.

Angelelli, I.

1967a. *Studies on Gottlob Frege and traditional philosophy*, Dordrecht (Reidel).

Aspray, W.

1991a. "Oswald Veblen and the origins of mathematical logic at Princeton", in Drucker *1991a*, 54–70.

Auroux, S.

1973a. *L'encyclopédie "grammaire" et "langue" au XVIIIe siècle*, [Paris] (Mame).

1982a. *L'illuminismo francese e la tradizione logica di Port-Royal*, Bologna (CLUEB).

Ayer, A. J. (1910–1989)

1935a. *Language, truth and logic*, London (Gollancz). [Various later eds.]

1937a. R of Scholz and Schweitzer *1935a*, *Mind n.s. 46*, 244–247.

1977a. *Part of my life*, London (Collins).

Babbage, C. (1792–1871)

1820a. *Examples of the solutions of functional equations*, Cambridge ([no publisher]). [F: Gergonne *1821a*.]

1827a. "On the influence of signs in mathematical reasoning", *Trans. Cambridge Phil. Soc. 2*, 325–378 ≈ *Works 1*, 1989, London (Pickering), 371–408.

Bachmann, F. (1909–1982) *See also* Scholz and Bachmann

1934a. *Untersuchungen zur Grundlegung der Arithmetik, mit besondere Beziehung auf Dedekind, Frege und Russell*, Leipzig (Meiner).

Baire, R. L. (1874–1932) *See also* Letters; Schönflies and Baire

Works. (ed. P. Lelong), *Oeuvres scientifiques*, Paris (Bordas).

1899a. "Sur les fonctions de variables réelles", *Ann. mat. pura appl. (3)3*, 1–123 = Milan (Bernardoni) = *Works*, 49–170.

1905a. Leçons sur les fonctions discontinues, Paris (GV) = *Works*, 195–327.

Ballue, L. E. (1863–1938)
1894a. "Le nombre entier considéré comme fondement de l'analyse mathématique", *RMM 2*, 317–328.

Bandmann, H.
1992a. Die Unendlichkeit des Seins. Cantors transfinite Mengenlehre und ihre metaphysischen Wurzeln, Frankfurt/Main (Lang).

Barrau, J. A. (1873–1946)
1910a. R of Brouwer *1907a*, *JFM 38* (1907), 81–84.

Becher, W. H.
1980a. "William Whewell and Cambridge mathematics", *Hist. stud. phys. sci. 11*, 3–48.

Becker, O. (1881–1964)
1927a. "Mathematische Existenz", *Jbch. Phil. phänom. Forschung 8*, 441–809‡ = Halle (Niemeyer).

Behmann, H. (1891–1970)
Papers: Philosophical Institute, Erlangen (Germany), University (see Haas *1981a*).
1927a. Mathematik und Logik, Leipzig and Berlin (Teubner). [R: Bennett *1930a*.]
1931a. "Zu den Widersprüchen der Logik und der Mengenlehre", *JDMV 40*, 37–48. [R: Schmidt *1931a*.]
1932a. "Zur Richtigstellung einer Kritik meiner Auflösung der logisch-mengetheoretischen Widersprüchen", *Erkenntnis 2*, 305–306.
1934a. "Sind die mathematischen Urteile analytisch oder synthetisch?", *Erkenntnis 4*, 1–27.
1937a. "The paradoxes of logic", *Mind n.s. 46*, 218–221.

Behrens, G. J. A. A. C. (b. 1892)
1918a. Die Prinzipien der mathematischen Logik bei Schröder, Russell und König, Hamburg (Berngruber & Henning). [Rare: Kiel University, *Dissertation* there.]

Beisswanger, P.
1966a. "Hermann Weyl and mathematical texts", *Ratio 8*, 25–45.

Belnap, N. D. Jr. and Massey, G. J.
1990a. "Semantic holism", *Studia logica 49*, 67–82.

Benacerraf, P. and Putnam, H.
1985a. (Eds.) *Philosophy of mathematics. Selected readings₂*, Oxford (Clarendon Press).

Bendixson, I. (1861–1935)
1883a. "Quelques théorèmes de la théorie des ensembles de points", *AM 2*, 415–429.

Benis-Sinaceur, H.
1987a. "Structure et concept dans l'épistemologie mathématique de Jean Cavaillès", *Rev. d'hist. sci. 40*, 5–30 [with letters to A. Lautmann on pp. 117–129].

Benjamin, A. C.
1927a. R of Burkamp *1927a*, *JP 24*, 385–387.

Bennett, A. A.
1930a. R of Behmann *1927a*, *Bull. AMS 36*, 615.

Bentham, G. (1800–1884)
1823a. Essai sur la nomenclature [. . .] *Ouvrage extrait du* Chrestomathia *de Jérémie Bentham*, Paris (Bossange).
1827a. Outline of a new system of logic, with a critical examination of Dr. Whately's "Elements of logic", London (Hunt and Clarke) = 1990, Bristol (Thoemmes). [R: Hamilton *1833a*.]

Bentley, A. F. (1870–1957)
Papers: Bloomington (Indiana), University.
1931a. "The linguistic structure of mathematical consistency", *Psyche 12*, no. 3, 78–91.
1932a. Linguistic analysis of mathematics, Bloomington (Principia Press).

Berlin, I. (1909–1997)
1978a. (Ed. H. Hardy), *Concepts and categories*, London (Hogarth).

Bernays, P. (1888–1977) *See also* Hilbert and Bernays
Papers: Zurich (Switzerland), Technical High School.
m1918a. "Beiträge zur axiomatischen Behandlung des Logik-Kalküls", *Dissertation*, Göttingen University. [Copies at University Library; and Bernays Papers, File 973:192.]
1922a. R of Hilbert *1918a*, *JFM 46* (1916–1918), 62–64.
1922b. R of Ziehen *1917a*, *JFM 46* (1916–1918), 65.
1926a. "Axiomatsche Untersuchungen des Aussagenkalkuls der „Principia Mathematica‴", *Math. Ztsch. 25*, 305–320.
1975a. R of Schröder *1966a*, *JSL 40*, 609–614.

Bernkopf, M.
1968a. "A history of infinite matrices", *AHES 4*, 308–358.

Bernstein, B. A. (1881–1964)
1926a. R of *PM₂*, *Bull. AMS 32*, 711–713.
1931a. "Whitehead and Russell's theory of deduction as a mathematical science", *Bull. AMS 37*, 480–488.

Bernstein, F. (1878–1956)
1901a. Untersuchungen aus der Mengenlehre, Göttingen. [*Dissertation*.]
1905a. "Untersuchungen aus der Mengenlehre", *MA 61*, 117–155. [Revision of *1901a*.]
1912a. "Über eine Anwendung der Mengenlehre auf ein aus der Theorie der säkularen Störungen herrührendes Problem", *MA 71*, 417–439.

Bettazzi, R. (1861–1941)
1890a. Teoria delle grandezze, Pisa (Spoeni) = *Ann. univ. Toscana 19* (1893), pt. 2, 1–181. [R: Vivanti *1891a*.]
1895a. "Gruppi finiti ed infiniti di enti", *Atti Accad. Torino 31*, 362–368.

Biermann, K.-R.
1969a. "Did Husserl take his doctor's degree under Weierstrass's supervision?", *Organon 9*, 261–264.

Black, M. (1909–1988)
1931a. "Note on Mr. Bentley's alleged refutation of Cantor", *Psyche 12*, no. 4, 77–79.
1933a. The nature of mathematics. A critical survey, London (KP). [R: Wisdom *1934a*.]
1935a. R of Quine *1934b*, *Mind n.s. 44*, 524–526.

Blackwell, K. M.
1973a. "Our knowledge of 'Our knowledge'", *Russell no. 12*, 11–13.
1985a. "Part 1 of *The principles of mathematics*", *Russell n.s. 4*, 271–288.

Blackwell, K. M. and Ruja, H.
1994a. A bibliography of Bertrand Russell, 3 vols., London (Routledge).

Blakey, R.
1847a. An essay on logic₂, London (Saunders).
1851a. Historical sketch of logic, from the earliest times to the present day, Edinburgh (Nichol).

Blumberg, H.
1920a. R of Hausdorff *1914a*, *Bull. AMS (2)27*, 116–129.

Blumenthal, O. (1876–1944)
1935a. "Lebensgeschichte", in Hilbert *Papers 3*, 388–435.

Bôcher, M. (1867–1918)
1904a. "The fundamental concepts and methods of mathematics", *Bull. AMS 11*, 115–135.

Bollinger, M.
1972a. "Geschichtliche Entwicklung des Homologiebegriffs", *AHES 9*, 94–166.

Bolzano, B. P. J. N. (1781–1848)
Works. Gesamtausgabe (in progress in three series), 1969–, Stuttgart (Frommann-Holzboog).
Writings. Gesammelte Schriften, 12 vols., 1882, Vienna (Braunmüller).
Mathematics. Early mathematical works, 1981, Prague (Academy of

Science). [Photoreprints, including of *1817a*.]

1817a. *Rein analytischer Beweis*, Prague (Haase) = *Abh. Gesell. Wiss. Prague* (*3*)5 (1814–1817, publ. 1818), no. 5. [Other issues and ≈ s; various Ts.]

1837a. *Wissenschaftslehre*, 4 vols., Sulzbach (Seidel) ≈ *Writings 7–10* ≈ *Works* (*1*)*11–14* [in progress]. [Also other issues and partial Ts.]

1851a. (Ed. F. Přihonský), *Paradoxien des Unendlichen*, Leipzig (Reclam). [Various eds. and Ts. E: *Paradoxes of the infinite*, 1950, London (RKP).]

Boole, G. (1815–1864)

Papers: London (England), Royal Society, ms. 782; some material at Cork (Ireland), University.

Studies. (Ed. R. Rhees), *Studies in logic and probability*, 1952, London and La Salle, Ill. (Open Court).

Manuscripts. (Eds. I. Grattan-Guinness and G. Bornet), *Selected manuscripts on logic and its philosophy*, 1997, Basel (Birkhäuser).

1844a. "On a general method in analysis", *Phil. trans. Royal Soc. London 134*, 225–282.

1847a. *The mathematical analysis of logic*, Cambridge (Macmillan) and London (Bell)‡ = 1948, Oxford (Blackwell) ≈ *Studies*, 49–124 ≈ Ewald *1996a*, 451–509.

1848a. "The calculus of logic", *Cambridge Dublin math. j. 3*, 183–198 ≈ *Studies*, 125–140.

1851a. "On the theory of probabilities", *Phil. mag.* (*4*)*1*, 521–530 ≈ *Studies*, 247–259‡.

1854a. *An investigation of the laws of thought*, Cambridge (MacMillan) and London (Walton and Maberly) = 1958, New York (Dover)‡ ≈ 1916, La Salle, Ill. (Open Court). [R: Ulrici *1855a*.]

1859a. *A treatise on differential equations*$_1$, Cambridge (MacMillan).

1860a. *A treatise on the calculus of finite differences*$_1$, Cambridge and London (MacMillan).

Boole–De Morgan

Letters. (Ed. G. C. Smith), *The Boole–De Morgan correspondence*, 1982, Oxford (Clarendon Press). [R: Corcoran *1986a*.]

Boole, M. E. (1832–1916)

1890a. *Logic taught by love*, London [private] ≈ *Collected works 1*, 1931, London (Daniel), 399–515‡.

Boolos, G.

1994a. "The advantages of honest toil over theft", in George *1994a*, 27–44.

Boos, W.

1985a. "'The true' in Gottlob Frege's 'Über die Grundlagen der Geometrie'", *AHES 34*, 141–192.

Borel, E. F. E. J. (1871–1956) *See also* Letters

Works. *Oeuvres*, 4 vols., 1972, Paris (GV). [Consecutively paginated.]

1894a. "Sur quelques points de la théorie des fonctions", *Ann. sci. Ecole Normale Sup.* (*3*)*12*, 9–55 ≈ *Selecta*, 1940, Paris (GV), 3–48 = *Works*, 239–286.

1898a. *Leçons sur la théorie des fonctions*$_1$, Paris (GV).

1907a. "La logique et l'intuition en mathématiques", *RMM 15*, 273–283 = *Works*, 2081–2091.

1928a. *Leçons sur la théorie des fonctions*$_3$, Paris (GV).

Borga, M.

1985a. "La logica, il metodo assiomatico e la problematica metateorica", in Borga and others *1985a*, 11–75.

Borga, M., Freguglia, P and Palladino, D.

1985a. *I contributi fondazionale della scuola di Peano*, Milan (Franco Angeli).

Borga, M. and Palladino, D.

1997a. *Oltre il mito della crisi*, Brescia (La Scuola).

Bostock, D.
1974a, 1979a. Logic and arithmetic, 2 vols., Oxford (Clarendon Press). [R: Resnik *1982a*.]

Boswell, T.
1988a. "On the textual authenticity of Kant's *Logic*", *HPL 9*, 193–203.

Bottazzini, U.
1985a. "Dall'analisi matematica al calcolo algebraico: origini delle prime ricerche di logica di Peano", *HPL 6*, 25–52.
1986a. The higher calculus, New York (Springer).
1991a. "Angelo Genocchi e i principi del calcolo", in A. Conte and L. Giacardi (eds.), *Angelo Genocchi e i suoi interlocutori scientifici*, Turin (Palazzo Congrano), 32–60.

Bouligand, G. (1889–1979)
1928a. "Ensembles impropres et nombre dimensionnel", *Bull. sci. math.* (2)*52*, 320–344, 361–376.
1931a. "Les courants de pensée Cantorienne et l'hydrodynamique", *Rev. gén. sci. pures appl. 42*, 103–110.
1932a. "Sur quelques applications de la théorie des ensembles à la géométrie infinitésimale", *Bull. Acad. Polon. Sci. Lett.*, *classe sci. math. natur.* (*A*), 1–13.

Boutroux, P. (1880–1922)
1904a. "Sur la notion de correspondance dans l'analyse mathématique", *RMM 12*, 909–920.
1905a. "Correspondance mathématique et relation logique", *RMM 13*, 620–637.
1905b. "Sur la notion de correspondance dans l'analyse mathématique", in Claparède *1905a*, 713–719. [Not identical to *1904a*.]
1914a. "Congrès International de Philosophie Mathématique. Discours d'ouverture", *RMM 22*, 571–580.

Bozzi, S. *See* Mangione and Bozzi

Bradley, F. H. (1846–1924)
1883a. The principles of logic$_1$, Oxford (Clarendon Press). [$_2$ 2 vols., 1922‡.]
1893a. Appearance and reality$_1$, Oxford (Clarendon Press). [$_2$ 1897.]

Braithwaite, R. B. (1900–1990)
1934a. R of Peirce *Papers 1–4*, *Mind n.s. 43*, 487–511.
1962a. "Introduction", in Gödel *1962a'*, 1–32.

Brent, J.
1993a. Charles Sanders Peirce. A life, Bloomington (Indiana UP).

Brieskorn, E.
1996a. (Ed.) *Felix Hausdorff zum Gedächtnis*, Braunschweig (Vieweg).

Brink, C. H.
1978a. "On Peirce's notation for the logic of relatives", *Trans. C. S. Peirce Soc. 14*, 285–304.

Broad, C. D. (1887–1971)
1915a. R of Russell *1914c*, *Mind n.s. 24*, 250–254.

Brock, W. H.
1967a. (Ed.) *The atomic debates*, Leicester (Leicester UP).

Bronstein, D. J.
1936a. R of Quine *1934b*, *PR 45*, 416–418.

Brouwer, L. E. J. (1881–1966)
Works. Collected works, 2 vols. 1975–1976, Amsterdam (North-Holland).
1907a. Over de grondslagen der wiskunde, Amsterdam (Mass and van Suchtelen). [Part E: *Works 1*, 11–101‡. R: Barrau *1910a*.]
1910a. "On the structure of perfect sets of points", *Proc. Kon. Akad. Wetens. Amsterdam 12*, 785–794 = *Works 2*, 341–351.
1914a. R of Schönflies *1913a*, *JDMV 23*, 78–83 = *Works 1*, 139–144.
1918–1919a. "Begründung der Mengenlehre unabhängig vom logischen Satz von ausgeschlossenen Dritten", *Verh. Kon. Akad. Wetens. Amsterdam sect. 1 12*, no. 5 (43

pp.), no. 12 (33 pp.) = *Works 1*, 150–221.

1925a. "Intuitionistische Zerlegung mathematischer Grundbegriffe", *JDMV 33*, 251–256 = *Works 1*, 275–280. [E: Mancosu *1998a*, 290–292.]

1925–1927a. "Zur Begründung der intuitionistischen Mathematik", *MA 93*, 244–257; *95*, 453–472; *96*, 451–488 = *Works 1*, 301–389.

1929a. "Mathematik, Wissenschaft und Sprache", *Monats. Math. Physik 36*, 153–164 = *Works 1*, 417–428. [E: Ewald *1996a*, 1170–1185.]

1930a. *Die Struktur des Kontinuums*, Vienna (Gistel) = *Works 1*, 429–440. [E: Ewald *1996a*, 1186–1197 ≈ Mancosu *1998a*, 54–63.]

1930b. R of Fraenkel *1927b*, *JDMV 39*, pt. 2, 10–11 = *Works 1*, 441–442.

Brunschvicg, L. (1869–1944)

1911a. "La notion moderne d'intuition et la philosophie des mathématiques", *RMM 19*, 145–176.

1912a. *Les étapes de la philosophie mathématique*, Paris (Alcan).

Brush, S. G.

1967a. "Foundations of statistical mechanics 1845–1915", *AHES 4*, 145–183 ≈ *The kind of motion we call heat*, 2 vols., 1976, Amsterdam (North-Holland), ch. 5.

Buickerood, J. G.

1985a. "The natural history of the mind: Locke and the rise of facultative logic in the eighteenth century", *HPL 6*, 157–190.

Bunn, R.

1977a. "Quantitative relations between infinite sets", *Ann. of sci. 34*, 177–191.

Burali-Forti, C. (1861–1931)

1894a. "Sulle classi ordinate ed i numeri trasfiniti", *Rend. Circolo Mat. Palermo 8*, 169–179.

1894b. *Logicamatematica₁*, Milan (Hoepli).

1896a. "Le classi finite", *Atti Accad. Sci. Torino* (2)32, 34–52. [F: C & G, 1–28.]

1897a. "Una questione sui numeri trasfiniti", *Rend. Circolo Mat. Palermo 11*, 154–164. [E: van Heijenoort *1967a*, 104–111. F: C & G, 41–49.]

1897b. "Sulle classi ben ordinate", *Rend. Circolo Mat. Palermo 11*, 260. [E: van Heijenoort *1967a*, 111–112. F: C & G, 53.]

1899a. "Les propriétés formales des operations logiques", *RdM 6*, 141–177.

1899b. "Sur l'égalité, et l'introduction des éléments dérivés dans les sciences", *L'ens. math.* (1)1, 246–261.

1901a. "Sur les différentes méthodes logiques pour la définition du nombre réel", in Congress *1901a*, 289–307.

1903a. "Sulla teoria generale delle grandezze e dei numeri", *Atti Accad. Sci. Torino* (2)39, 192–208.

1909a. "Sulle definizioni mediante «coppie»", *Boll. di mat.* (1)8, 237–242.

1912a. "Gli enti astratti definiti come enti relativi ad un campo di nozione", *Rend. Accad. Lincei* (5)21, pt. 2, 677–682.

1913a. "Sur les lois générales de l'algorithme des symboles de fonction et d'opération", in Hobson and Love *1913a 1*, 480–491.

1919a. *Logica matematica₂*, Milan (Hoepli).

1921a. "Polemica logica-matematica", *Per di mat.* (4)1, 354–359 = *Notizie di logica 7* (1988), no. 4, 21–26.

Burali-Forti, C. and Ramorino, A.

1898a. *Elementi di aritmetica razionale*, Turin (Petrini).

Burkamp, W. (1879–1939)
1927a. *Begriff und Beziehung*, Hamburg (Meiner). [Rs: Benjamin *1927a*, Wright *1928a*.]
1929a. *Die Struktur der Ganzheiten*, Berlin (Junker und Dünnhaupt).
1932a. *Logik*, Berlin (Mittler). [R: Ackermann *1938a*.]
Burke, J. B. B. (b. 1873)
1931a. *The emergence of life*, Cambridge (Cambridge UP).
Burke, T.
1994a. *Dewey's new logic—a reply to Russell*, Chicago (University of Chicago Press).
Burkhardt, H. and Smith, B.
1991a. (Eds.) *Handbook of ontology and metaphysics*, 2 vols., Munich (Philosophia)
Burn, R. P.
1992a. "Irrational numbers in English language textbooks, 1890–1915", *HM 19*, 158–176.
Byrd, M.
1987a. "Part II of *The principles of mathematics*", *Russell n.s. 7*, 60–70.
1989a. "Russell, logicism, and the choice of the logical constants", *Notre Dame j. formal logic 30*, 343–361.
1994a. "Part V of *The principles of mathematics*", *Russell n.s. 14*, 47–86.
1996a. "Parts III–IV of *The principles of mathematics*", *Russell n.s. 16*, 145–168.
1999a. "Part VI of *The principles of mathematics*", *Russell n.s. 19*, 29–61.

Cajori, F. (1859–1930)
1929a. *A history of mathematical notations 2*, La Salle, Ill. (Open Court).
Cantor, G. F. L. P. (1845–1918)
Here "ÜP" abbreviates "Über unendliche, lineare Punktmannichfaltigkeiten".
Papers: Göttingen (Germany), University.

Papers. (Ed. E. Zermelo), *Gesammelte Abhandlungen mathematischen und philosophischen Inhalts*, 1932, Berlin (Springer) = 1980 = 1966, Hildesheim (Olms). [Russian ed.: (Ed. F. A. Medvedev), *Trudi po teorii mnodjestv*, 1985, Moscow (Nauka).]
Letters. (Eds. H. Meschkowski and W. Nilson), *Briefe*, 1991, Berlin (Springer).
1869a. "Über die einfachen Zahlensysteme", *Ztsch. math. Physik 14*, 121–128 ≈ *Papers*, 35–42.
1872a. "Über die Ausdehnung eines Satzes der Theorie der trigonometrischen Reihen", *MA 5*, 122–132 ≈ *Papers*, 92–102‡. [F: *1883d*, 336–348.]
1874a. "Über eine Eigenschaft des Inbegriffes aller reellen algebraischen Zahlen", *J. rei. ang. Math. 77*, 258–262 ≈ *Papers*, 115–118. [F: *1883d*, 305–310. E: Ewald *1996a*, 839–843.]
1878a. "Ein Beitrag zur Mannigfaltigkeitslehre", *J. rei. ang. Math. 84*, 242–258 ≈ *Papers*, 119–133. [F: *1883d*, 311–328.]
1879a. "ÜP", pt. 1, *MA 15*, 1–7 ≈ *Papers*, 139–145‡. [F: *1883d*, 349–356.]
1880a. "ÜP", pt. 2, *MA 17*, 355–358 ≈ *Papers*, 145–148‡. [F: *1883d*, 357–360.]
1882a. "Über ein neues und allgemeines Kondensationsprinzip der Singularitäten von Funktionen", *MA 19*, 588–594 ≈ *Papers*, 107–113.
1882b. "ÜP", pt. 3, *MA 20*, 113–121 ≈ *Papers*, 149–157‡. [F: *1883d*, 361–371.]
1883a. "ÜP", pt. 4, *MA 21*, 51–58 ≈ *Papers*, 157–164‡. [F: *1883d*, 372–380.]

1883b. "ÜP", pt. 5, *MA 21*, 545–591 ≈ *1883c* ≈ *Papers*, 165–209‡. [Revised F: *1883d*, 381–408.]

1883c. Grundlagen einer allgemeinen Mannigfaltigkeitslehre, Leipzig (Teubner). [R: Simon *1883a*. Es: *The campaigner 9* (1976), 69–96; Ewald *1996a*, 881–920.]

1883d. F of various papers, *AM 2*, 305–408.

1883e. "Sur divers théorèmes de la théorie des ensembles", *AM 2*, 409–414 ≈ *Papers*, 247–251.

1884a. "De la puissance des ensembles parfaits des points", *AM 4*, 381–392 ≈ *Papers*, 252–260.

1884b. "ÜP", pt. 6, *MA 23*, 453–488 ≈ *Papers*, 210–246.

m1885a. "Principien einer Theorie der Ordungstypen. Erste Mittheilung", Cantor Papers; in my *1970b*, 82–101.

1885b. "Über verschiedene Theoreme aus der Theorie der Punktmengen in einem n-fach ausgedehnten stetigen Raume G_n. Zweite Mittheilung", *AM 7*, 105–124 ≈ *Papers*, 261–277.

1885c. R of Frege *1884b*, *Dtsch. Lit.-Zeit. 6*, cols. 728–729 ≈ *Papers*, 440–442.

1886a. "Über die verschiedenen Standpunkte in Bezug auf das Actuale Unendliche", *Ztsch. Phil. phil. Kritik 66*, 224–233 ≈ *Natur und Offenbarung 32* (1886), 46–49, 226–233 (= 1886, Halle/Saale (Beyer)) ≈ *1890a*, 1–10 ≈ *Papers*, 370–377.

1886b. "Über verschiedenen Ansichten in Bezug auf die Actualunendlichen Zahlen", *Bihang Kongl. Svenska Vet.-Akad. Handlingar 11*, no. 19 (10 pp.).

1887–1888a. "Mitteilungen zur Lehre vom Transfiniten", *Ztsch. Phil. phil. Kritik 91*, 81–125, 272–270; *92*, 240–265 ≈ *1890a*, 11–93 ≈ *Papers*, 378–439.

1890a. Zur Lehre vom Transfiniten, Halle/Saale (Pfeffer (Stricker)). [R: Frege *1892c*.]

1892a. "Über eine elementare Frage der Mannigfaltigkeitslehre", *JDMV 1* (1890–1891), 75–78 ≈ *Papers*, 278–281. [E: Ewald *1996a*, 920–922. I: *1892b*.]

1892b. I of *1892a*, *RdM 2*, 165–167.

1895a. "Sui numeri transfiniti", *RdM 5*, 104–109.

1895b. "Beiträge zur Begründung der transfiniten Mengenlehre", pt. 1, *MA 46*, 481–512 ≈ *Papers*, 282–311. [R: Vivanti *1898a*. I: *1895c*. E: *1915a*, 85–136.]

1895c. I of *1895b*, *RdM 5*, 129–162.

1897a. *1895b*, pt. 2, *MA 49*, 207–246. [R: Vivanti *1900a*. E: *1915a*, 137–201.]

1915a. (E and ed. P. E. B. Jourdain), *Contributions to the founding of the theory of transfinite numbers*, La Salle, Ill. (Open Court) = 1955, New York (Dover). [E of *1895b* and *1897a*.]

Cantor-Dedekind

Letters. (Eds. E. Noether and J. Cavaillès), *Briefwechsel Cantor-Dedekind*, 1937, Paris (Hermann). [Manuscripts: Technical High School, Braunschweig. F: Cavaillès *1962a*, 177–251.]

Carnap, R. (1891–1970)

Papers: Pittsburgh (Pennsylvania), University of Pittsburgh.

m1910–1913a. Notes of lecture courses given by Frege, Carnap Papers, 11–10. [(Ed. and int. by G. Gabriel), *HPL 17* (1996), iii–xvi, 1–48.]

1927a. "Eigentliche und uneigentliche Begriffe", *Symposion 1*, 355–374.

1928a. Der logische Aufbau der Welt. Versuch einer Konstitutionstheorie der Begriffe, Berlin (Welt-Kreis). [E: *1967a*.]

1929a. Abriss der Logistik, mit besondere Berücksichtigung der Relationstheorie und ihre Anwendungen, Vienna (Springer).

1930a. "Die alte und neue Logik", *Erkenntnis 1,* 12–26.

1931a. "Bericht über Untersuchungen zur allgemeinen Axiomatik", *Erkenntnis 1,* 303–307.

1931b. R of *PM*$_2$, *Erkenntnis 2,* 73–75.

1931c. "Die logizistiche Grundlegung der Mathematik", *Erkenntnis 2,* 91–105. [Rs: Gödel *1932a,* Hempel *1938a.* E: Pears *1972a,* 175–191 ≈ Benacerraf and Putnam *1985a,* 41–52.]

m1931d. "Metalogik", in Carnap Papers, 81-01-17 to -19 and in Vienna Circle Papers 186/9–10. [Ed. with int. by J. Padilla Gálvez in *Mathesis 11* (1995), 137–192; and by F. Stadler in Stadler *1997a,* 314–329.]

1931e. "Überwindung der Metaphysik durch logische Analyse der Sprache", *Erkenntnis 2,* 219–241.

1934a. "On the character of philosophical problems", *Phil. of sci. 1,* 5–15.

1934b. Logische Syntax der Sprache, Vienna (Springer). [Rs: Dubislav *1935a,* Nagel *1935a,* Stebbing *1935a.* E: *1937a.*]

1934c. "Die Antinomien und die Unvollständigkeit der Mathematik", *Monats. Math. Physik, 41,* 263–284. [E: in *1937a.*]

1934d. (E and int. by M. Black), *The unity of science,* London (KP) = 1996, Bristol (Thoemmes). [R: Stebbing *1935a.*]

1935a. Philosophy and logical syntax, London (KP) = 1996, Bristol (Thoemmes). [R: Stebbing *1935a.*]

1935b. "Ein Gültigkeitskriterium für die Sätze der klassischen Mathematik", *Monats. Math. Physik 42,* 163–190. [E: in *1937a.*]

1937a. The logical syntax of language, London (KP). [E of *1934b, d and 1935b* by Countess Zeppelin.]

1939a. Foundations of logic and mathematics, Chicago (University of Chicago Press).

1942a. Introduction to semantics, Cambridge, Mass. (Harvard UP).

1943a. Formalization of logic, Cambridge, Mass. (Harvard UP).

1963a. "Intellectual autobiography", in P. A. Schilpp (ed.), *The philosophy of Rudolf Carnap,* La Salle, Ill. (Open Court), 1–84.

1967a. The logical syntax of language, London (RKP). [E of *1928a* by R. George.]

Carnot, L. N. M. (1753–1823)

1803a. Géométrie de position, Paris (Duprat).

Carroll, L. [Dodgson, C. L.] (1832–1898)

1894a. "A logical paradox", *Mind n.s. 3,* 436–438.

1895a. "What the tortoise said to Achilles", *Mind n.s. 4,* 278–280. [Various ≈ s.]

Carus, P. C. (1852–1919)

1892a. R of Schröder *1890b, The monist 2,* 618–623.

Cassina, U. (1897–1964)

1933a. "Su la logica matematica di G. Peano", *Boll. Unione Mat. Ital.* (2)*12,* 57–65 ≈ *1961a,* 331–342.

1933b. "L'opera scientifica de Giuseppe Peano", *Rend. Sem. Mat. Fis. Milano 7,* 323–389 ≈ *1961b,* 397–468.

1933c. "L'oeuvre philosophique de G. Peano", *RMM 40,* 481–491.

1940a. "Sul teorema fondamentale della geometria proiettiva", *Per. di mat. (4)20,* 65–83 ≈ *1961a,* 402–424.

1948–1949a. "Sui fondamenti della geometria secondo Hilbert", *Rend. Ist. Lombardo, cl. sci. mat. fis. 81,* 71–84; *82,* 67–94 ≈ *1961a,* 425–496.

1952a. "Alcune lettere e documenti inediti sul trattato do calcolo di

Genocchi-Peano", *Rend. Ist. Lombardo, cl. sci. mat. fis. 85*, 337–362 ≈ *1961b*, 375–397.

1955a. "Storia ed analisi del «Formulario completo» di Peano", *Boll. Unione Mat. Ital.* (*3*)*10*, 244–265, 544–574 ≈ *1961b*, 469–535.

1955b. "Sul «Formulario completo» di Peano", in Terracini *1955a*, 71–102 ≈ Peano *Formulary*$_5$, 1960 reprint, v–xxxiii ≈ *1961a*, 371–401.

1961a. Critica dei principî della matematica e questione di logica, Rome (Cremonese).

1961b. Dalle geometria egiziana all matematica moderna, Rome (Cremonese).

Cassinet, J. (1925–1999) and Guillemot, M.

1983a. "L'axiome du choix dans les mathématiques de Cauchy (1821) à Gödel (1940)", 2 vols., Toulouse. [Double *Docteur d'état-sciences*, University Paul Sabatier. Vol. 2 comprises Fs of many primary writings.]

Cassirer, E. A. (1874–1945)

1907a. "Kant und die moderne Mathematik", *Kantstudien 12*, 1–49.

1910a. Substanzbegriff und Funktionbegriff, Berlin (B. Cassirer). [E: *Substance and function*, 1923, La Salle, Ill. (Open Court) = 1953, New York (Dover).]

1923a, 1929a. Philosophie der symbolischen Formen pts. 1, 3, Oxford (B. Cassirer). [E: *The philosophy of symbolic forms*, 1953, 1957, New Haven (Yale UP).]

Castelnuovo, G. (1865–1952)

1909a. (Ed.) *Atti del IV Congresso Internazionale dei Matematici*, 3 vols., Rome (Accademia dei Lincei) = 1967, Liechtenstein (Kraus).

Castrillo, P.

1997a. "Christine Ladd-Franklin y su puesto en la tradición algebraica de la lógica", *Mathesis 13*, 117–130.

Cauchy, A.-L. (1789–1857)

Works. (Eds. various), *Oeuvres complètes*, 12 + 15 vols., 1882–1974, Paris (GV).

1821a. Cours d'analyse, Paris (de Bure) ≈ *Works* (*2*)*3* = 1968, Darmstadt (Wissenschaftliche Buchgesellschaft) = 1992, Bologna (CLUEB) [with int. by U. Bottazzini].

1822a. "Sur le développement des fonctions en séries", *Bull. Soc. Philom. Paris*, 49–54 ≈ *Works* (*2*)*2*, 276–282.

1823a. Résumé des leçons données à l'Ecole Polytechnique sur le calcul infinitésimal, Paris (de Bure) ≈ *Works* (*2*)*4*, 5–261.

Cavaillès, Jean (1903–1944)

1935a. "Le cercle de Vienne au Congrès de Prague", *RMM 42*, 137–149.

1938a. Remarques sur la formation de la théorie abstraite des ensembles, 2 pts., Paris (Hermann) ≈ *1962a*, 23–176.

1938b. Méthode axiomatique et formalisme, 3 pts., Paris (Hermann). [Thesis at the University of Paris, 1937.]

1962a. Philosophie mathématique, Paris (Hermann).

Cayley, A. (1821–1895)

Papers. Collected mathematical papers, 14 vols., 1889–1898, Cambridge (Cambridge UP).

1854a. "On the theory of groups", *Phil. mag.* (*4*)*7*, 40–47 ≈ *Papers 2*, 123–130.

1864a. "On the notion and boundaries of algebra", *Quart. j. pure appl. maths.*, *6*, 382–384 ≈ *Papers 5*, 292–294.

Charraud, N.

1994a. Infini et inconscient. Essai sur Georg Cantor, Paris (Anthropos).

Chihara, C. S.

1973a. Ontology and the vicious circle principle, London and Ithaca (Cornell UP).

1980a. "Ramsey's theory of types: suggestions for a return to Fregean sources", in D. H. Mellor (ed.), *Prospects for pragmatism*, Cambridge (Cambridge UP), 21–47.

Christen, P. G. *See* Rahman and Christen

Church, A. (1903–1995)

1927a. "Alternatives to Zermelo's assumption", *Trans. AMS 29*, 178–208. [Ph.D., Princeton University.]

1928a. "On the law of excluded middle", *Bull. AMS 34*, 75–78.

1928b. R of PM_2 *2–3*, *Bull. AMS 34*, 237–240.

1932a. R of Ramsey *Essays*, *Amer. math. monthly 39*, 355–357.

1936a. "A bibliography of symbolic logic", *JSL 1*, 121–218.

1938a. "Additions and corrections" to *1936a*, *JSL 3*, 178–212.

1939a. "Schröder's anticipation of the simple theory of types", *Erkenntnis 9*, 149–152.

1944a, 1956a. Introduction to mathematical logic$_{1,2}$, Princeton (Princeton UP).

1974a. "Russellian simple type theory", *Proc. Amer. Phil. Soc. 47*, 21–33.

1984a. "Russell's theory of the identity of propositions", *Philos. natur. 21*, 513–522.

Chwistek, L. (1884–1944)

1921a. "Antynomje logiki formalnej", *Przeg. filoz. 24*, 164–171. [E: McCall *1967a*, 338–345.]

1922a. "Über die Antinomien der Prinzipien der Mathematik", *Math. Ztsch. 14*, 236–243.

1924a, 1925a. "The theory of constructive types", *Ann. Soc. Polon. Math. 2*, 9–48; *3*, 92–141.

1926a. "Über die Hypothesen der Mengenlehre", *Math. Ztsch. 25*, 439–473.

1929a. "Neue Grundlagen der Logik und Mengenlehre", pt. 1, *Math. Ztsch. 30*, 704–724.

1933a. "Die nominalistische Grundlegung der Mathematik", *Erkenntnis 3*, 367–388.

1935a. Granice nauki, Lvòv and Warsaw (Atlas). [E version: *1948a*.]

1939a. "A formal proof of Gödel's theorem", *JSL 4*, 61–68.

1948a. (E by H. Brodie), *The limits of science*, London (KP). [R: Myhill *1949a*.]

Claparède, E.

1905a. (Ed.) *Congrès International de Philosophie. IIme session*, Geneva (Droz) = 1967, Liechtenstein (Kraus).

Clark, G.

1997a. "New light on Peirce's iconic notation for the sixteen binary connectives", in Houser and others *1997a*, 304–333. [*See also* S. Zellweger on pp. 334–386.]

Clauberg, K. W. and Dubislav, W. *See also* Dubislav

1922a. Systematische Wörterbuch der Philosophie, Leipzig (Meiner).

Cleaver, F. L. *See* Anderson and Cleaver

Cocchiarella, N.

1987a. Logical studies in early analytic philosophy, Columbus (Ohio State UP).

Coffa, J. A.

1979a. "The humble origins of Russell's paradox", *Russell nos. 33–34*, 31–38.

1980a. "Russell and Kant", *Synthese 46*, 247–263.

1991a. The semantic tradition from Kant to Carnap, Cambridge (Cambridge UP).

Cohen, M. R. (1880–1947)

1912a. R of PM_1*1*, *PR 21*, 87–91.

1918a. "The subject matter of formal logic", *JP 15*, 673–688.

Cohen, M. R. and Nagel, E.

1934a. Introduction to logic and scientific method, New York (Harcourt Brace) and London (Routledge). [Abridged version 1939. Part 1 of original = 1962;$_2$ (ed. J. Corcoran), Indianapolis (Hackett).]

Condillac Etienne Bonnot, Abbé de (1714–1780)

1780a. La logique ou les premiers développmens de l'art de penser, Paris (L'Esprit and De Bure) = [with E and int. by W. R. Albury] 1980, New York (Abaris).

1798a. La langue des calculs, Paris (Gratiot etc.) = [ed. and int. S. Auroux and A.-M. Chouillet] 1981, Lille (PU).

Congress

1901a. Bibliothèque du Congrès International de Philosophie 3, Paris (Colin) = 1967, Liechtenstein (Kraus).

Contro, W.

1976a. "Von Pasch zu Hilbert", *AHES 15*, 283–295.

Corcoran, J.

1973a. "Gaps between logical theory and mathematical practise", in M. Bunge (ed.), *The methodological unity of science*, Dordrecht (Reidel), 23–50.

1980a. "On definitional equivalence and related topics", *HPL 1*, 231–234 [*see also* pp. 187–207].

1986a. R of Boole-De Morgan *Letters*, *HPL 7*, 65–75.

Corcoran, J. and Wood, S.

1980a. "Boole's criteria of validity and invalidity", *Notre Dame j. formal logic 21*, 609–638.

Cornish, K.

1998a. The Jew of Linz, London (Century).

Corry, L.

1997a. "David Hilbert and the axiomatisation of physics (1894–1905)", *AHES 51*, 83–198.

Costello, H. T.

1928a. R of *PM$_2$*, *JP 25*, 438–445.

Cournot, A.-A. (1801–1877)

1847a. De l'origine et des limites de la correspondance entre l'algèbre et la

géométrie, Paris and Algiers (Hachette).

Couturat, A.-L. (1868–1914)

1896a. De l'infini mathématique, Paris (Alcan) = 1969, New York (Franklin) = 1973, Paris (Blanchard) = 1975, Hildesheim (Olms). [R: Russell *1897a*.]

1898a. R of Russell *1897c*, *RMM 6*, 354–380.

1898b. "Sur les rapports du nombre et de grandeur", *RMM 6*, 422–427.

1899a. "La logique mathématique de M. Peano", *RMM 7*, 616–646.

1900a. "Sur une définition logique du nombre", *RMM 8*, 23–36.

1900b. "Sur la définition du continu", *RMM 8*, 157–168.

1900c. R of Schröder *1890b*, *1891a* and *1895a*, *Bull. des sci. math.* (2)*24*, 49–68, 83–102.

1900d. R of Whitehead *1898a*, *RMM 8*, 323–362.

1900e. "Congrès International de Philosophie. Séance générale. Logique et histoire des sciences", *RMM 8*, 538–547, 556–565, 589–598, 638–647, 670–678. [Attributed; full report is pp. 503–698.]

1900f. "Les mathématiques au Congrès de Philosophie", *L'ens. math.* (1)*2*, 397–410.

1901a. R of Peano *Formulary$_{1-3}$*, *Bull. des sci. math.* (2)*25*, 141–159.

1903a. Opuscules et fragments inédits de Leibniz, Paris (Alcan).

1904a. R of Russell *1903a*, *Bull. des sci. math.* (2)*28*, pt. 1, 129–147.

1904b. "La philosophie des mathématiques de Kant", *RMM 12*, 321–383 ≈ *1905b*, 235–306.

1904c. "IIme Congrès International de Philosophie, Génève", *RMM 12*, 1037–1077.

1904–1905a. "Les principes des mathématiques", *RMM 12*, 19–50, 211–240, 664–698, 810–844; *13*, 244–256 ≈ *1905b*, 1–218.

1905a. L'algèbre de la logique, Paris (GV) = 1969, Hildesheim (Olms) = 1980, Paris (Blanchard). [E: (Ed. P. E. B. Jourdain), *The algebra of logic*, 1914, Chicago and London (Open Court). Polish T: *1918a.*]

1905b. Les principes des mathématiques, Paris (Alcan) = 1965, Hildesheim (Olms). [G: *1908a.*]

1905c. "Sur l'utilité de la logique algorithmique", in Claparède *1905a*, 706–711.

1906a. "Pour la logistique (réponse à M. Poincaré)", *RMM 14*, 208–250. [Part E: *The monist 22* (1912), 283–523.]

1906b. "La logique et la philosophie contemporaine", *RMM 14*, 318–341 ≈ *1983a'*, 17–34.

1908a. Die philosophischen Prinzipien der Mathematik, Leipzig (Klinkhardt). [G of *1905b* by C. Siegel.]

1913a. "The principles of logic", in Windelband and Ruge *1913a*, 136–198.

1917a. "Sur les rapports logiques des concepts et des propositions", *RMM 24*, 15–58. [Posthumous.]

1918a. Algebra logiki, Warsaw (Wydawnictwo Kasy). [Trans. of *1905a* by B. Knaster. Not found: see Wolenski *1989a*, 333.]

1983a'. L'oeuvre de Louis Couturat, Paris (Presses de l'Ecole Normale Supérieure). [Conference proceedings.]

Couturat, L. and Ladd Franklin, C. *See also* Ladd

1902a. "Symbolic logic or algebra of logic", in J. M. Baldwin (ed.), *Dictionary of philosophy and psychology 2*, London and New York (MacMillan), 640–651.

Crowe, M. J.

1967a. A history of vector analysis, Notre Dame and London (Notre Dame UP).

Curry, H. B. (1900–1982) *See also* Ducasse and Curry

1930a. "Grundlagen der kombinatorischen Logik", *Amer. j. maths. 52*, 509–536, 789–834. [*Dissertation.*]

Czyż, J.

1994a. Paradoxes of measures and dimensions originating in Felix Hausdorff's ideas, Singapore (World Scientific).

De Amicis, E. (1846–1908)

1892a. "Dipendenza fra alcune proprietà notevoli delle relazioni fra enti di un medesimo sistema", *RdM 2*, 113–127.

de Laguna, T.

1906a. R of Royce *1905a*, *JP 3*, 357–361.

1915a. "The logico-analytic method in philosophy", *JP 12*, 449–462. [R of Russell *1914c* and other works.]

1916a. "On certain logical paradoxes", *PR 25*, 16–27.

De Morgan, A. (1806–1871) *See also* Boole–De Morgan Letters

Here *"TCPS"* cites the *Transactions of the Cambridge Philosophical Society*.

Logic. (Ed. P. Heath), *On the syllogism and other logical writings*, 1966, London (RKP).

1831a. On the study and difficulties of mathematics, London (Society for the Diffusion of Useful Knowledge) ≈ (Ed. T. J. McCormack), 1902, Chicago (Open Court).

1832a. "State of the mathematical and physical sciences in the University of Oxford", *Quart. j. educ. 4*, 191–208.

1833a. "On the methods of teaching the elements of geometry", *Quart. j. educ. 6*, 35–49, 237–251.

1835a. R of Peacock *1830a*, *Quart. j. educ. 9*, 91–110, 293–311.

1835b. The elements of algebra$_1$, London (Taylor and Walton).

1836a. "Calculus of functions", in *Encyclopaedia Metropolitana 2*,

305–392. [Date of offprint version; volume carries "1845".]

1839a. First notions of logic (*preparatory to the study of geometry*)$_1$, London (Taylor and Walton). [$_2$ 1840 ≈ *1847a*, ch. 1.]

1841a. "Relations (mathematical)", in *The penny cyclopaedia 19*, 372–374.

1842a. The differential and integral calculus, London (Taylor and Walton). [Published in parts from 1836.]

1842b. "The foundations of algebra", pt. 2, *TCPS 7*, 173–187‡ ≈ Ewald *1996a*, 336–348.

1846a. "On the syllogism", pt. 1, *TCPS 8*, 379–408. [Part ≈: *Logic*, 1–21‡.]

1847a. Formal logic, London (Walton and Maberly) ≈ (Ed. A. E. Taylor), 1926, La Salle, Ill. (Open Court).

1849a. "The foundations of algebra", pt. 4, *TCPS 8*, 139–142, 241–254.

1849b. Trigonometry and double algebra, London (Taylor, Walton and Maberly).

1850a. "On the syllogism", pt. 2, *TCPS 9*, 79–127 ≈ *Logic*, 22–68‡.

1858a. "On the syllogism", pt. 3, *TCPS 10*, 173–230 ≈ *Logic*, 74–146.

1860a. "On the syllogism", pt. 4, *TCPS 10*, 331–357, *355–*358 [*sic*] ≈ *Logic*, 208–246‡.

1860b. Syllabus of a proposed system of logic, London (Walton and Maberly). Part ≈ *Logic*, 147–207‡.

1860c. "Logic", in *English cyclopedia 5*, 150–164. Part ≈ *Logic*, 247–270‡.

1862a. "On the syllogism", pt. 5, *TCPS 10*, 428–487 ≈ *Logic*, 271–345.

1865a. "George Boole, F. R. S.", *Macmillan's mag. 11*, 279–280. [Attributed.]

1866a. "On infinity, and on the sign of equality", *TCPS 11*, 145–189.

1868a. R of J. M. Wilson, *Elementary geometry* (1868), *The Athenaeum 2* [for year], 71–73.

De Morgan, S. E. (1808–1892)

1882a. Memoir of Augustus De Morgan, London (Longmans, Green).

de Rouilhan, P.

1988a. Frege. Les paradoxes de la représentation, Paris (de Minuit).

1996a. Russell et le cercle des paradoxes, Paris (PUF).

Del Val, J. A.

1973a. "Los escritos de Ventura Reyes y Prósper (1863–1922)", *Teorema 3*, 313–328.

du Bois Reymond, P. D. G. (1831–1889)

1877a. "Über die Paradoxien des Infinitärkalküls", *MA 11*, 149–167.

1880a. "Der Beweis des Fundamentalsatzes der Integralrechnung", *MA 16*, 115–128.

Dahms, H.-J.

1985a. Philosophie, Wissenschaft, Aufklärung. Beiträge zur Geschichte und Wirkung des Wiener Kreises, Berlin (de Gruyter).

1985b. "Verbreitung und Emigration des Wiener-Kreises zwischen 1931 und 1940", in *1985a*, 307–365.

Dantzig, T. (1884–1956)

1930a. Number, the language of science$_1$, New York and London (A & U).

Darboux, G. (1842–1917)

1875a. "Sur les fonctions discontinues", *Ann. sci. Ecole Normale Sup.* (*2*)*4*, 57–112.

Dassen, C. C. (1873–1941)

1933a. "Réflexions sur quelques antinomies et sur la logique empiriste", *Anales Soc. Cient. Argentina 115*, 135–166, 199–232, 275–296. [Not found: R by A. Schmidt in *ZfM 9* (1934), 1–2.]

1939a. "Vida y obra de Louis Couturat", *Anales Acad. Nat. Ciencias Buenos Aires 4*, 73–204. [Not found: R by W. V. Quine in *JSL 5* (1940), 168–169.]

Dauben, J. W.
  1971a. "The trigonometric background to Georg Cantor's theory of
  sets", *AHES 7*, 181–216.
  1977a. "C. S. Peirce's philosophy of
  infinite sets", *Math. mag. 50*, 123–
  135.
  1979a. Georg Cantor, Cambridge,
  Mass. and London (Harvard UP)
  = 1990, Princeton (Princeton UP).
  1980a. "Mathematicians and World
  War I: the international diplomacy
  of G. H. Hardy and Gösta Mittag-
  Leffler", *HM 7*, 261–288.
  1995a. Abraham Robinson, Princeton
  (Princeton UP).
Davis, M.
  1965a. (Ed.) *The undecidable*, Hewlett,
  New York (Raven Press).
Dawson, J. W. Jr.
  1985a. "Completing the Gödel-
  Zermelo correspondence", *HM 12*,
  66–70.
  *1997a. Logical dilemmas. The life and
  work of Kurt Gödel*, Wellesley,
  Mass. (Peters).
Dedekind, J. W. R. (1831–1916) *See
  also* Cantor-Dedekind Letters
  Papers: Göttingen (Germany),
  University.
  *Works. Gesammelte mathematische
  Werke*, 3 vols., 1930–1932, Braun-
  schweig (Vieweg) = 1969, New
  York (Chelsea).
  m1854a. "Über die Einführung neuer
  Funktionen in der Mathematik", in
  Works 3, 447–449.
  m1862a. (Eds. M.-A. Knus and W.
  Scharlau), *Vorlesung über die
  Differential- und Integralrechnung
  1861/62*, Braunschweig (Vieweg).
  *1872a. Stetigkeit und irrationale
  Zahlen₁*, Braunschweig (Vieweg) =
  1892, 1905, 1912 ≈ *Works 3*, 315–
  334. [E: *1901a*, 1–27 ≈ Ewald
  1996a, 766–779. I: *1926a*, 119–153.]
  m1887a. "Ähnliche (deutliche) Abbildung und ähnliche Systeme.
  1887.7.11", in *Works 3*, 447–449.

  *1888a. Was sind und was sollen die
  Zahlen?*, Braunschweig (Vieweg) ≈
  Works 3, 335–391. [R: F. W. F.
  Meyer *1891a*. =s with new ints.
  1893, 1911. E: *1901a*, 29–115 ≈
  Ewald *1996a*, 790–833. Polish T:
  1914, Warsaw (Mianovski). I:
  1926a, 1–118.]
  1893a. "Vorwort zur zweiten
  Auflage", in *1872a*, second printing, 1x–xi.
  1897a. "Über Zerlegung von Zahlen
  durch ihre grössten gemeinsamen
  Teiler", in *Festschrift der Technische Hochschule zu Braunschweig*,
  Braunschweig (Vieweg), 1–40 ≈
  Works 2, 103–147.
  1900a. "Über die von drei Moduln
  erzeugte Dualgruppe", *MA 53*,
  371–403 ≈ *Works 2*, 236–271.
  *1901a. Essays on the theory of num-
  bers*, Chicago (Open Court) = 1963,
  New York (Dover). [E of *1872a*
  and *1888a* by W. W. Beman.]
  *1926a. Essenza e significato dei nu-
  meri*, Rome (Stock). [I of *1872a*
  and *1888a* by O. Zariski. Rare:
  University of Milan, Department of
  Mathematics.]
Dehn, M. (1878–1952)
  1905a. R of Hilbert *Geometry₂* (1903),
  JFM 34 (1903), 523–524.
  1905b. R of Frege *1903b* and Korselt
  1903a, *JFM 34* (1903), 525.
  1909a. R of Frege *1906a*, *JFM 37*
  (1906), 485.
Dejnožka, J.
  1990a. "The ontological foundation
  of Russell's theory of modality",
  Erkenntnis 32, 383–418.
  *1996a. The ontology of the analytic tra-
  dition and its origins*, Lanham, Md.
  (Littlefield, Adams).
  *1999a. Bertrand Russell on modality
  and logical relevance*, Aldershot
  (Ashgate).

Demopoulos, W.

1994a. "Frege, Hilbert and the conceptual structure of model theory", *HPL 15*, 211–225.

1995a. (Ed.) *Frege's philosophy of mathematics*, Cambridge, Mass. and London (Harvard UP).

Demos, R. (1892–1968)

1917a. "A discussion of a certain type of negative proposition", *Mind n.s. 26*, 188–196.

Detlefsen, M.

1986a. Hilbert's program, Dordrecht (Reidel).

1993a. "Logicism and the nature of mathematical reasoning", in Irvine and Wedeking *1993a*, 265–292.

Dewey, J. (1859–1949)

Papers: Carbondale and Edwardsville (Illinois), Southern Illinois University (not used).

Works. Works, 3 ser., 1969–1990, Carbondale (Illinois), Southern Illinois UP.

1903a. (Ed.) *Studies in formal logic*, Chicago (University of Chicago Press) ≈ *Works (1)2.*

1916a. Essays in experimental logic, Chicago (University of Chicago Press) ≈ *Works (2)10.* [R: Russell *1919a.*]

1938a. Logic: the theory of enquiry, New York (Holt) ≈ *Works (3)12.*

Dhombres, J.

1986a. "Quelques aspects de l'histoire des équations fonctionnelles", *AHES 36*, 91–181.

Dickson, L. E. (1874–1954)

1903a. "Definition of a linear associative algebra by independent postulates", *Trans. AMS 4*, 21–26 = *Mathematical papers 2*, 1975, New York (Chelsea), 109–116.

Dickstein, S. (1851–1939)

1899a. "Zur Geschichte der Prinzipien der Infinitesimalrechnung", *Abh. Gesch. Math. 9*, 65–79.

Dingler, H. A. E. H. (1881–1954)

1911a. Über die Bedeutung der Burali-Fortischen Antinomie für die Wohlordnungssätze der Mengenlehre, Munich (Ackermann). [*Dissertation.* Rare: Erlangen University.]

1912a. Über wohlgeordnete Mengen und zerstreute Mengen im allgemeinen, Munich (Ackermann). [*Habilitation.*]

1913a. "Über die logischen Paradoxien der Mengenlehre und eine paradoxiefreie Mengendefinition", *JDMV 22*, pt. 1, 305–315.

1915a. R of König *1914a*, *Arch. Math. Physik (3)24*, 153–159.

1915b. Das Prinzip der logischen Unabhängigkeit in der Mathematik zugleich als Einführung in die Axiomatik, Munich (Ackermann). [R: Löwenheim *1922a.*]

1923a. Die Grundlagen der Physik$_2$, Berlin and Leipzig (de Gruyter).

1931a. Philosophie der Logik und Arithmetik, Munich (Reinhardt). [R: Nagel *1932a.*]

Dini, U. (1845–1918)

1892a. Grundlagen für eine Theorie der Functionen einer veränderlichen reellen Grösse, Leipzig (Teubner). [G of Italian original (1878).]

Dipert, R.

1978a. "Development and crisis in late Boolean logic: the deductive logics of Peirce, Jevons and Schröder", Bloomington (Indiana University Ph.D.).

1983a. R of 1983 reprint of Peirce *1883b*, *HPL 5*, 225–232.

1991a. "The life and work of Ernst Schröder", *Modern logic 1*, 119–139.

1994a. "The life and logical contributions of O. H. Mitchell", *Trans. C. S. Peirce Soc. 30*, 515–542.

Dirichlet, J. P. G. Lejeune (1805–1859)

1829a. "Sur la convergence des séries trigonométriques", *J. rei. ang. Math. 4*, 157–169 ≈ *Gesammelte Werke 1*,

1889, Berlin (Reimer) = 1969, New York (Chelsea), 117–132.

Dreben, B.
1990a. "Quine", in R. B. Barrett and R. E. Gibson (eds.), *Perspectives on Quine*, Cambridge, Mass. and Oxford (Blackwell), 81–95. [Reply by Quine, pp. 96–97.]

Dreben, B. and Kanamori, A.
1997a. "Hilbert and set theory", *Erkenntnis 110*, 77–125.

Drucker, T.
1991a. (Ed.) *Perspectives on the history of mathematical logic*, Boston (Birkhäuser).

Dubislav, W. (1895–1937) *See also* Clauberg and Dubislav
1926a. "Über das Verhältnis der Logik zur Mathematik", *Ann. der Phil. 5*, 193–208.
1927a. Über die Definition₂, Berlin (Weiss).
1930a. "Über den sogenannten Gegenstand der Mathematik", *Erkenntnis 1*, 27–48.
1931a. Die Definition₃, Leipzig (Meiner: *Erkenntnis Beiheft 1* [and only?]) = 1981. [R: Schmidt *1932a.*]
1932a. R of Gödel *1931a, ZfM 2*, 1.
1932b. "Les recherches sur la philosophie des mathématiques en Allemagne", *Rech. phil. 1*, 299–311.
1932c. Die Philosophie der Mathematik in der Gegenwart, Berlin (Junker und Dünnhaupt). [Rs: Schmidt *1933a*, Scholz *1933b.*]
1935a. R of Carnap *1934b, JFM 60* (1934), 19–20.

Ducasse, C. J. (1881–1969) and Curry, H. B. *See also* Curry
1962a. "Early history of the Association for Symbolic Logic", *JSL 27*, 255–258.

Dudman, V. H.
1971a. "Peano's review of Frege's Grundgesetze", *Southern j. phil. 9*, 25–37. [Es of Peano *1895a* and *1896a*, and Frege *1896a.*]

Dufumier, H.
1909a. "Les théories logico-métaphysiques de MM. B. Russell et G. E. Moore", *RMM 17*, 620–653.
1911a. R of *PM₁1, Bull. des sci. math. (2)35*, pt. 1, 213–221.
1912a. "La philosophie des mathématiques de MM. Russell et Whitehead", *RMM 20*, 538–566.

Dugac, P.
1973a. "Eléments d'analyse de Karl Weierstrass", *AHES 10*, 41–176.
1976a. Richard Dedekind et les fondements des mathématiques, Paris (Vrin).
1976b. "Notes et documents sur la vie et l'oeuvre de René Baire", *AHES 15*, 298–383.

Dummett, M. E.
1991a. Frege: philosophy of mathematics, London (Duckworth) and Cambridge, Mass. (Harvard UP).

Duporcq, E.
1902a. (Ed.) *Compte rendus du Deuxième Congrès International de Mathématiciens*, Paris (GV) = 1967, Liechtenstein (Kraus).

Eaton, R. M. (1892–1932)
1931a. General logic. An introductory survey, Cambridge, Mass. (Harvard UP).

Eccarius, W.
1985a. "Georg Cantor und Kurd Lasswitz: Briefe zur Philosophie des Unendlichen", *Schr. Gesch. Naturwiss. Tech. Med. 22*, no. 1, 7–28.

Edwards, A. W. F.
1989a. "Venn diagrams for many sets", *The new scientist*, (7 January), 51–56.

Edwards, H. M.
1989a. "Kronecker's views on the foundations of mathematics", in Rowe and McCleary *1989a 1*, 67–77.

Edwards, H. M., Neumann, O. and
Purkert, W. *See also* Purkert
1982a. "Dedekinds 'Bunte Bemerkungen' zu Kroneckers
'Grundzüge'", *AHES 27*, 49–85.

Ellis, A. J. (1814–1890)
1873a. "On the algebraic analysis of
logical relations", *Proc. Roy. Soc.
London 21*, 497–498.

Ellis, R. L. (1817–1859)
m1863a. "Notes on Boole's Laws of
thought", in *The mathematical and
other writings*, Cambridge (Deighton, Bell), 391–394 ≈ *Rep. Brit. Ass.
Adv. Sci.* (1870, pb. 1871). 12–14.

Engel, F. (1861–1941)
1901a. R of Hilbert *Geometry*₁ (1899),
JFM 30 (1899), 424–426.
1905a. R of Russell *1903a, JFM 34*
(1903), 62–63.

Enriques, F. (1871–1946)
*1906a. Problemi della scienza*₁,
Bologna (Zanichelli). [G by K.
Grelling: *Probleme der Wissenschaft*, 2 vols., 1910, Leipzig and
Berlin (Teubner). E, ed. J. Royce:
Problems of science, 1914, Chicago
(Open Court).]
1921a. "Noterelle di logica matemat-
ica", *Per. di mat.* (4)*1*, 233–244 =
Notizie di logica 7 (1988), no. 3,
18–26.
1921b. Reply to Burali-Forti *1921a*,
Per. di mat. (4)*1*, 360–365 = *Notizie
di logica 7* (1988), no. 4, 27–32.
1922a. Per la storia della logica,
Bologna (Zanichelli). [F: *L'evolu-
tion de la logique*, 1926, Paris
(Chiron). G by L. Bieberbach: *Zur
Geschichte der Logik*, 1927, Leipzig
(Teubner). E: *The historical development of logic*, 1929, New York
(Holt) = 1968, New York (Russell
and Russell).]

Enros, P. J.
1983a. "The Analytical Society
(1812–1813)", *HM 10*, 24–47.

Erdmann, B. (1851–1921)
1892a. Logik 1, Halle/Saale
(Niemeyer).

Ewald, W. B.
1996a. (Ed.), *From Kant to Hilbert. A
source book in the foundations of
mathematics*, 2 vols., New York and
Oxford (Clarendon Press).

Färber, C. (1863–1912)
1905a. R of Frege *1903a, JFM 34*,
(1903), 71–72.

Feferman, S.
1988a. "Weyl vindicated: 'Das Kontinuum' 70 years later", in *Atti del
Congresso "Temi e prospettive della
logica e della filosofia della scienza
contemporanea" 1*, Bologna
(CLUEB), 60–93 ≈ *In the light of
logic*, 1998, New York (Oxford UP),
249–283.

Fehr, H. (1870–1954)
1904a. "Le 3ᵐᵉ congrès international
des mathématiciens, Heidelberg,
1904", *L'ens. math.* (*1*)6, 379–400.
1905a. "Sur la fusion progressive de
la logique et des mathématiques",
in Claparède *1905a*, 677–679.

Feibleman, J. (1904–1987)
1944a. "A reply to" Russell *1937a*, in
Schilpp *1944a*, 155–174.

Ferrari, M.
*1996a. Ernst Cassirer dalla scuola di
Marborgo alla filosofia della cultura*,
Florence (Olschki).

Ferreirós, J.
1993a. "On the relations between
Georg Cantor and Richard
Dedekind", *HM 20*, 343–363.
1997a. "Notes on types, sets and
logicism, 1930–1950", *Theoria 12*,
91–124.
*1999a. Labyrinth of thought. A history
of set theory and its role in modern
mathematics*, Basel (Birkhäuser).

Ferrières, G.
*1950a. Jean Cavaillès. Philosoph et
combatant*, Paris (PUF).

Feys, R. (1889–1961)

1924–1925a. "La transcription logistique du raisonnement", *Rev. néoscol. de phil. 26*, 299–324, 417–451; *27*, 61–86.

1926–1927a. "Le raisonnement en termes de faits dans la logistique russellienne", *Ibidem 29*, 393–421; *30*, 154–192, 257–274 ≈ 1927, Louvain.

Finsler, P. (1894–1970)

Essays. (Ed. G. Unger), *Aufsätze zur Mengenlehre*, 1975, Darmstadt (Wissenschaftliche Buchgesellschaft).

Sets. (Eds. D. Booth and R. Ziegler), *Finsler set theory. Platonism and circularity*, 1996, Basel (Birkhäuser).

1925a. "Gibt es Widersprüche in der Mathematik?", *JDMV 34*, 143–155 = *Essays*, 1–10. [E: *Sets*, 39–49.]

1926a. "Über die Grundlegung der Mengenlehre", *Math. Ztsch. 25*, 683–713 = *Essays*, 19–49. [E: *Sets*, 103–132.]

Fisher, G.

1981a. "The infinite and the infinitesimal quantities of Paul du Bois Reymond", *AHES 24*, 101–163.

Fitch, F. B.

1938a. "The consistency of the ramified Principia", *JSL 3*, 140–149.

1974a. "Towards proving the consistency of *Principia mathematica*", in Nakhnikian *1974a*, 1–17.

Forsyth, A. R. (1858–1942)

1893a. Theory of functions of a complex variable₁, Cambridge (Cambridge UP).

1935a. "Old Tripos days at Cambridge", *Math. gaz. 19*, 162–179.

Fraenkel, A. A. H. (1891–1965)

Papers: The Jewish National and University Library, Jerusalem (Israel).

1919a. Einleitung in der Mengenlehre. Eine gemeinverständliche Einführung in das Reich der unendlichen Grössen₁, Berlin (Springer).

1922a. "Der Begriff «definit» und die Unabhängigkeit des Auswahlsaxioms", *Sitz.-ber. Preuss. Akad. Wiss., phys. math. Kl.*, 253–257. [E: van Heijenoort *1967a*, 284–289.]

1923a. Einleitung in der Mengenlehre. Eine elementare...₂, Berlin (Springer).

1925a. R of Weyl *1921a*, *JFM 48* (1921–1922), 47–50.

1927a. R of Hölder *1924a*, *JFM 49* (1923), 24–28.

1927b. Zehn Vorlesungen über die Grundlegung der Mengenlehre, Leipzig and Berlin (Teubner). [R: Brouwer *1930b*.]

1928a. Einleitung in der Mengenlehre₃, Berlin (Springer).

1930a. "Georg Cantor", *JDMV 39*, 189–266 ≈ Leipzig (Teubner).

1931a. "Die heutige Gegensätze in der Grundlegung der Mathematik", *Erkenntnis 1*, 286–302.

1932a. "Das Leben Georg Cantors", in Cantor *Papers*, 452–483.

1938a. R of Gödel *1931a*, *JFM 57*, (1931), 54.

1953a. Abstract set theory₁, Amsterdam (North-Holland).

1968a. Lebenskreise. Aus dem Erinnerungen eines jüdischen Mathematikers, Stuttgart (Deutsche Verlags-Anstalt).

Franchella, M.

1994a. "Heyting's contribution to the change in research into the foundations of mathematics", *HPL 15*, 149–172.

Fraser, C.

1985a. "J. L. Lagrange's changing approach to the foundations of the calculus of variations", *AHES 32*, 151–191.

Fréchet, M. (1878–1956)

1906a. "Sur quelques points du calcul fonctionnel", *Rend. Circolo Mat. Palermo 22*, 1–74.

Frege, F. L. G. (1848–1925)
Here *"SBJG"* cites the *Sitzungs-
berichte der Jenaischen Gesellschaft
der Medizin und Naturwissenschaft-
en.* Editions of Frege' not listed.
Papers: Münster University.
Writings. (Ed. I. Angelelli), *Kleine
Schriften*, 1967, Hildesheim (Olms).
Manuscripts. (Eds. H. Hermes and
others), *Nachgelassene Schriften*₂,
1983, Hamburg (Meiner).
Letters. (Eds. H. Hermes and others),
Wissenschaftlicher Briefwechsel,
1976, Hamburg (Meiner).
*1874a. Rechnungsmethoden, der sich
auf eine Erweiterung des Grössen-
begriffes gründen*, Jena (Frommann)
≈ *Writings*, 50–84. [*Dissertation.*]
*1879a. Begriffsschrift, eine der arith-
metischen nachgebildete Formel-
sprache des reinen Denkens*,
Halle/Saale (Niebert). [Rs:
Michaelis *1880a* and *1881a*,
Schröder *1880a.* Various ≈ s. Es:
van Heijenoort *1967a*, 1–82; *1972a*,
101–203.]
1879b. "Anwendung der Begriffsschrift", *SBJG 13*, 29–33. [E: *1972a*,
204–208.]
m1880a. "Booles rechende Formel-
sprache und die Begriffsschrift", in
Manuscripts, 9–52.
1882a. "Über den Zweck der
Begriffsschrift", *SBJG 16*, 1–10.
[E: *1972a*, 90–100.]
m1882b. "Booles logische Formel-
sprache und meine Begriffsschrift",
in *Manuscripts*, 53–59.
1884a. "Geometrie der Punktpaare in
der Ebene", *SBJG 17*, 98–102 ≈
Writings, 94–98.
*1884b. Die Grundlagen der Arithmetik.
Eine logisch-mathematische Unter-
suchung über den Begriff der Zahl*,
Breslau (Köbner) = 1934 = 1964,
Hildesheim (Olms) ≈ *1986a.* [R:
Cantor *1885c.* E: *The foundations
of arithmetic*₂, 1953, Oxford
(Blackwell).]

1885a. "Über formale Theorien der
Arithmetik", *SBJG 19*, 94–104 ≈
Writings, 103–111. [E: *1971a*, 141–
153.]
1885b. "Erwiderung auf" Cantor
1885c, Dtsch. Lit.-Zeit. 6, col. 1030
≈ *Writings*, 144.
1891a. Funktion und Begriff, Jena
(Pohle) ≈ *Writings*, 125–142‡. [R:
Michaelis *1894a.* E: *1960a*, 21–41.]
1892a. "Über Sinn und Bedeutung",
Ztsch. Phil. phil. Kritik 100, 25–50
≈ *Writings*, 143–162‡. [E: *1960a*,
56–76.]
1892b. "Über Begriff und Gegenstand", *Vrtlj. wiss. Phil. 16*, 192–205
≈ *Writings*, 167–178‡. [E: *1960a*,
42–55.]
1892c. R of Cantor *1890a, Ztsch. Phil.
phil. Kritik 100*, 269–272 ≈ *Writings*,
163–166‡.
*1893a. Grundgesetze der Arithmetik,
begriffsschriftlich abgeleitet 1*, Jena
(Pohle) = 1962, Hildesheim
(Olms). [Rs: Michaelis *1896a*,
Peano *1895a.* Part E: *1964a.*]
1894a. R of Husserl *1891a, Ztsch.
Phil. phil. Kritik n.s. 3*, 313–332 ≈
Writings, 179–192‡. [E: *Mind n.s. 81*
(1972), 321–337.]
1895a. "Kritische Beleuchtung einiger
Punkte in E. Schröders Vorlesung-
en über die Algebra der Logik",
Arch. syst. Phil. 1, 433–456 ≈
Writings, 193–210‡. [E: *1960a*, 86–
106.]
1895b. "Le nombre entier", *RMM 3*,
73–78 ≈ *Writings*, 211–219 [with
G]. [E: *Mind n.s. 79* (1970), 481–
486.]
1896a. "Über die Begriffsschrift des
Herrn Peano und meine eigene",
*Ber. Verh. Königl. Sächs. Gesell.
Wiss. Leipzig, math. phys. Kl. 48*,
361–378 ≈ *Writings*, 220–233‡.
1896b. "Lettera dal Sig. G. Frege
all'Editore", *RdM 6*, 53–59 ≈
Writings, 234–239 ≈ Peano *Works
2*, 288–294‡. [E: Dudman *1971a.*]

1899a. Über die Zahlen des Herrn H. Schubert, Jena (Pohle) ≈ *Writings*, 240–261‡.

1903a. Grundgesetze der Arithmetik, begriffsschriftlich abgeleitet 2, Jena (Pohle) ≈ 1962, Hildesheim (Olms). [Rs: Färber *1905a*. Part E: *1964a*.]

1903b. "Über die Grundlagen der Geometrie", *JDMV 12*, 319–324, 368–375 ≈ *Writings*, 262–272‡. [R: Dehn *1905b*. E: *1971a*, 22–37.]

1906a. "Über die Grundlagen der Geometrie", *JDMV 15*, 293–309, 377–403, 423–430 ≈ *Writings*, 281–323‡. [R: Dehn *1909a*. E: *1971a*, 49–112.]

1906b. "Antwort auf die Ferienplauderei des Herrn Thomae", *JDMV 15*, 586–590 ≈ *Writings*, 324–328.

m1906c. "Einleitung in die Logik", in *Manuscripts*, 201–212.

1908a. "Die Unmöglichkeit der Thomaeschen formalen Arithmetik aufs Neue nachgewiesen", *JDMV 17*, 52–55 ≈ *Writings*, 329–333.

1918a. "Logische Untersuchungen. Erster Teil. Der Gedanke", *Beiträge Phil. dtsch. Idealismus 1*, 58–77 ≈ *Writings*, 342–362‡. [E: Klemke *1968a*, 507–535.]

m1919a "Aufzeichnungen für Ludwig Darmstaedter", in *Manuscripts*, 273–277.

1919b. "Logische Untersuchungen. Zweiter Teil. Die Verneinung", *Beiträge Phil. dtsch. Idealismus 1*, 143–157 ≈ *Writings*, 362–378‡. [E: *1960a*, 117–135.]

1923a. "Logische Untersuchungen. Dritter Teil: Gedankengefüge", *Ibidem 3*, 36–51 ≈ *Writings*, 378–394‡. [E: Klemke *1968a*, 537–558.]

m1924a. (Eds. G. Gabriel and W. Kienzler), "Gottlob Freges politisches Tagebuch", *Dtsch. Zeit. Phil. 42* (1996), 1057–1098. [Diary kept 1923–1924.]

m1924b?. "Logische Allgemeinheit", in *Manuscripts*, 278–281. [Date conjectured.]

m1924–1925a. "Erkenntnisquelle der Mathematik und der mathematischen Naturwissenschaften", in *Manuscripts*, 286–294. [See also pp. 282–285, 298–303.]

1953a. The foundations of arithmetic$_2$, Oxford (Blackwell). [E of *1884b* by J. L. Austin.]

1960a. (E and ed. P. Geach and M. Black), *Translations from the philosophical writings of Gottlob Frege$_2$*, Oxford (Blackwell).

1964a. The basic laws of arithmetic, Berkeley and Los Angeles (University of California Press). [Part E of *1893a* and *1903a*, with int., by M. Furth.]

1971a. (E and ed. E.-H. Kluge), *On the foundations of geometry and formal theories of arithmetic*, New Haven and London (Yale UP).

1972a. (E and ed. T. W. Bynum), *Conceptual notation and related articles*, Oxford (Clarendon Press).

1986a. (Ed. C. Thiel), *Die Grundlagen der Arithmetik. Centenarausgabe*, Hamburg (Meiner). [Of *1884b*. R: Schirn *1988a*.]

1989a. (Eds. A. Janik and C. P. Berger), "Gottlob Frege. Briefe an Ludwig Wittgenstein", *Grazer phil. Studien 33–34*, 5–34.

Frege, K. A. (b. 1809)

1862a. Hilfsbuch zum Unterrichte in der deutschen Sprache für Kinder von 9 bis 13 Jahren$_3$, Wismar and Ludwigsluft (Hinstorff). [Not found: *see* Kreiser *1995a*.]

Freguglia, P.

1985a "Il calcolo geometrico ed i fondamenti della geometria", in Borga and others *1985a*, 174–236.

Frei, G.

1985a. (Ed.) *Der Briefwechsel zwischen David Hilbert und Felix Klein*,

Göttingen (Vandenhoeck und Ruprecht).

Frewer, M.
1981a. "Felix Bernstein", *JDMV 83*, 84–95.

Friedman, M.
1996a. "Overcoming metaphysics: Carnap and Heidegger", in Giere and Richardson *1996a*, 45–79.

Furth, M.
1964a. "Editor's introduction", in Frege *1964a*, v–lx.

Gabriel, G.
1972a. *Definitionen und Interessen*, Stuttgart (Frommann-Holzboog).

Gabriel, G. and Kienzler, W. *See also* Frege *m1924a*.
1997a. (Ed.) *Frege in Jena*, Würzburg (Königshausen & Neumann).

Gätschenberger, R. (b. 1865)
1920a. *SUMBOLA. Anfangsgründer der Erkenntnistheorie*, Karlsruhe (Braun).

Galdeano, Z. G. de (1846–1924)
1891a. R of Schröder *1890b*, *El prog. mat. 1*, 139–142, 194–203.
1892a. R of Schröder *1891a*, *El prog. mat. 2*, 354–361.

Galison, P.
1996a. "Constructing modernism: the cultural location of *Aufbau*", in Giere and Richardson *1996a*, 17–44.

García Bacca, J. (1901–1992)
1933a. "Simbólica (lógica)", in *Enciclopedia universal illustrada 9* (*Apendice*), Madrid (Espasa-Calpe), 1326–1399. [Ref. in Church *1938a*, 190 inaccurate.]
1934a. *Introducció a la logistica amb aplicacions a la filosofia i a las matematicas*, 2 vols., Barcelona (Institut d'Estudis Catalans). [Not found.]
1936a. *Introducción a la logica moderna*, Barcelona (Labor). [Rare: Complutense University, Madrid.]

Garciadiego, A. *See also* Moore, G. H. and Garciadiego
1992a. *Bertrand Russell and the origins of the set-theoretic "paradoxes"*, Basel (Birkhäuser).

Gårding, L.
1998a. *Mathematics and mathematicians. Mathematics in Sweden before 1950*, [no place] (AMS and LMS).

Gehman, H. M.
1927a. R of Hausdorff *1927a*, *Bull. AMS* (*2*)*34*, 778–781.

Genocchi, A. (1817–1889)
1884a. *Calcolo differenziale e principii di calcolo integrale, pubblicato con aggiunte dal Dr Giuseppe Peano*, Turin (Bocca). [G: *1898–1899a.*]
1898–1899a. *Differentialrechnung und Grundzüge der Integralrechnung herausgegeben von G. Peano*, 2 vols., Leipzig (Teubner). [G of *1884a.*]

Gentzen, G. (1909–1945)
Papers. (E and ed. M. E. Szabo), *The collected papers*, 1969, Amsterdam (North-Holland).
1935a. "Untersuchungen über das logische Schliessen", *Math. Ztsch. 39*, 176–210, 405–431. [E: *Papers*, 68–101‡.]
1936a. "Die Widerspruchsfreiheit der Stufenlogik", *Math. Ztsch. 41*, 357–366. [E: *Papers*, 214–222.]

George, A.
1994a. (Ed.), *Mathematics and mind*, New York and London (Oxford UP).

Gergonne, J. D. (1771–1859)
1816a. "Essai de dialectique rationnelle", *Ann. math. pures appl. 7*, 189–228.
1817a. "De l'analyse et de la synthèse dans les sciences mathématiques", *Ibidem 7*, 345–372.
1818a. "Essai sur la théorie des définitions", *Ibidem 9*, 1–35.
1821a. "Des équations fonctionnelles", *Ibidem 12*, 73–103. [F of Babbage *1820a.*]

Gerlach, H.-M. and Sepp, H. R.
1994a. (Eds.) *Husserl in Halle*, Bern (Lang).

Geymonat, L. (1908–1991)
1932a. "Sul significato filosofico di alcuni recenti indirizzi di logica matematica", *Arch. di filos. 4*, 263–282.
1986a. "L'opera di Peano di fronte alla cultura Italiana", in Peano *1986a'*, 7–15.

Geyser, J.
1909a. "Logistik und Relationslogik", *Philos. Jahrbuch 22*, 123–143.

Giekie, A. (1835–1924)
1923a. "Sir Alfred Bray Kempe", *Proc. Roy. Soc. London* (*A*)*102*, i–x.

Giere, R. N. and Richardson, A. W.
1996a. (Eds.) *Origins of logical empiricism*, Minneapolis and London (University of Minnesota Press).

Gilman, B. I. (1852–1933)
1883a. "Operations in relative number with applications to the theory of probabilities", in Peirce *1883a*, 107–126.
1892a. "On the properties of a one-dimensional manifold", *Mind n.s. 1*, 518–526.

Gilson, L.
1955a. Méthode et métaphysique selon Franz Brentano, Paris (Vrin).

Goblot, L. L. E. (1858–1935)
1918a. Traité de logique, Paris (Colin).

Gödel, K. (1906–1978)
Papers: Princeton (New Jersey), University (used in Gödel *Works*).
Works. Collected works, 4 vols., 1986–, New York (Oxford UP). [Includes also *en face* Es of all items written in German.]
1930a. "Einige metamathematische Resultate über Entscheidungsdefinitheit und Widerspruchsfreiheit", *Anz. Akad. Wiss. Wien 67*, 214–215 ≈ *Works 1*, 140–143.
1930b. "Die Vollständigkeit der Axiome des logischen Funktionenkalküls", *Monats. Math. Physik 37*,

349–360 ≈ *Works 1*, 102–123. [E: van Heijenoort *1967a*, 582–591.]
1931a. "Über formal unentscheidbare Sätze der Principia Mathematica und verwandter Systeme", *Monats. Math. Physik 38*, 173–198‡ ≈ *Works 1*, 145–195. [Rs: Dubislav *1932a*, Fraenkel *1938a*. Es: *1962a'*; Davis *1965a*, 4–38; van Heijenoort *1967a*, 596–616.]
1931b. R of Hilbert *1931a*, *ZfM 1*, 260 ≈ *Works 1*, 212–213.
1932a. "Nachtrag", *Erkenntnis 2*, 149–151 ≈ *Works 1*, 202–205.
1932b. "Über Vollständigkeit und Widerspruchsfreiheit", *Ergebnisse eines Mathematischen Kolloquiums no. 3* (1930–31), 12–13 ≈ *Works 1*, 234–236 = Menger *1998a*, 168–169.
1932c. R of Carnap *1931c*, Heyting *1931a* and von Neumann *1931a*, *ZfM 2*, 321–322 ≈ *Works 1*, 242–249.
1944a. "Russell's mathematical logic", in Schilpp *1944a*, 123–153‡ ≈ *Works 2*, 119–143 ≈ Pears *1972a*, 192–226 ≈ Benacerraf and Putnam *1985a*, 447–469. [G: see Whitehead and Russell *1932a*.]
1947a. "What is Cantor's continuum problem?", *Amer. math. monthly 54*, 515–525. [Revised in Benacerraf and Putnam *1985a*, 470–485 ≈ *Works 2*, 176–188.]
1962a'. On formally undecidable propositions of PM, Edinburgh and London (Nelson). [Contains an E of *1931a*.]

Goldfarb, W.
1979a. "Logic in the twenties: the nature of the quantifier", *JSL 44*, 351–368.
1989a. "Russell's reasons for ramification", in C. W. Savage and C. A. Anderson (eds.), *Rereading Russell*, Minneapolis (University of Minnesota Press), 24–40.

Gonseth, F. (1890–1974)

Papers: Lausanne (Switzerland), University.

1926a. Les fondements des mathématiques, Paris (Blanchard).

Gordon, W. T.

1990a. C. K. Ogden: a bio-bibliographical study, Metuchen, New Jersey and London (Scarecrow).

Grassmann, H. G. (1809–1877)

Works. Gesammelte mathematische und physikalische Werke, 3 vols., each in 2 pts., 1894–1911, Leipzig (Teubner) = 1972, New York (Johnson).

1844a. Die lineale Ausdehnungslehre, Leipzig (Wiegand) ≈ *Works 1*, pt. 1. [E (by L. C. Kannenberg): *A new branch of mathematics*, 1995, Chicago and La Salle, Ill. (Open Court).]

1861a. Lehrbuch der Mathematik für höhere Lehranstalten pt. 1, Berlin (Enslin).

1862a. Die Ausdehnungslehre, vollständig und in strenger Form bearbeitet, Berlin (Enslin) ≈ *Works 1*, pt. 2.

1878a'. R. Sturm, E. Schröder and L. Sohnke, "Hermann Grassmann", *MA 14*, 1–45.

Grassmann, R. (1815–1901)

1872a. Die Formenlehre oder Mathematik, Stettin (Grassmann) = 1966, Hildesheim (Olms).

Gratry, A. J. A. (1805–1872)

1855a. Philosophie—Logique$_5$, Paris (Douniot). [E: *Logic* (E and int. by H. and M. Singer), 1944, La Salle, Ill. (Open Court).]

Grattan-Guinness, I.

1970a. The development of the foundations of mathematical analysis from Euler to Riemann, Cambridge, Mass. (MIT Press).

1970b. "An unpublished paper by Georg Cantor: *Principien einer Theorie der Ordnungstypen. Erste Mittheilung*", *AM 124*, 65–107. [Cantor *m1885a.*]

1971a. "The correspondence between Georg Cantor and Philip Jourdain", *JDMV 73*, pt. 1, 111–130.

1971b. "Materials for the history of mathematics in the Institut Mittag-Leffler", *Isis 62*, 363–374.

1971c. "Towards a biography of Georg Cantor", *Ann. of sci. 27*, 345–391.

1972a. "A mathematical union: William Henry and Grace Chisholm Young", *Ann. of sci. 29*, 105–186.

1973a. "Not from nowhere. History and philosophy behind mathematical education", *Int. j. math. educ. sci. tech. 4*, 421–453.

1974a. "Russell's home at Bagley Wood", *Russell no. 13*, 24–26.

1974b. "The rediscovery of the Cantor-Dedekind correspondence", *JDMV 76*, pt. 1, 104–139.

1974c. "Achilles is still running", *Trans. C. S. Peirce Soc. 10*, 8–16.

1974d. "The Russell Archives: some new light on Russell's logicism", *Ann. of sci. 31*, 387–406.

1975a. "Russell's election to a Fellowship of the Royal Society", *Russell no. 17*, 23–26.

1975b. "Wiener on the logics of Russell and Schröder. An account of his doctoral thesis, and of his subsequent discussion of it with Russell", *Ann. of sci. 32*, 103–132.

1975c. "The Royal Society's financial support of the publication of *PM*", *Notes rec. Roy. Soc. London 30*, 89–104.

1975d. "Preliminary notes on the historical significance of quantification and of the axioms of choice in the development of mathematical analysis", *HM 2*, 475–488.

1977a. "The Gergonne relations and the intuitive use of Euler and Venn diagrams", *Int. j. math. educ. sci. tech.* 8, 23–30.

1977b. Dear Russell–dear Jourdain, London (Duckworth) and New York (Columbia UP).

1978a. "How Bertrand Russell discovered his paradox", *HM 5*, 127–137.

1979a "In memoriam Kurt Gödel: his 1931 correspondence with Zermelo on his incompletability theorem", *HM 6*, 294–304.

1980a. "Georg Cantor's influence on Bertrand Russell", *HPL 1*, 61–93.

1981a. "Are there paradoxes of the set of all sets?", *Int. j. math. educ. sci. tech. 12*, 9–18.

1981b. "On the development of logics between the two world wars", *Amer. math. monthly 88*, 495–509.

1982a. "Psychology in the foundations of logic and mathematics: the cases of Boole, Cantor and Brouwer", *HPL 3*, 33–53 ≈ *Psicoanalisi e storia della scienza*, 1983, Florence (Olschki), 93–121.

1983a. "Psychical research and parapsychology; notes on the development of two disciplines", in W. Roll, J. Beloff and R. White (eds.), *Research in parapsychology 1982*, Metuchen, New Jersey (Scarecrow Press), 283–304.

1985a. "Mathematics and mathematical physics at Cambridge, 1815–40", in P. Harman (ed.), *Wranglers and physicists. Cambridge physics in the nineteenth century*, Manchester (Manchester UP), 84–111.

1985b. "Bertrand Russell's logical manuscripts: an apprehensive brief", *HPL 6*, 53–74.

1986a. "Discovering Whitehead", *Trans. C. S. Peirce Soc. 22*, 61–68.

1986b. "From Weierstrass to Russell: a Peano medley", in Peano *1986a'*, 17–31 ≈ *Rev. stor. sci. 2* (1985: publ. 1987), 1–16.

1986c. "Russell's logicism versus Oxbridge logics, 1890–1925", *Russell, n.s. 5*, 101–131.

1987a. "What was and what should be the calculus?", in (ed.), *History in mathematics education*, Paris (Belin), 116–135.

1988a. "*Grandes écoles, petite Université*: some puzzled remarks on higher education in mathematics in France, 1795–1840", *Hist. of univs. 7*, 197–225.

1988b. "Living together and living apart: on the interactions between mathematics and logics from the French Revolution to the First World War", *South African j. phil. 7*, no. 2, 73–82.

1990a. Convolutions in French mathematics, 1800–1840. From the calculus and mechanics to mathematical analysis and mathematical physics, 3 vols., Basel (Birkhäuser) and Berlin (Deutscher Verlag der Wissenschaften).

1990b. "Bertrand Russell (1872–1970) after twenty years", *Notes rec. Roy. Soc. London 44*, 280–306.

1991a. "The Hon. Bertrand Russell and *The educational times*", *Russell n.s. 11*, 86–91.

1991b. "The correspondence between George Boole and Stanley Jevons, 1863–1864", *HPL 12*, 15–35.

1992a. "Russell and G. H. Hardy: a study of their relationship", *Russell n.s. 11*, 165–179.

1992b. "Charles Babbage as an algorithmic thinker", *Ann. hist. computing 14*, no. 3, 34–48.

1994a. (Ed.), *Companion encyclopedia of the history and philosophy of the mathematical sciences*, 2 vols., London (Routledge).

1996a. " 'I never felt any bitterness': Alys Russell's interpretation of her separation from Bertie", *Russell n.s 16*, 37–44.

1997a. "Peirce between logic and mathematics", in Houser and others *1997a*, 23–42.

1997b. "How did Russell write *The principles of mathematics* (1903)?", *Russell n.s. 16*, 101–127.

1997c. "A retreat from holisms: Carnap's logical course, 1921–1943", *Ann. of sci. 54*, 407–421.

1997d. "Benjamin Peirce's *Linear Associative Algebra* (1870): new light on its preparation and 'publication'", *Ann. of sci. 54*, 597–606.

1998a. "Karl Popper for and against Bertrand Russell", *Russell, n.s. 18*, 25–42.

1998b. "Structural similarity or structuralism? Comments on" Priest *1994a, Mind n.s. 107*, 823–834.

Graves, R. P.
  1889a. Life of Sir William Rowan Hamilton 3, London (Hodges, Figges).

Gray, J. D.
  1994a. "Georg Cantor and transcendental numbers", *Amer. math. monthly 101*, 819–832.

Greenhill, A. G. (1847–1927)
  1923a. "Mathematics of reality and metamathematics", *Math. gaz. 11*, 358–367.

Greg, W. W. (1875–1959)
  1927a. The calculus of variants, Oxford (Clarendon Press).

  1932a. "Bibliography—an apologia", *The library 13*, 113–143 ≈ *Collected papers*, 1966, Oxford (Clarendon Press), 239–266‡.

Gregory, D. F. (1813–1844)
  1839a. "On the elementary principles of the application of algebraic symbols to geometry", *Cambridge math. j. 2*, 1–9 ≈ *The mathematical writings*, 1865, Cambridge (Deighton, Bell), 150–162.

  1841a. Examples of the processes of the differential and integral calculus$_1$, Cambridge (Deighton, Bell).

Grelling, K. (1886–1942)
  1910a. Die Axiome der Arithmetik mit besondere Berücksichtigung der Beziehungen zur Mengenlehre, Göttingen (Kaestner). [*Dissertation.*]

  1924a. Mengenlehre, Leipzig (Teubner).

  1928a. "Philosophy of the exact sciences: its present status in Germany", *The monist 38*, 97–119.

Grelling, K. and Nelson, L. (1882–1927)
  1908a. "Bemerkungen zu den Paradoxien von Russell und Burali-Forti" and three appendices, *Abh. Fries'schen Schule* (2)2, 301–334 ≈ Nelson, *Beiträge zur Philosophie der Logik und Mathematik*, Frankfurt (Öffentliches Leben), 55–86.

Griffin, N.
  1980a. "Russell on the nature of logic (1903–1913)", *Synthese 45*, 117–188.

  1985a. "Russell's multiple relation theory of judgement", *Phil. studies 47*, 213–247.

  1991a. Russell's idealist apprenticeship, Oxford (Clarendon Press).

  1996a. "Denoting concepts in *The principles of mathematics*", in Monk and Palmer *1996a*, 23–64.

Griffin, N. and Lewis, A. C. *See also* A. C. Lewis
  1990a. "Bertrand Russell's mathematical education", *Notes rec. Roy. Soc. London 44*, 51–71.

Grosche, G. *See* Kreiser and Grosche

Grunsky, H. (1904–1986)
  1935a. R of Weyl *1927b, JFM 52* (1926), 40–43.

Guillemot, M. *See* Cassinet and Guillemot

Gumbel, E. J.
  1924a. R of Russell *1919b, JFM 47* (1919–1920), 36–38.

Gutzmer, A. (1860–1924)
  1904a. "Geschichte der *DMV*", *JDMV 10*, pt. 1, 1–49 ≈ Leipzig (Teubner) [with contents of *1–10*].

Haack, S.
1993a. "Peirce and logicism: notes towards an exposition", *Trans. C. S. Peirce Soc. 29*, 33–56.

Haas, G. and E. Stemmler
1981a. Der Nachlass Heinrich Behmann (1891–1970). Gesamtverzeichnis, Aachen (Technical High School).

Hadamard, J. (1865–1963) *See also* Letters
Works. Oeuvres, 4 vols., 1968, Paris (CNRS).
1898a. "Sur certaines applications possibles de la théorie des ensembles", in Rudio *1898a,* 201–202 = *Works 1,* 311–312.

Hager, P.
1994a. Continuity and change in the development of Russell's philosophy, Dordrecht (Kluwer).

Hahn, H. (1879–1934)
Papers. (Eds. L. Schmetterer and K. Sigmund), *Gesammelte Abhandlungen 3,* 1997, Vienna and New York (Springer).

Hailperin, T.
1984a. "Boole's abandoned propositional logic", *HPL 5,* 39–48.
1986a. Boole's logic and probability$_2$, Amsterdam (North-Holland).

Hall, R.
1972a. "Unnoticed terms in logic", *Notes and queries 217,* 131–137, 165–171, 203–209.

Haller, R. and Stadler, F. *See also* Stadler
1993a. (Eds.) *Der Aufstieg der wissenschaftliche Philosophie,* Vienna (Hölder-Pichler-Tempsky).

Hallett, M.
1984a. Cantorian set theory and limitation of size, Oxford (Clarendon Press).
1994a. "Hilbert's axiomatic method and the laws of thought", in George *1994a,* 158–200.
1995a. "Hilbert and logic", in M. Marion and R. Cohen (eds.), *Que-*
bec studies in the philosophy of science, pt. 1, Dordrecht (Kluwer), 183–257.

Halsted, G. B. (1853–1922)
1878a. "Boole's logical method", *J. specul. phil. 12,* 81–91. [R: Ulrici *1878a.*]
1878b. "Professor Jevons's criticism of Boole's logical system", *Mind 3,* 134–137.

Hamilton, W. (1788–1856)
1833a. R of Bentham *1827a,* Whately *Logic$_3$* (1829) and other books on logic, *Edinburgh rev. 27,* 194–238 ≈ *Discussions on philosophy$_1$,* 1852, London and Edinburgh ([various]), 116–174 ≈ $_2$1853, 118–175.

Hankel, H. (1839–1873)
1870a. Untersuchungen über die unendlich oft oscillierenden und unstetigen Functionen, Tübingen [*Dissertation*] ≈ *MA 20* (1882), 63–112 ≈ (Ed. P. E. B. Jourdain), 1905, Leipzig (Engelsmann), 44–102.
1871a. "Grenze", in *Allgemeine Encyclopädie der Wissenschaften und Künste,* sect. 1, pt. 90, Leipzig, 185–211.

Hannequin, A. (1856–1905)
1895a. Essai critique sur l'hypothèse des atomes dans la science contemporaine, Paris (Masson) ≈ *Annales Univ. Lyon 7.* [R: Russell *1896b.*]

Hardwick, C. S.
1977a. (Ed. with J. Cook), *Semiotics and significs. The correspondence between Charles S. Peirce and Victoria Lady Welby,* Bloomington (Indiana UP).

Hardy, G. H. (1877–1947)
Papers. (Eds. various), *Collected papers 7,* 1979, Oxford (Clarendon Press).
1903a. "A theorem concerning the infinite cardinal numbers", *Quart. j. pure appl. maths. 35,* 87–94 = *Papers,* 427–434‡. [F: C & G, 265–276.]

1903b. R of Russell *1903a, Times literary suppl.*, 263 = *Papers*, 851–854. [Anonymous.]

1911a. R of *PM*₁ *1, Times literary suppl.*, 321–322 = *Papers*, 859–862. [Anonymous.]

1918a. "Sir George Stokes and the concept of uniform convergence", *Proc. Cambridge Phil. Soc. 19,* 148–156 = *Papers*, 505–513 ≈ Ewald *1996a*, 1235–1242.

*1924a. Orders of infinity*₂, Cambridge (Cambridge UP).

1929a. "Mathematical proof", *Mind n.s. 38,* 1–25 = *Papers*, 581–606 ≈ Ewald *1996a*, 1243–1263.

1942a. Bertrand Russell and Trinity: a college controversy of the last war, Cambridge [private] = 1970, Cambridge (Cambridge UP) = 1977, New York (Arno).

Harley, R. (1828–1910)

1866a. George Boole, F. R. S.", *British quart. rev. 41,* 141–181 ≈ Boole *Studies*, 425–472.

1867a. "Remarks on Boole's mathematical analysis of logic", *Rep. Brit. Ass. Adv. Sci.* (1866), pt. 2, 3–6.

1871a. "On Boole's 'Laws of thought'", *Ibidem* (1870), pt. 2, 14–15.

Harrell, M.

1988a. "Extension to geometry of *PM* and related systems II", *Russell n.s. 8,* 140–160.

Hasse, H. (1898–1979) and Scholz, H.
See also Scholz

1928a. "Die Grundlagenkrise der griechischen Mathematik", *Kantstudien 33,* 4–34.

Hausdorff, F. (1868–1942)

1904a. "Der Potenzbegreff in der Mengenlehre", *JDMV 13,* 569–571. [F: C & G, 445–450.]

1905a. R of Russell *1903a, Vrtlj. wiss. Philos. Soz. 29,* 119–124.

1906a, 1907a. "Untersuchungen über Ordnungstypen", *Ber. Verh. Königl.*

Sächs. Akad. Wiss. Leipzig 58, 106–169; *59,* 84–159.

1908a. "Grundzüge einer Theorie der geordneten Mengen", *MA 65,* 435–505.

1909a. "Die Graduierung nach dem Endverlauf", *Abh. Königl. Sächs. Akad. Wiss. Leipzig, math.-phys. Kl. 31,* 297–334.

1914a. Grundzüge der Mengenlehre, Leipzig (de Gruyter) = 1949, New York (Chelsea). [R: Blumberg *1920a.*]

1927a. Mengenlehre, Berlin and Leipzig (de Gruyter). [R: Gehman *1927a.* E: *Set theory*₁, 1957, New York (Chelsea).]

Hawkins, B. S.

1997a. "Peirce and Russell: the history of a neglected 'controversy'", in Houser and others *1997a,* 111–146.

Hawkins, T. W.

1970a. Lebesgue's theory of integration, Madison (University of Wisconsin Press) = 1975, New York (Chelsea).

Hawtrey, R. W. (1879–1974)
Papers: Cambridge (England), Churchill College.

Heck, R. G. Jr.

1993a. "The development of arithmetic in Frege's *Grundgesetze der Arithmetik*", *JSL 58,* 579–601 ≈ Demopoulos *1994a,* 257–294.

Hedrick, E. R. (1876–1943)

1933a. "Tendencies in the logic of mathematics", *Science 77,* 335–343.

Heidegger, M. (1889–1976)

1912a. "Neuere Forschungen über Logik", *Liter. Rundschau Kathol. Dtschld. no. 38,* cols. 465–472, 517–524, 565–570 ≈ *Gesamtausgabe 1,* 1978, Frankfurt/Main (Klostermann), 17–43.

Heine, E. H. (1821–1881)

1870a. "Über trigonometrische Reihen", *J. rei. ang. Math. 71,* 353–365.

1872a. "Die Elemente der Function-enlehre", *J. rei. ang. Math. 74*, 172–188.

Heinzmann, G.
1986a. (Ed.) *Poincaré, Russell, Zer-melo et Peano*, Paris (Blanchard). [Mostly photoreprints.]

Helmholtz, H. L. von (1821–1894)
Writings. (Eds. P. Hertz and M. Schlick, rev. R. S. Cohen and Y. Elkana and E by M. F. Lowe), *Epistemological writings*, 1977, Dordrecht (Reidel),
1878a. *Die Tatsachen in der Wahrnehmung*, Berlin (Hirschwald). [Many ≈ s and Ts. E with notes: *Writings*, 115–185‡ ≈ Ewald *1996a*, 689–727.]
1878b. "The origin and meaning of geometrical axioms", pt. 2, *Mind 3*, 212–225. [Part ≈ : Ewald *1996a*, 685–689. Another E, with notes: *Writings*, 1–38.]

Hempel, C. (1905–1997)
1933a. R of Kaufmann *1930a*, *JFM 56* (1930), 39–40.
1938a. R of Carnap *1931c*, Heyting *1931a*, von Neumann *1931a* and Vienna Circle *1931a*, *JFM 57* (1931), 52–54.

Henderson, H.
1993a. *Catalyst for controversy. Paul Carus of Open Court*, Carbondale, Ill. (Southern Illinois UP).

Henkin, L.
1962a. "Are logic and mathematics identical?", *Science 138*, 788–794.

Henry, D. P. *See* Mays and Henry
Herbertz, R. (b. 1878)
1912a. *Die philosophische Literatur. Ein Studienführer*, Stuttgart (Spermann).

Herbrand, J. (1908–1931)
Writings. (Ed. J. van Heijenoort), *Ecrits logiques*, Paris (PUF). [E: *1971a*.]
1930a. "Recherches sur la théorie de la démonstration", *Prace Tow. Nauk. Warsaw (3)*, no. 33 ≈

Writings, 35–153. [E: *1971a*, 44–202; part ≈ van Heijenoort *1967a*, 525–581.]
1930b. "Les bases de la logique Hilbertienne", *RMM 37*, 243–255 ≈ *Writings*, 155–166. [E: *1971a*, 302–324.]
1971a. (E and ed. W. Goldfarb), *Logical writings*, Dordrecht (Reidel).

Hermann, I.
1949a. "Denkpsychologische Betrachtungen im Gebiet der mathematischen Mengenlehre", *Schweiz. Ztsch. Psych. 8*, 189–231. [F: *Parallélismes*, 1980, Paris (Denoël), 227–242.]

Hermes, H. *See* Scholz and Hermes
Hesseling, D. E.
1999a. "Gnomes in the fog. The reception of Brower's intuitionism in the 1920s", Utrecht University doctorate. [To be published commercially.]

Hessenberg, G. (1874–1925)
1906a. "Grundbegriffe der Mengen-lehre", *Abh. Fries'schen Schule (2)1*, 479–706 ≈ Göttingen (Vandenhoeck und Ruprecht).
1922a. *Von Sinn der Zahlen*, Leipzig (Verlag der Neue Geist).

Heyting, A. (1898–1980)
1931a. "Die intuitionistische Grundlegung der Mathematik", *Erkenntnis 2*, 103–115. [Rs: Gödel *1932a*, Hempel *1938a*. E: Benacer-raf and Putnam *1985a*, 52–61.]
1934a. *Mathematische Grundlagen-forschung. Intuitionismus. Beweis-theorie*, Berlin (Springer).

Hilbert, D. (1862–1943)
Papers: Göttingen (Germany), University.
Papers. Gesammelte Abhandlungen, 3 vols., 1932–1935, Berlin (Springer) = 1970 = 1966, New York (Chelsea).
Geometry. Grundlagen der Geometrie₁, 1899, Leipzig (Teubner). [Many later eds., especially₇ (1930). F of

$_1$: *1900b*. E of $_1$: *1902a*. Rs of $_1$: Engel *1901a*, Veblen *1903a*. R of $_2$: Dehn *1906a*.]

1891a. "Über die stetige Abbildung einer Linie auf ein Flächenstück", *MA 38*, 359–360 ≈ *Papers 3*, 1–2.

1893a. R of Peano *1890a*, *JFM 22* (1890), 405–406.

1894a. "Über die gerade Linie als kürzeste Verbindung zwischen zweier Punkte", *MA 46*, 91–96 ≈ *Geometry*$_7$ (1930), 126–132 [from $_2$ (1903) onwards].

1900a. "Über den Zahlbegriff", *JDMV 8*, pt. 1, 180–184 ≈ *Geometry*$_7$ (1930), 241–246 [from $_3$ (1909) onwards].

1900b. "Les principes fondementaux de la géométrie", *Ann. sci. Ecole Normale Sup.* (*3*)*17*, 103–209. [F of *Geometry*$_1$ (1899).]

1900c. "Mathematische Probleme", *Nachr. Königl. Gesell. Wiss. Göttingen, math. phys. Kl.*, 253–297 ≈ *Archiv Math. Physik* (*3*)*1* (1901), 44–63, 213–237 [with some additions] ≈ *Papers 3*, 290–309. [E: *Bull. AMS 8* (1902), 437–479. F: Duporcq *1902a*, 58–114. Various other ≈ s and Ts.]

1902a. *The foundations of geometry*$_1$, Chicago (Open Court) = 1947. [E of *Geometry*$_1$ (1899). $_2$ 1971, La Salle, Ill. (Open Court).]

1902b. "Über die Grundlagen der Geometrie", *MA 56*, 381–422 ≈ *Geometry*$_7$ (1930), 178–230.

m1905a. "Logische Principien des mathematischen Denkens", versions in Papers and in Göttingen University Library‡. [Edition to be prepared.]

1905b. "Über die Grundlagen der Logik und Arithmetik", in Krazer *1905a*, 174–185 ≈ *Geometry*$_7$ (1930), 247–261 [from $_3$ (1909) onwards]. [Es: *The monist 15* (1905), 338–352; van Heijenoort *1967a*, 129–138.]

m1917a. "Mengenlehre", Mathematical Institute, Göttingen. [Lecture course.]

1917b. "Axiomatisches Denken", *Actes Soc. Helvét. Sci. Natur. 99*, pt. 2, 139–130. [F: *L'ens. math.* (*1*)*19*, 330–331. Full paper: *1918b*.]

m1918a. (Ed. P. Bernays), "Prinzipien der Mathematik", Mathematical Institute, Göttingen. [Lecture course.]

1918b. "Axiomatisches Denken", *MA 78*, 405–415 ≈ *Papers 3*, 146–156‡. [R: Bernays *1923a*. F: *L'ens. math.* (*1*)*20* (1918), 122–130. Es: *Phil. math. 7* (1971), 1–12; Ewald *1996a*, 1107–1115.]

m1921a. (Ed. P. Bernays), "Logik-Kalkül", Mathematical Institute, Göttingen. [Lecture course.]

1922a. "Neubegründung der Mathematik", *Abh. Math. Seminar Hamburg Univ. 1*, 157–177 ≈ *Papers 3*, 157–177 [*sic*]‡. [E: Ewald *1996a*, 1115–1134 ≈ Mancosu *1998a*, 198–214.]

m1922–1923a. "Logische Grundlagen der Mathematik", Mathematical Institute, Göttingen. [Lecture course.]

1923a. "Die logischen Grundlagen der Mathematik", *MA 88*, 151–165 ≈ *Papers 3*, 178–191‡. [E: Ewald *1996a*, 1136–1148.]

1926a. "Über das Unendliche", *MA 95*, 161–190. [Parts in *JDMV 36* (1927), pt. 1, 201–215; and *Geometry*$_7$ (1930), 262–288. F: *AM 48* (1926), 91–122. E: van Heijenoort *1967a*, 367–392.]

1929a. "Probleme der Grundlegung der Mathematik", *MA 102*, 1–9‡ ≈ *Geometry*$_7$ (1930), 313–323. [E: Mancosu *1998a*, 227–233.]

1930a. "Naturkennen und Logik", *Die Naturwiss. 18*, 959–963 ≈ *Papers 3*, 378–387. [F: *L'ens. math.* (*1*)*30* (1931), 22–33. E: Ewald *1996a*, 1157–1165.]

1931a. "Die Grundlegung der elementaren Zahlenlehre", *MA 104*, 485–494‡. [Part in *Papers 3*, 192–195. R: Gödel *1931b*. E: Ewald *1996a*, 1149–1157 ≈ Mancosu *1998a*, 266–274.]

Hilbert, D. and Ackermann, W. *See also* Ackermann

1928a. Grundzüge der theoretischen Logik₁, Berlin (Springer).

Hilbert, D. and Bernays, P. *See also* Bernays

1934a, 1939a. Grundlagen der Mathematik, 2 vols., Berlin (Springer).

Hill, C. O.

1991a. Word and object in Husserl, Frege and Russell, Athens, Ohio (Ohio UP).

1994a. "Frege's attack on Husserl and Cantor", *The monist 77*, 345–357.

1995a. "Husserl and Hilbert on completeness", in Hintikka *1995a*, 143–163.

1997a. "Did Georg Cantor influence Edmund Husserl?", *Synthese 113*, 145–170.

1997b. Rethinking identity and metaphysics, New Haven and London (Yale UP).

Hintikka, J.

1988a. "On the development of the model-theoretic viewpoint in logical theory", *Synthese 77*, 1–76.

1995a. (Ed.) *From Dedekind to Gödel*, Dordrecht (Kluwer).

1995b. "Standard vs. non-standard distinction: a watershed in the foundations of mathematics", in *1995a*, 21–44.

1996a. The principles of mathematics revisited, Cambridge (Cambridge UP).

Hobson, E. W. (1856–1933)

1905a. "On the general theory of transfinite numbers and order types", *Proc. LMS (2)3*, 170–188. [F: C & G, 277–300.]

Hobson, E. W. and Love, A. E. H. (1863–1940)

1913a. (Eds.) *Fifth International Congress of Mathematicians. Proceedings*, 2 vols., Cambridge (Cambridge UP) = 1967, Liechtenstein (Kraus).

Hodges, A.

1983a. Alan Turing. The enigma, London (Burnett).

Höfler, A. (1853–1922)

1922a. Logik₂, Vienna and Leipzig (Hölder-Pichler-Tempsky).

Hölder, L. O. (1859–1937)

1914a. Die Arithmetik in strenger Begründung, Leipzig (Teubner).

1924a. Die mathematische Methode, Berlin (Springer) = 1978. [Rs: Fraenkel *1927a*, A. R. Schweitzer *1926a*, Wrinch *1925a*.]

Houser, N. *See also* Anellis and Houser

1991a. "The Schröder-Peirce correspondence", *Modern logic 1*, 206–236.

1991b. "Peirce and the law of distribution", in Drucker *1991a*, 10–32.

1992a. "The fortune and misfortunes of the Peirce Papers", in M. Belat and J. Deledalle-Rhodes (eds.), *Signs of humanity*, Berlin (Mouton, de Gruyter), 1259–1268.

1993a. "Peirce and logicism: a response to" Haack *1993a*, *Trans. C. S. Peirce Soc. 29*, 57–68.

Houser, N., Roberts, D. and van Evra, J. *See also* Roberts; van Evra

1997a. (Eds.) *Studies in the logic of Charles S. Peirce*, Bloomington (Indiana UP).

Howard, D.

1996a. "Relativity, *Eindeutigkeit* and monomorphism", in Giere and Richardson *1996a*, 115–164.

Huntington, E. V. (1874–1952)

1901a. Über die Grund-Operationen an absoluten und complexen Grössen in geometrischen Behandlung, Braunschweig (Vieweg). [*Dissertation*, Strassburg University.]

1902a. "A complete set of postulates for the theory of absolute continuous magnitudes", *Trans. AMS 3*, 264–279.

1902b. "Complete sets of postulates for the theory of positive integral and positive rational numbers", *Ibidem 3*, 280–284.

1903a. "Complete sets of postulates for the theory of real quantities", *Ibidem 4*, 358–370.

1904a. "Sets of independent postulates for the algebra of logic", *Ibidem 5*, 288–309.

1905a, 1905b. "The continuum as a type of order", *Ann. maths.* (*2*)6, 151–184; (*2*)7, 15–43 ≈ 1905, Cambridge, Mass. (Harvard University Publications Office). [₂ *The continuum and other types of serial order. With an introduction to Cantor's transfinite numbers*, 1917, Cambridge, Mass. (Harvard UP) = 1955, New York (Dover).]

1933a. "New sets of independent postulates for the algebra of logic, with special reference to *PM*", *Trans. AMS 35*, 274–304.

Huntington, E. V. and Ladd Franklin, C. *See also* Ladd

1905a. "Logic, symbolic", in F. C. Beach (ed.), *The Americana 6*, 6 pp. (unpaginated) ≈ [new ed.] *17* (1934), 568–573.

Hurwitz, A. (1859–1919)
Papers: Göttingen (Germany), University.

1898a. "Über die Entwickelung der allgemeinen Theorie der analytischen Functionen in neuerer Zeit", in Rudio *1898a*, 91–112 ≈ *Mathematische Werke 1*, 1932, Basel (Birkhäuser) (= 1962), 461–480.

Husserl, E. (1859–1928)
Works. (Eds. various), *Husserliana*, 1950–, Den Haag (Nijhoff), then Dordrecht (Kluwer).

Letters. Briefwechsel, 10 vols., 1994, Dordrecht (Kluwer).

m1882a. "Beiträge zur Theorie der Variationsrechnung", in G. Scrimieri, *Analitica matematica e fenomenologica in Edmund Husserl*, Bari, 39–60. [*Dissertation*, Vienna University (Biermann *1969a*).]

1887a. Über den Begriff der Zahl. Psychologische Analyse, Halle/Saale (Heynemans) ≈ *Works 12*, 289–339‡. [*Habilitation.* E: *1981a*, 92–118.]

1891a. Philosophie der Arithmetik 1 [and only], Halle/Saale (Pfeffer) ≈ *Works 12*, 5–283‡. [Rs: Michaelis *1894b*, Tannery *1892a*.]

1891b. R of Schröder *1890a*, Göttingen gel. Anz., 243–278 ≈ *Works 22*, 3–43‡. [E: *1994a*, 52–91.]

1900a, 1901a. Logische Untersuchungen₁, 2 vols., Halle/Saale (Niemeyer). [₂ 1913–1921 ≈ *Works 18–19*. E: *1970a*.]

m1901b. "Das Imaginäre in der Mathematik", in *Works 12*, 430–451.

1919a. "Erinnerungen auf Franz Brentano", in O. Kraus (ed.), *Franz Brentano*, Munich (Beck), 151–167 ≈ *Works 25*, 304–315‡. [E: *1981a*, 342–349.]

1929a. "Formale und transzendentale Logik", *Jbch. Phil. phänom. Forschung 10*, 1–298 ≈ Halle/Saale (Niemeyer) ≈ *Works 17*, 1–339.

1970a. Logical investigations, London (RKP)‡. [E of *1900a*₂ and *1901a*₂ by J. N. Findlay.]

1981a. (E and eds. P. McCormack and F. A. Elliston), *Shorter works*, Notre Dame (University of Notre Dame Press).

1994a. (E and ed. D. Willard), *Early writings on the philosophy of logic and mathematics*, Dordrecht (Kluwer).

Ibragimoff, S. G.
  1966a. "O zabitikh rabotakh Ernsta Shrodera", *Ist.-mat. issled.* (*1*)17, 247–258.

Iglesias, T.
  1977a. "Russell's introduction to Wittgenstein's 'Tractatus'", *Russell nos. 25–28*, 21–38.
  1984a. "Russell's theory of knowledge and Wittgenstein's earliest writings", *Synthese 60*, 285–332.

Ilgauds, K.-H. *See also* Purkert and Ilgauds
  1982a. "Zur Biographie Georg Cantors: Georg Cantor und die Bacon-Shakespeare-Theorie", *Schr. Gesch. Naturwiss. Tech. Med. 19*, no. 2, 31–49.

Irvine, A. D. and Wedeking, G. A.
  1993a. (Eds.) *Russell and analytic philosophy*, Toronto (University of Toronto Press).

Jadacki, J. J.
  1986a. "Leon Chwistek—Bertrand Russell's correspondence", *Dialectic and humanism 13*, 240–263.

Jager, R.
  1960a. "Russell's denoting complex", *Analysis 20*, 53–62.

Jané, I.
  1995a. "The role of the absolute infinite in Cantor's conception of set", *Erkenntnis 42*, 375–402.

Jaśkowski, S. (1906–1965)
  1934a. "On the rules of suppositions in formal logic", *Studia logica 1*, 5–32 ≈ McCall *1967a*, 232–258.

Jevons, W. S. (1835–1882)
  Papers: Manchester (England), Manchester University.
  Letters. (Ed. H. A. Jevons), *Letters and journal*, 1886, London (MacMillan).
  Works. (Eds. R. Adamson and H. A. Jevons), *Pure logic and other minor works*, 1890, London (MacMillan) = 1991, Bristol (Thoemmes).

  1864a. Pure logic, London (Stanford) ≈ *Works*, 1–77.
  1866a. "On a logical abacus", *Proc. Manchester Phil. Soc. 5*, 161–165.
  1870a. "On the mechanical performance of logical inference", *Phil. trans. Roy. Soc. London 160*, 497–518 ≈ *Works*, 137–172.
  1873a. "Who discovered the quantification of the predicate?", *Contemp. rev. 21*, 821–824.
  1874a. The principles of science₁, London (MacMillan).
  1876a. Logic₁, London (MacMillan).
  1880a. Studies in deductive logic. A manual for students, London (MacMillan).
  1883a. The principles of science₂, London (MacMillan).

Johnson, D. M.
  1979a, 1981a. "The problem of the invariance of dimension in the growth of modern topology", *AHES 20*, 97–181; *25*, 85–267.

Johnson, W. E. (1858–1931)
  1892a. "The logical calculus", *Mind n.s. 1*, 3–30, 235–250, 340–357.
  1921a, 1922a, 1924a. Logic, 3 pts., Cambridge (Cambridge UP). [R of *1922a*: Ramsey *1922a*.]

Jones, E. E. C. (1848–1922)
  1890a. Elements of logic as a science of propositions, Edinburgh (Clark).

Jordan, M. E. C. (1838–1922)
  1887a, 1893a. Cours d'analyse₁, 3; ₂, *1*, Paris (Gauthier-Villars).

Jordan, Z. A.
  1945a. The development of mathematical logic and of logical positivism in Poland between the two wars, Oxford (Oxford UP). [Part on logic ≈ McCall *1967a*, 346–406.]

Jørgensen, J. J. F. T. (1894–1969)
  1931a. A treatise of formal logic, 3 vols., Copenhagen (Lewin and Munksgaard) and Oxford (Oxford UP) = 1962, New York (Russell and Russell). [Rs: Scholz *1993a*, Stebbing *1932a*.]

1932a. "Über die Ziele und Probleme der Logistik", *Erkenntnis 3*, 73–100.

Joseph, H. W. B. (1867–1943)

1928a. "Logic and mathematics", *J. phil. studies 3*, 3–14.

1932a. "A defence of free thinking in logistics", *Mind n.s. 41*, 424–440.

1933a. "A defence of free thinking in logistics resumed", *Mind n.s. 42*, 417–443.

Jourdain, P. E. B. (1879–1919)

Papers: Djursholm (Sweden), Institut Mittag-Leffler (only two letter-books: see my *1977b*).

History. (Ed. I. Grattan-Guinness), *Selected essays on the history of set theory and logics* (*1906–1918*), 1991, Bologna (CLUEB).

1904a. "On the transfinite cardinal numbers of well-ordered aggregates", *Phil. mag.* (*6*)*7*, 61–75.

1906a. "De infinito in matematica", *RdM 8*, 121–136.

1910a. "The development of the theories of mathematical logic and the principles of mathematics", pt. 1, *Qu. j. pure appl. maths. 41*, 324–352 = *History*, 101–132.

1911a. "The philosophy of Mr. B∗rtr∗nd R∗ss∗ll", pt. 1, *The monist 21*, 483–508.

1911b. R of Natorp *1910a*, *Mind n.s. 20*, 552–560.

1912a. 1910a, pt. 2, *Qu. j. pure appl. maths. 43*, 219–314 = *History*, 133–228.

1913a. "The development of the theory of transfinite numbers", pt. 4, *Archiv Math. Physik 22*, 1–21 = *History*, 79–99.

1913b. R of PM_1 2, *J. Indian Math. Soc. 5*, 20–23.

1913c. R of PM_1 1, *JFM 41* (1910), 83–84.

1913d. 1910a, pt. 3, *Qu. j. pure appl. maths. 44*, 113–128 = *History*, 229–244.

1913e. "A correction and some remarks", *The monist 23*, 145–148.

1913f. "Tales with philosophical morals", *Open court 27*, 310–315.

1914a. R of Russell *1914c*, *Math. gaz. 7*, 404–409.

1915a. R of PM_1 2, *JFM 43* (1912), 93–94.

1916a. 1911a, pt. 2, *The monist 26*, 24–62.

1918a. R of PM_1 3, *JFM 44* (1913), 68–71.

1918b. The philosophy of Mr. B ∗ rtr ∗ nd R ∗ ss ∗ ll, London (A & U) = *History*, 245–342.

Kanamori, A. *See* Dreben and Kanamori

Kaufmann, F. (1895–1949)

Papers: Waterloo (Ontario), Wilfred Laurier University (mostly after 1930 (*see* Reeder *1991a*); not used).

1930a. Das Unendliche in der Mathematik und seine Ausschaltung, Leipzig and Vienna (Deuticke) = 1968, Darmstadt (Wissenschaftliche Buchgesellschaft). [Rs: Hempel *1933a*, Nagel *1932a*. E: *1978a.* I: *L'infinito in matematica* (T and ed. by L. Abertazzi), 1990, Trento (Reverdito).]

1931a. "Bemerkungen zum Grundlagenstreit in Logik und Mathematik", *Erkenntnis 2*, 262–290. [E: *1978a*, 165–187.]

1978a. (Ed. B. F. McGuinness), *The infinite in mathematics*, Dordrecht (Reidel).

Kempe, A. B. (1849–1922)

Papers: Chichester (England), West Sussex Record Office.

1872a. How to draw a straight line. A lecture on linkages, London (MacMillan).

1885a. "On the application of Clifford's graphs to ordinary binary quantics", *Proc. LMS* (*1*)*17*, 107–121.

1886a. "A memoir on the theory of mathematical form", *Phil. trans. Roy. Soc. London 177*, 1–70.

1887a. "Note to a memoir on the theory of mathematical form", *Proc. Roy. Soc. London 42*, 193–196.

1890a. "On the relation between the logical theory of classes and the geometrical theory of points", *Proc. LMS 21*, 147–182.

1890b. "The subject-matter of exact thought", *Nature 43*, 156–162.

1894a. "Mathematics", *Proc. LMS 26*, 5–15.

1897a. "The theory of mathematical form. A correction and explanation", *The monist 7*, 453–458.

Kennedy, H. C.

1975a. "Nine letters from Giuseppe Peano to Bertrand Russell", *J. hist. phil. 13*, 205–220.

1980a. Peano, Dordrecht (Reidel). [I version: *Peano*, 1983, Turin (Boringhieri).]

Kerry, B. B. (1858–1889)

1885a. "Über G. Cantors Mannig-faltigkeitsuntersuchungen", *Vrtlj. wiss. Phil. 9*, 191–232.

1887a. "Über Anschauung und ihre psychische Verarbeitung", pt. 4, *Ibidem 11*, 249–307.

Kertesz, A. (1929–1974)

1983a. (Ed. M. Stern), *Georg Cantor*, Halle/Saale (Leopoldina).

Keynes, J. M. (1883–1946)

1921a. A treatise on probability, London and New York (Macmillan).

Keyser, C. J. (1862–1947)

Papers: New York (New York), Columbia University.

1901a. "Theorems concerning positive definitions of finite assemblage and infinite assemblage", *Bull. AMS 7*, 218–226.

1903a. "Concerning the axiom of infinity and mathematical induction", *Bull. AMS 9*, 424–434.

1904a. "The axiom of infinity: a new supposition of thought", *Hibbert j. 2*, 532–553 ≈ *The human worth of rigorous thinking*₁, 1916, New York (Columbia UP), ch. 7.

1905a. "The axiom of infinity", *Hibbert j. 3*, 380–383.

1907a. Mathematics, New York (Columbia UP).

1912a. R of *PM*₁ *1, Science n.s. 35*, 106–110.

1922a. Mathematical philosophy. A study of fate and freedom, New York (Dutton).

1923a. R of Wittgenstein *1922a, New York evening post*, (18 August), 409. [Part in *Bull. AMS 30* (1924), 179–180.]

Kienzler, W. *See* Gabriel and Kienzler

Kimberling, C.

1972a. "Emmy Noether", *Amer. math. monthly 79*, 136–149.

King, J.

m1984a. "A report on the manuscripts" Russell *m1898a, m1899a* and *m1899–1900a*, in RA.

Klein, C. F. (1849–1925)

Papers: Göttingen (Germany), University.

Klemke, E. D.

1968a. (Ed.) *Essays on Frege*, Urbana, Chicago and London (University of Illinois Press).

Klyce, S. (1879–1933)

Papers: Washington (D.C.), Library of Congress.

1924a. "Foundations of mathematics", *The monist 34*, 615–637.

1932a. Outline of basic mathematics, Winchester, Mass. (the author). [Rare: Harvard University.]

Köhler, E. *See also* Wolenski and Köhler

1991a. "Gödel and Carnap in Vienna", *Jhrb. Kurt-Gödel-Gesellschaft*, (1990), 54–62.

König, J. (1849–1913)

1905a. "Zum Kontinuum-Problem", in Krazer *1905a*, 144–147 ≈ *MA 60*, 177–180 [with correction on p. 462]. [F: C & G, 469–476.]

1905b. "Über die Grundlagen der Mengenlehre und das Kontinu-

umproblem", *MA 61*, 156–160. [E:
van Heijenoort *1967a*, 145–149.]
1906a. "Sur la théorie des
ensembles", *C. r. Acad. Sci. 143*,
110–112.
1914a. (Ed. D. König), *Neue Grundlagen der Logik, Arithmetik und Mengenlehre*, Leipzig (von Veit). [Rs:
Dingler *1915a*, Mirimanoff *1914a*.]

Kötter, E. (1859–1922)
1895a. R of G. Veronese, *Fundamenti
di geometria* (1891) and its review
by Peano, *JFM 24* (1892), 483–495.

Koppelman, E.
1971a. "The calculus of operations
and the rise of abstract algebra",
AHES 8, 155–242.

Korselt, A. (1864–1947)
1896–1897a. R of Schröder *1890b*,
Dtsch. Math. mathem. Unterr. 28,
578–599; *29*, 30–43.
1903a. "Über die Grundlagen der
Geometrie", *JDMV 12*, pt. 1, 402–
407. [R: Dehn *1905b*.]
1905a. "Über die Grundlagen der
Mathematik", *JDMV 14*, 365–389.
1906a. "Paradoxien der Mengenlehre", *JDMV 15*, 215–219.
1906b. "Über Logik und Mengenlehre", *JDMV 15*, 266–269.
1908a. "Über die Logik der Geometrie", *JDMV 17*, pt. 1, 98–124.
1911a. "Über einen Beweis der
Äquivalenzsates", *MA 70*, 294–296.
1911b. "Über mathematische Erkenntnis", *JDMV 20*, pt. 1, 364–380.

Kowalevski, G. (1876–1950)
1950a. *Bestand und Wandel*, Munich
(Oldenbourg).

Krazer, C. A. J. (1858–1926)
1905a. (Ed.) *Verhandlungen des dritten
Internationalen Mathematiker-
Kongresses*, Leipzig (Teubner) =
1967, Liechtenstein (Kraus).

Kreisel, G.
1968a. R of a paper by R. Suszko,
Math. rev. 39, 10.

Kreiser, L.
1979a. "W. Wundts Auffassung der
Mathematik—Briefe von G. Cantor an W. Wundt", *Wiss. Ztsch.
Karl-Marx Univ., Gesch. Sprachwiss.
Reihe 28*, 197–206.
1995a. "Freges ausserwissen-
schaftliche Quellen seines logischen Denkens", in Max and
Stelzner *1995a*, 219–225.

Kreiser, L. and Grosche, G.
1983a. "Anhang. Nachschrift einer
Vorlesung und Protokolle mathematischer Vorträge Freges", in
Frege *Manuscripts*, 325–388.

Krois, J. M.
1997a. "On editing Cassirer's unpublished papers", *Etudes et lettres*, nos.
1–2, 163–183.

Kronecker, L. (1823–1891)
Works. Werke, 5 vols., 1895–1931,
Leipzig (Teubner) = 1968, New
York (Chelsea).
1885a. "Die absolut kleinsten Reste
reeller Grössen", *Sitz.-Ber. Königl.
Preuss. Akad. Wiss. Berlin*, 383–396,
1045–1049 ≈ *Works 3 pt. 1*, 113–
136.
1887a. "Über den Zahlbegriff", *J. rei.
ang. Math., 101*, 337–355 ≈ *Works
3 pt. 1*, 249–274.
1891a. "Auszug aus einem Briefe von
L. Kronecker an Herrn G. Cantor",
JDMV 1 (1891–1892), 23–25 ≈
Works 5, 495–499‡.
1894a. (Ed. E. Netto), *Vorlesungen
über Mathematik 1*, Leipzig
(Teubner).

Kuklick, B. R.
1972a. *Josiah Royce: an intellectual
biography*, Indianapolis and New
York (Bobbs, Merrill).
1977a. *The rise of American philosophy: Cambridge, Massachusetts
1860–1930*, New Haven and
London (Yale UP).

Kuratowski, K. (1896–1980)
1921a. "Sur la notion de l'ordre dans

la théorie des ensembles", *Fund. math. 2*, 161–171.

1980a. A half century of Polish mathematics, Oxford (Pergamon) and Warsaw (Polish Scientific Publishers).

Łukasiewicz, J. (1878–1956)
Works. (Ed. J. Słupecki), *Z zagadień logiki i filozofii*, Warsaw (Polish Scientific Publishers). [E version: *1971a*.]

1910a. O zasadzie sprzeczności u Aristotelesa, Cracow (Academy of Sciences). [Not found; G summary in *Bull. Soc. Sci. Cracovie, cl. phil.*, (1910), 15–38 (rare: RA).]

1912a. "O twórczości w nauce", in *Ksiega pamiatkowa ku uczeniu 250 rczwicy zatażenia*, Lvov, 1–15 ≈ *Works*, 66–75. [E: *1971a*, 1–15‡.]

1920a. "O logice trówartościowej", *Ruch filoz. 5*, 170–171. [E: *1971a*, 87–88.]

1921a. "Logika dwuwartościowa", *Przeg. filoz. 13*, 189–205. [E: *1971a*, 89–108.]

1925a. "Démonstration de la compatibilité des axiomes de la théorie de la déduction", *Ann. Soc. Polon. Math. 3*, 149.

1930a. "Philosophische Bemerkungen zu mehrwertigen Systemen des Aussagenkalküls", *C. r. Soc. Sci. Lett. Varsovie 23*, cl. 3, 51–77. [E: McCall *1967a*, 40–65 ≈ *1971a*, 153–178.]

1936a. "Logistyka u filozofia", *Przeg. filoz. 39*, 115–131 ≈ *Works*, 195–209. [E: *1971a*, 218–235.]

1971a. (E and ed. L. Borkowski), *Selected works*, 1970, Amsterdam (North-Holland) and Warsaw (Polish Scientific Publishers).

Łukasiewicz, J. and Tarski, A. *See also* Tarski
1930a. "Untersuchungen über den Aussagenkalkul", *C. r. Soc. Sci. Lett. Varsovie 23*, cl. 3, 30–50 = Tarski *Papers 2*, 321–343. [E: Tarski *1956a*, 38–59 ≈ Łukasiewicz *1971a*, 131–152. Polish T: *Works*, 129–143.]

Łukasiewicz, R.
1990a. Letter to the editor, *Metalogicon 3*, no. 1, 54–55.

Lacroix, S. F. (1765–1843)
1799a. "De la méthode en mathématiques", in *Elémens de géométrie₁* 1799–₄ 1804, Paris (Duprat), preface; then in *Essais sur l'enseignement₁* 1805, Paris (Courcier)–₄ 1838, Paris (Bachelier), sect. 2, ch. 2. [E by J. Toplis: *Phil. mag.* (*1*)20 (1804–1805), 193–202.]

1816a. An elementary treatise on the differential and integral calculus, Cambridge (Deighton). [E by C. Babbage, G. Peacock and J. F. W. Herschel of F original (1802).]

1819a. Traité du calcul différentiel et du calcul intégral₂ 3, Paris (Courcier).

Ladd, C. (1847–1930) *See also* Couturat and Ladd Franklin; Huntington and Ladd-Franklin; Ladd(-)Franklin
Papers: New York (New York), Columbia University.
1880a. "On De Morgan's extension of the algebraic process", *Amer. j. maths. 3*, 210–225.

1883a. "On the algebra of logic", in Peirce *1883b*, 17–71.

Ladd(-)Franklin, C. (1847–1930)
1890a. "Some proposed reforms in common logic", *Mind 15*, 75–88.

1892a. R of Schröder *1890b*, *Mind n.s. 1*, 126–132.

1912a. "Implication and existence in logic", *Phil. rev. 21*, 641–655.

m1918a? "Bertrand Russell and symbol logic", in Papers, Box 10. [Two versions.]

1928a. "The antilogism", *Mind n.s. 37*, 532–534.

Lagrange, J. L. (1736–1813)
Works. Oeuvres, 14 vols. 1867–1892, Paris (GV) = 1968, Hildesheim (Olms).
*1788a. Méchanique analitique*₁, Paris (Desaint). [Not in *Works*.]
*1797a. Théorie des fonctions analytiques*₁, Paris (Imprimerie Impériale) = *J. Ecole Polyt.* (*1*)3, cah. 9 (1801), 1–277. [Not in *Works*.]
1806a. Leçons sur le calcul des fonctions, Paris (Courcier) ≈ *Works 10.* [Also other editions.]

Laita, L.
1977a. "The influence of Boole's search for a universal method in analysis on the creation of his logic", *Ann. of sci. 34*, 163–176.
1979a. "Influences on Boole's logic: the controversy between William Hamilton and Augustus De Morgan", *Ann. of sci. 36*, 45–65.
1980a. "Boolean algebra and its extra-logical sources: the testimony of Mary Everest Boole", *HPL 1*, 37–60.

Lalande, A. (1867–1963)
1914a. "L'oeuvre de Louis Couturat", *RMM 22*, 644–688.

Landau, E. J. (1877–1938)
1917a. "Richard Dedekind", *Nachr. Königl. Gesell. Wiss. Göttingen, Geschäft. Mitt.*, 50–70.
*1930a. Grundlagen der Analysis*₁, Leipzig (Teubner) = 1946, New York (Chelsea).

Landini, G.
1987a. "Russell's substitutional theory of classes and relations", *HPL 8*, 171–200.
1996a. "Will the real *PM* please stand up?", in Monk and Palmer *1996a*, 287–330.
1998a. " 'On denoting' against denoting", *Russell n.s. 18*, 43–80.
1998b. Russell's hidden substitutional theory, New York (Oxford UP).

Landis, E. H. *See* Richardson, R. P. and Landis

Langer, S. K. K. (1895–1985)
1926a. "Confusion of symbols and confusion of logical types", *Mind n.s. 35*, 222–229.
*1937a. An introduction to symbolic logic*₁, Boston (Houghton Mifflin) and London (A & U). [R: Nagel *1938a.* ₂ (small additions) 1953, New York (Dover)‡.]

Langford, C. H. (1895–1965) *See also* Lewis and Langford
1926–1927a. "Some theorems on deducibility", *Annals of mathematics* (*2*)28, 16–40, 459–471.
1927a. "On propositions belonging to logic", *Mind n.s. 36*, 342–346.
1928a. "Concerning logical principles", *Bull. AMS 34*, 573–582.
1928b. R of *PM*₂ 2–3, *Isis 10*, 513–519.
1929a. "General propositions", *Mind n.s. 38*, 436–457.

Lebesgue, H. L. (1875–1941) *See also* Letters
Works. (Eds. G. Chatelet and G. Choquet) *Oeuvres scientifiques*, 5 vols., 1972–1973, Geneva (Kundig).
1902a. "Intégrale, longueur, aire", *Ann. mat. pura ed appl.* (*3*)7, 231–359 = Milan (Bernardoni) = *Works 1*, 201–331. [*Thèse.*]
*1904a. Leçons sur l'intégration*₁ Paris (GV) = *Works 2*, 11–154. [₂ 1928.]
1905a. "Sur les fonctions représentables analytiquement", *J. math. pures appl.* (*6*)1, 139–216 = *Works 3*, 103–180.
1922a. "A propos d'une nouvelle périodique: Fundamenta mathematicae", *Bull. des sci. math.* (*2*)46, 35–48 = *Works 5*, 339–351.

Lennes, N. J. *See* Veblen and Lennes
Lenzen, V. (1890–1975)
1965a. "Reminiscences of a mission to Milford, Pennsylvania", *Trans. C. S. Peirce Soc. 1*, 3–11.
1971a. "Bertrand Russell at Harvard, 1914", *Russell no. 3*, 4–6.

1973a. "The contributions of Charles S. Peirce to linear algebra", in D. Riepe (ed.), *Phenomenology and natural existence*, Albany (State University of New York Press), 239–254.

Leśniewski, S. (1886–1939)
Works. Collected works, 2 vols., 1992, Warsaw (Polish Scientific Publishers) and Dordrecht (Kluwer). [Consecutively paginated. All Es.]
1914a. "Czy klasa klas, nie podporzadkowanych sobie, jest podporzadkowana sobie?", *Przeg. filoz.* *17*, 63–75. [E: *Works*, 115–128.]
1927–1931a. "O podstawach matematyki", *Przeg. filoz.* *30* (1927), 164–206; *31* (1928), 261–291; *32* (1929), 60–101; *33* (1930), 77–105; *34* (1931), 142–170. [E: *Works*, 174–382‡.]
1929a. "Grundzüge eines neuen Systems der Grundlagen der Mathematik", pt. 1, *Fund. math.* *14*, 1–81. [R: Skolem *1935a.* E: *Works*, 410–492‡.]
1930a. "Über die Grundlagen der Ontologie", *C. r. Soc. Sci. Lett. Varsovie 23*, cl. 3, 111–132. [E: *Works*, 606–628.]
1931a. "Über Definitionen in der sogenannten Theorie der Deduktion", *Ibidem 24*, cl. 3, 289–309. [E: McCall *1967a*, 170–187 ≈ *Works*, 629–648.]

Letters [by R. Baire, E. Borel, J. Hadamard and H. Lebesgue]
1905a. "Cinq lettres sur la théorie des ensembles", *Bull. Soc. Math. France 33*, 261–273 ≈ Borel, *Leçons sur la théorie des fonctions*₂, 1914, Paris (GV), 150–160 = Borel *Works*, 1253–1266 = Lebesgue *Works 3*, 82–94 = Hadamard *Works 1*, 335–347. [E: G. H. Moore *1982a*, 311–320 ≈ Ewald *1996a*, 1077–1086.]

Levi, B. (1875–1961)
1908a. "Antinomie logiche?", *Ann. mat. pura ed appl.* (*3*)*15*, 187–216.

Lewis, A. C. *See also* Griffin and Lewis
1977a. "H. Grassmann's *Ausdehnungslehre* and Schleiermacher's *Dialektik*", *Ann. of sci. 34*, 103–162.
1995a. "Hermann Grassmann and the algebraisation of arithmetic", in P. Schreiber (ed.), *Hermann Grassmann. Werk und Wirkung*, Greifswald (University), 47–58.

Lewis, C. I. (1883–1964)
Papers: Stanford (California), University (mostly from 1950s: not used).
Papers. (Eds. J. D. Cohen and J. L. Mothershead), *Collected papers*, 1970, Stanford (Stanford UP). [In fact a selection.]
1912a. "Implication and the algebra of logic", *Mind n.s. 21*, 522–531 ≈ *Papers*, 351–359.
1913a. "A new algebra of implications and some consequences", *JP 10*, 428–438.
1914a. "The calculus of strict implication", *Mind n.s. 23*, 240–247.
1914b. R of *PM*₁ 2, *JP 11*, 497–502.
1914c. "The matrix algebra for implications", *JP 11*, 589–600.
1916a. "Types of order and the system S", *PR 25*, 407–419 ≈ *Papers*, 360–370.
1918a. *A survey of symbolic logic*, Berkeley (University of California Press). [Part = 1960, New York (Dover).]
1922a. "La logique et la méthode mathématique", *RMM 29*, 455–474.
1926a. R of Smart *1925a*, *JP 23*, 220–222.
1928a. R of *PM*₂, *Amer. math. monthly 35*, 200–205 ≈ *Papers*, 394–399.
1932a. "Alternative systems of logic", *The monist 42*, 481–507.
1933a. "Note concerning many-valued logic systems", *JP 30*, 364 ≈

Lewis and Langford *1932a*, 1959 reprint, 234.

1934a. "Paul Weiss on alternative logics", *PR 43*, 70–74.

Lewis, C. I. and Langford, C. H. *See also* Langford

1932a. Symbolie logic, New York (Century) = 1959, New York (Dover). [R: Scholz *1935b*.]

Liard, L. (1846–1917)

1878a. Les logiciens anglais contemporains, Paris (Germer Baillère).

Linke, P.

1946a. "Gottlob Frege als Philosoph", *Ztsch. phil. Forschung 1*, 75–99. [E by C. Hill in R. Poli (ed.), *The Brentano puzzle*, 1998, Aldershot (Ashgate), 45–73.]

Locke, J. (1632–1704)

1690a. An essay concerning human understanding, London (Basset). [Numerous ≈ s.]

Löwenheim, L. (1878–1957)

1908a. "Über das Auflösungsproblem im logischen Klassenkalkül", *Sitz.-Ber. Berliner Math. Gesell. 7*, 89–94. [Publ. with *Archiv Math. Physik.*]

1915a. "Über Möglichkeiten im Relativkalkül", *MA 76*, 447–470. [E: van Heijenoort *1967a*, 228–251.]

1922a. R of Dingler *1915b*, *JFM 45* (1914–1915), 100–101.

Loria, G. (1862–1954)

1891a. R of Peano *1888a*, *JFM 20* (1888), 689–692.

1892a. R of Peano *1889a*, *JFM 21* (1889), 51–52.

1901a. R of Pieri *1899a*, *JFM 30* (1899), 426–428.

Love, A. E. H. *See* Hobson and Love

Lovett, E. O. (1871–1957)

1900a. "Mathematics at the international congress of philosophy, Paris, 1900", *Bull. AMS 7*, 157–183.

Lowe, V. (1907–1988)

1941a. "The development of Whitehead's philosophy", in Schilpp *1941a*, 15–124.

1962a. Understanding Whitehead, Baltimore and London (Johns Hopkins UP).

1975a. "A.N. Whitehead on his mathematical goals: a letter of 1912", *Ann. of sci. 32*, 85–101.

1985a, 1990a. Alfred North Whitehead. The man and his work, 2 vols. (vol. 2 ed. J. B. Schneewind), Baltimore (Johns Hopkins UP).

Lüroth, J. (1844–1910)

1903a. "Ernst Schröder", *JDMV 12*, 249–265 ≈ Schröder *1905a*, iii–xix.

1904a. "Aus der Algebra der Logik. (Nach [Schröder *1905a*])", *JDMV 13*, 73–111.

Luschei, E. C.

1962a. The logical systems of Leśniewski, Amsterdam (North-Holland).

MacColl, H. (1837–1909)

1877a, 1877b. "The calculus of equivalent statements and integration limits" and "second paper", *Proc. LMS (1)9*, 9–20, 177–186.

1880a. "Symbolic reasoning", *Mind 5*, 45–60.

1897a. "Symbolic reasoning. (II.)", *Mind n.s. 6*, 493–510.

1899a. R of Whitehead *1898a*, *Mind n.s. 8*, 108–113.

1902a. "Symbolic reasoning. (IV.)", *Mind n.s. 11*, 352–368.

1904a. "Symbolic logic VI. VII. VIII", *The Athenaeum 2* [for year], 149–151, 213–214, 879–880 [corrections, pp. 244 and 811].

1905a. "Symbolic reasoning. (VI.)", *Mind n.s. 14*, 74–81 ≈ Russell *Analysis*, 308–316.

1905b. "Existential import", *Mind n.s. 14*, 295–296 ≈ Russell *Analysis*, 317.

1905c. Reply to Russell *1905b*, *Mind n.s. 14*, 401–402. ≈ Russell *Analysis*, 317–319.

1906a. Symbolic logic and its applications, London (Longmans, Green).

1907a. "Symbolic logic (a reply)", *Mind n.s. 16*, 470–473.

1908a. " 'If' and 'imply' ", *Mind n.s. 17*, 151–152.

1908b. " 'If' and 'imply' ", *Mind n.s. 17*, 453–455.

MacFarlane, A. (1851–1913)

1879a. Principles of the algebra of logic, with examples, Edinburgh (Douglas). [R: Venn *1879a.*]

1879–1881a. "On a calculus of relationship", *Proc. Roy. Soc. Edinburgh 10*, 224–232; *11*, 5–13, 162–173.

1881a. "An analysis of relationship", *Phil. mag.* (*5*)*11*, 436–446. [Summary in *Math. qus. Educ. times*, 36 (1883), 78–81.]

1885a. "The logical spectrum", *Phil. mag.* (*5*)*19*, 286–290.

MacHale, D.

1985a. George Boole—his life and work, Dublin (Boole Press).

Mac Lane, S.

1934a. Abgekürzte Beweise in Logikkalkül, Göttingen (Hubert) = *Selected papers*, 1979, New York (Springer), 1–62.

1935a. "A logical analysis of mathematical structure", *The monist 5*, 118–130.

McCall, S.

1967a. (Ed.) *Polish logic 1920–1939*, Oxford (Clarendon Press). [Source book of Es.]

McCarty, D. C.

1995a. "The mysteries of Richard Dedekind", in Hintikka *1995a*, 53–96.

McColl, H. *See* MacColl

McCoy, R. E.

1987a. Open Court. A centennial bibliography 1887–1987, La Salle, Ill. (Open Court).

McGuinness, B. F.

1988a. Wittgenstein: a life, Berkeley (University of California Press).

McGuinness, B. F. and von Wright, G. H.

1991a. "Unpublished correspondence between Russell and Wittgenstein", *Russell n.s. 10*, 101–124.

McKinsey, J. C. C. (1908–1953)

1935a. "On a redundancy in PM", *Mind n.s. 44*, 270–271.

Maccaferri, E. (1870–1953)

1913a. "Le definizioni per astrazione e la classe di Russell", *Rend. Circolo Mat. Palermo 35*, 165–171.

Mace, C. A. (b. 1894)

1931a. R of Stebbing *1930a*, *Mind n.s. 40*, 354–364.

1933a. The principles of logic. An introductory survey, London (Longmans, Green).

Magnell, T.

1991a. "The extent of Russell's modal views", *Erkenntnis 34*, 171–185.

Majer, U.

1997a. "Husserl and Hilbert on completeness", *Synthese 110*, 37–56.

Mally, E. (1879–1944)

1912a. Gegendstheoretische Grundlagen der Logik und Logistik, Leipzig (*Ztsch. für Phil. 148*, suppl. vol.).

1914a. "Über die Unabhängigkeit der Gegendstände vom Denken", *Ztsch. Phil. phil. Kritik 155*, 37–52.

Mancosu. P.

1998a. (Ed.), *From Brouwer to Hilbert*, New York and Oxford (Oxford UP). [Source book.]

1999a. "Between Vienna and Berlin: the immediate reception of Gödel's incompleteness theorems", *HPL 20*, 33–45.

Mangione, C. and others

1990a. "Italian logic in the 19th century before Peano", *HPL 11*, 203–210.

Mangione, C. and Bozzi, S.

1993a. Storia della logica da Boole ai nostri giorni, [Milan] (Garzanti).

Mansel, H. L. (1820–1871)
  1851a. "Recent extensions of formal logic", *North British rev.*, 90–121.
Marchisotto, E. A.
  1995a. "In the shadow of giants: the work of Mario Pieri in the foundations of mathematics", *HPL 16*, 107–119.
Massey, G. J. *See* Belnap and Massey
Max, I. and Stelzner, W.
  1995a. (Eds.), *Frege-Kolloquium Jena 1993*, Berlin and New York (de Gruyter).
Mays, W. and Henry, D. P.
  1953a. "Jevons and logic", *Mind n.s.* *62*, 484–505.
Maz'ya, V. and Shaposhnikova, T.
  1998a. Jacques Hadamard, a universal mathematician, [no place] (AMS and LMS).
Medvedev, F. A. (1923–1993)
  1966a. "Rannyaya istoriya teorii ekvivalentnosti", *Ist.-mat. issled.* *17*, 229–246.
  1982a. Rannyaya istoriya aksiomi vibora, Moscow (Nauka).
  1991a. (E by R. Cooke), *Scenes from the history of real functions*, Basel (Birkhäuser).
Mehrtens, H.
  1979a. Die Entstehung der Verbandstheorie, Hildesheim (Gerstenberg).
  1990a. Moderne Sprache Mathematik, Frankfurt/Main (Suhrkamp).
Meinong, A. (1853–1920)
  Works. Gesamtausgabe, 7 vols., 1968–1978, Graz (Akademische Druck- und Verlagsanstalt).
  1910a. Über Annahme₂, Leipzig (Barth) = *Works 4*, xv–xxv, 1–384. [₁ 1902, part = pp. 387–489.]
  1916a. "Über emotionale Präsentationen", *Sitz.-Ber. Kaiserl. Akad. Wiss. Wien, phil.-hist. Kl. 183*, 181 pp.‡ = *Works 3*, 285–467.
Menger, K. (1902–1985)
  1979a. Selected papers in logic and foundations, didactics, economics, Dordrecht (Reidel).

  1994a. (Eds. various), *Reminiscences of the Vienna Circle and the Mathematical Colloquium*, Dordrecht (Kluwer).
  1998a. (Eds. E. Dierker and K. Sigmund), *Ergebnisse eines Mathematischen Kolloquiums*, Vienna and New York (Springer).
Merrill, D. D.
  1978a. "DeMorgan [*sic*], Peirce and the logic of relations", *Trans. C. S. Peirce Soc. 14*, 247–284.
  1990a. Augustus De Morgan and the logic of relations, Dordrecht (Kluwer).
Meyer, C.
  1916a. "La filosofía de la matemáticas y su evolución en el siglo XIX", *Rev. filos. cult. cienc. educ.* [Argentina] *4*, 11–51.
  1918a. "Filosofía logistica de las matemáticas", *Ibidem 6*, 8–45.
Meyer, Friedrich (1842–1898)
  1885a. Elemente der Arithmetik und Algebra₂, Halle/Saale (Schmidt). [Rare: Braunschweig University.]
Meyer, Friedrich W. F. (1856–1934)
  1891a. R of Dedekind *1888a*, *JFM 20* (1888), 49–52.
Meyerson, E. (1859–1933)
  1931a. Du cheminement de la pensée, 3 vols., Paris (Alcan). [Consecutively paginated.]
Michaelis, C. T.
  1880a. R of Frege *1879a*, *Ztsch. Völkerpsych. Sprachwiss. 12*, 232–240. [E: Frege *1972a*, 212–218‡.]
  1881a. R of Frege *1879a*, *JFM 11* (1879), 48–49.
  1882a. R of C. S. Peirce *1880a*, *JFM 12* (1880), 41–44.
  1883a. R of C. S. Peirce *1881a*, *JFM 13* (1881), 55.
  1894a. R of Frege *1891a*, *JFM 23* (1891), 53.
  1894b. R of Husserl *1891a*, *JFM 23* (1891), 58–59.

1896a. R of Frege *1893a, JFM 25* (1893–1894), 101–102.

Michell, T.

    1997a. "Bertrand Russell's 1897 critique of the traditional theory of measurement", *Synthese 110*, 257–276.

Mirimanoff, D. (1861–1945)

    1914a. R of König *1914a, L'ens. math.* (*1*)*16*, 399–402.

    1917a. "Les antinomies de Russell et de Burali-Forti", *L'ens. math.* (*1*)*19*, 37–52.

    1917b. "Remarques sur la théorie des ensembles et les antinomies cantoriennes", pt. 1, *L'ens. math.* (*1*)*19*, 209–217.

Mitchell, O. H. (1851–1889)

    1883a. "On a new algebra of logic", in Peirce *1883b*, 72–106.

Mittag-Leffler, M. G. (1846–1927)

    Papers: Sweden, Institut Mittag-Leffler.

    1923a. (Ed.) "Briefe von K. Weierstrass an P. du Bois-Reymond", *AM 39*, 99–125.

Monk, R.

    1996a. Bertrand Russell. The spirit of solitude, London (Cape).

Monk, R. and Palmer, A.

    1996a. (Eds.) *Bertrand Russell and the origins of analytic philosophy*, Bristol (Thoemmes).

Monro, C. J.

    1881a. R of Venn *1881a, Mind 6*, 574–581.

Moore, E. H. (1862–1932)

    Papers: Chicago (Illinois), University.

    1900a. "On certain crinkly curves", *Trans. AMS 1*, 72–90.

    1902a. "On the projective axioms of geometry", *Trans. AMS 3*, 142–158, 501 [correction].

    1903a. "On the foundations of mathematics", *Bull. AMS 9*, 402–424‡ ≈ *Science 17*, 410–416.

1910a. Introduction to a form of general analysis, New Haven (Yale UP).

Moore, G. E. (1873–1958)

    Papers: Cambridge (England), University.

    1899a. "The nature of judgement", *Mind n.s. 8*, 167–193.

    1899b. R of Russell *1897c, Mind n.s. 8*, 397–405.

    1920a. "External and internal relations", *Proc. Aristotelian Soc. n.s. 20*, 40–62 ≈ *Philosophical studies*, 1922, London (KP), 276–309.

    1927a. "Facts and propositions", *Proc. Aristotelian Soc. suppl. 7*, 171–206 ≈ *1959a*, 60–88.

    1954a, 1955a. "Wittgenstein's lectures in 1930–33", *Mind n.s. 63*, 1–15, 289–316; *64*, 1–27 ≈ *1959a*, 252–324‡.

    1959a. Philosophical papers, London (A & U).

Moore, G. H.

    1978a. "The origins of Zermelo's axiomatisation of set theory", *J. phil. logic 7*, 307–329.

    1980a. "Beyond first-order logic: the historical interplay between mathematical logic and axiomatic set theory", *HPL 1*, 95–137.

    1982a. Zermelo's axiom of choice, New York (Springer).

    1983a. "Lebesgue's measure problem and Zermelo's axiom of choice", *Ann. New York Acad. Sci. 412*, 129–154.

    1989a. "Towards a history of Cantor's continuum problem", in Rowe and McCleary *1989a 1*, 79–121.

    1997a. "Hilbert and the emergence of modern mathematical logic", *Theoria 12*, 65–90.

Moore, G. H. and Garciadiego, A. *See also* Garciadiego

    1981a. "Burali-Forti's paradox: a reappraisal of its origins", *HM 8*, 319–350.

Morris, C. W. (1901–1979)
Papers: Indianapolis (Indiana), University of Indiana, Peirce Project.
1937a. Logical positivism, pragmatism and scientific empiricism, Paris (Hermann).

Moss, J. M. B.
1972a. "Some B. Russell's sprouts (1903–1908)", in *Conference in mathematical logic—London '72*, Berlin (Springer), 211–250.

Mostowski, A. (1913–1975)
1966a. Thirty years of foundational studies, Oxford (Blackwell).

Mulligan, K. *See* Smith, B. and Mulligan

Murawski, R.
1998a. "Undefinability of truth. The problem of priority. Tarski vs Gödel", *HPL 19*, 153–160.

Murphey, M.
1961a. The development of Peirce's philosophy, Cambridge, Mass. (Harvard UP) = 1993, Philadelphia (Hackett).

Myhill, J. R. (1923–1987)
1949a. R of Chwistek *1948a*, *JSL 14*, 119–125.
1974a. "The undefinability of the set of natural numbers in the ramified *Principia*", in Nakhnikian *1974a*, 19–27.
1979a. "A refutation of an unjustified attack on the axiom of reducibility", in Andrews *1979a*, 81–90.

Nagel, E. (1901–1985) *See also* Cohen and Nagel
1932a. R of Kaufmann *1930a* and Dingler *1931a*, *JP 29*, 401–409.
1935a. R of Carnap *1934b*, *JP 32*, 49–52.
1935b. " 'Impossible numbers': a chapter in the history of science", *Stud. hist. ideas 3*, 429–474 ≈ *1979a*, 166–194, 322–328.
1936a. "Impressions and appraisals of analytic philosophy in Europe", *JP 33*, 7–24, 29–53.

1938a. R of Langer *1937a*, *JP 35*, 613–614.
1939a. The formation of modern conceptions of modern logic in the development of geometry", *Osiris 7*, 142–224 ≈ *1979a*, 195–259, 328–339.
1979a. Teleology revisited, New York (Columbia UP).

Nakhnikian, G.
1974a. (Ed.) *Bertrand Russell's philosophy*, London (Duckworth).

Natorp, P. G. (1854–1924)
1901a. "Zu den logischen Grundlagen der neueren Mathematik", *Arch. system. Phil. 7*, 177–208, 372–384.
1910a. Die logischen Grundlagen der exakten Wissenschaften, Leipzig and Berlin (Teubner) = 1923.
[R: Jourdain *1911a*.]

Neil, S. (1825–1901)
1865a. "Modern logicians—the late George Boole", *The British controversialist and lit. mag.* (3)*13*, 81–94, 161–174.
1872a. "Augustus De Morgan", *Ibidem* (3)*28*, 1–21.

Nelson, L. *See* Grelling and Nelson

Neumann, O. *See* Edwards, Neumann and Purkert

Neurath, O. (1882–1945)
Papers: Haarlem (The Netherlands), State Archives of North Holland.
1938a. (Ed.) *Encyclopedia and unified science*, Chicago (University of Chicago Press).

Newman, M. H. A. (1897–1984)
m1923a. "The foundations of mathematics from the standpoint of physics", Cambridge, Saint John's College Library.
1928a. "Mr. Russell's 'causal theory of perception' ", *Mind n.s. 37*, 137–148.

Nicod, J. G. P. (1893–1924)
1917a. "A reduction in the number of primitive propositions of logic",

Proc. Cambridge Phil. Soc. 19, 32–41.

1922a. "Les tendances philosophiques de M. Bertrand Russell", *RMM 29*, 77–84.

1922b. "Mathematical logic and the foundations of mathematics", in *Encyclopaedia Britannica$_{12}$ 32*, 874–876.

1930a. Foundations of geometry and induction, London (KP).

1969a. Geometry and induction, London (RKP).

Nidditch, P.

1963a. "Peano and the recognition of Frege", *Mind n.s. 72*, 103–100.

Novak, G.

1989a. "Riemann's *Habilitationsvortrag* and the synthetic *a priori* status of geometry", in Rowe and McCleary *1989a 1*, 17–46.

O'Briant, W. H.

1984a. "Russell on Leibniz", *Studia Leibnitiana 11*, 160–222.

O'Gorman, F. P.

1977a. "Poincaré's conventionalism of applied geometry", *Stud. hist. phil. sci. 8*, 303–340.

Ogden, C. K. (1889–1957)

Papers: Hamilton (Ontario), McMaster University.

Ogden, C. K. and Richards, I. A.

1923a. The meaning of meaning$_1$, London (KP) = (Ed. W. Terrence Gordon), 1994, London (RKP). [Rs: Ramsey *1924a*, Russell *1926a*. $_2$ 1926; later eds. follow this version.]

Open Court Publishing Company

Papers: Carbondale (Illinois), Southern Illinois University.

Otte, M. and Panza, M.

1997a. (Eds.), *Analysis and synthesis in mathematics. History and philosophy*, Dordrecht (Kluwer).

Padilla Galvez, J.

1995a. "La metalógica en la propuesta de R. Carnap", *Mathesis 11*, 113–136. [Int. to ed. of Carnap *m1931b.*]

Padoa, A. (1868–1937)

m1896a?. "Saggio di una teoria di proposizioni", in *1968a'*, 322–336.

m1897a. "Interpretazione aritmetica della logica matematica", in *1968a'*, 317–321.

1898a. Conférences sur la logique mathématique, Brussels (Larcier). [Lithograph. Rare: University of Pavia, Department of Mathematics, Berzolari Collection.]

1899a. "Note di logica matematica", *RdM 6*, 105–121.

1900a. Rassiunto delle conferenze su l'algebra e la geometria quali teorie deduttive, [Rome (University)]. [Lithograph. Rare: as for *1898a.*]

1901a. "Essai sur une théorie algébrique des nombres entiers, precédé d'une introduction logique à une théorie déductive quelconque", in Congress *1901a*, 309–365. [Part E: van Heijenoort *1967a*, 118–123.]

1901b. "Numeri interi relativi", *RdM 7*, 73–84.

1902a. "Théorie des nombres entiers absolus", *RdM 8*, 45–54.

1902b. "Un nouveau système irreductible de postulats pour l'algèbre", in Duporcq *1902a*, 249–256.

1902c. "Un nouveau système de définitions pour la géométrie euclidienne", in Duporcq *1902a*, 353–363.

1902d. Logica matematica e matematica elementare, Livorno (Giusti). [Pamphlet. Rare: University of Bologna.]

1906a. "Ideografica logica", *Ateneo Veneto 29*, no. 1, 323–340. [Rare: University of Bologna.]

1911–1912a. "La logique déductive dans sa dernière phase de développement", *RMM 19*, 828–883; *20*, 48–67, 207–231 ≈ 1912, Paris (GV). [Lectures at Geneva.]

1913a. "La valeur et les rôles du principe d'induction mathématique", in Hobson and Love *1913a*, 471–479.

1930a. "Logica", in (eds. L. Berzolari and others), *Enciclopedia delle matematiche elementari 1*, pt. 1, Milan (Hoepli), 1–79.

1936a. "Ce qui la logique doit à Peano", in *Actes du Congrès International de Philosophie Scientifique*, Paris (Hermann), pt. 8, 31–37.

1968a'. (Ed. A. Giannattassio), "Due inediti di Alessandro Padoa", *Physis 10*, 309–336.

Palladino, D. *See also* Borga and Palladino

1985a. "I fondamenti della teoria dei numeri e dell'analisi", in Borga and others *1985a*, 76–173.

Palmer, A. *See* Monk and Palmer

Panteki, M.

1992a. "Relationships between algebra, differential equations and logic in England: 1800–1860", Ph.D., C.N.A.A. (London).

1993a. "Thomas Solly (1816–1875): an unknown pioneer of the mathematicization of logic in England, 1839", *HPL 14*, 133–169.

Panza, M. *See* Otte and Panza

Parry, W. T.

1968a. "The logic of C. I. Lewis", in P. A. Schilpp (ed.), *The philosophy of C. I. Lewis*, La Salle, Ill. (Open Court), 15–54.

Parshall, K. H.

1998a. James Joseph Sylvester. Life and work in letters, Oxford (Clarendon Press).

Parshall, K. H. and Rowe, D. *See also* Rowe and McCleary

1994a. The emergence of the American mathematical research community,

Providence, Rhode Island (AMS) and London (LMS).

Parsons, C.

1976a. "Some remarks on Frege's conception of extension", in M. Schirn (ed.), *Studien zu Frege I*, Stuttgart (Frommann-Holzboog), 265–277.

Pasch, Moritz (1843–1930)

1882a. Einleitung in die Differential- und Integralrechnung, Leipzig (Teubner).

1882b. Vorlesungen über neuere Geometrie, Leipzig (Teubner).

1924a. Mathematik und Logik$_2$, Leipzig (Engelsmann).

Peacock, G. (1791–1858)

1830a. A treatise on algebra, Cambridge (Deighton). [R: De Morgan *1835a*.]

1834a. "Report on the recent progress and actual state of certain branches of analysis", *Rep. Brit. Ass. Adv. Sci.*, (1833), 185–332.

Peano, G. (1858–1932)

Here "T(B)" abbreviates the publisher "Turin (Bocca)".

Works. (Eds. Unione Matematica Italiana), *Opere scelte*, 3 vols., 1957–1959, Rome (Cremonese).

Selection. (E and ed. H. C. Kennedy), *Selected works of Giuseppe Peano*, 1973, Toronto and London (Toronto UP).

Formulary. (Ed.) *Formulario di mat(h)ematico*, [then] *Formulaire (de) mathématiques*, $_1$ 1895, T(B); $_2$ 3 pts., 1897–1899, T(B); $_3$ Paris (Carré et Naud); $_4$ 1902–1903, T(B); $_5$ 1908 T(B) = (ed. U. Cassina), 1960, Rome (Cremonese). [R of $_{1-3}$: Couturat *1901a*.]

Letters. (Ed. G. Osimo), *Lettere di Giuseppe Peano a Giovanni Vacca*, 1992?, Milan (University Bocconi).

1884a. "Annotazioni", in Genocchi *1884a*, vii–xxxii‡ ≈ *Works 1*, 47–75.

1887a. Applicazioni geometriche del calcolo infinitesimale, T(B).

1888a. Calcolo geometrico secondo [. . .] *Grassmann*, T(B). Part in *Works 2*, 3–19. [R: Loria *1891a*. Part E: *Selection*, 75–100. E: *Geometric calculus*, 2000, Boston (Birkhäuser).]

1889a. Arithmetices principia, novo methodo exposita, T(B) ≈ *Works 2*, 20–55‡. [R: Loria *1892a*. Es: *Selection*, 101–134; part in van Heijenoort *1967a*, 83–97.]

1889b. I principii di geometria logicamente esposti, T(B) ≈ *Works, 2*, 56–91‡.

1890a. "Sur une courbe, qui remplit toute une aire plane", *MA 36*, 157–160 ≈ *Works 1*, 110–114. [R: Hilbert *1893a*. E: *Selection*, 143–148.]

1890b. "Démonstration de l'intégrabilité des équations différentielles ordinaires", *MA 37*, 182–228 ≈ *Works 1*, 119–170‡.

1891a. "Principii di logica matematica", *RdM 1*, 1–10 ≈ *Works 2*, 92–101. [Spanish T: *1892b*. E: *Selection*, 153–161.]

1891b. "Formole di logica matematica", *RdM 1*, 24–31, 182–184 ≈ *Works 2*, 102–113‡.

1891c. "Sul concetto di numero. Nota I. Nota II", *RdM 1*, 87–102, 256–267 ≈ *Works 3*, 80–109‡.

1891d. R of Schröder *1890b* and *1891a*, *RdM 1*, 164–170 ≈ *Works 2*, 114–121‡.

1892a. "Dimostrazione dell'impossibilità di segmenti infinitesimi costanti", *RdM 2*, 58–62 ≈ *Works 3*, 110–114. [R: Vivanti *1895a*.]

1892b. "Principios de lógica matemática", *El prog. mat. 2*, 20–23, 49–53. [Spanish T of *1891a*.]

1894a. "Sur la définition de la limite d'une fonction. Exercice de logique mathématique", *Amer. j. math. 17*, 38–68 ≈ *Works 1*, 228–257‡.

1894b. Notations de logique mathématique. Introduction au Formulaire de mathématiques, Turin ([Guadagnini?]) ≈ *Works 2*, 123–176.

1894c. "Sui fondamenti della geometria", *RdM 4*, 51–90 ≈ *Works 3*, 115–157.

1895a. R of Frege *1893a*, *RdM 5*, 122–128 ≈ *Works 2*, 189–195. [E: Dudman *1971a*.]

1895b. "Logique mathématique", in *Formulary₁*, 1–7, 127–129. Part ≈ *Works 2*, 177–188.

1896a. "Riposta" to Frege *1896a*, *RdM 6*, 60–61 ≈ *Works 2*, 295–296. [E: Dudman *1971a*.]

1897a. "Sul §2 del Formulario, t. II: aritmetica", *RdM 6*, 75–89 ≈ *Works 3*, 232–248‡.

1897b. "Logique mathématique", in *Formulary₂*, no. 1, 63 pp. ≈ *Works 2*, 218–287‡.

1897c. "Studii di logica matematica", *Atti Reale Accad. Sci. Torino 32*, 565–583 ≈ *Works 2*, 201–217. [E: *Selection*, 190–205.]

1898a. R of Schröder *1898a*, *RdM 6*, 95–101 ≈ *Works 2*, 297–303‡.

1898b. Text on the foundations of arithmetic, in *Formulary₂*, no. 2, vii + 60 pp. Part ≈ *Works 3*, 215–231.

1899a. "Sui numeri irrazionali", *RdM 6*, 126–140 ≈ *Works 3*, 249–267.

1900a. "Formules de logique mathématique", *RdM 7*, 1–41 ≈ *Works 2*, 304–361‡.

1901a. "Les définitions mathématiques", in *Congress 1901a*, 279–288 ≈ *Works 2*, 362–368‡.

1901b. "Dizionario di matematica. Parte prima. Logica matematica", *RdM 7*, 160–172 ≈ *2*, 369–383‡.

1902a. Arithmetica generale e algebra elementare, Turin (Paravia). [Rare: University of Milan, Department of Mathematics.]

1904a. "Sur les principes de la géométrie selon M. Pieri. (Rapport.)", in Vasiliev *1904a*, 92–95.

1906a. "Super theorema de Cantor-Bernstein", *Rend. Circolo Mat. Palermo 8*, 136–143 ≈ *RdM 8*, 360–366 ≈ *2*, 337–344.

1906b. "Addizione" to *1906a*, *RdM 8*, 143–157 ≈ *Works 1*, 344–358 = Heinzmann *1986a*, 106–120. [E: *Selection*, 206–218. F: C & G, 83–96.]

1912a. "Préface", in Padoa *1911–1912a*, book version, 3–4.

1913a. R of *PM*₁, *1–2*, *Boll. bibl. stor. sci. mat. 15*, 45–53, 75–81 ≈ *Works 2*, 389–401.

1913b. "Della proposizioni esistenziali", in Hobson and Love *1913a 2*, 497–500 ≈ *Works 2*, 384–388.

1916a. "L'esercuzione tipografia delle formule matematiche", *Atti Reale Accad. Sci. Torino 52*, 279–286 ≈ *Works 3*, 397–404.

1923a. Review of A. Natucci, *Il concetto di numero* (1922), *Archeion. Arch. stor. sci. 4*, 382–383.

1986a'. Celebrazioni in memoria di Giuseppe Peano, Turin (Department of Mathematics, University).

Pears, D. F.
1972a. (Ed.), *Bertrand Russell*, New York (Doubleday).

Pearson, K. (1857–1936)
1936a. "Old Tripos days at Cambridge, as seen from another viewpoint", *Math. gaz. 20*, 27–36.

Peckhaus, V.
1990a. Hilbertprogramm und Kritische Philosophie. Das Göttinger Modell interdisziplinärer Zusammenarbeit zwischen Mathematik und Philosophie, Göttingen (Vandenhoeck und Ruprecht).

1991a. "Ernst Schröder und die 'Pasigraphischen Systeme' von Peano und Peirce", *Modern logic 1*, 174–205.

1992a. "Hilbert, Zermelo und die Institutionalisierung der mathematischen Logik in Deutschland", *Ber. Wissengesch. 15*, 27–38.

1994a. "Benno Kerry: Beiträge zu seiner Biographie", *HPL 15*, 1–8.

1994b. "Logic in transition: the logical calculi of Hilbert (1905) and Zermelo (1908)", in D. Prawitz and D. Westerståhl (eds.), *Logic and philosophy of science in Uppsala*, Dordrecht (Kluwer), 311–323.

1995a. Hermann Ulrici (1806–1884). Der Hallesche Philosoph und die englische Algebra der Logik, Halle/Saale (Hallescher Verlag).

1995b. "The genesis of Grelling's paradox", in Max and Stelzner *1995a*, 269–280.

1995c. "Ruestow's thesis on Russell's paradox", *Modern logic 5*, 165–167.

1996a. "The influence of Hermann Günther Grassmann and Robert Grassmann on Ernst Schröder's algebra of logic", in Schubring *1996a*, 217–227.

1997a. Logik, Mathesis universalis und allgemeine Wissenschaft. Leibniz und die Wiederentdeckung der formalen Logik im 19. Jahrhundert, Berlin (Akademie Verlag).

Peirce, B. (1809–1880)
Papers: Cambridge (Mass.), Harvard University, Houghton Library.

1870a. Linear associative algebra, Washington (lithograph) ≈ (Ed. C. S. Peirce), *Amer. j. math. 4* (1881), 97–229 = in I. B. Cohen (ed.), *Benjamin Peirce: "Father of pure mathematics" in America*, 1980, New York (Arno).

Peirce, C. S. S. (1838–1914)
Here "AAAS" stands for "American Academy of Arts and Sciences"
Papers: Cambridge (Mass.), Harvard University, Houghton Library; copies at Indianapolis (Indiana), University, Peirce Project.

Papers. Collected papers, 6 vols. (eds. C. Hartshorne and P. Weiss) 1931–1935, and vols. 7–8 (ed. A. Burks) 1958, Cambridge, Mass. (Harvard UP).

Writings. (Eds. various) *Writings of C. S. Peirce. A chronological edition,* about 30 vols., 1982–, Bloomington (Indiana UP).

Elements. (Ed. C. Eisele), *The new elements of mathematics,* 4 vols., 1976, The Hague (Mouton) and Atlantic Highlands, New Jersey (Humanities Press).

Essential. (Eds. N. Houser and C. Kloesel), *The essential Peirce 1,* 1992, Bloomington (Indiana UP).

m1865a. "On the logic of relatives", in *Writings 1,* 162–302.

m1866a. "The logic of science, or, induction and hypothesis", in *Writings 1,* 358–504.

1868a. "On an improvement in Boole's calculus of logic", *Proc. AAAS 7,* 250–261 ≈ *Papers 3,* 3–15 ≈ *Writings 2,* 12–22.

1868b. "On a new list of categories", *Proc. AAAS 7,* 287–298 ≈ *Papers 1,* 287–299 ≈ *Writings 2,* 49–59 ≈ *Essential,* 1–10.

1868c. "Upon the logic of mathematics", *Proc. AAAS 7,* 402–412 ≈ *Papers 3,* 16–26 ≈ *Writings 2,* 59–69.

1870a. "Description of a notation for the logic of relatives", *Mem. AAAS 9* (1873), 317–378 ≈ *Papers 1,* 27–98 ≈ *Writings 2,* 359–429‡. [Offprints printed 1870.]

1875a. "On the application of logical analysis to multiple algebra", *Proc. AAAS 10,* 392–394 ≈ *Papers 3,* 99–101 ≈ *Writings 3,* 177–179.

1880a. "On the algebra of logic", *Amer. j. math. 3,* 15–57 ≈ *Papers 3,* 104–157 ≈ *Writings 4,* 163–209‡. [Part in *Essential,* 200–209. R: Michaelis *1882a.*]

m1880b. "A Boolean algebra with one constant", in *Papers 4,* 13–18 ≈ *Writings 4,* 218–221.

1881a. "On the logic of number", *Amer. j. math. 4,* 85–95 ≈ *Papers 3,* 158–170 ≈ *Writings 4,* 299–309‡

≈ Ewald *1996a,* 596–607. [R: Michaelis *1883a.*]

1882a. "On the relative forms of the algebra", *Amer. j. math. 4,* 221–225 ≈ *Papers 3,* 171–175 ≈ *Writings 4,* 319–322.

1883a. [Ed.] *Studies in logic. By members of the Johns Hopkins University,* Boston (Little, Brown) = 1983, Amsterdam and Philadelphia (Benjamins). [R: Venn *1883a.*]

1883b. "A theory of probable inference", in *1883a,* 126–181 ≈ *Papers 2,* 433–477 ≈ *Writings 4,* 408–450.

1883c. "Note B", in *1883a,* 187–203 ≈ *Papers 3,* 195–209 ≈ *Writings 4,* 453–466.

1885a. "On the algebra of logic: a contribution to the philosophy of notation", *Amer. j. math. 7,* 180–202 ≈ *Papers 3,* 210–238 ≈ *Writings 5,* 162–190‡ ≈ Ewald *1996a,* 608–631. [Part in *Essential,* 225–228.]

m1890a. "A guess at the riddle", in *Papers 1,* 181–226. [Date conjectured.]

1896a. "The regenerated logic", *The monist 7,* 19–40 ≈ *Papers 3,* 266–287‡.

1897a. "The logic of relatives", *The monist 7,* 161–217 ≈ *Papers 3,* 288–345‡. [Sort of R of Schröder *1895a.*]

m1897b. "Multitude and number", in *Papers 4,* 145–188.

1898a. "The logic of mathematics in relation to education", *Educ. review,* 209–216 ≈ *Papers 3,* 346–359 ≈ Ewald *1996a,* 632–637.

1900a. "Infinitesimals", *Science 2,* 430–433 ≈ *Papers 3,* 360–365.

1903a. R of Russell *1903a* and Lady Welby, *What is meaning?* (1903), *The nation 77,* 308–309 ≈ *Papers 8,* 130.

m1903b. "Lowell lectures, 1903", in *Elements 3,* 39–402.

Perron, O. (1880–1975)

1907a. "Was sind und was sollen die irrationalen Zahlen?", *JDMV 16*, 140–155.

1939a. Irrationalzahlen₂, Berlin (de Gruyter) = 1951, New York (Chelsea).

Piaget, J. (1896–1982)

1923a. Le jugement et le raisonnement chez l'enfant₁, Geneva (Delachaux et Niestlé). [E: *Judgement and reasoning in the child*, 1928, London (KP).]

1941a. La génèse du nombre chez l'enfant, Neuchâtel (Delachaux et Niestlé). [E: *The child's conception of number*, 1952, London (RKP).]

1942a. Nombres, classes et relations, Paris (Vrin). [Rare: RA.]

Picardi, E.

1994a. "Kerry and Frege über Begriff und Gegenstand", *HPL 15*, 9–32.

1994b. La chimica di concetti: linguaggio, logica, psicologia, Bologna (Il Mulino).

Pieri, M. (1860–1913)

Works. Opere sulle fondamenti della matematica, Rome (Cremonese).

Letters. (Ed. G. Arrighi), *Lettere a Mario Pieri (1884–1913)*, 1997, Milan (Universita Bocconi).

1898a. "I principii della geometria di posizione, composta in un sistema logico-deduttivo", *Mem. Accad. Sci. Torino* (2)*48*, 1–62 = *Works*, 101–162.

1899a. "Della geometria elementare come sistema ipotetico deduttivo", *Ibidem* (2)*49*, 173–222 = *Works*, 183–233. [R: Loria *1901a.*]

1901a. "Sur la géométrie envisagée comme un système purement logique", in Congress *1901a*, 367–404 = *Works*, 235–272.

1906a. "Sur la compatibilité des axiomes de l'arithmétique", *RMM 14*, 196–207 = *Works*, 377–388.

1906b. "Uno sguardo al nuovo indirizzo logico-matematico delle scienze deduttive", *Annu. Univ. Catania*, 21–82 = *Works*, 389–448‡.

Poincaré, J. H. (1854–1912)

Works. (Eds. various), *Oeuvres*, 11 vols., 1916–1956, Paris (GV).

1894a. "Sur la nature de raisonnement mathématique", *RMM 2*, 371–384 ≈ *1902a*, ch. 1.

1899a. "Des fondements de la géométrie: à propos d'un livre de M. Russell", *RMM 7*, 251–279. [On Russell *1897a.*]

1900a. "Sur les principes de la géométrie: réponse à M. Russell", *RMM 8*, 73–86.

1902a. La science et l'hypothèse, Paris (Flammarion). [Es: *1905a*; *1913a*, 9–197.]

1904a. "Rapport sur les travaux de M. Hilbert", in Vasiliev *1904a*, 11–48.

1905a. Science and hypothesis, London and Newcastle-upon-Tyne (Scott) = 1952, New York (Dover). [E of *1902a*. R: Russell *1905c.*]

1905b. Letter to the editor, *Mind n.s. 15*, 141–143.

1905c. "Les mathématiques et la logique", pt. 1, *RMM 13*, 815–835‡ ≈ *1908b*, ch. 3.

1906a. 1905c, pt. 2, *RMM 14*, 17–34‡ ≈ *1908b*, ch. 4.

1906b. 1905c, pt. 3, *RMM 14*, 294–317‡ ≈ *1908b*, ch. 5. [R: Sheldon *1906a.*]

1906c. "A propos de la logistique", *RMM 14*, 866–868 = Heinzmann *1986a*, 145–147.

1908a. "L'avenir des mathématiques", *Rev. gén. sci. pures appl.*, *19*, 930–939 ≈ *Scientia 4*, 1–23 ≈ *Bull. des sci. math.* (2)*32*, pt. 1, 168–190 ≈ *Rend. Circ. Mat. Palermo 26*, 152–168 ≈ *1908b*, ch. 1 ≈ Castelnuovo *1909a 1*, 167–182. [Parts in *Works 5*, 19–23; *Scientia 110* (1975), 357–368.]

1908b. Science et méthode, Paris (Flammarion). [Es: *Science and method* (E by F. Maitland), 1914, London and Edinburgh (Nelson) = 1952, New York (Dover); *1913a*, 357–546.]

1909a. "Réflexions sur les deux notes précédentes", *AM 32*, 195–200 = Heinzmann *1986a*, 224–229 ≈ *Works 11*, 114–119.

1909b. "La logique de l'infini", *RMM 17*, 461–482 = Heinzmann *1986a*, 235–256 ≈ *Dernières pensées*, 1913, Paris (Flammarion), ch. 4.

1910a. "Über transfinite Zahlen", in *Sechs Vorträge über ausgewählte Gegenstände*, Leipzig and Berlin (Teubner), 45–48 = Heinzmann *1986a*, 231–234 ≈ *Works 5*, 120–124.

1913a. The foundations of science Lancaster, Pa. (Science Press). [E by G. B. Halsted of *1902a, 1908b* and another book.]

Popper, K. R. (1902–1994)
1935a. Logik der Forschung$_1$, Vienna (Springer). [E: *The logic of scientific discovery*, 1959, London (Hutchinson).]

Post, E. L. (1897–1954)
1921a. "Introduction to a general theory of propositions", *Amer. j. math. 43*, 163–185 = (Ed. M. Davis), *Solvability, provability, definability. The collected works*, 1994, Boston (Birkhäuser), 21–43 ≈ van Heijenoort *1967a*, 264–283.

Posy, C. J.
1992a. (Ed.) *Kant's philosophy of mathematics*, Dordrecht (Kluwer).

Price, M.
1994a. Mathematics for the multitude? A history of the Mathematical Association, Leicester (Mathematical Association).

Priest, G.
1994a. "The structure of the paradoxes of self-reference", *Mind n.s. 103*, 27–34.

Pringsheim, A. (1850–1941)
1898a. "Irrationalzahlen und Konvergenz unendlicher Prozessen", in *EMW 1*, sect. 1, 47–146 (article IA3).

1899a. "Grundlagen der allgemeinen Funktionentheorie", in *EMW 2*, sect. 1, 1–53 (article IIA1).

Prior, A. N. (1914–1969)
1949a. "Categoricals and hypotheticals in George Boole and his successors", *Australian j. phil. 27*, 171–196.

1965a. "Existence in Lesniewski and Russell", in J. N. Crossley and M. E. Dummett (eds.), *Formal systems and recursive functions*, Amsterdam (North-Holland), 149–155.

Pulkkinen, J.
1994a. The threat of logical mathematism, Frankfurt/Main (Lang).

Purkert, W. *See also* Edwards, Neumann and Purkert
1986a. "Georg Cantor und die Antinomien der Mengenlehre", *Bull. Soc. Math. Belgique* (*A*)38, 313–327.

1989a. "Cantor's views on the foundations of mathematics", in Rowe and McCleary *1989a 1*, 49–65.

1995a. Felix Hausdorff Findbuch, Bonn (Mathematisches Institut).

Purkert, W. and Ilgauds, H. J. *See also* Ilgauds
1987a. Georg Cantor 1845–1918, Basel (Birkhäuser).

Putnam. H. *See* Benacerraf and Putnam

Pycior, H. M.
1981a. "George Peacock and the British origins of symbolical algebra", *HM 8*, 23–45.

1983a. "Augustus de Morgan's algebraic work: the three stages", *Isis 74*, 211–226.

Quine, W. V.
m1932a. "The logic of sequences. A generalisation of *PM*", Cambridge,

Mass. (Harvard University Ph.D.)
= 1989, New York and London
(Garland).

1932b. "A note on Nicod's postulate",
Mind n.s. 43, 472–476.

1934a. Summary of lecture related to
Whitehead *1934a*, *Amer. math.
monthly 41*, 129–131.

1934b. *A system of logistic*, Cam-
bridge, Mass. (Harvard UP). [Rs:
Ackermann *1938b*, Black *1935a*,
Bronstein *1936a*.]

1934c. "Ontological remarks on the
propositional calculus", *Mind n.s.
43*, 472–476.

1935a. R of Carnap *1934b*, *PR 44*,
394–397.

1936a. "A theory of classes presupposing no canons of logic", *Proc.
Nat. Acad. Sci. 22*, 320–326.

1936b. "On the axiom of reducibility",
Mind n.s. 45, 498–500.

1940a. *Mathematical logic*$_1$, Cambridge, Mass. (Harvard UP).
[$_2$ 1951.]

1941a. "Whitehead and the rise of
modern logic", in Schilpp *1941a*,
125–163.

1955a. "On Frege's way out", *Mind
n.s. 64*, 145–159 ≈ *Selected logic
papers*, 1966, New York (Random
House), 146–158.

1962a. "Paradox", *Scientific American
206*, no. 4, 84–96 ≈ *The ways of
paradox*, 1966, New York (Random
House), 1–20 ≈ (no ed.), *Mathematics in the modern world*, 1968,
San Francisco (Freeman), 200–208.

1969a. *Set theory and its logic*$_2$, Cambridge, Mass. (Harvard UP).

1985a. *The time of my life. An autobiography*, Cambridge, Mass. (MIT
Press).

1986a. "Peano as logician", in Peano
1986a′, 33–43 ≈ *HPL 8*, 15–24
[with bibliography by me].

1989a. "Preface", in *1932a*, 1989
photoreprint.

Rademacher, H. A. (1892–1969)

1923a. R of Weyl *1918a*, *JFM 46*
(1916–1918), 56–59.

Rado, R. (1906–1989)

1975a. "The cardinal module and
some theorems on families of sets",
Ann. mat. pura ed appl. (*4*)*102*,
135–154.

Rahman, S.

1997a. "Hugh MacColl: eine bib-
liographische Erschliessung seiner
Hauptwerke und Notizen zu ihrer
Rezeptionsgeschichte", *HPL 18*,
165–183.

Rahman, S. and Christen, P. G.

1997a. "Hugh MacColls Begriff der
hypothetischen Aussage", *Philos.
scient. 2*, cah. 4, 95–138.

Ramorino, A. *See* Burali-Forti and
Ramorino

Ramsey, F. P. (1903–1930)

Papers: Pittsburgh (Pennsylvania),
University of Pittsburgh.

Essays. (Ed. R. B. Braithwaite), *The
foundations of mathematics and
other logical essays*, 1931, London
(KP). [Rs: Church *1932a*, Russell
1931a. Further incomplete eds.
1978 and 1980.]

Notes. (Ed. M. C. Galavotti), *Notes on
philosophy, probability and mathematics*, Naples (Bibliopolis).

1922a. R of Johnson *1992a*, *The new
statesman 19*, 469–470.

1923a. R of Wittgenstein *1922a*, *Mind
n.s. 32*, 465–478 ≈ *Essays*, 270–
286‡.

1924a. R of Ogden and Richards
1923a, *Mind n.s. 33*, 108–109.

1925a. R of *PM*$_2$ *1*, *Nature 116*,
127–128. [Attributed.]

1925b. R of *PM*$_2$ *1*, *Mind n.s. 34*,
506–507.

1926a. "Mathematical logic", *Math.
gaz. 13*, 185–194 ≈ *Essays*, 62–81‡.

1926b. "The foundations of mathematics", *Proc. LMS* (*2*)*25*, 338–
384 ≈ *Essays*, 1–61‡.

1927a. "Facts and propositions", *Proc. Aristotelian Soc.*, *suppl. 7*, 153–170 ≈ *Essays*, 138–155.

1930a. "On a problem of formal logic", *Proc. LMS* (*2*)*30*, 264–286 ≈ *Essays*, 82–111.

Rang, B. and Thomas, W.

1981a. "Zermelo's discovery of the 'Russell paradox'", *HM 8*, 15–22.

Raspa, V.

1999a. *In contradizzione. Il principio di contraddizione alle origini della nuova logica*, Trieste (Parnaso).

Reeder, H. P.

1991a. *The work of Felix Kaufmann*, Washington (University Presses of America).

Reich, K.

1985a. "Aurel Voss", in M. Folkerts and U. Lindgren (eds.), *Mathemata. Festschrift für Helmut Gericke*, Munich (Steiner), 674–699.

Reichenbach, H. (1891–1953)

Papers: Pittsburgh (Pennsylvania), University of Pittsburgh.

1930a. "Tagung für Erkenntnislehre der exakten Wissenschaften in Königsberg", *Die Naturwiss. 18*, 1093–1094.

1936a. "Logistic empiricism in Germany and the present state of its problems", *JP 33*, 141–160.

1944a. "Bertrand Russell's logic", in Schilpp *1944a*, 21–54.

Reisch, G.

1994a. "Planning science: Otto Neurath and the *Encyclopedia of Unified Science*", *Brit. j. hist. sci. 27*, 153–176.

1995a. "A history of the *Encyclopedia of Unified Science*", Chicago (University of Chicago Ph.D.).

Rescher, N.

1969a. *Many-valued logic*, New York (McGraw-Hill).

1974a. "Bertrand Russell and modal logic", in (ed. with others), *Studies in modality*, Oxford (Blackwell),

85–96 ≈ in Andrews *1979a*, 139–149.

Resnik, M.

1982a. R of Bostock *1974a* and *1979a*, *JSL 47*, 708–713.

Reyes y Prósper, V. de los (1863–1922)

Here *"EPM"* cites the journal *El progreso matemático*.

Writings. Reprint of some papers following Del Val *1973a*, *Teorema 3* (1973), 329–354.

1891a. "Cristina Ladd Franklin", *EPM 1*, 297–300 ≈ *Writings*, 336–339.

1892a. "Ernesto Schroeder", *EPM 2*, 33–36 ≈ *Writings*, 344–348.

1892b. "Charles Santiago Peirce y Oscar Honward [*sic*] Mitchell", *EPM 2*, 170–173 ≈ *Writings*, 340–343.

1892c. "Projecto de clasificación de los escritos lógico-simbólicos especialmente de los post-Boolianos", *EPM 2*, 229–232 ≈ *Writings*, 333–335.

1893a. "La lógica simbólica en Italia", *EPM 3*, 41–43 ≈ *Writings*, 349–351.

Reymond, A.-F. (1874–1958)

1908a. *Logique et mathématiques*, Saint-Blaise (Foyer Solidariste). [Rs: Russell *1909a*, Sheffer *1910a*.]

1909a. "Note sur la théorie d'existence des nombres entiers et sur la définition logistique du zéro", *RMM 17*, 237–279.

1914a. "Le premier congrès de philosophie mathématique. Paris 6–8 avril 1914", *L'ens. math.* (*1*)*16*, 370–378.

1917a. "Les ordinaux transfinis de Cantor et leur définition logique", *RMM 24*, 693–709.

1932a. *Les principes de la logique et la critique contemporaine*, Paris (Boivin).

Rice, A.

1997a. "Inspiration or desperation? Augustus de Morgan's appointment to the chair of mathematics at London University in 1828", *Brit. j. hist. sci. 30*, 257–274.

Richard, J. (1862–1956)

1903a. Sur la philosophie des mathématiques, Paris (GV).

1905a. "Les principes des mathématiques", *Rev. gen. sci. pures appl. 16*, 541. [E: van Heijenoort *1967a*, 142–144.]

Richards, I. A. *See* Ogden and Richards

Richards, Joan

1980a. "The art and the science of British algebra", *HM 7*, 343–365.

1988a. Mathematical visions: the pursuit of geometry in Victorian England, San Diego (Academic Press).

Richards, John

1980a. "Boole and Mill: differing perspectives on logical psychologism", *HPL 1*, 19–36.

1980b. "Propositional functions and Russell's philosophy of language, 1903–14", *The phil. forum 11*, 315–339.

Richardson, A. W. *See* Giere and Richardson

Richardson, R. P. (b. 1876) and Landis, E. H. (b. 1876)

1915a. "Numbers, variables, and Mr. Russell's philosophy", *The monist 25*, 321–364 = 1915, Chicago (Open Court).

1916a. Fundamental conceptions of modern mathematics, Chicago and London (Open Court).

Rickey, V. F.

1975a. "Creative definitions in propositional calculi", *Notre Dame j. formal logic 16*, 273–294.

Rider, R.

1990a. "Measure of ideas, rule of language: mathematics and language in the 18th century", in T. Frängsmyr, J. L. Heilbron and R. Rider (eds.), *The quantifying spirit in the 18th century*, Berkeley and Los Angeles (University of California Press), 113–142.

Riemann, G. F. B. (1826–1866)

Works. Gesammelte mathematische Werke$_2$, 1892, Leipzig (Teubner) = 1990, Berlin (Springer) [with much additional apparatus].

1867a. "Über die Darstellbarkeit einer Function durch eine trigonometrische Reihe", in *Abh. Königl. Gesell. Wiss. Göttingen 13*, 87–132 ≈ *Works*, 227–271.

1867b. "Über die Hypothesen, welche der Geometrie zu Grunde liegen", in *Ibidem*, 133–152 ≈ *Works*, 272–287.

Risse, W. (1931–1998)

1979a. Bibliographia logica 3, Hildesheim and New York (Olms).

Roberts, D. *See also* Houser, Roberts and van Evra

1973a. The existential graphs of Charles S. Peirce, The Hague and Paris (Mouton).

Rodriguez-Consuegra, F. A.

1987a. "Russell's logicist definitions of numbers 1899–1913: chronology and significance", *HPL 8*, 141–167.

1988a. "Russell's theory of types, 1901–1910: its complex origins in the unpublished manuscripts", *HPL 9*, 131–164.

1988b. "Elementos logisticas en la obra de Peano y su escuela", *Mathesis 4*, 221–299.

1991a. The mathematical philosophy of Bertrand Russell: origins and development, Basel (Birkhäuser).

1997a. "Nominal definitions and logical consequence in the Peano school", *Theoria 12*, 125–137.

Rosinger, K. E.

1929a. "The formalization of implication", *The monist 39*, 273–280.

1930a. "Concerning the symbols ϕx and $\phi \hat{x}$", *Ann. maths. (2)31*, 181–184.

1938a. R of Woodger *1937a*, *JP 35*, 273–274.

Rosser, J. B. (1907–1989)

1936a. "Extensions of some theorems of Gödel and Church", *JSL 1*, 87–91.

Rossi-Landi, F.

1953a. Charles Morris, Rome and Milan (Bocca).

Rougier, L. P. A. (b. 1889)

1916a. "La démonstration géométrique et le raisonnement déductif", *RMM 23*, 809–858.

1920a. Le philosophie géométrique de Henri Poincaré, Paris (Alcan).

Rowe, D. E. and McCleary, J. *See also* Parshall and Rowe

1989a. (Eds.) *The history of modern mathematics*, 2 vols., Boston (Academic Press).

Royce, J. (1855–1916)

Papers: Cambridge (Massachusetts), Harvard University, Pusey Library.

Essays. (Ed. D. S. Robinson), *Logical essays*, 1951, Dubuque (Brown).

Letters. (Ed. J. Clendenning), *Letters*, 1970, Chicago and London (University of Chicago Press). [Only letters from him.]

1881a. Primer of logical analysis for the use of composition students, San Francisco (Bancroft).

1902a. "The concept of the infinite", *Hibbert j. 1*, 21–45.

1905a. "The relations of the principles of logic to the foundations of geometry", *Trans. AMS 6*, 353–415‡ = *Essays*, 379–441. [R: de Laguna *1906a.*]

1913a. "The principles of logic", in Windelband and Ruge *1913a*, 67–135 = *Essays*, 310–378.

Rudio, F. (1856–1929)

1898a. (Ed.) *Verhandlungen des I. Internationalen Mathematiker-Kongresses*, Leipzig (Teubner) = 1967, Liechtenstein (Kraus).

Ruge, A. *See* Windelband and Ruge

Ruja, H. *See* Blackwell and Ruja

Russell, B. A. W. (1872–1970) *See also* Whitehead and Russell

Papers: Hamilton (Ontario), McMaster University, Russell Archives (RA).

Papers. (Eds. various), *Collected papers*, about 30 vols., 1983–, London ([now] Routledge).

Letters 1. (Ed. N. Griffin), *The selected letters of Bertrand Russell 1*, 1992, London (Allen Lane).

Logic. (Ed. R. C. Marsh), *Logic and knowledge*, 1956, London (A & U).

Analysis. (Ed. D. P. Lackey), *Essays in analysis*, 1973, London (A & U) and New York (Braziller).

1896a. "The logic of geometry", *Mind n.s. 5*, 1–23 ≈ *Papers 1*, 266–286.

1896b. R of Hannequin *1895a*, *Mind n.s. 5*, 410–417 ≈ *Papers 2*, 35–43‡.

1897a. R of Couturat *1896a*, *Mind n.s. 6*, 112–119 ≈ *Papers 2*, 59–67‡.

1897b. "On the relations of number and quantity", *Mind n.s. 6*, 326–341 ≈ *Papers 2*, 68–90‡.

1897c. An essay on the foundations of geometry, Cambridge (Cambridge UP) = 1956, New York (Dover). [Rs: Couturat *1897a*, G. E. Moore *1899b*, Wilson *1903a*. F: *1901f.*]

m1898a. "An analysis of mathematical reasoning", in *Papers 2*, 155–242.

m1898b. "On the principles of arithmetic", in *Papers 2*, 245–260. [Date conjectured.]

m1898c. "Note on order", in *Papers 2*, 339–358.

1898d. "Les axiomes propres à Euclide, sont-ils empiriques?", *RMM 6*, 759–776 ≈ *Papers 2*, 420–433 [and E on pp. 322–338‡].

m1899a. "The classification of relations", in *Papers 2*, 136–146.

m1899b. "The fundamental ideas and axioms of mathematics", in *Papers 2*, 261–305.

1899c. "Sur les axiomes de la géométrie", *RMM* 7, 684–707 ≈ *Papers 2*, 434–451 [and original and part E on pp. 390–415‡].

m1899–1900a. "Principles of mathematics", in *Papers 3*, 9–180.

1900a. R of J. Schultz, *Psychologie der Axiome* (1900), *Mind n.s. 9*, 120–121 ≈ *Papers 3*, 508–510.

1900b. A critical exposition of the philosophy of Leibniz with an appendix of leading passages, Cambridge (Cambridge UP) = 1937, London (A & U) [with a new int.].

m1900c. "On the logic of relations with applications to arithmetic and the theory of series", in *Papers 3*, 590–612.

m1900d. "Recent Italian work on the foundations of mathematics", in *Papers 3*, 350–379.

1901a. "On the notion of order", *Mind, n.s. 10*, 30–51 ≈ *Papers 3*, 287–309‡.

1901b. "Sur la logique des relations avec des applications à la théorie des séries", *RdM 7*, 115–136, 137–148. [Parts in *Papers 3*, 618–627; E on pp. 310–349‡, from *Logic*, 3–38.]

m1901c. "Part I. The variable", in *Papers 3*, 181–206.

1901d. "Recent work on the principles of mathematics", *Int. monthly 4*, 83–101 ≈ *1918a*, ch. 5 ≈ *Papers 3*, 363–369‡.

1901e. "Is position in time and space absolute or relative?", *Mind n.s. 10*, 293–317 ≈ *Papers 3*, 259–282 [and draft on pp. 219–233].

1901f. Essai sur les fondements de la géométrie, Paris (GV). [F of *1897c* by A. Cadenat, with notes by Russell and Couturat.]

1901g. "L'idée d'ordre et la position absolue dans l'espace et le temps", in *Congress 1901a*, 241–277 ≈ *Papers 3*, 570–588 [and E on pp. 234–258‡].

1902a. "Théorie générale des séries bien-ordonnées", *RdM 8*, 12–16, 17–43. [Parts in *Papers 3*, 661–673; E on pp. 384–421‡.]

1902b. "The teaching of Euclid", *Math. gaz. 2*, 165–167 ≈ *55* (1971), 143–145 ≈ *Papers 3*, 465–469.

1902c. "Geometry, non-Euclidean", in *Encyclopaedia Britannica*₁₀ suppl. 4, 664–674 ≈ *Papers 3*, 470–504.

1902d. "On finite and infinite cardinal numbers", Whitehead *1902a*, 378–383 ≈ *Papers 3*, 422–430.

1903a. The principles of mathematics, Cambridge (Cambridge UP) = 1937, London (A & U) [with new int. *1937a*]. [Rs: Couturat *1904a*, Engel *1905a*, Hardy *1903b*, Hausdorff *1905a*, Peirce *1903a*, Shearman *1907a*, Wilson *1903a*.]

1903b. "Recent work on the philosophy of Leibniz", *Mind n.s. 12*, 177–201 ≈ *Papers 3*, 537–561.

m1903c. Sequence of manuscripts on classes, relations and functions, in *Papers 4*, 3–73.

m1903d. "On the meaning and denotation of phrases", in *Papers 4*, 283–296.

m1904a. "Outline of symbolic logic", in *Papers 4*, 77–84.

m1904b. "On functions, classes and relations", in *Papers 4*, 85–95.

m1904c. "On functions", in *Papers 4*, 96–110.

m1904d. [Fundamental notions], in *Papers 4*, 111–259. [Title conjectured; first folio lost.]

1904e. "Meinong's theory of complexes and assumptions", *Mind n.s. 18*, 204–219, 336–354, 509–524 ≈ *Analysis*, 21–76 ≈ *Papers 4*, 431–474.

1904f. "The axiom of infinity", *Hibbert j. 2*, 809–812 ≈ *Analysis*, 256–259 ≈ *Papers 4*, 475–478.

1904g. "Non-Euclidean geometry", *The Athenaeum 2* [for year], 592–593 ≈ *Papers 4*, 482–485.

m1905a. "On fundamentals", in *Papers 4*, 359–413.

1905b. "The existential import of propositions", *Mind n.s. 14*, 398–401 ≈ *Analysis*, 98–102 ≈ *Papers 4*, 486–489.

1905c. R of Poincaré *1905a*, *Mind n.s. 14*, 412–418 ≈ *Philosophical essays₂*, 1966, London (A & U), 70–78 ≈ *Papers 4*, 589–594.

1905d. "On denoting", *Mind n.s. 14*, 479–493 ≈ *Logic*, 41–56 ≈ *Analysis*, 103–119 ≈ *Papers 4*, 414–427‡. [Various other ≈ s and = s, and Ts.]

m1905e. "On substitution", in *Papers 5*, to appear. [Distinct from *m1906f.*]

m1905f. "Necessity and possibility", in *Papers 4*, 507–520.

1905g. "Sur la relation des mathématiques à la logistique", *RMM 13*, 906–917 ≈ *Papers 4*, 622–631. [R: Sheldon *1906a*. Es: *Analysis*, 260–271; *Papers 4*, 521–532.]

1906a. "On some difficulties in the theory of transfinite numbers and order types", *Proc. LMS (2)4*, 29–53 = Heinzmann *1986a*, 54–78 ≈ *Analysis*, 135–164 ≈ *Papers 5*, to appear. [Part F: C & G, 323–336.]

1906b. "The theory of implication", *Amer. j. maths. 28*, 159–202 ≈ *Papers 5*, to appear.

1906c. R of MacColl *1906a*, *Mind n.s. 15*, 255–260 ≈ *Papers 5*, to appear.

m1906d. On the substitutional theory of classes and relations", in *Analysis*, 165–189‡ ≈ *Papers 5*, to appear.

m1906e. "Logic in which propositions are not entities", in *Papers 5*, to appear.

m1906f. "On substitution", in *Papers 5*, to appear. [Distinct from *m1905e.*]

1906g. "On the nature of truth", *Proc. Aristotelian Soc. n.s. 7*, 28–49 ≈ *Papers 5*, to appear.

1906h. "Les paradoxes de la logique", *RMM 14*, 627–650‡ = Heinzmann *1986a*, 121–144 ≈ *Papers 5*, to appear. [E: *Analysis*, 190–214.]

m1906i. "The paradox of the liar", in *Papers 5*, to appear.

1906j. "The nature of truth", *Mind n.s. 15*, 528–533 ≈ *Papers 5*, to appear.

m1906k. "Multiplicative axiom", in *Papers 5*, to appear.

m1907a. "The regressive method of discovering the premises of mathematics", in *Analysis*, 272–283 ≈ *Papers 5*, to appear.

m1907b. "Fundamentals", in *Papers 5*, to appear.

m1907c. Manuscripts on functions (Files 230.030920, 22 and 24), in *Papers 5*, to appear?. [Date conjectured.]

1907d. "The study of mathematics", *New quart. 1*, 29–44 ≈ *1910a*, ch. 3 ≈ *1918a*, ch. 4 ≈ *Papers 12*, 83–93‡.

1908a. R of MacColl *1906a*, *The Athenaeum 1* [for year], 396–397 ≈ *Papers 5*, to appear.

1908b. "'If' and 'imply', a reply to Mr. MacColl", *Mind n.s. 17*, 300–301 ≈ *Papers 5*, to appear.

1908c. "Mathematical logic as based on the theory of types", *Amer. j. maths. 30*, 222–262 ≈ *Logic*, 59–102 ≈ van Heijenoort *1967a*, 150–182 ≈ *Papers 5*, to appear. [Part in Heinzmann *1986a*, 200–223.]

1909a. R of Reymond *1908a*, *Mind n.s. 18*, 299–301 ≈ *Papers 6*, 60–63.

1910a. *Philosophical essays₁*, London (Longmans, Green). [₂ 1966, London (A & U).]

1910b. "La théorie des types logiques", *RMM 18*, 263–301 = Heinzmann *1986a*, 257–295. [E: *Analysis*, 215–252 ≈ *Papers 6*, 3–31.]

1910c. "On the nature of truth and falsehood", in *1910a*, ch. 7 ≈ *Papers 6*, 115–124.

1911a. "Le réalisme analytique", *Bull. Soc. Franç. Phil. 11*, 53–82 ≈ *Papers 6*, 410–432 [and E on pp. 132–146]. [Part =: Heinzmann *1986a*, 296–304.]

1911b. "L'importance philosophique de la logistique", *RMM 19*, 282–291. [Not in *Papers 6*. E on pp. 32–40, based on that by P. E. B. Jourdain in *The monist 23* (1913), 481–493 ≈ *Analysis*, 284–294.]

1911c. "Knowledge by acquaintance and knowledge by description", *Proc. Aristotelian Soc. n.s. 11*, 209–232 ≈ *1918a*, ch. 10 ≈ *Papers 6*, 147–161‡.

1911d. "Sur les axiomes de l'infini et du transfini", *C. r. seances Soc. Math. France*, no. 2, 22–35 = *Bull. Soc. Math. France 39* (1911), 1967 reprint, 488–501 ≈ *Papers 6*, 398–408. [Es: pp. 41–53‡; my *1977b*, 162–174.]

1911e. "On the relation of universals and particulars", *Proc. Aristotelian Soc. n.s. 12*, 1–24 ≈ *Logic*, 105–124 ≈ *Papers 6*, 162–182.

1912a. The problems of philosophy, London (William & Norgate). [Many ≈ s. G: *1926c*.]

m1912b. "What is logic?", in *Papers 6*, 54–56.

m1913a. Draft book on epistemology, in *Papers 6*, 1–178. [Some chs. published in *The monist 24* (1914), 1–16, 161–187, 435–453 (≈ *Logic*, 125–174), 582–593; *25* (1915), 28–44, 212–233.]

1913b. Opening Chairman's remarks at a Congress session, in Hobson and Love *1913a 1*, 53 ≈ *Papers 6*, 444–449.

1914a. R of Windelband and Ruge *1913a*, *The nation 14*, 771–772 ≈ *Papers 8*, 93-96 [and another review on pp. 97–98].

1914b. "The relation of sense-data to physics", *Scientia*, 1–27 ≈ *1918a*, ch. 8 ≈ *Papers 8*, 3–26.

1914c. Our knowledge of the external world as a field for scientific method in philosophy, London (A & U) and Chicago (Open Court). [Rs: Broad *1915a*, de Laguna *1915a*, Jourdain *1914a*. G: *1926d*. Different rev. eds.: 1926, London (A & U); 1929, New York (Norton). All eds. various ≈ s, not always correctly identified.]

1914d. On scientific method in philosophy, Oxford (Oxford UP) ≈ *1918a*, ch. 6 ≈ *Papers 8*, 55–73.

1918a. Mysticism and logic, London (Longmans, Green). [Various ≈ s.]

1918–1919a. "The philosophy of logical atomism", *The monist 28*, 495–527; *29*, 32–63, 190–222, 345–380 ≈ *Logic*, 177–281‡ ≈ *Papers 8*, 157–243.

1919a. R of Dewey *1916a*, *JP 16*, 5–26 ≈ *Papers 8*, 132–154.

1919b. Introduction to mathematical philosophy, London (A & U) and New York (Macmillan). [R: Gumbel *1924a*. G: *1923a*. F: *1928a*.]

1919c. "On propositions: what they are and how they mean", *Proc. Aristotelian Soc. suppl. 2*, 1–43 ≈ *Logic*, 283–320‡ ≈ *Papers 8*, 276–306.

1921a. The analysis of mind, London (A & U).

1921b. "Vorwort" to Wittgenstein *1921a*, *Ann. Naturphil. 14*, 186–198. [Not in *Papers*.]

1922a. "Introduction", in Wittgenstein *1922a*, 7–23 ≈ 1961 ed., ix–xxii‡ ≈ *Papers 9*, 96–112.

1923a. Einführung in die mathematische Philosophie, Munich (Drei Masken) = 1953, Darmstadt and Geneva (Wissenschaftliche Buchgesellschaft). [G of *1919b* by E. J. Gumbel and W. Gordon.]

1924a. "Logical atomism", in (Ed. J. H. Muirhead), *Contemporary*

British philosophy, London (A & U), 356–383 ≈ *Logic*, 321–344‡ ≈ *Papers 9*, 160–179.

1925a. "Introduction to the second edition", in *PM₂ 1*, xiii–xlvi.

1926a. On education, especially in early childhood, London (A & U) and New York (Boni and Liveright).

1926b. R of Ogden and Richards *1923a, The dial 81*, August, 114–121 ≈ *Papers 9*, 138–144 [and another review on pp. 135–137].

1926c. Die Probleme der Philosophie, Berlin (Welt-Kreis). [G of *1912a* by P. Hertz. R: Schlick *1927a*.]

1926d. Unser Wissen von der Aussenwelt, Leipzig (Meiner). [G of *1914c* by W. Rothstock.]

1927a. The analysis of matter, London (KP).

1928a. Introduction à la philosophie mathématique, Paris (Payot). [F of *1919b* by G. Moreau.]

1930a. "Preface" to Nicod *1930a*, 5–9 (≈ *Papers 9*, 303–307) ≈ Nicod *1961a*, xiii–xvi (≈ *Papers 11*, 213–216).

1931a. R of Ramsey *Essays, Mind n.s. 40*, 476–482 ≈ *Papers 10*, 106–114 [and another review on pp. 114–117].

1932a. Letter to S. Pankhurst, in *Schola et vita 7*, 102. [Published in Esperanto.]

1936a. "On order in time", *Proc. Cambridge Phil. Soc. 32*, 216–228 ≈ *Logic*, 345–363 ≈ *Papers 10*, 122–137.

1936b. "Auto-obituary. The last survivor of a dead epoch", *The listener 16*, 289 ≈ *Unpopular essays*, 1950, London (A & U), 173–175. [Various other ≈ s.]

1937a. "Introduction to the second edition", in *1903a*, 1937 reprint, v–xiv. [R of edition: Scholz *1938a*.]

1938a. "On the importance of logical form", in Neurath *1938a*, 39–41 ≈ *Papers 10*, 138–140.

1939a. "Dewey's new *Logic*", in P. A.

Schilpp (ed.), *The philosophy of John Dewey*, New York (Tudor), 137–156 ≈ *Papers 10*, 141–160.

1940a. An enquiry into meaning and truth, London (A & U).

1944a. "My mental development", in Schilpp *1944a*, 3–20.

1944b "Reply to criticisms", in Schilpp *1944a*, 681–741.

1948a. "A turning point in my life", in L. S. Russell (ed.), *The Saturday book—8*, London (Hutchinson), 142–147.

1948b. "Whitehead and *PM*", *Mind n.s. 57*, 137–138 ≈ *Papers 11*, 190–191.

1950a. "Logical positivism", *Rev. int. phil. 4*, 3–19 ≈ *Logic*, 365–382‡ ≈ *Papers 11*, 155–167.

1956a. Portraits from memory and other essays, London (A & U).

1959a. My philosophical development, London (A & U).

1967a, 1968a, 1969a. The autobiography of Bertrand Russell, 3 vols., London (A & U). [Various ≈ s and Ts.]

1967b. "False and true", *The observer*, 12 March, 33.

1971a. "Addendum to" *1944a*, in Schilpp *1944a₄* (1971), xvii–xx.

Russell, J. J. (1923–1975)

1984a. Analysis and dialectic, The Hague (Nijhoff).

Sanchez Valeria, V.

1997a. "Head or tail? De Morgan on the bounds of traditional logic", *HPL 18*, 123–138.

Sanzo, U.

1976a. "Significato epistemologico della polemica Poincaré-Couturat", *Scientia 110*, 369–396 [and E on pp. 397–418].

Sarkar, S.

1997a. " 'The boundless ocean of unlimited possibilities': logic in Carnap's *Logical syntax of language*", *Synthese 93*, 191–237.

Sarton, G. A. L. (1884–1956)
1936a. The study of the history of mathematics, Cambridge, Mass. (Harvard UP) = 1957, New York (Dover).

Scanlan, M. J.
1991a. "Who were the American postulate theorists?", *JSL 56*, 981–1002.

Schiller, F. C. S. (1864–1937)
1935a. "Many-valued logics—and others", *Mind n.s. 44*, 467–483.

Schilpp, P. A. (1897–1993)
1941a. (Ed.) *The philosophy of Alfred North Whitehead*, New York (Tudor). [$_2$ 1951.]
1944a. (Ed.) *The philosophy of Bertrand Russell*, New York (Tudor). [Later eds.]
1963a. (Ed.) *The philosophy of Rudolf Carnap*, La Salle, Ill. (Open Court).

Schirn, M.
1983a. "Begriff und Begriffsumfang. Zu Freges Anzahldefinition in den Grundlagen der Arithmetik", *HPL 4*, 117–143.
1988a. R of Frege *1986a*, *JSL 53*, 993–999.
1996a. (Ed.) *Frege: importance and legacy*, Berlin and New York (de Gruyter).

Schlegel, S. F. V. (1843–1905)
1893a. R of Schröder *1890b*, *JFM 22* (1890), 73–78.
1898a. R of Schröder *1895a* and *1895b*, *JFM 26* (1895), 74–80.

Schlick, F. A. M. (1882–1936)
Papers: Haarlem (The Netherlands), State Archives of North Holland.
1910a. "Das Wesen der Wahrheit nach der modernen Logik", *Vrtlj. wiss. Phil. Soz. 34*, 386–477.
1918a. Allgemeine Erkenntnislehre$_1$, Berlin (Springer). [R: Weyl *1923a*. $_2$ 1925.]
1927a. R of Russell *1926c*, *Die Natur-wiss. 15*, 626.

Schmid, A.-F.
1978a. Une philosophie de savant. Henri Poincaré et la logique mathématique, Paris (Maspéro).
1983a. "La correspondance inédite entre Bertrand Russell et Louis Couturat", *Dialectica 37*, 75–109.

Schmidt, H. A. (1902–1967)
1931a. R of Behmann *1931a*, *ZfM 1*, 50.
1932a. R of Dubislav *1931a*, *ZfM 2*, 1–2.
1933a. R of Dubislav *1932c*, *ZfM 5*, 145.

Schnippenkötter, J.
1910a. "Die Bedeutung der mathematischen Untersuchungen Couturats für die Logik", *Philos. Jahrbuch 23*, 447–460.

Schönflies, A. M. (1853–1928)
1898a. "Mengenlehre", in *EMW 1*, sect. 1, 184–207 (article IA5).
1900a. "Die Entwickelung der Lehre von den Punktmannigfaltigkeiten", pt. 1, *JDMV 8*, pt. 2 = Leipzig (Teubner). [Rs: Tannery *1900a*, Vivanti *1902a*.]
1906a. "Über die logische Paradoxien der Mengenlehre", *JDMV 15*, 19–25.
1908a. "Die Entwickelung der Lehre von den Punktmannigfaltigkeiten", pt. 2, *JDMV suppl. vol. 2*, Leipzig (Teubner).
1909a. "Über eine vermeintliche Antinomie der Mengenlehre", *AM 32*, 177–184.
1910a. "Über die Stellung der Definition in der Axiomatik", *Schr. Phys.-ökon. Gesell. Königsberg 51*, pt. 1, 260–293 ≈ *JDMV 20* (1911), 222–255‡.
1913a. Entwickelung der Mengenlehre und ihrer Anwendungen, Leipzig and Berlin (Teubner). [R: Brouwer *1914a*.]
1927a. "Die Krisis in Cantors mathematischen Schaffen", *AM 50*, 1–23.

Schönflies, A. M. and Baire, R. L. *See also* Baire

1909a. "Théorie des ensembles", in *Encyclopédie des sciences mathématiques 1*, Paris (GV), vol. 1, 489–531 = Baire *Works*, 478–520.

Scholz, H. (1884–1956) *See also* Hasse and Scholz

Papers: Münster (Germany), University (mostly after 1933: not used).

1931a. Geschichte der Logik, Freiburg and Munich (Alber) = 1959. [E: *Concise history of logic*, 1961, New York (Philosophical Library).]

1933a. R of Jørgensen *1931a*, *JDMV 43*, pt. 2, 84–88.

1933b. R of Dubislav *1932c*, *JDMV 43*, pt. 2, 88–90.

1935a. "Das theologische Element im Beruf des logistischen Logikers", in (no ed.), *Christliche Verwirklichung*, Rotenfels am Main, 138–159 ≈ (Eds. H. Hermes and others), Scholz, *Mathesis universalis₂*, 1969, Darmstadt (Wissenschaftliche Buchgesellschaft), 324–340.

1935b. R of Lewis and Langford *1932a*, *JDMV 45*, pt. 2, 88–91.

1937a. R of Tarski *1936a*, *Dtsch. Lit.-Zeit. 58*, cols. 1914–1917.

1938a. R of Russell *1903a*, 1937 reprint, *Dtsch. Lit.-Zeit. 59*, cols. 465–468.

Scholz, H. and Bachmann, F. *See also* Bachmann

1936a. "Der wissenschaftliche Nachlass von Gottlob Frege", in *Actes du Congrès International de Philosophie Scientifique Sorbonne Paris 1935*, Paris (Hermann), pt. 8, 24–30.

Scholz, H. and Hermes, H.

1952a. "Mathematische Logik", in *EMW₂*, pt. 1, no. 1, 82 pp.

Scholz, H. and Schweitzer, H.

1935a. Die sogenannte Definition durch Abstraktion, Leipzig (Meiner). [R: Ayer *1937a*.]

Schrecker, P.

1946a. "On the infinite number of the infinite orders", in A. Montagu (ed.), *Studies and essays in the history of science and learning*, New York (Schuman), 359–373.

Schröder, F. W. K. E. (1841–1902) *See also* Grassmann *1878a'*

Here *"VAL"* abbreviates *"Vorlesungen über die Algebra der Logik (exakte Logik)"*.

1873a. Lehrbuch der Arithmetik und Algebra für Lehrer und Studirende, Leipzig (Teubner).

1874a. "Über die formalen Elemente der absoluten Algebra", in *Beilage zum Programm des Pro- und Real-Gymnasiums in Baden-Baden für 1873–74*, Stuttgart (Schweizbart). [Not found: see Peckhaus *1991a*.]

1877a. Der Operationskreis des Logikkalküls, Leipzig (Teubner) = 1966, Darmstadt (Wissenschaftliche Buchgesellschaft). [R: Adamson *1878a*.]

1880a. R of Frege *1879a*, *Ztsch. Math. Physik 25*, 81–94.

1890a. Über das Zeichen, Leipzig (Fock). [E: *1892a*.]

1890b. VAL 1, Leipzig (Teubner). [Rs: Couturat *1900c*, Carus *1892a*, Korselt *1896a* and *1897a*, Ladd(-)Franklin *1892a*, Peano *1891d*, Schlegel *1893a*.]

1891a. VAL 2 pt. 1, Leipzig (Teubner). [Rs: Couturat *1900c*, Husserl *1891b*, Peano *1891d*.]

1892a. "Signs and symbols", *Open court 6*, 3431–3434, 3441–3444, 3463–3466. [E of *1890a*.]

1895a. VAL 3, Leipzig (Teubner). [Rs: Couturat *1900c*, Peirce *1897a*.]

1895b. "Note über die Algebra der binären Relative", *MA 46*, 144–158.

1898a. "Über Pasigraphie", in Rudio *1898a*, 147–162. [R: Peano *1898a*. E: *1898b*.]

1898b. "On pasigraphy", *The monist 9*, 44–62 [corrigenda p. 320]. [E of *1898a.*]

1898c. "Ueber zwei Definitionen der Endlichkeit und G. Cantor'sche Sätze", *Abh. Kaiserl. Leop.-Car. Akad. Naturf. 71*, 301–362.

1898d. "Die selbstständige Definition der Mächtigkeiten 0, 1, 2, 3, und die explizite Gleichzahligkeitsbedingung", *Ibidem*, 363–378.

1901a. "Sur une extension de l'idée d'ordre", in Congress *1901a*, 235–240.

1905a. (Ed. E. Müller), *VAL 2 pt. 2*, Leipzig (Teubner).

1909a, 1910a. (Ed. E. Müller), *Abriss der Algebra der Logik*, 2 pts., Leipzig (Teubner) = *1966a 3*, 651–819.

1966a. Photoreprint of *VAL*, *1909a* and *1910a*, 3 vols., New York (Chelsea). [Complete but slightly reorganised. R: Bernays *1975a*.]

Schubert, H. C. H. (1848–1911)
1898a. "Grundlagen der Arithmetik", in *EMW 1*, sect. 1, 1–27 (article IA1).

Schubring, G.
1996a. (Ed.) *Hermann Günther Grassmann (1809–1877)*, Dordrecht (Kluwer).

1998a. "An unknown part of Weierstrass's *Nachlass*", *HM 25*, 423.

Schuhmann, K.
1977a. *Husserl Chronik*, The Hague (Nijhoff).

Schur, F. H. (1856–1932)
1898a. "Über den Fundamentalsatz der projectiven Geometrie", *MA 51*, 401–409.

Schwarz, K. H. A. (1843–1921)
1872a. "Zur Integration der partiellen Differentialgleichung", *J. rei. ang. Math. 74*, 218–253 ≈ *Gesammelte mathematische Abhandlungen 2*, 1890, Berlin (Springer), 175–210.

Schweitzer, A. R.
1926a. R of Hölder *1924a*, *Amer. math. monthly 33*, 147–150.

Schweitzer, H. *See* Scholz and Schweitzer

Sebestik, J.
1992a. *Logique et mathématique chez Bernard Bolzano*, Paris (Vrin).

Sepp, H. R. *See* Gerlach and Sepp

Servois, F. J. (1767–1847)
1814a. "Essai sur un nouveau mode d'exposition des principes du calcul différentiel", *Ann. math. pures appl. 5*, 93–140‡ ≈ *Essai sur un nouveau mode d'exposition des principes du calcul différentiel*, Nismes (Blachier Belle), 3–50.

Shaposhnikova, T. *See* Maz'ya and Shaposhnikova

Shaw, J. B. (b. 1866)
1907a. *Synopsis of linear associative algebra*, Washington (Carnegie Institution).

1912a. R of *PM*$_1$ *1*, *Bull. AMS 18*, 386–411.

1916a. "Logistic and the reduction of mathematics to logic", *The monist 26*, 397–414 ≈ *1918a*, ch. 5.

1918a. *Lectures on the philosophy of mathematics*, Chicago and London (Open Court).

Shearman, A. T. (1866–1937)
1905a. "Some controverted points in symbolic logic", *Proc. Aristotelian Soc. 5*, 74–105.

1906a. *The development of symbolic logic. A critico-historical study of the logical calculus*, London (Williams and Norgate) = 1990, Bristol (Thoemmes).

1907a. R of Russell *1903a*, *Mind n.s. 16*, 254–265.

1911a. *The scope of formal logic*, London (University of London Press).

Sheffer, H. M. (1882–1964)
Papers: Cambridge (Massachusetts), Harvard University, Houghton Library.

1910a. R of Reymond *1908a*, *PR 19*, 89–90.

1913a. "A set of five independent postulates for Boolean algebras", *Trans. AMS 14*, 481–488.

m1921a. "The general theory of notational relativity", in Sheffer Papers. [Copy also at RA.]

1926a. R of PM_2 *1*, *Isis 8*, 226–231.

1927a. "Notational relativity", in E. S. Brightman (ed.), *Proceedings of the Sixth International Congress of Philosophy*, New York (Longmans, Green) = 1968, Liechtenstein (Kraus), 348–351.

Sheldon, W. H.
1906a. R of Russell *1905g* and Poincaré *1906b*, *JP 3*, 296–298.

Shen, E.
1927a. "The Ladd-Franklin formula in logic: the antilogism", *Mind n.s. 36*, 54–60.

Shin, S.-J.
1994a. *The logical status of diagrams*, Cambridge (Cambridge UP).

Shosky, J.
1997a. "Russell's use of truth tables", *Russell n.s. 17*, 11–26.

Sieg, W.
1990a. "Relative consistency and accessible domains", *Synthese 84*, 259–297.

1999a. "Hilbert's programs: 1917–1922", *Bull. symb. logic 5*, 1–44.

Siegmund-Schulze, R.
1983a. "Die Anfänge der Functionalanalysis", *AHES, 26*, 13–71.

1994a. *Mathematische Berichterstattung in Hitlerdeutschland. Der Niedergang des „Jahrbuch über die Fortschritte der Mathematik"*, Göttingen (Vandenhoeck und Ruprecht).

1998a. "Eliakim Hastings Moore's 'General analysis' ", *AHES 52*, 51–89.

Sierpinski, W. (1882–1969)
1918a. "L'axiome de M. Zermelo et son rôle dans la théorie des ensem-bles et l'analyse", *Bull. Acad. Sci. Cracovie, cl. sci. math. nat.* (A), 97–152‡ = *Oeuvres choisis 2*, 1975, Warsaw (Polish Scientific Publishers), 208–255.

Silva da Silva, C. M.
1997a. "Manuel Amoroso Costa e a filosofia da matematica no Brasil", in *VI Seminário Nacional de História da Ciência e da Tecnologia. Anais*, Rio de Janiero (Sociedade Brasileira de História da Ciência), 158–165.

Simon, M. (1844–1918)
1883a. R of Cantor *1883c*, *Dtsch. Lit.-Zeit. no. 18*, 642–643.

Simons, P. M.
1987a. "Frege's theory of real numbers", *HPL 8*, 25–44.

Sinaceur, M.-A.
1971a. "Appartenance et inclusion. Un inédit de Richard Dedekind", *Rev. d'hist. sci. 24*, 247–254.

1974a. "L'infini et les nombres. Commentaires de R. Dedekind", *Ibidem 27*, 251–278.

1979a. "La méthode mathématique de Dedekind", *Ibidem 32*, 107–142.

Sinisi, V. F.
1976a. "Lesniewski's analysis of Russell's antinomy", *Notre Dame j. formal logic 17*, 19–34.

Skolem, T. (1887–1962)
1923a. "Begründung der elementaren Arithmetik durch die rekurriende Denkweise", *Videns. skr. I. mat. naturw. kl.*, no. 6, 38 pp. = (Ed. J. E. Fenstad), *Selected works*, 1970, Oslo (Universitetsforlaget), 153–188.

1935a. R of Leśniewski *1929a*, *JFM 55* (1929), 626–627.

Słupecki, J. (1904–1987)
1953a. "St. Lesniewski's prototheric", *Studia logica 1*, 44–112.

1955a. "St. Lesniewski's calculus of names", *Studia logica 3*, 7–76.

Sleeper, R. W.

1986a. The necessity of pragmatism. John Dewey's conception of philosophy, New Haven and London (Yale UP).

Smart, H. R.

1925a. The philosophical presuppositions of mathematical logic, New York (Longmans, Green). [R: C. I. Lewis *1926a*.]

1926a. "On mathematical logic", *JP* 23, 296–300.

1931a. The logic of science, New York and London (Appleton).

1949a. "Cassirer's theory of mathematical concepts", in P.A. Schilpp (ed.), *The philosophy of Ernst Cassirer*, La Salle, Ill. (Open Court), 239–267.

Smith, B.

1994a. Austrian philosophy. The legacy of Franz Brentano, Chicago and La Salle, Ill. (Open Court).

Smith, B. and Mulligan, K.

1982a. (Eds.) *Parts and moments. Studies in logic and formal ontology*, Munich (Philosophia).

Smith, B. and Woodruff Smith, D.

1995a. (Eds.) *The Cambridge companion to Husserl*, Cambridge (Cambridge UP).

Smith, G. C.

1980a. "De Morgan and the transition from infinitesimals to limits", *Australian Math. Soc. gaz. 7*, 46–52.

1983a. "Boole's annotations on 'A mathematical analysis of logic'", *HPL 4*, 27–38.

Smith, H. J. S. (1826–1883)

1875a. "On the integration of discontinuous functions", *Proc. LMS (1)6*, 140–153 ≈ *Collected mathematical papers 2*, 1894, Oxford (Clarendon Press) = 1965, New York (Chelsea), 86–100.

Smith, J. Farrell

1985a. "The Russell-Meinong debate", *Phil. phenom. res. 45*, 305–350.

Sobociński, B.

1949–1950a. "L'analyse de l'antinomie Russellienne par Lesniewski", *Methodos 1*, 94–107, 220–228, 308–316; 2, 237–257.

Sofroniou, S.

1979a. "Tarski's analysis of objective truth in type theory", in *Epethpis. Filosofikon pararthma*, Leukosia, Crete (Centre for Scientific Investigation), 233–303.

Spaier, A. (1883–1934)

1927a. La pensée concrète. Essai sur le symbolisme intellectuel, Paris (Alcan).

1927b. La pensée et la quantité. Essai sur la signification et la réalité des grandeurs, Paris (Alcan).

Spencer Brown, G.

1969a. The laws of form, London (A & U). [Various ≈ s.]

Spiegelberg, H.

1981a. The phenomenological movement. A historical introduction, The Hague (Nijhoff).

Stäckel, P. (1862–1919)

1905a. "Elementare Dynamik der Punktsysteme und starren Körper", in *EMW 4*, sect. 1, 435–684 (article IV6).

Stadler, F. *See also* Haller and Stadler

1997a. Studien zum Wiener Kreis. Ursprung, Entwicklung und Wirkung des Logischen Empirismus im Kontext, Frankfurt/Main (Suhrkamp).

Stamm, E. (d. 1940)

1911a. "Beitrag zur Algebra der Logik", *Monats. Math. Physik 22*, 137–149.

Stammler, G. (b. 1898)

1928a. Begriff Urteil Schluss. Untersuchungen über Grundlagen und Aufbau der Logik, Halle/Saale (Niemeyer). [Rare: Göttingen University Library.]

Stebbing, L. S. (1885–1943)

Papers: Englefield Green (England), Royal Holloway College (career files).

1929a. "Logistic", in *Encyclopaedia Britannica$_{14}$* *14*, 330–334.

1930a. A modern introduction to logic$_1$, London (Methuen). [R: Mace *1931a.*]

1932a. R of Jørgensen *1931a*, *Mind n.s. 41*, 236–241.

1933a. 1930a$_2$, London (Methuen).

1933b. "Mr. Joseph's defence of free thinking in logistics", *Mind n.s. 42*, 338–351.

1935a. R of Carnap *1934b*, *1934d*, *1935a* and another work, *Mind n.s. 44*, 499–511.

Steiner, J. (1796–1863)

1867a. (Ed. H. Schroter), *Vorlesungen über synthetischen Geometrie pt. 2$_1$*, Leipzig (Teubner).

Stelzner, W. *See* Max and Stelzner

Sternfeld, R.

1966a. Frege's logical theory, Carbondale and Edwardsville, Ill. (Southern Illinois UP).

Stolz, O. (1842–1905)

1882a. "B. Bolzano's Bedeutung in der Geschichte der Infinitesimalrechnung", *MA 18*, 255–279.

1885a. Vorlesungen über allgemeine Arithmetik 1, Leipzig (Teubner).

1888a. "Über zwei Arten von unendlich Kleinen und von unendlich grossen Grössen", *MA 31*, 601–604.

Strachey, J. (1886–1927)

1911a. R of *PM$_1$ 1*, *The spectator 107*, 142–143.

Strachey, O. (1874–1960)

1915a. "Mr. Russell and some recent criticisms of his work", *Mind n.s. 24*, 16–28.

Study, C. H. E. (1862–1930)

1928a. Denken und Darstellung in Mathematik und Naturwissenschaften$_2$, Braunschweig (Vieweg).

Stump, D.

1997a. "Reconstructing the unity of mathematics circa 1900", *Perspectives on science 5*, 383–417.

Styazhkin, N. I.

*1969a. From Leibniz to Peano: a con-*cise history of mathematical logic, Cambridge, Mass. (MIT Press).

Sylvester, J. J. (1814–1897)

1884a. "Lectures on the principles of universal algebra", *Amer. j. maths. 6*, 270–286 ≈ Collected mathematical papers 4, 1912, Cambridge (Cambridge UP) = 1965, New York (Chelsea), 208–224.

Szaniawski, K.

1989a. (Ed.) *The Vienna Circle and the Lvov-Warsaw school*, Dordrecht (Kluwer).

Tannery, J. (1848–1910)

1884a. R of *AM 1–2*, *Bull. des sci math. (2)8*, pt. 2, 136–171.

1886a. Introduction à l'étude des fonctions d'une variable$_1$, Paris (GV). [$_2$ 1904.]

1892a. R of Husserl *1891a*, *Bull. des sci math. (2)16*, pt. 1, 239–245.

1900a. R of Schönflies *1900a*, *Ibidem (2)24*, pt. 1, 239–245.

Tannery, S. P. (1843–1904)

1934a. Mémoires scientifiques 13, Toulouse (Privat) and Paris (GV).

Tarski, A. Tajtelbaum (1902–1983) *See also* Łukasiewicz and Tarski

Papers. Collected papers, 4 vols., 1986, Basel (Birkhäuser).

Semantics. Logic, semantics, metamathematics, $_1$ 1956, Oxford (Clarendon Press); $_2$ (Ed. J. Corcoran) 1983, Philadelphia (Hackett).

1923a. "Sur le terme primitif de la logistique", *Fund. math. 4*, 197–200 = *Papers 1*, 15–19‡. [E, with another paper: *Semantics*, 1–23.]

1924a. "Sur les 'truth-functions' au sens de MM. Russel [*sic*] et Whitehead", *Fund. math. 5*, 59–74 = *Papers 1*, 21–38‡.

1930a. "Über einige fundamentale Begriffe der Metamathematik", *C. r. Soc. Sci. Lett. Varsovie 23*, 22–29 = *Papers 1*, 311–320. [E: *Semantics*, 30–37.]

1932a. "Der Wahrheitsbegriff in den Sprachen der deduktiven Disziplinen", *Akad. Wiss. Wien, math.-naturwiss. Kl., akad. Anz. 69*, 23–25 = *Papers 1*, 613–617.

1933a. "Einige Betrachtungen über die Begriffe ω-Widerspruchsfreiheit und der ω-Vollständigkeit", *Monats. Math. Physik 40*, 97–112 = *Papers 1*, 619–636. [E: *Semantics*, 279–295.]

1935a. "Einige methodologische Untersuchungen über die Definierbarkeit der Begriffe", *Erkenntnis 5*, 80–100 = *Papers 1*, 637–659. [E: *Semantics*, 296–319.]

1936a. "Der Wahrheitsbegriff in den formalisirten Sprachen", *Studia phil. 1*, 261–405 = *Papers 2*, 51–198. [R: Scholz *1937a*. E: *Semantics*, 152–278‡.]

1937a. "Appendix E", in Woodger *1937a*, 161–172 = *Papers 2*, 335–350.

1941a. Einführung in der mathematischen Logik, Berlin (Springer). [E version: *Introduction to logic*, 1941, New York (Norton).]

1944a. "The semantic conception of truth and the foundations of semantics", *Phil. phenom. res. 4*, 341–375 = *Papers 2*, 661–699.

Taussky-Todd, O. (1906–1995)

1987a. "Reminiscences of Kurt Gödel", in P. Weingartner and L. Schmetterer (eds.), *Gödel remembered*, Naples (Bibliopolis), 29–41.

Taylor, A. E. (1869–1945)

1920a. R of Ziehen *1920a*, *Mind n.s. 29*, 488–490.

Terracini, A. (1889–1968)

1955a. (Ed.) *In memoria di Giuseppe Peano*, Cuneo (Liceo Scientifico Statale).

Thiel, C.

1965a. Sinn und Bedeutung in der Logik Gottlob Freges, Meisenheim am Glan (Hain). [E: *Sense and ref-*

erence in Frege's logic, 1968, Dordrecht (Reidel).]

1975a. "Leben und Werk Leopold Löwenheims (1878–1957)", *JDMV 77*, pt. 1, 1–9.

1977a. "Leopold Löwenheim: life, work, and early influence", in R. O. Gandy and J. M. E. Hyland (eds.), *Logic colloquium 76*, Amsterdam (North-Holland), 235–252.

1982a. "From Leibniz to Frege: mathematical logic between 1679 and 1879", in *Logic, methodology and philosophy of science*, Amsterdam (North-Holland), 755–770.

1993a. "Carnap und die wissenschaftliche Philosophie auf der Erlanger Tagung 1923", in Haller and Stadler *1993a*, 218–223.

Thomae, J. K. (1840–1921)

1880a, 1898a. Elementare Theorie der analytischen Functionen einer complexen Veränderlichen$_{1,2}$, Halle/Saale (Niebert).

1906a. "Gedankenlose Denker. (Eine Ferienplauderei.)", *JDMV 15*, 434–438.

1906b. "Erklärung", *JDMV 15*, 590–592.

Thomas, W. *See* Rang and Thomas

Thomson, W. (1819–1890)

1842a. Outline of the laws of thought, London (Pickering) and Oxford (Graham). [Anonymous. $_2$, named *An outline of the necessary laws ...* (1849).]

Toepell, M. M.

1986a. Über die Entstehung von David Hilberts "Grundlagen der Geometrie", Göttingen (Vandenhoeck und Ruprecht).

1991a. (Ed.) *Mitgliedergesamtverzeichnis der DMV 1890–1990*, Munich (DMV).

Tricot, J.

1930a. Traité de logique formelle, Paris (Vrin).

Turnbull, H. W. (1885–1961)
m1907a. "Principles of mathematics", in Cambridge University Library, Turnbull Papers, Box 3, File 3. [Notes of lecture course by Whitehead.]

Twardowski, K. (1866–1938)
m1927a. (Eds. J. Wolenski and T. Binder), "Selbstdarstellung", *Grazer phil. Studien 39*, 1–26.

Tymoczko, T.
1986a. (Ed.) *New directions in the philosophy of mathematics*, Boston (Birkhäuser).

Ulrici, H. (1806–1884)
1855a. R of Boole *1854a*, *Ztsch. Phil. philos. Kritik n.s. 27*, 273–291 ≈ Peckhaus *1995a*, 87–104‡.
1878a. R of Halsted *1878a*, *Ibidem n.s. 73*, 314–316 ≈ Peckhaus *1995a*, 105–107.

Urbach, B.
1910a. "Über das Wesen der logischen Paradoxa", *Ibidem n.s. 140*, 81–108.

Urquhart, A.
1995a. "G. F. Stout and the theory of descriptions", *Russell n.s. 14*, 163–171.

van Dalen, D.
1978a. "Brouwer: the genesis of his intuitionism", *Dialectica 32*, 291–303.
1990a. "The war of the frogs and the mice, or the crisis of Mathematische Annalen", *Math. intell. 12*, no. 4, 17–31.
1999a. Mystic, geometer, and intuitionist. The life of L. E. J. Brouwer 1, Oxford (Oxford UP).

Van Evra, J. *See also* Houser, Roberts and van Evra
1977a. "A reassessment of George Boole's theory of logic", *Notre Dame j. formal logic 18*, 363–377.
1984a. "Richard Whately and the rise of modern logic", *HPL 5*, 1–18.

van Heijenoort, J. (1912–1986)
1967a. [Ed.] *From Frege to Gödel. A source book in mathematical logic*, Cambridge, Mass. (Harvard UP). [Much editorial material.]

Van Horn, C. E. (b. 1884)
1920a. "An axiom in symbolic logic", *Proc. Cambridge Phil. Soc. 19* (1916–1919), 22–31.

van Vleck, E. B. (1863–1943)
1915a. "The role of the point-set theory in geometry and dynamics", *Bull. AMS (2)21*, 321–341.

von Neumann, J. (1903–1957)
Papers: Washington (D.C.), Library of Congress (mostly after 1940).
Works. (Ed. A. H. Taub), *Collected works*, 6 vols., Oxford (Pergamon).
1923a. "Zur Einführung der transfiniten Zahlen", *Acta Litt. Sci. Szeged 1*, 199–208 = *Works 1*, 24–33. [E: van Heijenoort *1967a*, 346–354.]
1925a. "Eine Axiomatisierung der Mengenlehre", *J. rei. ang. Math. 154*, 219–240 = *Works 1*, 34–56. [E: van Heijenoort *1967a*, 393–413.]
1927a. "Zur Hilbert'schen Beweistheorie", *Math. Ztsch. 26*, 1–46 = *Works 1*, 256–300.
1931a. "Die formalistische Grundlegung der Mathematik", *Erkenntnis 2*, 116–121 = *Works 2*, 234–239. [Rs: Gödel *1932a*, Hempel *1938a*. E: Benacerraf and Putnam *1985a*, 61–65.]

Vailati, G. (1863–1909)
Works₁. (Eds. various), *Scritti*, 1911, Leipzig and Florence (Barth und Seeber).
Works₂. (Ed. M. Quaranto), *Scritti*, 3 vols., 1987, [no place] (Forini).
Letters. (Ed. G. Lanaro), *Epistolario 1891–1909*, 1971, Turin (Einaudi).
1892a. "Dipendenza fra le proprietà delle relazioni", *RdM 2*, 161–164 ≈ *Works₁*, 14–17 ≈ *Works₂ 2*, 331–334.

1898a. Il metodo deduttivo come strumento di ricerca, Turin (Frassati) ≈ *Works*₁, 118–148 ≈ *Works*₂ *2*, 18–48. [F: *1898b*.]

1898b. "La méthode déductive comme instrument de recherche", *RMM 8*, 667–703. [F of *1898a*.]

1899a. "La logique mathématique et sa nouvelle phase de développement dans les écrits de M. J. Peano", *RMM 7*, 86–102 ≈ *Works*₁, 229–242 ≈ *Works*₂ *2*, 172–185.

1904a. "La più recente definizione della matematica", *Leonardo 2* ≈ *Works*₁, 528–534 ≈ *Works*₂ *1*, 7–12.

Vasiliev, A. V. (1853–1929)

1904a. (Ed.) "Tret'e prisuzhdenie premii N. I. Lobacheskogo 1903", *Obshchestva Fiziko-Matematicheskago Kazan' Universiteta (2)14*, 1–100.

Veblen, O. (1880–1960)

Papers: Washington (D.C.), Library of Congress.

1903a. R of Hilbert *Geometry*₁ (1899), *The monist 13*, 303–309.

1904a. "A system of axioms for geometry", *Trans. AMS 5*, 343–384.

Veblen, O. and Lennes, N. J.

1907a. *Introduction to infinitesimal analysis. Functions of one real variable*, New York (Wiley) = 1935, New York (Stechert).

Vega Reñon, L.

1996a. Una guía de historia de la lógica, Madrid (Universidad Nacional de Educación a Distancia).

Venn, J. (1834–1923)

Papers: Cambridge (England), Gonville and Caius College.

1879a. R of MacFarlane *1879a*, *Mind 4*, 580–581.

1880a. "On the diagrammatic and mechanistic representation of propositions and reasonings", *Phil. mag. (5)10*, 1–18.

1881a. "On the various notations adopted for expressing the common propositions of logic", *Proc. Cambridge Phil. Soc. 4*, 36–47.

1881b. "On the employment of geometrical diagrams for the sensible representation of logical propositions", *Ibidem 4*, 47–59.

*1881c. Symbolic logic*₁, London (Macmillan). [R: Monro *1881a*.]

1883a. R of Peirce *1883a*, *Mind 8*, 594–603.

*1894a. Symbolic logic*₂, London (Macmillan) = 1970, New York (Chelsea).

Vercelloni, L.

1989a. Filosofia delle strutture, Florence (Nuovo Italia).

Vienna Circle

Papers: Haarlem (The Netherlands), State Archives of North Holland (files of protocols).

1931a. "Diskussion zur Grundlegung der Mathematik am Sonntag, dem 7. Sept. 1930", *Erkenntnis 2*, 135–155 [with bibliography]. [E of text: *HPL 5* (1984), 111–129. R: Hempel *1938a*.]

Vilkko, R.

1998a. "On the reception of Frege's *Begriffsschrift*", *HM 25*, 412–422.

Vitali, G. (1875–1932)

1905a. Sul probleme della misura dei gruppi di punti di una retta, Bologna (Gamberini and Parmeggiani). [Rare pamphlet. F: C & G, 73–74.]

Vivanti, G. (1859–1949)

1891a. R of Bettazzi *1890a*, *Bull. sci. math. (2)15*, pt. 1, 53–68.

1895a. R of Peano *1892a*, *JFM 24* (1892), 68–69.

1898a. R of Cantor *1895b*, *JFM 26* (1895), 81–82.

1900a. R of Cantor *1897a*, *JFM 28* (1897), 61–62.

1902a. R of Schönflies *1900a*, *JFM 31* (1900), 70–74.

Voss, A. E. (1845–1931)

1901a. "Die Prinzipien der rationellen Mechanik", in *EMW 4*, sect. 1, 3–121 (article IV1).

1914a. Über die mathematische Erkenntnis, Leipzig and Berlin (Teubner).

Vuillemin, J.
1968a. Leçons sur la première philosophie de Russell, Paris (Colin).

Waismann, F. (1896–1969)
1936a. Einführung in das mathematische Denken$_1$, Vienna (Springer). [E: *Introduction to mathematical thinking*, 1951, New York (Unger) = 1959, New York (Harper).]
1976a. (Eds. various) *Logik, Sprache, Philosophie*, Stuttgart (Reclam).

Walsh, A.
1997a. "Differentiation and infinitesimal relatives in Peirce's 1870 paper on logic: a new interpretation", *HPL 18*, 61–78.

Wang, H. (1921–1995)
1957a. "The axiomatization of arithmetic", *JSL 22*, 145–158.
1968a. "Russell and his logic", *Ratio 7*, 1–34.
1987a. Reflections on Kurt Gödel, Cambridge, Mass. (MIT Press).
1996a. A logical journey. From Gödel to philosophy, Cambridge, Mass. (MIT Press).

Warlow, W.
1850a. Letter on Bentham knowing quantification of the predicate, *The Athenaeum*, (21 December), 1351.

Webb, J.
1980a. Mechanism, mentalism, and metamathematics, Dordrecht (Reidel).

Weber, H. (1842–1913)
1893a. "Leopold Kronecker", *JDMV 2*, 5–31‡ ≈ *MA 45*, 1–25.

Wedeking, G. A. *See* Irvine and Wedeking

Weierstrass, K. T. W. (1815–1897)
Works. (Ed. R. Rothe), *Mathematische Werke 7*, 1927, Berlin (Springer) = 1967, Hildesheim (Olms) and New York (Johnson).

Weinberg, J. R.
1965a. Abstraction, relation, and induction, Madison and Milwaukee (University of Wisconsin Press).

Weiss, P.
1928a. "The theory of types", *Mind n.s. 37*, 338–348.
1928b. "Relativity in logic", *The monist 38*, 536–548.
1929a. "The nature of systems", *The monist 39*, 281–319, 440–472 = Chicago (Open Court).
1932a. "The metaphysics and logic of classes", *The monist 42*, 112–154.
1933a. "On alternative logics", *PR 42*, 520–525.
1936a. Report of the opening session of the Association for Symbolic Logic, *JSL 1*, 120.

Weyl, C. H. H. (1885–1955)
Papers: Zurich (Switzerland), Technical High School.
Papers. (Ed. K. Chandrasekhran), *Gesammelte Abhandlungen*, 4 vols., 1968, Berlin (Springer).
1910a. "Über die Definitionen der mathematischen Grundbegriffe", *Math.-naturw. Blätter 7*, 93–95, 109–113 = *Papers 1*, 298–304.
1918a. "Der *circulus vitiosus* in der heutigen Begründung der Analysis", *JDMV 28*, pt. 1, 85–92 = *Papers 2*, 43–50‡.
1918b. Das Kontinuum, Leipzig (Veit) = no year, New York (Chelsea). [R: Rademacher *1923a.*]
1921a. "Über die neue Grundlagenkrise der Mathematik", *Math. Ztsch. 10*, 39–79 ≈ *Papers 2*, 143–180. [R: Fraenkel *1925a.* E: Mancosu *1998a*, 86–118.]
1923a. R of Schlick *1918a*, *JFM 46* (1916–1918), 59–62.
1927a. "Die heutige Erkenntnislage in der Mathematik", *Symposion 1*, 1–32 = *Papers 2*, 511–542. [E: Mancosu *1998a*, 123–142.]
1927b. Philosophie der Mathematik und Naturwissenschaften, Munich

(Oldenbourg). [R: Grunsky *1935a*. Revised E: *Philosophy of mathematics and natural science*, 1952, Princeton (Princeton UP).]

1946a. "Mathematics and logic", *Amer. math. monthly 53*, 2–13 = *Papers 4*, 268–279‡.

Whately, R. (1787–1863)

Logic. Elements of logic$_1$, 1826, London (Mawman) = (Ed. and int. by P. Dessì), Bologna (CLUEB). [Many later eds. R of $_3$ (1829): Hamilton *1833a*.]

1823a. "Logic", in *Encyclopaedia metropolitana 1*, 193–240.

Whitehead, A. N. (1861–1947)

1898a. A treatise on universal algebra with applications, Cambridge (Cambridge UP) = 1960, New York (Hafner). [Rs: Couturat *1900d*, MacColl *1899a*.]

1901a. "Memoir on the algebra of symbolic logic", *Amer. j. maths. 23*, 139–165, 297–316.

1902a. "On cardinal numbers", *Ibidem 24*, 367–384.

1903a. "The logic of relations, logical substitution, groups, and cardinal numbers", *Ibidem 25*, 157–178.

1904a. "Theorems on cardinal numbers", *Ibidem 26*, 31–32.

1905a. "Note" to Russell *1905g*, *RMM 13*, 916–917.

1906a. The axioms of projective geometry, Cambridge (Cambridge UP). [Part F: *1907a*.]

1906b. "On mathematical concepts of the material world", *Phil. trans. Roy. Soc. London (A) 205*, 465–525 ≈ (Eds. F. S. C. Northrop and M. W. Gross), *An anthology*, 1953, Cambridge (Cambridge UP), 7–82.

1907a. "Introduction logistique à la géométrie", *RMM 15*, 34–39. [F of part of *1906a*.]

1907b. The axioms of descriptive geometry, Cambridge (Cambridge UP).

1911a. "Axioms of geometry", in *Encyclopaedia Britannica*$_{11}$ *11*, 730–736 ≈ *1948a*, 178–194 ≈ *1948b*, 243–268.

1911b. "Mathematics", in *Encyclopaedia Britannica*$_{11}$ *17*, 878–883 ≈ *1948a*, 195–208 ≈ *1948b*, 269–288.

1911c. An introduction to mathematics, London (Williams and Norgate). [Various ≈ s.]

1913a. "The mathematical curriculum", *Math. gaz. 7*, 87–94 ≈ *1917b*, ch. 4 ≈ *1929b*, ch. 6.

1913b. "The principles of mathematics in relation to elementary teaching", in Hobson and Love *1913a 2*, 449–454 ≈ *1917b*, ch. 5.

1916a. "The aims of education—a plea for reform", *Math. gaz. 8*, 191–203 ≈ *1917b*, ch. 1 ≈ *1929b*, ch. 1.

1916b. "La théorie relationniste de l'espace", *RMM 23*, 423–454.

1917a. "The organisation of thought", *Rep. Brit. Ass. Adv. Sci.* (1916), 355–363 ≈ *1917b*, ch. 6 ≈ *1929b*, ch. 8.

1917b. The organisation of thought educational and scientific, London (Williams and Norgate) and Philadelphia (Lippincott). [Various ≈ s.]

1918a. "Graphical solution for high-angle fire", *Proc. Roy. Soc. London (A)94*, 301–307.

1926a. "PM. To the editor of 'Mind' ", *Mind n.s. 35*, 130.

1929a. Process and reality, Cambridge (Cambridge UP) and New York (Macmillan).

1929b. The aims of education and other essays, New York (Macmillan) and London (Williams and Norgate).

1933a. Adventures of ideas, Cambridge (Cambridge UP).

1934a. "Indication, classes, number, validation", *Mind n.s. 43*, 281–297‡, 543 [corrigenda] ≈ *1948a*, 227–240 ≈ *1948b*, 313–330. [R: Ackermann *1938c*.]

1934b. "Foreword", in Quine *1934b*, ix–x.

1948a, 1948b. [Ed. D. Runes], *Essays in science and philosophy*, New York (Philosophical Library); London (Rider).

Whitehead, A. N. and Russell, B. A. W. *See also* Russell

PM₁ Principia mathematica₁, 3 vols., 1910, 1912, 1913, Cambridge (Cambridge UP). [Rs: Cohen *1912a*, Dufumier *1911a*, Hardy *1911a*, Jourdain *1913b, 1913c, 1915a* and *1918a*, Keyser *1912a*, C. I. Lewis *1914b*, Peano *1913a*, Shaw *1912a*, Strachey *1911a*.]

PM₂ Principia mathematica₂, 3 vols., 1925, 1927, 1927, Cambridge (Cambridge UP) = 1955[?], Taipei (Rainbow Bridge) [in octavo format]. Parts of *1* = 1962, Cambridge (Cambridge UP). [Rs: B. A. Bernstein *1926a*, Carnap *1931b*, Church *1928b*, Costello *1928a*, Langford *1928b*, C. I. Lewis *1928a*, Ramsey *1925a* and *1925b*, Sheffer *1926a*.]

1911a. "Geometry. VI. Non-Euclidean geometry", in *Encyclopaedia Britannica₁₁ 11*, 724–730 ≈ Whitehead *1948a*, 209–226 ≈ Whitehead *1948b*, 289–312.

1932a. Einführung in die mathematischen Logik, Munich and Berlin (Drei Masken) = 1984, Vienna (Medusa). [G by H. Mokre of introductions to *PM₁* and *PM₂*, and also of Gödel *1944a* in reprint.]

Wiener, H. L. G. (1857–1939)

1892a. "Ueber Grundlagen und Aufbau der Geometrie", *JDMV 1* (1890–1891), 45–48.

Wiener, N. (1894–1964)

Papers: Cambridge (Massachusetts), MIT.

Works. (Ed. P. Masani), *Collected works 1*, 1976, Cambridge, Mass. (MIT Press).

m1913a. "A comparison between the treatment of the algebra of relatives by Schröder and that by Whitehead and Russell", Cambridge, Mass. (Harvard University Ph.D.). [Parts in my *1975b*.]

1914a. "A simplification of the logic of relations", *Proc. Cambridge Phil. Soc. 17*, 387–390 = *Works*, 29–32 ≈ van Heijenoort *1967a*, 224–227.

1914b. "A contribution to the theory of relative position", *Ibidem 17*, 441–449 = *Works*, 34–42.

1914c. "Studies in synthetic logic", *Ibidem 18*, 14–28 = *Works*, 43–57.

1916a. "Mr. Lewis and implication", *JP 13*, 656–662 = *Works*, 226–232.

1919a. "A new theory of measurement: a study in the logic of mathematics", *Proc. LMS* (2)*19*, 181–205 = *Works*, 58–82.

Willard, D.

1984a. Logic and the objectivity of knowledge: a study in Husserl's early philosophy, Athens (Ohio UP).

Wilson, E. B. (1879–1964)

1903a. "The so-called foundations of geometry", *Archiv Math. Physik 6*, 104–122.

1904a. R of Russell *1897c* and *1903a*, *Bull. AMS 11*, 74–93.

Windelband, W. and Ruge, A.

1913a. (Ed.) *Encyclopaedia of the philosophical sciences 1* [and only] *Logic*, London (Macmillan). [R: Russell *1914a*. G original: 1912, Tübingen (Mohr (Siebeck)).]

Winter, E. (1896–1982)

1969a. Bernard Bolzano. Ein Lebensbild, Stuttgart (Frommann-Holzboog).

Winter, M.

1905a. "Métaphysique et logique mathématique", *RMM 13*, 589–619.

1911a. La méthode dans la philosophie des mathématiques, Paris (Alcan).

Wisdom, A. J. T. (1904–1993)

1934a. R of Black *1933a*, *Mind n.s. 43*, 529–531.

Wittgenstein, L. (1889–1951)
  Letters. (Eds. G. H. von Wright with
  B. F. McGuinness), *Letters to
  Russell, Keynes and Moore*, 1974,
  Oxford (Blackwell) and Ithaca
  (Cornell UP).
  Ogden. (Ed. G. H. von Wright), *Let-
  ters to C. K. Ogden*, 1973, Oxford
  (Blackwell) and London and Boston
  (RKP).
  1921a. "Logische-philosophische
  Abhandlung", *Ann. Naturphil. 14*,
  198–262.
  1922a. Tractatus logico-philosophicus,
  London (KP). [≈ and E by F.
  Ramsey and C. K. Ogden of *1921a*.
  Rs: Keyser *1923a*, Ramsey *1923a*.
  New E: 1961, London (RKP).]
Wolenski, J.
  *1989a. Logic and philosophy in the
  Lvov-Warsaw school*, Dordrecht
  (Kluwer).
Wolenski, J. and Köhler, E.
  1999a. (Eds.) *Alfred Tarski and the
  Vienna Circle*, Dordrecht (Kluwer).
Wood, S. *See* Corcoran and Wood
Woodger, J. H. (1894–1981)
  *1929a. Biological principles. A critical
  study*, London (KP).
  *1937a. The axiomatic method in biol-
  ogy*, Cambridge (Cambridge UP).
  [R: Rosinger *1938a*.]
  1958a. "Formalization in biology",
  Log. et anal. 1, 97–104.
Woodhouse, R. (1773–1827)
  1801a. "On the necessary truth of
  certain conclusions obtained by
  means of imaginary quantities",
  Phil. trans. Roy. Soc. London 91,
  89–119.
Woodruff Smith, D. *See* Smith and
  Woodruff Smith
Wright, J. N.
  1928a. R of Burkamp *1927a*, *Mind
  n.s. 37*, 377–378.
Wrinch, D. M. (1894–1976)
  Papers: Northampton (Massachusetts), Smith College.

1919a. "On the nature of judgement",
  Mind n.s. 28, 319–329.
1920a. "On the nature of memory",
  Mind n.s. 29, 46–61.
1923a. "On mediate cardinals", *Amer.
  j. maths. 45*, 87–92.
1925a. R of Hölder *1924a*, *Mind n.s.
  34*, 507–508.
Wundt, W. (1832–1920)
  1910a. "Psychologismus und Logizismus", in *Kleine Schriften 1*, Leipzig
  (Engelsmann), 511–634.
Wussing, H.
  *1984a. The genesis of the abstract group
  concept*, Cambridge, Mass (MIT
  Press). [German original: 1969,
  Berlin.]

Young, W. H. (1863–1942)
  1905a. "On the general theory of
  integration", *Phil. trans. Roy. Soc.
  London (A)* 204, 221–254.
Young, W. H. and Young, G. C.
  (1868–1944)
  1906a. The theory of sets of points,
  Cambridge (Cambridge UP) =
  (Eds. R. C. H. Tanner and I. Grattan-Guinness), 1972, New York
  (Chelsea) [with additional material].
  1929a. R of E. W. Hobson, *The theory
  of functions of a real variable*₂
  (1921–1926), *Math. gaz. 14*, 98–104.

Zaitsev, E. A.
  1989a. "Dj. Peano o ponyatii «tot,
  kotorii» i o vozmotnosti ego elim-
  initsatsii iz teorii", *Ist. i metod. es-
  testv. nauk 36*, 50–58.
Zaremba, S. (1863–1942)
  1926a. La logique des mathématiques,
  Paris (GV).
Zariski, O. (1899–1986)
  1926a. "Note", in Dedekind *1926a*,
  157–300.
Zermelo, E. F. F. (1871–1953)
  Papers: Freiburg (Germany),
  University.
  1901a. "Über die Addition trans-
  finiter Kardinalzahlen", *Nachr.*

*Königl. Gesell. Wiss. Göttingen,
math.-phys. Kl.*, 34–38. [F: C & G,
435–443.]

1904a. "Beweis, dass jede Menge
wohlgeordnet werden kann", *MA
59*, 514–516. [E: van Heijenoort
1967a, 139–141. F: C & G, 451–
457.]

1908a. "Neuer Beweis für die
Möglichkeit einer Wohlordnung",
MA 65, 107–128. [E: van Heijenoort *1967a*, 183–198. F: C & G,
523–555.]

1908b. "Untersuchungen über die
Grundlagen der Mengenlehre. I",
MA 65, 261–281. [No successor. E:
van Heijenoort *1967a*, 199–215. F:
C & G, 557–593.]

1909a. "Sur les ensembles finis et le
principe d'induction complète" and
"Supplément", *AM 32*, 185–193 =
Heinzmann *1986a*, 148–156, 230.

1909b. "Über die Grundlagen der
Arithmetik", in Castelnuovo *1909a
2*, 8–11.

1913a. "Über eine Anwendung der
Mengenlehre auf die Theorie des
Schachspiels", in Hobson and Love
1913a 2, 501–504.

1930a. "Über Grenzzahlen und Mengenlehre", *Fund. Math. 16*, 29–47.
[E: Ewald *1996a*, 1219–1233.]

1932a. "Über Stufen der Quantifika-
tion und die Logik des Unendlich-
en", *JDMV 41*, pt. 2, 86–88.

Ziehen, T. (1862–1950)

*1917a. Das Verhältnis der Logik zur
Mengenlehre*, Berlin (Reuther und
Reichard). [R: Bernays *1923b*.]

*1920a. Lehrbuch der Logik auf
positivistische Grundlage mit
Berücksichtigung der Geschichte der
Logik*, Bonn (Marcus und Webers).
[R: Taylor *1920a*.]

READERS who intend to use this index substantially should browse in it first in order to appreciate its organisation; only a few of the many cross-references are given (when '/' divides entry from sub-entry). Attention is drawn especially to the entries 'Cardinal(s)', 'Numbers', 'Ordinal(s)', and 'Order'; 'Proof-methods'; and the groups beginning 'Algebras', 'Axiom', 'Logic(s)', 'Paradox(es), and 'Philosophy'. The group beginning 'Class(es)' refer only to the Cantorian theory; otherwise consult 'Part-whole theory'.

The principle of optimal approximation has been used, so that sub-(sub-)entries often contain more citations than the parent entries. Topics are often placed under main headings; for example, 'Sheffer stroke' comes only under 'Logical connectives'.

Sub-entries under names of figures are confined to their main notions and publications and principal details of their career. They also cover 'relationships' between figures, intellectual as well as personal, when the sub-entry is placed under *each* entry.

Simple citations of items listed in the bibliography are rarely indexed, and the tables not at all. The symbol '~' is used in page references 'm~n' to indicate that the (sub-)entry in question appears on *most* pages between m and n and is relevant throughout; 'm-n' cites *all* pages as usual.